Membrane Structure and Mechanisms of Biological Energy Transduction

Membrane Structure and Mechanisms of Biological Energy Transduction

Edited by

J. Avery

Department of Chemistry, H.C. Ørsted Institute
University of Copenhagen, Denmark

PLENUM PRESS · LONDON AND NEW YORK

Plenum Publishing Company Ltd.
42 Lower John Street
London W1R 3PD
Tel. 01-437 1408

U.S. Edition published by
Plenum Publishing Corporation
227 West 17th Street
New York New York 10011

Reprinted from *Journal of Bioenergetics* Volumes 3 and 4

1st Printing 1973
2nd Printing 1974

ISBN 0-306-30718-9
Library of Congress Catalog Card Number: 72-95064

Printed in Great Britain by
William Clowes & Sons Limited
London, Colchester and Beccles

Preface

Ten years ago, Calvin proposed that the chloroplast thylakoid membrane may play an important role in separating photochemically produced oxidizing and reducing agents, thus preventing the back-reactions in photosynthesis. Similarly, in the study of oxidative phosphorylation, Mitchell's chemiosmotic hypothesis has focused attention on the role of electrochemical gradients across the mitochondrial membranes. Much knowledge of membrane structure and function in these systems has been gained recently by the use of freeze-etch and negative staining electron microscopy techniques, by low-angle X-ray diffraction, and by the study of lipid-depleted membranes. The models of membrane structure put forward by Vanderkooi, Green, Packer, Sjöstrand, Lenaz, Benson and others, challenge the classical Danielli-Davson lipid bilayers model. This book contains a collection of papers reporting these new developments reprinted from two special issues of the *Journal of Bioenergetics* devoted to membrane structure and function. I am extremely grateful to Professor D. E. Green for organizing the contributions of Mechanisms of Biological Energy Transduction and to Professor S. Papa for editorial advice during the preparation of the section on Membrane Structure. Sincere thanks are also due to the distinguished scientists who have contributed to this volume.

John Avery
H.C. Ørsted Institute
University of Copenhagen

Contents

Part A: Mechanisms of Biological Energy Transduction

Part B: Membrane Structure

Part A

Mechanisms of Biological Energy Transduction

Reprinted from *Journal of Bioenergetics* Volume 3

The Development of Bioenergetics

Albert Szent-Gyorgyi

Marine Biological Laboratory, Woods Hole, Massachusetts

Living systems, being built of matter, and being driven by energy, their analysis can be approached from two sides: that of matter or from the side of energy. This division is analogous to the division into structure and function, or, anatomy and physiology. This division may be very convenient but is artificial, and exists only in our mind, not in nature. Structure and function, essentially, are identical, two sides of the same coin. Structure generates function and function generates structure. A deeper understanding can be achieved only by a fusion of the two.

The basic blueprint of bioenergetics is laid down, and progress can be expected rather in the development in details than in the discovery of new principles. What I would like to see is a more philosophical attitude. "Philosophy" means the search for deeper and wider relations. Needless to say, a philosophical approach presupposes the knowledge of the details, and so it is no wonder that we should have lost ourselves in details. All the same, the loss of a philosophical attitude is regrettable and is characteristic for the contemporary biology. The word "Science", with its present meaning, is of recent origin. In the last century "natural philosophy" was used instead, which indicates a different attitude.

If I were a newcomer in Bioenergetics, I would start my research with trying to clear my mind about the meaning of the word "energy". Living systems consist of atoms, and atoms are, essentially, a cloud of electrons held together by nuclei. So energy can be only electronic energy. Kinetic energy plays no major role in biology except in activation. The source of our energies are the foodstuffs, and their energy can be only electronic. So my first question would be how an electron can release energy? Strictly speaking, electrons have no energy, so how can they release any?

In trying to answer this question I would take my refuge to the table of atoms of Mendeleef. It contains all the atoms out of which living systems are built. On its left side are the atoms which tend to give off electrons to approach the noble gas structure, while on the right are the atoms which tend to take up electrons. A "trend to give off electrons" involves that the electrons are loosely held, given off easily, that is have a low ionization potential, are on a high energy level. The "trend to take up electrons" means that there are low energy levels capable of accommodating electrons. It follows that energy can be released by electrons of an element of the left side by interacting with an element of the right. In such interaction the electrons of the former could go to a lower energy level of the latter and doing so, would release energy.

What would probably strike me when taking a second look at this table, is the extraordinary nature and position of H. It is a group by itself. This atom is the only one which has the qualities needed to support life. It is small, it parts easily with its electron and the proton, left behind, can merge with the proton pool of the general solvent, water. From this pool an electron can easily pick up a proton and form an H with it. This would lead me to the conclusion that H had to be the fuel of life. Looking over the elements of the right side I would soon come to the conclusion that for many reasons oxygen only is fit to act as general H- or electron-acceptor. However, the acceptor qualities are not as specific as the donor ones, and in absence of O_2 various other elements may also act as acceptors, and, evidently, had to do so before O_2 appeared in the atmosphere. About the possibilities of storage of H, the Mendeleef table would tell me that only C, with its position in the middle, is fit to build up the complex molecules which could hold the H's.

If energy production only demands the interaction of an element from the left of Mendeleef table with an element of the right, then why do not all atoms of the left side simply unite with the atoms on the right side? Why do not all thermodynamically possible reactions take place spontaneously? Should this happen then the whole biosphere would burn up and there would be no place left for life, since no life can exist without energy. Whatever the living machinery does, it has to pay for it in terms of energy. Thus why do all thermodynamically possible reactions not take place? Where is the "hang-up" which makes life possible, and how can life make the reactions go which do not take place spontaneously?

The hang-up is the activation energy, the fact that most chemical reactions are, in a way, similar to a rock on the mountain side. By rolling down, the rock would liberate great amounts of energy. If it does not do so this is simply because in order to roll down it would have to be loosened up, which demands a relatively small quantity of energy which has to be invested. This small initial investment is which holds reactions up and makes life possible. What life does is to "loosen up" H atoms, decrease the activation energy to a level at which the energy of heat agitation is sufficient to make the reaction go. This loosening up, this decrease of activation energy, is done by enzymes, dehydrogenases, or "H-activation".

With these factors in mind I would try to follow the road on which life has built its machinery. Life builds gradually and does not reject what it has built but builds on top. Consequently, the present cell is comparable to an archeological excavation site where one can find all successive layers on top of one another, the oldest one the deepest. It is like a pyramid. The oldest ones can be expected to be most widely spread and anchored most firmly. By this I mean that the newer a process, the more fancy its chemical equipment, the easier it will be discarded, while the older and simpler ones will be retained more tenaciously. For example: the most complete and newest energy producing machinery is that of biological oxidation. It demands an involved chemical structure. If the cell divides it has to liquify and has to give up its complex solid structures. So it gives up biological oxidation and reverts for energy production to older and simpler processes.

If we try to go back to the very beginning, the prebiotic period, there we find a "primordial soup" with organic molecules in it. When life appeared it had to organize part of these molecules into living systems, while the others were used as "foodstuffs", a source of energy. If we imagined two molecules, side by side, a "living macromolecule" and a foodstuff molecule, the question would come up how could the former take energy out of the latter?

According to what has been said before I can see one way only: taking H atoms (or electrons) off the foodstuff molecule. In order to do this the binding of these H's would have to be loosened up, their activation energy would have to be decreased. The H transfer involves the "dehydrogenation" of the foodstuff molecule. Dehydrogenation being the oldest process we find dehydrogenases widely spread in present living systems. They are actually the foundation of our metabolism.

This simple scheme had to be complicated by the problem of storage. Life needs energy all the time, but foodstuffs are accessible only occasionally. The H's or electrons taken off the foodstuff had thus to be stored to be used later according to need. There had to be an H-, or electron pool. As I have pointed out in my book on "The Living State" (now in press)* one of the most important building blocks of the energy household is an H- or electron pool which plays an important role even today and is, probably, the most

* Academic Press, New York.

important energy source of the cell during division. In the interphase it is replenished by the dehydrogenases.

Structural organization has three grades. The basic level is that of "constitution", by which we mean how the atoms are put together to a molecule, and how simple molecules are put together to make more complex ones. This putting together results in fibrous protein molecules. On the second level of organization this fiber is folded up in a meaningful way and a globule is formed. This second step of organization can be destroyed by heating. For complete destruction mostly heating to 100°C at pH 4 is needed. The third level of organization, which I call "coordination", consists of putting together the single globules in a meaningful way so that chain reactions can smoothly take place. This subtlest and highest level of organization can be disturbed most easily, even by moderate heating. It is remarkable that at this highest level of molecular organization the units are put together two-dimensionally, into membranes. It seems likely that only in a two-dimensional structure can the sequence of chemical reactions be controlled and determined by structure.

The history of bioenergetics probably consists of three periods, based on three discoveries of Nature. In the first of them H's were taken off from the foodstuff molecules, to be coupled subsequently to elements, or atomic groups, of the right side of the Mendeleef table (e.g. NO_3). This archaic method involved, possibly, only conformation. The second period of bioenergetics may have been based on the discovery that energy can be derived from another molecule by reshaping its constitution to a form in which it contained less energy, the energy difference being stabilized and transferred by means of "high-energy-phosphate bonds". This led to fermentation. The third discovery which opened the way to the formation of the more complex forms which demanded more energy was the discovery, made by Nature, that the energy of solar radiation could be used to separate H from O, splitting water, the H being used as energy source, and O being sent back as O_2 into the atmosphere. The energy of H was released gradually in a complex chain of reaction, being transduced into $\sim P$, or being stored in the form of foodstuffs. This method demanded a highest level of organization which I called "coordination". This method of energy release and storage became possible only after O_2 appeared in the atmosphere and could be used as final H- or electron acceptor.

Anaerobic bacteria may be left-overs of the first period.

Chemiosmotic Coupling in Energy Transduction: A Logical Development of Biochemical Knowledge

Peter Mitchell

Glynn Research Laboratories, Bodmin, Cornwall, England

Introduction

At the invitation of the Editors, this paper gives a summary sketch of my position regarding some metabolic aspects of energy transduction and describes some present and anticipated perspectives from my point of view. To maintain as broad a horizon as possible, however, I have used this opportunity to describe how my views, and the rationale that I have developed to express them, have been derived from accepted or acceptable physicochemical theory and biochemical knowledge stemming from the creative and painstaking observations of my progenitors and colleagues.

My interest in the conceptual and functional relationships between chemical and osmotic reactions was first seriously stimulated when I was studying the specific exchange and uptake of inorganic phosphate and arsenate through the plasma-membrane of staphylococci.[1,2,3] The remarkably high specificity of the phosphate translocation reaction, its susceptibility to specific inhibitors including SH-reactors, its high entropy of activation which indicated a large conformational change in the translocator system, and the tight coupling of phosphate translocation against arsenate translocation,[1,2,3] indicated how closely osmotic translocation reactions could resemble (or be functionally related to) enzyme catalysed group-transfer reactions. Further, the observation that the plasma-membrane material isolated from staphylococci contained the cytochrome system and associated enzyme activities[4,5] suggested that certain of the group-transfer reactions catalysed by the enzyme and catalytic carrier systems in the plasma-membrane might actually be vectorial group-translocation reactions because of the spatial orientation of the catalytic systems. These were the circumstances that led me to remark at a symposium nineteen years ago (see ref. 2) that "in complex biochemical systems, such as those carrying out oxidative phosphorylation (e.g. Slater &

Cleland, 1953),[6] the osmotic and enzymic specificities appear to be equally important and may be practically synonymous". The same circumstances drew my attention to certain related conceptual and factual knowledge from the fields of enzymology and membrane transport which, at that time, tended to be pursued by separate schools of thought. In this paper I show how the bringing together of this knowledge opened up some new perspectives and provided a logical foundation for the development of the chemiosmotic coupling concept.

Primary Chemical Coupling

The general principle of primary chemical coupling was already recognized in the nineteen thirties when Green, Stickland and Tarr[7] described the coupling between the oxidation of one substrate (AH_2) and the reduction of another (B) by a solution of the appropriate AH_2 and BH_2 dehydrogenases in the presence of the specific coupling factor (NAD) which was alternately reduced and oxidized in the following type of reaction (see also Green[8]):

$$\begin{matrix} AH_2 & \searrow\swarrow & NAD^+ & \nwarrow\nearrow & BH_2 \\ A & \swarrow\searrow & NADH + H^+ & \nearrow\searrow & B \end{matrix} \qquad (1)$$

According to the elegant rationale and terminology subsequently introduced by Lipmann,[9] this type of system effectively coupled the group-transfer reactions,

$$\begin{matrix} AH_2 & \searrow & \\ A & \swarrow\searrow & 2H \end{matrix} \qquad (2)$$

and

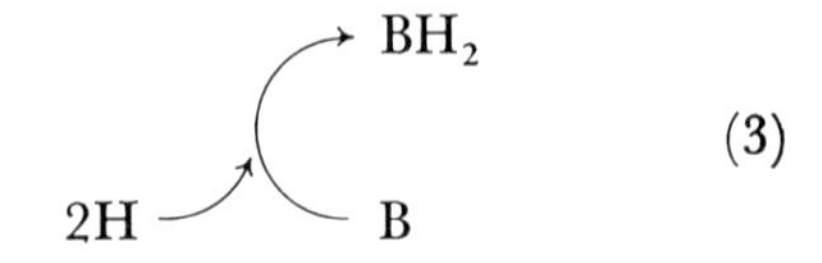

by facilitating the transfer of the hydrogen groups, and equilibrating the hydrogen group potential, between the donor and acceptor redox couples. Thus, "energy-coupling" actually corresponded to the material transfer of hydrogen groups via the substrate-specific enzymes and the NAD coenzyme mediating between the chemical components A and B; and Lipmann's view enabled one to imagine the reaction going forward because of a "thrust" transmitted down the group-potential gradient—although, of course, there was no *net* vector component of this "thrust" in aqueous "homogeneous" enzyme solutions.

In the case of the soluble enzyme systems catalysing substrate-level phosphorylation by a coupling between redox reactions and phosphorylation or dehydration reactions, a more complex but fundamentally similar coupling scheme was found to be required. It was difficult to see, at first, how the "driving" redox reaction could involve a component in common with the "driven" phosphorylation or dehydration reaction so that the "thrust" of a group-potential gradient could be transmitted between the two reactions. But this apparent difficulty was overcome when it was shown by Warburg[10] and by Racker[11] that the oxidation of 3-phosphoglyceraldehyde (PG) to 3-phosphoglycerate (PGA) could be coupled to ADP phosphorylation through an appropriate series of group transfers because both the hydrogen and phosphoryl group transfer reactions were channelled through the intermediate 1,3-diphosphoglycerate (PGAP), as summarised in the following equation:

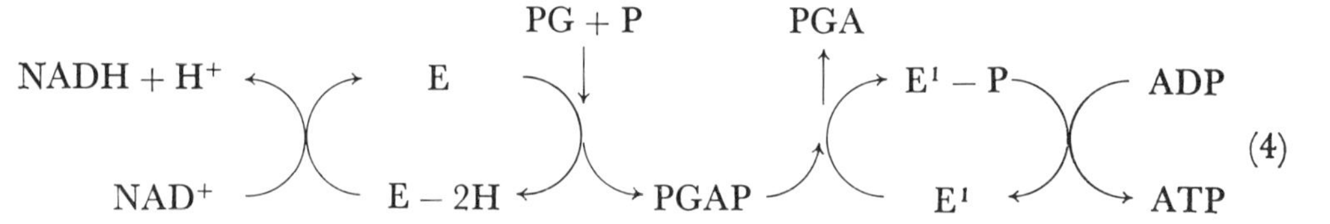

where E and E' stand for 3-phosphoglyceraldehyde dehydrogenase and 3-phosphoglycerate kinase enzyme complexes respectively.

Thus, in the nineteen fifties it was evident that the transfer of one type of chemical group could be "energetically coupled" to the transfer of another type of chemical group by means of a series of appropriate chemical intermediates that would enable the thermodynamic driving force to be transmitted through the corresponding group-potential gradients—as described most eloquently by Lipmann.[12] This type of mechanism implicitly attributed the transmission of the thermodynamic force through the system to interaction of chemical groups across covalent bonds.

The success and elegance of this "substrate-level" type of coupling mechanism led to the belief that it would be generally applicable, and that even in the particulate systems catalysing oxidative and photosynthetic phosphorylation, the coupling mechanism would be explained in this way when the pathways of group transfer were elucidated by isolating and identifying the covalent "energy-rich" chemical intermediates.

The emphasis placed on the covalent intermediates in "substrate level" metabolic coupling tended to obscure the fact that, while the changes of group potential accompanying covalent bond interchanges between certain standard states could properly be regarded as the source of metabolic energy, the transmission of this energy through a given region of a metabolic pathway by the "thrust" of chemical particles diffusing through it should be attributed not only to forces along covalent bonds but also to concentration-dependent osmotic forces and to forces along ionic and secondary bonds. The relevance of this important fact to "energy coupling" in metabolism was indicated by Glasstone, Laidler and Eyring's "theory of absolute reaction rates" in which chemical transformation and viscous fluid flow were treated as proceeding through fundamentally similar thermally activated rearrangements of chemical particles via transitional states of appropriately low free energy.[13] Likewise, the suggestion by Pauling[14] that enzymes have a higher affinity for the transitional configurations of their substrates than for the normal reactant and resultant species—and thus catalyse group transfer by lowering the free energy (and increasing the probability) of the transition state—illustrated the role of secondary bonding relationships in catalysing or channelling the transfer of chemical groups from donors to acceptors, and therefore in coupling group transfer processes. In the case of the transfer of a given type of chemical group between acceptor and donor systems, as illustrated by equations 1–3, it was evident that the coupling should be attributed to non-covalent bonding interactions inasmuch as these were involved in catalysing the flow of the specific group through the prescribed pathway, and minimizing energy loss through side reactions. In the case of systems like that of equation (4), where the transfer of one type of chemical group is coupled to that of another, it followed that the osmotic and non-covalent bonding interactions might play a major part in linking the two different group-transfer processes as well as being involved in the direct catalytic role described above. This is illustrated by the following equation, representing the overall process of oxidative phosphorylation:

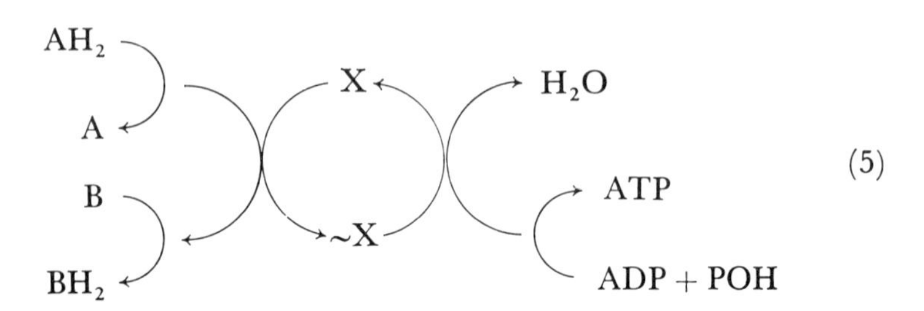

(5)

In this system, the intermediate process, represented as the transition between X and ~X, might mainly represent a change of covalent bonding of X—as in the orthodox chemical coupling hypothesis—but might, equally well, mainly represent a change of osmotic status or ionic or secondary bonding of X—providing a basis for alternative coupling hypotheses.

Electron Translocation and Group Translocation

The idea of spatially orientated chemical reactions in biochemistry stemmed from the work of Lundegardh[15] who suggested that cytochrome pigments might provide an electron-conducting pathway across plant cell membranes so that oxygen could be reduced (and H^+ ions consumed) on one side while hydrogenated substrates were oxidized (and H^+ ions were produced) on the other side, as represented by the following equation:

$$\begin{matrix} AH_2 & \searrow & & \swarrow & \frac{1}{2}O_2 + 2H^+ \\ &) & \xrightarrow{2e^-} & (& \\ 2H^+ + A & \swarrow & & \searrow & H_2O \end{matrix} \qquad (6)$$

This idea of electron conduction or translocation, accompanied by redox and acid-base changes across natural membranes, was further developed, notably by Davies and Ogston,[16] by Conway,[17] and by Robertson[18] (see Robertson[19]).

Since Curie[20] pointed out that effects cannot be less symmetric than their causes, it obviously followed that, when a (vectorial) transport or osmotic process across a membrane was caused by a chemical reaction, the group-potential gradient acting as the driving force[9] must be directed in space. The beautifully simple thermodynamic treatment of chemicomotive cells and circuits by Guggenheim[21] likewise showed how the transport of a chemical component is determined by gradients of chemical potential representing what he called the "chemicomotive forces" directed across (artificial) membranes or phase boundaries; and a somewhat similar treatment of biological transport was also given by Rosenberg.[22] Thus, it was logical to introduce the concept of group translocation[23,24] as a generalization of the idea of electron translocation, so that one could more readily recognize and describe, not only electromotive forces, but also other chemicomotive forces (such as phosphorylmotive forces or protonmotive forces) as prime movers in metabolically-coupled translocation reactions. In this way it was possible to consider all biological transport reactions as arising from the thermally activated diffusion of solutes and chemical groups down the (vectorial or higher order tensorial) electrochemical potential gradients of their various compounds and complexes in the local aqueous and lipid-membrane media, including conformationally mobile "active centre" and "carrier centre" regions of enzymes and catalytic carriers.[2,25,26]

As shown by my mobile enzyme model for phosphoryl translocation,[25] I was initially encouraged to regard the idea of specific conformational flexibilities or articulations in the condensed complex of the enzymes or catalytic carriers catalysing group translocation as acceptable by analogy with the notion of conformational mobility in membranes, which originated from Langmuir's elegant idea of the turning over of lipid molecules in artificial monolayers.[27,28] This view implied that such group-translocation proteins or protein complexes (including, for example, prosthetic groups or lipid) would undergo cyclic conformational changes synchronous with the group transfer reaction; and it also followed that in such three-dimensional articulated complexes the translocation of the group undergoing chemical transfer along one pathway might be tightly coupled to the translocation of the acceptor or donor substrate species or of other solute species in the same or in different directions in space.[29] The point to be stressed is that according to this view "energy coupling" is attributed to the transmission of "thrusts" along interacting trains of chemical particles including the macromolecular components. In this type of system, the osmotic "thrust" or potential gradient of a small molecular weight solute can generally be exerted only along the translational degree of freedom across an osmotic barrier or into a group-transfer reaction centre, but the osmotic "thrust" of a group translocation catalyst is exerted through the complex degrees of freedom defined by the permitted articulations, as in a macroscopic mechanical engine. This development of ideas was consistent with contemporary developments in the fields of classical enzymology and protein structure, such as the "induced fit" interpretation of enzyme kinetic data by Koshland,[30] the suggestion by Hammes[31] that the activation energy for group transfer may be lowered by the balancing of stress-strain relationships in mobile enzyme-substrate complexes, and the observations of Muirhead and Perutz[32] on conformational changes in haemoglobin molecules during oxygenation. Subsequent experimental work on translocation catalysis has served to confirm my view that the conformational mobility of the translocation catalysts is associated with their normal group-translocation and solute-translocation functions.[33–36]

Proton Translocating Respiratory Chain and Photoredox Chain

The development of my working hypothesis for the chemiosmotic type of coupling mechanism in oxidative and photosynthetic phosphorylation[37] depended on a logical sophistication of the concept of group translocation, because, in order to account for "energy coupling" by the transmission of an osmotic "thrust" through a train of protons between the redox chain and reversible ATPase systems, it was necessary to formulate group-translocation mechanisms by which the electron and hydrogen transfer reactions through the redox chain, and the H_2O transfer through the ATPase system, could each result in effective proton translocation.

Proton translocation in the redox chain was attributed to paired electron translocation and hydrogen atom (or hydride ion) translocation reactions[37–39] which could conveniently be represented by so-called redox loops of type I or type II, according to whether one or two H^+ ions were translocated respectively per divalent redox equivalent transferred, as represented in the following equations:

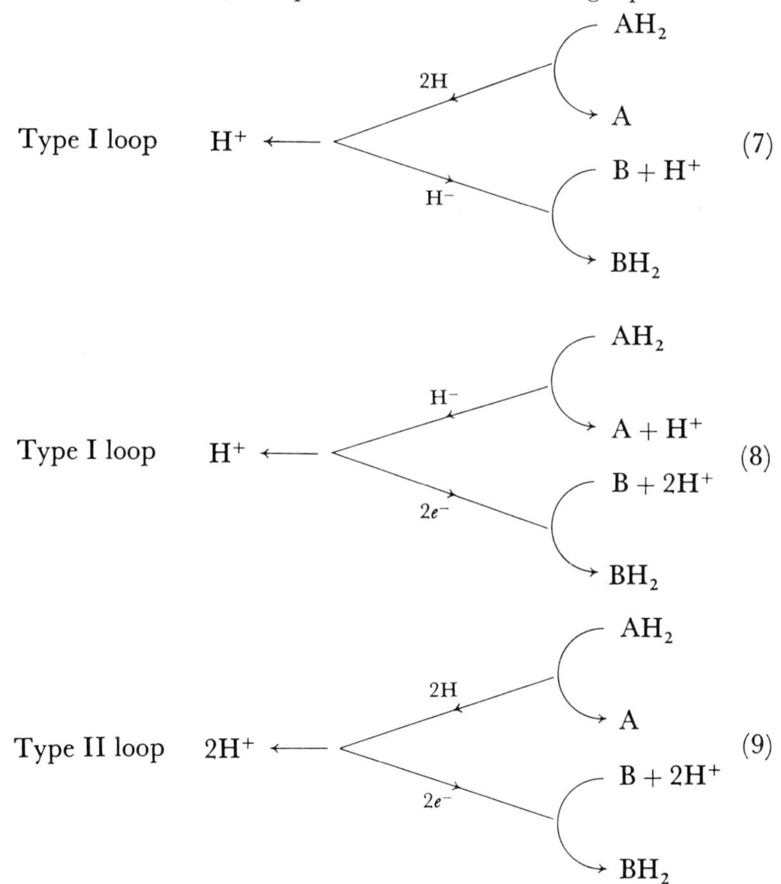

The essential point to be noted is that these proposals depicted proton translocation as occurring, not by the movement of protons as such, but by the stoichiometrically coupled translocation of hydrogen atoms to the left and of electrons to the right through the duplex catalytic carrier system represented abstractly by the redox loop. Thus, this redox loop type of group translocation complex was considered to transform the (scalar) redox potential difference between the A/AH_2 and B/BH_2 couples present on the right into a hydrogenmotive force to the left and an electromotive force to the right, adding up to a protonmotive force (and a corresponding *net* flow of protons) to the left.[38–41]

In the case of the photoredox chain, it was assumed that the electron-carrying arm of the redox loop could include a photoelectric reaction so that the photoelectric potential could be transformed into the protonmotive force.[38–40]

There was a danger that the need to present the concept of the redox loop in a somewhat abstract form for the sake of explicitness and simplicity in defining general principles might detract from the realization of the functional detail involved. Therefore it was emphasized that, in accordance with fundamental observations by Chance, Holmes, Higgins and Connelly[42] on the bimolecular characteristics of respiratory carrier interactions, it was necessary to attribute both hydrogen and electron translocation largely to the specific mobilities of hydrogen-carrying and electron-carrying groups in the catalytic carrier complexes.[38,39]

The idea of specific mobility in the group-translocation complex was related to the requirement that the coupling between the redox reaction and the translocation process in a redox loop must be specified on the one hand by the uniqueness of the primary oxido-reduction changes, and on the other hand by the uniqueness of the translocation reactions, channelled by the specific packing and secondary articulations in or between the enzymic and/or catalytic-carrier components of the redox loop complex.[38,39] For example, it was pointed out that the redox couple $NAD^+/(NADH + H^+)$ could act as a carrier of hydrogen groups (2H) or of hydride groups (H^-), depending upon whether the essentially cyclic translocation process catalysed was that of ($NADH + H^+$ minus NAD^+) translocation, as follows:

$$2H \longrightarrow \left(\begin{array}{c} \nearrow (NADH + H^+) \searrow \\ \\ \nwarrow NAD^+ \swarrow \end{array} \right) \longrightarrow 2H \qquad (10)$$

or (NADH minus NAD^+) translocation, as follows:

$$H^- \longrightarrow \left(\begin{array}{c} \nearrow NADH \searrow \\ \\ \nwarrow NAD^+ \swarrow \end{array} \right) \longrightarrow H^- \qquad (11)$$

Likewise, it was pointed out[38,39] that a dithiol (R_1–SH, R_2–SH) could act as an electron carrier if the translocational specificities were selective for the disulphide and deprotonated dithiol species only, as follows:

$$2e^- \longrightarrow \left(\begin{array}{c} \nearrow (R_1\text{—}S^-, R_2\text{—}S^-) \searrow \\ \\ \nwarrow (R_1\text{—}S\text{—}S\text{—}R_2) \swarrow \end{array} \right) \longrightarrow 2e^- \qquad (12)$$

Proton Translocating ATPase

In the ATPase systems of mitochondria, chloroplasts and bacteria, the effective translocation of protons was attributed to the specific translocation of OH^- or O^{2-} groups.[37–40] It was suggested that this could occur by the following type of cyclic translocation reaction involving cyclically mobile groups X–I and XH + IO^- or X^- + IO^- in the translocation system, which would come reversibly into equilibrium with water on the left but with ATP (and not with water) via a kinase on the right:

$$\text{Type I} \qquad \begin{array}{ccccc} H^+ & \leftarrow & XH + IO^- & \rightarrow & ATP + H^+ \\ & \times & & \times & \\ H_2O & \rightarrow & X - I & \leftarrow & ADP + POH \end{array} \qquad (13)$$

$$\text{Type II} \qquad \begin{array}{ccccc} 2H^+ & \leftarrow & X^- + IO^- & \rightarrow & ATP + 2H^+ \\ & \times & & \times & \\ H_2O & \rightarrow & X - I & \leftarrow & ADP + POH \end{array} \qquad (14)$$

By analogy with the case of the redox loops, the proposed ATPase systems could be of types I or II, translocating one or two protons respectively per ATP hydrolysed, depending on the specificity of the translocation channel for XH + IO^- (corresponding to OH^- translocation) or for X^- + IO^- (corresponding to O^{2-} translocation).

Chemiosmotic Coupling

Considerations of the conditions required to enable enzyme-catalysed group translocations to produce thermodynamically macroscopic metabolic effects led to the recognition of two possible types of chemiosmotic coupling, one applying to molecular-level enzyme associations, and the other applying to associations of enzymes in subcellular vesicles.[24] In the molecular-level (or microscopic) type of chemiosmotic coupling, described by Dr. Moyle and me,[24] enzyme molecules or sub-units catalysing two consecutive reactions were supposed to be associated in pairs so that a microscopic internal phase was enclosed between them. A product of the first reaction was supposed to be translocated through the active centre of the first enzyme (or sub-unit) into the internal phase so that it could gain access to and react via the active centre of the second enzyme (or sub-unit) faster than it escaped into the neighbouring medium. In this case, chemiosmotic coupling would be achieved through the osmotic "thrust" of the common intermediate sequestered in the microscopic internal phase—as we suggested might, for example, be the case in the coupling between isocitrate oxidation to oxalosuccinate and oxalosuccinate decarboxylation to ketoglutarate in isocitrate dehydrogenase.[24]

In the subcellular vesicle (or macroscopic) type of chemiosmotic coupling, proposed as the direct means of coupling redox reactions to dehydration reactions in oxidative and photosynthetic phosphorylation systems,[37] the topological arrangement was supposed to be fundamentally similar to that of the molecular-level type, but on a larger scale. Thus, the membrane-bound respiratory chain systems of mitochondria and bacteria, and the membrane-bound photoredox chain systems of chloroplasts and bacteria, were supposed to consist of assemblies of proton-translocating redox loops; and several such assemblies, present in the topologically closed membrane vesicles of mitochondrial cristae, bacterial protoplasts, chloroplast grana or derived "particles" were taken to contribute to the establishment of the total effective proton flux and to the protonmotive force across the so-called coupling membrane of the vesicles. The resulting "thrust" of protons through the reversible proton-translocating ATPase assemblies, also present in the coupling membrane, was supposed to give rise to ATP synthesis by reversal of the ATPase reaction.[37–40]

The effect of scale in the macroscopic type of chemiosmotic coupling introduced a more fundamental difference from the microscopic type than might have been expected, because it meant that there need be no direct physical (or chemical) contact between the redox chain assemblies and the ATPase assemblies in systems catalysing oxidative and photosynthetic phosphorylation. It also meant that the coupling membrane, of low proton conductance, in which the proton-translocating assemblies were assumed to be situated, would be an essential component of the normal coupled oxidative and photosynthetic phosphorylation systems.[37] Further, it followed that specific proton-coupled solute-translocation systems would be required in the coupling membrane to maintain osmotic stability[37] and to facilitate the entry and exit of appropriate substrates.[29, 38, 41]

In this brief paper it is not possible to give a meaningful survey of the experimental work that has been carried out with the object of defending or refuting one or another of the proposed mechanisms of coupling in oxidative and photosynthetic phosphorylation systems; but the review by the late G. D. Greville[43] provides an excellent main source of information, which can readily be supplemented by more recent reviews.[44–49] The wealth of detail has tended, however, to divert attention from some rather general considerations about the biochemical relationships between the parts of the different feasible coupling mechanisms that should, I believe, make it rather easier than has commonly been supposed to distinguish the macroscopic chemiosmotic coupling mechanism from other possible mechanisms.

Comparisons between Chemical, Conformational and Chemiosmotic Coupling Mechanisms

Referring back to equation (5), the fundamental question of so-called "energy transduction" can be answered in biochemically realistic terms only when we can specify how the flow of redox particles (electrons and hydrogen atoms) in the redox chain system is connected by actual physicochemical interactions with the flow of the elements of water (electrons, hydrogen atoms and oxygen atoms) through the reversible ATPase system. In other words, what are X and $\sim$X in equation (5), and how do they participate in the redox-chain and reversible ATPase systems?

As illustrated by equation (4), in substrate-level oxidative phosphorylation, the flow of redox particles through the redox system on the left is coupled to the flow of phosphoryl into (or H_2O out of) the phosphorylation system on the right because the phosphorylated intermediate (PGAP) that passes between the two systems contains both the redox functional acyl group and the phosphoryl group linked together covalently in the same small molecule. The pattern of events is not quite the same as that illustrated by equation (5), because the dehydration and oxidoreduction systems overlap in equation (4); but the fundamental coupling principle, depending on the actual coupling between the flows of the chemical particles, is identical.

It is noteworthy that in substrate-level phosphorylation, the redox system and the ADP phosphorylation system, which are linked by the diffusion of the intermediate substrate, are physically separate from one another, although they are both present in the same aqueous phase. Likewise, in the macroscopic type of chemiosmotically coupled oxidative phosphorylation, the proton-translocating redox chain or photoredox chain system, and the proton-translocating ATPase system, which are linked by the diffusion of the intermediate H^+ ion, are physically separate from one another, although they are both present in the same non-aqueous coupling membrane between the two aqueous phases. In this case, X and $\sim$X of equation (5) correspond to the H^+ ion at different electrochemical potentials in the two aqueous phases.

In the other conceivable types of coupling mechanism that have seriously been advocated, a direct interaction is required, at the molecular level of dimensions, between certain of the redox chain or photoredox chain components and components of the reversible ATPase. For example, in the microscopic chemiosmotic mechanism[24] advocated by Williams,[50] protons, trapped in microscopic domains, are assumed to mediate between redox activity and ADP phosphorylation at various sites along the respiratory chain. Similarly, in versions of the chemical coupling hypothesis, where it is assumed that non-diffusable derivatives of catalytic components of the redox chain act as intermediates between redox activity and ADP phosphorylation,[51] direct contact is required between the appropriate redox chain components and components of the reversible ATPase. Likewise, according to the conformational coupling hypothesis originally conceived by Boyer[52] and by Green,[53] and developed in different ways by Slater[54] and by Green,[55] segments of the respiratory chain and components of the reversible ATPase are supposed to undergo related conformational and local electrical changes during redox and reversible ATPase activity, and coupling is attributed to the direct transmission of forces and flows by intimate contact between the respiratory chain segments and the reversible ATPase components.

In all the "direct interaction" mechanisms of coupling, the cyclic transition between X and $\sim$X in equation (5) is represented by the cyclic change of chemical or physical state and configuration in the specific domain of contact assumed to exist between components of the redox (or photoredox) chain and the reversible ATPase. Thus, according to these mechanisms, the respiratory chain (and photoredox chain) should be chemically duplex, and in view of the intimate association assumed between the redox and reversible ATPase systems, it would be expected that normal biochemical fractionation methods should provide evidence for well-defined structural associations and functional interrelationships between the redox (or photoredox) and ATPase components. However, as I have pointed out before,[56] and re-emphasized in a recent review,[57] Keilin's chemically simple concept of the respiratory chain, as a hydrogen- and electron-carrying system, was almost universally rejected in the nineteen fifties in favour of a chemically duplex type of system, although there was never any experimental support for this remarkable swing of opinion. Despite earlier claims (see ref. 56), there is no experimental evidence for a *direct* influence of the redox state of components of the respiratory chain (or photoredox chain) on the activity of the ATPase, and there is no evidence for a *direct* influence of the poise of the ATPase on redox activity through the chain. Nor is there any evidence of *direct* associations between components of the redox chain (or photoredox chain) and components of the reversible ATPase system, such as would be required by the "direct interaction" type of coupling mechanism; even though there were strong incentives for finding such associations, as illustrated, for example, by the work of Racker and co-workers.[58–62]

It is now generally acknowledged as a fact (see refs. 43–49) that coupling between the redox (or photoredox) chain system and the reversible ATPase system is dependent on the osmotic intactness of the topologically closed coupling membrane in which the two systems are situated, and that the classical uncoupling agents, such as 2,4-dinitrophenol, act by specifically equilibrating the electrochemical potential of H^+ ions across the coupling membrane—in accordance with a chemiosmotic type of coupling mechanism.[37] Slater[63] has sought to reconcile the "direct interaction" type of coupling mechanism with these facts by assuming the existence of a proton pump, separate from the redox and ATPase systems, but actuated by an energy-rich intermediate in equilibrium with the redox and ATPase systems (see Chappell and Crofts[64] and Lardy, Connelly and Johnson[65]). Thus, in complete oxidative or photosynthetic phosphorylation systems, the energy-rich intermediate (or state) would be discharged by activity of this superimposed proton pump if the coupling membrane were broken or made specifically permeable to H^+ ions. However, one would expect that, by the use of suitable inhibitors or by partial resolution of the system, it would be possible to eliminate the activity of the superimposed proton pump and thus obtain coupled systems that would be independent of the osmotic integrity of the coupling membrane and unaffected by the proton-conducting uncoupling agents.

At all events, one finds it hard to accept that it would have been beyond the wit of all the biochemists engaged in studying the coupling mechanism in oxidative and photosynthetic phosphorylation during the last forty years to demonstrate a single relevant *direct* functional interaction or structural complexation between components of the redox (or photoredox) and ATPase systems, if such direct interactions and complexations are actually responsible for "energy coupling" in the natural systems.

Some Perspectives on Future Developments: Design Principles for Redox Loops and Cation-Translocating ATPase Systems

At the end of their recent review on energy conservation in mitochondria, van Dam and Meyer[46] have remarked that it may be appropriate to caution the proponents of the different hypotheses with the words of the Greek philosopher Myson: "It should not be our aim to explain facts in the light of arguments, but to argue on the basis of facts." I would wholeheartedly agree with this sentiment as long as it is remembered that facts are not the same as randomly selected experiments, and that facts are formulated in words and other symbols and emerge from the raw experimental data only in the light of arguments. It is in this spirit that I have thought it useful to present the following comments on some possible interpretative applications of the chemiosmotic coupling rationale.

In my attempts to assign known carriers of the respiratory chain and photoredox chain to appropriate locations in the redox loop systems, I have begun with the assumption that natural electron acceptors, such as haem groups of cytochromes and iron–sulphide groups of non-haem iron proteins, would probably function as electron translocators, and that natural hydrogen acceptors such as ubiquinone, plastoquinone and the flavin groups of flavoproteins, would probably function as hydrogen translocators.[37–40] However, as indicated by equations (10) to (12), it was also explicitly recognized[38,39] that the translocational specificities could affect the overall translocation reaction by determining the degree of protonation of the species undergoing translocation; and that, as in equation (10), functional groups of the carrier (such as the phosphate groups of NAD), which were not directly involved in the redox changes, could participate in proton translocation. Thus, my suggestion for one possible mechanism for Loop I of the mitochondrial respiratory chain[38,39] may be represented as follows:

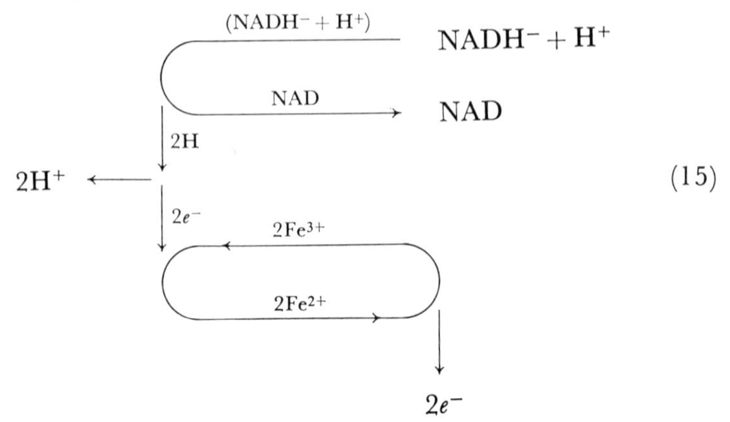

where $NADH^-$ and NAD represent the actual ionization states of the nicotinamide nucleotides in water near pH 7, $(NADH^- + H^+)$ represents $NADH^-$ with the phosphate group fully protonated in the translocation complex, and Fe stands for the iron atom of a non-haem iron protein in the NADH dehydrogenase.

It was not possible to account for proton translocation by the "energy-linked" nicotinamide nucleotide transhydrogenase (Loop O) in terms of a normal type II loop (see equation 9), because this was not compatible with the observed transfer of tritium between the nicotinamide nucleotides without dilution in the water hydrogen-pool.[38, 39] However, I pointed out at a meeting of the Biochemical Society in 1969, during the discussion of a paper on the "energy-linked" transhydrogenase,[66] that the principle which I employed to account for translocation of 2H atoms by NAD in Loop I provided the basis for a mechanism of proton translocation in Loop O, according to the following type of scheme:

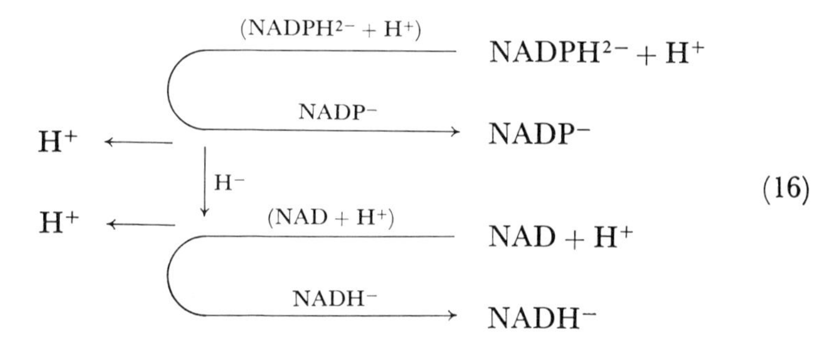

(16)

where $NADPH^{2-}$, $NADP^-$, $NADH^-$ and NAD represent the actual ionization states of the nucleotides in water near pH 7 and the bracketed forms represent the protonated species in the specific translocation complexes. The same general principle as that used above was also used by Skulachev[67] to suggest a proton-translocating transhydrogenase mechanism which was fundamentally similar to that of equation (16), but rather more complicated in detail. Incidentally, it is noteworthy[68] that the transhydrogenase system may contain a flavin nucleotide (FAD) and that this might possibly be included as shown in the scheme of equation (17) (see top of p. 20).

As suggested previously,[41] hydrogen translocation in Loop I might be mediated by FMN rather than by NAD, and likewise FAD could conceivably be involved rather than NAD or NADP in the type of Loop O mechanism represented by equation (16). The general principle of the proton-translocating mechanism would be the same.

Referring to equation (14), representing the proton-translocating ATPase of type II, since there are no known anhydrides formed prior

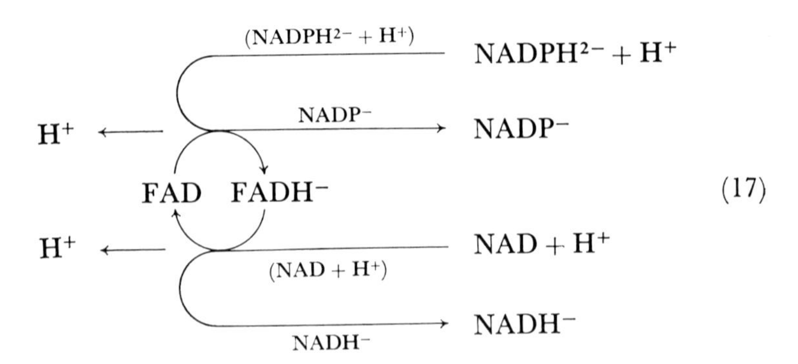

to ATP in oxidative and photosynthetic phosphorylation systems, and by analogy with the suggestions mentioned above for proton-translocating redox loops, it is possible that X–I, X^- and IO^- might be represented by ATP, ADP and inorganic phosphate. Accordingly, equation 14 might involve the adenine nucleotide and phosphate in proton translocation in some such way as the following:

$2H^+$ ←	MgH_2ATP ←	$ATP^{4-} + Mg^{2+} + 2H^+$	
	$H_2PO_4^-$ →	$HPO_4^{2-} + H^+$	(18)
H_2O →	$MgADP^-$ →	$ADP^{3-} + Mg^{2+}$	

where ATP^{4-}, ADP^{3-} and HPO_4^{2-} represent the actual states of ionization of ATP, ADP and phosphate in the right hand aqueous phase near pH 8 and MgH_2ATP and $MgADP^-$ represent the corresponding magnesium salts in the specific translocation complexes.

Figure 1A illustrates the suggested functions [in the process represented by equation (18)] of the biochemical components of the ATPase known as F_0 and F_1 in the terminology of Racker.[59] The The component F_1 is here represented as the natural ATPase, accessible to H^+ (and H_2O) in the complex with F_0, only from the left hand phase; and the region shown crudely as a hole through F_0 represents a specific proton-conducting pathway that can be blocked by oligomycin. Figure 1B illustrates, for comparison, my previous suggestion[38–40] according to which F_1 is an XI synthetase which only shows (artificial) ATPase activity when accessible to H_2O (in place of XI) on the left after dissociation from the F_0 component, as illustrated in Fig. 2.

It is interesting to compare equation (18) with a corresponding type of mechanism speculatively suggested for the $(Na^+ + K^+)$-activated

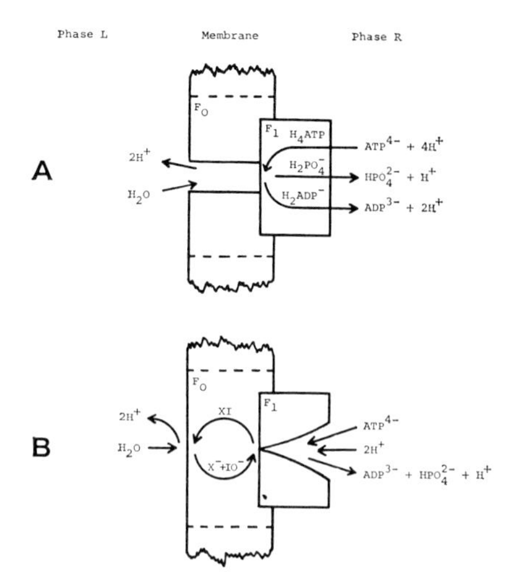

Figure 1. Diagrams of proton-translocating ATPase II: A, in which F_1 is represented as a natural ATPase catalysing proton translocation by the translocation of the adenine nucleotides and phosphate in appropriate protonation states; and F_0 is a locator of F_1, having a specific proton-conducting pathway through it. B, in which F_1 is represented as an XI synthetase and F_0 is a locator of F_1 having XI hydrolase activity and providing specific translocation channels for the functional groups XI and $X^- + IO^-$.

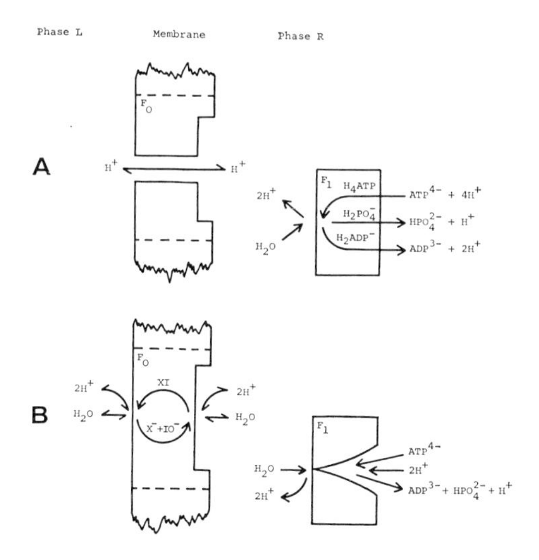

Figure 2. Diagrams of proton-translocating ATPase II, as in Fig. 1, but dissociated into separate F_0 and F_1 components, to illustrate the oligomycin-sensitive proton-conducting activity of F_0, and the natural (A) or artificial (B) ATPase activity of F_1 (see ref. 40).

ATPase, involving the adenine nucleotides directly in Na^+ and K^+ translocation,[29] in some such way as the following:

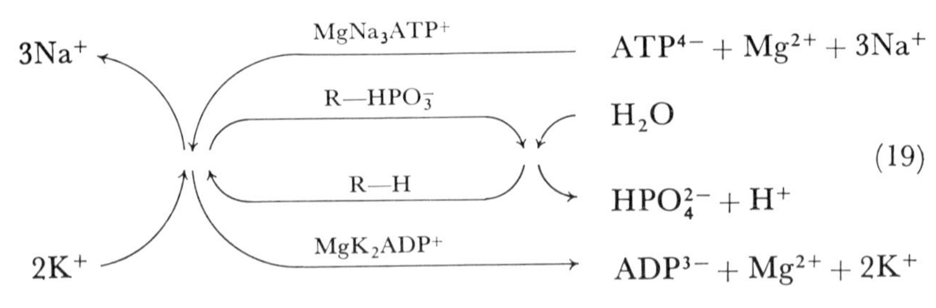

where R–HPO_3^- represents the phosphorylated intermediate in the enzyme and the other conventions are as before.

The above comments, concerning the possible involvement of some nucleotide coenzymes in proton-translocation reactions, cannot be discussed fully here; but it is hoped that the general principles illustrated by the specific examples that I have given will help to provide new models for the interpretation of known experimental data, and also that they will help to enable new experiments to be designed so as to establish new unequivocal facts. In this context it is relevant to draw attention to the fact that the catalytic complexes in the systems catalysing oxidative and photosynthetic phosphorylation, which we may legitimately regard as "miniature engines", have actually been designed by the evolutionary process of natural selection.[35, 36, 42] Therefore, this brief discussion of some of the design principles that may possibly be involved in the redox loop and ATPase systems is not merely an academic exercise, but, if correct, should be related to the actual biochemical potentialities exploited by natural selection.

Acknowledgements

I would like to thank my research colleagues Dr. Jennifer Moyle and Dr. Ian West for helpful discussion and criticism, and Miss Stephanie Phillips for expert secretarial assistance, during the preparation of this paper. I am also indebted to Glynn Research Ltd. for general financial support.

References

1. P. Mitchell, *J. Gen. Microbiol.*, **9** (1953) 273.
2. P. Mitchell, *Soc. Exp. Biol. Symp.*, **8** (1954) 254.
3. P. Mitchell, *J. Gen. Microbiol.*, **11** (1954) 73.
4. P. Mitchell, *J. Gen. Microbiol.*, **11** (1954) x.
5. P. Mitchell and J. Moyle, *Biochem. J.*, **64** (1956) 19P.
6. E. C. Slater and K. W. Cleland, *Biochem. J.*, **53** (1953) 557.
7. D. E. Green, L. H. Stickland and H. L. Tarr, *Biochem. J.*, **28** (1934) 1812.
8. D. E. Green, in: *Perspectives in Biochemistry*, J. Needham and D. E. Green (eds.), University Press, Cambridge, 1937, p. 175.
9. F. Lipmann, *Advan. Enzymol.*, **1** (1941) 99.

10. O. Warburg and W. Christian, *Biochem. Z.*, **303** (1939) 40.
11. E. Racker and I. Krimsky, *J. Biol. Chem.*, **198** (1952) 731.
12. F. Lipmann, in: *Currents in Biochemical Research*, D. E. Green (ed.), Interscience Publishers Inc., New York, 1946, p. 137.
13. S. Glasstone, K. J. Laidler and H. Eyring, *The Theory of Rate Processes*, McGraw-Hill, New York and London, 1941.
14. L. Pauling, *Ann. Rep. Smithsonian Institute*, (1950) 225.
15. H. Lundegardh, *Ark. Bot.*, **32A** (1945) 1.
16. R. E. Davies and A. G. Ogston, *Biochem. J.*, **46** (1950) 324.
17. E. J. Conway, *Intern. Rev. Cytol.*, **2** (1953) 419.
18. R. N. Robertson, *Biol. Rev.*, **35** (1960) 231.
19. R. N. Robertson, *Protons, Electrons, Phosphorylation and Active Transport*, Cambridge University Press, Cambridge, 1968.
20. P. Curie, *J. Phys.*, *3ème Ser.*, (1894) 393.
21. E. A. Guggenheim, *Modern Thermodynamics by the Methods of Willard Gibbs*, Methuen, London, 1933.
22. T. Rosenberg, *Acta Chem. Scand.*, **2** (1948) 14.
23. P. Mitchell, *Nature*, **180** (1957) 134.
24. P. Mitchell and J. Moyle, *Nature*, **182** (1958) 372.
25. P. Mitchell, *Biochem. Soc. Symp.*, **16** (1959) 73.
26. P. Mitchell, *Ann. Rev. Microbiol.*, **13** (1959) 407.
27. I. Langmuir, *Science*, **87** (1938) 493.
28. I. Langmuir, *Proc. Roy. Soc. A*, **170** (1939) 1.
29. P. Mitchell, *Biochem. Soc. Symp.*, **22** (1963) 142.
30. D. E. Koshland, *Advan. Enzymol.*, **22** (1960) 45.
31. G. G. Hammes, *Nature*, **204** (1964) 342.
32. H. Muirhead and M. F. Perutz, *Nature*, **199** (1963) 633.
33. P. Mitchell, in: *Comprehensive Biochemistry*, M. Florkin and E. H. Stotz (eds.), Vol. 22, Elsevier, Amsterdam, 1967, p. 167.
34. P. Mitchell, *Advan. Enzymol.*, **29** (1967) 33.
35. P. Mitchell, in: *Membranes and Ion Transport*, E. E. Bittar (ed.), Vol. 1, Wiley, London, 1970, p. 192.
36. P. Mitchell, *Symp. Soc. Gen. Microbiol.*, **20** (1970) 123.
37. P. Mitchell, *Nature*, **191** (1961) 144.
38. P. Mitchell, *Chemiosmotic Coupling in Oxidative and Photosynthetic Phosphorylation*, Glynn Research, Bodmin, 1966.
39. P. Mitchell, *Biol. Rev.*, **41** (1966) 445.
40. P. Mitchell, *Federation Proc.*, **26** (1967) 1370.
41. P. Mitchell, *Chemiosmotic Coupling and Energy Transduction*, Glynn Research, Bodmin, 1968.
42. B. Chance, W. Holmes, J. Higgins and C. M. Connelly, *Nature*, **182** (1958) 1190.
43. G. D. Greville, in: *Current Topics in Bioenergetics*, D. R. Sanadi (ed.), Vol. 3, Academic Press, New York, 1969, p. 1.
44. L. Packer, S. Murakami and C. W. Mehard, *Ann. Rev. Plant Physiol.*, **21** (1970) 271.
45. D. A. Walker and A. R. Crofts, *Ann. Rev. Biochem.*, **39** (1970) 389.
46. K. van Dam and A. J. Meyer, *Ann. Rev. Biochem.*, **40** (1971) 115.
47. B. Chance and M. Montal, in: *Current Topics in Membranes and Transport*, F. Bronner and A. Kleinzeller (eds.), Vol. 2, Academic Press, New York and London, 1971, p. 99.
48. V. P. Skulachev, in: *Current Topics in Bioenergetics*, D. R. Sanadi (ed.), Vol. 4, Academic Press, New York, 1971, p. 127.
49. R. A. Dilley, in: *Current Topics in Bioenergetics*, D. R. Sanadi (ed.), Vol. 4, Academic Press, New York, 1971, p. 237.
50. R. J. P. Williams, in: *Current Topics in Bioenergetics*, D. R. Sanadi (ed.), Vol. 3, Academic Press, New York, 1969, p. 79.
51. B. T. Storey, *J. Theoret. Biol.*, **28** (1970) 233.
52. P. D. Boyer, in: *Oxidases and Related Redox Systems*, T. E. King, H. S. Mason and M. Morrison (eds.), Vol. 2, Wiley, New York, 1965, p. 994.
53. D. E. Green, in: *Oxidases and Related Redox Systems*. T. E. King, H. S. Mason and M. Morrison (eds.), Vol. 2, Wiley, New York, 1965, p. 1012.
54. E. C. Slater, C. P. Lee, J. A. Berden and H. J. Wegdam, *Biochim. Biophys. Acta*, **223** (1970) 354.
55. D. E. Green, *J. Bioenergetics*, **3** (1972) 159.
56. P. Mitchell, *Chemiosmotic Coupling in Oxidative and Photosynthetic Phosphorylation*, Glynn Research, Bodmin, 1966, p. 150.
57. P. Mitchell, in: *Energy Transduction in Respiration and Photosynthesis*, E. Quagliariello, S. Papa and C. S. Rossi (eds.), Adriatica Editrice, Bari, 1971, p. 123.

58. E. Racker, *Biochem. Biophys. Res. Communs*, **14** (1964) 75.
59. E. Racker, *Mechanisms in Bioenergetics*, Academic Press, New York, 1965.
60. E. Racker, *Federation Proc.*, **26** (1967) 1335.
61. E. Racker, C. Burstein, A. Loyter and R. O. Christiansen, in: *Electron Transport and Energy Conservation*, J. M. Tager, S. Papa, E. Quagliariello and E. C. Slater (eds.), Adriatica Editrice, Bari, 1970, p. 235.
62. A. F. Knowles, R. J. Guillory and E. Racker, *J. Biol. Chem.*, **246** (1971) 2672.
63. E. C. Slater, *European J. Biochem.*, **1** (1967) 317.
64. J. B. Chappell and A. R. Crofts, in: *Regulation of Metabolic Processes in Mitochondria*, J. M. Tager, S. Papa, E. Quagliariello and E. C. Slater (eds.), Elsevier, Amsterdam, 1966, p. 293.
65. H. A. Lardy, J. L. Connelly and D. Johnson, *Biochemistry*, **3** (1964) 1961.
66. J. Rydström, A. Teixeira da Cruz and L. Ernster, *Biochem. J.*, **116** (1970) 12P.
67. V. P. Skulachev, *FEBS Letters*, **11** (1970) 301.
68. P. T. Cohen and N. O. Kaplan, *J. Biol. Chem.*, **245** (1970) 2825.

Solution of the Problem of Energy Coupling in Terms of Chemiosmotic Theory

V. P. Skulachev

Department of Bioenergetics, Laboratory of Bioorganic Chemistry, Moscow State University, Moscow, USSR

Summary

The present state of the chemiosmotic hypothesis of oxidative phosphorylation is considered. It is pointed out that the available data testify to the validity of the following postulates of this hypothesis:

(1) Energization of coupling membranes results in formation of a transmembrane electric potential and/or a pH difference whose values prove to be of the same order of magnitude as standard free energy of ATP hydrolysis.

(2) The redox chain can generate a membrane potential independently of whether or not high-energy intermediates are formed.

(3) ATPase can generate a membrane potential independently of whether or not mechanisms of electron transfer via coupling sites are operative.

(4) Energy accumulated in the form of transmembrane electric and osmotic gradients can be utilized for ATP synthesis ("ion transfer phosphorylation").

The observations summarized in these items are sufficient to conclude that electron transfer and phosphorylation can be coupled by a membrane potential, as was postulated by the chemiosmotic theory.

It is noted that a number of consequences of Mitchell's principle of energy coupling are also experimentally proved. It was shown, in particular, that

(a) an increase in electric conductance and ion permeability, initially very low for coupling membranes, results in uncoupling of oxidative phosphorylation;

(b) electron (hydrogen) transfer in some segment(s) of the respiratory chain is directed across the membrane;

(c) energy-linked transhydrogenase represents reverse electron transfer via the additional (fourth) site of the redox chain energy coupling, etc.

Thus, chemiosmotic theory of oxidative phosphorylation seems to be acceptable as working hypothesis for the further study of the mechanism of oxidative phosphorylation.

Introduction

The problem of electron transfer phosphorylation was first formulated in 1939 by Belitser and Tsibakova[1] who introduced the "P:O" criterion and showed that this ratio is more than 1. The authors noted that such a high value of the phosphorylation coefficient suggests a fundamental difference in the mechanisms of ATP formation coupled with respiration, and glycolysis, since in the latter case, the amount of the ATP synthesized is equal to that of the substrate utilized.

A lot of hypothetical schemes were put forward to explain the nature of coupling between electron transfer and phosphorylation, but none of them solved the problem. Only quite recently, one hypothetical scheme of energy coupling, viz. Mitchell's chemiosmotic concept,[2, 3] was supported by experimental data which allow us to prefer it to alternative possibilities. In this paper, I shall try to substantiate the statement that oxidation and phosphorylation can be coupled via a membrane potential as was postulated by Mitchell.

Membrane Potential and Some Other Postulates of Chemiosmotic Theory

Formation of a membrane potential in coupling membranes is confirmed by the following independent lines of evidence:

(1) Energization of coupling membranes results in the appearance of transmembrane fluxes of penetrating ions, cations and anions being transferred in the opposite directions. The only requirement for the ion to be involved in the respiration or ATP-dependent translocation is the ability to penetrate through the membrane. If the ion does not penetrate the membrane, its energy-dependent transport can be induced by any treatment increasing the specific membrane permeability towards this ion.

The most demonstrative in this respect are experiments with synthetic unnatural ions readily penetrating through phospholipid membranes such as phenyl dicarbaundecaborane (PCB^-),* tetraphenyl boron, picrate anions and dibenzyl dimethyl ammonium, tetrabutylammonium, triphenyl methyl phosphonium cations. As it was found by Dr. E. A. Liberman's and our groups, the anions mentioned are extruded from mitochondria and are taken up by "inside out" submitochondrial particles in an energy-dependent manner, whereas cations are moved in the opposite direction.[4, 5] This observation defined as "transmembrane electrophoresis"[6] was confirmed by Papa *et al.*[7] using the penetrating anion of thiocyanate, and by Chance and co-workers[8] who applied tetraphenyl boron anion and tetraphenyl arsonium cation. It is very probable, that the list of synthetic ions

* Abbreviations: *p*-trifluoromethoxycarbonylcyanide phenylhydrazone, FCCP; phenyl dicarbaundecaborane, PCB^-; anilinonaphthalenesulfonate, ANS^-.

which can be involved in the energy-dependent transport may be supplemented by such compounds as ANS^-, auramine O, ethidium bromide,[9–11] pyrenesulfonate[12] and some others.

The list of foreign ionophores inducing energy-dependent movement of non-penetrating ions includes now, besides antibiotics of the valinomycin type, dipentafluorophenyl mercury and diborenyl mercury for iodide anion (as was shown by Liberman *et al.*[13]), tetraphenyl boron and PCB^-—for tetraethyl- and tetramethyl ammonium (according to the data of this laboratory[14]) and some others. The characteristics of the process of charge-specific transport of ions of different structure strongly suggests its electrophoretic nature.

(2) The diffusion potential of penetrating ions induces the same changes in the coupling membrane parameters as energization.

The most important observation along this line is the discovery of "ion transfer phosphorylation". The phenomenon in question consists in energization of the membrane and ATP synthesis coupled with diffusion of ions down their concentration gradients. Ion transfer phosphorylation was found to be coupled with the H^+ efflux from chloroplasts,[15] K^+ efflux from mitochondria,[16] Na^+ influx into and K^+ efflux from erythrocytes,[17] and Ca^{2+} efflux from sarcoplasmic reticulum.[18]

As was found in our laboratory by Jasaitis and Kuliene[19] and independently in Italy by Azzi and co-workers,[10] diffusion potential of K^+ ions induces the same changes in ANS^- fluorescence in mitochondria and particles as does energization. The linear relationship between values of the membrane potential and the ANS^- responses was revealed. The same effect was found in this group[19] when a membrane potential was generated by H^+ diffusion down pH gradient in the presence of a protonophorous uncoupler. The suggestion that the ANS^- probe indicates the formation of a membrane potential was directly confirmed recently by Changeux and associates[21] who showed that the ANS^- fluorescence depends linearly upon the electric potential difference generated across the cell membrane of *Electrophorus electricus* by the microelectrode current-clamp technique.

According to Jackson and Crofts,[20] the effects of diffusion potentials of K^+ and H^+ imitate those of energization on the carotenoid spectrum in chromatophores of photosynthetic bacteria. Dr. V. D. Samuilov in our laboratory observed similar relationships for spectral changes of bacteriochlorophyll. In Slater's laboratory, the red shift of the cytochrome *b* spectrum observed previously under energization was found to be induced by the diffusion potential of SCN^-, NO_3^- and I^- anions in the appropriate direction which was opposite in mitochondria and "inside-out" particles.[22]

(3) Agents discharging the membrane potential de-energize coupling membranes. This effect was demonstrated with various

detergents, penetrating ions, ionophorous antibiotics and a wide range of synthetic compounds increasing proton conductance of membranes. The uncoupling efficiency of these compounds was found to depend linearly on their efficiency as proton conductors in artificial phospholipid membranes.* [23, 24]

These three lines of evidence were obtained for all types of coupling membranes, namely for inner mitochondrial membrane, chromatophore membrane, membrane of aerobic bacteria and thylakoid membrane of chloroplasts (for reviews see refs. 24, 26). In the latter case, a quite independent line of evidence for a membrane potential was furnished by Witt, Junge and coworkers measuring "electrochromic" chances in spectra of some pigments and dyes.[27–31]

All these data indicate that a membrane potential can be formed in coupling membranes and its formation is essential for energy coupling as was postulated by Mitchell.[3]

A number of Mitchell's postulates which have been confirmed experimentally are listed below:

(1) *Coupling membranes can transduce chemical energy into electric form.* Accepting this postulate we ought to agree with Mitchell that coupling membranes can perform, besides chemical and osmotic work, one more, third type of energy transduction, namely electric work.

(2) *Formation of the electrochemical potential of H^+ ions ($\Delta\bar{\mu}_{H^+}$) is the primary event in the performance of the osmotic work by coupling membranes.* This conclusion is based, in particular, on observations of pH changes accompanying electrophoretic transport of synthetic and natural ions across coupling membranes. In terms of this concept we can explain the mechanism of "active" transport of ions in mitochondria, chloroplasts and bacteria (for reviews see refs. 3, 14, 24, 26).

(3) *In coupling membranes transduction of chemical energy into electric and osmotic forms is reversible.* This postulate is confirmed by observations of the ion transfer phosphorylation, demonstrating a new principle of bioenergetics, namely reversibility of some energy transductions in the cell. It is noteworthy that all previously described biological processes of energy transductions were irreversible (e.g. transduction of chemical energy of ATP into mechanical energy of actomyosin contraction,

* It is noteworthy that such a relationship can be shown only if phospholipids from mitochondria are used for the formation of the artificial membrane. For example, dicoumarol did not, and dinitrophenol did increase, but too slightly, the proton conductance of the membranes prepared from the total fraction of brain phospholipids; both compounds affect membranes made of mitochondrial phospholipids at concentrations corresponding to their uncoupling activity.[23] Another condition important for studying uncouplers of high lipid solubility is to measure their concentration in the water phase of the incubation mixture with mitochondria. Calculation of the uncoupling efficiency proceeding from the total concentration of uncoupler added to the sample may lead to mistakes since the amount of lipid in sample with mitochondria is much higher than in that with artificial membranes.[23] These two conditions were disregarded in recent work by Ting *et al.*[25] who confirmed the observation that uncouplers can increase the conductance of phospholipid membranes but failed to obtain the quantitative correlation between the uncoupling and conducting activities.

into electric energy of nerve excitation, or into light energy in luciferin–luciferase system).

(4) *The redox chain of coupling membranes involves stages of transmembrane electron and/or hydrogen movement.* This property was demonstrated firstly for the middle part of mitochondrial respiratory chain: it was found that flavins and cytochrome *c* are localized on the opposite sides of the membrane, so reducing equivalents flowing from flavins to cytochrome *c* must traverse the membrane (for reviews see refs. 3, 14, 24, 32, 33).

(5) *Redox chains of mitochondria and chromatophores include not three, but four energy coupling sites, the fourth site being localized between NADPH and NAD⁺.* This postulate was supported by Dr. E. A. Liberman's and our observations demonstrating formation of membrane potential under transhydrogenase reaction.[4,6,13,14,24,33a–35] Recently Van Dam and others obtained an indication of ATP formation coupled with oxidation of NADPH by NAD^+.[36]

(6) *Protonophores uncouple oxidation and phosphorylation.* Confirmation of several of Mitchell's postulates served as important evidence in favour of the chemiosmotic hypothesis. However, it was insufficient to accept this hypothesis *in toto* if one takes into account very high level of complexity of studied experimental objects which represent closed membrane vesicles containing multienzyme assemblies.

As a matter of fact, all the data mentioned above could be explained in terms of, for example, the scheme of Chappell and Crofts[37] postulating the formation of the membrane potential as a special energy-linked mitochondrial function which is alternative to phosphorylation.

Four possible versions of the relationships of redox chain, membrane potential and high-energy intermediates (~) are given below:[24]

redox chain ⟷ $\Delta\bar{\mu}_{H^+}$ ⟷ ~ (I)

redox chain ⟷ ~ ; $\Delta\bar{\mu}_{H^+}$ ⟷ ~ (II)

redox chain ⟵ ~ ; redox chain ⟷ $\Delta\bar{\mu}_{H^+}$ (III)

redox chain ⟷ ~ ; redox chain ⟷ $\Delta\bar{\mu}_{H^+}$ ⟷ ~ (IV)

According to the chemiosmotic theory (scheme I), electron transfer and squiggle formation are coupled via the electrochemical potential of H^+, $\Delta\bar{\mu}_{H^+}$. Scheme II ("hydrolytic proton pump") postulates the squiggle as an intermediate between electron transfer and $\Delta\bar{\mu}_{H^+}$. Scheme III ("redox proton pump") proposes $\Delta\bar{\mu}_{H^+}$ connected with squiggle via the redox chain. Scheme IV represents the combination of any two of the schemes mentioned, so the redox chain, $\Delta\bar{\mu}_{H^+}$: and squiggle are connected in the cyclic manner. To exclude version II it would be enough to demonstrate generation of a membrane potential when no high-energy intermediates are formed. To exclude scheme III it would be enough to show a membrane potential formation when electron transfer in the coupling sites of the redox chain is impossible. Both these questions were recently analysed experimentally and solved in terms of the chemiosmotic theory.

Oxidation-supported Formation of a Membrane Potential with no High-energy Intermediates Involved

Such a possibility was analysed in our group when studying the fourth (transhydrogenase) coupling site. Redox potentials of the electron donor and acceptor operating at this site (NADP and NAD) are similar differing by less than 0·005 V. So, ΔG of the transhydrogenase reaction is determined solely by the ratio of concentrations of substrates and products, i.e. NADPH, NAD^+, $NADP^+$ and NADH. Varying this ratio one can change ΔG within a wide range of values. Keeping the substrate-to-product ratio at a low level it is possible to create conditions unfavourable for synthesis of any high-energy compounds. Under such conditions, transhydrogenase-supported formation of a membrane potential should be hindered if high energy intermediates are involved in this process (scheme II). On the contrary, in terms of schemes I and III a membrane potential might be formed at any substrate-to-product ratio but 1, the direction of potential being changed when this ratio decreases below 1 (reversal of the membrane potential-producing transhydrogenase, reaction 1 from right to left).

Transhydrogenase-supported Formation of $\Delta\mu_{H^+}$

(a) *In terms of the chemiosmotic hypothesis:*

$$NADPH + NAD^+ + 2H^+_{out} \leftrightarrow NADP^+ + NADH + 2H^+_{in} \quad (1)$$

where "out" and "in" mean outside and inside of sonicated submitochondrial particles, respectively.

(b) *In terms of the hydrolytic proton pump hypothesis (scheme II):*

$$NADPH + NAD^+ + XH + YOH \leftrightarrow NADP^+ + NADH + X \sim Y + H_2O \quad (2)$$

$$X \sim Y + H_2O + 2H^+_{out} \leftrightarrow XH + YOH + 2H^+_{in} \quad (3)$$

where $X \sim Y$ a hypothetical high-energy intermediate.

Several years ago Mitchell and Moyle[38] mentioned that the reverse transhydrogenase reaction is accompanied by a pH shift opposite to that coupled with respiration or ATP hydrolysis. However, this preliminary study was not developed because the pH changes were rather small. We studied the same problem using such a sensitive test for membrane potential as the PCB^- probe.[6, 34, 35, 39]

It was found that treatments of submitochondrial particles with substrates of both the forward ($NADPH + NAD^+$) and reverse ($NADH + NADP^+$) transhydrogenase reaction result in the appearance of the PCB^- flows in the particle membrane. Direction of PCB^- movement was determined by that of the transhydrogenase reaction: influx, if the reaction was $NADPH \rightarrow NAD^+$ directed, and efflux, if it was $NADH \rightarrow NADP^+$ directed. In both cases equalization of the concentration of substrates and products as well as addition of uncouplers returned the PCB^- level to the initial one.

In other experiments, we measured the PCB^- responses coupled with the forward transhydrogenase reaction under different substrate-to-product ratios which vary from 2 up to 300. It was observed that the lowering of this ratio results in a decrease in both the rate and amplitude of the PCB^- responses. Nevertheless, measurable PCB^- uptake could be obtained at a substrate-to-product ratio as low as 2. The straight line relationship between the substrate-to-product ratio and the PCB^- response has been revealed. It is important for this discussion that there is no threshold for the PCB^- response. A linear transhydrogenase energy yield dependence of the PCB^- response could be demonstrated at so low values of energy production as a half of kcal/mole of NADPH oxidized. Similar results were obtained when PCB^- was substituted by another penetrating anion, tetraphenyl boron.

It is highly improbable that traces of $X \sim Y$ which might be present in the non-energized particles could be sufficient to observe a measurable rate of reaction (2) running from right to left when $NADP^+$ and NADH are added.* It means that formation of the membrane potential of the opposite polarity coupled with the reverse transhydrogenase reaction ($NADH \rightarrow NADP^+$) must be greatly impeded if the second mechanism is operative.

The second scheme meets similar kinetic difficulties, if dealing with the forward transhydrogenase reaction ($NADPH \rightarrow NAD^+$) producing the membrane potential at low substrate-to-product ratios

* The following calculation gives a rough idea of the $X \sim Y$ concentration under non-energized conditions. When no energy source is available, $X \sim Y$ should be equilibrated with XH and YOH according to the formula:

$$\Delta G^\circ = RT \ln \frac{[XH] \times [YOH]}{[X \sim Y]},$$

where ΔG° is the standard energy of the $X \sim Y$ hydrolysis. ΔG° values for high-energy compounds lie between 7 and 16 kcal/mole. Assuming, e.g., that $\Delta G^\circ = 7$ kcal/mole, and $XH = YOH = 10^{-3}$ M, we obtain, from the above formula, $X \sim Y$ of about 10^{-11} M. If ΔG° is assumed as equal to 16 kcal/mole, $X \sim Y$ should be 10^{-18} M. The $X \sim Y$ concentration per mg of mitochondrial protein proved to be 10^{-17} moles in the former, and 10^{-24} moles in the latter case, values which are far below the level of mitochondrial enzymes (10^{-9}–10^{-10} moles per mg protein).

insufficient to increase significantly the initial, very low level of $X \sim Y$.

So the data obtained testify against the involvement of any high-energy intermediates in the transhydrogenase-linked formation of the membrane potential. It means that transhydrogenase can be a precedent of the redox chain coupling site producing the membrane potential with no high-energy intermediates involved.

A quite different line of evidence testifying against the hydrolysis proton pump hypothesis (scheme II) is connected with the study of the third (cytochrome oxidase) coupling site.

As was found by Hinkle and Mitchell,[40] alkalinization of the incubation medium with CO-treated mitochondria previously equilibrated with the mixture of ferri- and ferrocyanides induces oxidation of cytochrome *a*, while acidification induces its reduction. Both effects required a protonophore (10^{-6} M FCCP was used).

According to scheme I the oxidation of cytochrome *a*, induced by alkalinization of the extramitochondrial space should be explained as a result of the formation of a diffusion potential of H^+ ions which go out from mitochondria down the gradient of H^+ concentration. The sign "plus" of this potential must be outside mitochondria since positive charges (H^+) leave the mitochondrial interior. Electrons from cytochrome *a* situated, after Mitchell, in the middle part of the membrane are transferred to the outer surface of the membrane where cytochrome *c* is localized. In general, this process can be defined as symport of H^+ ions and electrons in the direction from inside to outside the mitochondria.

Such an explanation is not valid if one deals with scheme II. In this case the electron transfer must be directed along the membrane being controlled by the electrochemical potential of H^+ via a system of the $X \sim Y$ hydrolyzing proton pump. It means that electron transfer can be affected by the total $\Delta\bar{\mu}_{H^+}$, or proton motive force, including its both components: electric membrane potential ($\Delta\psi$) and ΔpH. In the presence of uncoupler, changes in pH of the incubation mixture do not result in the formation of any proton motive force since $\Delta\psi$ and ΔpH being energetically equal and oppositely directed are mutually annihilated. For example, addition of alkali leads to the formation of the ΔpH favourable for H^+ efflux. However, such an H^+ efflux must give rise to the appearance of $\Delta\psi$ ("plus" outside mitochondria) which is favourable for H^+ influx. As a result, the total proton motive force should be equal to zero. Therefore, any effects mediated by the proton pump are impossible if a protonophorous uncoupler is present. Since there was no change in the cytochrome *a* reduction when FCCP was omitted, Hinkle and Mitchell concluded that oxidoreduction of cytochrome *a* is affected by $\Delta\psi$ but not by ΔpH, nor by total $\Delta\bar{\mu}_{H^+}$, as was predicted by chemiosmotic theory. This conclusion was confirmed by the observation that the redox potential of cytochrome *a* in

FCCP-treated mitochondria represents a linear function of $\Delta\psi$ calculated from the transmembrane pH difference. Formation of a $\Delta\psi$ equal to 20 mV proved to be necessary for the shift of the cytochrome *a* redox potential by 10 mV as if cytochrome *a* were situated half-way between two membrane surfaces.[40]

The results of Hinkle and Mitchell clearly show that electron transfer between cytochromes *c* and *a* can be an electrogenic process without the assistance of any hydrolytic pumps, utilizing high-energy intermediates. If cytochrome a_3 is localized on the side opposite to that of cytochrome *c*, the terminal part of the respiratory chain could form a membrane potential by the electron transfer oriented across the membrane, as it was postulated by Mitchell.[3]

The data testifying against scheme II were obtained also when studying formation of membrane potentials in photosynthetic systems. Jackson and Crofts[41] reported that photo-induced changes in light absorption in the region of 520 nm which can be observed at liquid nitrogen temperature, may be induced also by creation of a K^+ (or H^+) diffusion potential across the chromatophore membrane. The rate of the light-induced absorption changes was faster than $2 \cdot 10^{-8}$ sec. Both the temperature and time characteristics of the 520 nm response exclude involvement of chemical reactions of synthesis and hydrolysis of high-energy intermediates.

Photoinduced spectral changes in the same region taking less than $2 \cdot 10^{-8}$ sec. were also observed in chloroplasts.[27–31] Unfortunately, no direct evidence was furnished allowing us to decide whether formation of an electric field in the membrane is indispensable for the spectral shift studied.

ATP-supported Formation of a Membrane Potential with no Electron Transfer Involved

The results summarized in the preceding section are hardly compatible with the hydrolytic proton pump hypothesis (scheme II) which is considered usually as an alternative to the chemiosmotic concept (scheme I). However, these data could be explained in terms of another alternative scheme, III. This version requires electron transfer via redox chain loops to be the only mechanism for membrane potential generation. According to such a scheme, there are two redox chains in the coupling membrane: one, oriented along the membrane which forms ATP, and another one, oriented across the membrane which generates the membrane potential. It is proposed that ATP energy is utilized for charging the membrane by the cyclic oxidoreductions involving reverse electron transfer in the first chain followed by forward electron transfer in the second chain. If this were the case, the ATP-supported formation of a membrane potential would be possible only as long as at least one energy coupling site of the redox chain was operative.

This possibility was analysed in our group[42] when studying respiratory chain-deficient particles of beef heart mitochondria (E-particles prepared according to the method of Arion and Racker, see ref. 43). Low temperature spectrophotometry showed that the concentration of cytochrome oxidase in E-particles is negligible, while amounts of cytochromes b, c_1 and c per mg of protein are decreased by factor two or three.

As experiments with PCB^- and ANS^- showed, E-particles cannot form any membrane potential in the first, third and fourth sites of energy coupling of the redox chain. At the same time, ATP hydrolysis produces a membrane potential. Thus, E-particles represent the unique type of submitochondrial fragments which retain an ATP-dependent mechanism of membrane potential generation while three of the four energy coupling sites of the redox chain are absent. Arresting the second site by antimycin, one can obtain a system where none of the coupling sites can catalyse the cyclic electron transfer required for ATP-supported charging of the membrane according to scheme III. Nevertheless, as it was found in the experiments, addition of ATP to the antimycin-treated E-particles induced PCB^- and ANS^- responses testifying to the generation of a membrane potential. Complete reduction or oxidation of redox carriers still present in E-particles, was also without effect on the ATP-supported mechanism of membrane potential production.[42] These data make scheme III highly improbable, if not excluded.

Another line of evidence for this conclusion was obtained in experiments with very different objects, namely promitochondria from anaerobic yeasts. As was shown by Groot, Kovač and Schatz,[44] promitochondria respond by a decrease in ANS^- fluorescence to the addition of ATP. Subsequent treatment with an uncoupler reverses the ATP effect. Valinomycin + K^+ as well as uncouplers inhibit P_i-ATP exchange. All these membrane potential-dependent responses could be demonstrated in spite of the lack of many respiratory chain carriers such as cytochromes b, c_1, c, a, a_3, CoQ, and nonheme iron proteins of $g = 1{\cdot}94$. Promitochondria contained some dehydrogenases and a cytochrome b_5-like pigment, the latter being localized, apparently, in the outer membrane of the organelle. The authors did not determine directly whether a membrane potential might be formed by these redox catalysts. However, this possibility seems to be hardly probable since all the known electron carriers involved in the respiratory chain coupling sites are absent in promitochondria.

Hence, one can conclude that electron transfer via coupling sites is not required for membrane potential formation supported by ATP.

The Proof of the Basic Theorem of the Chemiosmotic Hypothesis

The data presented in the two preceding sections seem to be sufficient to conclude that ATPase and the redox chain are two separate mechanisms for charging the mitochondrial membrane. Taking into account this fact and keeping in mind observations on ion transfer phosphorylation we can point out that basic theorem of the chemiosmotic hypothesis postulating electron transfer and ATP synthesis to be coupled via a membrane potential is experimentally proved.

(1) Coupling membranes can transduce redox chain or ATP energy into membrane potential:

$$\text{redox chain and/or} \sim \rightarrow \Delta\bar{\mu}_{H^+} \tag{4}$$

This item is confirmed, first of all, by the data with penetrating synthetic ions and ANS^-.

(2) The redox chain-supported $\Delta\bar{\mu}_{H^+}$ formation does not require $\sim$:

$$\text{redox chain} \rightarrow \Delta\bar{\mu}_{H^+} \tag{5}$$

as follows from the data on membrane potential formation coupled with reverse transhydrogenase reaction, as well as with the forward one when proceeding under conditions of low energy yields. Another line of evidence of this item originates from Hinkle and Mitchell's experiments on FCCP-required reverse and forward electron-transfers in the cytochrome oxidase region of the chain under alkali or acid bath conditions, respectively.

(3) ATP-supported $\Delta\bar{\mu}_{H^+}$ formation does not require the redox chain:

$$\sim \rightarrow \Delta\bar{\mu}_{H^+} \tag{6}$$

This point is confirmed by the data on antimycin-treated E-particles and yeast promitochondria.

(4) The redox chain and ATPase are localized in the same membrane. Consequently, $\Delta\bar{\mu}_{H^+}$ is their common product:

$$\text{redox chain} \rightarrow \Delta\bar{\mu}_{H^+} \leftarrow \sim \tag{7}$$

(5) Taking into account the reversibility of $\Delta\bar{\mu}_{H^+} \leftrightarrow \sim$ energy transduction (ion transfer phosphorylation) we can modify the latter equation as follows:

$$\text{redox chain} \rightarrow \Delta\bar{\mu}_{H^+} \rightarrow \sim \tag{8}$$

Hence, electron transfer and phosphorylation can be coupled by the membrane potential as was postulated by Mitchell.[3]

Questions Which Should be Settled

The final solution of the problem of chemiosmotic coupling requires two main questions to be settled:

(1) How redox chain charges the membrane?

(2) How ATP-synthetase discharges the membrane in a manner resulting in formation of ATP?

Both these problems remain obscure, making difficult immediate acceptance of Mitchell's theory *in toto*. However, this cannot be a reason to call in question the phenomenon of chemiosmotic coupling of oxidation and phosphorylation just as the existence of a metabolic pathway cannot be denied solely because there are some uncertainties in the action mechanism of enzymes catalyzing partial reactions of the pathway.

Critical analysis of the chemiosmotic theory carried out recently by Slater[45] clearly showed that most of the objections to this concept originate from scarcity of information about details of enzymatic mechanisms of electron and energy transfers in coupling membranes. There have not yet been discovered any facts inconsistent with the idea that oxidation and phosphorylation can be coupled via a membrane potential.

Mitchell's opponents consider the results of calculations of the energetics and stoichiometry of oxidative phosphorylation and ion transport in mitochondria to be the main arguments against the chemiosmotic theory. It was found, in particular, that the equilibrium conditions of ATP synthesis in state 4 mitochondria require the proton motive force to be as high as 370 mV (instead of values of 200–250 mV measured experimentally) if the $P:2\bar{e}$ ratio is assumed to be equal to 2 for each coupling site of respiratory chain.[45] Discussing this question we should keep in mind that (1) $P:2\bar{e}$ is measured in state 3 mitochondria while the value of 370 mV was calculated for state 4, and (2) there are mechanisms of ADP^{3-}–ATP^{4-} antiport down the $\Delta\psi$ gradient and H_3PO_4 transport down the ΔpH gradient in mitochondria; these mechanisms can provide additional energy for phosphorylation of external ADP by external inorganic phosphate. Operation of these systems can decrease the $P:2\bar{e}$ ratio and allow us to explain how a proton motive force of about 200–250 mV maintains the high $[ATP]/[ADP] \times [P_i]$ ratio observed in experiments with state 4 mitochondria.[24, 46] As it was pointed out by Klingenberg,[46] the ADP–ATP antiport system can be transformed, under some conditions, from the electrogenic porter mechanism to an electroneutral one. Switching over these mechanisms it might be possible to regulate the $P:2\bar{e}$ ratio. In any case, the statement, that the proton motive force must be 370 mV under the conditions of $P:2\bar{e} = 2$, seems not to be rigorously substantiated.

Similar uncertainties arise when dealing with calculations of the K^+:ATP ratio in mitochondria accumulating K^+ at the cost of ATP energy or synthesizing ATP under conditions of downhill K^+ efflux.[47]

Ratios of $H^+:\bar{e}$, $K^+:\bar{e}$, and $Ca^{2+}:\bar{e}$ in mitochondria taking up K^+ or Ca^{2+} in a respiration-dependent fashion, were found to be, under some conditions, higher than those to be expected by Mitchell.[24, 32, 45, 47] However, again, these observations are not conclusive, if one tries to use them for disproving the chemiosmotic theory since the nature and action mechanism of the different ion carriers of mitochondrial membrane remain obscure. Movement of charged species, other than H^+, K^+, and Ca^{2+} measured in these experiments and of substrates of oxidation and endogeneous anions of mitochondria, can also hinder the unequivocal explanation of the mentioned data.

Summarizing the present state of the study of oxidative phosphorylation we can conclude that there is now a linear system of logical reasoning supporting the chemiosmotic principle of energy coupling which excludes the alternative schemes of the hydrolytic proton pump and the redox proton pump. If, nevertheless, one would still like to confine oneself to, e.g., the chemical concept of energy coupling, one will be compelled to assume that (1) neither the penetration anion probe in submitochondrial particles, (2) nor 520 nm responses in chloroplasts and chromatophores indicate electric field formation, and (3) electrogenic oxidoreduction between cytochromes *c* and *a* in mitochondrial membranes is not involved in membrane potential generation. These postulates together seem highly improbable.

At the same time it is impossible now to exclude that a chemical mechanism of energy coupling coexists with the chemiosmotic one (scheme IV mentioned above). However, like any compromise this scheme should be considered only when possibilities of the orthodox (chemiosmotic) version have been exhausted.

Acknowledgements

The author cordially thanks Professor P. Mitchell, Drs. L. L. Grinius, A. A. Jasaitis, E. A. Liberman, I. I. Severina and L. S. Yaguzhinsky for useful advice, stimulating discussion and help. He is grateful to Miss T. I. Kheifets for correcting the English version of the paper.

References

1. V. A. Belitser and E. T. Tsibakova, *Biokhimiya USSR*, **4** (1939) 516.
2. P. Mitchell, *Nature*, **191** (1961) 144.
3. P. Mitchell, *Chemiosmotic Coupling in Oxidative and Photosynthetic Phosphorylation*, Glynn Res. Ltd., Bodmin, 1966.
4. L. L. Grinius, A. A. Jasaitis, Yu. P. Kadziaskas, E. A. Liberman, V. P. Skulachev, V. P. Topali, L. M. Tsofina and M. A. Vladimirova, *Biochim. Biophys. Acta*, **216** (1970) 1.
5. L. E. Bakeeva, L. L. Grinius, A. A. Jasaitis, V. V. Kuliene, D. O. Levitsky, E. A. Liberman, I. I. Severina and V. P. Skulachev, *Biochim. Biophys. Acta*, **216** (1970) 13.

6. V. P. Skulachev, *FEBS Letters*, **11** (1970) 301.
7. S. Papa, F. Guerrieri, M. Lorusso and E. Quagliariello, *FEBS Letters*, **10** (1970) 295.
8. M. Montal, B. Chance and C.-P. Lee, *J. Membrane Biol.*, **2** (1970) 201.
9. A. Azzi, *Biochem. Biophys. Res. Commun.*, **37** (1969) 254.
10. A. Azzi, P. Cherardini and M. Santato, *J. Biol. Chem.*, **246** (1971) 2035.
11. A. Azzi, A. Tamburro, E. Gobbi and M. Santato, *7th FEBS Meeting Abstracts*, Varna, 1971, p. 51.
12. J. R. Brocklehurst, R. B. Freedman, D. J. Hancock, and G. K. Radda, *Biochem. J.*, **116** (1970) 721.
13. P. I. Isaev, E. A. Liberman, V. D. Samuilov, V. P. Skulachev and L. M. Tsofina, *Biochim. Biophys. Acta*, **216** (1970) 22.
14. V. P. Skulachev, *Energy Transformations in Biomembranes*, Nauka Press, 1972.
15. A. T. Jagendorf and E. Uribe, *Proc. Natl. Acad. Sci. U.S.*, **55** (1966) 170.
16. R. S. Cockrell, E. J. Harris and B. C. Pressman, *Nature*, **215** (1967) 1487.
17. P. J. Garrahan and I. M. Glynn, *Nature*, **211** (1966) 1414.
18. M. Makinose, *FEBS Letters*, **5** (1971) 269.
19. A. A. Jasaitis, V. V. Kuliene and V. P. Skulachev, *Biochim. Biophys. Acta*, **234** (1971) 177.
20. J. B. Jackson and A. R. Crofts, *FEBS Letters*, **4** (1969) 185.
21. J. Patrick, B. Valeur, L. Monnerie and J.-P. Changeux, *J. Membrane Biol.*, **5** (1971) 102.
22. H. Haaker, I. A. Berden and K. Van Dam, *Biochim. Biophys. Acta*, in press.
23. E. A. Liberman, V. P. Topali, L. M. Tsofina, A. A. Jasaitis and V. P. Skulachev, *Nature*, **222** (1969) 1076.
24. V. P. Skulachev, *Current Topics in Bioenergetics*, **4** (1971) 127.
25. H. P. Ting, D. F. Wilson and B. Chance, *Arch. Biochem. Biophys.*, **141** (1970) 141.
26. E. A. Liberman and V. P. Skulachev, *Biochim. Biophys. Acta*, **216** (1970) 30.
27. W. Junge and H. T. Witt, *Z. Naturforsch.*, **23** (1968) 244.
28. W. Schliephake, W. Junge and H. T. Witt, *Z. Naturforsch.*, **23** (1968) 1571.
29. Ch. Wolff, H.-E. Buchwald, H. Rüppel, K. Witt and H. T. Witt, *Z. Naturforsch.*, **24** (1969) 1041.
30. H. M. Emrich, W. Junge and H. T. Witt, *Z. Naturforsch.*, **24** (1969) 1144.
31. W. Junge, B. Rumberg and H. Schröder, *Eur. J. Biochem.*, **14** (1970) 575.
32. G. D. Greville, *Current Topics in Bioenergetics*, **3** (1969) 1.
33. L. L. Grinius, M. A. Il'ina, V. P. Skulachev and G. V. Tikhonova, *Biochim. Biophys. Acta* (submitted).
33a. E. A. Liberman and L. M. Tsofina, *Biofisika USSR*, **14** (1969) 1017.
34. L. L. Grinius and V. P. Skulachev, *Biokhimiya USSR*, **36** (1971) 430.
35. A. E. Dontsov, L. L. Grinius, A. A. Jasaitis, I. I. Severina and V. P. Skulachev, *J. Bioenergetics* (in press).
36. R. J. Van de Stadt, F. J. R. M. Nienwenhuis, K. Van Dam, *Biochim. Biophys. Acta*, **234** (1971) 173.
37. J. B. Chappell and A. R. Crofts, *Biochem. J.*, **95** (1965) 393.
38. P. Mitchell and J. Moyle, *Nature*, **208** (1965) 1205.
39. V. P. Skulachev, in: *Energy Transduction in Respiration and Photosynthesis*, Bary, Adriatica Editrice, 1971, p. 99.
40. P. Hinkle and P. Mitchell, *J. Bioenergetics*, **1** (1970) 45.
41. J. B. Jackson and A. R. Crofts, *Eur. J. Biochem.*, **18** (1971) 120.
42. A. A. Jasaitis, I. I. Severina, V. P. Skulachev and S. M. Smirnova, *J. Bioenergetics* (in press).
43. W. J. Arion and E. Racker, *J. Biol. Chem.*, **245** (1970) 5186.
44. G. S. P. Groot, L. Kovač and G. Schatz, *Proc. Natl. Acad. Sci. U.S.*, **68** (1971) 308.
45. E. C. Slater, *Quarterly Reviews of Biophysics*, **4** (1971) 35.
46. M. Klingenberg, *Essays in Biochem.*, **6** (1970) 119.
47. E. Rossi and G. F. Azzone, *Europ. J. Biochem.*, **12** (1970) 319.

A Model of Membrane Biogenesis

Alexander Tzagoloff

The Public Health Research Institute of the City of New York, Inc.
New York, N.Y.

Fundamental questions dealing with the site of synthesis of membrane proteins and phospholipids, their integration into functional enzyme complexes and ultimately into membranes, remain largely unanswered at present. On the other hand, considerable progress has been made in elucidating the structure and organization of membranes and this new information has provided a conceptual framework for postulating possible mechanisms of membrane assembly. In this communication a model of membrane biogenesis is described which is based on studies of the structure and biosynthesis of constitutive enzymes of the mitochondrial inner membrane. Although the proposed mechanism will be discussed in terms of the biogenesis of this particular membrane, it could apply equally well to other biological membranes.

Because the biosynthetic processes involved in membrane biogenesis are still poorly understood, the model presented here is to a large extent an intuitive one. It must also be pointed out that a critical review of the existent literature in this area has not been attempted. Such a search would undoubtedly have uncovered plentiful evidence both pro and con. The model contains four essential points each of which has been intentionally specified in sufficient detail to allow experimental verification.

Organization of the Mitochondrial Inner Membrane

The mitochondrion is a highly specialized organelle—being concerned almost exclusively with the utilization of the energy of oxidation of NADH and succinate for the synthesis of ATP. The macromolecular assembly which carries out this function is localized in the inner membrane of the mitochondrion and is composed of four electron transfer complexes and of the oligomycin-sensitive ATPase complex. Although these five complexes account for the bulk of the membrane mass,[1] there are probably numerous additional enzymes present in the membrane in lower concentrations. The complexes have been found in every instance to be high molecular weight enzymes (200,000 to 500,000)[2,3] consisting of multiple species of proteins and of phospholipids.

An important advance in understanding the structure of the membrane has emerged from the finding that the complexes, even in the most highly purified form, retain the capacity to spontaneously assemble into membranes.[4,5] This finding strongly implies that in addition to their specific catalytic roles, the enzyme complexes also function as the structural units of the membrane. According to this view, the membrane is a mosaic of its component enzymes. Although the mosaic model avoids the need of postulating any other organizational principle such as structural protein, it does not deny the possibility that there are structural proteins present within the domains of the enzyme complexes which are concerned with the protein-protein and protein-lipid interactions responsible for maintaining the structure of the membrane.

Site of Synthesis of the Membrane Proteins

Assuming the mosaic model of the membrane, the mechanism of biogenesis can be experimentally approached at the simpler level of the biosynthesis of the individual constitutive enzyme complexes of the membrane. This approach has been recently explored in a number of laboratories. The aim of these studies up to now has been primarily to define the site of synthesis of the various polypeptides of known constitutive enzymes of the inner membrane. From such studies it is now evident that both cytochrome oxidase[6–9] and the oligomycin-sensitive ATPase[10,11] contain some subunit proteins which are synthesized by the cytoribosomal system and some that are synthesized by the mitochondrial system of protein synthesis. The ATPase complex is so far the better documented example. This enzyme complex contains ten subunit proteins.[12] Of the ten subunits, four have been shown to be mitochondrial products[13] and six to be products of the cytoribosomal system of protein synthesis.[14,15]

Are there any characteristics which distinguish the mitochondrial and cytoplasmic products? Studies on the ATPase complex indicate that the more soluble proteins of this enzyme are made by the cytoribosomal system. The five subunit proteins of the oligomycin-insensitive ATPase, F_1, as well as the oligomycin sensitivity conferring protein (OSCP) are all synthesized by the cytoribosomal system.[10,14,15] Both F_1 and OSCP can be extracted under conditions which do not disrupt the basic structure of the membrane. Once separated from the membrane, these proteins behave as classical water soluble proteins. The mitochondrial products, however, are highly insoluble species which are associated with phospholipid and can only be solubilized with surfactant reagents which cause a dissolution of the membrane.[13]

Green[16] has recently proposed that membranes are composed of two classes of proteins—*intrinsic* hydrophobic proteins which are bonded to phospholipids and form the basic structure of the membrane and *extrinsic* proteins which are water soluble species which are also tightly associated with the membrane but are not essential for the maintainance of the membrane structure.

The protein components of the ATPase complex can be operationally separated into these two categories of proteins. The cytoplasmic products exhibit properties of extrinsic proteins while the mitochondrial products behave as intrinsic proteins. The first idea of the model proposed here is that all intrinsic proteins of the membrane are synthesized by the mitochondrion and all extrinsic proteins are synthesized by the cytoribosomal system.

Chemical Nature of the Mitochondrial Products

It has been known for some time that mitochondria contain substantial amounts of proteolipids,[17] i.e. proteins which are soluble in mixtures of chloroform and methanol. The function of these proteins is still not clear. Some recent studies suggest that the complexes of the mitochondrial inner membrane contain low molecular weight proteins which exhibit solubility properties characteristic of proteolipids. The ATPase complex of yeast and of beef mitochondria contain subunit proteins which are soluble in chloroform:methanol.[18] Cattell *et al.*[19] have recently reported that a protein which binds dicyclohexylcarbodiimide, a potent inhibitor of the mitochondrial ATPase, has the properties of a proteolipid. Studies on cytochrome oxidase have also revealed that several subunit proteins of this complex are soluble in acidic chloroform:methanol.[18] That proteolipids may be integral components of many membrane complexes is also supported by the finding that a highly purified preparation of the Ca^{2+} ATPase of sarcoplasmic reticulum contains a low molecular weight protein which is soluble in acidic chloroform:methanol.[20]

Although the chemistry of proteolipids is still obscure, some work on a brain proteolipid isolated by Stoffyn and Folch-Pi[21] suggests that the hydrophobic properties of this protein may be due to the presence of esterified fatty acids. Thus, the purified brain proteolipid has been shown to contain esterified fatty acids which could not be accounted for by contaminating phospholipids. Stoffyn and Folch-Pi[21] have suggested that the fatty acids of the brain proteolipid may be esterified directly to some amino acids of the protein moiety. The proteolipid purified from the ATPase of sarcoplasmic reticulum has also been found to contain saponifiable fatty acids which are in excess to the phospholipid content of the preparation.[20]

A question which may be raised is whether there is any relationship of the mitochondrially synthesized products (intrinsic proteins) to the membrane proteolipids. Here again the evidence at present is only fragmentary, but there is some indication from studies on the ATPase complex of yeast mitochondria that at least two of the mitochondrially synthesized subunits of this complex are soluble in chloroform: methanol.[18] Recently Kadenbach[22] has reported that a low molecular weight peptide synthesized by rat liver mitochondria has the properties of a proteolipid. The second idea of the model is that the mitochondrial products or intrinsic proteins are proteolipids with esterified fatty acids.

An extension of this notion is that (1) all membrane complexes contain as integral components one or more subunits which are proteolipids, (2) the hydrophobic properties of this class of proteins is due in part if not entirely to the presence of protein esterified fatty acids, (3) the proteolipids can combine with phospholipids through hydrophobic interactions of the fatty acids of the proteins and of phospholipids, and (4) the ability of the proteolipids to combine with phospholipids determines the capacity of membrane enzyme complexes to spontaneously assemble into membrane structures.

Cytoplasmic Control of Mitochondrial Protein Synthesis

Studies with yeast have shown that the synthesis of certain extrinsic membrane proteins by the cytoplasmic system is independent of mitochondrial protein synthesis. Thus, the synthesis of inner membrane proteins such as cytochrome c,[23] F_1,[14] and OSCP[15] proceeds normally even when mitochondrial protein synthesis is inhibited by chloramphenicol. An analogous situation is found in "petite" mutants of yeast which have a defective mitochondrial protein synthesizing system. Such mutants have been shown to synthesize cytochrome c[24] and F_1.[10,25] The converse does not hold—at least not in the case of the ATPase complex. Very little synthesis of the four intrinsic proteins of the ATPase complex occurs when the cytoribosomal system is inhibited by cycloheximide.[11] These mitochondrial products of the ATPase, however, are synthesized in the presence of cycloheximide, provided the cells are first incubated in chloramphenicol.[11] Moreover, it was found that the chloramphenicol preincubation caused a marked stimulation of overall mitochondrial protein synthesis during the subsequent incubation in cycloheximide. These results indicate that during the initial incubation in chloramphenicol some cytoplasmic products accumulate which are necessary for mitochondrial protein synthesis. In a normally growing culture of cells, presumably the synthesis of both mitochondrial and cytoplasmic products is a synchronized process and the latter do not accumulate.

There are two obvious mechanisms which could explain these observations. The first is that products of the cytoribosomal system act as activators of the mitochondrial protein synthesizing system. In the case of the ATPase complex, these activators could be F_1, OSCP or even some other protein(s) which are not intrinsic components of the ATPase. An alternative explanation of this phenomenon is that the intrinsic proteins of the membrane are made jointly by the mitochondrial and cytoribosomal systems. According to this mechanism, the cytoplasmically derived part of the protein is a water soluble peptide which is then converted to an intrinsic membrane protein by the addition of a hydrophobic peptide (handle) which is synthesized by the mitochondrion. This mechanism could explain the observed cytoplasmic control of mitochondrial protein synthesis if the synthesis of handles were dependent on the availability of the cytoplasmically made peptides. The third idea of the model is that the mitochondrial system of protein synthesis makes only part of the intrinsic proteins of the membrane. The handles of the intrinsic proteins could be small peptides with esterified fatty acids which confer upon the cytoplasmic peptides proteolipid properties.

Can such a mechanism be reconciled with what is known about the mitochondrial machinery of protein synthesis? It has become evident from recent work that the mitochondrion is responsible for the synthesis of a large number of membrane proteins.[26–28] These observations are difficult to explain in view of the small amount of DNA in the mitochondrion which in addition to coding for messenger RNA also codes for transfer RNA (see ref. 29 for a review of this topic). The mechanism proposed here could help to explain how a large number of distinct proteins may be coded for by mitochondrial DNA if one assumes that only a limited number of different handles are made by the mitochondrion. In fact it is only necessary to postulate a single species of handle since the enzymatic and functional specificity of the intrinsic proteins could reside in that portion of the polypeptides which are coded for by the nuclear DNA.

Synthesis and Assembly of Membrane Enzymes

A question which is paramount to understanding membrane biogenesis is whether the constitutive enzymes are synthesized outside of the membrane and are subsequently incorporated as preassembled units or whether the synthesis is an intramembrane process. There is no evidence at present that membrane enzymes can be found in free form. The exception to this are proteins such as cytochrome *c* or enzymes such as F_1 which are composed entirely of extrinsic proteins that are synthesized independently of the mitochondrial system.[14] A reasonable guess at present is that the central events in the assembly of membrane complexes occur within the membrane itself. The fourth idea of the model proposed here is that the membrane contains growing points which consist of the whole protein synthesizing machinery of the mitochondrion. This machinery synthesizes the hydrophobic handles—low molecular weight peptides which combine with phospholipids to form the basic substructure of the complexes and also of the membrane mosaic. The cytoplasmic counterparts of the intrinsic proteins are then enzymatically condensed to the handles possibly through peptide bond formation.

Although some membrane complexes could be composed exclusively of intrinsic proteins, most, probably contain extrinsic protein components as well, e.g. the ATPase complex. The incorporation of the extrinsic protein components into the membrane complexes could be a self-assembly process. There are currently a number of examples of extrinsic proteins of the mitochondrial inner membrane which can spontaneously interact with their intrinsic protein counterparts to reconstitute functional enzyme complexes.[30, 31]

Biogenesis of Other Cellular Membranes

With the exception of mitochondria and chloroplasts, no other cellular membranes have been shown to synthesize their own proteins. It must be borne in mind, however, that *in vitro* protein synthesis by mitochondria and chloroplasts is very low when compared to that of the classical cytoribosomal system.[29] The absence of a detectable protein synthesizing activity in an isolated membrane fraction could merely reflect a stringent control by the cytoplasmic system for that particular membrane. In principle, therefore, all membranes could have their own protein synthesizing machinery designed to make the hydrophobic handles of their intrinsic proteins.

Acknowledgments

I wish to thank Dr. Maynard E. Pullman for many stimulating discussions on the topic of this paper and for his help in the preparation of the manuscript.

References

1. D. H. MacLennan, in: *Current topics in membranes and transport*, F. Bronner and A. Kleinzseller (eds.), Academic Press, New York, 1970, p. 177.
2. Y. Hatefi, in: *The Enzymes*, P. D. Boyer, H. Lardy and K. Myrbäck (eds.), Academic Press, New York, 1963, Volume 7, p. 495.
3. D. E. Green and A. Tzagoloff, *Arch. Biochem. Biophys.* **116** (1966) 293.

4. D. E. Green, D. W. Allmann, E. Bachmann, H. Baum, K. Kopaczyk, E. F. Korman, S. Lipton, D. H. MacLennan, D. G. McConnell, J. F. Perdue, J. S. Rieske and A. Tzagoloff, *Arch. Biochem. Biophys.*, **119** (1967) 312.
5. Y. Kagawa and E. Racker, *J. Biol. Chem.*, **241** (1966) 2475.
6. W. L. Chen and F. C. Charalampous, *J. Biol. Chem.*, **244** (1969) 2767.
7. G. D. Birkmayer, *Eur. J. Biochem.*, **21** (1971) 258.
8. H. Weiss, W. Sebald and T. Bücher, *Eur. J. Biochem.*, **22** (1971) 19.
9. H. R. Mahler, in: *Probes of Structure and Function of Macromolecules and Membranes*, Volume I, Academic Press, New York, 1971, p. 411.
10. G. Schatz, *J. Biol. Chem.*, **243** (1968) 2192.
11. A. Tzagoloff, *J. Biol. Chem.*, **246** (1971) 3050.
12. A. Tzagoloff and P. Meagher, *J. Biol. Chem.*, **246** (1971) 7328.
13. A. Tzagoloff and P. Meagher, *J. Biol. Chem.*, **247** (1972) 594.
14. A. Tzagoloff, *J. Biol. Chem.*, **245** (1970) 1545.
15. A. Tzagoloff, *J. Biol. Chem.*, **244** (1969) 5027.
16. D. E. Green, *Proc. N.Y. Acad. Sciences*, in press.
17. C. D. Joel, M. L. Karnovsky, E. G. Ball and O. Cooper, *J. Biol. Chem.*, **233** (1958) 1565.
18. M. Sierra and A. Tzagoloff, unpublished observations.
19. K. J. Cattell, C. R. Lindop, I. G. Knight and R. B. Beechey, *Biochem. J.*, **125** (1971) 66P.
20. D. H. MacLennan, private communication.
21. P. Stoffyn and J. Folch-Pi, *Biochem. Biophys. Res. Commun.*, **44** (1971) 157.
22. B. Kadenbach, *Biochem. Biophys. Res. Commun.*, **44** (1971) 724.
23. A. W. Linnane, in: *Oxidases and related redox systems*, T. E. King, H. S. Mason and M. Morrison, (eds.), John Wiley, New York, 1965, p. 1102.
24. P. P. Slonimski, in: *La formation des enzymes respiratoires chez la levure*, Masson, Paris, 1953.
25. L. Kováč and K. Weissová, *Biochim. Biophys. Acta*, **153** (1968) 55.
26. S. Yang and R. S. Criddle, *Biochem.*, **9** (1970) 3063.
27. P. O. Weislogel and R. A. Butow, *J. Biol. Chem.* **246** (1971) 5113.
28. D. Y. Thomas and D. H. Williamson, *Nature*, **233** (1971) 196.
29. M. Ashwell and T. S. Work, *Ann. Rev. Biochem.*, **39** (1970) 251.
30. Y. Kagawa and E. Racker, *J. Biol. Chem.* **241** (1966) 2467.
31. T. E. King and S. Takemori, *J. Biol. Chem.*, **239** (1964) 3559.

Energy Transduction in the Functional Membrane of Photosynthesis

Results by Pulse Spectroscopic Methods

H. T. Witt

Max-Volmer-Institut 1. Institut für Physikalische Chemie
Technische Universität Berlin

1. *Methods*

The functional membrane of photosynthesis performs molecular events which are also realized in membranes of photoreceptors, mitochondria, nerves and muscles. These events are light quanta phenomena, electron transfer, electrical field generation, ion translocation, ATP synthesis and hydrolysis respectively. Therefore photosynthesis provides a good example for studying principles of molecular dynamics and energetics in biomembranes in general.

The molecular events in photosynthesis are in action between the beginning of light absorption and the end of the overall process which is terminated by the production of NADPH and ATP. With NADPH and ATP absorbed CO_2 can be reduced into sugar and "everything else".[1]

Valuable information on the molecular events in photosynthesis has been obtained by *optical indication* of the events and by excitation and registration of the indicated events with the *repetitive pulse spectroscopic method*.[2]

In this way three improvements have been realized:

(a) The extremely high sensitivity of the repetitive techniques permits an analysis at conditions of *one single turnover* of the molecular machinery. In this way the interrelationships between the different events became most transparent.

(b) The time resolution of the repetitive technique exceeds the hitherto range by six orders of magnitude and works down to 10 *nanoseconds*. Thereby the analysis is open to all phases of the events between excited singlet states and the formation of the endproducts.

(c) The special optical indications made it possible to extend the analysis—formerly most restricted to redox-reactions—to *six other events*.

After a short introduction to the types of events which have been measured the further presentation is focussed to the electrically energized membrane and to the mechanism of phosphorylation. Subsequently the results are discussed in respect to the current hypothesis of bioenergetics. The report is restricted to results by pulse spectroscopic techniques.

Details on the methods are published in ref. 2. Details on the results are outlined in ref. 3. Because this report is an abstract of what is presented in ref. 3, the reader may be referred in respect of the detailed literature to ref. 3.

2. *Events*

In toto eight events have been analysed in the above explained way.

1. *Metastable states* of carotenoids
2. *Light reactions* of chlorophylls
3. *Redox reactions* during the transfer of electrons.

These three events are indicated—as *in vitro*—by intrinsic absorption changes.

4. 1966—an analysis of the mechanism of ATP generation was opened with the observation of specific absorption changes which are coupled with a *high energy state* of phosphorylation[4]:

1967—these spectral changes have been identified as indicating an *electrical field* across a membrane.[5] A first report on this result has been published in ref. 6. The mechanism of the field indication occurs by electrochromism (shift of absorptbaion nds in a field), see refs. 7 and 7a.

This interpretation has been proved by different independent lines of evidences (see ref. 3):

(i) by the observation that all bulk pigments within the membrane response optically to the event;
(ii) by the shape of the spectral changes which are typical for being induced by a field;
(iii) by the sensitivity of these changes to artificial ion translocators;
(iv) by agreement of this spectrum with that induced by fields on artificial pigment layers;
(v) by artificially embedded dyes as probes for the field.

An electrical field has recently been established also across the inner membrane of photobacteria.[7b]

The field indicating absorption changes ΔA can be used as a molecular voltmeter which can be calibrated in electrical units[8]:

$$\Delta\phi \approx 50\ \text{mV} . \Delta A / {}^{1}\Delta A$$

$\Delta A / {}^{1}\Delta A$ = absorption changes in relation to that produced in a saturated single turnover flash.

5. The field drives ions across the membrane. These *ion fluxes* i change the extent of the field in time. Therefore the time course of the field indicating absorption changes can be used as a molecular ammeter:

$$i \approx 1\ \mu F . 50\ mV . d\Delta A/dt . 1/{}^{1}\Delta A$$

6. *Proton transfer* ΔH^+ across membranes has been measured in the *outer* phase by artificial indicators. During the transfer of H^+ into the inner phase the indicator is deprotonated. This can be measured in the case of the indicator umbelliferon (UBF) by a change from a non-fluorescing into a fluorescing state of UBF.[9]

7. *Proton accumulation* H^+_{in} in the inner phase has been measured by an intrinsic indicator for H^+_{in} changes. As such an indicator the response of Chlorophyll-a_I-700 has been used.[10]

8. *Phosphorylation* has been measured by the indicator UBF, too. During the ATP generation H^+ is consumed and can be measured by the irreversible fluorescence changes of UBF.[11]

Some results and relationships which have been evaluated between the cited eight events are the following.

3. *Electrically Energized Membrane*

(a) Within one electron transport chain two *light reaction* centres have been spectroscopically observed which drive the transport of electrons from H_2O to $NADP^+$. These centres are an excited Chlorophyll-a_I-700[12] and Chlorophyll-a_{II}-680.[13] These centres have been recognized spectroscopically as coupled in series.[14,15,16] This has been proposed in ref. 16a.

(b) The analysis of the *electron transfers* Δe gave inter alia evidence that a pool of plastoquinone PQ ($\sim$ 10 PQ) is the link between the two light centres.[17] Some 100 of these arrangements are covering the surface of one thylakoid. The light driven electron transfer within this network from H_2O to $NADP^+$ converts light energy in form of the reducing power of NADPH.

(c) The observed *electrical potential change* $\Delta\phi$ indicates a primary electron transfer perpendicular to a membrane (charge separation with $\boxed{-}$ outside and $\boxed{+}$ inside).[5]

The generation of $\Delta\phi$ takes place in <20 ns simultaneously with the photoact.[18]

This leads to the idea that the electrically charged membrane ($\frac{1}{2}.C\Delta\phi^2$) is a further state in which light energy is converted.

(d) The electrical potential change $\Delta\phi$ has been measured as to be proportional to the amount of *moved electrons* Δe[19,20]:

$$\Delta e \sim \Delta\phi$$

In a single turn-over flash it is ${}^{1}\Delta\phi \approx 50$ mV and in a steady-state light $\Delta\phi_{ss} \approx 100$ mV.

(e) The amount of *inward translocated protons* ΔH^+ which is not balanced by counter ion movement is proportional to the previously set up potential change (exchange of $\boxed{-}$ against OH^-, and $\boxed{+}$ against H^+)[8, 20]:

$$\Delta\phi \sim \Delta H^+$$

The amount of inward translocated protons ΔH^+ which is balanced by counter ion movement give rise at steady state light conditions and $pH_{out} = 8$ to a maximal change in the inner phase of the membrane from pH_{in} 8 to pH_{in} 5,[10] i.e. to a *chemical potential gradient* of

$$\Delta pH_{ss} \approx 3$$

(f) For the $\Delta\phi$ generation and ΔH^+ translocation two *light generators hν* per electron chain have been identified[8]:

$$h\nu_I \sim \tfrac{1}{2}\Delta\phi \sim \tfrac{1}{2}\Delta H^+ \qquad h\nu_{II} \sim \tfrac{1}{2}\Delta\phi \sim \tfrac{1}{2}\Delta H^+$$

This supports in an independent way the concept of two light centres reported in (a).

(g) The *discharging* of the electrically energized membrane occurs by a field driven ion flux i ($i \sim \Delta\phi$). Because the $\Delta\phi$ decay is accelerated 20-fold with the increase from pH_{in} 8 to pH_{in} 5 in the inner phase of the membrane it is assumed that the discharging occurs predominantly by field driven *proton* effluxes[21]:

$$\Delta\dot{\phi} = \Delta\dot{\phi}_{H^+} = f(H^+_{in})$$

(h) Only one artificial ion translocator molecule as gramicidine GMCD per 10^5 chlorophyll molecules works already as a shunt for the intrinsic decay. About 10^5 chlorophyll molecules cover an area of about (5000 Å)^2. This area has the same order of size as that of one thylakoid. Therefore it is proposed that the *functional unit* of the electrical events is the membrane of 1 thylakoid[5]:

$$\Delta\phi\text{-unit} = 1 \text{ Thylakoid}$$

4. *Phosphorylation*

For the investigation of phosphorylation the system has firstly been restricted to electrical events by excluding the energetical involvement of ΔpH. This is possible because in a single turnover flash the produced potential change is $\Delta\phi \approx 50$ mV but the pH gradient only $\Delta pH \ll 0{\cdot}1$.

(a) Under the described conditions ATP generation takes place, i.e. phosphorylation is possible in the presence of an electrical potential difference only, without an energetical contribution of a pH gradient.

Whether $\Delta\phi$ and the thereby energized membrane is necessary for

ATP generation or whether $\Delta\phi$ is only a parallel event is answered by the following results.

(b) The extent of the potential $\Delta\phi$ can be changed from ${}^{1}\Delta\phi \approx 50$ mV up to $\Delta\phi_{max} \approx 200$ mV by using longer flashes. The *yield of ATP* per flash increases thereby linearly with the potential change[11]:*

$$\Delta\phi \sim \mathrm{ATP}$$

(c) Under phosphorylation conditions the rate of $\Delta\phi$ decay is accelerated 2-5-fold. From this and the following results it is supposed that phosphorylation is coupled to a field driven proton flux chanelled through a special *ATPase pathway* within the membrane.[22]

(d) According to the interpretation in (c) it is expected that the rate of phosphorylation should be increased when the field driven proton flux is increased. In fact the *rate of ATP* generation increases 20-fold when the intrinsic $\Delta\phi$ decay is accelerated 20-fold by a change from pH_{in} 8 to pH_{in} 5,[23] i.e.

$$\Delta\dot{\phi}_{H^+} \sim \mathrm{A\dot{T}P}$$

(e) From the result in (d) it is evident that phosphorylation is predominantly coupled to electrically driven *proton* effluxes across a ATPase pathway.

(f) The number of H^+ which passes across an ATPase coupled pathway can be read out from the difference in the decay of $\Delta\phi$ with and without phosphorylation (see (c)), because according to section 2.5 the absolute value of ion fluxes can be measured. For the *stoichiometry* H^+:ATP it results that the generation of one ATP is coupled to three field driven H^+[24]:

$$3H^+ \sim 1\mathrm{ATP}$$

(g) A consequence of these results is that *deactivation* of phosphorylation should take place as soon as the intrinsic field driven H^+ flux is outrun by e.g. alcaline ions. This has been proved by addition of alcaline translocators as valinomycine (VMC) [24] and gramicidine (GMCD)[23]:

$$\Delta\dot{\phi}_{K^+} \rightarrow \mathrm{A\dot{T}P} = 0$$

(h) The results in (b–g) indicate that obviously the discharging of the electrically energized membrane by protons is coupled with the generation of ATP. This is lastly strongly supported by the result that when only one artificial ion translocator GMCD per one thylakoid acts as a shunt for the intrinsic discharging of the membrane (see section 3 (h)), phosphorylation is deactivated, too. The *function unit* of phosphorylation is therefore based by the field on one thylakoid[25]:

$$\text{ATP-unit} = 1\ \text{Thylakoid}$$

* A nonlinear dependence has been reported in ref. 24. This deviation is discussed in ref. 3.

5. *Conclusions*

The results are in principle consistent with the electrochemical hypothesis of Mitchell.[26] However the initial generation of the electrical potential by an orientated shift of 2 electrons and the stoichiometry of $3H^+$ per 1 ATP are different from his original concept.

The demonstrated coupling between the electrical events and ATP generation can be written in a short hand drive as

$$\Delta e \rightleftarrows \Delta\phi \rightleftarrows \text{ATP}$$

Two other hypotheses on the coupling between electron transfer and ATP-generation—the chemical[27] and conformational[28] hypotheses—can be written briefly as

$$(\Delta e \rightleftarrows [\sim] \rightleftarrows \text{ATP})$$

The squiggle [~] represents an unknown chemical intermediate in which energy is conserved as covalent bond energy or it represents conformational changes which conserve energy in the folding of numerous protein molecules. The dotted arrows symbolize the subsequent physical and chemical steps by which ATP is synthesized.

The reported results support the acceptance of a coupling of electron transfer and phosphorylation by $\Delta\phi$. The coupling by a squiggle is clearly in contradiction to our results.*

One can argue that the demonstrated coupling of $\Delta\phi$ with phosphorylation takes place from a side path in which $\Delta\phi$ is equilibrated with a squiggle in the mainpath:

$$\begin{array}{c}(\Delta e \rightleftarrows [\sim] \rightleftarrows \text{ATP}) \\ \upharpoonleft\!\downharpoonright \\ \Delta\phi\end{array}$$

The as yet available data do, however, also not support this modified squiggle phosphorylation:

(i) The squiggle has as yet never been shown to exist.

(ii) Phosphorylation should be possible without $\Delta\phi$ in suspension without closed vesicles. Treatments in this direction were as yet without success.

(iii) Peculiar properties of the squiggle have to be assumed to explain the observed stoichiometry of $3H^+$ per 1 ATP.

(iv) Generation of the squiggle as well as the subsequent equilibration with $\Delta\phi$ must occur in <20 ns because $\Delta\phi$ generation has been measured to take place in this time (see above). The synthesis of the squiggle must be therefore faster than any electron transfer event apart from that during the photo act. This is a very improbable assumption.

* $\Delta\phi$ can of course induce conformational changes which in turn may support the synthesis of ATP. This step is a result of charging and therefore different from the conformational change hypothesis.

In permanent light H^+_{in} is dumped up and it results at steady state conditions a gradient of $\Delta pH_{ss} \approx 3$ (see above). Evidence for an energetical contribution of a pH gradient to ATP generation was first demonstrated by Jagendorf *et al.*[29] Further detailed proofs have been given in refs. 30 to 33. Also at steady-state conditions a stoichiometry of $3H^+$ per 1 ATP has been observed.[31] This value satisfies, together with $\Delta\phi_{ss} \approx 100$ mV and $\Delta pH_{ss} \approx 3$, the energetical requirement for phosphorylation.[3]

Extending the electrochemical concept for ΔpH contribution it would follow analog to the arguments at single turnover conditions:

$$\Delta e \rightleftarrows (\Delta\phi, \Delta pH) \rightleftarrows ATP$$

The presented lines of evidences for a electrochemical coupling of electron transfer and ATP generation will need further elaboration. For instance the linear dependence of ATP formation with $\Delta\phi$ and the stoichiometry of $3H^+$ per ATP has to be motivated. This is in close connection with the open question for the chemical steps (symbolized by dotted arrows) by which ATP is synthesized.

References

1. M. Calvin and A. A. Benson, *Science*, **107** (1948) 767.
2. H. Rüppel and H. T. Witt, in: *Methods in Enzymology*, **16**, "Fast Reactions", Academic Press New York, (1969) pp. 316–380.
3. H. T. Witt, *Quarterl. Review of Biophysics*, **4** (1971) No. 4.
4. H. T. Witt, G. Döring, B. Rumberg, P. Schmidt-Mende, U. Siggel and H. H. Stiehl, in: *Energy Conversion by the Photosynthetic Apparatus*, Brookhaven Symposia in Biology No. 19, Upton, New York 1-67-500, (1966) p. 161.

4a. B. Rumberg, R. Schmidt-Mende, U. Siggel and H. T. Witt, *Ang. Chem.* **5** (1966) 522 (Intern. ed. in English).

5. W. Junge und H. T. Witt, *Z. Naturforsch.* **23b** (1968) 244.
6. H. T. Witt, in: *Fast Reactions and Primary Processes in Chemical Kinetics* (Nobel Symposium V), S. Claesson (ed.), Almqvist & Wiksell, Stockholm; Interscience Publ. New York, London, Sydney, 1967, p. 261.
7. H. M. Emrich, W. Junge and H. T. Witt, *Z. Naturforsch.* **24b** (1969) 1144.

7a. S. Schmidt, R. Reich and H. T. Witt, *Naturwiss.* **8** (1971) 414.

7b. J. B. Jackson and A. R. Crofts, *FEBS Lett.* **4** (1969) 185.

8. W. Schliephake, W. Junge and H. T. Witt, *Z. Naturforsch.* **23b** (1968) 1571.
9. H. H. Grünhagen and H. T. Witt, *Z. Naturforsch.* **25b** (1970) 373.
10. B. Rumberg and U. Siggel, *Naturwiss.* **56** (1969) 130.
11. M. Boeck and H. T. Witt, *Eur. J. Biochem.* (1972a) (in the press), previous report in (3).
12. B. Kok, *Biophys. Acta* **48** (1961) 527.
13. G. Döring, G. Renger, J. Vater and H. T. Witt, *Z. Naturforsch.* **24b** (1969) 1139.
14. B. Kok and G. Hoch, in: *Light and Life*, McElroy & B. Glass (eds.), Johns Hopkins Press, Baltimore, 1961, p. 397.
15. L. N. M. Duysens, J. Amesz and B. M. Kamp, *Nature*, **190** (1961) No. 4775 510–511.
16. H. T. Witt, A. Müller und B. Rumberg, *Nature* **191** (1961) 194.

16a. R. Hill & F. Bendall, *Nature*, **186** (1960) 136–137.

17. H. H. Stiehl und H. T. Witt, *Z. Naturforsch.* **23b** (1968) 220; **24b** (1969) 1588.
18. Ch. Wolff, H. E. Buchwald, H. Rüppel, K. Witt and H. T. Witt, *Z. Naturforsch.* **24b** (1969) 1038.
19. H. T. Witt, B. Rumberg, P. Schmidt-Mende, U. Siggel, B. Skerra, J. Vater und J. Weikard, *Angew. Chem.*, Intern. Ed. in English **4** (1965) 799.
20. E. Reinwald, H. H. Stiehl and B. Rumberg, *Z. Naturforsch.* **23b** (1968) 1616.
21. M. Boeck and H. T. Witt, *Eur. J. Biochem.* (1972b) (in the press), previous report in (3).
22. B. Rumberg und U. Siggel, *Z. Naturforsch.* **23b** (1968) 239.
23. M. Boeck and H. T. Witt, *Eur. J. Biochem.* (1972c) (in the press), previous report in (3).

24. W. Junge, B. Rumberg and H. Schröder, *Eur. J. Biochem.* **14** (1970) 575.
25. M. Boeck and H. T. Witt, *Eur. J. Biochem.* (1972d) (in the press), previous report in (3).
26. P. Mitchell, *Biol. Rev.* **41** (1966) 445.
27. E. C. Slater, *Nature,* **172** (1953) 975.
28. P. D. Boyer, in: *Oxidases and related redox systems,* T. E. King, H. S. Mason and M. Morrison (eds.), John Wiley, New York, 1965, Vol. 2, p. 994.
29. A. T. Jagendorf and E. Uribe, *Brookhaven Symp. Biol.* **14** (1966) 215.
30. A. R. Crofts, *Biochem. Biophys. Res. Comm.* **24** (1966) 127.
31. H. Schröder, U. Siggel, H. Muhle and B. Rumberg, in: *II. Intern. Congr. Photosynthesis Res., Stresa,* G. Forti (ed.), Naples (1971) (in press).
32. M. Schwartz, *Nature,* London, **219** (1968) 915.
33. C. Carmeli, *FEBS Lett.* **7** (1970) 297.

Toward a Theory of Muscle Contraction*

Gustavo Viniegra-Gonzalez† and Manuel F. Morales‡

Cardiovascular Research Institute, University of California, San Francisco, California 94122

The general chemistry of muscle contraction was elucidated early on from test-tube reconstitutions of function:[1,2] Myosin and actin form an ATP-dissociable complex; myosin is an actin-modifiable ATPase, and the free energy of ATP hydrolysis "pays for" any work performed in contraction. The central structural feature of contraction was discovered later:[3,4] In the vertebrate sarcomere, thick (myosin) and thin (actin + control proteins) filaments interdigitate to produce shortening without change in filament length. Finally came the macroscopic "law of force":[5,6] The "tetanic" shortening force, and/or the rate of chemical energy dissipation, is proportional to the number of interactions between (myosin) cross-bridges and actin. Even together these great discoveries are not an explanation of how muscle works because it remains unknown how myosin, actin, and ATP, located as they are, produce a force at the microscopic level, or how this force is directed and timed to produce the behavior expressed by the macroscopic law. On the other hand, it is nowadays pointless to theorize about these unknown processes unless the theory springs clearly and naturally from these firmly-established central facts.

The direction that *molecular* theory should take was suggested in far-seeing papers by W. O. Fenn,[7] and by A. F. Huxley[8]: Macroscopic shortening, or tension must result from the *cycling* of molecular elements; the cycling must be coupled to free energy dissipation (ATP hydrolysis). For participants in the molecular cycle the obvious candidates are a myosin cross bridge and an adjacent stretch of actin filament. Perhaps during some portion of the kinematic cycle the cross bridge makes impulsive contact with (transfers momentum to) the actin. The foregoing suggestions are reasonable, but they remain unproven; for example, no one has "seen" a myosin cross bridge executing a kinematic cycle, or "felt" actin recoiling locally from a

* A summary of this work was presented as an invited paper at the 15th Meeting of the Biophysical Society, February 1971. This article also includes portions of a Dissertation submitted by G. V.-G. in partial fulfillment of the requirements for the Doctorate of Philosophy at the University of California.

† Senior Fellow of the Bay Area Heart Research Committee.

‡ Career Investigator of the American Heart Association.

myosin impulse. Furthermore, there is now much new physicochemical information about myosin, actin, etc. that has to be fitted together in order to fill out a picture of molecular events. At this juncture it seemed to us worthwhile speculating on how a cycling mechanism might work at the molecular level. Our speculation of course has both logical and informational gaps—gaps which we have tried to stuff with suggestions for experiment.

We believe the transfer of momentum from myosin to actin to be somewhat different from that postulated by A. F. Huxley and embraced by many subsequent authors,[9, 10] and it is necessary first to explain our dissent and state some of its implications.

Because individual cross-bridges—i.e., projections of individual myosin molecules—are discernible in electron micrographs,[11, 12] it was simplest to assume that these structures moved on actin as would the oars of a boat on water. There are now reasons for disbelieving this simplest assumption. The projection of a myosin molecule is now known to be structurally and functionally *duplex*.[13, 14, 15] In the model actomyosin contraction known as "superprecipitation" only molecules with *both* myosin halves functional participate.[16] In its interaction with myosin, the activity-composition relationships of actin copolymerized from native and inactive monomers suggests that the monomers participate by *twos*.[17]

In still simpler systems, in the presence of excess globular actin, the two "S-1" moieties of myosin when attached to one another (as in "HMM") bind four actin monomers, but separate they bind only two.[18] Finally, a system consisting of monomeric myosin (S-1), and monomeric (G) actin shows no actin-activated ATPase,[19] suggesting that interaction requires S-1 to S-1 communication through actin. These various observations lead us to suggest that in the transfer of momentum the two S-1 moieties of a single myosin cross bridge alternately bind on, and exert force at, two different monomers of the actin filament, much as one would use one's hands in continuously drawing in or playing out a rope. As in this human analogy the movements of the S-1 moieties would have to be *coordinated* in order to be effective. This requirement will be considered after we discuss sequential transformations of a single S-1 moiety.

We assume, on the basis of Barany's work,[20] that an individual S-1 moiety has two distinct combining sites—one for polyphosphates, e.g., ATP, and one for actin monomer. It will be further assumed that when one type of ligand binds at its site there is produced a structural distortion which makes the other site unattractive to its ligand, i.e., binding of the two types of ligand to different sites will be assumed to be "competitive". On the other hand, we will assume that the reactions of hydrolysis compel a temporal sequence of appearance of polyphosphates, e.g., "ATP, then ADP". Because the affinities of these sequentially-appearing substances for the polyphosphate binding site are different from one another, the S-1 moiety will pass through a succession of states, sometimes binding a polyphosphate, sometimes binding actin. With the "right" disposition of the molecules, these circumstances will cause an S-1 moiety to execute a kinematic cycle, which we now specify.

Competitive inhibition experiments suggested to Blum[21] that in hydrolysis of a bound triphosphate the orthophosphate fragment is first ejected, and that the rate of its desorption limits the overall rate of hydrolysis. Therefore, for catalysis by myosin the minimal chemical cycle would be,

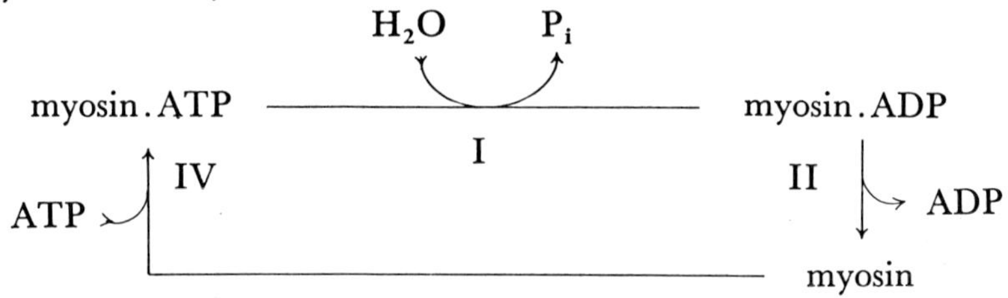

If the affinities are in the order, ATP-myosin > actin-myosin > ADP-myosin, then we write the chemical cycle in the presence of actin as,

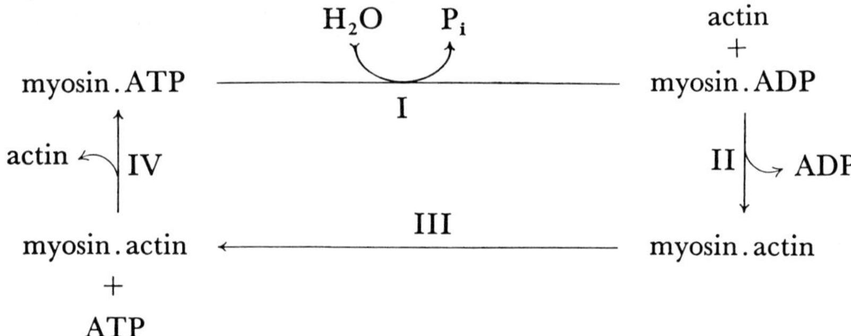

Because externally-added ADP does not affect the transients expected from steps I and II,[22] and because muscular *activity* stimulates a catalyzed P-ATP isotope exchange[23] it is also worth considering that step I in the actomyosin cycle is actually,

myosin.ATP

H_2O → I-A

P_i ←

myosin*ADP —I-B→ myosin.ADP (with * above)

In this elaborated Step I, "myosin*ADP" and "$\overset{*}{\text{myosin}}$.ADP" are both supposed to be complexes of ADP with myosin, but physically different from one another, for instance in the conformation of the protein. The asterisks emphasize that energy is transferred from one set of modes to another within the same complex. The foregoing chemical cycles are written so as to suggest a corresponding kinematic cycle. Thinking that in the relative translation of thick and thin filaments the "root" (axial coordinate of zero force, left side) of an S-1 moiety is stationary with respect to the thick filament, we envision the tip of the S-1 moiety as moving out to remote actin (right side), and pulling (or pushing) the actin back to a location opposite the root. Repetition of these processes of course results in continuous dissipation of ATP, and continuous translation of actin past the location of the root. As the schemes are written energy transduction is supposed to occur in Step I of the simple actomyosin cycle, or in Step I-B of the elaborated cycle, i.e., during this step the chemical transformation in some way induces the S-1 structure to "reach out". Having outlined the cycles we can return to explain what we mean by "communication" and "coordination" between the two S-1 moieties.

In isolated myosin molecules there is evidence that (as regards substrate binding) one S-1 moiety does not "know" what the other is doing.[15] We suggest, however, that the information necessary for coordination is transmitted directionally from one moiety to the other via polymerized actin. Through his demonstration that myosin "arrowheads" always point away from the *z*-membrane in isolated "double-brush" structures, H. E. Huxley[24] established that the F-actin filament has a polarity; therefore it is not unreasonable to think that transmission of information (e.g., a distortion) along the filament has a direction. We suggest that the binding of one S-1 moiety of a myosin cross bridge to a monomer of the actin filament produces in the filament a distortion which makes a particular "downstream" monomer expecially attractive to the other S-1 moiety of the same cross bridge. If this is the effect that explains the aforementioned binding of 4 actin monomers to the two S-1 moieties of a single myosin,[18] then we would surmise that, parallel to the thin filament axis, the distance spanned by the two S-1 moieties (on the same strand) is ca. 150 A. It is interesting that this same estimate of the span is obtained from reconciling shortening velocity with maximal acto-myosin ATPase,[25] and from anlyzing velocity transients.[26] Recently, Doppler-shift broadening of the spectrum of scattered light,[27] and X-ray diffraction,[28] have begun to provide evidence that the F-actin helix is deformable under "normal" circumstances. Possibly the increase in flexibility induced in F-actin as the (bound myosin)/(actin monomer) ratio rises from 0 toward 1/5 reflects an increase in the number of micro-distortions of the sort just postulated; possibly with further increases in the ratio the flexibility starts to decrease because the "other ends" of the distortions begin to be stabilized by the additional S-1 moieties that become available for binding. Whatever be the details of coordination, a specification of the temporal relations between S-1 moieties can be made in terms of the aforementioned kinematic cycle of an individual moiety. Myosin and actin are *unattached* during Step I, and they are *attached* during Step III (For the present purposes steps II and IV are disregarded as briefer transients). If we designate the two moieties by prime (′) and double prime (″), then coordination consists in having I′ coincide with III″, and I″ with III′. This is just the temporal arrangement which allows the bound S-1 in step III″ to direct the free S-1 in step I′ to the right actin monomer, and conversely for III′ and I″. Because the location of the myosin-binding site on actin is unknown, and because we assume that binding of one S-1 moiety distorts the actin helix so as to make a particular actin monomer attractive to the free S-1 moiety, we cannot specify the lattice of actin sites to which S-1 moieties may attach. The most reasonable expectation is that along the thin filament sites would occur alternately on one actin strand or the other; in that case the two troponin-tropomyosin "controller strands" lying in the grooves of the actin strands[29] could function by widening or narrowing the *transverse* distance between successive sites of the lattice.

Having set forth the bare outline of a kinematic cycle, we attempt to assimilate into the scheme what current knowledge we can muster. In this effort we note that there should be some connection between steps of myosin catalysis and actomyosin catalysis, and between these and the "relaxation" state of the organized fiber.

As regards steps II and IV the following observations are cogent. First, is that ATP-binding and actin-binding deteriorate at different rates[20]; as we have said, this is the basis for assuming that the binding sites are distinct. Second, is the evidence that myosin suffers a conformational change on binding ATP—F. Morita's[30] report of a change in the UV-difference spectrum of myosin. H. C. Cheung's[44] report of a change in tryptophane → bound ANS energy transfer, and J. Seidel *et al.*'s[31] report of a change in the EPR spectrum of a myosin-bound spin label. Third, is the evidence that myosin suffers a conformational change upon binding actin, viz., T. Tokiwa's[32] report that actin-myosin binding immobilizes a myosin-bound spin label. Last is the evidence that actin and ATP compete with each other for myosin. That ATP dissociates actomyosin is obviously consistent with competition; evidence of the reverse effect is not so plentiful. However, it is noteworthy that according to T. Tokiwa[32] actin-binding and

ATP-binding produce qualitatively opposite changes in the EPR-spectrum (immobilization and mobilization, respectively) of a myosin-bound spin label. In step II actin is the displacer, and in step IV it is displaced; this contrast is consistent with the fact that under similar ionic conditions myosin prefers ATP over ADP by a factor between five and ten.[15, 33] The species, "myosin*ADP" is hypothetical, and its stability is unknown. We believe (but it remains to be proved) that when myosin is affinity-labelled with "SH-ATP"[34] the nucleotide which remains blocking the site is actually the diphosphate; if this be true, then affinity-labelled myosin should be a model of "myosin*ADP", since in this species the dinucleotide should be positioned as it is at the instant that it is formed from the ATP. Tokiwa[32] has found that affinity-labelled myosin does not bind actin at all. Tentatively then, we will think that in the competition with actin "*ADP" is more like ATP than ADP.

In step II it is supposed that some agency (e.g., actin) produces in myosin some structural change which facilitates the desorption of ADP (thereby speeding up the overall hydrolysis, according to Blum's hypothesis). Agencies other than actin will also do this, and the generalized phenomenon is called "modification". Modification has been studied most extensively in 0·5–0·6 M KCl at 25°C, in the presence of 0·005–0·010 M $CaCl_2$. It is important, however, that as recently shown by Stone and Prevost[35] modification also occurs under contraction conditions, and that under these conditions actin behaves qualitatively like other modifiers. Under such conditions prior application of a variety of modifiers strongly activates the *A*TPase activity of myosin, at pH near 7. In his hypothesis Blum[21] suggested that the bond which allowed the diphosphate product (e.g., ADP) to cling to myosin after cleavage (beyond step I) involved both the 6-substituent of the purine ring and a Mg^{2+} tightly held by the myosin. Kielley and Kalckar[36] added that perhaps the Mg^{2+} chelation includes a special SH group of myosin (actually of the S-1 moiety). These suggestions would explain why either a chelating agent such as EDTA, or a sulfhydryl reagent such as PCMB are able to activate ATPase. A cardinal postulate in Blum's hypothesis—that hydrolysis is rate-limited by ADP-desorption—has recently been shown directly by E. W. Taylor;[37] however, much remains to be elucidated. As it happens, there are several other agencies that modify, e.g., certain organic solvents, changing from 25° to 0°C, substituting the purine ring of ITP for that of ATP, etc. This diversity led Rainford *et al.*,[38] and independently, T. Sekine and W. W. Kielley,[39] to suggest that underlying the effects of all these agencies there was a common change in conformation. This idea has been confirmed, and the change has been characterized by several of our colleagues, using non-kinetic physical methods, viz., anion variation,[40] ANS fluorescence,[41] spin-labelling,[42] tryptophane → ANS energy transfer,[43,44] and pre-steady state transients.[22] To the extent that these simpler modifications are genuine analogs of actin modification the observations cited in this paragraph justify ascribing to step II the conformational nature discussed above. Pending proof of the analogy, we accept it as a working hypothesis. The mechanism of step IV would be just that of step II acting in reverse because myosin binds ATP more tightly than ADP. Parenthetically, we note that these ideas must apply as well to contractants other than ATP, e.g. ITP. ITP is also a good contractant, but only if more Mg^{2+} is made available to it than is required by ATP. This extra requirement is interesting because ITP and IDP are held less tightly than ATP and ADP (argumentation based on K_M and K_I, respectively), no doubt accounting for the fact that overall ITPase is faster than overall ATPase. Seemingly, in order to achieve cycling a balance must be struck between having enough ring binding to achieve certain states, e.g., myosin.NDP but not so much that the complex is stabilized to the point where actin is unable to displace NDP.

In our scheme step III is the "power stroke", and may, in some structural respects be a reversal of step I (not necessarily, of course, since ATP and actin are quite different ligands). Although motion itself is an incontrovertible indicator that something like step III occurs, there are at present no methods capable of directly detecting step III, either in the fiber or in a model system such as precipitation.

We turn finally to step I, which in the fiber is the "displacement" stroke. As already mentioned step I (or I-A, I-B) should be observable *and comparable*, both in the fiber and in solution. As regards solution, very little knowledge has accumulated. Spin labels attached to myosin show environmental differences depending on whether the bound ligand is nothing, ATP, or ADP, but these differences can as yet not be related to kinematic properties of the S-1 moieties; in our community attempts are being made to get at these kinematic properties from studying the depolarization of fluorescence emitted by fluorophores attached to the S-1 moieties. Our information about the organized fiber is actually a little better. The polarization of tryptophane fluorescence when the fiber is excited by polarized UV light vibrating in a plane perpendicular to the fiber axis ($P_\perp$) appears to reflect orientation of S-1 moieties.[45,46] Dos Remedios *et al.*[46] have now shown that the terminus of the displacement stroke ($P_\perp$ of relaxation) is different depending on whether the relaxant is PP (or externally added ADP), or whether it is ATP (or a daughter substance such as *ADP). Further measurements of $P_\perp$, in the presence of certain analog substances should clarify the nature of step I. Experiments in collaboration with R. G. Yount, employing "unsplittable ATP" (actually

ARPPNP)[47] should show whether the terminus (that cannot be achieved with PP) is achieved with ATP itself. As already explained, affinity-labelling with SH-ATP should be equivalent to binding *ADP, so the $P_{\perp}$ of a fiber treated with SH-ATP should reveal whether the terminus is achieved with *ADP. Naturally the analog experiments suggested here could also be done for X-ray diffraction analysis.

Before closing this paper we must consider the energetics of the kinematic cycle which we have proposed, particularly in the light of McClare's[48] penetrating critique of various models of contraction. For this purpose it is convenient to begin with step IV. In this step we assume that myosin "prefers" ATP over actin. The individual energy decreases when either ligand binds to myosin are undoubtedly large, but we can expect the *net* decrease to be small; nevertheless this net has to be subtracted from the useful work which the overall hydrolysis might otherwise perform. During step I we assume that, in order to hydrolyze the bound ATP, the S-1 moiety has to be displaced; in other words, the energy is stored in some deformation of the S-1 moiety. Supposedly, this deformation has moved the moiety near a new position along the actin filament. It is assumed that myosin prefers actin over the bound ADP which is now attached, so that a bond forms between the S-1 moiety and a new location on the actin filament (step II). In this exchange of partners net energy is once again lost. The energy of the hydrolysis minus the sum of the losses in steps IV and II would thus be available for performing useful work during step III. There is however, a proviso, namely, that step I be practically "adiabatic"; for example, we must assume that the motion of the S-1 moiety is not arrested by collisions, and that the nucleotide it bears (either ATP or *ADP) is not exchanged for, say, ADP from solution. In practice then conditions may be achieved by having the truly *displacement* step sufficiently rapid ("chemical" phases, e.g., water attack, could be slow). In the foregoing we have also suggested as a basis for coordination that binding of an S-1 moiety may "prepare" a downstream actin site for receiving the other S-1 moiety. This creates no special problem for our energetic accounting; mechanically, it merely means that not only is one actin molecule deformed (in binding an S-1 moiety) but that a region of actin filament is deformed, as in the propagation of a defect in a crystal.

In this paper there has been proposed a kinematic cycle (and its underlying chemistry) which could be responsible for the cyclic impulses that an S-1 moiety delivers to the actin filament in muscle contraction. It has also been suggested that the activities of the two S-1 moieties of a single myosin molecule are closely coordinated. Finally, an attempt has been made to show that much modern data can be fitted into our hypothesis.

Acknowledgements

The authors deeply appreciate many critical discussions with fellow members of the Muscle Biochemistry Group, of the Cardiovascular Research Institute, University of California.

This research was supported by grant C. I.-8 of the American Heart Association, grant GB 24992 of the National Science Foundation, and grant HE 13649 of the National Heart and Lung Institute.

References

1. V. A. Engelhardt and M. N. Ljubimova, *Nature*, **144** (1939) 669.
2. A. Szent-Gyorgyi, *Muscular Contraction*, Academic Press, New York, 1947.
3. A. F. Huxley and R. Niedergerke, *Nature*, **173** (1954) 971.
4. H. E. Huxley and J. Hanson, *Nature*, **173** (1954) 979.
5. A. M. Gordon, A. F. Huxley and F. J. Julian, *J. Physiol.*, **184** (1966) 170.
6. C. J. Ward, C. Edwards and E. S. Benson, *Proc. Nat. Acad. Sci.* **53** (1965) 1377.
7. W. O. Fenn, in: *Physical Chemistry of Cells and Tissues*, R. Höber (ed.), Chap. 33, Blakiston, Philadelphia, 1945.
8. A. F. Huxley, *Progr. Biophys. Biophys. Chem.*, **7** (1957) 257.
9. R. E. Davies, *Nature*, **199** (1962) 1068.
10. W. F. Harrington, *Proc. Nat. Acad. Sci.*, **68** (1971) 685.
11. H. E. Huxley, *J. Biophys. Biochem. Cytol.*, **3** (1957) 631.
12. M. K. Reedy, *J. Mol. Biol.*, **31** (1968) 155.
13. C. Cohen and K. Holmes, *J. Mol. Biol.*, **6** (1963) 423.
14. H. S. Slater and S. Lowey, *Proc. Nat. Acad. Sci.*, **58** (1967) 1611.
15. A. J. Murphy and M. F. Morales, *Biochemistry*, **9** (1970) 1528.
16. T. Tokiwa and M. F. Morales, *Biochemistry*, **10** (1971) 1722.
17. K. Tawada and F. Oosawa, *J. Mol. Biol.*, **44** (1969) 309.
18. R. Cooke and M. F. Morales, *J. Mol. Biol.*, in press.
19. W. Beschorner and A. Weber, *Biophys. Soc. Abstracts*, **11** (1971) 106a.
20. M. Barany, K. Barany and H. Oppenheimer, *Nature*, **199** (1963) 694.
21. J. J. Blum, *Arch. Biochem. Biophys.*, **55** (1955) 486.
22. K. Imamura, J. A. Duke and M. F. Morales, *Arch. Biochem. Biophys.*, **136** (1970) 2.
23. M. Ulbrecht and J. C. Ruegg, *Experientia*, **27** (1971) 45.
24. H. E. Huxley, *J. Mol. Biol.*, **7** (1963) 281.
25. M. F. Morales, *Rev. Mod. Phys.*, **31** (1959) 426.
26. R. J. Podolsky and A. C. Nolan, *Symposium of the Society of General Physiologists*, R. J. Podolsky (ed.), Prentice-Hall, New York, 1971.
27. S. Fujime and S. Ishiwata, *J. Phys. Soc. Japan*, **29** (1970) 1651.
28. H. E. Huxley, *Biophys. Soc. Abstr.*, **11** (1971) 235a.
29. J. A. Spudich and H. E. Huxley, *Biophys. Soc. Abstr.*, **11** (1971) 220a.
30. F. Morita and K. Yagi, *Biochem. Biophys. Res. Comm.*, **22** (1966) 297.
31. J. C. Seidel, M. Chopek, and J. Gergely, *Biochemistry*, **9** (1970) 3265.
32. T. Tokiwa, *Biochem. Biophys. Res. Comm.*, **44** (1971) 471.
33. S. Lowey and S. M. Luck, *Biochemistry*, **8** (1969) 3195.
34. A. J. Murphy, J. A. Duke and L. Stowring, *Arch. Biochem. Biophys.*, **137** (1970) 297.
35. D. B. Stone and S. Prevost, *Biochim. Biophys. Acta* (1971), in press.
36. W. W. Kielley, H. Kalckar and L. B. Bradley, *J. Biol. Chem.*, **219** (1956) 95.
37. E. W. Taylor, R. W. Lymn and G. Moll, *Biochemistry*, **9** (1970) 2984.
38. P. Rainford, K. Hotta and M. F. Morales, *Biochemistry*, **3** (1964) 227.
39. T. Sekine and W. W. Kielley, *Biochim. Biophys. Acta*, **81** (1964) 2.
40. J. C. Warren, L. Stowing and M. F. Morales, *J. Biol. Chem.*, **241** (1966) 309.
41. J. A. Duke, R. McKay and J. Botts, *Biochim. Biophys. Acta*, **126** (1966) 600.
42. J. Quinlivan, H. M. McConnell, R. Cooke and M. F. Morales, *Biochemistry*, **8** (1969) 3188.
43. H. C. Cheung and M. F. Morales, *Biochemistry*, **8** (1969) 2177.
44. H. C. Cheung, *Biochim. Biophys. Acta*, **194** (1969) 478.
45. J. F. Aronson and M. F. Morales, *Biochemistry*, **8** (1969) 4517.
46. C. G. Dos Remedios, R. Millikan and M. F. Morales, *J. Gen. Physiol.* (1971), in press.
47. R. G. Yount, D. Babcock, W. Ballantyne and D. Ojala, *Biochemistry*, **10** (1971) 2484.

48. W. F. McClare, *J. Theoret. Biol.*, **30** (1971) 1.
49. J. C. Seidel and J. Gergely, *Biochem. Biophys. Res. Comm.*, **44** (1971) 826.

In conversation, one of us (M. F. M.) learned that, independently, Dr. John Gergely and his colleagues have also been considering the existence of "myosin*ADP". After this manuscript was completed there appeared a paper[49] reporting their own interesting evidence for this concept.

Dr. Joseph Duke has now shown by direct experiments that after affinity labelling with SH–ATP the nucleotide actually remaining attached is the diphosphate; this result confirms the assumption on p. 60.

Bioenergetics of Nerve Excitation

Ichiji Tasaki and Mark Hallett

Laboratory of Neurobiology
National Institute of Mental Health,
Bethesda, Maryland

Abstract

The process of action potential production is analyzed in relation to the problem of energy transduction in the nerve. Describing the conditions required for the maintenance of excitability, the indispensability of divalent cations and the dispensability of univalent cations in the external medium are emphasized. Univalent cations with a strong tendency toward hydration enhance the action potential amplitude when added to the external Ca-salt solution. Experimental facts are described in consonance with the macromolecular interpretation of nerve excitation which postulates a transition of the negatively charged membrane macromolecules from a hydrophobic (resting) state to a hydrophilic (excited) state. Thermodynamic implications are discussed in relation to changes in enthalpy and volume accompanied by action potential production. Difficulties associated with analyses of excitation processes on a molecular basis are stressed.

Introduction

From a physicochemical and biochemical point of view, nervous tissue is peculiar, difficult material to investigate. The physicochemical processes underlying production of an action potential occur within an extremely thin membrane structure and progress at a disturbingly high rate. Most of the chemists' standard tools are totally inadequate to follow such rapid processes involving such a limited quantity of chemical substance in labile, "living" tissue. Only electronic devices employed by communication engineers have had the sensitivity and the rapidity to respond to the signs of physicochemical events taking place in the nerve membrane. For this reason, "axonology" has developed almost as a branch of applied electronic engineering.

In spite of this past and present trend, many "axonologists" are keenly aware of the fact that precise measurements of electrical quantities alone do not yield meaningful information about what is happening in the nerve membrane. It is now widely recognized that the goal and the destiny of "axonology" is toward harmony and amalgamation with the branch of science known as molecular biology. New instruments and techniques introduced in recent years to study excitation processes may be regarded as products of a painstaking struggle to achieve this goal.

In this article, we make an attempt to review the experimental findings obtained by using these new instruments and techniques. We clarify, in the first place, the conditions which internal and external media have to satisfy in order to maintain excitability. Then, we describe the experimental data concerning ion fluxes across the membrane. Next, we discuss our present day knowledge about the thermochemical and metabolic processes in axons. Attempts will be made, whenever possible, to point out thermodynamic implications of the experimental findings described.

Conditions Necessary for Maintenance of Excitability

Soon after Hodgkin and Katz[1] confirmed and extended Overton's[2] finding demonstrating the importance of the role played by sodium ion in nerve excitation, many electrophysiologists believed that the presence of Na^+ in the external medium (or the existence of an electrochemical potential gradient for Na^+) was an indispensable factor for the maintenance of axon excitability. Later, however, Lorente de No and his associates[3] found that the ability of frog nerve fibers to develop action potentials could be maintained in a medium containing no Na^+, namely after replacing Na^+ in the medium with guanidinium and other univalent cations (cf. Hille,[4] for more recent data).

Figure 1 shows two examples of observations demonstrating the dispensability of Na^+ in squid giant axons.[5] The axons were internally perfused with a Rb-salt solution and were immersed in a medium containing either hydrazinium or guanidinium chloride. There was neither Na^+ nor K^+ on either side of the membrane. It is important to note, however, that a calcium-salt was present in the external medium in addition to the salt of nitrogenous univalent cations. It is seen in these records that the all-or-none action potentials evoked were accompanied by a simultaneous reduction in the membrane impedance. The voltage clamp technique applied to axons under these conditions yielded an N-shaped curve which was qualitatively similar to those obtained with Na^+ (and Ca^{2+}) externally and K^+ internally.

An extensive study was made concerning the ability of polyatomic univalent cations in the external medium to enhance the action potential amplitude. Ammonium-ions can, to some extent, replace Na^+ in the external medium and maintain axon excitability. When one H-atom on NH_4^+ was replaced with CH_3, the Na-substituting capability of the resultant cation was decreased considerably. With two or more H-atoms substituted with nonpolar groups, such as methyl or ethyl, the resultant cation was no longer capable of replacing Na^+. When, however, the H-atoms were replaced with polar groups, such as amino or hydroxyl, the resultant cation was capable of

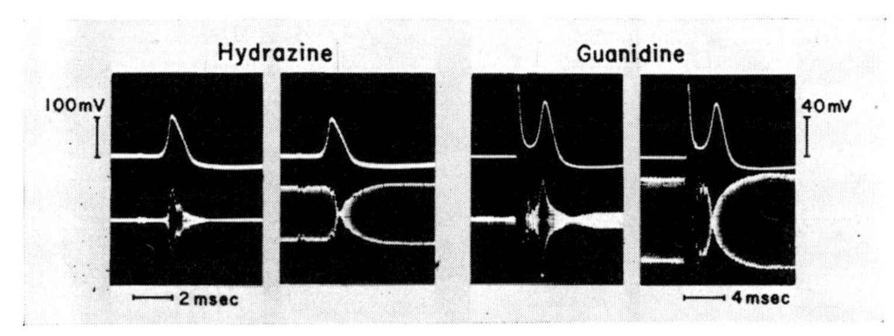

Figure 1. Impedance and potential variation recorded from squid giant axons immersed in sodium-free external media. In each record, potential variations are indicated by upper trace; simultaneous impedance variations are indicated by lower trace. In each pair of records, the record at left was obtained when the impedance bridge was balanced initially for the resting state. Record at right was obtained when the bridge was unbalanced in the resting state, so that best balance could be obtained in the excited state. *Left.* The external medium contained 0·3 M hydrazinium chloride and 0·2 M $CaCl_2$; the internal perfusion solution contained 0·1 M RbF. *Right.* The external medium contained 0·1 M guanidinium chloride, 0·2 M TMA-Cl and 0·2 M $CaCl_2$; the internal solution contained 0·1 M RbF. (From I. Tasaki, I. Singer and A. Watanabe, *Amer. J. Physiol.*, **211** (1966) 746.)

sustaining large action potentials. These findings can be summarized in the following manner:

$$CH_3 < H < OH \sim NH_2$$

where the inequality signs describe the relative magnitude of the amplitude-augmenting effect on the action potential.[6]

Soon after the dispensability of Na^+ in squid giant axons was established, there was another step forward in our understanding of the ionic requirement of the axon membrane.[7, 8] It was found that, under intracellular perfusion with a dilute solution of favorable salts, large action potentials could be obtained in the complete absence of univalent cation in the external medium. The first demonstration of action potentials of this type was made under maintained internal perfusion with a solution containing a mixture of CsF and Cs-phosphate at pH 7·3. The external medium contained only $CaCl_2$ and a trace of Tris-buffer. (Glycerol or sucrose was added to the media on both sides of the membrane to maintain the tonicity.)

The Cs-salts used in those initial experiments were not the only favorable internal electrolytes.[9] Tetramethylammonium (TMA) tetraethylammonium (TEA), guanidium, aminoguanidinium, Na^+, Li^+, etc., were found to be usable instead of Cs^+ in this type of experiment. The action potentials observed with the salt of a univalent cation internally and the salt of a divalent cation externally are often called "bi-ionic action potentials". The amplitude of a bi-ionic action potential was found to vary with the chemical species of the internal cation. Among the nitrogenous cations mentioned above, TMA and TEA gave rise to very large bi-ionic action potentials. The salt of guanidinium or aminoguanidinium ion were found to produce all-or-none, but small action potentials. In general, the cations that are favorable as Na-substitutes (externally) gave rise to small bi-ionic

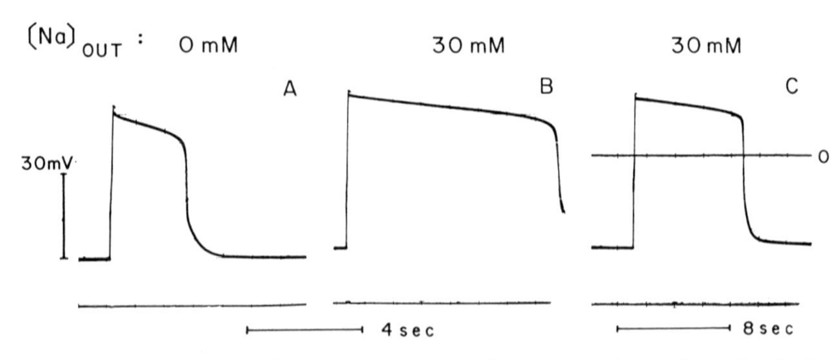

Figure 2. Resting and action potentials recorded from an axon internally perfused with a 30 mM sodium solution. The external media contained 100 mM $CaCl_2$ for all the records. The external sodium concentrations were varied as indicated. The internal anion was 1:1 mixture of fluoride and phosphate in this case. Stimuli used were 100 msec in duration and approximately 1·5 $\mu a/cm^2$ in intensity (indicated by the lower trace). Note that record C was taken at a slower sweep speed. In record C, the potential recorded when the internal recording electrode was withdrawn and placed in the external medium was superposed on the action potential trace. Axon diameter: approximately 450 μ. 20°C. (From A. Watanabe, I. Tasaki and L. Lerman, *Proc. Nat. Acad. Sci.*, **58** (1967) 2246.)

action potentials, and vice versa. In other words, the sequence of substituted ammonium-ions in affecting the action potential amplitude determined by internal application is roughly opposite to that established by external application. The significance of these findings will be clarified later.

The amplitude of a bi-ionic action potential is affected to some extent by the divalent cation concentration in the external medium, a higher concentration leading to a larger amplitude. Addition of the salt of Na^+ or its substitutes increases the amplitude also. In the external medium, Sr^{2+} or Ba^{2+} (but not Mg^{2+}) can be used to sustain bi-ionic action potentials. It is important to note that in the internal medium the divalent cation concentration has to be kept at an extremely low level, for the axon membrane would otherwise be irreversibly damaged. Figure 2 shows an example of the action potential records obtained under continuous intracellular perfusion with a dilute solution of a Na-salt.

We may summarize the experimental facts described above in the following manner. To maintain the ability of a squid giant axon to develop all-or-none action potentials, the presence of the salt of a divalent cation (Ca, Sr, or Ba) externally and of the salt of a univalent cation internally is required. Addition to the external medium of the salt of a univalent cation with a strong tendency toward hydration increases the action potential amplitude. Univalent cations with nonpolar (hydrophobic) side-groups are more favorable as the internal cations than those with polar side-groups. (Information concerning the effects of different anions may be found elsewhere.[10])

Ion Fluxes Across Axon Membrane

The electric resistance (or impedance) of the axon membrane is known to fall drastically when the membrane develops an action potential.[11] Reflecting this situation, there is, during an action potential, an enormous increase in the fluxes of cations across the membranes; this can be demonstrated readily by the use of the radio-tracer technique.[12, 13] Unfortunately, the time resolution of the radio-tracer technique is quite limited. Furthermore, for ion species which have finite concentrations on both sides of the membrane, estimation of the (net) ion flux is difficult (see Kedem and Essig[14]).

A squid giant axon immersed in artificial sea water remains highly excitable under continuous intracellular perfusion with a K-phosphate solution. The influx of Na^+ can be traced without ambiguity, under these conditions, by labelling the external Na-ions with Na^{22} or Na^{24}. Complete removal of the K^+ in the artificial sea water changes neither the resting potential nor the action potential of the axon significantly. The efflux of K^+ can then be traced without ambiguity by labelling the internal perfusion fluid with K^{42}. Because of the poor time resolution of the radio-tracer technique, however, no information can be obtained by this technique concerning the time course of the fluxes during different phases of an action potential.

Figure 3 in reference 15 shows an example of the results obtained with radioactive K^+. The average level of K-efflux during one sampling period (5 min) is seen to show a definite increase when the axon was stimulated repetitively at a rate of 25/sec. The extra K-efflux was roughly doubled when the frequency of stimulation was doubled. From these measurements, the amount of K-ion transferred across the membrane associated with production of an action potential was estimated to be about 15 pmole/cm^2 per nerve impulse.[15] This value is not very different from those obtained by previous investigators (see below).

Transport of one species of cation across the membrane should be accompanied by transfer of net electric charge. The amount of 15 pmole carries approximately $1{\cdot}5 \times 10^{-6}$ coulomb which would alter the potential difference across the membrane capacity of 1 $\mu F/cm^2$ by 1·5 V. Actually, such charge transfer does not take place, because this K-efflux is accompanied by simultaneous influx of Na^+. The total amount of Na^+ transferred during one action potential is known to be very close to the value for K^+. (Note, however, that previous ion-flux measurements were always complicated by the presence of the non-radioactive counterpart of the labelled species on both sides of the axon membrane.) There is also a drastic increase in Ca-influx during excitation;[16,7] but the total charge transferred by Ca^{2+} is far smaller than the value mentioned above.

Had the axon membrane no capacitative properties, the axon interior would be expected to satisfy the conditions of electroneutrality. The apparent capacity of the membrane is about 1 $\mu F/cm^2$; and this capacity is charged at the peak of the action potential to about 0·1 v above its resting potential level. This process involves transfer of charge of about 10^{-7} coulomb/cm², or about 1 pmole/cm² of univalent cations. This quantity is far smaller than (and is equal to only 7% of) the amount of univalent cation transported across the membrane during an action potential. It follows from these considerations that the deviation from the condition of electroneutrality of the axon interior is less than about 7% and, consequently, that the fluxes of Na- and K-ions are roughly equal in magnitude and opposite in sign during the entire course of an action potential. It is important to note that this requirement is independent of the difference in intramembrane mobility between the two cations involved.

It is of some interest to consider, in passing, the amount of free energy dissipated as the result of cation interdiffusion during an action potential. We denote the difference in chemical potential of ion species i across the membrane by $\Delta\mu_i$ and the quantity of this species transferred across the membrane by n_i. Then, the free energy dissipated during one complete cycle of an action potential is given by the sum of products $n_i \, \Delta\mu_i$ for all the ion species involved. Under the conditions of internal perfusion mentioned above, n's are known; but $\Delta\mu$'s are either positive or negative infinity. Since ions are moving downhill along their chemical potential gradient, this dissipation of free energy is very large. Obviously, this energy derives from the difference in the ion concentrations maintained by perfusion. Similarly, there is a very large free energy dissipation associated with production of a bi-ionic action potential.

In an intact axon immersed in normal sea water $\Delta\mu$'s are finite. It is difficult, however, to estimate $\Delta\mu$'s and n's accurately under these conditions. Assuming that the net flux of Na^+ and K^+ to be about 5 pmole/cm^2 per impulse and the Na- and K-ion activity ratio across the membrane to be 10:1 and 1:10, respectively, the amount of free energy dissipated (in the form of entropy production) is found to be roughly 3×10^{-9} cal/cm^2 per impulse. [The free energy ($\frac{1}{2}CV^2$) lost (or gained) by charging or discharging the membrane capacity (C) when the voltage (V) involved is ± 50 mV is about 10 times less than the value stated above.] The contribution of the divalent cation fluxes to energy dissipation is also considered to be very small. The difference in ion activities (concentrations) in this case is maintained by metabolism (see below).

Macromolecular Interpretation of Nerve Excitation

The electric resistance of the membrane of an unperfused axon is known to fall, at the peak of excitation, to about 1/200 of its value in the resting state.[11] Under bi-ionic conditions,[9] the ratio of the membrane resistance at the peak of an action potential to that at rest is known to be 1/20. This large fall indicates that there is a drastic change in the structure of the membrane during excitation.

Most biological macromolecules have their isoelectric points in the acidic range of pH; therefore, at physiological pH, they are negatively charged. There is ample evidence to suppose that the macromolecules of which the axon membrane is composed are negatively charged[6,7] and, as the consequence, that the membrane exhibits properties of a cation-exchanger membrane.[17] In inanimate ion-exchanger membranes, the membrane conductance varies directly with its water content.[18,19] It is therefore highly probable that, during nerve excitation, the membrane macromolecules undergo a drastic conformational change which leads to an increase in the membrane water content.[6]

The possibility that the membrane macromolecules may undergo a conformational change during excitation presents a new problem concerning energy transduction in axons. When the membrane is thrown into a state with a low electric resistance, the cations start to flow rapidly in the direction of negative concentration gradient. The observed increase in cation fluxes during excitation is then nothing more than a reflection of high cation mobilities in the low resistance state of the membrane; and, the direction of ion flow is determined simply by the distribution of the cation species chosen by the experimenter (see e.g., Fig. 2) or adopted by nature. Thus, this line of argument leads us to believe that radio-tracer measurements as well as resistance measurements tell us something about the low resistance (i.e., excited) state of the membrane but yield no information about how such a state is produced by a stimulus. Thus, a different approach has to be adopted to elucidate the mechanism whereby the membrane macromolecules are transformed from their "resting" conformation to their "excited" conformation.

Many electrophysiologists visualize the axon membrane as consisting of "ion-pores" which can be opened or closed by application of an electric field.[20,21,4] This popular viewpoint is vividly illustrated in a figure published by Baker in *Scientific American*.[21] In the resting state, the membrane is assumed to possess pores suited for transport of only K^+. On application of an electric field, a new type of pores, capable of transporting Na^+, are opened. Then, after a certain delay, these Na-pores are automatically closed and more K-pores are opened. Obviously, this scheme is a pictorial representation of the equivalent circuit theory of Hodgkin and Huxley.[22] One shortcoming of this scheme is that it does not answer the question of how ion-pores with such sizes and complex time-dependent characteristics can be constructed with proteins and phospholipids.

An alternative approach to the problem of nerve excitation is to rely on the knowledge about the behavior of various bio-colloids and macromolecules. The axon membrane, consisting of negatively charged macromolecules, is immersed in a medium usually containing the salts of divalent and univalent cations. If the charge density is high, the electroselectivity of the membrane (see Helfferich[23]) is expected to selectively bring divalent cations to the negatively charged sites. There is good evidence that the charge density in the squid axon membrane is much higher than the anion concentration in the external medium.[6] In physical chemistry, many instances are known[24-26] in which synthetic polyacids make a hydrophobic complex with Ca^{2+} or Sr^{2+} but not with Na^+ or K^+. It is also known that a cation exchanger with a rigid crystalline structure sometimes undergoes a phase-transition when one kind of counter ion is replaced with another.[27,28] Since the axon membrane has a rigid, crystalline structure,[29] it is possible that the membrane macromolecules undergo some kind of phase-transition when the chemical composition of the external medium is varied. A series of experiments carried out on internally perfused axons[30] indicate that such a transition actually occurs (Fig. 3).

An axon, immersed in a $CaCl_2$ solution, was internally perfused with a dilute Cs-salt solution. The axon in this circumstance is capable of responding to electric stimuli with all-or-none action potentials; but no electric stimulus was delivered during the following observation. When a small amount of KCl solution was added to the external $CaCl_2$ solution, no significant change in the membrane potential was observed. When the external K^+ concentration was raised to 10 mequiv/l, a large, abrupt rise in the membrane (intracellular) potential was observed after a long delay. This abrupt potential rise was accompanied by a simultaneous fall in the membrane resistance. Following production of an abrupt potential rise, a further continuous increase in the external KCl concentration changed the membrane potential continuously. A decrease in the KCl concentration or an increase in $CaCl_2$ concentration frequently produced an abrupt fall in the membrane potential.

This phenomenon of abrupt depolarization was first described by Osterhout and Hill[31] who analyzed the effect of KCl on single plant cells of Nitella. These investigators regarded this potential jump as generation of an action potential with an infinitely long depolarization period. We recognize striking similarities between this phenomenon

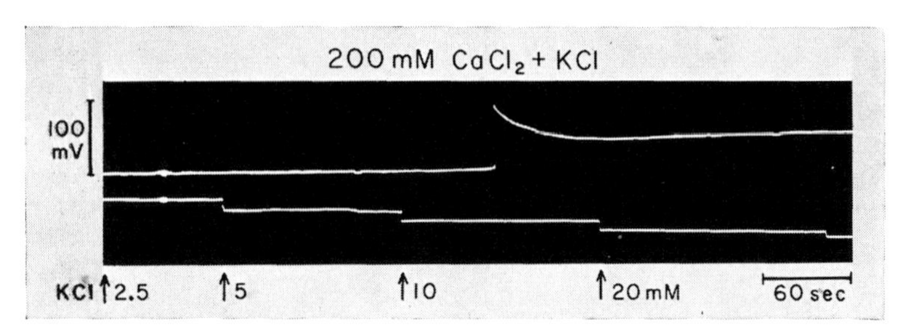

Figure 3. Abrupt depolarization of squid axon membrane induced by external application of KCl (upper oscillograph trace). The KCl concentration in the rapidly flowing external fluid medium was raised by a factor of 2 at the times marked by the lower oscillograph trace. The axon was perfused internally with a CsF perfusion fluid. No electric stimuli were delivered to the axon during the experiment. The potential jump produced by 10 mM KCl was 83 mV initially, and was followed by a gradual potential fall of approximately 40 mV. 16°C. (From I. Tasaki, T. Tekenaka and S. Yamagishi, *Amer. J. Physiol.*, **215** (1968) 152.)

and the phenomenon of phase-transition in various synthetic and natural macromolecules.[25,26,32] As one might expect from those *in vitro* experiments, abrupt depolarization in axons with prolonged action potentials can be produced not only with the salt of K^+ but also with those of Rb^+, Cs^+, or Na^+. The existence of a hysteresis loop has been demonstrated in axons undergoing abrupt depolarization and repolarization.[30]

In axons which develop action potentials of a short duration, it is practically impossible to raise the external K^+-concentration uniformly during the time involved in production of an action potential. This is the reason why abrupt depolarization has not been observed in intact axons.

The natural termination of the action potential (repolarization), in a manner opposite to depolarization, might be characterized as an exchange of Ca^{2+} for K^+. An experiment on repolarization similar to K^+-depolarization would be to raise the external Ca^{2+}-concentration during the action potential, which is also clearly impossible in axons which develop action potentials of short duration. In unperfused axons immersed in normal sea water, there is an enormous increase in cation fluxes during excitation. This increase is expected to bring about a gradual change in the ionic composition (including Ca^{2+}) in and on both sides of the axon membrane. There seems little doubt that the rapidity of the potential fall from the peak of an action potential is directly related to the rate at which the ionic composition changes during excitation.[6]

Thermochemical Studies of Axons

In inanimate cation-exchangers, exchange of K^+ or Na^+ for Ca^{2+} is known to be exothermic; the change in enthalpy associated with such exchange is of the order of 2·5 kcal per equivalent of exchanger.[33–35] In consonance with the notion that a cation-exchanger process is operative in axons, the initiation of an action potential in axons is known to be exothermic.[36, 37] The macromolecular interpretation of nerve excitation mentioned above demands that the termination of an action potential involves exchange of divalent cations for univalent cations. In agreement with this interpretation, the process of action potential termination is endothermic.[36, 37]

The positive initial heat produced in the rabbit vagal nerve[37] is about 24 μcal/g. With an estimated membrane area of 6000 cm^2/g, the heat production in the membrane is roughly 0·004 $\mu cal/cm^2$. One explanation is that all of this heat derives from ion-exchange processes. If exchange of 2 Na^+ for Ca^{2+} at the negatively charged membrane sites is taken as generating 2·5 kcal/equiv, then the density of the sites in the membrane is roughly $1{\cdot}6 \times 10^{-12}$ equiv/cm^2 or 10^{12} sites/cm^2. A somewhat larger estimate for the number of sites was obtained from measurements of hydrophobic probe binding to squid axons (unpublished). Estimation based on the effect of tetrodotoxin[38] is very different; this discrepancy may be resolved by assuming that TTX affects only trigger sites for a cooperative process in the axon membrane (see below).

LeChatelier-Braun's law states that an endothermic process is promoted by a rise in temperature. Hence, it is expected that exchange of Ca^{2+} for univalent cations in the axon membrane is encouraged by a temperature rise. The fact that a brief heat pulse (as well as a Ca-pulse) applied during the plateau of a prolonged action potential can bring about abrupt repolarization[39] is therefore in agreement with the macromolecular interpretation. It is well-known that the action potential duration is reduced by a rise in temperature; this fact is also consistent with the notion that repolarization of the axon membrane is an endothermic process.

The fact that the process of action potential termination is endothermic poses an interesting thermodynamic problem. Since the free energy of the system (G) must not increase during a spontaneously progressing process, there must be a simultaneous increase in the entropy ($\Delta S > 0$) large enough to overcome the increase in the enthalpy ($\Delta H > 0$). (Note that $\Delta G = \Delta H - T\Delta S \geqslant 0$, where T is the absolute temperature.) It was surmised that the expected increase in entropy is associated with rearrangement of water molecules. In the excited state of the axon, water molecules are tightly fixed around the negatively charged sites of the membrane macromolecules; when this "special hydration region" (see Oosawa[40]) in the membrane is destroyed by invasion of divalent cations, the entropy of the system is expected to increase.

The notion that water molecules are involved in nerve excitation received experimental support when the physicochemical properties of favorable univalent cations in- and outside the axon were examined. As we have discussed already, cations of a hydrophilic nature (Na^+, Li^+, hydrazinium, guanidinium, etc.) are capable of sustaining large action potentials when applied externally. In the internal solution, polyatomic cations with hydrophobic side-groups (TMA, TEA, choline, etc.) are favorable. These findings can readily be explained on the assumption that nerve excitation involves a transition of the membrane from a hydrophobic (resting) state to a hydrophilic (excited) state. Additional support of this notion was obtained from the studies of changes in fluorescent light emitted by a "hydrophobic probe". In axons injected with 2-*p*-toluidinylnaphthalene-*b*-sulfonate (TNS), it was found that the wave-length of emission maximum shortens and the quantum yield falls when an action potential is produced.[29] These changes are typical of the effects associated with a rise in the polarity (dielectric constant) of the microenvironment of TNS molecules.

The density of water molecules in the "special hydration region" is higher than that in ordinary water.[40] Hence, a small decrease in the total volume of a nerve (including the surrounding medium) is expected to occur when an action potential is produced and restoration of the volume to take place on termination of an action potential. Although no direct detection of a volume change has been made, the existence of such a volume change has been inferred from studies of the effect of a high hydrostatic pressure on frog myelinated nerve fibers and squid axons.[41] It was found that high pressure increased the action potential duration reversibly. The effect of heavy water on excitation processes[42, 43] also indicates the importance of water in the process of action potential production.

From the thermodynamic considerations stated above, it may be argued that the hydrophobic-hydrophilic transition in the axon membrane exhibits many characteristics of a phase-transition of the first-order. In a true first order phase-transition, the change in the total free energy (ΔG) is zero; in the axon membrane the change in the free energy associated with rapid, reversible transition between the two membrane states (ΔG) is undoubtedly very small as compared with the changes in enthalpy (ΔH) and in entropy ($T\Delta S$).

In this connection, it is important to note that the energy delivered to the membrane by a brief electric stimulus is extremely small. In threshold stimulation, an action potential is produced when the potential difference across the axon membrane is altered by 12–25 mV. Since kT/e (where k is the Boltzmann constant and e is the electronic charge) is approximately 25 mV, and since the thermal energy is $\frac{1}{2}\,kT$ per degree of freedom, the energy delivered by a stimulus is, by itself, insufficient to drastically alter the thermal motion of the ions, water or of side-groups of macromolecules in the membrane. It is, by itself, far smaller than the value needed to break the weakest chemical bonds such as H-bonds. (Note that $\frac{1}{2}\,kT$ corresponds to about 0·3 kcal/mole.) We introduce at this point the concept of thermodynamic instability (see e.g., Guggenheim,[44] p. 37). The system consisting of the axon membrane and the two contiguous aqueous phases is stable against very small perturbations;[6] but, the system is so close to an unstable state that large perturbations are not required to trigger a large change in the system cooperatively. The nature of the forces involved in this cooperative change of the axon membrane is not well understood at present. It has been pointed out, however, that both short-range forces (action between neighboring sites in the membrane) and long-range forces (mediated by electric currents) are operative in the axon membrane.[6]

Active Transport and Metabolism

It is generally believed that active transport and metabolism in the nerve are only indirectly related to the process of action potential production. In order to maintain excitability, it is essential that the normal structure of the membrane macromolecules be kept unaltered and the chemical composition of the axon interior be maintained within a certain limit. Particularly, a rise in the internal Ca^{2+} concentration is detrimental to the maintenance of excitability. A rise in the internal Na^{+} concentration tends also to diminish the action potential amplitude. Active transport and metabolism is regarded as playing a role of maintaining a high K^{+}, a low Na^{+} and a low Ca^{2+} concentration in the axon interior.

Investigations of active transport in axons rely almost exclusively on measurements of radio-tracers of the ion species involved. Radio-tracer fluxes across biological membranes have been analyzed on the basis of thermodynamics of irreversible processes.[14, 45] From this standpoint, it is in general very difficult to decide whether an observed tracer flux is taking place through the active transport channel or through the passive channel.[45] Partly because of this difficulty and partly because the unfamiliarity of the present authors with the vast literature dealing with this subject, the problem of active transport and metabolism is treated in this article only superficially.

It seems clear from extensive studies of cation transport in squid giant axons[46] that there are two distinct processes influencing the efflux of the radio-tracer (and hence the cold species) of Na^+. The first process is considered to account for 50–90% of the Na-efflux in normal axons. The Na-efflux by this pump is coupled to the K-influx and the required energy derives from the metabolism of high energy phosphate compounds, particularly from ATP.[47,48] This pump is inhibited by cyanide which blocks ATP synthesis and by ouabain which works in an unknown fashion (without seemingly interfering with metabolism). Many biologists believe that this type of Na-pump can be electrogenic, that is to say, it alters the membrane potential when the K-influx and Na-efflux are not strictly equal (see e.g., Rapoport[49]). The second kind of Na-efflux process involves a Na–Ca exchange.[46] The Na-efflux by this process is independent of the external K^+ concentration and ouabain, but is dependent on the external Ca^{2+} concentration. This pump can also be blocked by cyanide. Recent studies[50] indicate that this process in reverse, Ca-efflux coupled to Na-influx, is present in nerve; this is the only pump found so far in nerve which deals with Ca-extrusion.

The metabolic activity of nervous tissues has been studied using squid axons,[51,52] crab nerves,[53] mammalian non-myelinated nerve fibers,[54] sympathetic ganglia,[55] and crayfish stretch receptor neurons.[56] The conclusions drawn from these studies are that, in the presence of oxygen, glucose is metabolized through glycolysis, the Kerbs cycle and oxidative phosphorylation and the end product is ATP. Repetitive stimulation is known to enhance metabolism of the nerve.

Current Trend of Research Activities

Most of the current research can be divided into the following four categories: (1) precise measurements of various electrical properties of the axon membrane;[57,58] (2) measurements of non-electrical phenomena during nerve excitation including changes in turbidity,[59] birefringence,[60] and extrinsic fluorescence;[29] (3) experimental studies of model membranes;[61,62] and (4) theoretical studies of excitation processes on a mathematical, physical or physicochemical basis.[63–68] It is evident that most investigators are interested in elucidating the excitation process on a molecular basis.

Theoretical approaches are expected to be quite valuable, but are (and always will be) limited by the quality and quantity of the experimental results available. Model membrane studies have been exciting because the underlying molecular events are relatively well defined and amenable to analysis; their limitation lies in the uncertainty as to the similarity between model and real membranes. Further analyses of electrical properties are expected to reveal shortcomings of previous experimental data. It is our view that studies of non-electrical properties of the axon membrane should lead to deeper insights into the molecular events in the nerve membrane. With this new information, our understanding of excitation processes may reach in the near future a level far beyond what has been described in this article.

References

1. A. L. Hodgkin and B. Katz, *J. Physiol.*, **108** (1949) 37.
2. E. Overton, *Pflügers Arch. f. ges. Physiol.*, **92** (1902) 346.
3. R. Lorente de No, F. Vidal and L. M. H. Larramendi, *Nature*, **179** (1957) 737.
4. B. Hille, *Proc. Nat. Acad. Sci.*, **68** (1971) 280.
5. I. Tasaki, I. Singer and A. Watanabe, *Amer. J. Physiol.*, **211** (1966) 746.
6. I. Tasaki, *Nerve Excitation, A Macromolecular Approach*, Charles C. Thomas, Springfield, Illinois, 1968.
7. I. Tasaki, A. Watanabe and L. Lerman, *Amer. J. Physiol.* **213** (1967) 1465.
8. A. Watanabe, I. Tasaki and L. Lerman, *Proc. Nat. Acad. Sci.*, **58** (1967) 2246.
9. I. Tasaki, L. Lerman and A. Watanabe, *Amer. J. Physiol.*, **216** (1969) 130.
10. I. Tasaki, I. Singer and T. Takenaka, *J. Gen. Physiol.*, **48** (1965) 1095.
11. K. S. Cole and H. J. Curtis, *J. Gen. Physiol.*, **22** (1939) 649.
12. R. D. Keynes, *J. Physiol.*, **114** (1951) 119.
13. A. L. Hodgkin and R. D. Keynes, *J. Physiol.*, **119** (1953) 513.
14. O. Kedem and A. Essig, *J. Gen. Physiol.*, **48** (1965) 1047.
15. I. Tasaki, I. Singer, and A. Watanabe, *J. Gen. Physiol.*, **50** (1967) 988.
16. A. L. Hodgkin and R. D. Keynes, *J. Physiol.*, **138** (1957) 253.
17. T. Teorell, *Progr. Biophys.*, **3** (1953) 305.
18. K. Sollner, *J. Macromol. Sci. Chem.*, **A3** (1969) 1.
19. J. A. Kitchner, in: *Modern Aspects of Electrochemistry*, J. O. Brockris (ed.), Vol. II, Academic Press, New York, 1959, p. 87.
20. L. Mullins, *J. Gen. Physiol.*, **43** (1960) 105.
21. P. F. Baker, *Sci. Amer.*, **214** (1966) 74.
22. A. L. Hodgkin and A. F. Huxley, *J. Physiol.*, **117** (1952) 500.
23. F. Helfferich, *Ion Exchange*, McGraw-Hill, New York, 1962.
24. A. Ikegami and N. Imai, *J. Polymer Sci.*, **56** (1962) 133.
25. O. Smidsrod and A. Haug, *J. Polymer Sci. C.* **16** (1967) 1587.
26. E. Matijevic, J. Leja and R. Nemeth, *J. Colloid and Interface Sci.*, **22** (1966) 419.
27. R. M. Barrer and J. D. Falconer, *Proc. Roy. Soc. A*, **236** (1956) 227.
28. D. H. Olson and H. S. Sherry, *J. Phys. Chem.*, **72**, (1968) 4095.
29. I. Tasaki, A. Watanabe and M. Hallett, *Proc. Nat. Acad. Sci.*, **68** (1971) 938.
30. I. Tasaki, T. Takenaka and S. Yamagishi, *Amer. J. Physiol.*, **215** (1968) 152.
31. W. J. V. Osterhout and S. E. Hill, *J. Gen. Physiol.*, **22** (1938) 139.
32. F. T. Wall and J. W. Drenan, *J. Polymer Sci.*, **7** (1951) 83.
33. N. T. Coleman, *Soil Sci.*, **74** (1952) 115.
34. R. M. Barrer, L. V. C. Rees and D. J. Ward, *Proc. Roy. Soc. A*, **273** (1963) 180.
35. H. S. Sherry and H. F. Walton, *J. Phys. Chem.*, **71** (1967) 1457.
36. B. C. Abbott, A. V. Hill and J. V. Howarth, *Proc. Roy. Soc. B*, **148** (1958) 149.
37. J. V. Howarth, R. D. Keynes and J. M. Richie, *J. Physiol.*, **194** (1968) 745.
38. J. W. Moore, T. Narahashi and T. I. Shaw, *J. Physiol.*, **188** (1967) 99.
39. C. S. Spyropoulos, *Amer. J. Physiol.*, **200** (1961) 2064.
40. F. Oosawa, *Polyelectrolytes*, Marcel Dekker, Inc., New York, 1971.
41. C. S. Spyropoulos, *J. Gen. Physiol.*, **40** (1957) 849.

42. C. S. Spyropoulos and E. M. Ezzy, *Amer. J. Physiol.*, **197** (1959) 808.
43. F. Conti and G. Palmieri, *Biophysik*, **5** (1968) 71.
44. E. A. Guggenheim, *Thermodynamics*, 3rd Ed., Interscience, New York, 1957.
45. A. Essig, *Biophys. J.*, **8** (1968) 53.
46. P. F. Baker, M. P. Blaustein, A. L. Hodgkin and R. A. Steinhardt, *J. Physiol.*, **200** (1969) 431.
47. P. C. Caldwell, *J. Physiol.*, **152** (1960) 545.
48. F. J. Brinley and L. J. Mullins, *J. Gen. Physiol.*, **52** (1968) 181.
49. S. I. Rapoport, *Biophys. J.*, **11** (1971) 631.
50. M. P. Blaustein and A. L. Hodgkin, *J. Physiol.*, **200** (1969) 497.
51. C. M. Connelly, *Biol. Bull.*, **103** (1952) 315.
52. M. G. Doane, *J. Gen. Physiol.*, **50** (1967) 2603.
53. P. F. Baker, *J. Physiol.*, **180** (1965) 383.
54. J. M. Richie, *J. Physiol.*, **188** (1967) 309.
55. M. G. Larrabee, *Progr. in Brain Res.*, **31** (1969) 95.
56. E. Giacobini, *Protoplasma*, **63** (1967) 52.
57. L. A. Cuervo and W. J. Adelman, *J. Gen. Physiol.*, **55** (1970) 309.
58. H. N. Fishman, *Biophys. J.*, **10** (1970) 799.
59. L. B. Cohen and R. D. Keynes, *J. Physiol.*, **212** (1971) 259.
60. L. B. Cohen, B. Hille and R. D. Keynes, *J. Physiol.*, **211** (1970) 495.
61. P. Mueller and D. O. Rudin, *J. Theor. Biol.*, **18** (1968) 222.
62. G. Ehrenstein, H. Lecar and R. Nossal, *J. Gen. Physiol.*, **55** (1970) 119.
63. D. E. Goldman, *Biophys. J.*, **4** (1964) 167.
64. L. Bass and W. J. Moore, in: *Structural Chemistry and Molecular Biology*, A. Rich and N. Davidson (eds.), W. H. Freeman and Co., San Francisco, 1968, p. 356.
65. G. Adam, *Z. Naturforsch.*, **23b** (1968) 181.
66. J. H. Wang, *Proc. Nat. Acad. Sci.*, **67** (1970) 916.
67. L. Y. Wei, *Math. Biophys.*, **33** (1971) 187.
68. T. L. Hill and Y.-D. Chen, *Proc. Nat. Acad. Sci.*, **68** (1971) 1711.

An Analytical Appraisal of Energy Transduction Mechanisms

R. J. P. Williams

Wadham College, Oxford

Abstract

The nature of mechanisms and energy profiles for reactions in biological systems is examined. Simple molecular and complex "phase" intermediates are described. The conceptual problems in energy transduction, related to "phase" intermediates are made evident by reference to different facets of the problem—stereochemical changes, charge fluxes, E° changes, and changes in the activities of H_2O and H^+.

The problem to which I have been asked to address myself is the mechanism of energy transduction in biology. Before inspecting the proposed solutions to the problem it is as well to have in our minds what it is that we are looking for. In other words what is the sense in which the word *mechanism* is being used. Reaction pathways can be broken down to individual steps—elementary reactions—and conventional mechanisms are descriptions of the way in which these step processes take place. Very much of our thinking about such steps is based on the consideration of gas phase reactions such as

$$H^* + HH \rightleftharpoons H + H^*H$$

where * identifies one hydrogen atom in the exchange. From the considerations of such gas reactions a mechanism has come to mean a molecular sequence with a *molecular* intermediate here H*HH, and an associated energy profile with identifiable molecular species at the minima, Fig. 1. Starting from this point of view biochemists have assumed that mechanisms of biological reactions can also be written down as energy contour diagrams in which they see a sequence of *atom movements in small* (*substrate*) *molecules*. Is this a true or necessary view of such reactions? What happens when we turn to larger assemblies of atoms not in the gas phase?

Consider a problem involving a condensed phase—for example the evaporation of an NaCl molecule, a well defined species in the gas phase, from a crystal of sodium chloride. The equilibrium vapour pressure is a measure of the relative free energy of NaCl in the gas phase and its free energy per mole in the lattice. In the gas phase the molecular entity is NaCl and we can ignore interactions between molecules although molecular concentration adjusts the free energy through the entropy. In the condensed phase NaCl no longer exists (i.e. can not be identified) as a molecule and the energy per mole of an infinite crystal of sodium chloride is the sum to infinity of all the Na^+ and Cl^- interactions in the whole of the solid phase. How does this fact, that long-range interactions dominate the crystal, alter the

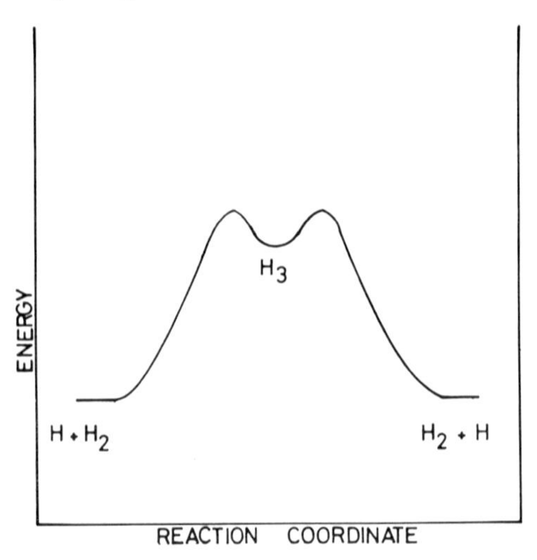

Figure 1. The energy profile for hydrogen exchange in the gas phase—a molecular intermediate.

picture of a contour diagram, Fig. 1, for a reaction—say that of the evaporation of an NaCl molecule from sodium chloride crystal surface. Clearly a potential energy diagram of the type of Fig. 1, see Fig. 2, can be drawn but while the right-hand side gives the energy of a defect lattice plus an NaCl molecule, the left-hand side gives the cooperative energy of a whole phase per mole and its description requires the whole lattice to be described (an infinite number of "molecules"). Thus a molecular mechanism for this reaction, requiring an intermediate to be drawn on a molecular scale, is impossible as the intermediate involves some degree of incipient NaCl molecule formation on the surface plus a cooperative change in the whole lattice.

Organic chemists do not often meet this lattice problem and they define reactions in terms of molecules. Their systems usually approximate to gas phase reactions and their solvents can be treated as inert matrices. They are then entitled to write contour diagrams invoking simple chemical intermediates, Fig. 1. Does such a mechanistic view apply to biological systems or is the situation closer to that of a continuous phase problem?

The name "molecule" is sensibly applied to a condensed system of a small number of atoms linked by relatively apolar bonds. It has no ultimate justification in the description of highly polar liquids or solids, e.g. sodium chloride crystals, or in a solid where the energy

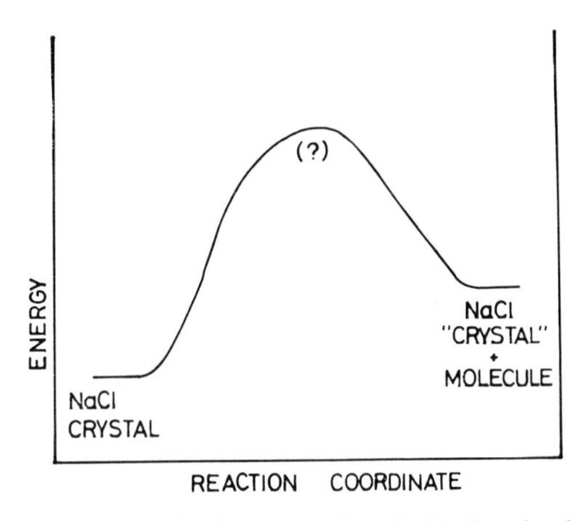

Figure 2. The energy profile for the evaporation of a NaCl molecule from a crystal of sodium chloride. No energy minimum is shown for it is quite uncertain how this should be represented even though it is known to occur. This is a case of a "phase" intermediate.

states result from any strong *cooperative* interaction over three dimensions e.g. diamond or sodium metal. What about a liquid such as water? Clearly a molecular description, H_2O, is a useful approximation for some properties but equally clearly there is considerable long-range interaction over a large number of neighbours. Evaporation of a water molecule is therefore a problem requiring consideration of large numbers of water molecules. It follows that mechanisms of reactions and their energy profiles in water cannot be fully described by the same treatments as apply to the hydrogen exchange reaction in the gas phase or to reactions of organic molecules in hydrocarbon solvents for energy is distributed into the *bulk* medium. How then should we look upon the reactions of proteins in water? Examples illustrate the type of description of the protein that must be invoked.

An obvious starting example is the reaction of myoglobin with oxygen in water for the crystal structure of the protein is known and NMR allows us to follow some of its properties in solution.

$$Mb + O_2 \rightleftharpoons MbO_2$$

Before describing the intermediates we must describe the final states $Mb + O_2$ and MbO_2. O_2 is a gas molecule of no polarity and we can feel justified in treating it as a small separate molecule. Mb is a large partly polar unit cross-linked by hydrogen-bonds and ionic-links and interacting strongly with the bulk liquid. Any alteration that is made at one point in the Mb unit must alter to some degree (1) the whole of the interactions in the chain, (2) the solvent interaction. Thus a molecular description localized upon the active site may be quite inappropriate even when comparing reactant and product and therefore to the intermediates of the reaction with oxygen. In fact such molecular mechanisms could only be said to be a valid approximation if the major part (say 90%) of the energetics of the reaction were strongly localized. We see that energy profile diagrams of enzyme reactions can be misleading if they are taken to refer to the interaction of substrate and *localized* enzyme regions for in principle a very large part of an enzyme can be involved in a reaction path. The recognition of this feature of proteins is being developed slowly as the significance of remote protein sidechains upon enzyme reactions is felt.

Cooperativity in haemoglobin can be used as an example of increased protein complexity. The binding of the α and β chains together alters the oxygen affinity or redox potentials of the active site iron of the α and β chains separately by a very general rearrangement of the protein which includes minor movements of even α-helical regions. Oxygen binding to the $\alpha_2\beta_2$ unit stimulates further rearrangement which is cooperative or anti-cooperative with proton binding (at several points), carbon dioxide binding, and anion binding in the effector site—very many anions have some effect. No description of interaction between haem units which keeps small regions of the protein in mind can be of value in the discussion of the energetics for a general problem of long-range and short-range interactions over the whole of four protein sub-units is involved. Points of interest in the mechanism of oxygen uptake, such as changes at the iron and the salt-bridges are worthy of special note but it is quite improper to partition energy between the bits.

In myoglobin and haemoglobin then there are ramifications at many remote groups of the proteins due to the movement of the FG part of the back-bone chain with oxygenation. In chymotrypsin there is a chain of interacting residues glu-his-ser and it is their interaction, and with other groups, which produces the unexplained anomolous serine reactivity; in carboxypeptidase the movement of the distant tyrosine assists attack; in ribonuclease the role of the distant arginine in the attack is not readily described; in lysozyme the anomalous pK_a of a carboxylate group and the role of tryptophans is uncertain, and so on. In all these cases a general bulk description of the protein may be necessary if all features of the energy profile are to be appreciated. In order to avoid discussion of cooperative energies covering the whole protein it is conventional to try to reduce these problems to molecular dimensions by referring to pictorial concepts such as "induced fit", and "anomalous environment". Perhaps it would be better to aim at clarifying the energetics of such processes before great store is placed on the validity of these concepts. For example many of the above enzymes bind *charged* substrates. The charge on the substrate usually binds far from the attacking groups but might it not alter the whole protein both through bond and through space? Is that an "induced fit" or an "anomalous environment"? As we turn from the above, usually extracellular, proteins to intracellular proteins it becomes apparent, as in haemoglobin, that the protein reflects almost any change of its environment—salts, pH, phosphate effectors, etc., far from the active site. Other *intracellular* proteins respond to large numbers of so-called effectors, e.g. glutamate synthatase responds to about fifty known effectors. There are general as well as specific allosteric factors. Are we being driven to realize that each *soluble* intracellular protein is reflecting the solution conditions of the whole solution phase of the cell—is it an osmometer, an electroscope, and a selective reagent for a large number of chemicals? An energy profile, Fig. 1, must then come to reflect the state of the cell as a whole as well as the molecular events recorded by looking at the substrate, intermediate, and product, for the activation energy will be changing with cell conditions.

With the above conceptual problems in mind I am able to state why I think that there is so much confusion about the mechanism of energy transduction. When a biochemist wishes to write an intermediate for this "reaction" he is tempted to postulate a molecular equation (as in the gas phase)

$$E + S \rightleftharpoons ES \rightarrow \text{products}$$

Useful though such molecular descriptions are we need to remember their limitations and to note that we have no proof that *energy transduction* can be written with any form of molecular ES complex as an intermediate for although the final reactants and products can be reasonably well written in molecular form.

$$\text{energy source: } h\nu + \text{chlorophyll} \rightarrow \text{energy},$$
$$\text{or } O_2 + 2RH_2 \rightarrow 2R + 2H_2O + \text{energy};$$
$$\text{energy sink: energy} + P + ADP \rightarrow ATP,$$

there is much evidence now that molecular intermediates cannot be so drawn for the energy transduction itself although there may be many simultaneous local changes which can be so drawn. We may have to be content with a more general extended view over a relatively large bulk of material, a *phase intermediate*, much as is true for the evaporation of NaCl molecules from a sodium chloride lattice

$$\begin{matrix}\text{perfect sodium} \\ \text{chloride } \textit{crystal}\end{matrix} \rightleftharpoons \text{NaCl } \textit{molecule} + \begin{matrix}\text{defective sodium} \\ \text{chloride } \textit{crystal}\end{matrix}$$

Energy can be so spread out as to belong to a phase.

Now the nature of the evidence about intermediates in such a situation is bound to be very complicated and the more so when the phase is very complicated. Firstly the appearance of a "phase" intermediate as opposed to a "molecular" intermediate could be observed in two ways in the course of reaction:

(a) *General* redistribution of charge instead of very *local* charge changes.
(b) *General* redistribution of atomic positions instead of very *local* changes in atom positions.

There may or may not be changes in covalent bonds so typical of the ES complexes.

As changes of charge and atomic position are often easily followed we can tackle reactions of such complicated units as mitochondria and chloroplasts, or their fragments, immediately we have defined the phase to which we are referring. (In the case of haemoglobin we refer to the total $\alpha_2\beta_2$ subunit and ignore the water in the first instance.)

In all energy-conserving reactions there is the added complication that we have to consider at first a multiphase system, one membrane and two aqueous phases at least. It is a general truth that those phases which are *in equilibrium* will rapidly return to balance reflecting changes initially brought about in one of them so that all three phases should be examined, apparently. Fortunately biological phases are not in equilibrium so that there can be "effective" or "semi-effective" discontinuity between action in one and action in any other phase. The reactions which give rise to ATP formation do not equilibrate with the aqueous phase of the cell and in the first instance we shall treat the membrane phase in which ATP is produced separately from the aqueous phase. (*Note.* Chemiosmosis does not allow us to do this.)

The Membrane Phase: Charge Distribution

The mitochondrial and chloroplast inner membranes or the regions of the outer membranes of bacteria which carry out either oxidative phosphorylation or photophosphorylation are composed of a large group of chemicals including lipids, quinones, electron-transfer proteins, ATP-ases. What do we know of the charge distribution amongst these chemicals? As yet the chemicals are so ill-defined that we can hardly make any quantitative statement about their absolute nature. However if we go from a state in which oxidative phosphorylation is occurring, state III, to one, state IV, in which it is not, then it is obvious that there is a *relatively large switch* to higher oxidation, i.e. increase in positive charge. The charge redistribution is gross and involves upwards of twenty proteins and many small molecules. It is our desire to find a connection between the changes in these charged entities and the ATP-ase(s) which are apparently unaffected in the first instance by this charge redistribution in that they contain no redox centres. It is also clear that in these reactions there is no need for us to discuss the original source of the change of charge for no matter which substrates provide the reducing or oxidizing equivalents the oxidation in the phase is that of bound hydrogen or metal ions in the first instance. *The substrates are irrelevant.* Moreover there is no experimental reason why we should try to analyse particular oxidative phosphorylation stages—I, II or III in mitochondria or the stages of photophosphorylation—for we have only discovered *one* type of ATP-ase protein which must respond to all the stages. Finally as the membrane phase is surrounded by aqueous phases there is no reason to doubt that membrane charge, surface potentials, will change with the degree of oxidation of the membrane and it is bound to do so asymmetrically as the reactions are dislocated[1] or translocated.[2] Inevitably there is a connection with phases surrounding the membrane but I shall maintain that this is secondary to the major events of oxidative phosphorylation and is not necessarily brought into *equilibrium* with the membrane phase processes. (This is in strict contrast with chemi-osmotic theory.)

Changes in Atomic Positions

There is a danger that changes in atomic position, conformation changes will be over interpreted. The limits of our knowledge are (1) mitochondrial and chloroplast membranes alter their properties *as a whole* from the energized to the de-energized state. (2) *All* redox enzymes cycle through conformational states with the changes of oxidation state no matter which membrane state we are discussing. This is known for cytochromes a_3, b, c, cc^1, flavoproteins, ferredoxins, rubredoxin. In some cases the evidence is from redox potential data, E° changes, and in others it is from structural or spectroscopic studies. It is also self-evident that conformational changes and redistribution of charge are inevitably linked in protein systems because of the nature of proteins as polar molecules, see earlier discussion of NaCl, myoglobin, and haemoglobin. The evidence for a phase mechanism of initial oxidative energy distribution rather than an active site mechanism is strong, on the basis of these conformational and charge fluctuations, although we cannot indicate the way in which the energy is partitioned in the different parts of the membrane phase.

Sources of Charge

The obvious source of charge is the removal of electrons by oxygen. There is then a flow of negative charge from the hydrogen of substrates into the space of accumulated positively charged centres. The change in net charge causes the switch of the mitochondrial state; the flux of charge in the energized state carries out oxidative phosphorylation. The flux is not of substrate hydrogen but of electrons, though in

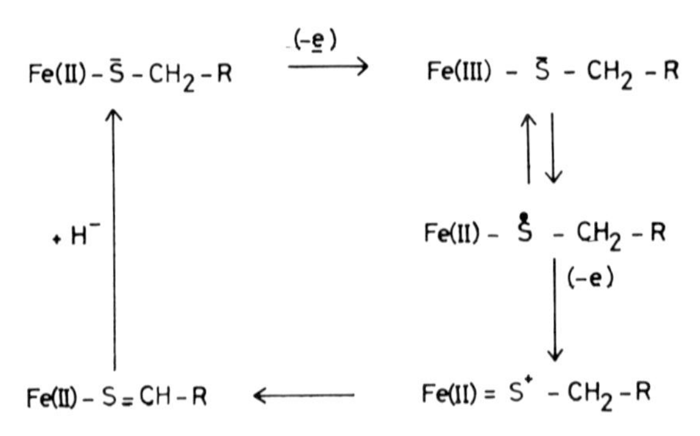

Figure 3. A suggestion as to the all-important role of the iron–sulphur proteins. They are the fundamental catalysts of the reaction $H^+ \rightleftharpoons H^- - 2e$. It could be that copper and molybdenum can carry out this reaction too for all these sulphides are active as inorganic hydrogenation catalysts. Electron transfer could be mediated by a similar series of steps.

different regions of the path of the electron, coenzyme and enzyme protons are first picked up and then rejected. These protons are those that are bound by such coenzymes as NADH, reduced flavin, and reduced quinones and by protein –SH and imidazole. The catalysis of the oxidation of these moieties is brought about by iron–sulphur and possibly copper–sulphur and molybdenum–sulphur proteins. A possible mechanism for this step can be drawn from our knowledge of the iron–sulphur protein structures and involves a metal–sulphide double bond, Fig. 3. Loss of the metal–sulphur proteins will cause oxidative phosphorylation to disappear. As the electrons flow through the membrane phase protons are cycling on to the coenzyme (and protein?) carriers, reduction step, and are leaving the membrane for the aqueous phase as H_3O^+, oxidation step, Fig. 4. In this sense the electron flux can act as a pump for lowering the water activity in the membrane. In 1956 I pointed out that this cycle was a possible way of generating ATP by generating protons in an organic phase.[3] As we have pointed out subsequently there is a compulsory, concomitant, conformational change.[1,4]

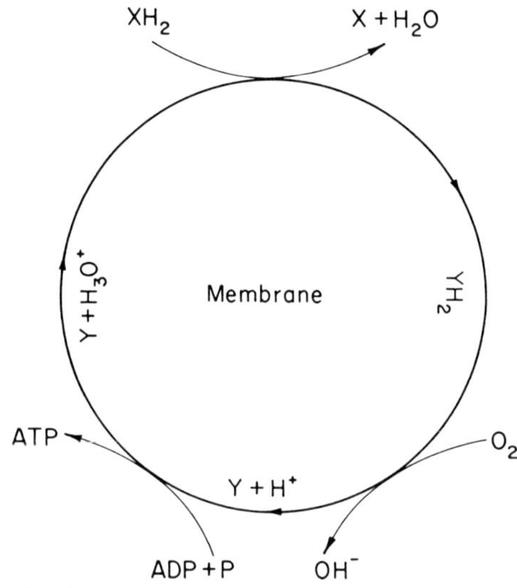

Figure 4. The dehydration cycle in oxidative phosphorylation.

The Problem of Standard State E° Values

In 1969 I drew attention to the fact that all the $E°$ values of redox carriers must be expected to change in the different states of mitochondria.[1,4] (An interesting side effect is that this undermined cross-over theory.) The evidence for such changes was then slight. Subsequently Slater and coworkers[5] and Chance and coworkers[6] claimed that specific changes occurred but these authors have gone on and attempted to connect the $E°$ changes directly with coupling. I fail to see any justification for this for I am convinced that *all* the very many proteins of the cytochrome chain will prove to show such changes. All the changes will be kinetically competent if judged by rate data—they have to be for the reaction is coupled—but in the absence of a *demonstrated physical or chemical connection* between one or another of the carriers the direct relevance to coupling is minimal. However this switch of $E°$ is a direct pointer to the changed ambient conditions of the membrane phase—its changed conformation and charge, its changed effective H^+ activity, its changed surface potential and so on. All are changes of condition of a phase.

Uncoupling Problem

It is my opinion that uncoupling will not prove to be one type of reaction. This follows from the nature of the connection with "phase-intermediates". Removal of charge from the membrane, e.g. by phenolates, or prevention of conformational change can uncouple—compare the effect of organic phosphates, protons, CO_2, salts on the Hill coefficient for oxygen uptake by haemoglobin.

The ATP-ase Problem

There is the distinct possibility that all *membrane* ATP-ases can be run forwards or backwards, i.e. to ATP degradation or formation, in a highly selective way. ATP formation in a membrane can be coupled to salt *gradients*, to *oxidative* or to *photo-energy*. ATP hydrolysis can be coupled to *reduction*, *light production*, or reversed salt *gradients*. However ATP hydrolysis uncoupled to energy conserving processes does *NOT* normally occur in membranes. A catalyst by itself cannot direct reactions in this way and therefore the ATP-ase must be switched on or off in a controlled way. In mitochondria or chloroplasts there are two situations (a) a situation in which the ATP-ase governs an *equilibrium in the membrane phase which is unrelated to the ATP equilibrium in the aqueous phase*. (b) a situation in which the ATP-ase is switched off. Uncoupling is a chemical effect operating on the link between the phase intermediate and the ATP-ase.

The Switched-on Problem

The reaction of concern in the membrane phase is

$$[ADP]_0 + [P]_0 \rightleftharpoons [ATP]_0 + [H_2O]_0$$

We know that $[ADP]_0$, $[P]_0$ and $[ATP]_0$ relate closely to aqueous phase concentrations for the energy calculation ratios used by Slater balance $[ATP]_w/[ADP]_w \cdot [P]_w$ against the flux of redox equivalents. As the system does not equilibrate in the aqueous phase we know therefore that $[H_2O]_w$ cannot equilibrate with the

$$[ATP]_0/[ADP]_0[P]_0$$

quotient. Thus the ATP-ase cannot have access to bulk water. The bulk components of the organic and aqueous phases are not connected.

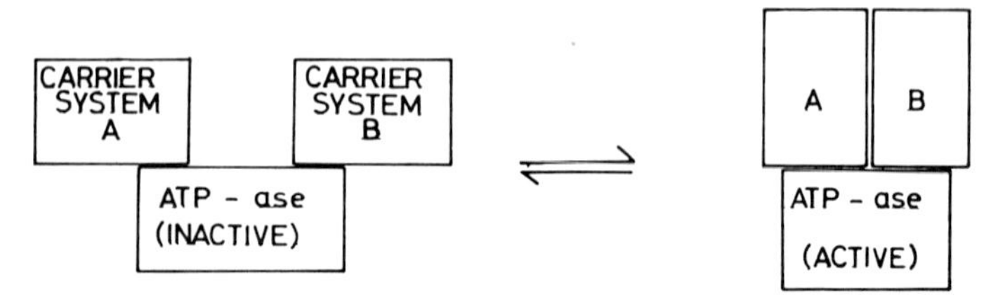

Figure 5. The switch for the ATP-ase.

The above diagram, Fig. 5, is an attempt to illustrate the switched-off $\rightleftharpoons$ switched-on relationship indicating that gross conformation changes in carriers and ATP-ase which affect the membrane.

The Charge Flux Problem

A major mystery in the energy conservation steps is the nature of the charge carriers, electrons or protons, and the sites, traps, which they visit. Soon the detailed structures of four or five electron-transfer type proteins will be known, cytochromes *c* and b_5, flavodoxin, high potential iron protein (HIPIP), and rubredoxin. We know already that minor or major conformation changes occur on redox change in very many (all?) such proteins so that there is an inevitable conformational and charge flux in the coupled state. For many of the carriers tunnelling of electrons cannot be the only charge-carrying process for this cannot be coupled. Again NADH, flavin, quinone, must undergo protonic changes. There is then a mobile pool of charge in the membrane as well as one associated with the fixed proteins so that there is a phase potential—a proton potential—connected directly with the redox potential. This potential does *not* equilibrate rapidly with the aqueous phase as H^+ is not freely mobile across the phase boundary. *This potential must not be linked to osmosis which is inevitably linked to the aqueous water activity*. The iron–sulphur proteins could be the essential $H^+ \rightleftharpoons H^- - 2e$ catalysts of the membrane which are absent from, and out of contact with, the aqueous phase.

A Specific Example

We shall consider the series of proteins associated with succinate dehydrogenase

$$\text{Succinate} \rightarrow \text{Flavin } Fe_4S_4 \rightarrow Fe_4S_4 \rightarrow \text{cyt. } b \rightarrow \text{cyt. } c \text{ cyt } c_1$$

Succinate is a source of hydride to the flavin Fe_4S_4 unit. This unit passes electrons to the iron sulphur protein Fe_4S_4 at a potential around 0·0 volts. Thus this step of the reaction is very like that in many simpler dehydrogenases, e.g. in detoxification

$$\text{NADH} \rightarrow \text{Flavin} \rightarrow Fe_2S_2 \rightarrow \text{cyt. P450}$$

Now in this detoxification system the flavin/Fe_2S_2 is able to hold electrons until the substrates for P450 are bound i.e. O_2 and sterol. The electron transfer is triggered by the presence of substrates but equally substrate/substrate reaction is triggered by the presence of electrons in an overall four electron redox reaction

$$2e + O_2 + SH \xrightarrow{2H^+} H_2O + SOH$$

The mutual reaction is controlled by conformation changes and this must be the case in the succinate chain. The substrate for the Fe_4S_4/cyt. *b* system is a quinone present in large excess so that the electrons are passed next to the quinone which must pick up protons from the surrounding medium. This is a two-electron step. The hydroquinone diffuses or transfers protons to the interior of the system where it reacts with oxidized cytochrome *b*/cytochrome *c* and must give $2H^+$ and two electrons. Now there is a pool of potentially poised couples Flavin.Fe_4S_4, Fe_4S_4, cyt. *b*, cyt. *c*, cyt. c_1, and perhaps other Fe_4S_4 proteins, as well as the major redox buffer QH_2/Q, H^+. All these proteins change conformation on redox reaction and their phase interaction is controlled by their potential poising and vice versa, Fig. 6. Thus they all change $E°$ values. In this coupled state electrons

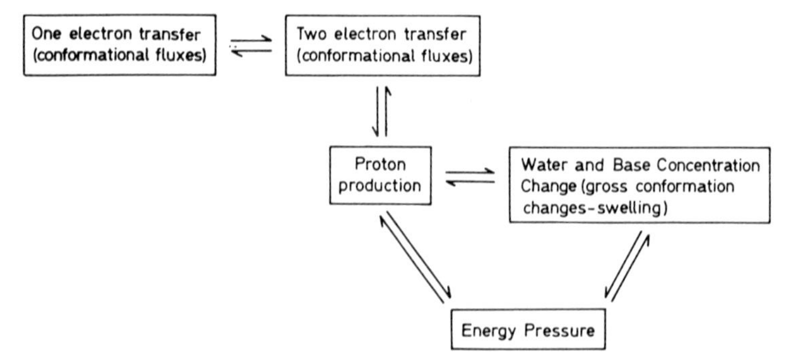

Figure 6. The relationship between the conformation changes in the chloroplast and mitochondrial membranes. Note all proteins change their conformation both with redox and proton flux and with changes of steady state concentration of charge.

flow, effectively two at a time, due to the controlled quinone reaction. It is necessary to remove H^+ from inside the membrane and to pick it up from the aqueous phase at each turn of the cycle, Fig. 4, and the internally generated H^+ can remove water from the interior of the membrane, Fig. 4. The water is replaced at a steady rate by $ADP + P \rightarrow ATP$ condensation. The system is just a logical extension from haemoglobin via mixed function oxidases (and in general multi-protein dehydrogenases coupled to oxygen), to dehydration by carrying out the dehydrogenation and dehydration steps in a non-aqueous phase.

Conclusion

My conclusion is as follows. Energy transduction has been examined with an incorrect view of mechanism taken over from small molecule chemistry. Energy transduction occurs through the generation of a charged/conformational store of energy which does not equilibrate with the aqueous phase though it can equilibrate internally in the membrane phase with [ATP], [P], [ADP], H^-, H^+ and $[H_2O]_0$. This occurs in the switched-on geometry. The switched-on geometry is in steady state balance with the switched-off geometry through the supply of substrates. The switched-on geometry is asymmetric and it enters into charge exchange with and transport of ions from the bulk aqueous medium. This exchange is in competition with ATP formation and equilibrates more slowly than ATP formation. The examination of many external and internal probes, e.g. $[H^+]$, $[Ca^{2+}]$, $[K^+]$, $[Na^+]$, carotenoid absorption, fluorescence probes, etc., will reflect this phase intermediate. The energetics of the process are then best represented by a description of the local activity of water at the ATP-ase site. The water activity is driven to a low value by the flux of charge, probably protons, through the site. The only "molecular" intermediates are those of the ATP-ase. The experimental test of such a theory rests in the measurement of local H_2O or H^+ concentrations—a very difficult problem. In my opinion none of the present experimental approaches adequately tackle this problem.

References

1. R. J. P. Williams, *J. Theoret. Biol.*, **1** (1961) 1; **3** (1962) 209; also see *Current topics in bioenergetics*, D. R. Sanadi (ed.), Vol. 3, Academic Press, New York, 1969, p. 79.
2. P. Mitchell, *Nature*, **191** (1961) 194.
3. R. J. P. Williams, *Chem. Rev.*, **56** (1956) 515.
4. R. J. P. Williams, in: *Electron Transport and Energy Conservation*, J. M. Tager, S. Papa, E. Quagliariello and E. C. Slater (eds.), Adriatica Editrice, Bari, 1970, pp. 7, 373.
5. E. C. Slater, in: *Electron Transport and Energy Conservation*, J. M. Tager, S. Papa, E. Quagliariello and E. C. Slater (eds.), Adriatica Editrice, Bari, 1970, p. 533.
6. B. Chance, in: *Electron Transport and Energy Conservation*, J. M. Tager, S. Papa, E. Quagliariello and E. C. Slater (eds.), Adriatica Editrice, Bari, 1970, p. 551.

Oxidative Phosphorylation, A History of Unsuccessful Attempts: Is It Only An Experimental Problem?

Giovanni Felice Azzone

CNR Unit for Physiology of Mitochondria and Institute of General Pathology, University of Padova, Italy

It is due to the predominance of the logic based on the inductive reasoning the commonly held view that the purpose of a scientist is that of making observations, of verifying theories through observations and of proposing theories through the generalization of observations. There is an opposite view, however, which starts from the assumption that empirical sciences are systems of theories.[1] In this type of logic, theories are conjectures proposed by the scientist to explain and predict natural processes, and cannot be inferred from observations, although they must be compatible with them. Empirical sciences are thought as including two types of statements: universal and specific. The formers are hypotheses with the character of natural laws, the latters assert the occurrence of single events.[1] The validity of both statements is decided empirically. However there is an asymmetric relationship between the two. The universal statements can never be verified but only refuted by specific statements (experimental observations). On the other hand a specific statement can be verified through another observation. A theory can be disproven if there exists at least one class of events forbidden for the theory. A theory which has a larger number (in comparison to another) of potential falsifiers, and can therefore more easily be tested and refuted, has also a larger empirical content. In the following, I shall try to show the impact of this logic of scientific knowledge on current hypotheses and future theories of energy conservation.

Comments on the Chemical or Conformational Hypotheses

"Can you describe any possible observations which, if they are actually made, would refute your theory? If you cannot, then your theory has clearly not the character of an empirical theory; for if all conceivable observations agree with your theory then you are not entitled to claim of any particular observation that it gives empirical support to your theory".[2]

The "chemical" hypothesis has had a considerable influence for almost twenty years in the field of oxidative phosphorylation, and it has been the first attempt of symbolic analysis of the process.

A symbolic formulation of electron transport coupled phosphorylation is:

$$AH_2 + B + ADP + P_i \rightleftharpoons A + BH_2 + ATP \quad (1)$$

The "chemical" hypothesis[3, 4] has been recently formulated as[5]:

$$AH_2 + B + C\ (I) \rightleftharpoons A \sim C\ (I) + BH_2 \quad (2)$$

$$A \sim C\ (I) + ADP + P_i \rightleftharpoons A + C + ATP \quad (3)$$

Sum

$$AH_2 + B + ADP + P_i \rightleftharpoons A + BH_2 + ATP \quad (4)$$

$A \sim C$ (I) indicates a high energy intermediate where the redox energy is conserved in a covalent bond between one of the products of the redox reaction and a ligand C. The "conformational" hypothesis[6–8] has been written[5] as:

$$AH_2 + B \rightleftharpoons A^* + BH_2 \quad (5)$$

$$A^* + ADP + P_i \rightleftharpoons A + ATP \quad (6)$$

Sum

$$AH_2 + B + ADP + P_i \rightleftharpoons A + BH_2 + ATP \quad (7)$$

A* indicates the conservation of the redox energy into a rearrangement of the protein structure of a redox carrier.[5] The chemical and conformational hypotheses have been considered as "variants of a single hypothesis", otherwise denoted as the C hypothesis.[5]

All scientific theories exclude a specific class of events.[1] A theory speaks about empirical reality only in so far as it sets limits to it. Every theory can thus be put into the form "such and such does not happen".[2] The observation of the event forbidden by the theory constitutes a refutation of the theory. By using this type of analysis, the empirical content of equations (2–7) can be formulated as such: "There is no conversion of redox into chemical energy (say ATP) which does not involve: (a) (chemical)-a covalent bond with a redox carrier or; (b) (conformational)-a conformational rearrangement of a redox carrier". A disproof of these hypotheses requires the observation of a process of energy transduction without either the formation of a covalent bond or a conformational change at the level of the respiratory chain. I maintain that the class of the potential falsifiers of these hypotheses, as they stand in equations (2–7) is practically empty. Until they are expressed in such general terms it is impossible to make specific predictions leading to falsifying experiments. Hence, their empirical content is negligible and they are of little help to the experimentalist. This criticism is not directed to a chemical or conformational hypothesis in general, but to the view of considering equations (2–7) *as* a hypothesis.

Attempts have been made to propose specific mechanisms. Although some of them are of considerable interest, an analytical discussion is out of place in the present context. On the other hand I shall briefly consider the view that shifts of the α band of cytochrome b and of the midpoint potentials of cytochromes b and a upon energization[9,10] are in support of the formulation given in equations (2–7). Two problems arise here, one is experimental, the other theoretical. The experimental question concerns the amount of information on membrane conformation which can be obtained from the red shift as compared, for example, to other extrinsic probes such as those used in many other laboratories. That the importance of the red shift derives from its being "primary" because of its faster kinetics, in respect to the other dyes, is irrelevant because this may be due to an additional diffusion step present in the latter but not in the former case. On the other hand it might be easily predicted that the red shift will be a difficult source of structural information because (a) it is difficult to decide whether the source of the shift is an intramolecular rearrangement[9,10] or it concerns the environment of the chromophore,[11] (b) there are no known models to which compare the "energy state" of cytochrome b. Point (b) is a criticism also to the molecular interpretation of the shift of the midpoint potentials. From a theoretical standpoint, the observations are "in accord" with the chemical or conformational hypotheses simply because equations (2–7) are compatible with "any observation". However they can be equally accounted for by the chemiosmotic hypothesis as far as this hypothesis is equally elusive. The cytochrome's shifts cannot be used to refute, or therefore to support, a particular mechanism of oxidative phosphorylation.

Comments on the Chemiosmotic Hypothesis

This hypothesis[12,13] is a mixture of both precise and of more general statements. In the former case it has a high empirical content and it has been put under stringent experimental test whereas in the latter it suffers the same objections as the chemical-conformational hypotheses. Some of its basic formulations can be stated as follows.

"The conversion of redox into chemical energy does not require high energy intermediates of the A $\sim$ C or C $\sim$ I type". The disproof of this statement requires the isolation of the high energy intermediates. Since these intermediates are however not defined in chemical terms, the class of the potential falsifiers of this statement is empty.

"The process of electron transport coupled phosphorylation does not occur without (a) a precise arrangement of the electron and hydrogen carriers, (b) a certain stoicheometric transfer of protons, (c) a membrane potential of a certain magnitude and (d) an impermeability of the mitochondrial membrane to diffusion of charged species."

Hypotheses (a) and (d) have been most fruitful. Although I do not think that the arrangement of the respiratory loops proposed by Mitchell, is feasible, it must be conceded that this proposal has given a great impetus to studies of the topography of the membrane. The requirement for a restriction to charge translocation has led not only to a new approach for studying the mechanism of action of uncouplers[14–16] but also to the discovery of the anion carriers[17] (the proposal for which was also included in the hypothesis).

On the other hand strong experimental evidence has been provided that hypotheses (b) and (c) are not correct. The stoicheometry of the proton translocation per energy rich bond during the conversion of chemical into osmotic energy or vice versa is 4 and not 2 as predicted by the chemiosmotic hypothesis.[18–23] Indeed Cockrell *et al.*[18] have reported a $K^+/\sim$ ratio of 3 and 7, for the respiration and ATP driven uptake, respectively. We have obtained a ratio of about 4 in both cases.[19–23] The stoichiometry of the H^+/K^+ exchange during active transport is variable and dependent on the internal and external concentrations of H^+ and K^+ whereas in the chemiosmotic hypothesis it should be constant and independent of the H^+ and K^+ concentrations.[21, 24, 25] Finally the osmotically driven ATP synthesis takes place in the presence of an osmotic potential (as calculated from the H^+ and K^+ concentration gradients) which is half that predicted by the chemiosmotic hypothesis.[22, 24–26]

Also, experiments on the saturation kinetics[27] and the dependence of the $K/\sim$ ratio on the H^+ and K^+ concentrations[21] strongly oppose an electrophoretic ion diffusion down an electrical gradient.[25] However the question of the transmural potential will presumably remain open for a certain time due to the difficulty of performing conclusive refuting experiments. On the other hand the Occam's razor here applies: what is needed for a membrane potential in active ion uptake once its role in ATP synthesis is questioned? My opinion is that the chemiosmotic hypothesis has had more success in predicting some general properties of the mitochondria than in devising a specific mechanism for oxidative phosphorylation.

The Basic Properties of the Mitochondrial Membrane

I propose that these be individuated as follows:

1. *Coupling of transport to metabolism and transformation of scalar into vectorial reactions.* The property is shared by the Na/K^+ ATPase reaction of the plasma membrane. According to the Curie's theorem "the effects cannot be more asymmetric than the causes". If one assumes that the forces of chemical affinity are scalar, a dilemma arises as to the conversion of a scalar into a vectorial reaction. It has been pointed out however that the enzyme catalyzed group transfer reaction occurs in an anisotropic microscopic non-aqueous phase as to show in reality a vectorial character.[28] The orientation of all the enzyme molecules within a membrane is required in order to avoid a random distribution in a homogeneous phase which renders isotropic the chemical reaction. A fundamental difference between our and Mitchell's concept for active transport in mitochondria, is that the former requires a specific geometric arrangement of membrane groups which are uniquely involved in ion translocation, whereas in the latter the vectorial character concerns only the operation of the respiratory chain and of the ATPase. Our concept would be refuted by the observation that active transport entails an electrophoretic diffusion of cations down an electrical gradient.

2. *Propagation of high energy state in membrane structure.* This property is unique to the mitochondrial membrane although it may be compared with the conformational change proposed to accompany the action potential in the nerve membrane. The redox energy is primarily conserved in a particular conformation of the membrane structure rather than in a covalent bond or a conformational change involving an electron carrier. This concept, may be formalized as follows:

$$AH_2 + B + M \rightleftharpoons A + BH_2 + M^* \quad (8)$$

$$M^* + P_i + ADP \rightleftharpoons M + ATP \quad (9)$$

Sum

$$AH_2 + B + P_i + ADP \rightleftharpoons A + BH_2 + ATP \quad (10)$$

Where M and M* indicate a low and high energy state of the membrane. Whatever the energy conserving site, the high energy state can be propagated along the membrane without formation of covalent bonds or diffusion of low molecular weight high energy intermediates. The concept expressed in equations (8–9) will be refuted by one of the following observations.

(a) The occurrence of site specificity for reaction of ADP, uncouplers, and inhibitors of oxidative phosphorylation. These effects are predicted in equations (2–7) since the compounds $A \sim C$ (I) or A* vary according to the site in the respiratory chain. On the other hand they are excluded in equations (8–9) where M* is common for the three sites.

(b) Presence of energy conservation and respiratory control in soluble electron transport systems, devoid of membrane organization. In equations (2–7) the respiratory control is due to bringing a respiratory carrier into an inhibited form $A \sim C$ (I) or A*. In equations (8–9) the respiratory control is due to the fact that the transfer of reducing equivalents from A to B, at the site where energy is conserved, must occur across the membrane M, and is inhibited when the membrane is in the energized state M*. Equations (8–9) therefore imply that the high energy state is a property of the environment and becomes manifest only when electron transport takes place in a lipid-rich multimolecular assembly with specific geometric and structural requirements. However equations (8–9) do not require an alternation of hydrogen and electron carrier as in the chemiosmotic hypothesis.

It may be argued that by writing down equations (8–9) I have exposed myself to the same criticism of elusiveness used for equations (2–7). My answer is: (a) the equations (8–9) are not a hypothesis for oxidative phosphorylation; they are simply a symbolic means for expressing the alternative concept that energy is not conserved in respiratory chain components and does not involve the existence of low molecular weight high energy intermediates; (b) although equations (8–9) are vague, I have proposed some observations through which my concept may be refuted; I expect the proponents of equations (2–7) to suggest similar falsifying experiments for their concepts.

Membrane Structure, Carrier "Gating" and Nucleophilic Sites

The growing of a theory for oxidative phosphorylation necessarily involves a multistep process. This means to start by proposing *very narrow range hypotheses* which have a precise empirical content and lead to definite predictions and experimental tests. After suitable controls broader hypotheses can be proposed which should be tested also by utilizing the concepts derived from the simpler hypotheses. And so on until the general principles underlying the mechanism of energy transduction may be included in a theory of oxidative phosphorylation.

The present proposals of our laboratory for active transport and oxidative phosphorylation are based on three general concepts.

1. *The mitochondrial membrane is formed by a core of polypeptide chains buried in hydrocarbon regions.*[25, 27, 29] The polar sites of the membrane proteins are partly exposed or "bare" (in respect to the aqueous phase) and partly unexposed; i.e., not accessible to hydrophilic ions.

Transport across the membrane and coupling of ion fluxes involves interaction with these sites. Specific transport mechanisms are required (for example the presence of extrinsic ionophores) when the interaction involves the unexposed polar groups (carriers). This proposal implies a membrane structure of the type envisaged by Lenard and Singer[30] and Wallach and Zahler,[31] rather than the classical lipid bilayer. Furthermore this proposal is alternative to those implying electrophoretic, electrogenic ion diffusion across the membrane and coupling of fluxes through membrane potential. Instead we assume that the electrophoretic step is in series with another involving an electrostatic interaction with a fixed negative charge system. The term electroneutral exchange[21,27] indicates that the coupling of the various forms of energy, including the ion fluxes which constitute the osmotic energy, does not occur through a transmembrane electrical potential, but rather involve common chemical reactions within the membrane. Electroneutrality is also a stringent thermodynamic requirement for an electrostatic interaction in a hydrophobic environment.

2. *The accessibility of membrane hydrophobic and polar groups for the aqueous phase is modified when the membrane undergoes an energy-linked rearrangement.* The geometric properties of the groups which undergo an energy-linked protonation and therefore operate as proton carriers, are dependent on the metabolic state of the membrane ("gating" of the carrier).[22,25] This proposal is alternative to those implying either a free diffusion of protons across the membrane or a translocation through respiratory loops and ATPase and explains the high and low rates of proton translocation in the energized and deenergized membrane, respectively.

3. *The energy-linked conformational change involves the formation of highly nucleophilic sites.*[22,23,25,26,29,32,33,35,36] The proton donating and acceptor properties of membrane groups are markedly modified during transition from the deenergized to energized state. Such changes of apparent pK_a may be due to: (a) enhanced hydrophobicity[34] with restriction to ionization of ion pairs and/or increased activity of protons or (b) increased strength of specific groups due to intramolecular rearrangement of the electron configuration. The formation of superacid or superbasic regions, where substrates are more easily protonated or deprotonated, can be utilized for acid-base catalysis (ATP synthesis) and operation of the proton pump. Our proposal gives an alternative explanation to a number of recent observations. The shifts of the absorption band and of the midpoint potentials are not due to the formation of high energy forms of the cytochromes but rather to pK_a shifts in the cytochrome environments. The increased uptake of organic anions to energized submitochondrial

particles is not due to diffusion down an electrical gradient (positive inside) or attraction to positive charges, but rather to an entropy driven process accompanying a decrease of electrostatic repulsion in certain membrane regions.[37] Finally uncouplers act by transferring protons in the lipid phase. However the uncoupling effect is not due to the H^+ movement across the membrane to collapse an electrical potential but to the interaction, within the membrane, of the protons with the energy-linked nucleophiles or to an increase of H^+ ions activity in certain membrane regions.[26,29] The sink for the protons is due to the increase of electron density in some groups of the membrane proteins, rather than to the low electrochemical potential of protons in the inner aqueous mitochondrial phase. We are at present actively engaged in deriving from these very general ideas simple experimental questions to put under stringent test. A number of observations have been made which appear to be in accord with the general formulation of the problem.

It may be argued that these proposals are still far from the "great" problems of oxidative phosphorylation. On the other hand they have the advantage to be possibly expressed in chemical terms and lead to definite questions. The answer to these questions may permit to move the preliminary steps toward a structural analysis of the membrane during energy transduction.

Acknowledgements

I am indebted to my collaborators Stefano Massari, Raffaele Colonna and Paolo Dell'Antone, who have contributed to the development of many of the concepts expressed in this article.

References

1. K. R. Popper, *The logic of scientific discovery*, Einaudi, Torino, 1970.
2. K. R. Popper, *Fed. Proc. Symposia*, **22** (1963) 961.
3. E. C. Slater, *Nature*, **172** (1953) 975.
4. B. Chance and G. B. Williams, *Adv. Enzym.*, **17** (1956) 65.
5. E. C. Slater, *Quarterly Review Biophys.*, **4** (1971) 35.
6. P. D. Boyer, in: *Oxidase and related redox systems*, T. E. King, H. S. Mason and M. Morrison (eds.), John Wiley, New York, 1965, p. 994.
7. C. R. Hackenbrock, *J. Cell. Biol.*, **30** (1966) 269.
8. D. E. Green, J. Asai, R. A. Harris and J. T. Penniston, *Arch. Biochem. Biophys.*, **125** (1969) 684.
9. E. C. Slater, C. P. Lee, J. A. Berdem and H. J. Wedgam, *Nature*, **226** (1970) 1248.
10. B. Chance, D. F. Wilson, P. L. Dutton and M. Erecinska, *Proc. Natl. Acad. Sci. USA*, **66** (1970) 1175.
11. M. Wikstrom, in: *Biochemistry and Biophysics of Mitochondrial Membranes*, G. F. Azzone, E. Carafoli, A. L. Lehninger, E. Quagliariello and N. Siliprandi (eds.), Academic Press, New York, in press.
12. P. Mitchell, *Nature*, **191** (1961) 144.
13. P. Mitchell, *Biol. Rev.*, **41** (1966) 445.
14. V. Hopfer, A. L. Lehninger and F. E. Thompson, *Proc. Natl. Acad. Sci. USA*, **59** (1968) 484.
15. E. A. Liberman, V. P. Topali, L. M. Tsofina, A. A. Jasaitis and V. P. Skulachev., *Nature*, **222** (1969) 1076.

16. P. Mitchell and J. Moyle, *Biochem. J.*, **104** (1967) 588.
17. J. B. Chappell and K. N. Haarhoff, in: *Biochemistry of Mitochondria*, E. C. Slater, Z. Kaniuga and L. Wojtczak (eds.), Academic Press, London, 1967, p. 75.
18. R. S. Cockrell, E. J. Harris and B. C. Pressman, *Biochemistry*, **5** (1966) 2326.
19. E. Rossi and G. F. Azzone, *Europ. J. Biochem.*, **7** (1969) 418.
20. E. Rossi and G. F. Azzone, *Europ. J. Biochem.*, **12** (1970) 319.
21. G. F. Azzone and S. Massari, *Europ. J. Biochem.*, **19** (1971) 97.
22. G. F. Azzone and S. Massari, in: *Membrane Bound Enzymes*, G. Porcellati and Di Jeso (eds.), Plenum Press, New York, 1971, p. 19.
23. G. F. Azzone, R. Colonna, P. Dell'Antone and S. Massari, in: *Energy Transduction in Respiration and Photosynthesis*, E. Quagliariello, S. Papa and C. S. Rossi (eds.), Adriatica, Ed., in press.
24. S. Massari and G. F. Azzone, in: *Biochemistry and Biophysics of Mitochondrial Membranes*, G. F. Azzone, E. Carafoli, A. L. Lehninger, E. Quagliariello and N. Siliprandi (eds.), Academic Press, New York, in press.
25. G. F. Azzone and S. Massari, in: *Treatise on Electron and Coupled Energy Transfer in Biological Systems*, T. E. King and M. Klingenberg (eds.), in press.
26. S. Massari and G. F. Azzone, *Europ. J. Biochem.*, **12** (1970) 309.
27. S. Massari and G. F. Azzone, *Europ. J. Biochem.*, **12** (1970) 301.
28. P. Mitchell, in: *Comprehensive Biochemistry*, M. Florkin and E. H. Stotz (eds.), Elsevier, Amsterdam, 1967, p. 167.
29. G. F. Azzone, R. Colonna and P. Dell'Antone, in: *Biochemistry and Biophysics of Mitochondrial Membranes*, G. F. Azzone, E. Carafoli, A. L. Lehninger, E. Quagliariello and N. Siliprandi (eds.), Academic Press, New York, in press.
30. J. Lenard and S. J. Singer, *Proc. Natl. Acad. Sci. USA*, **56** (1966) 1828.
31. D. F. H. Wallach and P. H. Zahler, *Proc. Natl. Sci. USA*, **56** (1966) 1557.
32. R. Colonna, P. Dell'Antone and G. F. Azzone, *FEBS Letters*, **10** (1970) 13.
33. P. Dell'Antone, R. Colonna and G. F. Azzone, *Biochim. Biophys. Acta*, **234** (1971) 541.
34. J. R. Brocklehurst, R. B. Freedman, D. J. Hancock and G. K. Radda, *Biochem. J.*, **116** (1970) 721.
35. P. Dell'Antone, R. Colonna and G. F. Azzone, *Europ. J. Biochem.*, **24** (1972) 553.
36. P. Dell'Antone, R. Colonna and G. F. Azzone, *Europ. J. Biochem.*, **24** (1972) 566.
37. G. F. Azzone, P. Dell'Antone and R. Colonna, *7th FEBS meeting*, Varna 1971, p. 51.

On the Coupling of Electron Transport to Phosphorylation*

Jui H. Wang

Kline Chemistry Laboratory, Yale University, New Haven, Connecticut 06520

Abstract

After a general thermodynamic discussion of the coupling of oxidation to phosphorylation, quantitative treatments of free energy transduction based upon the proton gradient model, the charged membrane model and the chemical model respectively are summarized and compared with experimental data. The relationship between energy transduction and respiratory control is reexamined.

Introduction

Since the dawn of history man has been wondering—what is it that keeps him alive? His soul or his respiration or something else? Should his attempted answer be developed from the conservation of souls (reincarnation), or the conservation of mass and energy, or the non-conservation of free energy, or some other hypothesis?

The free energy liberated by respiratory or photosynthetic electron transport is often first converted to and stored in a certain form prior to the formation of ATP.[1-3] The various forms of this stored free energy suggested in the literature include energy-rich intermediates,[1] concentration gradients,[4,5] changes in macromolecular conformation and membrane structure[6,7] and electric potential energy.[8-10] In this paper, we shall develop these suggestions quantitatively and examine their validity by comparison with experimental data.

Thermodynamic Considerations

Let us consider an oxidation process represented by

$$\sum_k \nu_k A_k \rightarrow \sum_l \nu_l A_l$$

* Supported by a research grant (GM 04483) from the National Institute of General Medical Sciences, National Institutes of Health, Department of Health, Education and Welfare.

where ν_1 moles of molecular species A_1, ν_2 moles of A_2, etc. react to form ν_n moles of A_n, ν_{n+1} moles of A_{n+1}, etc. The summation $\sum_k$ is over all reactants and the summation $\sum_l$ is over all products of this oxidation reaction. At constant temperature T and pressure P, the decrease of free energy due to this reaction is given by

$$-\Delta G_0 = \sum_k \nu_k \mu_k - \sum_l \nu_l \mu_l' \geqslant 0 \tag{1}$$

where μ_k, μ_l' represent the electrochemical potentials of A_k, A_l at their respective locations in the system, and the equality sign applies to the limiting case in which the system is already at equilibrium. If the oxidation-reduction system is not at equilibrium, the free energy released may be utilized to drive a thermodynamically unfavorable reaction represented by

$$\sum_i \nu_i A_i \rightarrow \sum_j \nu_j A_j$$

resulting in the storage of an amount of free energy equal to

$$\Delta G_s = \sum_j \nu_j \mu_j' - \sum_i \nu_i \mu_i > 0 \tag{2}$$

Thermodynamics requires

$$-\Delta G_0 > -(\Delta G_0 + \Delta G_s) \geqslant 0 \tag{3}$$

The electrochemical potential μ_i of the reactant molecular species A_i of charge number Z_i at a location with electric potential Ψ in the system may be written as

$$\mu_i = \mu_i^\circ + Z_i \mathscr{F} \Psi + RT \ln m_i + RT \ln \gamma_i \tag{4}$$

where m_i is the molal concentration, γ_i the activity coefficient of A_i, $\mathscr{F}$ the faraday, R the gas constant, and the standard potential μ_i° is characteristic of A_i but for a given set of arbitrarily chosen standard states is independent of Ψ and m_i. Likewise, the electrochemical potential μ_j' of the product molecular species A_j at another location in the system with electric potential Ψ', concentration m_j' and activity coefficient γ_j' may be written as

$$\mu_j' = \mu_j^\circ + Z_j \mathscr{F} \Psi' + RT \ln m_j' + RT \ln \gamma_j' \tag{5}$$

Substitution of equations (4) and (5) into (2) gives

$$\begin{aligned} \Delta G_s = & \left(\sum \nu_j \mu_j^\circ - \sum \nu_i \mu_i^\circ\right) + \mathscr{F}\left(\Psi' \sum \nu_j Z_j - \Psi \sum \nu_i Z_i\right) \\ & + RT\left(\sum \nu_j \ln m_j' - \sum \nu_i \ln m_i\right) \\ & + RT\left(\sum \nu_j \ln \gamma_j' - \sum \nu_i \ln \gamma_i\right) \end{aligned} \tag{6}$$

The first through the fourth terms on the righthand side of equation (6) have been used loosely to represent the free energies converted by the energy transduction process and stored in the form of energy-rich chemical intermediates, electric potential energy, concentration gradients and additional molecular interactions respectively. This arbitrary representation is misleading, because in a biological system each of these four effects contributes to two or more terms on the righthand side of equation (6). For example, the formation of chemical intermediates may affect all four terms on the righthand side of equation (6) for the following reasons: (a) Molecules in general react at concentrations m_i, m'_j ..., with activity coefficients γ_i, γ'_j..., instead of at their standard states; (b) it is well-known since Volta's experiments in 1800 and Faraday's experiments in 1833 that chemical reactions may cause electric potential changes and local concentration changes. Therefore the experimental detection of concentration gradients generated by energy transducing processes in chloroplasts[3,11–13] and mitochondria[14] does not support the proton gradient or chemiosmotic model in preference to the chemical intermediates hypothesis. For the same reason, the observation of membrane electric potential and structure changes does not necessarily support the charged membrane model and the conformation change model respectively in preference to the competing hypotheses. On the other side, the detection of a phosphorylated intermediate[15] does not necessarily support the chemical coupling mechanism in preference to the chemiosmotic and electric potential models either, because it is thermodynamically possible to drive the generation of chemical intermediates with the free energy stored in concentration gradients or electrically charged membranes.

The successful coupling of phosphorylation to electron transfer in homogeneous model systems[16,17] does demonstrate the feasibility and potential efficiency of the chemical coupling mechanism, because macroscopic homogeneity precludes concentration gradient, electric potential gradient and membrane structural changes. But results obtained from artificial model systems cannot be used as direct experimental evidence for the coupling mechanism in chloropolasts and mitochondria.

Proton Gradient Models

The central assumption of proton gradient models[4,5] is that the first two terms on the righthand side of equation (6) are negligible and hence the free energy stored prior to ATP synthesis is given by

$$\Delta G_s \approx RT(\sum \nu_j \ln m'_j - \sum \nu_i \ln m_i) + RT(\sum \nu_j \ln \gamma'_j - \sum \nu_i \ln \gamma_i) \quad (7)$$

To simplify the theoretical treatment, let us avoid pH changes due to net oxidation–reduction reactions and consider the efficiency of photosynthetic energy conversion under conditions of cyclic electron transport, as measured by Z^* or X_E, by a chloroplast in a medium buffered by $B^- + BH$. If the ionic strength of the medium is maintained constant by a higher concentration of sodium chloride, the free energy increase due to the light-driven transfer of δ mole of H^+ (or H_3O^+), with accompanying counter ions, from Side 1 to Side 2 of the chloroplast membrane at constant temperature is given by

$$\Delta G_s = RT \int_0^\delta \ln\left(\frac{[H^+]_2}{[H^+]_1}\right) d\delta \tag{8}$$

where $[H^+]_1$ and $[H^+]_2$ represent the concentration of H^+ (or H_3O^+) on Side 1 and Side 2 of the membrane respectively. Let us assume that the chloroplast has been immersed in the buffered medium in the dark for a long time so that

$$[BH]_1/[B^-]_1 = r = [BH]_2/[B^-]_2 \tag{9}$$

According to the proton gradient models, the number of moles of H^+ translocated is proportional to the number of moles of electrons transported. By using intense light for a very short illumination period, it is possible to translocate a sufficiently large number, δ, of moles of H^+ from one side of the chloroplast membrane to the other without appreciable amounts of B^- and BH diffusing across simultaneously. Under such experimental conditions, the concentrations of H^+ on the two sides immediately after the light-driven proton translocation are given by

$$[H^+]_2 = K'_a\left\{\frac{[BH]_2}{[B^-]_2}\right\} = K'_a\left\{\frac{a_2 r/(1+r) + \delta}{a_2/(1+r) - \delta}\right\}, \tag{10}$$

$$[H^+]_1 = K'_a\left\{\frac{[BH]_1}{[B^-]_1}\right\} = K'_a\left\{\frac{a_1 r/(1+r) - \delta}{a_1/(1+r) + \delta}\right\} \tag{11}$$

where a_2 and a_1 represent the total number of moles of buffer, B^- plus BH, on Side 2 and Side 1 of the membrane respectively, $K'_a = [B^-][H^+]/[BH]$, and the illumination is sufficiently short so that $\delta < a_2 r/(1+r) < a_1 r/(1+r)$ and $\delta < a_2/(1+r) < a_1/(1+r)$.

Substituting equations (10) and (11) into equation (8) and integrating, we get[18]

$$\Delta G_s \approx \frac{(1+r)^2}{2r}\left(\frac{1}{a_1} + \frac{1}{a_2}\right) RT\delta^2 \tag{12}$$

Equation (12) shows that for a given number, δ, of moles of proton translocated, the free energy stored should decrease as the buffer concentration is increased. Since experimental data[18] clearly show that ΔG_s increases rapidly as the buffer concentration is raised, we can only conclude that most of the Z^* or X_E produced by photosynthetic energy conversion under conditions of cyclic electron transport is not of proton gradient nature.

Attempts have been made to improve the proton gradient theory by postulating that protons are not translocated from Side 1 to Side 2, but from the exterior to the interior of the chloroplast membrane. This type of proton translocation would produce a much larger proton gradient for the same number of protons translocated. But the improved theory also contradicts the observed rapid increase of ΔG_s as the buffer concentration is raised.

Charged Membrane Models

These models assume that the free energy of electron transport is first used to charge the thylakoid or inner mitochondrial membrane and stored as electric potential energy. This energy is later utilized to translocate ions and phosphorylate ADP.[8–10] Accordingly, Z^* or X_E is equal to the work required to charge a membrane condenser by redistributing the ions in the chloroplastic or mitochondrial system. The work required to charge a condenser of constant capacity C to a potential V is equal to $CV^2/2$. But since practically all the electric charges in a biological system reside on ions and since the capacity of a membrane condenser also depends on the distribution of ions, we expect the free energy stored by charging such a membrane condenser to vary with the concentration of the principal ionic species, say Na^+ and Cl^-, but to be practically independent of the concentration of the dilute buffer, $[B^-] + [BH]$, which does not contribute significantly to the total ionic strength.

For simplicity, let us consider the free energy stored by charging an infinite planar membrane by bringing sodium ions to it from the sodium chloride solution of molar concentration C_s on one side of the membrane until the membrane potential becomes Ψ_0. For $C_s < 0{\cdot}05$ M and $\Psi_0 < 1$ V, theoretical considerations[18] show that the free energy stored per unit area of the membrane is equal to

$$\Delta G_s = \sqrt{\frac{2DC_sRT}{1000\pi}}\left\{\Psi_0 \sinh\left(\frac{\mathscr{F}\Psi_0}{2RT}\right) - \frac{2RT}{Z\mathscr{F}}\left[\cosh\left(\frac{\mathscr{F}\Psi_0}{2RT}\right) - 1\right]\right\} \quad (13)$$

where D is the dielectric constant of the solvent. Therefore in the range of experimental conditions for which equation (13) is applicable, ΔG_s at a given sodium chloride concentration should be independent of buffer concentration which does not contribute significantly to the total ionic strength.

Since this deduction contradicts the experimental observation[18] that at approximately constant ionic strength ΔG_s increases rapidly as the buffer concentration is raised and that at a given buffer concentration ΔG_s does not increase as C_s changes from 10 to 50 mM, we conclude that most of the Z^* or X_E produced by photosynthetic energy conversion under conditions of cyclic electron transport is not in the form of electrically charged thylakoid membrane. The experimental observation also precludes a combination of proton gradient and electric potential energy as the principal form in which ΔG_s is stored.

Chemical Coupling Mechanisms

Chemical mechanisms include all four terms on the righthand side of equation (6) and hence are thermocynamidally more general than the proton gradient and charged membrane theories.

For the convenience of discussion, let us tentatively adopt as a working hypothesis the radical coupling mechanism discovered in model systems[16, 17] for the cyclic photophosphorylation in chloroplasts, and assume that under illumination an imidazole group of a cytochrome molecule is first photo-oxidized to a substituted imidazolyl radical. If this substituted imidazolyl radical is adjacent to a suitable phospholipid or phosphoprotein, $(RO)(R'O)PO_2^-$, of the chloroplast membrane, it can rapidly react with the latter to form a substituted phosphoimidazolyl radical which can subsequently be reduced to substituted orthophosphoimidazole. The unstable substituted orthophosphoimidazole can either decompose to the original reactants or capture a proton to form water and substituted 1-phosphoimidazole. The latter can then transfer its substituted phosphoryl group to an acceptor group A^- or AH on the membrane to form the intermediate (RO)(R'O)POA. Both the substituted 1-phosphoimidazole and (RO)(R'O)POA are "non-phosphorylated" energy-rich intermediates[19] (customarily represented by $X \sim I$, Z^* or X_E) and either of them can subsequently react with inorganic phosphate P_i and ADP to form ATP and regenerate the original reactants.

For short illumination periods, an approximate steady-state treatment of photosynthetic energy conversion by this radical coupling mechanism under conditions of cyclic electron transport gives the following result[18]:

$$\Delta G_s \approx B\left\{st - \frac{1}{4}\left[1 + \frac{2m(1+r)^2 a}{(1+m)\, rc}\right] s^2 t^2\right\} \tag{14}$$

where $c = [B^-] + [BH] \neq 0$, $r = [BH]/[B^-]$ before the illumination, t is the illumination time, and B, s, m, a are constants. Although the washed chloroplasts must still contain a small amount of endogenous buffer, we expect from equation (14) that the additional free energy stored due to the added buffer will increase with the buffer concentration c at low c when

$$\frac{2m(1+r)^2}{(1+m)\, rc} \geqslant 1, \qquad c \neq 0.$$

But at very high c, we expect this additional ΔG_s to become independent of c when

$$\frac{2m(1+r)^2 a}{(1+m)\, rc} \ll 1.$$

The available experimental data[18] are consistent with this inference at low c. At $c = 10$ mM, the yield of $X \sim I$ has not yet reached its saturation value.

The net reactions leading to $X \sim I$ in this particular chemical coupling mechanism are

$$\text{HN}\!\!\diagup\!\!\text{N (imidazole)} + {}^{-}\text{O-}\overset{\text{O}}{\overset{\|}{\text{P}}}\text{-OR} \,(\text{OR}') + \text{H}^+ \rightleftharpoons \text{N}\!\!\diagup\!\!\text{N-}\overset{\text{O}}{\overset{\|}{\text{P}}}\text{-OR}\,(\text{OR}') + \text{H}_2\text{O} \qquad (15)$$

and

$$\text{N}\!\!\diagup\!\!\text{N-}\overset{\text{O}}{\overset{\|}{\text{P}}}\text{-OR}\,(\text{OR}') + \text{A}^- + \text{H}^+ \rightleftharpoons \text{HN}\!\!\diagup\!\!\text{N} + \text{O=P(A)(OR}')\text{(OR)} \qquad (16)$$

The net reactions for the formation of ATP from $X \sim I$ are

$$\text{N}\!\!\diagup\!\!\text{N-}\overset{\text{O}}{\overset{\|}{\text{P}}}\text{-OR}\,(\text{OR}') + \text{ADP} + \text{P}_i \rightleftharpoons \text{HN}\!\!\diagup\!\!\text{N} + \text{O=P(O}^-)\text{(OR}')\text{(OR)} + \text{ATP} + \nu\text{H}^+ \qquad (17)$$

or

$$\text{O=P(A)(OR}')\text{(OR)} + \text{ADP} + \text{P}_i \rightleftharpoons \text{O=P(O}^-)\text{(OR}')\text{(OR)} + \text{A} + \text{ATP} + \nu'\text{H}^+ \qquad (18)$$

Since Reactions 15 and 16 involve the uptake of protons whereas Reactions 17 and 18 involve the release of protons, we can drive the first two reactions to the right by lowering the pH and drive the last two reactions to the right by raising the pH. Consequently according to this chemical coupling mechanism it should be possible to make ATP from ADP and P_i by incubating the mixture with chloroplasts in the dark at low pH and then suddenly raise the pH to a high value, but it should not be possible to do the same by incubating with chloroplasts in the dark at high pH and then suddenly decrease the pH to a low value. These inferences are quite consistent with the well-known experimental work of Jagendorf and Uribe.[11] Similarly when a chloroplast suspension is illuminated under the conditions of cyclic electron transport in the absence of ADP and P_i, the external pH is expected to rise initially due to Reactions 15 and 16, and gradually to reach a steady-state value when the rate of production of $X \sim I$ equals the rate of its hydrolysis. After the light is turned off, a drop in external pH should follow the hydrolysis of $X \sim I$. Whether or not this radical coupling mechanism is applicable to mitochondria and chloroplasts has yet to be shown. But the above discussion shows that all chemical mechanisms which involve proton uptake during the formation of $X \sim I$ and proton release during the reaction of $X \sim I$ with ADP and P_i to form ATP are consistent with these experimental data.

The formation of $X \sim I$ may cause conformation changes in either or both of the macromolecular moieties (X or I) and consequently trigger changes in the gross structure of the membrane.[20] These conformational and membrane structural changes consume some of the converted free energy, since they follow spontaneously the formation of the energy-rich bond. (If these secondary processes had not taken place, the energy-rich bond may be energy-richer, although its formation would be slower for the same amount of free energy input from electron transport.) Although these conformational and membrane structural changes do not by themselves represent free energy storage, they may be very important to the related physiological functions of the organelle, such as permeability regulation and respiratory control.

The observed dependence of respiration rate on the concentrations of P_i, ADP and ATP[21,22] has been attributed to the effect of free energy storage on the midpoint reduction potentials of the electron carriers at the control sites.[23,20] Qualitatively this mechanism of respiratory control may be described as follows. At steady-state, the rate of oxidation of the reduced form of every carrier in an electron transport chain is equal to the rate of reduction of its oxidized form. Therefore in order for the electron transport chain to function efficiently in the kinetic sense, every carrier must maintain a healthy ratio of its oxidized and reduced forms. If the midpoint reduction potential of a particular carrier is raised substantially as a result of the energy storage prior to the formation of ATP, then a very large fraction of this carrier will automatically go to its reduced form. Consequently, the steady-state rate of electron transfer to it from its neighboring carriers on the substrate side of the respiratory chain will become very slow and hence rate-limiting. Conversely, if the midpoint reduction potential of this carrier is lowered substantially as a result of the energy storage, a very large fraction of this carrier will automatically go to its oxidized form. Consequently, the steady-state rate of electron transfer from it to its neighboring carrier on the oxygen side of the respiratory chain will become very slow and hence rate-limiting. Therefore unless the primary form of stored free energy is being continually used up for the phosphorylation of ADP or discharged by the action of uncouplers etc., electron transport in the normal direction will be severely retarded.

Evidence of quite large changes of the *in vivo* midpoint reduction potentials of cytochrome b_T and of heme a_3 in cytochrome oxidase by free energy storage has been obtained from spectroscopic data.[24]

In recent years a number of investigators have treated the coupling of electron transport to phosphorylation as an ideal, thermodynamically reversible process. Indeed, by completely inhibiting cytochrome oxidase with CN^-, it is even possible to use ATP to reverse the direction of electron transport through a part of the respiratory chain.[25,26] But inasmuch as no one has yet succeeded in preparing the ideal mitochondria which utilize the hydrolysis free energy of ATP to sustain the steady-state oxidation of water by $NADP^+$, the basic assumption of this convenient approach is open to question.

The observed coupling of electron transfer to phosphorylation in homogeneous solutions[16,17] shows clearly that in general respiratory control is not necessary for energy transduction, although energy transduction is necessary for respiratory control. For these reasons, it may be more realistic conceptually and more fruitful experimentally to treat the coupling of electron transport to phosphorylation as a steady-state rate process with optimum characteristics than as an ideal, thermodynamically reversible process. After all, life itself is irreversible!

References

1. E. C. Slater, *Nature,* **172** (1953) 975.
2. Y. K. Shen and G. M. Shen, *Scientia Sinica,* **11** (1962) 1097.
3. G. Hind and A. T. Jagendorf, *Proc. Natl. Acad. Sci. USA,* **49** (1963) 715.
4. P. Mitchell, *Nature,* **191** (1961) 144; P. Mitchell, in: *Regulation of Metabolic Processes in Mitochondria,* J. M. Tager, S. Papa, E. Quangliariello and E. C. Slater (eds.), Elsevier, Amsterdam, 1966, p. 65.
5. R. J. P. Williams, *J. Theoret. Biol.,* **1** (1961) 1.
6. P. D. Boyer, in: *Oxidases and Related Redox Systems,* T. E. King, H. S. Mason and M. Morrison (eds.), Vol. 2, Wiley, New York, 1965, p. 994.
7. E. F. Korman, A. D. F. Addink, T. Wakabayashi and D. E. Green, *J. Bioenergetics,* **1** (1970) 9.
8. H. H. Grünhagen and H. T. Witt, *Zeit. Naturforsch.,* **25b** (1970) 373.
9. W. Junge, *Eur. J. Biochem.,* **14** (1970) 582.
10. E. A. Liberman and V. P. Skulachev, *Biochim. Biophys. Acta,* **216** (1970) 30.
11. A. T. Jagendorf and E. Uribe, *Proc. Natl. Acad. Sci. USA,* **55** (1966) 170.
12. D. W. Deamer, A. R. Crofts and L. Packer, *Biochim. Biophys. Acta,* **131** (1967) 81.
13. S. Izawa, *Biochim. Biophys. Acta,* **223** (1971) 165.
14. P. Mitchell and J. Moyle, *Biochem. J.,* **105** (1967) 1147; *European J. Biochem.,* **4** (1968) 530.
15. R. L. Cross, B. A. Cross and Jui H. Wang, *Biochem. Biophys. Res. Commun.,* **40** (1970) 1155.
16. W. S. Brinigar, D. B. Knaff and Jui H. Wang, *Biochemistry,* **6** (1967) 36.
17. S. I. Tu and Jui H. Wang, *Biochemistry,* **9** (1970) 4505.

18. Jui H. Wang, C. S. Yang and S. I. Tu, *Biochemistry*, **10** (1971) 4922.
19. C. P. Lee and L. Ernster, *Eur. J. Biochem.*, **3** (1968) 385.
20. Jui H. Wang, *Science*, **167** (1970) 25.
21. H. A. Lardy and H. Wellman, *J. Biol. Chem.*, **195** (1952) 215.
22. B. Chance and G. R. Williams, *J. Biol. Chem.*, **217** (1955) 383.
23. Jui H. Wang, *Proc. Natl. Acad. Sci. USA*, **58** (1967) 37.
24. D. E. Wilson and P. L. Dutton, *Arch. Biochem. Biophys.*, **136** (1970) 583.
25. B. Chance and B. Hagihara, *Proc. Intern. Congr. Biochem., 5th, Moscow*, **5** (1961) 3.
26. M. Klingenberg and P. Schollmeyer, *Proc. Intern. Congr. Biochem., 5th* (1961) 46.

Functional Organization of Intramembrane Particles of Mitochondrial Inner Membranes

Lester Packer

Department of Physiology-Anatomy, University of California, Berkeley, California 94720

It is generally assumed that the high efficiency of electron transport and energy coupling in primary energy transducing membranes is the result of a special spatial arrangement of the interacting components within the membrane. Indeed, in recent years much evidence has been accumulated for an assymmetrical organization of the electron transport components of the inner mitochondrial membrane. Thus, cytochrome *c* appears to be localized at the outside surface of the inner membrane; ATPase at the inside surface; and cytochrome oxidase and cytochrome *b* within the hydrophobic membrane center. These concepts accord with the known sidedness of cytochrome *c* and ATPase reactivity and extractability; and with the disruption of the membrane structure upon extraction of cytochrome oxidase or cytochrome *b* (cf. ref. 1).

Based on current knowledge of biochemical and morphological studies a conceptual view of the possible orientation of mitochondrial membrane components was constructed as illustrated in Fig. 1. This shows the membrane to have distinct hydrophobic and hydrophilic regions. In certain areas the lipids are oriented in a bilayer structure, but in other areas the bilayer structure is perturbed by proteins that protrude into the center of the membrane. This is especially true for the large ATPase molecule (molecular weight 355,000 with its subunits) and for the lipoprotein electron transport components such as cytochrome $a + a_3$ (molecular weight 70,000–80,000), and cytochrome *b* (molecular weight 41,000). Permeability studies have shown that the dehydrogenases are located on the matrix side of the inner membrane.[2] Hence, it is necessary for substrates to permeate the membrane before they can donate electrons to the respiratory chain. The arrows in Fig. 1 show diagrammatically the possible direction of substrate translocation and electron flow, H^+ transport and ATP synthesis. The diagram does not imply any specific "boats and/or bridges" hypothesis for the carrier mediated translocation of metabolites, inorganic ions, and adenine nucleotides, although these considerations must also be taken into account in developing a fuller comprehension of the functional organization of the inner mitochondrial membrane.

An important question is, how is it possible to determine the

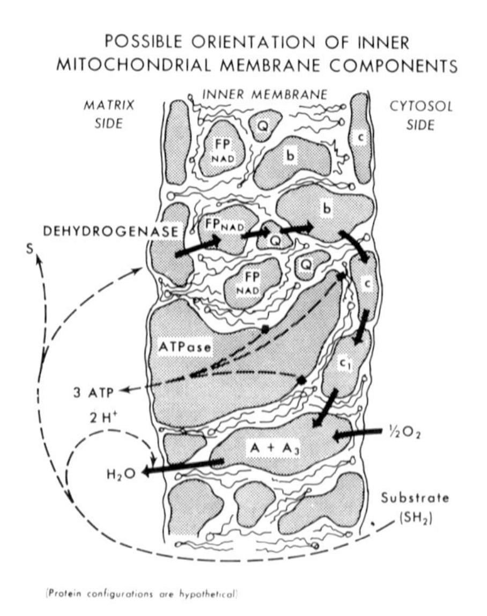

Figure 1. Illustration of possible orientation of inner mitochondrial membrane components. The membrane is visualized as combining the features of the lipid bilayer and subunit hypotheses of membrane structure (cf. Stoeckenius and Engelman[14]). Components which interact with the membrane predominantly by electrostatic forces (viz. cytochrome *c*) are surface components. Components organized in the center of the membrane are predominantly bound by strong hydrophobic forces. The large ATPase molecule, with its numerous subunits, is visualized to be bound by both hydrophobic and electrostatic forces each of which predominate in a different region of this large molecule. Excepting for the Q component the dark areas are for proteins and the light areas are lipid domains. The dimensions of the lipid and protein domains are hypothetical, drawn to emphasize functional organization (cf., Fig. 4, *bottom*).

components of the inner mitochondrial membrane by a direct experimental approach?

An approach to this problem was suggested in 1970 by Wrigglesworth, Packer and Branton,[3] who discovered the existence of particles in mitochondrial membranes. This has prompted our laboratory to carry out a detailed study of the factors which affect the distribution and dispersal, and identification of these particles in the membrane. Electron microscopy after freeze-cleavage is the only technique presently available which reveals the presence of structural components in the membrane interior.

I. *Interpretation of Electron Micrographs After Freeze Cleavage and Etching*

Freeze-cleavage and etching electron microscopy reveal components of the hydrophobic interior and membrane surface respectively. The possible orientation of lipids and proteins and their relationship to the fracture pattern and etch face are shown diagrammatically in Fig. 2. Lipids are generally thought to be present as bilayer structures. X-ray

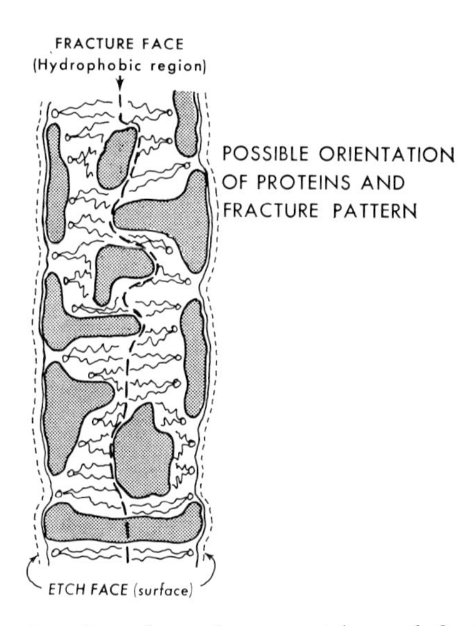

Figure 2. Possible orientation of membrane proteins and fracture patterns after freeze-cleavage. The dashed line shows the possible fracture face that might arise as a result of splitting the membrane down its hydrophobic center. This would cause exposure of components, which are revealed after the platinum shadowing to form a replica. In the case where proteins might entirely traverse the membrane the route that the fracture plane would take is uncertain; the fracture plane may either be around the component or it may itself break. The dashed line near the membrane surface is the etch face that would be revealed after sublimation of ice. Components on the etch face are revealed after platinum shadowing.

data of membranes suggest that the bilayer regions extend over at least a 400 Å field. In mitochondrial membranes fatty acids vary from about C-16 to C-22. This would vary the chain length from 19–27 Å respectively. It may be that shorter fatty acid chains preferentially are organized in those regions where proteins protrude more deeply into the membrane center. Accordingly, the fracture faces reveal protein components protruding into the hydrophobic, central region of the membrane; these are replicated by the platinum shadowing process. This interpretation, originally proposed by Branton,[4] has recently been established unequivocally by examination of complementary replicas of freeze-fracture specimens.[5] After fracturing, if the specimen is warmed from liquid nitrogen temperature to −100°C, ice sublimes from the membrane surface at a rate of about 1000 Å/min. This process, known as etching, exposes the membrane surface.

II. *Structural Organization of Inner and Outer Mitochondrial Membranes Observed by Electron Microscopy After Freeze-Cleavage*

In collaboration with Dr. R. Melnick[6] outer and inner membranes of rat liver mitochondria have been separated and their structure examined by electron microscopy after freeze-cleavage. The photographs in Fig. 3 show concave and convex fracture faces observed on these membrane preparations. Note that the replica of the rat liver inner mitochondrial membrane preparation shows a large area of a concave fracture (left side) and a convex fracture face (right side). On the other hand the replica of the outer membrane preparation shows a large area of a concave fracture face (lower portion) and a smaller area of a convex fracture face (upper left).

The histograms prepared from these photographs (Fig. 4) show the distribution of particle sizes. On the average, inner membranes contain about twice the number of particles in convex as compared with concave fracture faces, while in outer membranes this ratio is about 1:4. The diagram (lower part) illustrates how the particles in the half membranes may be distributed after fracturing. The data suggest that most of the particles face either the cytoplasmic or the matrix compartments in the cell and that fewer of the particles are distributed in the areas facing the intermembrane space. Note that others have found that in plasma membrane the majority of particles are distributed on the side facing the inside of the cell. Secondly, the results accord with the idea that the proteins of the outer mitochondrial membrane are derived from a membrane of the eucaryotic cell and that the proteins of the inner mitochondrial membrane are vestiges of an early infection of a primitive nucleated cell by a procaryotic organism according to the endosymbiont hypothesis. The fatty acid components of the membranes are likely characteristic of the host because of the known ease with which environmental conditions lead to fatty acid exchange.

III. *Intramembrane Particles of Mitochondrial Inner Membranes—Dynamic State of Organization*

The experiments described above show that the outer and inner mitochondrial membranes contain a densely and a lightly particulated fracture face, designated "A" and "B": faces respectively. Such particle distribution, frequently seen in other membranes, represents

Figure 3. Membrane fracture faces observed after freeze-cleavage of inner and outer rat liver mitochondrial membranes; after Melnick and Packer.[6]

the distribution of particles in the half membranes. Particle-particle interaction may contribute to membrane cohesion, and may reflect the forces which bind the membrane together. The presence of two densely staining areas, i.e. "Railroad Tracks" in the "unit membrane" visualized by conventional electron microscopy after positive staining

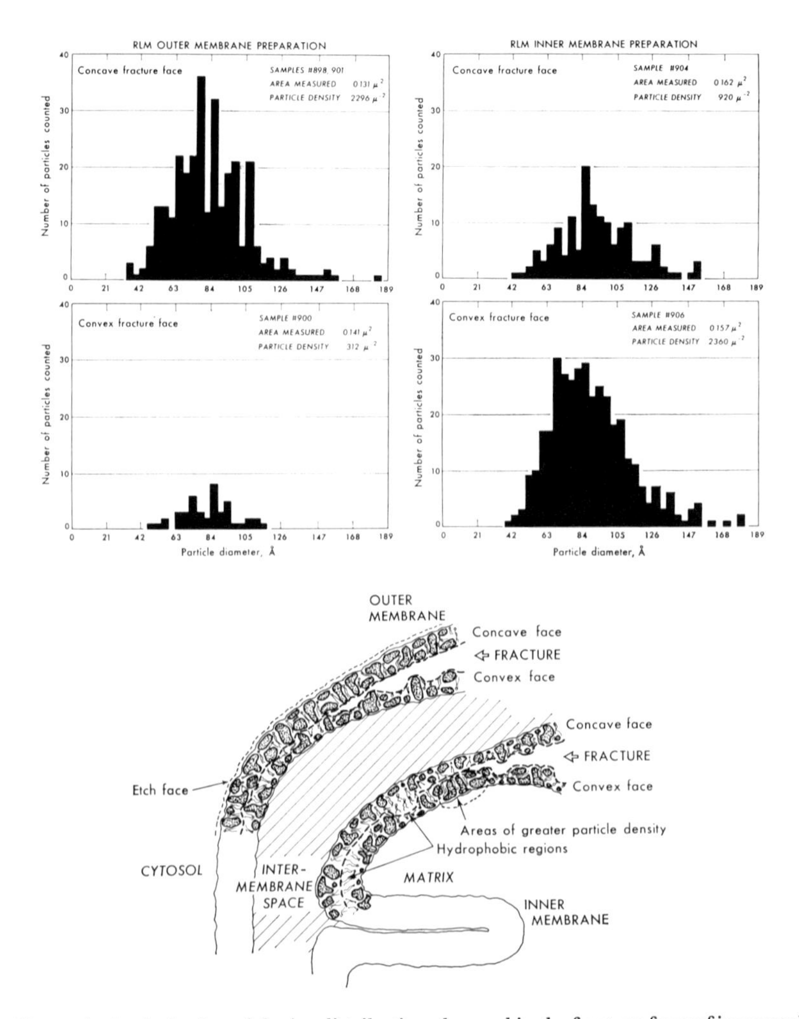

Figure 4. Analysis of particle size distribution observed in the fracture faces of inner and outer mitochondrial membranes and suggested distribution in these membranes. The measured particle sizes observed in the fracture faces after freeze-cleavage electron microscopy have been tabulated as a function of their frequency. In constructing these histograms no correction has been made for the thickness of the particles caused by the deposition of platinum during the shadowing process. An estimated correction is 20–30 Å for particles in this size range. Therefore, the actual particle diameters are 20–30 Å smaller in diameter. In the lower portion of the figure the diagram illustrates the possible organization of components in the half membranes that arise from inner and outer mitochondrial membranes.

with heavy metals suggests that the protein components predominate at the periphery. Since energy coupling requires a precise spatial relationship of membrane components, a vectorial dispersal of membrane components in the hydrophobic membrane phase may play a crucial role in the efficiency of energy coupling. It is important to identify what factors govern the properties of the particles in these "half membrane" structures, e.g., are the particles fixed within the membranes or do they change their orientation?

A. *Membrane Dehydration Experiments*

We have recently observed that particle distribution in the half membranes varies with water content of the membranes (biological membranes generally have about 15% water content). These changes were noted after dehydrating the membranes using 2 M sucrose (and also by changing the relative humidity). Some results are shown in Table I. At 0·33 M sucrose the inner membrane particle density

TABLE I. Influence of sucrose concentration on the particle distribution observed in fracture faces of inner mitochondrial membranes

		Particle analysis						
				Particle clusters (% of total)				
Inner membrane preparation	Area analyzed ($\times 10^{-1}$ μ^2)	Total No.	Density $\times 10^{-1}$ μ^2	1	2	3	4	5
0·33 M sucrose, pH 7·2								
Convex	19·35	2675	1452	57·0	9·3	10·6	9·5	12·7
Concave	7·49	418	558	65·3	8·7	6·2	6·8	10·3
2·0 M sucrose, pH 7·2								
Convex	6·60	1525	2310	65·4	9·7	8·5	9·5	7·4
Concave	2·03	534	2630	78·9	9·0	7·4	3·2	0·0

Inner membranes were prepared as previously described and incubated under the conditions indicated in the Table before freezing in liquid nitrogen.

shows typical A and B faces, i.e. heavy and light distribution of particles in the convex and concave fracture faces. In contrast, at 2·0 M sucrose the density of particles in the hydrophobic center of the membrane is 60% greater. Also, the "typical" A face or B face are no longer apparent and both half membranes have an approximately equal distribution of particles.

It is noteworthy that under similar conditions of high sucrose concentration, Packer, Pollack, Munn and Greville[7] have found that the energy coupling is inhibited and that upon lowering the sucrose concentration to the isotonic range, normal electron energy transfer activity is restored. It remains to be shown, however, that the structural modifications seen by freeze-cleavage are also reversible.

Moreover, we have noted, as shown in Table II and in other experiments, that in the concave fracture face more single particles occur and are less aggregated with other particles than in the convex face. This approach may eventually lead to a better understanding of the factors which cause the components in the membrane to aggregate; a problem which is of particular relevance to the organization of the respiratory chain.

B. *Lipid Depletion Studies*

The above experiments show that by modifying the water content of the membranes we can change the particle distribution in the half

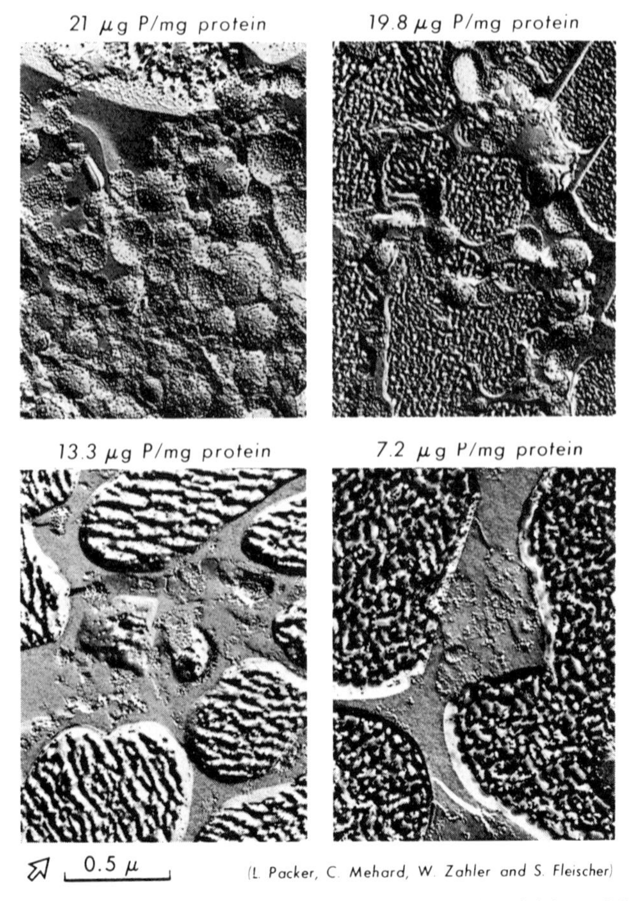

Figure 5. Effect of lipid depletion of beef heart submitochondrial particles on the particle distribution observed by electron microscopy after freeze-cleavage. Experiment carried out in collaboration with Dr. C. Mehard, Dr. W. Zahler, and Dr. S. Fleischer.

membranes. It was also of interest to determine the effect of lipid depletion.

In collaboration with Dr. S. Fleischer, Dr. C. Mehard and Dr. W. Zahler heart submitochondrial vesicles are being extracted in a stepwise manner with organic solvents. The extracted particles, of known lipid composition, are then examined by electron microscopy. Typical results are shown in Fig. 5. It has been established that after extensive extraction of lipid the membrane particles are disorganized and only an amorphous distribution pattern may be discerned. If lipids are removed gradually, then the smooth fracture faces disappear and only clusters of particles are deposited, and these are observed on the replica by such experiments. It should be possible to obtain evidence on the number of aggregated components in membrane vesicles. It is clear from these results that lipid plays a crucial role in establishing the proper conditions for dispersal of components in the membrane.

C. *Unsaturated Fatty Acid Depletion Studies*

To study effects of alteration *in vivo* of fatty acid composition, experiments were performed in which the unsaturated fatty acid composition of rat liver mitochondria was varied by dietary treatment.

The fatty acid composition of mitochondrial membrane is altered by growing rats on diets completely deficient in the essential unsaturated fatty acids.[8] In collaboration with Dr. M. A. Williams we have found (Table II) that the inner membrane preparations from

TABLE II. Particle densities in membrane fracture faces of mitochondria—Influence of unsaturated fatty acid composition

Unsaturated fatty acid composition	Sucrose in suspending medium mM	Fracture face	Area analyzed $\times 10^{-1}$	Particles No.	Particles Density/μ^2	Particles A/B
EFA deficient	70	A	1·696	220	1333	1·62
		B	3·084	230	821	
	300	A	3·277	507	1317	1·14
		B	2·22	264	1170	
+ Linolenic acid	70	A	4·547	658	1491	2·18
		B	1·61	110	684	
	300	A	3·88	678	1742	1·86
		B	5·597	521	939	

Rat liver mitochondria were isolated from animals grown on diets deficient in essential unsaturated fatty acids or supplemented with linolenic acid for periods up to four weeks.[15] The mitochondria were suspended in the sucrose concentrations indicated in the Table prior to freezing in liquid nitrogen for preparation of samples for freeze fracturing.

fatty acid deficient, and linolenic acid fed rats show differences in the distribution of particles in the half membranes (A and B faces). The value for the ratio, particles in A face/particles in B face, is higher in the membranes from linolenic fed animals whose mitochondria also show a high degree of energy coupling as compared with the fatty acid deficient preparations. These two types of mitochondria have remarkably different capacities for energy coupling as observed by

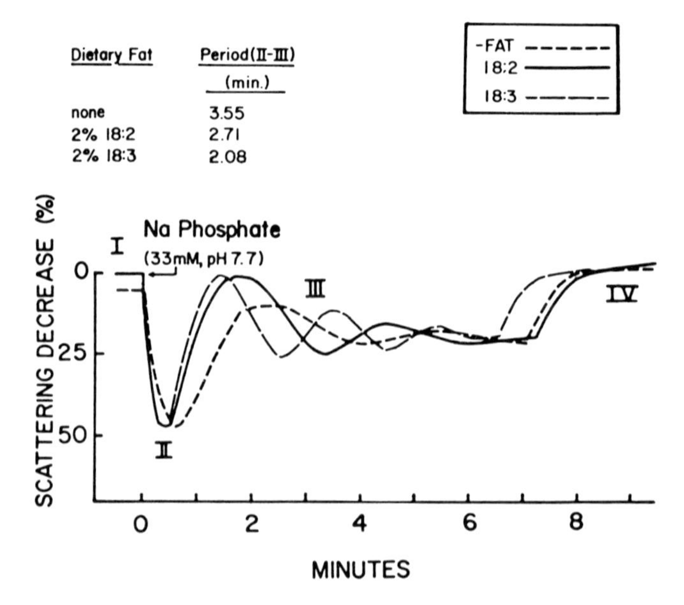

Figure 6. Effect of unsaturated fatty acids on the phase relationships of mitochondrial oscillations. Rat liver mitochondria were obtained from animals fed for a period of four weeks on a diet completely deficient in unsaturated fatty acids or supplemented with fatty acids of the type indicated in the Figure. Mitochondria were isolated as previously described,[15] suspended at pH 7·6 in an aerobic sucrose-EDTA test medium fortified with sodium succinate, the oxidizable substrate, at 25°C. Oscillations were started by adding a high concentration of permeant ions (33 mM sodium phosphate); experiments carried out in collaboration with M. B. Williams and R. C. Stancliff.

the period of the oscillatory state of ion transport (Fig. 6). It can be seen that the period of the oscillation is considerably lengthened in the mitochondria from the unsaturated fatty acid deficient animals. Under these conditions, the proportion of total unsaturated fatty acids in mitochondrial membranes does not change, but the unsaturated fatty acids contain fewer double bonds (i.e., their unsaturation index changes).[9] These experiments may provide direct evidence that the fluidity of the membrane is changing, and that this may be reflected by a changed pattern of distribution of particles in the half membranes that arise after freeze fracture. In this regard,

X-ray diffraction patterns suggest that membrane lipids are in bilayers with spatial distances of 4·8–5·0 Å. Therefore, the phospholipid molecules in the membranes must be packed closely together. Such tight packing suggests that the lipids would move in a coordinated manner.

Hence, the change in the hydrophobic center of the membrane after alteration of the unsaturated fatty acid content affects the pattern of membrane splitting during freeze-fracturing and can be detected by functional criteria, e.g., by measuring the period of the oscillatory state. From these results, together with the lipid depletion studies, it would appear that the inner mitochondrial membrane is in a liquid crystalline state where the proteins in the membrane center are in a dynamic state. These proteins can change their orientation in the membrane as the cohesive factors that bind them together, mainly hydrophobic forces, are modified by even small changes in the unsaturation index.

These interpretations have received support[10] from studies of mutant yeast mitochondria where the proportion of unsaturated fatty acids were lowered. When the unsaturated fatty acid content fell below 30% of the total fatty acids in the membrane, ATP synthesis was inhibited. Under these conditions, cytochrome content and ATPase activity were unchanged. Thus, these results suggest that the unsaturated fatty acid content of the membrane is important for energy coupling. The results can be rationalized by suggesting that the dispersal and orientation of the protein components are modified by changing the environment, and the flexibility inherent in the liquid crystalline state, which property is conferred upon the membrane by the presence of long chain unsaturated fatty acids which provide a milieu *par excellence* for protein dispersal.

D. *pH Dependence of Intraparticle Interactions*

Previous studies from this laboratory by Wrigglesworth and Packer[11] and House and Packer[12] with rat liver submitochondrial particles (SMP) have established that changes in pH markedly affect energy coupling and molecular conformations.

These studies revealed that, as the pH falls to 5·8, the following changes occur:

(a) Passive H^+ permeability of SMP decreases.

(b) The active H^+ uptake is increased.

(c) H^+/O ratio reaches a maximum of about 2 for choline or succinate oxidation.

(d) Respiration decreases.

These observations could be explained by changes in the structure of SMP. The main pH-dependent structural changes observed as the pH fell to 5·8 were:

(a) Volume changes seen by microscopy and packed volume.
(b) Increases in light scattering and ANS fluorescence.
(c) Changes in circular dichroism, suggesting alterations in quaternary protein structure, i.e., reversible particle aggregation.

In view of this it was of interest to determine if direct evidence could be obtained by electron microscopy after freeze-cleavage for reversible pH dependent intraparticle aggregation in the membrane. Indeed, we have observed large reversible aggregation effects of membrane preparation as the pH is lowered. Similar studies by Da Silva[13] on human erythrocyte ghosts have also demonstrated a reversible pH-dependent aggregation of membrane components.

IV. *Conclusions—Energy Coupling and Membrane Structure*

The above results indicate that respiration-dependent proton translocation markedly affects the molecular conformational state of functional proteins in the membrane by altering their mutual interaction. Since such changes are closely correlated with the degree to which the components are coupled to energetic processes, it seems that the dynamic organization of components in the membrane plays a crucial role in energy coupling in the inner mitochondrial membrane. These results are of interest with respect to the chemiosmotic and conformational hypotheses for energy coupling which state that the primary energetic event is the establishment of a H^+ gradient and/or conformational change. This is so because a change in H^+ gradient induced by electron transport would be anticipated to modify the aggregation state of the components involved in the tight coupling of energy from electron transport to oxidative phosphorylation or other energy-linked processes. At the molecular level of organization both are early events associated with energy coupling. Thus, a proton gradient can be viewed as causing a conformational change which brings the interacting proteins closer together, permitting an efficient chemical transfer of primary energy transduction into a form which can be conserved.

Other environmentally induced changes can also be considered as exerting a regulatory role on the energy coupling process by causing conformational changes. For example, changes in the concentration of bivalent cations such as calcium or magnesium would be expected to affect the aggregation state and particle interactions. These might be due to direct effects upon protein, or, in the case of calcium, to stabilization of the phospholipid bilayer structure, thus reducing the mobility of protein components in the hydrophobic phase. Likewise, changes in the unsaturated fatty acid composition of the membrane brought about by dietary means or by changed fatty acid metabolism could modify energy coupling by affecting structure at the molecular level by affecting intramembrane particle interactions. Similarly, modifications in the protein components of the membrane, brought about by a changed pattern of biosynthesis and/or genetic regulation may also modify energy coupling through deletion and/or through a change in the structure of components of the integrated electron transfer and energy coupling system.

References

1. B. Chance and C. P. Lee (eds.), *Probes of Structure and Function of Macromolecules and Membranes*, Vol. 1, Academic Press, New York and London, 1971.
2. M. Klingenberg, *Eur. J. Biochem.*, **13** (1970) 247.
3. J. M. Wrigglesworth, L. Packer and D. Branton, *Biochim. Biophys. Acta*, **205** (1970) 125.
4. D. Branton, *PNAS*, **55** (1966) 1048.
5. J. Wehrli, K. Muhlethaler and H. Moor, *Experimental Cell Res.*, **59** (1970) 336.
6. R. L. Melnick and L. Packer, *Biochim. Biophys. Acta*, in press.
7. L. Packer, J. K. Pollak, E. A. Munn and G. D. Greville, *J. Bioenergetics*, in press.
8. M. M. Guarnieri and R. M. Johnson, *Adv. Lip. Res.*, **8**, (1970) 115.
9. J. M. Lyons and C. M. Asmundson, *J. Amer. Oil Chem. Soc.*, **42** (1965) 1056.
10. J. W. Proudlock, J. M. Haslam and A. W. Linnane, *BBRC*, **37** (1969) 847.
11. J. M. Wrigglesworth and L. Packer, *J. Bioenergetics*, **1** (1970) 33.
12. D. R. House and L. Packer, *J. Bioenergetics*, **1** (1970) 273.
13. P. P. Da Silva, *J. Cell Biol.*, in press.
14. W. Stoeckenius and D. M. Engelman, *J. Cell Biol.*, **42** (1969) 613.
15. R. C. Stancliff, M. A. Williams, K. Utsumi and L. Packer, *Arch. Biochem. Biophys.*, **131** (1969) 629.

On Energy Conservation and Transfer in Mitochondria*

Y. Hatefi and W. G. Hanstein†

Department of Biochemistry, Scripps Clinic and Research Foundation, La Jolla, California 92037, U.S.A.

Abstract

Structural and functional features of energy conservation and transfer in mitochondria have been examined in the light of recent developments, and hypothetical schemes for energy coupling and transfer have been presented.

The following is in response to the invitation of this journal for a conceptual paper in the field of bioenergetics. The ideas presented were elicited by the significant findings of several laboratories (see references). Since this article is intended mainly for workers in the field, we have considered it proper to refrain from detailed explanations and enumeration of conclusions, which we feel will be rather obvious to our colleagues.

In mitochondria, electron transfer complexes I, III and IV include the three sites of energy conservation.[1–5] Each complex appears to contain an electron carrier, which undergoes a redox potential change upon energization of mitochondria by ATP,[5–9] plus other carriers whose redox potentials remain unaltered. Furthermore, each complex is located between two redox pools as shown in Table I. These pools make it possible for each complex to function independently of the others. In addition, the cytochrome *c* pool solves the problem of the unstoichiometric relationship of a and a_3 with the components of the other complexes.‡ The Q and *c* pools, along with most of the electron carriers, comprise two redox plateaus of short

* Supported by USPHS grant AM-08126 to Y.H. and San Diego County Heart Association Grant-in-Aid 97-71 to W.G.H.

† Recipient of USPHS career Development Award 1-K4-GM38291.

‡ Abbreviations: a, a_3, b, c_1, c; cytochromes a, a_3, b, c_1 and c respectively; b_T, transducing cytochrome b (see ref. 9); $b_{559 \cdot 5}$, cytochrome b of complex III with α band at 559·5 nm at 77°K; FeS2, iron–sulfur center 2 of complex I;[18] NADH deh., NADH dehydrogenase; Q, ubiquinone (coenzyme Q); OSCP, oligomycin sensitivity conferring protein; DOCA, deoxycholate.

TABLE I.

Correlations Among Coupling Regions I, II and III

Electron carriers	ΔE_m (mV) Non-energized	Energized	Redox pool Electron donor	Electron acceptor
FeS2-NADH deh.	+, large	±, small	NADH	Q
$c_1 - b_T$	+260	−15	Q	c
$a_3 - a$	+130	+35	c	O_2

potential spans. The former encompasses a span of not more than 150 mv, and includes, in the nonenergized state of mitochondria, the b cytochromes, iron–sulfur center 2 of complex I and Q.* The latter covers a span of only about 100 mv, and includes c_1, c, a, Cu, and the iron–sulfur protein of complex III (Fig. 1).

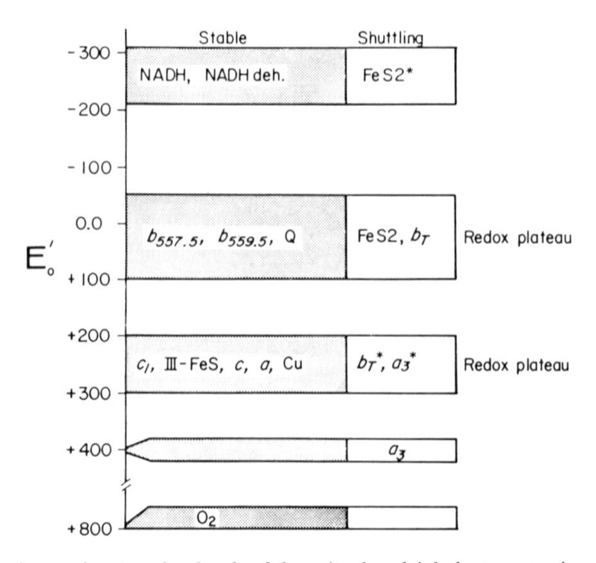

Figure 1. Approximate redox levels of the mitochondrial electron carriers. The redox-stable carriers in the shaded areas are those which do not appear to undergo potential change upon energization of mitochondria by ATP. The redox-shuttling carriers are those which appear to undergo potential change upon such treatment. These carriers in the energized state of mitochondria are identified with asterisks. For the redox potentials of Cu and III-FeS (complex III iron–sulfur protein) see refs. 14 and 15. On the basis of the data of refs. 16–18, the redox potential of FeS2 in the nonenergized state is assumed to be in the region shown. Bovine heart mitochondria contain 3 b-type cytochromes with 77°K α peaks at 557·5, 559·5 (b_K?) and 562·5 (b_T) nm. Cytochrome $b_{557 \cdot 5}$ is located in complex II, and the other two are located in complex III.[50]

* Succinate dehydrogenase and other electron tributaries of the respiratory chain will not be discussed in this article, because their function is not germane to the mechanism of energy conservation.

In the energized state, the composition of these plateaus is altered (Fig. 1). The carriers which change potential in complexes I, III and IV are considered to be respectively iron–sulfur center 2, b_T and a_3 (see, however, ref. 7). Upon energization, the potential of iron–sulfur center 2 approaches that of NADH dehydrogenase, which is an iron–sulfur flavoprotein,[10] and the potentials of b_T and a_3 approach those of c_1 and a respectively.* Thus, nearly all of the electron carriers on the oxygen side of Q congregate in the redox plateau at the level of cytochrome c, and iron–sulfur center 2 forms another region of low potential gradient with NADH and NADH dehydrogenase.† Figure 2 depicts the potential transitions of iron–sulfur center 2, b_T and a_3 in complexes I, III and IV respectively. It is seen that each complex

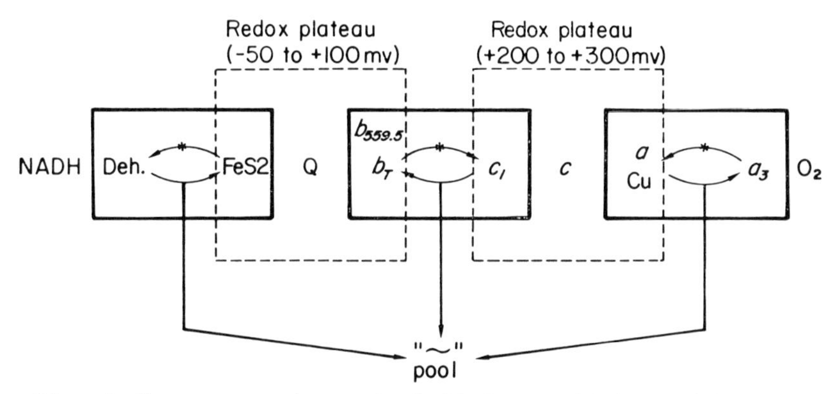

Figure 2. Energy conservation associated with the potential change of the shuttling carriers iron–sulfur center 2, cytochrome b_T and cytochrome a_3 in complexes I, III and IV respectively. Curved arrows show the redox potential changes of the above carriers; asterisks indicate energization; and the dashed squares show the redox plateaus at the pool levels of Q and c.

contains at least one component with a fixed potential at the level of the redox pool on its substrate side, and one component which is capable of potential change and redox-shuttling between the two pools which surround that complex.

The redox changes of iron–sulfur center 2, b_T and a_3 suggest that they participate in energy conservation when their potential is altered, and that they return to their original potential when the conserved energy is either transferred away from the respiratory chain or dissipated under uncoupling conditions. Thus, the redox potential shuttling property of these carriers and the closing of potential gaps as

* See, however, ref. 11 for critical comments regarding the redox potential measurements of refs. 6 and 8.

† The redox potential of NADH dehydrogenase appears to be close to those of NAD/NADH and their acetylpyridine analogues, because this enzyme catalyzes a transhydrogenation from NADH to acetylpyridine adenine dinucleotide.[10] The difference between the E_0' of the two nucleotides is 72 mv.[13]

$$a''' \underset{}{\overset{+e^-}{\rightleftharpoons}} a'' \tag{1A}$$

$$a'' + a_3''' \rightleftharpoons a''' + a_3'' \tag{2A}$$

$$a''' + a_3'' \overset{-e^-}{\rightleftharpoons} a'''\text{—}a_3''' \quad \text{(low potential } a_3\text{)} \tag{3A}$$

$$a'''\text{—}a_3''' \overset{+e^-}{\rightleftharpoons} a''\text{—}a_3''' \tag{4A}$$

$$a'' + a_3''' \rightleftharpoons a'''\text{—}a_3'' \tag{5A}$$

$$a'''\text{—}a_3'' \overset{-e^-}{\rightleftharpoons} a''' \sim a_3''' \tag{6A}$$

$$a''' \sim a_3''' + x \rightleftharpoons a''' + a_3''' \sim x \tag{7A}$$

$$a''' + a_3''' \sim x \rightleftharpoons a''' + a_3''' + {\sim}x \quad \text{(high potential } a_3\text{)} \tag{8A}$$

Scheme A. Hypothetical mechanism for energy coupling at site III. A variation of reactions 3A and 4A might be as follows:

$$a''' + a_3'' \overset{+e^-}{\rightleftharpoons} a'' + a_3''$$

$$a'' + a_3'' \overset{-e^-}{\rightleftharpoons} a'' - a'''$$

The high energy state formally represented by $\sim$ in this scheme could be with respect to hydrolysis, or coulombic or conformational strain.

$$b_{559\cdot5}''' \overset{+e^-}{\rightleftharpoons} b_{559\cdot5}'' \tag{1B}$$

$$b_{559\cdot5}'' + b_T''' + c_1''' \rightleftharpoons b_{559\cdot5}''' + b_T''' - c_1'' \quad \text{(high potential } b_T\text{)} \tag{2B}$$

$$b_T''' - c_1'' \overset{-e^-}{\rightleftharpoons} b_T''' - c_1''' \tag{3B}$$

$$b_T''' - c_1''' \overset{+e^-}{\rightleftharpoons} b_T'' \sim c_1''' \tag{4B}$$

$$b_T'' \sim c_1''' \rightleftharpoons b_T''' \sim c_1'' \tag{5B}$$

$$b_T''' \sim c_1'' \overset{-e^-}{\rightleftharpoons} b_T''' \sim c_1''' \tag{6B}$$

$$b_T''' \sim c_1''' + x \rightleftharpoons c_1''' + b_T''' \sim x \tag{7B}$$

$$b_T''' \sim x + c_1''' \rightleftharpoons b_T''' + c_1''' + {\sim}x \quad \text{(low potential } b_T\text{)} \tag{8B}$$

Scheme B. Hypothetical mechanism for energy coupling at Site II. A possible mechanism of electron transfer in reaction 2B, which would conform with a single site of antimycin inhibitiin, would be electron transfer from $b_{559\cdot5}$ to c_1 through low-potential b_T as an electron siphon.

shown in Table I appear to be phenomena which are intimately related to energy conservation.[8,9,19–21] These characteristics are summarized for better visualization in schemes A and B, which depict a unified hypothetical mechanism for energy conservation at sites III and II.* The main features of these mechanisms are:

1. Stepwise energy conservation, as a result of two cycles of single electron transfer, in carrier pairs, each of which is composed of a redox-stable and a redox-shuttling carrier.
2. Redox potential change of the shuttling carrier upon an energy yielding reduction or oxidation reaction, or upon energization by ATP.
3. Controlled respiration in state 4 because electron transfer through the coupled carriers could be slow.
4. Rapid electron transfer (characteristic of state 3) through the wide potential gap of each complex after transfer (or dissipation by uncouplers) of the "high energy" state and concomitant return of the shuttling carrier to its noncoupled state.

It might be added that in these hypothetical mechanisms a number of experimental observations have also been taken into consideration. Among these are:

a. The fact that b_T is not reduced without energy input and unless c_1 is in the oxidized state.[9,22,23]
b. The observations that the inhibitory effects of loosely bound compounds such as amytal, 2-heptyl-4-hydroxyquinoline-*N*-oxide,† and azide, are reversed by uncouplers.[25–28] This is because uncouplers might be expected to act near the site of action of these respiratory inhibitors.
c. The fact that antimycin A treatment leads to a potential change of b_T.[22,23] Several pieces of evidence indicate that antimycin A results in a stronger structural association of b and c_1,[29,30] which might be analogous to the coupling of b_T and c_1 in the energized state.

Although not included in the above hypothetical schemes, iron–sulfur centers 3 and 4 of complex I[18] and the iron–sulfur protein of complex III[14] might be involved respectively in electronic communication between the carrier pairs of each complex, whereas copper in complex IV might be required, in addition, for the 4 electron reduction of oxygen through the rapid sequential oxidation of a_3'', 2 Cu′, a''.[31,32]

The work of Lee and Ernster[33] indicates that the three coupling

* The mechanism for site I could be analogous to scheme A with the replacement of a and a_3 respectively by NADH dehydrogenase and iron–sulfur center 2 of complex I.

† Relative to the binding of antimycin A.[24]

sites communicate through a "high energy" pool. Other studies have shown that this pool is, in turn, in communication with the adenine nucleotide (ATP/ADP) pool through a set of factors (see reviews 34 and 35) capable of isoenergetic energy transfer. Thus, the mitochondrial energy conservation-transfer apparatus may be conceived of as 3 energy "transducing" systems (complexes I, III and IV) and one energy transfer system, each of which communicates with its neighbor by way of an energy (potential or "high-energy") pool (Figure 3).

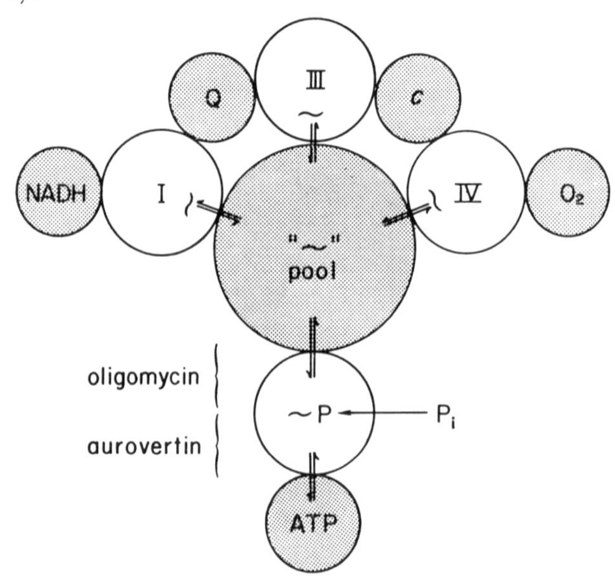

Figure 3. Schematic representation of the "transducing" complexes I, III and IV, the "high-energy" transfer complex (unshaded circles), and the energy pools (shaded circles).

The "high-energy" pool of Lee and Ernster suggests the possible existence of a "high-energy" carrier operating at this level. This carrier might be identical with component X in schemes A and B. Sanadi's factor B appears to be a possible candidate for this "high-energy" carrier. It is water soluble, has a small molecular weight (29,200), and is believed to operate near the respiratory chain.[36,37] The "high-energy" transfer system appears to involve the sites of inhibition by oligomycin and aurovertin.[38,39] The differential effect of these inhibitors on stimulation of state 4 respiration by arsenate has suggested that they act at separate sites, and that phosphate enters the phosphorylation sequence prior to ADP.[40] Wang and his colleagues have used the differential effect of these two inhibitors to trap membrane bound phosphate.[49] These results suggest that the energy transfer system is composed of at least 2 components, OSCP and F_1 or factor A (see review 34).

During isolation of the electron transfer complexes I, II, III and IV from submitochondrial particles, OSCP and F_1 accumulate in a fifth fraction, which is also particulate[42] (see also refs. 43 and 44) (Fig. 4). As in the case of the components of electron transfer complexes,

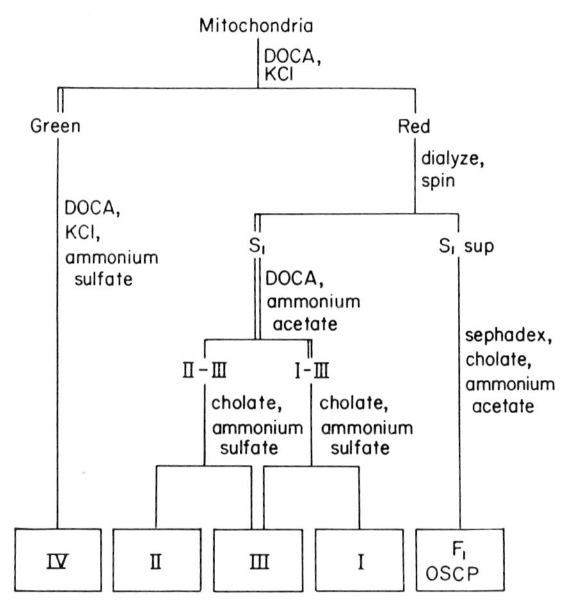

Figure 4. Resolution of mitochondria into the four electron transfer complexes and a fifth particulate fraction containing F_1 and OSCP.

F_1 and OSCP can be separated from one another by chaotropes.[29, 44–47] Therefore, it is possible that, similar to the "transducing" complexes, the "high-energy" transfer components of mitochondria also form a hydrophobic complex capable of protecting "∼" from hydrolysis. In addition to OSCP and factor A, such a complex might contain the Pullman-Monroy inhibitor[48] and embody the properties of Sanadi's factor A.D (ATPase synthetase).[49]

References

1. Y. Hatefi, *Comprehensive Biochemistry*, **14** (1966) 199.
2. B. Chance and G. R. Williams, *Advances in Enzymology*, **17** (1956) 65.
3. B. Chance and M. Erecińska, *Arch. Biochem. Biophys.*, **143** (1971) 675.
4. P. C. Hinkle, R. A. Butow and E. Racker, *J. Biol. Chem.*, **242** (1967) 5169.
5. M. Gutman, T. P. Singer and H. Beinert, *Biochem. Biophys. Res. Commun.*, **44** (1971) 1572.
6. D. F. Wilson and P. L. Dutton, *Arch. Biochem. Biophys.*, **136** (1970) 583.
7. P. Hinkle and P. Mitchell, *J. Bioenergetics*, **1** (1970) 45.

8. D. F. Wilson and P. L. Dutton, *Biochem. Biophys. Res. Commun.*, **39** (1970) 59.
9. B. Chance, D. F. Wilson, P. L. Dutton and M. Erecińska, *Proc. Nat. Acad. Sci. U.S.A.*, **66** (1970) 1175.
10. Y. Hatefi and K. E. Stempel, *J. Biol. Chem.*, **244** (1969) 2350.
11. A. H. Caswell, *Arch. Biochem. Biophys.*, **144** (1971) 445.
12. M. Gutman, T. P. Singer and H. Beinert, *J. Biol. Chem.*, **240** (1965) 475.
13. B. M. Anderson, C. J. Ciotti and N. O. Kaplan, *J. Biol. Chem.*, **221** (1956) 823.
14. J. S. Rieske, *Non-Heme Iron Proteins*, A. San Pietro (ed.), The Antioch Press, 1965, p. 461.
15. M. Erecińska, B. Chance and D. F. Wilson, *FEBS Letters*, **16** (1971) 284.
16. Y. Hatefi and K. E. Stempel, *Biochem. Biophys. Res. Commun.*, **26** (1967) 301.
17. C. I. Ragan and P. B. Garland, *European J. Biochem.*, **10** (1969) 399.
18. N. R. Orme-Johnson, W. H. Orme-Johnson, R. E. Hansen, H. Beinert and Y. Hatefi, *Biochem. Biophys. Res. Commun.*, **44** (1971) 446.
19. J. H. Wang, *Science*, **167** (1970) 25.
20. E. C. Slater, C.-P. Lee, J. A. Berden and H. J. Wegdam, *Biochim. Biophys. Acta*, **223** (1970) 354.
21. D. DeVault, *Biochim. Biophys. Acta*, **226** (1971) 193.
22. D. F. Wilson, M. Koppelman, M. Eercińska and P. L. Dutton, *Biochem. Biophys. Res. Commun.*, **44** (1971) 759.
23. J. S. Rieske, *Arch. Biochem. Biophys.*, **145** (1971) 179.
24. Y. Hatefi, A. G. Haavik and P. Jurtshuk, *Biochim. Biophys. Acta*, **52** (1961) 106.
25. B. Chance and G. Hollunger, *J. Biol. Chem.*, **238** (1963) 418, 432.
26. D. F. Wilson and B. Chance, *Biochem. Biophys. Res. Commun.*, **23** (1966) 751.
27. D. F. Wilson and B. Chance, *Biochim. Biophys. Acta*, **131** (1967) 421.
28. J. L. Howland, *Biochim. Biophys. Acta*, **73** (1963) 665.
29. J. S. Rieske, H. Baum, C. D. Stoner and S. H. Lipton, *J. Biol. Chem.* ,**242** (1967) 4854.
30. Y. Hatefi and W. G. Hanstein, *Arch. Biochem. Biophys.*, **138** (1970) 73.
31. Q. H. Gibson and C. Greenwood, *J. Biol. Chem.*, **240** (1965) 2694.
32. C. Greenwood and Q. H. Gibson, *J. Biol. Chem.*, **242** (1967) 1782.
33. C.-P. Lee and L. Ernster, *European J. Biochem.*, **3** (1968) 385.
34. H. A. Lardy and S. M. Ferguson, *Ann. Rev. Biochem.*, **38** (1969) 991.
35. K. van Dam and A. J. Meyer, *Ann. Rev. Biochem.*, **40** (1971) 115.
36. D. R. Sanadi, K. W. Lam and C. K. R. Kurup, *Proc. Nat. Acad. Sci. U.S.A.*, **61** (1968) 277.
37. D. R. Sanadi, K. W. Lam, B. P. Sani and J. C. Chen, *8th Int. Congress of Biochemistry, Switzerland, Abstracts*, p. 143.
38. H. A. Lardy, D. Johnson and W. C. McMurray, *Arch. Biochem. Biophys.*, **78** (1958) 587.
39. H. A. Lardy, J. L. Connelly and D. Johnson, *Biochemistry*, **3** (1964) 1961.
40. R. L. Cross and J. H. Wang, *Biochem. Biophys. Res. Commun.*, **38** (1970) 848.
41. R. L. Cross, B. A. Cross and J. H. Wang, *Biochem. Biophys. Res. Commun.*, **40** (1970) 1155.
42. W. G. Hanstein and Y. Hatefi (unpublished).
43. A. Tzagoloff, K. H. Byington and D. H. MacLennan, *J. Biol. Chem.*, **243** (1968) 2405.
44. D. H. MacLennan and A. Tzagoloff, *Biochemistry*, **7** (1968) 1603.
45. Y. Hatefi and W. G. Hanstein, *Proc. Nat. Acad. Sci. U.S.A.*, **62** (1969) 1129.
46. K. A. Davis and Y. Hatefi, *Biochemistry*, **8** (1969) 3355.
47. K. A. Davis and Y. Hatefi, *Biochemistry*, **10** (1971) 2509.
48. K. Asami, K. Juntti and L. Ernster, *Biochim. Biophys. Acta*, **205** (1970) 307.
49. R. J. Fisher, J. C. Chen, B. P. Sani, S. S. Kaplay and D. R. Sanadi, *Proc. Nat. Acad. Sci. U.S.A.*, **68** (1971) 2181.
50. K. A. Davis, Y. Hatefi, K. L. Pofl and W. L. Butler, *Biochem. Biophys. Res. Commun.*, in press.

An Enzymological Approach to Mitochondrial Energy Transduction

John H. Young*

Institute for Enzyme Research, The University of Wisconsin, Madison, Wisconsin 53706

Abstract

An approach to the problem of mitochondrial energy transduction is outlined. The approach is based on the fundamental assumption that there is an intimate relation between the mechanisms of enzyme catalysis and energy transduction. The implications of this assumption for the coupling of two chemical reactions and the coupling of a chemical reaction to an ion flux are discussed.

The basic physical problem of the mitochondrion is the mechanism of energy transduction. Examples of mitochondrial transductions are the coupling of an oxido-reduction to a bond rearrangement reaction (oxidative phosphorylation) and the coupling of either type of chemical reaction to an ion flux. The apparent requirement of a common intermediate for coupling these different physical and chemical processes is both restrictive and suggestive. Among the various models of mitochondrial function that have been proposed to date the conformational model[1-4] has for me the greatest intuitive appeal. Unfortunately, since none of these models can yet claim a solid experimental foundation, one is forced at this stage to rely on his intuition, however unreliable that might be.

One of the fundamental concepts in the conformational model is the coupling of a chemical reaction and a conformational transition. It is very possible, and perhaps even likely that enzyme catalysis may generally involve conformational changes on the part of the enzyme as an intrinsic feature of enzyme catalysis.[5-7] If this should prove to be the case, then the mere demonstration of conformational changes in the enzymes associated with oxidative phosphorylation would not provide conclusive evidence for a conformational model of oxidative phosphorylation. Consequently, a conformational model would be very difficult to directly verify experimentally. Nonetheless, the very

* Present address: School of Pharmacy, University of Wisconsin, Madison, Wisconsin.

likelihood of conformational changes being an intrinsic feature of enzyme catalysis suggests an "enzymological" approach to the problem of energy transduction. This merely means that one tries to construct a model of energy transduction drawing only upon the familiar properties of enzymes. This may seem like a rather trivial statement, but, as I will attempt to demonstrate, it has decidedly nontrivial consequences.

Formulation of Problem

The mitochondrion has the capacity to reversibly couple a number of different processes. It is instructive to briefly review from a purely phenomenological point of view what the requirements are that acceptable models of mitochondrial function must meet.

The principal coupled process is oxidative phosphorylation. This is essentially the coupling of two chemical reactions, substrate oxidation through the respiratory chain and ATP synthesis. The thermodynamic driving force for two coupled chemical reactions is the difference in driving forces of the two partial reactions.[8, 8a] Thus if the driving force for substrate oxidation is greater than for ATP hydrolysis, oxidative phosphorylation ensues. Conversely, if the driving force for ATP hydrolysis is greater than for substrate oxidation, reversed electron transfer ensues, at least through the first two coupling sites. If the driving forces for the two partial reactions are equal, then a tightly coupled system is in equilibrium despite the fact that neither of the two subsystems considered separately is in equilibrium. Under these conditions (State 4) both respiration and ATPase activity are severely inhibited in well coupled mitochondria.[9, 10]

Another coupled process in mitochondria is active transport. Both respiration and ATP hydrolysis can be coupled to a flux of ions across the inner mitochondrial membrane. I will consider some of the mechanistic details later, but here I merely wish to note that the ion gradients established across the inner membrane constitute a form of free energy. There are a number of reports in the literature on the coupling of potassium efflux and proton influx to ATP synthesis so that this coupled process too appears to be reversible.[11–14] This is as one would expect from the principle of microscopic reversibility.

Still another system is the transhydrogenation reaction. The reduction of $NADP^+$ by NADH can be coupled to either respiration or ATP hydrolysis.[15] Conversely, van de Stadt *et al.* have recently shown that the oxidation of NADPH by NAD^+ can drive ATP synthesis.[16] Grinius *et al.* have also shown that the oxidation of NADPH by NAD^+ can drive ion translocation.[17]

The overall picture which seems to emerge therefore is that of a number of different free energy forms (redox, bond energy, ion gradients) which can all be reversibly transformed—one into another. Each of these free energy forms might be considered to constitute a thermodynamic subsystem. In a truly tightly coupled system, which may be only an idealization, the total system may be at equilibrium despite the fact that each of the subsystems taken separately is far from equilibrium. This is the case when the thermodynamic driving forces of the subsystems are all nonzero but equal to one another. If the total system is perturbed from equilibrium by changing the driving force of one of the subsystems then the total system relaxes to a new equilibrium state. Thus, for example, if ADP is added to a system in State 4, a certain number of moles of substrate are oxidized and a stoichiometrically related number of moles of ATP are synthesized until a new equilibrium state is attained.

Still on the phenomenological level one might now ask about the manner in which these various free energy forms are coupled to one another. A *necessary precondition* for coupling these various free energy forms together is that each subsystem be constrained from independently approaching equilibrium. The simplest way to achieve this constraint is to reversibly couple each of the subsystems to a common intermediate free energy form. At equilibrium the intermediate itself would be in equilibrium with each of the subsystems. Thus the various subsystems would be both constrained from independently approaching equilibrium and coupled together by one and the same mechanism.

A common intermediate free energy form is supported by the fact that a given reaction (e.g. ATP hydrolysis) can be catalyzed by just one enzyme (e.g. oligomycin-sensitive ATPase) to drive reversed electron transfer, ion translocation, and transhydrogenation. This type of inhibitor result provides strong support for the concept that each of these coupled processes share in part a common pathway, as required if there is a common intermediate.[18]

Further support for a common intermediate is provided by the fact that all of these processes can be uncoupled from one another by the so-called classical uncouplers.[4, 18] This could be readily explained if the uncouplers act so as to decrease the stability of the intermediate. If the relaxation time for spontaneous decay of the intermediate becomes much less than the relaxation times for the various coupled processes, then the intermediate relaxes in an uncoupled manner before it can relax in a coupled manner. The constraint preventing each subsystem from independently approaching equilibrium is thereby removed.

If the rate of spontaneous relaxation is not zero, then there exists no true state of equilibrium apart from the state in which each of the subsystems is at equilibrium (a state of no interest for mitochondriology). The essential point to recognize, however, is that the constraint preventing each of the subsystems from independently approaching equilibrium need be only relative in the sense that tight coupling merely requires that *relative* to the rates of coupled relaxation the rate of spontaneous, uncoupled relaxation be slow, not zero.

The picture of the mitochondrion that emerges from this discussion is that of a variety of free energy forms all reversibly coupled to one another by being reversibly coupled to a common intermediate free energy form.[4, 18] We are now ready to ask about the physical and chemical nature of this intermediate.

Before proceeding I might note here that the manner of coupling the various subsystems to one another is actually more complex than that described here because I have ignored the concentration gradients across the inner membrane of the reactants and products of the various chemical reactions.[19] Rather than attempting to deal with the integrated mitochondrial system, I have merely tried to describe a kind of first approximation as a starting point for dealing with the problem of energy transduction.

Coupling Catalyst

In this section, I would like to outline in somewhat more detail what is meant by an enzymological approach to energy transduction. Basically it means that there must be an intimate relation between the mechanisms of enzyme catalysis and energy transduction. This may appear to be inconsistent with the conclusion of the previous section that a necessary precondition for coupling is the constraint preventing each subsystem from independently approaching equilibrium. By definition a catalyst allows a chemical reaction to reach equilibrium more rapidly. However, a moment's reflection reveals that an enzyme could be a "coupling catalyst" if two conditions are met: (1) the enzyme catalyzed reaction is *inhibited* and (2) the inhibition is relieved by the transfer of free energy from one subsystem to another.

A schematic representation of an inhibited ATPase is given in Fig. 1. The reaction takes place in four steps: (1) the attractive enzyme-substrate interactions stabilize both the enzyme and the reactant in higher energy states such that the perturbed reactant lies energetically closer to the appropriate transition state, (2) the enzyme-bound reactant is transformed to the enzyme-bound products, (3) the products dissociate from the enzyme leaving it in the higher energy conformation, and (4) the enzyme spontaneously relaxes from $\mathscr{E}$ to E so that the cycle can begin again. If conformation $\mathscr{E}$ is a long lived or metastable conformation, then the rate limiting step in the reaction is the slow spontaneous decay of conformation $\mathscr{E}$. Thus the dissociation of products from an enzyme in a metastable conformation provides a mechanism of inhibiting an enzyme catalyzed reaction.

Although the enzyme represented in Fig. 1 would have a low ATPase activity it would tend to have a high $^{32}P_i \rightleftharpoons ATP$ and ^{14}C-ADP $\rightleftharpoons$ ATP exchange activity if the standard free energy difference between conformations $\mathscr{E}$ and E is sufficiently large. Under these conditions the enzyme would act as a free energy reservoir and thereby couple the exergonic hydrolysis of unlabelled ATP to the endergonic synthesis of labelled ATP. If the labelled and unlabelled species are considered as separate subsystems, then the addition of $^{32}P_i$ or ^{14}C-ADP to a medium containing ATP, P_i or ADP, and the enzyme of Fig. 1 leads to a transfer of free energy from the unlabelled to the labelled subsystem until equilibrium is attained between the two.

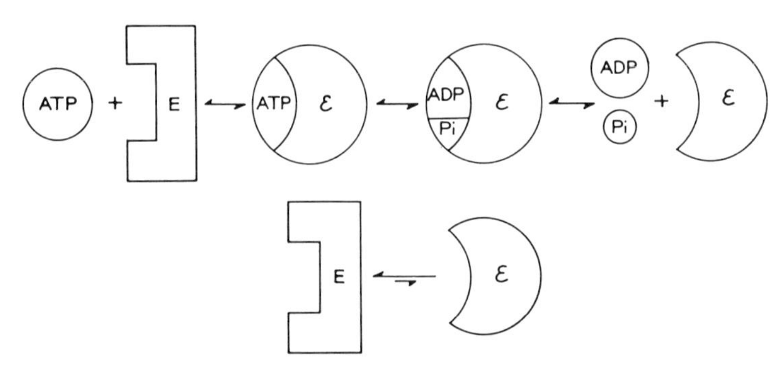

Figure 1. Schematic representation of inhibited ATPase. Perturbations of both enzyme and substrates are represented by geometric distortions. The spontaneous relaxation from conformation ϵ to conformation E is assumed to be slow, i.e., ϵ is a metastable conformation.

The enzyme of Fig. 1 therefore satisfies the two requirements of a coupling catalyst: (1) the enzyme catalyzed reaction is inhibited, and (2) the inhibition is relieved by the transfer of free energy from one subsystem to another. The two subsystems are constrained from independently approaching equilibrium by the very mechanism that couples them together. This is the essence of a coupling catalyst. (This concept will be developed more fully elsewhere.)[20]

This model for coupling together two *chemically equivalent* reactions (i.e. differing only by isotopic substitution) can be readily generalized to coupling two *chemically different* reactions if one postulates the existence of a coupling catalyst with *two* active sites, one active site to catalyze one partial reaction and another active site to catalyze the other partial reaction. Thus, for example, if electron transfer could also be reversibly coupled to the transition $E \rightleftharpoons \mathscr{E}$, respiration would be inhibited unless a mode of rapid relaxation is available for the metastable conformation $\mathscr{E}$. In the presence of sufficient levels of ADP and P_i, ATP synthesis would provide such a mode of relaxation, as illustrated in Fig. 1. Thus the release of respiratory control would be achieved by the same mechanism that couples oxidation to phosphorylation. In such a model tight coupling for oxidative phosphorylation would require a high activity for the ATP-linked isotopic exchange reactions discussed above.[21]

However, to postulate a coupling catalyst with two active sites provides only a formal, not a mechanistic solution to the problem of oxidative phosphorylation. Such an approach does not address itself to the fundamental question of the mechanism by which the respiratory and ATPase enzyme complexes are coupled together. To get at this question one might try to characterize the physical nature of the hypothesized conformational transitions and then ask what it is about these transitions that lends itself to the coupling of different enzyme complexes.

Characterization of Conformational Transition

One can partially characterize the hypothesized conformational transitions by drawing upon the reasonably well characterized passive transport properties of the inner membrane and then simply seeking a self-consistent model of active transport. We recently adopted this approach in developing a model that effectively incorporates basic features of the chemiosmotic model[22] into the conformational model.[23,24] To introduce this model I will try and place the reversible coupling of a chemical reaction and an ion flux within the context of the reversible coupling of two thermodynamic subsystems previously discussed.

As previously mentioned, a necessary precondition for coupling various subsystems together is that each subsystem be constrained from independently approaching equilibrium. If the subsystem is a concentration gradient across a membrane, then it must be a concentration gradient of a relatively *impermeant* species in order to be stable against spontaneous decay. The requirement for an impermeant species is analogous to the requirement for an *inhibited* enzyme catalyzed chemical reaction discussed in the previous section.

One can distinguish two categories of active *salt* transport in mitochondria, one facilitated by the neutral ionophores and one facilitated by the monocarboxylate ionophores. A common denominator of both types of active salt transport appears to be that *one and only one* of the ions of the salt crosses the membrane via an electrogenic mechanism while the other ion crosses via an electrically neutral mechanism either with a proton or in exchange for a proton. Given these passive transport properties, a necessary and sufficient condition for such a salt to be relatively impermeant is that the proton and hydroxyl conductances of the membrane be small. That is, the protons and hydroxyl ions must be constrained from being passively transported across the membrane via an *electrogenic* mechanism. Although both ions of the salt are readily exchangeable, the stability of the proton gradient in such a system insures the stability of the salt gradient.

Let us now consider the mechanistic problem of coupling a flux of an impermeant salt, as defined here, to a conformational cycle which is itself coupled to a chemical reaction as discussed in the previous section. If one accepts that the passive transport properties outlined here and discussed more fully elsewhere[24] are a valid first approximation to those of the inner mitochondrial membrane, then self-consistency requires that the conformational cycle involves an effective transmembrane transport of protons. This is the only way to obtain a *transmembrane* flux of an *impermeant* salt.

We have recently presented a model that satisfies this requirement by invoking: (1) pK changes of ionizable groups on opposite sides of the membrane (asymmetric membrane Bohr effect) and (2) a mechanism of facilitated proton conductance to allow for rapid equilibration between these two sets of ionizable groups. The conformational cycle thereby incorporates the proton generated membrane potential and pH gradient of the chemiosmotic model as schematically represented in Fig. 2. The mechanistic details of these hypothesized maneuvers are admittedly hazy at this point, but something along these lines is required simply for self-consistency.

Excitation in Fig. 2 corresponds to the conformational transition $E \rightarrow \mathscr{E}$. It is important to recognize, however, that although relaxation must correspond in a *net* sense to the conformational transition $\mathscr{E} \rightarrow E$, the enzyme is relaxing via a *different* pathway than that by which it was excited. In this model relaxation releases the inhibition of respiratory or ATPase activity, and as discussed previously this release must be the consequence of transferring free energy from the chemical subsystem to the concentration gradient subsystem. This transfer takes place in two steps in the present model. First, there is the electrogenic transport of ions in response to the proton generated membrane potential. Then, subsequent to the electrogenic transport of ions, there is the generation of the pH gradient and the subsequent transport of those species that cross the membrane either with a proton or in exchange for a proton. Only upon completion of the proton cycle is relaxation complete.

The fact that in salt transport the enzyme relaxes via a pathway different than that by which it was excited has important implications for characterizing the conformation $\mathscr{E}$ as an intermediate in any of the coupled chemical reactions of the mitochondrion. Electrogenic transport of ions, which initiates relaxation in active salt transport, is presumably *not* on the pathway of any of the coupled chemical reactions. Thus in the present model the pH gradient, which occurs only subsequent to the electrogenic transport of ions, would also *not* be on the pathway of any of the coupled chemical reactions. A membrane potential, however, would be on the pathway of all the coupled chemical reactions in this model. Thus a partial characterization of the hypothesized conformational transition has been achieved.

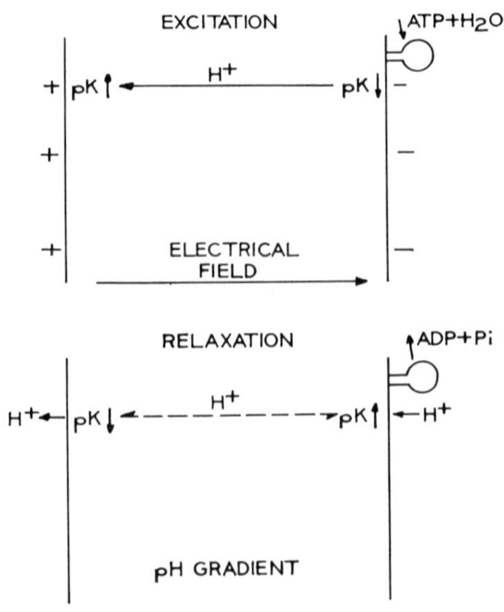

Figure 2. Conformationally induced proton translocation as the basis for the generation of the membrane potential and pH differential. The solid line connecting the two sets of ionizable groups in the excitation phase indicates high proton conductivity whereas the dotted line in the relaxation phase indicates low proton conductivity. The coupling of the conformational transition with a chemical reaction is represented by the adsorption of ATP and H_2O in the excitation phase and the desorption of ADP and inorganic phosphate (P_i) in the relaxation phase. (Reprinted by permission of Gordon and Breach, Science Publishers Inc.)

Although the present model incorporates both the membrane potential and the pH gradient of the chemiosmotic model in accounting for mitochondrial active transport, it incorporates only the membrane potential and not the pH gradient as an intermediate in the various coupled chemical reactions. This is a fundamental difference between the two models.

Conclusion

At the present stage of development the model outlined here provides only an approach, not a solution to the problem of mitochondrial function. This model is still vague and poorly defined concerning the central question of the mechanism of oxidative phosphorylation. Nonetheless, the approach of seeking in the properties of the enzymes the mechanisms of energy transduction as well as enzyme catalysis appears to be a promising one. To seek the mechanisms of energy transduction in the properties of the enzymes rather than in the properties of the reactions they catalyze is the fundamental distinction between this approach and the approach of both the chemical and the chemiosmotic models. The present approach is essentially just an extension of a view of enzyme catalysis in which the enzyme itself can play the role of a temporary free energy reservoir.[3,5,6]

The conclusion that a proton generated membrane potential is an intrinsic component of the common intermediate implies that in significant measure the form of the intermediate free energy is electrostatic. We might now attempt to formulate somewhat more precisely the fundamental question raised earlier: How could electrostatic free energy lend itself to coupling different enzyme systems? If this approach is on the right path, then this question is undoubtedly related to another fundamental question: Why does oxidative phosphorylation occur in a membrane system? As is well known, biological membranes have the character of parallel plate capacitors and are therefore well suited for the storage of electrostatic free energy. However, not only the storage but also the transfer of electrostatic free energy appears to be involved. As Liberman and Skulachev have already observed, if electrostatic free energy could be transmitted along the membrane, then the individual enzyme complexes of the inner mitochondrial membrane could be effectively united into a common system.[25] This observation does not in itself provide answers to the fundamental questions raised above, but it does identify what appears to be a promising area for seeking the answers to these questions.

Obviously, one cannot develop this model further without making some assumptions about the structure and arrangement of the various enzyme complexes within the inner membrane. However, an excursion into membrane structure and its relation to the present model is beyond the scope of the present paper. Hopefully, an understanding of the structure-function relation in this membrane system will soon be forthcoming.

Acknowledgements

This work was supported by a Public Health Service Research Career Development Award (K4-GM-21,171) from the Institute of General Medical Sciences.

References

1. D. E. Green, *Proc. Nat. Acad. Sci. U.S.*, **67** (1970) 544.
2. D. E. Green and J. H. Young, *Amer. Sci.*, **59** (1971) 92.
3. R. Lumry, in: *A Treatise on Electron and Coupled Energy Transfer in Biological Systems*, T. King and M. Klingenberg (eds.), Dekker, New York, 1969.
4. P. D. Boyer, in: *Biological Oxidations*, T. P. Singer (ed.), Interscience, New York, 1968, p. 193.
5. R. Lumry and R. Biltonen, in: *Structure and Stability of Biological Macromolecules*, S. N. Timasheff, G. D. Fasman, (eds.), Dekker, New York, 1969, p. 65.
6. W. Jencks, in: *Current Aspects of Biochemical Energetics: Fritz Lipman Dedicatory Volume*, N. Kaplan and E. Kennedy (eds.), Academic Press, New York, 1966, p. 272.
7. D. Koshland, Jr., *Proc. Nat. Acad. Sci. U.S.*, **44** (1958) 98; *Cold Spring Harbor Symp. Quat. Biol.*, **28** (1963) 473.
8. I. Prigogine, *Introduction to Thermodynamics of Irreversible Processes*, 3rd Ed., Interscience, New York, 1967, p. 25.

8a. H. Rottenberg, S. R. Caplan and A. Essig, in: *Membranes and Ion Transport*, Vol. 1, E. Bittar (ed.), Interscience, New York, 1970, p. 165.

9. M. Klingenberg and P. Schollmeyer, *Proc. 5th Intern. Congr. Biochem., Moscow 1961*, Vol. 5, Pergamon, London, 1963, p. 46.
10. S. Muraoka and E. C. Slater, *Biochim. Biophys. Acta*, **180** (1969) 221.
11. G. F. Azzone and S. Massari, *Eur. J. Biochem.*, **19** (1971) 97.
12. E. Rossi and G. F. Azzone, *Eur. J. Biochem.*, **12** (1970) 319.
13. R. S. Cockrell, E. J. Harris and R. Pressman, *Nature*, **215** (1967) 1487.
14. R. A. Reid, J. Moyle and P. Mitchell, *Nature*, **212** (1966) 257.
15. L. Danielson and L. Ernster, in: *Energy-Linked Functions of Mitochondria*, B. Chance (ed.), Academic Press, New York, 1963, p. 157.
16. R. J. van de Stadt, F. J. R. M. Nienwenhuis and K. van Dam, *Biochim. Biophys. Acta*, **234** (1971) 173.
17. L. L. Grinius, A. A. Jasaitis, Y. P. Kadziauskas, E. A. Liberman, V. P. Skulachev, V. P. Topali, L. M. Tsofina and M. A. Vladimirova, *Biochim. Biophys. Acta*, **216** (1970) 1.
18. L. Ernster and C. P. Lee, *Ann. Rev. Biochem.*, **33** (1964) 729.
19. M. Klingenberg, *Essays Biochem.*, **6** (1970) 119.
20. J. H. Young (manuscript in preparation).
21. P. D. Boyer, in: *Current Topics in Bioenergetics*, Vol. II, D. R. Sanadi (ed.), Academic Press, New York, 1967, p. 99.
22. P. Mitchell, *Biol. Rev. Cambridge Phil. Soc.*, **41** (1966) 445.
23. J. H. Young, G. A. Blondin and D. E. Green, in: *Physical Principles of Neuronal and Organismic Behavior*, Gordon and Breach, New York, in press.
24. J. H. Young, G. A. Blondin and D. E. Green, *Proc. Nat. Acad. Sci. USA*, **68** (1971) 1364.
25. E. A. Liberman and V. P. Skulachev, *Biochim. Biophys. Acta*, **216** (1970) 30.

ATP Synthesis in Oxidative Phosphorylation: A Direct-Union Stereochemical Reaction Mechanism

Ephraim F. Korman and Jerome McLick

Institute for Enzyme Research, University of Wisconsin, Madison, Wisconsin 53706

The Experimental Foundation for Formulating a Direct-Union Reaction Mechanism for ATP Synthesis

A fundamental understanding of oxidative phosphorylation will involve *chemical reaction mechanisms*. Since chemical reaction mechanisms describe in detail the making and/or breaking of chemical bonds, before a meaningful reaction mechanism can be formulated for a given reaction, the actual bonds involved must be known. This applies to all chemical reactions, including the chemical reactions in oxidative phosphorylation.

There is virtually universal agreement that in oxidative phosphorylation it is substrate inorganic phosphate (P_i) which forms a covalent chemical bond with some "first" polyatomic partner (of either high or low molecular weight) to give a phosphorylated entity. It is agreed that an atom which is a constituent of P_i engages an atom which is a constituent of some polyatomic partner to form a covalent bond. There is no question *whether* substrate P_i reacts to form a covalent bond with a "first" partner, but there is a fundamental question as to the *identity* of the "first" partner. In certain formulations of what is known as the "chemical hypothesis",[1] the "first" partner is *other than ADP*, and is arbitrarily designated "X". The covalent bond between "X" and substrate P_i could involve either the phosphorous atom or one of the oxygen atoms which are constituent to P_i. If such formulations were correct, ADP would clearly *not* be the "first" polyatomic chemical partner to react with P_i, but would be instead a later partner reacting with a covalent chemical "derivative" of P_i in a subsequent reaction to form the pertinent P–O bond in ATP. The various forms of the "chemical hypothesis" represent attempts, in terms of classical chemistry, to reckon, on the one hand, with the thermodynamically "downhill" energetics of respiratory oxidation-reduction, and on the other hand, with the thermodynamically "uphill" energetics of ATP synthesis. In the various formulations, the covalent chemical derivatives of P_i are intrinsically (or, upon oxidation, become) "high-energy" species with which ADP can react spontaneously to form ATP.

However, in the overall phosphorylation chemistry of oxidative phosphorylation, what has been actually experimentally observable over many years of investigation is given simply by equation (1). It is

$$\text{Ad—Ri—O—}\overset{\overset{\displaystyle O}{\|}}{\underset{\underset{\displaystyle O^-}{|}}{P}}\overset{\alpha}{—}\text{O—}\overset{\overset{\displaystyle O}{\|}}{\underset{\underset{\displaystyle O^-}{|}}{P}}\overset{\beta}{—}\mathbf{O}^- + {}^-\text{O—}\overset{\overset{\displaystyle O}{\|}}{\underset{\underset{\displaystyle O^-}{|}}{P}}\text{—}OH + H^+$$

$$= \text{Ad—Ri—O—}\overset{\overset{\displaystyle O}{\|}}{\underset{\underset{\displaystyle O^-}{|}}{P}}\overset{\alpha}{—}\text{O—}\overset{\overset{\displaystyle O}{\|}}{\underset{\underset{\displaystyle O^-}{|}}{P}}\overset{\beta}{—}\mathbf{O}\text{—}\overset{\overset{\displaystyle O}{\|}}{\underset{\underset{\displaystyle O^-}{|}}{P}}\overset{\gamma}{—}O^- + H_2O \qquad \text{Equation (1)}$$

specifically an oxygen atom which is a constituent of the β-phosphate group of ADP which ultimately engages in a covalent P–O bond with the phosphorous atom which is constituent to substrate P_i.[2] This P–O bond is the pertinent bond formed in ATP. Also, it is specifically an oxygen atom from P_i which ultimately ends up in H_2O.[2] These observations are the observations which are experimentally definitive at the level of bond making/breaking in the overall phosphorylation chemistry of oxidative phosphorylation.

The above experimentally definitive observations are insufficient of themselves to mechanistically implicate any polyatomic chemical partner reacting with P_i *other than ADP itself*. This does not mean that the data definitively rule out the possibility that some partner "X" covalently participates with P_i prior to ADP. What it does mean is that the data are simply insufficient to mechanistically implicate any such participation of "X". Indeed, for "X" to ever become mechanistically implicated, additional positive experimental data would have to demonstrate the existence of a covalent bond in a discrete "X"-derivative of P_i. The "X"-derivative of P_i would clearly have to be demonstrated as participating in oxidative phosphorylation proper.

In spite of many intensive investigations over a period of two decades, there is no experimentally definitive evidence for any chemical derivative of P_i other than the ADP-derivative of P_i, which is ATP itself, in the phosphorylation chemistry of oxidative phosphorylation.[3] The fact that the long-term search for an "X"-derivative of P_i has been unsuccessful clearly supports the very real possibility that there are in fact *no* "X"-derivatives of P_i participating in ATP synthesis and that ATP synthesis is occurring simply by a *direct-union* of ADP and P_i giving ATP plus H_2O. Indeed such a formulation is the *minimum speculative* formulation in terms of the actual experimental data at the level of bond making/breaking. In its strictly literal sense, equation (1) states that an oxygen atom which is a constituent of the β-phosphate group of ADP directly forms a covalent bond with the phosphorous atom of substrate P_i (where P_i is enzyme-bound as inorganic phosphate), while one of the oxygen atoms which is a constituent of substrate P_i breaks its covalent bond with phosphorous to directly give rise to an H_2O molecule. This strictly literal description, which is in terms of P–O bond-making and bond-breaking specifics, permits formulation of a chemical reaction mechanism for ATP synthesis in terms which are mechanistically classical. To us, a nucleophilic substitution reaction at the substrate inorganic phosphate phosphorous center is strongly indicated for the direct-union reaction. This would mean that the ATP synthesis reaction is amenable to mechanistic analysis using well-established fundamental principles applicable to nucleophilic substitution reactions at phosphate phosphorous centers in general.

Mechanistic Problems Engendered by the Isotopic Exchange Data

Historically, workers in oxidative phosphorylation have had the difficult problem of formulating an enzyme-catalyzed reaction mechanism to achieve ATP synthesis, while at the same time explaining the well-known isotopic exchanges which accompany ATP synthesis.[4,5] The isotopic exchange are:

1. ATP–$^{32}P_i$ } Rates approximate the net ATP synthesis rate
2. ATP–ADP(^{14}C) }
3. ATP–$H_2{}^{18}O$ } Rates much faster than net ATP synthesis rate
4. P_i–$H_2{}^{18}O$ }

All of these isotopic exchanges are inhibited by oligomycin and by uncouplers of oxidative phosphorylation and are therefore recognized as being intimately related to the mechanism of ATP synthesis in oxidative phosphorylation.[4,5] The relatively rapid ATP–$H_2{}^{18}O$ oxygen exchange has a special prominence, because it, unlike the other isotopic exchanges, is unique to oxidative phosphorylation systems, occurring rapidly in no other type of system.

The ATP–$^{32}P_i$ and ATP–ADP(^{14}C) exchanges and their relative rates can be quite simply explained by overall dynamic reversibility of ATP synthesis. As seen in equation (1) viewed in its overall sense, ATP cleaves (either by a direct hydrolysis or by some indirect cleavage) to ultimately give ADP and P_i. These unlabelled species can be replaced on the enzyme by ADP(^{14}C) and/or $^{32}P_i$. When ATP re-forms, ^{14}C and/or ^{32}P will be incorporated into ATP. This simple explanation applies irrespective of whether a direct-union reaction mechanism is involved in ATP synthesis or not. It depends solely on overall dynamic reversibility of ATP synthesis, whatever the mechanism. For this reason, the ^{14}C and ^{32}P exchanges have never posed a problem in the phosphorylation chemistry of oxidative phosphorylation. For the very same reason, however, the ^{14}C and ^{32}P exchanges have not provided a fruitful insight into the mechanism proper of ATP synthesis in oxidative phosphorylation.

By contrast to the ^{14}C and ^{32}P exchanges, the rapid ^{18}O exchanges cannot in any simple way be explained solely by dynamic reversibility of overall ATP synthesis. The ATP synthesis mechanism proper becomes critical here, including the obvious question of whether or not a direct-union reaction mechanism is involved. When viewed in its strictly literal sense, equation (1) would indicate that when $H_2{}^{18}O$ reacts to directly hydrolyze ATP to give ADP plus P_i, ^{18}O would be directly incorporated into P_i. If the ^{18}O-labelled P_i were to remain stereospecifically bound to the enzyme, resynthesis of ATP (the microscopic reverse of hydrolysis of ATP) would presumably expel ^{18}O from the ^{18}O-labelled P_i, and thus no ^{18}O would be incorporated into ATP. On the other hand, if the ^{18}O-labelled P_i fully dissociates from the enzyme and rebinds in a new steric orientation, resynthesis would incorporate ^{18}O into ATP. However, this latter sequence is a type of ATP-P_i exchange, and could thus occur no faster than the observed ATP-$^{32}P_i$ exchange.[6] The ^{18}O exchange from $H_2{}^{18}O$ into ATP actually can occur at a rate more than ten times as fast as the rate of the ATP-$^{32}P_i$ exchange.[6, 7]

Perhaps in part because of the seeming inability of a direct-union reaction mechanism for ATP synthesis to account for the rapid ATP-$H_2{}^{18}O$ oxygen exchange, some workers have continued to entertain a belief that ATP synthesis may not be occurring by a direct-union reaction mechanism but instead by a complex non-direct-union mechanism involving one, or perhaps a series of, covalently bonded phosphorylated intermediates. The rapid ATP-$H_2{}^{18}O$ oxygen exchange, although still unexplained in any simple way, may be, according to such workers, an expression ultimately of the complex mechanism of ATP synthesis whose mechanistic details remain obscure.

However, as we have strived to make clear, the experimentally definitive data are insufficient to implicate any phosphorylated entity other than ATP in oxidative phosphorylation. From this consideration we conclude that the resolution of the seeming "dilemma" of the rapid ATP-$H_2{}^{18}O$ oxygen exchange requires not a rejection of a direct-union reaction mechanism and the espousal of a more complex mechanism leaving the required explanation still obscure, but instead it requires adherence to the only mechanism for ATP synthesis in oxidative phosphorylation which is sufficiently experimentally founded. To us, a satisfactory mechanistic explanation of the rapid ATP-$H_2{}^{18}O$ oxygen exchange will consequently be in terms of a direct-union reaction mechanism for ATP synthesis in oxidative phosphorylation. Rather than finding the ATP-$H_2{}^{18}O$ oxygen exchange a "dilemma," we have found it to be the most revealing feature of our direct-union reaction mechanism. The significance of the ATP-$H_2{}^{18}O$ oxygen exchange emerges only within the framework of fundamental phosphorous reaction stereochemistry and reaction mechanisms. We deal with these matters of fundamental phosphorous chemistry in the next section as a prelude to the presentation of our proposed direct-union reaction mechanism for ATP synthesis in oxidative phosphorylation.

Fundamental Phosphorous Stereochemistry Relating to Reaction Mechanisms

A direct-union reaction mechanism in which ADP and P_i give ATP plus H_2O implicates to us a nucleophilic substitution reaction at the inorganic phosphate phosphorous center. It is generally accepted that nucleophilic substitution reactions at phosphate phosphorous centers can proceed by way of unstable pentacovalent reaction intermediates having trigonal bipyramidal geometry.[8,9] A classical trigonal bipyramid structure is depicted in Fig. 1 and described in the accompanying legend, including features such as apical and equatorial bonds. We formulate the ATP synthesis reaction mechanism in terms of such classical trigonal bipyramidal unstable reaction intermediates. In nucleophilic substitution reactions at phosphate phosphorous centers, a nucleophile classically attacks the tetrahedral phosphorus atom to give a trigonal bipyramidal structure in which the attacking nucleophile is apical. Exit of the leaving group is also classically from an apical position. In other words, groups both enter into and exit from trigonal bipyramidal unstable reaction intermediates classically by way of the long, weak, relatively reactive *apical* bonds.

Unstable trigonal bipyramidal phosphorous reaction intermediates can have structural non-rigidity, i.e., capacity for pseudorotation.[9,10] Pseudorotation of a trigonal bipyramid is depicted in Figure 1 and described in the legend.

Pseudorotation has been found to be governed by certain electronic rules[11]:

1. Electron-withdrawing groups prefer apical bonding.
2. Electron-releasing groups prefer equatorial bonding.

Some examples of these rules for oxygen groups bonded to pentacovalent phosphorous are:

–OH_2^+ (oxonium) groups prefer apical bonding.
–O^- groups prefer equatorial bonding.
–OH groups have relatively no bond orientation preference.

These fundamental matters concerning phosphorous stereochemistry and reaction mechanisms constitute a critical part of our proposed direct-union reaction mechanism for ATP synthesis in oxidative phosphorylation.

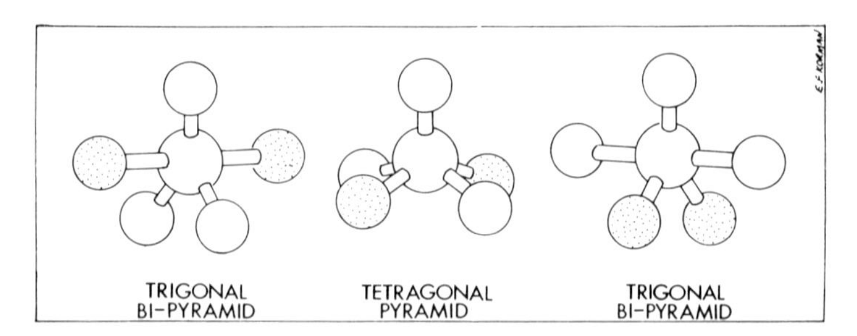

Figure 1. Classical trigonal bipyramidal geometry (see the structure on the extreme left).

The central atom has five atoms (or groups) bonded to it. Two of the atoms (shaded) are each bonded by a long, relatively weak *apical* bond and together are linear with the central atom. Three of the atoms (unshaded) are each bonded by a short, relatively strong *equatorial* bond and they form, together with the central atom, the equatorial plane. The apical bonds are perpendicular to the equatorial plane.

Pseudorotation of a trigonal bipyramid structure. In the structure at the extreme left, the pair of shaded atoms each start out apically bonded. By a process of bond-bending the shaded atoms become, in the structure at the extreme right, equatorially bonded. At the same time, in the original structure at the extreme left, the lower pair of equatorially bonded atoms (unshaded) become by the bond-bending process each apically bonded in the structure at the extreme right. The structure in the center is a tetragonal (square-base) pyramid and is the transition state halfway between the two trigonal bipyramidal structures. In structures having pairs of atoms (or groups) indistinguishable from each other (as pairs), the trigonal bipyramidal structure seemingly undergoes a 90° rotation, i.e., pseudorotates.

Description of the Enzyme-Catalyzed ATP Synthesis Reaction Mechanism

We are now prepared to formulate our proposed direct-union chemical reaction mechanism for the ATP synthesis reaction in oxidative phosphorylation.[12] In Fig. 2, which will describe the dynamic stereochemical reaction mechanism pictorially, we will designate substrate inorganic phosphate as HPO_4^{-2} since that protonation state is predominant in free solution at physiological pH. We will designate ADP as $ADPO^-$ to focus attention upon the pertinent oxygen atom of ADP directly involved in the reaction mechanism. The substrate oxygen atoms will be numbered for reference purposes.

In our ball-and-stick diagram, the formal double bond of P_i will be represented by one stick (σ-bond) and a pair of "sausage" π-clouds. In the enzyme active site we will invoke Mg^{2+} and proton-donor groups. We will not depict the chemical structures of the proton-donor groups (which we suggest could be imidazole groups[12]) but will depict only the pertinent protons. All substrates are considered to be tightly and stereospecifically bound to the enzyme during the entire catalysis.[13] The legend to Fig. 2 outlines some of the details of the enzyme-catalyzed chemical reaction mechanism. The proposed reaction mechanism encompasses features such as the direct-union of ADP and P_i to give ATP plus H_2O, fundamental phosphorous stereochemistry, tight stereospecific substrate binding during catalysis etc., which have been introduced in this and in previous sections as a prelude to its formulation.

Essence of the Reaction Mechanism

Pseudorotation, which commences with ADPO being apically bonded in reaction intermediate (I), results in ADPO becoming equatorially bonded in reaction intermediate (II). Since equatorial bonds are shorter and stronger than apical bonds, the pseudorotation constitutes *equatorial capture of ADPO* which is the essence of the proposed reaction mechanism.

Role of the Pair of Oxonium groups in the Pseudorotation

$ADPO^-$ is an excellent group to apically *leave* from the pentacovalent phosphorous center. Thus, ADPO has an inherently high preference for apical bond orientation at such a center. For ADPO to be equatorially captured, a pair of groups originally equatorially bonded must have a preference for apical bonding at least nearly equal to that of ADPO. In such an originally equatorially bonded pair, a single oxonium group (paired with some non-oxonium oxygen group) probably does not itself have sufficient preference for apical bonding for pseudorotation to be permitted. However, *two* oxonium groups, *acting together as a pair*, undoubtedly would have a combined preference for apical bonding which is sufficient for pseudorotation to be permitted, and thus for equatorial capture to be effected.

The Rate-Limiting Step

Since $ADPO^-$ is an excellent leaving group from the pentacovalent phosphorous center, it is conversely a poor entering group (i.e., it is a relatively weak nucleophile) at such a center. From that consideration, attainment of reaction intermediate (I) is probably the rate-limiting step in the chemical reaction mechanism.

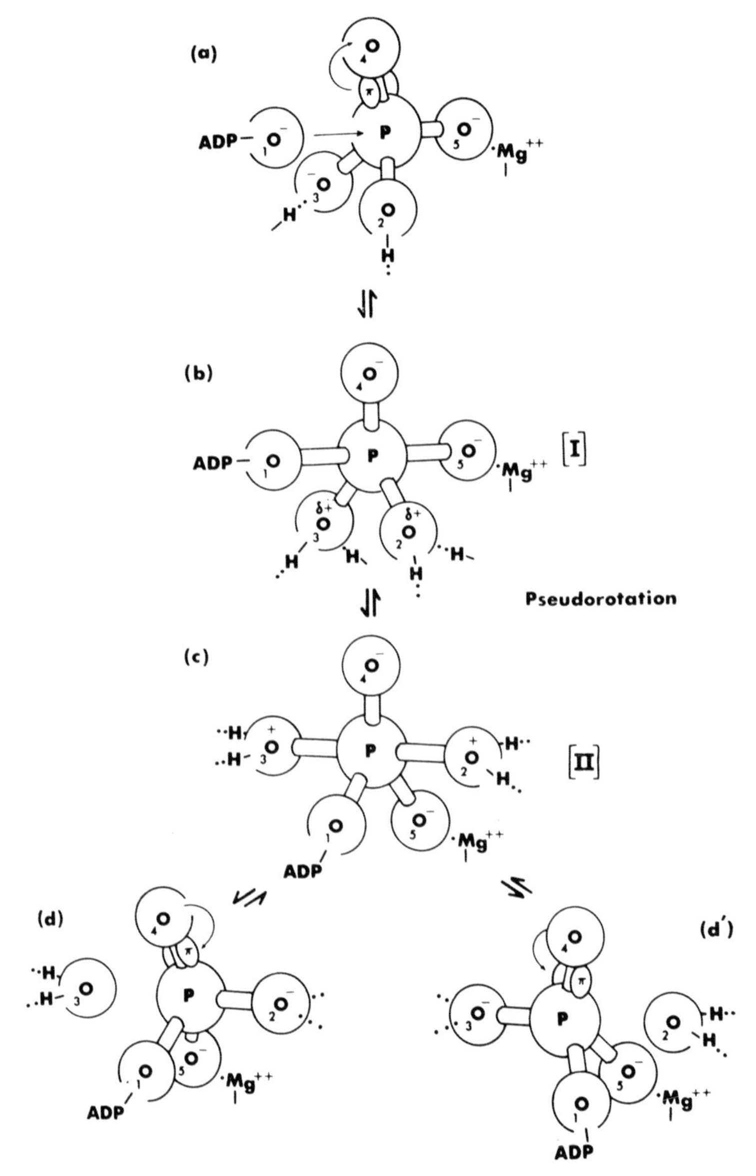

Figure 2. A proposed stereochemical reaction mechanism for ATP synthesis in oxidative phosphorylation.

In stage (a), tetrahedral HPO_4^{-2} binds tightly to the enzyme active site by a 3-point attachment: (1) the H of ($_2$O–H) H-bonds to an atom in the active site; (2) $_3O^-$ binds to a proton-donor group via a H-bond; and (3) $_5O^-$ coordinates Mg^{2+}. $_4$O is depicted as doubly bonded to the P center of P_i and no binding function is specified for it.

The tight binding (unspecified) of $ADPO^-$ together with P_i geometrically positions $_1O^-$ to interact with the P center of P_i. Any prohibitive charge-charge repulsion is removed by substrate binding to the enzyme. The interaction of $_1O^-$ and P constitutes a classically apical nucleophilic "attack" by $ADPO^-$ upon the P center of P_i. Tetrahedral P_i transforms to unstable trigonal bipyramidal reaction intermediate (I), depicted in stage (b). $_2$O, $_3$O,

Reaction Mechanism Stereochemistry

In the proposed reaction mechanism, the entering ADPO group has a 90° geometrical relationship with the leaving H_2O group at the phosphorous reaction center, which constitutes a *retention* stereochemistry path at that center.[14] The retention stereochemistry path is absolutely fundamental to the equatorial capture-apical dehydration reaction mechanism for ATP synthesis.

Mechanistic Explanation of the Isotopic Exchanges

We can now explain the isotopic exchanges which accompany the ATP synthesis reaction in oxidative phosphorylation in terms which are rigorously *mechanistic*. We explain the ATP-$^{32}P_i$ and ATP-ADP(^{14}C) exchanges in terms of the dynamic reversibility of the overall ATP synthesis mechanism, as described earlier above. By contrast, we explain the rapid ATP-$H_2{}^{18}O$ oxygen exchange as occurring simply and directly via the "double-oxonium" reaction intermediate (II) which is on the mechanistic path proper to ATP and which is at the heart of the equatorial capture of ADPO in that mechanistic path. If $H_2{}^{18}O$ apically attacks ATP (see Fig. 2, d or d′) leading to the "double-oxonium" reaction intermediate (II), in which one of the two coapical oxonium groups is now ^{18}O-labelled, followed by a Walden *inversion* type of expulsion of the unlabelled apical oxonium group, ^{18}O is directly incorporated into ATP. In this direct invertive exchange, ATP is never cleaved, i.e., the pertinent P–O bond formed in ATP synthesis is never broken.[15] The ATP-$H_2{}^{18}O$ oxygen exchange does not involve overall reversal of the ATP synthesis reaction mechanism and thus does not traverse the rate-limiting step. Therefore, the ATP-$H_2{}^{18}O$ oxygen exchange would clearly be expected to occur at a rate significantly faster than the rates of the ATP-$^{32}P_i$ and ATP-ADP(^{14}C) exchanges.

---

and $_4O$ assume equatorial bonding in going to (I). $_2O$ carries the H it brought into the active site, while $_3O$ acquires the H it was initially H-bonded to. At the same time, $_4O$ develops a negative charge.

In the active site, the trigonal bipyramidal geometry of (I) positions the equatorial $_2OH$ and $_3OH$ groups each into proximity to a proton-donor group. The proton-donor groups are geometrically disposed and sufficiently acidic to strongly H-bond to the $_2OH$ and $_3OH$ groups imparting partial oxonium character ($-OH_2^{\delta+}$) to those latter groups. (I) is now structured to permit pseudorotation to give reaction intermediate (II) depicted in stage (c). As (II) develops from (I) and the pair of partial oxonium groups become apically bonded, they develop essentially full oxonium character ($-OH_2^+$) in (II). (II) is called the "double-oxonium" reaction intermediate.

(II) can favorably "collapse" by classically apically losing either one (see stage d) or the other (see stage d′) of the oxonium groups as H_2O, while "returning" catalytic protons from the non-leaving oxonium group back to the enzyme active site. The loss of the H_2O transforms (II) to a tetrahedral phosphate group. The phosphate group thus formed has a π-bond between $_4O$ and the P center, and a full σ-bond between the P center and the $_1O$ oxygen atom. The pertinent P–O bond is formed and a molecule of ATP is achieved.

The inversion stereochemical reaction mechanism which we have deduced for the rapid ATP-$H_2{}^{18}O$ oxygen exchange is mechanistically classical. In principle, such an invertive exchange reaction mechanism would lead to racemization of a population of asymmetric reaction centers. Of course, the γ-phosphate phosphorous reaction centers in the ATP molecules are not normally asymmetric, and thus racemization cannot readily be measured. Nevertheless, in terms of its ability to catalyze the invertive ATP-$H_2{}^{18}O$ oxygen exchange reaction, the ATP synthesis enzyme has a "racemase" character. At the heart of the "racemizing" activity is the "double-oxonium" reaction intermediate (II), which is indispensable to effect equatorial capture of ADPO.

We explain the rapid P_i-$H_2{}^{18}O$ oxygen exchange with an inversion reaction mechanism analogous to that for the ATP-$H_2{}^{18}O$ oxygen exchange. In the P_i-$H_2{}^{18}O$ case, P_i acts as an analog of ATP, to give, upon attack of $H_2{}^{18}O$, a reaction intermediate having "double-oxonium" geometry analogous to reaction intermediate (II).

Implications Concerning Overall Oxidative Phosphorylation

The prominent ATP-$H_2{}^{18}O$ oxygen exchange, which heretofore seemed to be a "dilemma," emerges upon our analysis as exceptionally "meaningful" in that it arises directly from a reaction intermediate which is at the heart of equatorial capture of ADPO and thus indispensably on the mechanistic path proper to ATP. Since the ATP-$H_2{}^{18}O$ oxygen exchange can now be accounted for in *meaningful mechanistic terms* within the framework of a direct-union nucleophilic substitution reaction mechanism for ATP synthesis, an impediment to acceptance of such a direct-union reaction mechanism has been removed. The granting of such a direct-union nucleophilic substitution reaction mechanism for ATP synthesis has some very profound implications concerning the overall mechanism and energetics of oxidative phosphorylation. It immediately means that the ATP synthesis reaction is mechanistically "self-contained" and "separate" from the oxidation-reduction chemistry. Although "separate" in the sense that each has its own reaction mechanism identity (i.e., nucleophilic substitution as distinct from oxidation-reduction), the phosphorylation chemistry and the oxidation-reduction chemistry are nevertheless "coupled." The "coupling" must perforce be by way of the protein in its role as catalyst to both chemistries.

In the ATP synthesis reaction mechanism we alluded to the enzyme in its catalytic role. Proton-donor groups in the enzyme effect protonation of oxygen atoms of substrate P_i as a principal feature of the enzyme catalysis. The "double-oxonium" reaction intermediate

(II), which is at the mechanistic heart of both equatorial capture of ADPO and also of the ATP-$H_2{}^{18}O$ exchange reaction mechanism, involves four protons. The protonic events during catalysis are dependent upon proper proton geometric positioning and acidity. Our direct-union formulation implies that the requisite proton geometric positioning and acidity arise in the ATP synthesis active site upon respiratory electron transfer towards O_2 in active sites elsewhere in the enzyme architecture. Although we have not here included a mechanistic picture for the oxidation-reduction chemistry, we expect that chemistry to also involve protons. Protonic events very likely are salient mechanistic features of the oxidation-reduction chemistry.[16] Just as the proper geometric positioning and acidity of protons needed to catalyze the ATP synthesis reaction in the ATP synthesis site arise upon electron transfer towards O_2 in the oxidation-reduction sites, so, too, the proper geometric positioning and acidity of protons needed to catalyze reversed electron transfer in the oxidation-reduction sites arise upon reverse of ATP synthesis, i.e., upon ATP hydrolysis, in the ATP reaction site. The role of the enzyme is to provide a molecular architecture containing both the ATP synthesis site and the oxidation-reduction sites. As molecular architecture, the enzyme, in its state of tight binding to the substrates, will perforce undergo geometric changes in ensemble with the substrates as the latter obviously undergo geometric changes in transforming from reactants to products during catalysis, as is depicted, for example, in our stereochemical reaction mechanism. The geometric changes in the enzyme molecular architecture (conformational changes[17]), concomitant with the catalysis in one active site of a chemical reaction which is thermodynamically overall "downhill", geometrically positions protons of proton-donor groups (among other parameters) in the other active site to catalyze a chemical reaction which is thermodynamically overall "uphill". Thus, the enzyme, acting as a conformational system, weds the two chemistries, each of which nevertheless retains its separate mechanistic identity. In conformationally marrying the two chemistries, the enzyme "couples" their overall thermodynamic parameters. In terms of our view, a fundamental understanding of oxidative phosphorylation will involve *dynamic structural chemistry* applying not only to the substrates in the two chemistries but also to the enzyme to which the substrates are tightly bound.

Acknowledgements

This work was supported in part by Program Project Grant No. GM-12,847 and Training Grant GM-88 from the National Institute of General Medical Sciences (USPHS).

References

1. For a recent review of oxidative phosphorylation see: H. A. Lardy and S. M. Ferguson, *Ann. Rev. Biochem.*, **38** (1969) 991.
2. R. D. Hill and P. D. Boyer, *J. Biol. Chem.*, **242** (1967) 4320.
3. D. H. Jones and P. D. Boyer, *J. Biol. Chem.*, **244** (1969) 5767.
4. P. D. Boyer, in: *Biological Oxidations*, T. P. Singer (ed.), Interscience, New York, 1968, p. 193.
5. P. D. Boyer, in: *Current Topics in Bioenergetics*, D. R. Sanadi (ed.), Vol. 2, Academic Press, New York, 1967, p. 99.
6. P. C. Chan, A. L. Lehninger and T. Enns, *J. Biol. Chem.*, **235** (1960) 1790.
7. R. A. Mitchell, R. D. Hill and P. D. Boyer, *J. Biol. Chem.*, **242** (1967) 1793.
8. R. Kluger, F. Covitz, E. Dennis, L. D. Williams and F. H. Westheimer, *J. Amer. Chem. Soc.*, **91** (1969) 6066.
9. F. H. Westheimer, *Accounts Chem. Res.*, **1** (1968) 70.
10. E. L. Muetterties, *Accounts Chem. Res.*, **3** (1970) 266.
11. E. L. Muetterties, W. Mahler and R. Schmutzler, *Inorg. Chem.*, **2** (1963) 613.
12. For the original presentation of this reaction mechanism, see: E. F. Korman and J. McLick, *Proc. Nat. Acad. Sci.* (*U.S.*), **67** (1970) 1130.
13. For a treatment of the very important concept of *tight* substrate binding during enzyme catalysis, especially of reaction intermediates and transition states, see: R. Wolfenden, *Accounts Chem. Res.*, **5** (1972) 10.
14. For a useful pictorial introduction to general reaction stereochemistry at tetrahedral substrates see: L. H. Sommer, *Stereochemistry, Mechanism and Silicon*, Chap. 11, McGraw-Hill, New York, 1965.
15. We predict that an analog of ATP in which the terminal P–O–P oxygen bridge atom is replaced by a methylene bridge (P–CH_2–P) inert to enzymatic hydrolysis could, nevertheless, incorporate ^{18}O from $H_2{}^{18}O$ into the terminal phosphorus center.
16. For an introduction to the important concept of *proton* involvement in biological oxidation-reduction mechanisms see: G. A. Hamilton in *Progress in Bioorganic Chemistry*, E. T. Kaiser and F. J. Kezdy, (eds.), Vol. 1, Interscience, New York, 1970, p. 83.
17. R. A. Harris, J. T. Pennington, J. Asai, and D. E. Green, *Proc. Nat. Acad. Sci.* (*U.S.*), **59** (1968) 830.

The Electromechanochemical Model of Mitochondrial Structure and Function

David E. Green and Sungchul Ji

Institute for Enzyme Research, University of Wisconsin, Madison, Wisconsin 53706

The gross configurational changes which mitochondria undergo during energization and deenergization were the experimental foundation stones for the conformational model which we first proposed in 1967.[1–3] Out of these studies emerged the notion of the energized state of the transducing unit and the notion of coupling the relaxation of the energized unit either to the synthesis of ATP or to active transport.[4] Implicit in the conformational model is the conservation of energy in a metastable state[4,5] of the transducing unit rather than in a high-energy covalent intermediate.

Three major tasks faced us in the further development and extension of the conformational model: (1) to define the nature of the metastable energized state; (2) to define the mitochondrial structures implicated in the generation and relaxation of the energized state; and (3) to deduce how these structures could serve as the molecular instruments for implementing the physical principles underlying energy transduction. The model of mitochondrial structure and function to be described in the present communication embodies the essence of our current views on these questions

Before developing the model systematically, a few general comments about our efforts at model building might be appropriate. The model is rooted in the experimentally established structural attributes of the basic mitochondrial systems, namely the transducing units, the membranes, and the protein network systems in the two mitochondrial spaces. The present model is a product of a long process of fitting experiment with theory. The structural features of the mitochondrion have been our invariant guide to the selection of the different physical possibilities. But the model in its present form goes beyond the experiments on which it was based. Experiment provided the basis for the initial, inductive, developmental phase and theory provided the basis for the final, deductive, emergent phase.

A working model of mitochondrial structure and function fulfills three major roles: (1) it organizes and unifies the known phenomena of mitochondrial coupling on the basis of a few principles; (2) it can lead to the prediction of new phenomena; and (3) it sets a new standard for progress. A more acceptable model must at least rationalize the same wide range of phenomena rationalized by our model and encompass the same structural and functional parameters, only more accurately. Our view is that no general, all-encompassing model of mitochondrial structure and function can be constructed which does not mirror accurately the most fundamental attributes of energy transduction. The very generality of the model provides the most powerful support for its validity in principle.

Finally we would like to submit that biological problems of great magnitude such as the structure of DNA,[6] the structure of the membrane[7, 8] or the mechanism of muscular contraction[9] can only be approached via model building. The one-experiment-one-conclusion approach is inapplicable to problems of such dimension. But while experiment alone is clearly insufficient for meaningful progress, experiment in conjunction with model building can provide a combination equal to the requirements for solving problems of great complexity.

I. *The Basic Tenets of the Model*

A. *Electromechanochemical coupling*

The central concept in the development of the conformational model has been the "energized state"[4, 5] of the inner membrane repeating unit in which the component proteins assume metastable conformations. The discharge of these energized, metastable conformations was assumed to be coupled to work performances of the mitochondrion. In analyzing the nature of the energized state of the repeating unit, we came to the conclusion that the repeating unit in the energized state can be characterized by three fundamental properties—chemical, mechanical, and electrical. Chemical because the repeating units themselves are sites of bond formation or bond breaking (catalysis); mechanical because energization induces conformational rearrangements of the component proteins and leads to mechanical strain; and finally electrical not only because fixed charges are integral parts of protein structures in general but also because of the postulated existence of uncompensated charges in the interior of the energy transducing unit during energization. These conceptually distinct parameters are so intimately intertwined within the membranous macromolecular framework that it is no longer possible to separate the chemical, mechanical and electrical components involved in conformational changes accompanying energy transductions. Just as the impossibility of separating the electrical and chemical parts in the energetics of ion transports in solution led Guggenheim[10] to invoke the concept of an "electrochemical" potential, so our theoretical considerations require invoking an "electromechanochemical" potential. In addition to chemical activities, Coulombic charges and electric field intensities which appear in the mathematical expression for the electrochemical potential,[11] the quantitative expression for the electromechanochemical potential would involve information about the microscopic structure of the macromolecular energy transducing unit in the energized state.

In terms of this new concept, the ground and energized states of the repeating unit differ not only in respect to the content of electromechanochemical potential energy* but also in respect to the relative contributions of the three component potential energy forms to the total potential energy of the repeating units. The electrical and mechanical contributions are predominant in the energized state whereas the chemical contribution is predominant in the ground state. The metastable character of the energized state may be ascribed to the fact that the repeating unit becomes kinetically unstable when the system is strained electrically and mechanically.

Implicit in the argument is the assumption that the three different forms of potential energy are readily interconvertible within the repeating unit. Man-made energy transducing devices are known which illustrate the interconvertability of these three forms of energy. Fuel cells and batteries, for instance, are structurally designed to achieve the conversion of chemical potential energy into electrical potential energy. Electrolysis cells facilitate the opposite energy conversion, namely the conversion of electrical potential energy into chemical potential energy. These devices may be considered as examples of two-component transducers because they only involve two different types of potential energies (in this respect, we are not regarding thermal energy as potential energy). As an example of a three-component energy transducer, we may consider a mechanical spring connected on both ends to two spherical masses with opposite electrical charges. At the equilibrium state, the net forces acting on the two spheres will be zero because the electrostatic force will be equal and opposite to the mechanical force of the spring. In this resting state, the total potential energy of the system can be expressed as the sum of the mechanical and electrical potential energies.

*At the time when the manuscript was submitted, we were not certain whether the "potential energy" of the system in the sense we were using it could be considered equivalent to "free energy." Largely as the result of discussions with Dr. J. H. Young, we have concluded that the two terms are equivalent in the context of the present communication. Wherever the phrase "potential energy" is found in the text, the reader may assume that the concept of "free energy" is applicable. We are defining chemical potential energy as the free energy associated with changes in the electronic state of valence electrons.

However, if the system is perturbed by stretching the spring and then releasing it, the system will oscillate around the equilibrium position. The total potential energy of the system is now expressible in terms of three component energy forms—the electrical, mechanical, and kinetic energies (the electromechanical system). It can be readily shown that during any period of oscillation of the electromechanical system, the three different forms of energy undergo interconversion in accordance with the law of conservation of energy. These simple examples reveal the important observation that interconversions of different forms of potential energy are the inevitable consequences of the structural attributes of energy transducing devices.

Although there is no established precedent for a three-component energy transducer capable of interconverting the electrical, mechanical and chemical potential energies, we are postulating that the mitochondrion is structurally capable of performing this function. If we express the total potential energy of the repeating unit as ψ and the electrical, mechanical and chemical components of the potential energy as E, C and M respectively, we can write

$$\psi_g = E_g + M_g + C_g \tag{1}$$

and

$$\psi_e = E_e + M_e + C_e \tag{2}$$

where subscript g refers to the ground state and subscript e to the energized state. The potential energy difference between the two states is then

$$\Delta\psi = \psi_e - \psi_g = (E_e - E_g) + (M_e - M_g) + (C_e - C_g) \tag{3}$$

If the transition of the repeating unit from the ground state to the energized state is isoenergetic, equation (3) becomes

$$(E_e - E_g) + (M_e - M_g) + (C_e - C_g) = 0 \tag{4}$$

which shows that the electrical, mechanical and chemical potential energies are interconvertible. When the repeating unit undergoes a transition from the ground state to the energized state by virtue of a spontaneous chemical reaction, the chemical potential energy decreases, and the quantity $(C_g - C_e)$ becomes more positive. This means, according to equation (4), that the electrical potential energy difference $(E_g - E_e)$ and/or the mechanical potential energy difference $(M_g - M_e)$ must be negative. In other words, the chemically induced energized state of the repeating unit possesses higher electrical and/or mechanical potential energies than does the ground state. Thus, we can associate an electric field and conformational strain with the energized state.

The phenomenon of interconversion of the electrical, mechanical and chemical potential energies will be referred to as "electromechanochemical coupling". The enzyme system capable of catalyzing such energy conversion will be called the "electromechanochemical coupling system". It is our fundamental postulate that the repeating unit of the inner mitochondrial membrane is an electromechanochemical coupling system.

B. *The Mechanistic Principles of Mitochondrial Energy Transduction*

In the previous section, we have postulated that the mitochondrion is an electromechanochemical energy transducer in the sense that the structural attributes of the organelle enable facile interconversion of the electrical, mechanical and chemical potential energies. Since, the mitochondrion contains many thousands of repeating units,[4] we will define the smallest operational unit capable of catalyzing energy conversion as the "supermolecule" of mitochondrial energy transduction. We wish to distinguish the supermolecule from the classical tripartite repeating unit which was originally defined more or less as a structural unit and not as an operational unit.[12] Unlike the tripartite repeating unit which consists of the headpiece, the stalk and the basepiece, the structure of the supermolecule must be variable depending on the nature of work performance executed by the mitochondrion. Although the detailed macromolecular structure of the supermolecule for a particular work performance is still unknown, we can nevertheless recognize two major structural components of the energy transducing unit: (1) the component participating in the energization process (the energization structural component); and (2) the component involved in the deenergization process (the deenergization structural component). Clearly, the basepieces (in which the electron transfer complexes are localized[13]) as well as the headpieces (in which the ATPase is localized[13]) can be identified as the energization structural components. It is the identity of the deenergization structural component of the supermolecule which determines the nature of the work performance—the headpiece for oxidative phosphorylation,[14] transprotonase for active transport, the transhydrogenase for energized transhydrogenation, and the basepiece for reversed electron flow. It is interesting to note that both the headpiece and the basepiece can act either as the energization structural component or as the deenergization structural component of the supermolecule.

Having described the overall structural requirements for mitochondrial energy transduction, we are now in a position to inquire into the possible mechanism by which the supermolecule implements the principle of electromechanochemical coupling. We are of the opinion that whatever the mechanistic principles operating for a particular work function, the same principles must be applicable to all other work functions of the mitochondrion. On the basis of the biochemical and structural data presently available on oxidative phosphorylation and also on the basis of theoretical investigations in our and other laboratories, we have been led to the following conclusions: (1) the unit of mitochondrial energy transduction is a supermolecule in which electromechanochemical potential energy is manipulated; (2) when the supermolecule is energized either by substrate oxidation or by ATP hydrolysis, there is a vectorial charge separation in the energization structural component of the supermolecule—a consequence either of the separation of electrons and protons in the basepiece, or of the rearrangement of the structural dipoles in the headpiece (see below); (3) the energization of the supermolecule is the consequence of the enzyme-catalyzed conversion of the chemical potential energy of the bound substrate into the potential energy of the supermolecule (including the bound substrates) which is now predominantly in the electromechanical form; (4) the local electric field created as the result of the energization process (substrate oxidation or ATP hydrolysis) *induces* a complementary electric field in the deenergization structural components probably through an electric field effect; (5) the electrostatic interaction between the electrically polarized energization and deenergization structural components leads to a stabilization of the energized supermolecule (the metastable state); (6) when the polarized deenergization structural component is *depolarized* by a chemical reaction in the active site, the supermolecule is deenergized and relaxes to the ground state with concomitant conversion of the predominantly electromechanical potential energy either to kinetically stable molecules [ATP or TPNH] or to kinetically labile species with high electrochemical activities (for example accumulated ions); and finally (7) the local conformational perturbations at the active site where the chemical reactions actually take place, trigger the depolarization process.

The essential features of the above conclusions may be schematically represented as in Fig. 1. The system under consideration is a supermolecule (depicted as a regular rectangle or a ballooned-out rectangle) in which are contained the reduced electron carrier AH_2, the oxidized electron carrier B, and bound ADP and P_i in the ground state. When AH_2 is oxidized by B, the chemical potential energy decreases with concomitant increase in the electrical and mechanical potential energy of the "strained" supermolecule (depicted as a ballooned-out rectangle). We postulate that the strain in the metastable supermolecule is primarily due to a chemically induced charge separation. The electric field generated in the local region of the oxidoreductive active site is potentially available for electrostatic interaction with any charged groups or polarizable groups within the supermolecule. However, under the conditions of oxidative phosphorylation, the deenergization structural component binding ADP and P_i (i.e., the headpiece) is preferentially polarized. Now, the initiation P–O bond formation between ADP and P_i triggers the depolarization of the supermolecule with the conversion of the predominantly electromechanical potential energy inherent in the metastable supermolecule into the chemical potential energy of ATP.

Again, it is instructive to compare the present model with well known energy converters such as the hydrogen–oxygen fuel cell.[15] In this cell, the anode is electrically polarized by the oxidation reaction which converts hydrogen molecules into protons. The electron flow from the anode to the cathode now induces an electrical polarization of the latter electrode, which is in turn depolarized by the reduction

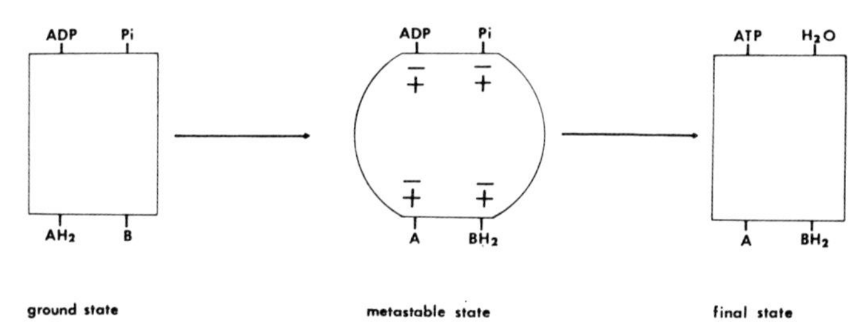

Figure 1. Schematic representation of the electromechanochemical model of oxidative phosphorylation.

of oxygen molecules into hydroxide ions.[16] In a sense, the anode–cathode system is analogous to the supermolecule in our model, with the important difference that whereas the electrode system is virtually free from charge-induced mechanical deformations, the supermolecule is extremely sensitive to charge-induced conformational deformations as schematically shown in Fig. 1. Just as we can regard the hydrogen molecule as a polarizer and the oxygen molecule as a depolarizer, we may consider the substrates of oxidoreduction as polarizers and the combination of ADP and P_i as a depolarizer. As is common with any analogy, there are many differences between the supermolecule and a fuel cell. In addition to the difference in structural rigidity between the two energy transducers already pointed out above, there are a few more points of difference: (1) both of the chemical reactions involved in polarization and depolarization of electrodes in fuel cells are oxidoreduction reactions, whereas in the supermolecule only the reaction in the basepiece and not the reaction in the headpiece is an oxidoreduction reaction; (2) both the polarization and the depolarization reactions are energetically down-hill in fuel cells, whereas in the supermolecule the polarization reaction is exergonic and the depolarization is endergonic; and finally (3) the induced polarization in fuel cells is mediated by actual movement of electrons from the anode to the cathode through a metallic conductor whereas in the supermolecule the induced polarization is achieved by an electric field effect without involving any movement of electrons between the two polarizable structural components.

We have considered in some detail the basic notions embodied in the electromechanochemical model of mitochondrial structure and function. It is our view that the mechanistic principles deduced from oxidative phosphorylation apply to all other energized processes. Accordingly, we may represent the work performances of the mitochondrion in terms of the following equations:

a. *Oxidative phosphorylation*

$$\begin{pmatrix}\text{Polarized}\\ \text{ATPase}\end{pmatrix} + \text{ADP} + \text{P}_i \rightarrow \begin{pmatrix}\text{Depolarized}\\ \text{ATPase}\end{pmatrix} + \text{ATP} + \text{H}_2\text{O} \quad (5)$$

b. *Active transport*

$$\begin{pmatrix}\text{Polarized}\\ \text{transprotonase}\end{pmatrix} + \text{H}^+_{\text{in}} + \text{K}^+_{\text{out}} \longrightarrow \begin{pmatrix}\text{Depolarized}\\ \text{transprotonase}\end{pmatrix} + \text{H}^+_{\text{out}} + \text{K}^+_{\text{in}} \quad (6)$$

$$\text{H}^+_{\text{out}} + \text{Ac}^-_{\text{out}} \longrightarrow \text{H}^+_{\text{in}} + \text{Ac}^-_{\text{in}} \quad (7)$$

c. *Energized transhydrogenation*

$$\begin{pmatrix}\text{Polarized}\\ \text{transhydrogenase}\end{pmatrix} + \text{DPNH} + \text{TPN}^+ \rightarrow \begin{pmatrix}\text{Depolarized}\\ \text{transhydrogenase}\end{pmatrix} + \text{DPN}^+ + \text{TPNH} \quad (8)$$

C. *Charge Separation and the Generation of Membrane Potential*

When an electron transfer complex is reduced by its appropriate substrate, simultaneously three events take place: (1) the complex becomes energized; (2) a pair of electrons is transferred to the electron transfer chain[4]; and (3) a membrane potential is generated.[17] These three events are clearly interrelated but the nature of this interrelationship has hitherto been an enigma. Two developments were crucial in deducing the nature of the interrelationship. First, the electron transfer complex by virtue of being an intrinsic repeating unit of the inner membrane most likely has a double-tiered structure (four proteins in each of the two tiers).[18] Second, Complex IV of the electron transfer chain can be resolved into two fractions equal in weight—one containing the oxidation–reduction components of the complex and the other devoid of these components.[19] From the splitting of the complex into two qualitatively different halves, we have deduced that in one of the two tiers, a set of four proteins is concerned with the transfer of electrons while in the other of the two tiers, a set of four proteins is concerned with the transfer of protons

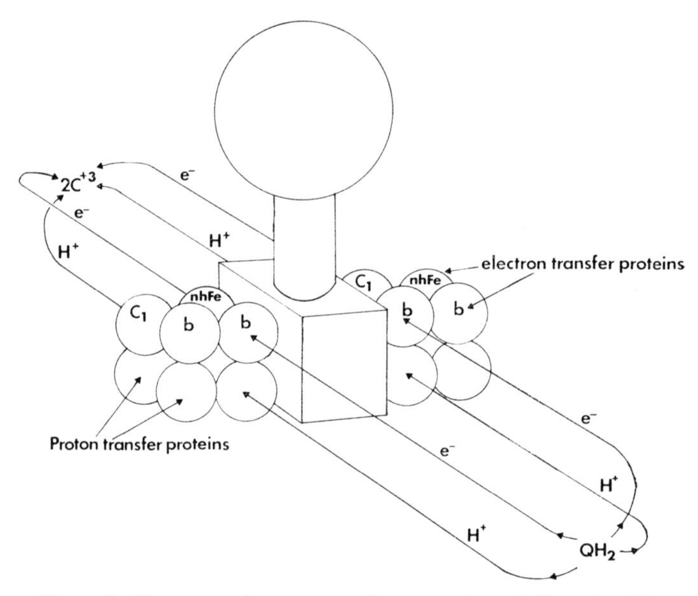

Figure 2. Electron and proton transfer routes through Complex III.

(Fig. 2). The substrate in effect transfers a molecule of H_2 to the complex. In this process, a pair of electrons ends up in the proteins of one tier of the complex (the electron transfer proteins) and a pair of protons end up in the proteins of the other tier of the complex (the proton transfer proteins).

The separation of charge which accompanies the reduction of the complex depends upon a simultaneous conformational transition the exact nature of which remains to be determined. If the electrons and protons were in cavities in the dead center of their respective proteins the distance of charge separation would be about 30 Å.

It is known that the electron can be localized within a deep channel in oxidoreduction proteins such as cytochrome *c*.[20] The electron that enters one of the empty 3d orbitals of heme iron[21] upon reduction can be treated as an uncompensated charge. We are postulating that the uncompensated charge of a proton can also be localized on a specific acceptor group within a deep channel in the proteins of the "proton transfer chain". The electron and proton are buried deep within the protein interior and are thus isolated from the aqueous milieu which surrounds the membrane. Penetration by the solvent into the protected internal cavities in which the uncompensated charges are localized would lead to the erosion of the electric potential.

The inner membrane when energized develops a membrane potential, positive on the intracristal side and negative on the matrix side.[17] Therefore, it is necessary for us to postulate that the electron transfer proteins of a complex are in the tier on the matrix side whereas the proton transfer proteins are in the tier on the intracristal side as shown in Fig. 2. In the interaction of DPNH with Complex I, or of reduced coenzyme Q with Complex III, or of reduced cytochrome *c* with Complex IV,[22] we could conceive of these interactions as the equivalent of the abstraction of 2 hydrogen atoms from the substrate and transfer of these hydrogen atoms to the complex. It is to be noted that no distinction is made between substrates such as Q_{H_2} which provide these 2 hydrogen atoms directly or substrates such as DPNH and cytochrome *c* which must be protonated as a preliminary to hydrogen abstraction. The point at issue is that a proton and an electron together are abstracted from the substrate by the appropriate dehydrogenating enzyme and that they are fed into respective acceptor sites in the proteins arranged in a double-tiered pattern—one set accepting the negative charge, and the other set accepting the positive charge. Thus, we view the charge separation in the electron transfer chain as the direct consequence of the catalysis of oxido-reduction reaction in the inner membrane. In this way, we avoid the necessity for invoking any hypothetical "proton pump" driven by substrate oxidation in order to achieve charge separation.

D. *Structure of the ATPase Coupling Unit*

The studies of Tzagoloff[23] suggest that the ATPase coupling unit consists of a headpiece which is attached by a cylindrical stalk to a structure in the membrane, as yet unidentified. There is apparently a set of associated proteins in the membrane (molecular weight of about 100,000[23]) in which the stalk is anchored. Let us describe the combination of the stalk sector and this membrane-anchoring structure as the "stalk complex" (see Fig. 3).

Two component elements of the mitochondrial energy transducing unit are well characterized morphologically, namely the headpiece (90 Å sphere) and the stalk (30 × 50 Å cylinder).[12] Experiments with ETP_H[24] have provided evidence that the headpiece may assume two distinct geometric relations with respect to the inner membrane depending on the coupling capability of ETP_H and on the presence or absence of ATP in the medium. In other words, the headpieces in tightly coupled ETP_H have been found to be on or partially in the membrane continuum in the presence of ATP (the apposed configuration; see Fig. 3a), whereas under the same experimental condition the headpieces in uncoupled submitochondrial particles project away from the membrane surface and are connected to the membrane by the 50 Å stalk sector (the extended configuration; see Fig. 3b). In the apposed configuration, the stalk is probably pushed deep into the membrane while in the extended configuration, the

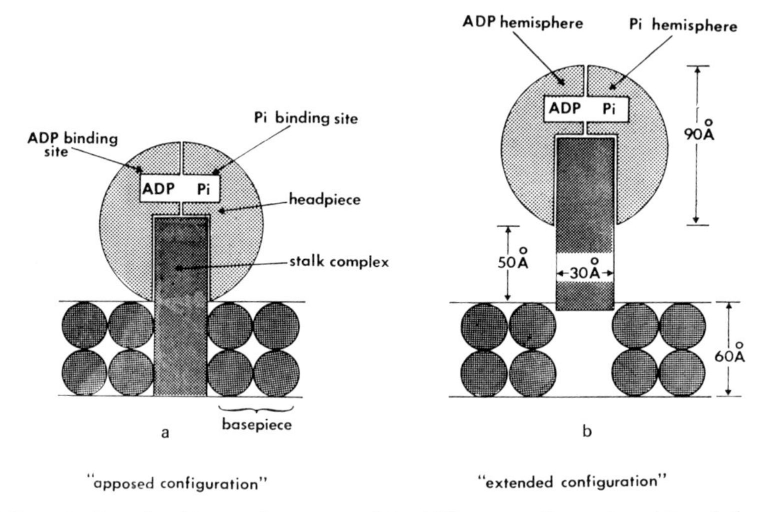

Figure 3. Postulated internal structure of the ATPase coupling unit and its relation to the electron transfer complexes.

stalk is pulled out of the membrane as shown in Fig. 3. Since the extended configuration is characteristic of uncoupled mitochondria or submitochondrial particles and the apposed configuration is associated with phosphorylating ETP_H,[24] we have deduced that the apposed configuration of the headpiece and stalk is a prerequisite for oxidative phosphorylation and active transport in mitochondria. Coupling requires the insertion of the stalk into the membrane continuum and the contact of the headpiece and membrane. Whenever these conditions are not fulfilled as in the orthodox configuration,* then electron transfer or ATP hydrolysis cannot be coupled to any work performance.

* It has not been generally appreciated that in mitochondria the headpiece-stalk projections are visualized only in the orthodox configuration.[12] The headpiece, but not the stalk, can be visualized in the aggregated and twisted configurations.

The evidence that supports the above structural interpretations may be summarized as follows. The ATPase coupling unit has been isolated as an oligomycin-sensitive complex from both beef heart and yeast mitochondria.[23,25] The unit interacts with phospholipids to form membranous arrays the periodicity of which is identical with the periodicity of headpieces in the intact mitochondrial membrane.[13] Under coupling conditions, the headpieces, but not the stalk, can be visualized by positive staining in the cristae of mitochondria.[26] However, in the orthodox configuration, both the stalk and the headpieces are visualized by negative staining and both project from the membrane.[12] Finally, as alluded to above, the headpiece in ETP_H extends away from the membrane via the stalk when ETP_H is uncoupled.[24]

We may consider the stalk as a dielectric for effective transmission of the electric field from the electron transfer complex to the headpiece of the ATPase coupling unit. When the stalk is extended from the membrane as in the orthodox configuration, the means for transmission of the electric field may be lacking. This would account for the loss of coupling capability when the inner membrane is stabilized in the orthodox configuration.[27]

E. *Induction of an Electric Field in the ATPase Coupling Unit*

To simplify visualization of the process by which an electric field is generated in the headpiece, we may consider the headpiece as a sphere divided longitudinally (perpendicular to the plane of the membrane) into two hemispheres, each corresponding to a multimeric unit of globular proteins (Fig. 3). The cavity into which the stalk complex is inserted is located between the two hemispheres. Furthermore, the active sites for ATP synthesis and hydrolysis are assumed to spread over the two hemispheres. ADP and P_i probably bind to separate hemispheres ("ADP hemisphere" and "P_i hemisphere" in Fig. 3). It is known that ATP hydrolysis leads to the generation of a membrane potential.[17] In principle there appear to be three possible ways of generating a membrane potential linked to ATP hydrolysis: (1) ATP hydrolysis drives a proton pump as in the chemiosmotic hypothesis[28]; (2) the conformational changes associated with ATP hydrolysis in the headpiece propagate through the stalk to the basepiece which undergoes a conformational transition leading to a membrane potential[29]; and (3) ATP hydrolysis conformationally polarizes the headpiece, and the electric field of the headpiece induces a polarization of the basepiece. For theoretical reasons (see above), we have chosen the third alternative as the preferred mechanism for generating a membrane potential. In our view, the electric field effect is singularly suited for coupling the ATP hydrolysis event to the polarization of the electron transfer chain over a relatively long distance (the center to center distance between the headpiece and the basepiece is about 70 Å[12]).

Since it is unlikely that ATP hydrolysis itself will lead to any net charge separation at the active site, we have deduced that ATP hydrolysis must be coupled to the perturbation of structural dipoles as a means for generating an electric field. To fulfill this role, the dipoles must be arranged on a plane parallel to the inner membrane surface (the *xy*-plane), so that in the unperturbed state of the headpiece there is no net dipole moment in the direction perpendicular to

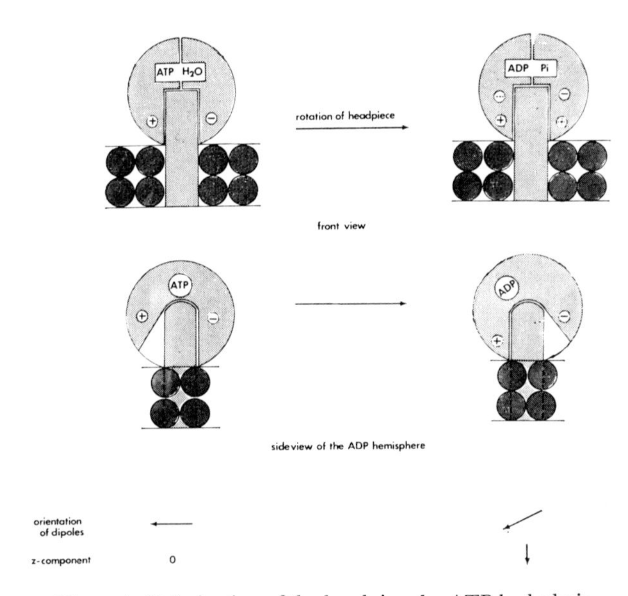

Figure 4. Polarization of the headpiece by ATP hydrolysis.

the plane of the inner membrane (the *z*-direction; see Fig. 4). When ATP hydrolysis takes place, the accompanying conformational perturbation compels the headpiece to undergo a state transition. In the perturbed state, the pair of structural dipoles are forced out of the planar arrangement and a net dipole movement in the *z*-direction results. This is schematically shown in Fig. 4 where the ATP-induced state transition of the headpiece is represented as an anti-clockwise rotation of the hemispheres. The important steps involved in linking ATP hydrolysis to the generation of the electric field in the headpiece are summarized in Fig. 5.

The conformational transition of the headpiece from the unperturbed to the perturbed state may be induced either by ATP hydrolysis as described above or by electron transfer in the basepiece. In the latter case, the electric field associated with the basepiece acts on the structural dipoles of the headpiece (charge–dipole interaction).

We have not specified the polarity of the immediate environment that surrounds the structural dipoles in Fig. 4. If the environment of the dipoles is hydrophobic, there would be a relatively strong dipole–dipole interaction even before the headpiece is conformationally perturbed, leading to an electromechanochemically strained headpiece. This electromechanochemical strain of the headpiece in the ground state is neither necessary from the point of view of our model

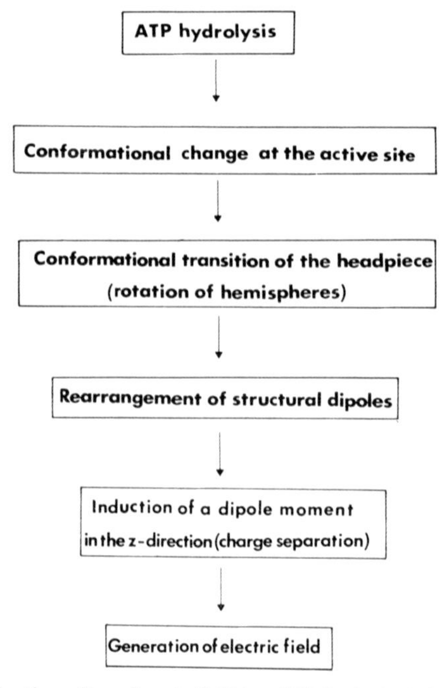

Figure 5. Induction of an electric field by ATP hydrolysis in the headpiece.

nor likely in view of the fact that protein systems (except cytochromes) rarely accommodate uncompensated charges in their hydrophobic interior under normal conditions. This consideration led us to entertain the notion that the dielectric environment of the structural dipoles undergoes a decrease in polarity simultaneously with the headpiece transition from the ground to the perturbed state; the polar environment of the dipoles would attenuate the electrostatic interaction between the dipoles in the ground state, but the decreased polarity of the dipole environment in the perturbed state would still induce a relatively strong electric field in the z-direction. Thus we postulate that the conformational perturbation of the headpiece involves not only the geometric rearrangement of the structural dipoles but also associated variations in the dielectric constant of the dipole environment in order to diminish the dipole–dipole interaction in the ground state and to enhance the intensity of the induced electric field in the perturbed state of the headpiece. Such a conformationally controlled variation of the polarity of protein microenvironment has been suggested by X-ray studies of the structure of cytochrome *c*.[30]

We have invoked above a full-blown fixed charge system in the interior of the headpiece. This does not mean that the surface charge redistribution and the changes in the chemical and mechanical states of the headpiece will not attenuate the effective magnitude of the fixed charges. For our purpose, all that is required is the presence of a fixed charge system which exerts a net electric field in the z-direction outside the headpiece when the headpiece undergoes conformational perturbation.

F. *Electric Field Effects*

The electric field in the inner membrane is established within a multimeric protein complex about 60 Å in thickness.[31] It is generated by the separation of two protons and two electrons per coupling site—the protons being localized near the intracristal side and the electrons toward the matrix side. The energy required to separate electrons and protons derived from the substrate molecule and to transfer these separated charges to the respective acceptor sites is obviously of paramount importance. A precondition for the acceptability of the mechanism we have proposed for generating a membrane potential is that the energy must be less than the energy derivable from the oxidoreduction reaction which takes place within a single coupling site. Unfortunately, the exact value of the energy required for charge separation in the inner membrane cannot be determined in the absence of more accurate structural and chemical information than is presently available. However, an order-of-magnitude estimation can be made by considering the major steps involved in the charge separation. There are five such steps: (1) breaking of the carbon–hydrogen bond of the substrate; (2) ionization of the abstracted hydrogen atom into an electron and a proton; (3) separation of the electron and proton from the initial intercharge distance after ionization (say 2 Å) to the final intercharge distance of about 30 Å; (4) chemical interactions of the electron and proton with their respective acceptor groups; and finally (5) the conformational rearrangements of polypeptide chains accompanying charge separation. The net energy change due to the charge separation in an electron transfer complex is then the algebraic sum of all the energies produced or consumed by the above elementary steps. It is interesting to note here that the energy required to carry out step (3) above (purely electrostatic work involved in increasing the intercharge distance from 2 to 30 Å) is about 15 Kcal/mole assuming a dielectric constant of 10. This energy is at least an order of magnitude smaller than the other energy terms. The important point to be emphasized is the fact that the electrostatic work required to separate an electron and a proton to a desired distance is not prohibitively large but is within the range of the energy derivable by oxidoreduction in the electron transfer complex.

The next item of importance is the strength of the electric field generated by charge separation in the basepiece. If we assume that the intercharge distance between electron and proton is 30 Å and the average dielectric constant of the medium separating the charges is 10, the electric field intensity at a point 70 Å away from the center of the basepiece is about 3×10^4 volts/cm. This electric field intensity is of the same order of magnitude as is known to induce conformational transitions in DNA molecules[32] and as is experimentally measured across the thylakoid membrane in chloroplast.[33] We are assuming that this electric field at the headpiece is of the right order of magnitude to drive the synthesis of ATP by direct union of ADP and P_i.[34]

The electric field generated in the electron transfer complexes has multiple effects on the component systems in the mitochondrion. Let us consider three such effects: (1) the asymmetric perturbations of the proteins in the inner membrane; (2) the perturbation of the matrix system; and (3) the induction of an electric polarization of the headpiece.

The electric field asymmetrically perturbs the relation of the bimodal proteins to the water–lipid interface at which these proteins are poised. If on one side of the membrane (the matrix side) the proteins sink below the interface due to the membrane electric field, then on the other side (the intracristal side) the proteins must rise above the interface due to the vectorial nature inherent in an electric field. If on the intracristal side, groups previously buried are exposed and ionize, leading to the generation of free protons, then on the matrix side, the opposite will take place, namely the burying of ionizable groups leading to the uptake of protons from the medium. Since the numbers of groups exposed on one side and buried on the other will be comparable, the net free energy change may be quite small even though the number of protons released or taken up is considerable.

The energized inner membrane is negatively polarized on the matrix side and positively polarized on the intracristal side.[17] The matrix network is enclosed between two cristae generally separated by relatively short distance, e.g. <300 Å in beef heart mitochondria. Thus the bulk of matrix proteins may exist within short distances from the inner membrane, probably in the order of one Debye length.[35] The intense cathodic electric field on the membrane surfaces which bound the matrix space will, therefore, inevitably affect the state of matrix proteins; positively charged proteins will migrate toward the membrane surfaces (electrophoretic effect); electrically neutral proteins with high dielectric constant will move to a region of high field intensity (dielectrophoretic effect[36]); and the p*K*a's of the ionizable groups on protein surfaces will decrease (the second Wien effect[37]) causing pH changes in the matrix space. In addition, the state of matrix water may undergo pronounced changes due to the fact that the water molecule is dipolar.[38] Several authors have speculated on the possibility of cooperative configurational transitions of protein systems induced by electric field.[39] It is probable that similar cooperative, electrostatic interactions exist between the inner membrane and the matrix system.

The electric field in the energized electron transfer complex can induce a complementary field in the ATPase coupling unit under the conditions of oxidative phosphorylation (see below). Similarly the electric field in the energized ATPase coupling unit can induce a complementary field in the electron transfer complex under the conditions of ATP-driven reversed electron flow and active transport. This induction of complementary field provides a mechanism for partitioning potential energy within different sectors of essentially one supramolecular unit (the electron transfer complex plus the ATPase coupling unit).

G. *De facto Unit of Mitochondrial Control*

The control of mitochondrial coupling depends upon the interplay of three basic systems: the inner membrane, the protein network systems in the matrix and the intracristal spaces. These three systems are so closely intertwined structurally and functionally that they can be considered as a single operational unit which we describe as the *de facto* unit of mitochondrial control. The matrix network can exist in a dispersed or condensed state[40]; and in the condensed state the matrix can exist either in a form with high ion binding capability or in a form with low ion binding capability. There are thus at least three alternative states of the matrix network system. The proteins in the intracristal space in general appear electron microscopically transparent. However, under certain conditions the intracristal space proteins can assume a lattice structure.[41] This again suggests that the intracristal network can also exist in a dispersed or condensed state. The matrix system is highly concentrated in the mitochondrion (between 10 and 50 grams protein per 100 ml of matrix fluid) and accounts for the bulk of the solubilizable protein of the mitochondrion. Moreover, the work of Hackenbrock clearly shows that the matrix proteins have a high degree of structural organization.[40]

The inner membrane contains two closely associated but independent repeat structures—the electron transfer complexes[22] and the ATPase coupling unit.[13,25] There is a 1:1 correspondence between the number of ATPase coupling units and electron transfer complexes. There are three alternative conformational states of the ATPase coupling unit: (1) the state in which the proteins of both the headpiece and stalk are in the membrane (fused regions of the coalesced configuration); (2) the state in which the headpiece projects from the membrane but the stalk is buried in the membrane (the twisted and paired configurations*); and finally (3) the state in which both the headpiece and stalk project from the membrane (the orthodox configuration).[12] The inner membrane can exist in at least four different states—nonenergized or energized with the headpiece-stalk unit projecting from the membrane (the orthodox configuration), nonenergized with the proteins of the headpiece-stalk unit buried in the membrane (fused regions of the coalesced configuration), and energized with the headpiece projecting from and the stalk in the membrane (the paired or the twisted configuration).

The configuration of the mitochondrion as visualized electron microscopically reflects the relationship of the three interdigitating systems and preponderantly the relationship of the inner membrane to the matrix system. Each of the four configurational states of the mitochondrion (orthodox, twisted, paired and coalesced) represents a given state of the matrix, a given state of the intracristal system and a given state of the inner membrane. These interrelationships are different for each of the four configurational states. Thus in a particular configuration such as the twisted, we can specify that the matrix system is condensed, the intracristal system is dispersed, the inner membrane has a helical twist, and the headpiece sector of the ATPase coupling unit is projecting into the matrix space while the stalk is embedded in the membrane.

There is a feedback relationship between the three mitochondrial systems. The state of the matrix affects the state of the inner membrane and the state of the intracristal system. The energization of the inner

* The configuration previously described as "aggregated" embraces two similar configurations—the "coalesced" which obtains under nonenergizing conditions in 0·25 M sucrose (the NEagg configuration according to Penniston *et al.*[1]), and the "paired" configuration which obtains under energizing conditions (the Eagg configuration according to Penniston *et al.*[1]). The purpose of introducing the new terms "coalesced" and "paired" is to describe the inner membrane configuration in purely morphological terms devoid of any reference to the energy state of the mitochondrion. This is necessary because of recent findings that the so-called "energized" configurations (e.g. E or ET) could be induced by nonenergized means.[45]

membrane affects the state of the matrix system which in turn affects the conformation of the inner membrane, etc. We are thus dealing with a delicately poised set of systems which collectively respond to changes in the ionic composition of the medium, to energization or deenergization, to osmotic pressures and to reagents which affect any one of the three interdigitating systems.

When the mitochondrion is exposed to energizing conditions in presence of the reagents required for oxidative phosphorylation, the *de facto* unit undergoes a rearrangement which leads to the establishment of a steady-state configuration (a mixture of the paired and coalesced configurations). When energized in presence of reagents that are required for active transport of K^+, it assumes yet another coupling configuration (the twisted configuration[42]). Thus for each coupling mode there is a unique configurational state of the *de facto* unit which defines the states of the three basic systems in relation to one another and ultimately the state of the supermolecule (see above). These unique configurations are maintained as long as energizing conditions are present. When energizing conditions are no longer available, the configuration of the *de facto* unit returns to the configuration that obtained prior to energization. Energization is therefore one of the crucial determinants of the coupling configuration. Many reagents are known (e.g. fluorescein mercuric acetate[43]) which can prevent configurational adjustment of the *de facto* unit and thereby interdict the induction of the configuration required for either oxidative phosphorylation or active transport.

The *de facto* unit can undergo ion-induced configurational rearrangements which are electron microscopically indistinguishable from those induced by energizing conditions.[44–47] Such nonenergized rearrangements of mitochondrial configurations are, however, stable and are not influenced by the reagents that block energized configurational changes. A major control feature of mitochondria is thus the regulation of the configurational adjustment of the *de facto* unit. Ions such as Ca^{2+} and Mg^{2+} play key roles in this regulation.[27]

Central to the notion of the *de facto* unit is the recognition that the macromolecular systems (the matrix, inner membrane and intracristal proteins) assume closely interacting relationships because of the fact that they are compelled to exist in a relatively small space enclosed by the outer membrane. Unlike the space outside the outer membrane, the intramitochondrial space is highly heterogeneous in material composition as well as in local electrostatic environment. Since it is clear that the configuration of the *de facto* unit will determine the state of the macromolecular systems inside the mitochondrion, it would follow that the configuration of the *de facto* unit also determines the conformational state of the supermolecule and hence the nature of mitochondrial work performance. In our view, therefore, the mechanism by which mitochondrial function is controlled is inextricably tied into the general causal sequence shown in Fig. 6. We may regard the *de facto* unit as the operational unit for mitochondrial control and the supermolecule as the operational unit for energy transduction.

The basic structural difference between ETP_H and intact mitochondria is that ETP_H lacks one of the two membranes (the outer membrane), one of the two spaces (the matrix space) and one of the two protein networks (the matrix network).

While ETP_H is incomplete in control functions, it nevertheless is capable of performing energy transductions—a token that ETP_H resembles more a collection of supermolecules. From the enzymological point of view, therefore, the intact mitochondrion represents an operational unit complete in terms of control as well as of energy transduction, whereas ETP_H represents a functional unit which is

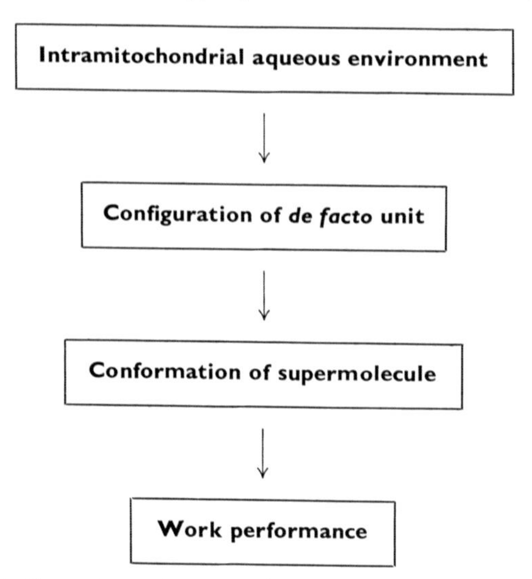

Figure 6. General causal sequence involved in mitochondrial control.

capable only of certain energy transductions with incomplete control capability.

It cannot be a happenstance that the various sets of ions which play a predominant role in determining configuration are the very ions which are required for the different work functions, e.g. $P_i + ADP$, $DPN^+ + TPNH$, $Ca^{2+} + P_i$, K^+ + weak acid anions. By this tactic, the *de facto* unit responds to the ions required for a particular work function by undergoing a transition to a configuration which is uniquely required for the exercise of that work function.

H. *Octet and Bimodal Themes for the Construction of the Electron Transfer Complexes and the Headpiece of the ATPase Complex*

The membrane model proposed by Vanderkooi and Green[7, 18] is based on two basic assumptions: (1) the membrane proteins (intrinsic proteins[49]) are globular; and (2) the surface properties of these intrinsic proteins are bimodal. Recently, Vanderkooi and Capaldi have shown that the intrinsic membrane proteins have polarities well below those of the majority of soluble proteins,[49] which is in line with the second assumption. Given the validity of these two assumptions, the simplest membrane model that is compatible with experimental facts would be the double-tiered model of Vanderkooi *et al.*[18] There is suggestive evidence that the active site of ATP synthesis in the headpiece as well as the active sites of oxidoreduction in the basepiece is (are) located within the hydrophobic interior of these units. Since

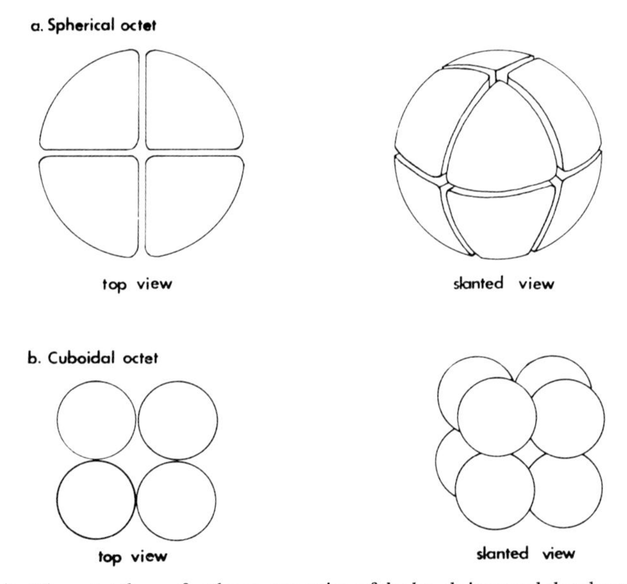

Figure 7. The octet theme for the construction of the headpiece and the electron transfer complexes.

the substrates for these active sites are either electrically charged molecules or potential charge generators, it is logical to postulate that these active sites consist of polar cavities located within the hydrophobic interior of globular protein systems, which are connected to the external medium through substrate-specific channels. Again, the simplest geometric arrangement of globular proteins which will afford the central cavity and substrate channels appears to be the octet arrangement (see Figs. 7 and 8).

The two transducing centers—the electron transfer complex which is part of the inner membrane, and the headpiece of the ATPase coupling unit which is a projection from the membrane—are probably built on the same constructional themes, namely the octet and bimodal themes. Let us first consider the structure of the electron transfer complex which is the easier of the two to visualize. There are four bimodal proteins on one tier apposing four bimodal proteins on the opposite tier of a double-tiered structure. Each set of four proteins will be arranged two in one tier and two in a tier immediately behind (Fig. 7). Top and bottom surfaces of the proteins of the octet will be polar whereas the interior surfaces of these proteins will be nonpolar. There is pairing of the hydrophobic surfaces of the proteins on the upper tier with the hydrophobic surfaces of the proteins on the lower tier. Moreover, there is also lateral juxtaposition of the hydrophobic

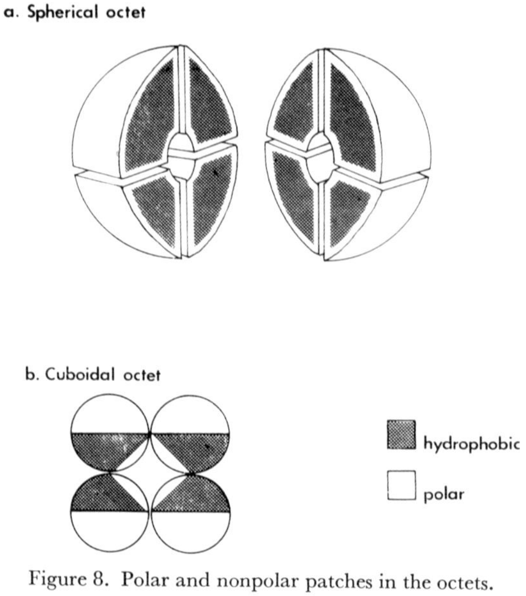

a. Spherical octet

b. Cuboidal octet

hydrophobic

polar

Figure 8. Polar and nonpolar patches in the octets.

surfaces of the proteins on the same tier. Thus, each protein makes hydrophobic links with its two neighboring proteins on the same tier. It is these multiple hydrophobic associations which endow the complex with a high degree of impenetrability to water. The complex is presumed to enclose a polar cavity which is in the hydrophobic center of the octet (Fig. 8b). This cavity in the interior is accessible from the exterior of the octet by selective channels into which reductant and/or oxidant can enter and leave. The stability of the polar region in the central cavity is achieved by the packing of the eight proteins in such a fashion that each protein contributes a polar "patch" to the wall of the cavity. The summation by appropriate packing of the eight polar "patches" leads to the formation of a polar lining for the cavity. A cavity formed in this fashion could be stable if polar residues in the hydrophobic interior of an octet are so constructed that they could not reach a more polar environment. The channels which lead into the central cavity from the exterior surface of the octet can be lined with either polar or nonpolar residues. In Complex III, there would be a nonpolar channel for coenzyme Q and a polar channel for cytochrome *c*; in Complex I, a polar channel for DPN^+ and a nonpolar channel for coenzyme Q, etc. The channels are selective for the specific reductant or oxidant. The polar channels would exit on the polar surface of the octet and the nonpolar channels would exit on the hydrophobic face of the octet. The same thermodynamic considerations that have been invoked for predicting the stability of the polar cavity would apply with equal force to the stability of the polar channels in hydrophobic regions of the octet.

The octet premise for the structure of the electron transfer complexes is compatible with the data on the molecular weights of the complexes[50] and the molecular weights of the individual proteins.[22] The assumption of eight proteins to each complex is yet to be proven but it corresponds to the simplest fit. The bimodal principle of membrane construction may be a general principle applicable to all biological membranes.[51] The concept of an internal cavity with channels leading into the cavity has been well established for multimeric units such as hemoglobin[52] and for monomeric units such as cytochrome *c*.[20]

If we bear in mind that the headpiece is surrounded by the aqueous phase, then the octet and bimodal premises would require that the bimodal proteins in the headpiece be rotated 45° with respect to the orientation of the bimodal proteins in the membrane (Fig. 8a). That is to say, the polar surface of the proteins would form the external surface of the headpiece whereas the nonpolar surface of the same proteins would be oriented interiorly. The same double tier arrangement would apply to the headpiece octet as for the electron transfer complex, and the concept of a central polar cavity with feeder channels would also apply. The active groups of the headpiece, i.e., the groups concerned with synthesis or hydrolysis of ATP are assumed to be localized in an interior cavity. Since ADP and P_i would probably be bound to the active sites by more than one ligand, it would appear that each of several globular proteins in the octet should contribute to the active sites. Hence no one protein alone of the headpiece would show ATPase activity in agreement with experimental findings.[53]

The evidence for the bimodality of the headpiece proteins rests on the polarity and solubility properties of these proteins as well as on the fact that the headpiece can be collapsed into the membrane as evidenced by the existence of fused regions of cristal membrane in the coalesced configuration. The assignment of eight molecules per headpiece is an approximation based on available data.

II. *Description of the Model*

The model which will be developed in this section is based on a set of physical principles and on the structure of the systems involved in mitochondrial transduction. It is our view that within the framework of these principles and this structural information, all the necessary ingredients are present for a complete description of mitochondrial coupling. The particular model we are presenting is a model and not necessarily *the* model. But *the* model will be, we suspect, a variation of the themes implicit in the present model. Thus the emphasis should be placed primarily on the strategic aspects of the model rather than the exact details.

A. *Oxidative Phosphorylation*

We have already considered some of the principles which underlie the mechanism of the coupling of electron transfer to the synthesis of ATP. Before we discuss oxidative phosphorylation in detail, it will be helpful to present a scheme for visualizing the mechanism of oxidative phosphorylation as we formulate it. Figure 9 depicts three major steps involved in oxidative phosphorylation at coupling site 2. The energy transducing unit is composed of the headpiece, the stalk complex and the electron transfer complexes. The substrates include reduced Coenzyme Q (QH_2), oxidized cytochrome *c* (c^{3+}), ADP and P_i. When QH_2 is oxidized by Complex III, the abstracted hydrogen atoms are separated into electrons and protons which are stabilized in the two tiers of the globular proteins as described earlier (Fig. 2). The electric field generated by the charge separation in the inner membrane simultaneously polarizes the pair of dipoles embedded in the interior of the headpiece giving rise to a metastable state of the energy transducing unit (Figure 9b). We postulate that the chemical potential energy decrease accompanying the oxidation of QH_2 is conserved predominantly as electrical (charge separation) and mechanical (mechanical strain) potential energies of a metastable state which has a finite life time. In the metastable intermediate state, ADP and P_i are bound to respective binding sites in such a way that one of the terminal oxygen atoms of ADP is in a precise orientation with respect to the phosphorus atom of inorganic phosphate favoring the formation of the P–O bond between ADP and P_i by the direct-union mechanism discussed by Korman and McLick.[34, 54] It is conceivable that in the metastable state both ADP and P_i are chemically activated via conformationally induced increase in the negative charge on the ADP oxygen atom and similarly induced increase in the positive charge on the phosphorus atom of P_i so that the activation energy barrier for the P–O bond formation is minimized. According to Korman and McLick,[54] the initially formed pentacovalent

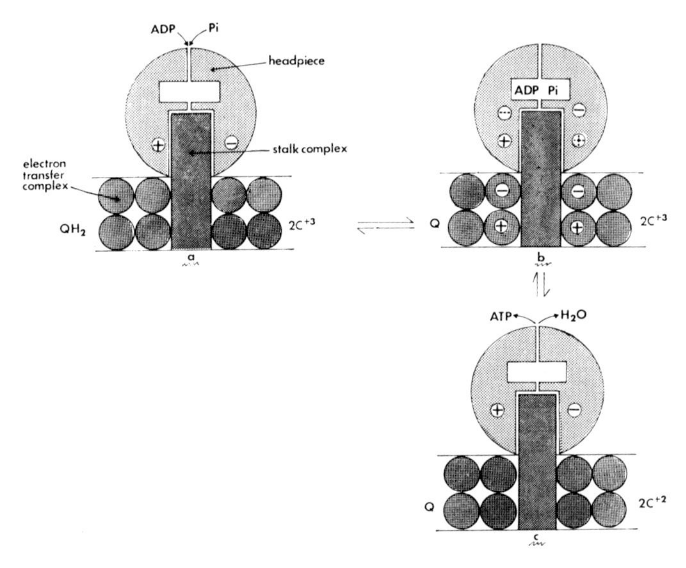

Figure 9. Mechanism of oxidative phosphorylation at coupling site 2.

phosphorus intermediate undergoes pseudorotation followed by dehydration before ATP synthesis is completed. Regardless of the precise molecular mechanism involved, it is likely that both the phosphorylation of ADP and the subsequent dehydration reaction proceed during the life time of the metastable state. If the chemical potential energy of ATP + H_2O is greater than that of ADP + P_i, the law of conservation of energy requires that ATP synthesis be accompanied by decreases in the electrical and mechanical potential energies of the enzymic system. In other words, as the covalent bond is formed between ADP and P_i, the supermolecule must relax electromechanically so that electrons and protons which were immobilized in the metastable state can now flow into the next mobile carrier and the original conformation can be restored (Fig. 9c).

An important observation to be noted here is that the electron transfer reaction is inhibited by the metastable state of the repeating unit (respiratory control). The charge-separated state of the electron transfer complexes are electrostatically stabilized by the polarized headpiece. The electron transfer in the basepiece can only proceed when the headpiece is depolarized via ATP formation. The conformational changes induced by the P–O bond formation between ADP and P_i act as a triggering mechanism for the relaxation of the headpiece and hence the collapse of the membrane potential. This triggering mechanism is not too surprising if we remember that the conformational changes at the active site *physically* affect the conformational state of the headpiece, and the headpiece and the electron

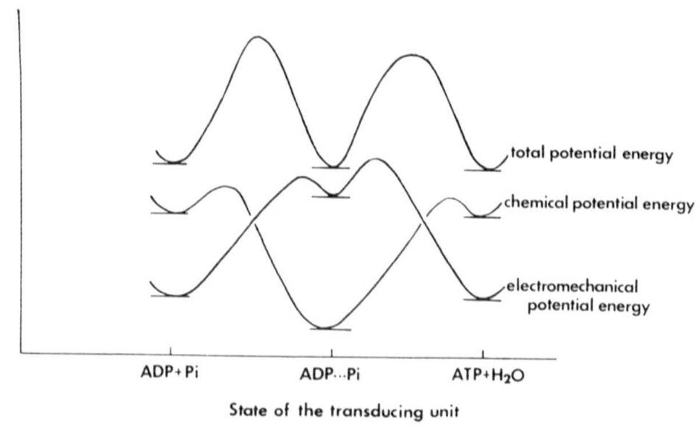

Figure 10. Potential energy diagram for oxidative phosphorylation.

transfer complexes are intimately linked by the electrostatic field effect.

If we assume that the oxidative phosphorylation reaction proceeds with 100% thermodynamic efficiency, the various types of potential energies of the energy transducing unit undergo fluctuations as shown in Fig. 10. It is clear from this diagram that the chemical potential energy is converted predominantly into the electrical and mechanical potential energies in the metastable state and this electromechanical potential energy is reconverted into predominantly chemical potential energy at the end of ATP synthesis. The difference between the initial and the final states of the energy transducing unit is that in the initial state the chemical potential energy resides mainly in the substrates of oxidoreduction whereas in the final state the chemical potential energy is accumulated, so to speak, in ATP. The interconversion of the different forms of potential energies during oxidative phosphorylation is summarized in Fig. 11. The sequence of potential energy changes chemical → electromechanical → chemical, represents in simplest form the principle of electromechanochemical coupling discussed earlier.

It is interesting to compare the salient features of the present model with the chemiosmotic hypothesis of P. Mitchell.[28] In the chemiosmotic model the membrane potential is generated secondarily as the result of separation of protons "out into" the intracristal space, and of hydroxide ions "out into" the matrix space, by maneuvering the movements of electron and hydrogen-atom carriers in the inner membrane. In the electromechanochemical model, the membrane

Predominant potential energy forms in

Initial state		Metastable state		Final state
chemical	→	electromechanical	→	chemical

Figure 11. Variation of potential energy forms during oxidative phosphorylation.

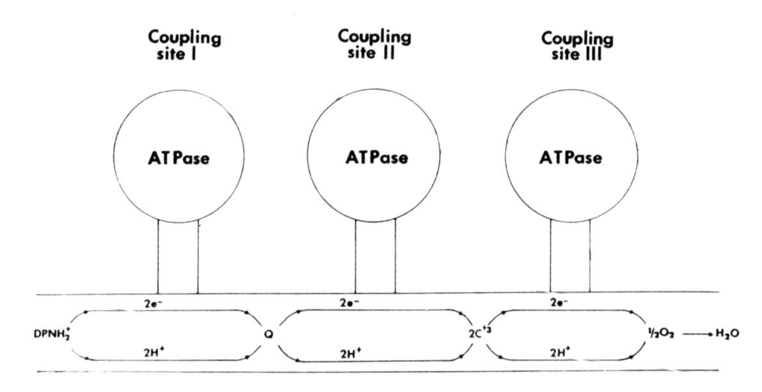

Figure 12. Arrangement of the proton and electron transfer pathways in the inner membrane.

potential is generated as a direct consequence of the chemistry of the oxidoreduction in the electron transport chain. We thus ascribe the charge-separating capability of the electron transport system to the localization of electron carriers in one tier of the double-tiered inner membrane and "proton carriers" in the other tier. Another important difference is that in the electromechanochemical model the protons separated from an oxidizable substrate are never allowed to leave the inner-membrane phase but are constrained to move within the hydrophobic environment (Figs. 12 and 13) thereby conserving the free energy of the oxidoreduction reaction within the inner membrane. Since there will be extensive conformational rearrangements of proteins in the vicinity of separated charges in the inner membrane, the pH of the external medium and the matrix space will be influenced by the conformational changes which are induced by the "internal" charge separation in the electron transport chain.

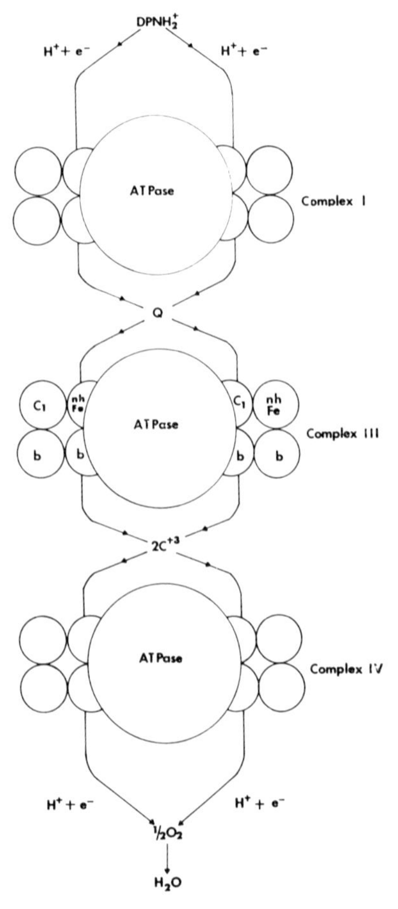

Figure 13. Surface view of the electron and proton transfer pathways in the inner membrane.

The electromechanochemical model of oxidative phosphorylation as presented above provides a unique approach to the mode of action of inhibitors and uncouplers.[55] There are four major steps involved in oxidative phosphorylation as evident in Fig. 14: (1) electron transfer; (2) electrical polarization of the basepiece (separation of e^- and H^+); (3) electrical polarization of the headpiece; and finally (4) bond formation between ADP and P_i. Any reagent or condition that prevent one or more of the above four steps can uncouple or inhibit oxidative phosphorylation. Thus, rotenone,[56] antimycin A[57] and cyanide[58] inhibit electron transport by interacting with Complexes I, III and IV respectively. Lipid-soluble organic acids such as 2,4-dinitrophenol (DNP)[59] and *m*-chlorocarbonyl cyanide phenylhydrazone (*m*-ClCCP)[60] may uncouple oxidative phosphorylation by

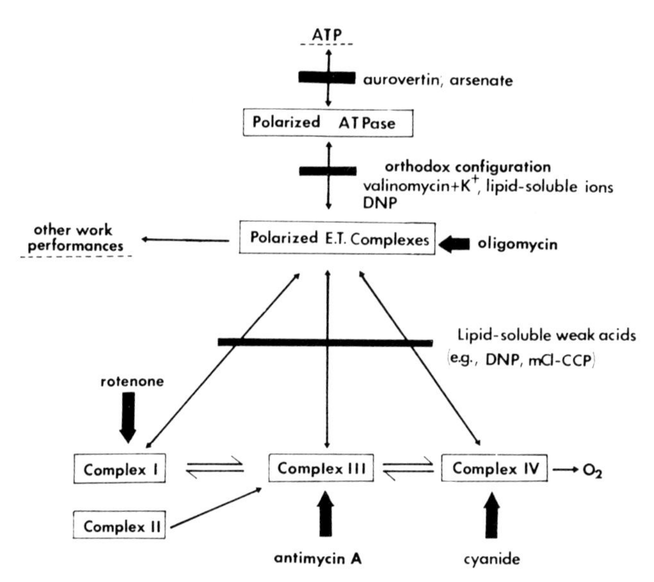

Figure 14. Proposed scheme for the sites of action of inhibitors and uncouplers of oxidative phosphorylation.

protonating the negative charge center and providing negative counter ions (conjugate bases) to the positive charge center in the inner membrane, thereby effectively depolarizing the basepiece. Since the basepiece is now disengaged in respect to electrostatic interaction with the headpiece, electron transfer in the basepiece is no longer inhibited by the electrical polarization of the headpiece and hence the electron transfer rate will be enhanced. Dinitrophenol can also act at the level of the headpiece by serving as the source of protons and dinitrophenoxide anion which can not only attenuate the electrostatic interaction between the headpiece and the basepiece but also prevent the electrostatic attraction between the two structural dipoles when the headpiece is in the perturbed state (Fig. 4). The elimination of the strong attraction between these two dipoles in the perturbed state will relieve the ATPase active site of the conformational constraints and will facilitate hydrolysis of ATP. This may account for the observation that dinitrophenol stimulates the ATPase activity of F_1 preparations.[61] Furthermore, if dinitrophenol acts at the level of the electron transport chain (the basepiece) with a greater affinity than at the level of the headpiece, one would expect that an insufficient amount of dinitrophenol could selectively uncouple oxidative phosphorylation without inducing the ATPase activity of the headpiece, in agreement with experimental findings.[62] It is clear from these discussions that the postulated existence of a pair of structural dipoles within the interior of the headpiece is completely in line with the dinitrophenol phenomena. In fact we are inclined to suggest that the dinitrophenol data strongly support our postulate about the internal structure of the headpiece (Fig. 4).

According to the present model, it is possible to uncouple oxidative phosphorylation by preventing the polarized basepiece from inducing the polarization of the headpiece, or *vice versa*. In the orthodox configuration, the headpiece and the basepiece are separated from each other 50 Å further than in the paired configuration by the intervening stalk sector.[12,24] This increase in the distance between the headpiece and the basepiece will weaken the electrostatic interaction between them to such an extent that electrostatic coupling is abolished. The same result could be achieved by introducing highly polarizable molecules such as water in a critical region between the headpiece and the basepiece. Therefore, various conditions (aging, swelling[63]) and reagents ($CaCl_2$,[27] thyroxine[64]) which induce the orthodox configurations could uncouple oxidative phosphorylation either by their effect on the ultrastructural organization of the energy transducing unit or by their ability to introduce a highly polarizable dielectric between the headpiece and the basepiece.

Since oligomycin inhibits the ATPase activity not only of tightly coupled mitochondria (i.e., ATP synthetase) but also of dinitrophenol-treated or arsenate-treated mitochondria,[65] it is most likely that this antibiotic prevents both P–O bond-breaking during ATP hydrolysis and P–O bond-formation during ATP synthesis by preventing the rotation of the two hemispheres in the headpiece (Fig. 4). Oligomycin may effectuate this result by acting as a sort of "hydrophobic glue" between the two hemispheres, thereby increasing the rotational energy barrier of the hemispheres. In the absence of oligomycin, the electrostatic attraction between the two headpiece dipoles presumably serves as a rotational barrier. The rotational barrier due to the structural dipoles but not the barrier imposed by oligomycin can be removed by the depolarizer, dinitrophenol.

The headpiece when detached from the inner membrane (F_1) acquires oligomycin-insensitive ATPase activity.[66] This observation raises two important questions: (1) why does isolation of the headpiece induce ATPase activity? and (2) why is the ATPase activity of F_1 insensitive to oligomycin? A simple explanation that can provide an answer to both of these questions is suggested by our model of the headpiece. It appears that in tightly coupled mitochondria there exists a hydrophobic region in the headpiece across which the two structural dipoles interact electrostatically and that this region is protected from the aqueous milieu outside the headpiece by the presence of the stalk sector. Therefore, as long as the stalk sector is attached to the headpiece, the hydrophobic region which is required for binding of oligomycin is available. Bound oligomycin then can prevent the quarternary structural changes (the rotation of the hemispheres) in the headpiece, and thereby inhibit the ATPase activity. On the other hand, if the stalk sector is removed from the headpiece, the oligomycin-binding site (or cavity) may be exposed to a more polar environment; oligomycin can no longer bind and thereby inhibit ATPase activity. There is a second sequella of removing the stalk sector from the headpiece. The removal of the stalk sector may enable a medium of high dielectric constant to intervene between the two structural dipoles in the perturbed state of the headpiece and thereby eliminate the electrostatic barrier to the rotation of the hemispheres. Thus the ATPase activity of F_1 is enhanced.

There have been several schemes proposed based on the chemical intermediate hypothesis[67, 68] or the chemiosmotic model[69] which attempt to explain the large body of experimental observations relating to uncouplers and inhibitors of oxidative phosphorylation. Unfortunately, all of the classical schemes that are available now are based on two or more "high-energy intermediates" such as $X \sim I$ or $X \sim P$ whose existence in mitochondrial systems has never been established experimentally.[70] Unlike these models, the scheme we have presented is based on experimentally established structural facts and biochemical states of the structures involved. The only postulate which needs experimental proof is the notion of the structural dipoles within the headpiece. Given the validity of this postulate, our scheme appears to be capable of accommodating the major portion of the data on uncouplers and inhibitors without invoking any unproven intermediates.

B. *Active Transport*

Most of the published data on mitochondrial active transport can be readily explained if we assume the existence of an enzyme system in the inner membrane which can catalyze a transmembrane proton movement. We will hereafter refer to this postulated enzyme system as "transprotonase." The idea that the transmembrane proton movement is fundamental in mitochondrial active transport was first elucidated by P. Mitchell in his chemiosmotic hypothesis[71] and was subsequently incorporated into the conformational model of active transport by Young, Blondin and Green.[72] It will be shown in the following section that although the electromechanochemical (EMC) model of active transport is formally indistinguishable from the chemiosmotic and the conformational models, the EMC model incorporates the basic structural features of biological membranes and in addition provides what we believe to be a more viable mechanistic linkage between the electron transfer chain (or the ATPase) and the active transport apparatus.

The principle of electromechanochemical energy transduction as formulated in the present paper and the molecular principle of biological membrane construction proposed by Vanderkooi and Green[7,18,51] provide a basis for deducing some of the essential structural and functional features of the transprotonase.

The transprotonase is most likely not a single protein but rather a set of intrinsic proteins.[49] This deduction is reasonable if one takes into account the distance (~ 60Å) across which the enzyme system transports protons. The electron transfer complexes which transport electrons over a similar distance are known to be composed of multiple proteins.[22] Since the transprotonase serves as the depolarizing structural component of the supermolecule for active transport, the enzyme system must be structured so as to undergo a precisely coordinated polarization–depolarization cycle coupled to the electron transfer reaction or to ATP hydrolysis. When the transprotonase is uncoupled from these polarizing reactions, however, the enzyme system must serve simply as a device for passively equilibrating protons across the inner membrane, since it is known that the inner membrane is permeable to protons to a limited extent under nonenergizing conditions.[73] Thus, when the transprotonase is coupled to a polarizing structural component it catalyzes a unidirectional flow of protons across the inner membrane (from the matrix to the intracristal side), and when the transprotonase is not coupled to any polarizing structural component, it facilitates the transmembrane equilibration of protons. We will speak of the coupled mode of transprotonase reaction as the "energized mode" and the uncoupled mode as the "nonenergized mode." Finally, we deduced from considerations of symmetry that one transprotonase complex transports two protons per turnover.

In Fig. 15a, a schematic diagram is shown in which the salient features of the transprotonase as described above are depicted. The structural details are perforce of secondary significance. There may be numerous ways of visualizing the same set of basic ingredients. We will represent the transprotonase as an octet of intrinsic proteins arranged in a double tier pattern in the phospholipid bilayer. In the interest of clarity the diagram in Fig. 15a shows only one-half of the transprotonase; the electron transfer complexes associated with transprotonase are omitted from the diagram. Each globular protein in the transprotonase complex is assumed to contribute one polar end of a dipole. In the ground state (i.e., the non-energized state), the permanent dipoles are arranged in a perfectly compensating configuration so that there is no net dipole moment associated with the octet. Because of the symmetric structural features, the transprotonase is capable of translocating protons in either direction across the inner membrane depending on the direction of the pH gradient. However upon charge separation in the electron transfer complexes (see the electrical signs in Fig. 15b), the resultant electric field imparts directionality to the proton flow catalyzed by the transprotonase. This

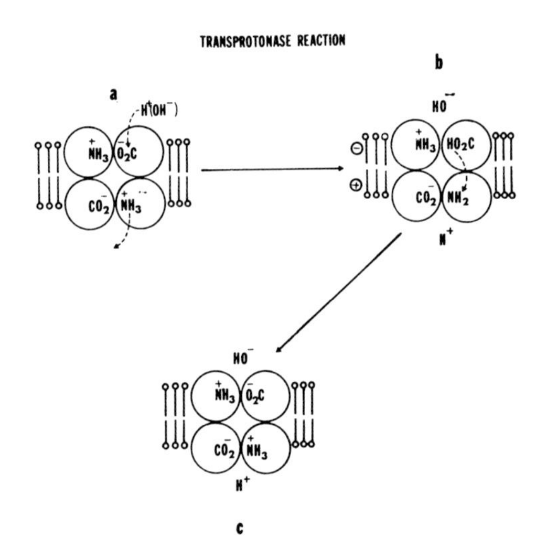

Figure 15. Mechanism of mitochondrial active transport.

directionality is imposed on the transprotonase probably through the field-induced asymmetric ionization of the polar groups within the octet. More specifically, the charge separation in the electron transfer complexes induces a complementary dipole moment in each of the two quartets of the transprotonase by protonating a basic group on the matrix side of the quartet and by deprotonating an acidic group on the intracristal side of the quartet as shown in Fig. 15b. For convenience we will refer to this phenomenon as "the field-induced protonation–deprotonation reaction." As the result of this field-induced protonation–deprotonation reaction, there is generated a pH differential (ΔpH) between the two compartments separated by the inner membrane (alkaline in the matrix space and acidic in the intracristal space). Most of the free energy derived from substrate oxidation in the case of electron transfer-driven active transport is still stored within the polarized supermolecule (the electron transfer complex + the transprotonase) in the form of electromechanochemical free energy. It is this energy which is available for driving the electrogenic proton movement across the inner membrane unidirectionally. The precise molecular mechanism involved in the transmembrane proton migration is unknown at the present time, but it is almost certain that the proton movement is mediated by a series of bond-forming and bond-breaking chemical reactions (i.e., a series of transprotonation reactions) involving a set of amino acid residues which form what may be called the "transmembrane proton relay system." We may represent one of such elementary transprotonation reactions as follows:

$$H-\overset{\oplus}{X}_i + X_{i+1} \rightarrow X_i + H-\overset{\oplus}{X}_{i+1} \qquad (9)$$

where X_i and X_{i+1} represent two adjacent proton-stabilizing amino acid residues of the transmembrane proton relay system. The subscript i varies from 1 to $(n-1)$ where n is the number of the proton-stabilizing amino acid residues constituting the relay system. If we assume that as the result of Reaction (9), a proton is translocated across the inner membrane say by 10 Å, there would be at most six proton-stabilizing residues per relay system, or $n = 6$.

It is assumed that regardless of whether the transprotonase is operating in the energized mode or in the nonenergized mode each elementary transprotonation reaction is accompanied by a compensating movement of membrane-permeable cations and/or anions across the inner membrane in appropriate directions so as to prevent any significant development of separated charges. In the absence of such compensating ion movement the transmembrane proton movement is probably suppressed.

When the transprotonase operates in the nonenergized mode, the net free energy change accompanying the transmembrane proton movement may be close to zero given the compensating charge migration so as to prevent charge separation (i.e., $\sum_i \Delta Y_i = O$). However, when the transprotonase operates in the energized mode, the unidirectional proton flow becomes an endergonic process even if accompanied by compensating charge movement and the corresponding free energy change will be positive (i.e., $\sum_i \Delta Y_i > O$). Therefore, under energizing conditions, Reaction (9) is to active transport what the reaction between ADP and P_i is to oxidative phosphorylation. Just as oxidative phosphorylation involves the conversion of electromechanochemical free energy into the chemical free energy of the P—O bond in ATP, so the transprotonation reaction involves the transduction of electromechanochemical free energy into the electrochemical free energy of the $H—\overset{\oplus}{X}_{i+1}$ bond according to Equation (9).

In this way it can be shown that both oxidative phosphorylation and energized transprotonation reaction are expressions of a common principle, namely the principle of electromechanochemical energy transduction.

Having described the fundamental features of the EMC model of active transport, we are now in a position to discuss specific examples. The mechanism of active transport of potassium acetate is shown in Fig. 16. The supermolecule involved in this particular work performance is composed of the electron transfer complex and transprotonase, both located in the inner membrane. The transprotonase is assumed to possess two acidic groups on the intracristal side and two basic groups on the matrix side. As electrons and protons are

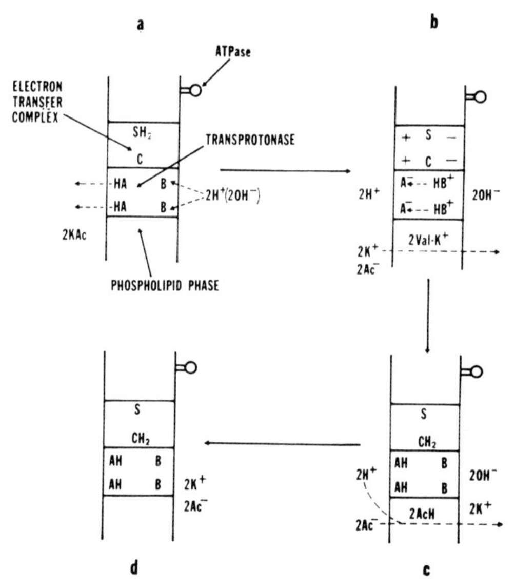

Figure 16. Depolarization of the metastable state induced by ion binding.

separated in the electron transfer complex, the ensuing electric field induces the non-electrogenic protonation and deprotonation reactions as shown in Fig. 16b leading to the generation of ΔpH. The transmembrane migration of protons can occur only when there is concomitant depolarization of the supermolecule. Valinomycin permits the rapid equilibration of K^+ in response to the proton movement. At the completion of the movement of protons across the inner membrane, two protons have left the matrix space in exchange for two potassium ions. The resulting ΔpH then drives the inward flow of acetate ion. The net result is that two molecules of K^+Ac^- are accumulated in each cycle of oxidoreduction involving the generalized substrate SH_2 and the generalized mobile electron carrier C. Our formulation predicts a $K^+/2e^-$ ratio of two per coupling site. The experimentally measured $K^+/2e^-$ ratio appears to vary widely.[74] This variability of $K^+/2e^-$ ratio may be due in part to the non-ideal behaviour of the K^+-specific electrode and to the possible uncertainty about the rate of oxygen uptake owing to the slowness of response of the oxygen electrode. We predict that when due corrections are applied, the observed $K^+/2e^-$ ratio should approach the theoretical value of two.

In the absence of weak acid anions, the H^+/K^+ ratio should be about unity according to our formulation. The experimental data agree closely with this theoretical ratio (75).

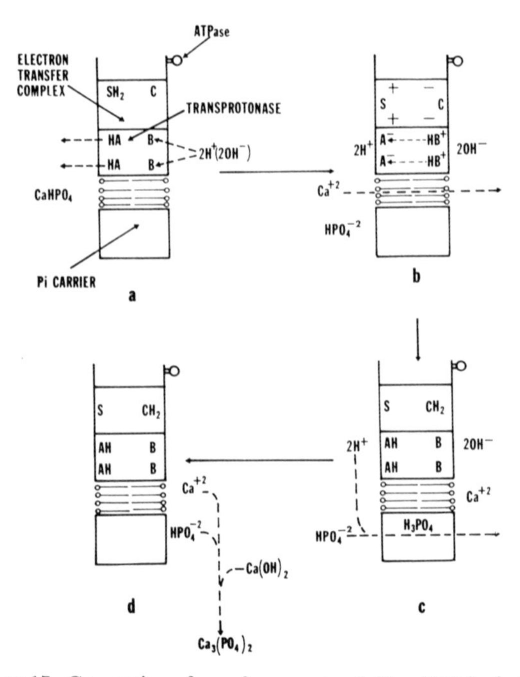

Figure 17. Generation of membrane potential by ATP hydrolysis.

The active transport of tricalcium phosphate $[Ca_3(PO_4)_2]$ can be accounted for in a similar manner as presented in Fig. 17. In this case, the calcium ion permeation is facilitated probably by an endogenous calcium ionophore while the permeation of inorganic phosphate is effectuated by the well-established phosphate carrier.[76,77] As developed in Fig. 17, charge separation in the electron transfer complex induces a ΔpH and polarizes the supermolecule. Concomitant with the depolarization of the supermolecule one calcium ion is translocated into the matrix space. If the influx of P_i driven by the pH gradient is slower than the rate of calcium ion uptake, the matrix space will be maintained alkaline throughout the duration of active transport. When a sufficient concentration of Ca^{++} and P_i accumulate in the alkaline matrix space, a spontaneous reaction between $CaHPO_4$ and $Ca(OH)_2$ will take place to give the $Ca_3(PO_4)_2$ precipitate and water (see Fig. 17d). Again the $Ca^+/2e^-$ ratio per coupling site predicted by the present formulation is unity in contrast to some of the higher values reported in the literature.[78] It is quite probable that $Ca^+/2e^-$ ratios higher than one are due to the non-energized binding of Ca^{++} to mitochondrial membrane surfaces. In the absence of permeant anions, the present scheme predicts a value of 2 for the H^+/Ca^{++} ratio.

Montal, Chance and Lee (79) reported the interesting observation that submitochondrial particles can take up both H^+ and K^+ to the extent of about 10 nmoles per mg protein in an energy-linked fashion in the absence of any added ionophore. The direction in which protons flow is normal in the sense that it is the direction predicted by most of the available models of active transport (the chemiosmotic,[71] the conformational[72] and the EMC models). However, the potassium movement is abnormal because it is in the direction opposite to the direction determined by the H^+/K^+ exchange diffusion process. The EMC model suggests a simple explanation for this apparent dilemma. As described above, charge separation in the electron transfer complexes generates a ΔpH via the field-induced protonation–deprotonation reaction, leading to an acidification of the intravesicular aqueous phase, and an alkalinization of the extravesicular phase. A part of the intravesicular free protons can then exchange with external K^+ ions—a process mediated by transprotonase in the non-energized mode and the endogenous potassium ionophore. Due to the non-identity of the two phases involved, the number of protons released into the intravesicular phase would probably be less than the number of protons taken up by the external surface of the vesicle. If this were the case, the complete exchange of the internal protons for K^+ would lead to the apparently simultaneous uptake of H^+ and K^+. On the other hand, if the number of protons released and taken up are identical, incomplete replacement of internal protons with external potassium ions would account for the same result. The essence of the present explanation is that the abnormal direction of the K^+ ion movement is due to the fact that the K^+ movement is driven not by the electromechanochemical free energy of the polarized supermolecule but by the ΔpH alone.

By now there is ample evidence that various substrate anions permeate the inner membrane through specific carrier systems analogous to the phosphate carrier.[76,77] We have incorporated these carrier systems into our EMC model. There appears to be at least two general carrier systems in the inner membrane—one for dicarboxylate anions and another for tricarboxylate anions as depicted in Fig. 18. It is to be noted that these carboxylate anions are assumed to pass through the respective carrier systems as un-ionized neutral species in either direction. That is, these carrier systems may be regarded as passive equilibrating systems for appropriate anions. There is no evidence that these anion carriers can be directly coupled to the primary energy source. That is to say, these carriers cannot serve as depolarizing structural components. Because of the non-electrogenic nature of the

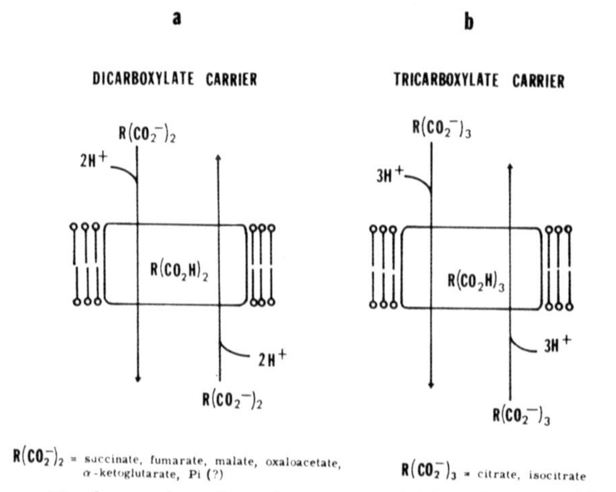

Figure 18. Generation of membrane potential by substrate oxidation.

translocation of these anions, the direction of the movement of these ions will be determined only by the electrochemical activities of H^+ and carboxylate anions. Consequently the transmembrane equilibration of substrate anions would be highly sensitive to the ΔpH across the inner membrane, and, conversely, the ΔpH would be influenced by the concentration gradient of these ions. As the result of the pH dependency of the transmembrane anion equilibration, various anion movements can be chemically linked.

A change in the electrochemical activity of any one of the numerous chemical species (i.e., H^+ and anions), therefore, would affect the equilibrating tendencies of all the other species. However, due to the kinetic parameters involved, it is conceivable that only a selected few of the ionic species would respond to a given electrochemical perturbation. This may underlie the phenomena of the numerous anion–anion antiports often reported in the literature.[80]

C. *Energized Transhydrogenation*

Mitochondrial transhydrogenase is an inner membrane system which during isolation is found to be associated with Complex I

(DPNH—Coenzyme Q reductase).[81] It may be considered as a basepiece of the inner membrane. Transhydrogenase catalyzes the transfer of the equivalent of a hydride ion from DPNH to TPN^+. Under energizing conditions, the equilibrium of the transhydrogenation reaction (9) shifts far to the right, the equilibrium constant for the reaction being approximately 500[82]:

$$DPNH + TPN^+ \rightleftharpoons DPN^+ + TPNH \qquad (9)$$

This phenomenon of energized transhydrogenation has been difficult to rationalize, since the redox couples DPN^+/DPNH and TPN^+/TPNH are known to be isoenergetic.[83] The energy transduction inherent in energized transhydrogenation represents yet another variation of mitochondrial coupling which any satisfactory general model of mitochondrial structure and function must rationalize. As we have indicated earlier, the electromechanochemical principle which satisfactorily rationalized the phenomena of oxidative phosphorylation and active transport also provides a useful mechanistic framework for accommodating the structural and biochemical data available on energized transhydrogenation.[84]

Let us first consider the components of the supermolecule involved in this work performance. The minimum requirement for the supermolecule implicated in transhydrogenation is met by a combination of transhydrogenase and one of the complexes of the electron transfer chain in the case of oxidoreductive transhydrogenation. This conclusion is based on the fact that decapitated submitochondrial particles are sufficient to catalyze energized transhydrogenation.[85] In ATP-dependent transhydrogenation, the supermolecule consists of the headpiece, the electron transfer chain and the transhydrogenase and we are assuming that these three entities are arranged as shown schematically in Fig. 19.

The transhydrogenase portion of the supermolecule is assumed to be constructed according to the Vanderkooi–Green membrane model (see Fig. 20[7,18,51]). The following postulates are required to account for the energized[86] as well as the nonenergized[17] transhydrogenation reaction within the framework of the electromechanochemical energy transduction principle: (1) there are two active sites per molecule of the transhydrogenase, one specific for the DPN^+/DPNH couple and the other specific for the TPN^+/TPNH couple; (2) the flavin coenzyme for the DPN^+/DPNH couple (F_D) is located in the intracristal tier of the DPN^+/DPNH couple (F_D) is located in the intracristal tier of the transprotonase and the flavin coenzyme for the TPN^+/TPNH couple (F_T) is located in the matrix tier of the transprotonase; (3) there exists a pair of globular proteins which are tightly associated with the F_D and F_T proteins and provide a proton transfer route accompanying the hydride transfer between the coenzymes F_D and F_T (see the right-hand tier of the quartet in Fig. 20); (4) the chemical reactions between the pyridine nucleotides and their respective flavin moieties are rapid and hence the rate-limiting step may involve either one or both of the two intramembrane charge migration processes—the hydride and the proton transfer reactions.

A possible mechanism for the energized transhydrogenation is presented in Fig. 20. In consequence of charge separation in the electron transfer complexes (see the encircled electrical signs in Fig. 20b), the transprotonase undergoes an electrical polarization via the hydride shift from DPNH to F_D and the binding of TPN^+ to F_T on the one hand and the intramembrane proton transfer from CO_2H to NH_2 on the other as indicated by the dotted arrows in Fig. 20a. The polarized supermolecule at this point can drive the energy-requiring hydride

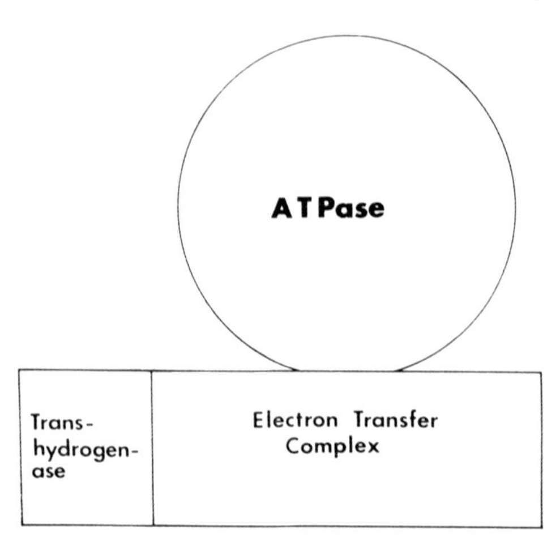

Figure 19. The supermolecule for energized transhydrogenase.

transfer from F_D to F_T. At the completion of this hydride transfer, the supermolecule becomes depolarized and TPNH and H^+ dissociate from the transhydrogenase leading to the state depicted in Fig. 20c. The net result is the conversion of the system DPNH + TPN^+ into the system DPN^+ + TPNH and the creation of ΔpH, acidic on the matrix side and alkaline on the intracristal side. The ATP hydrolysis reaction can achieve the same result because of the fact that the ATPase can induce the polarization of the electron transfer complexes. The predicted pH changes have been observed experimentally by Mitchell and Moyle.[87]

Under non-energizing conditions, the transhydrogenase acts independently of the electron transfer complexes and equilibrates H^- and H^+ across the enzyme system in response to the thermodynamic driving force imposed by Reaction (9). If the rate of the proton migration through the transprotonase is slower than the rate of the hydride migration, the non-energized transhydrogenation will lead to the intramembrane separation of H^+ and H^- and to a membrane potential —minus on the matrix side and plus on the intracristal side when the concentration of ($DPNH + TPN^+$) is greater than the concentration of ($DPN^+ + TPNH$), and the opposite polarity (i.e., plus on the matrix side and minus on the intracristal side) when the relative magnitude of the concentrations are reversed.

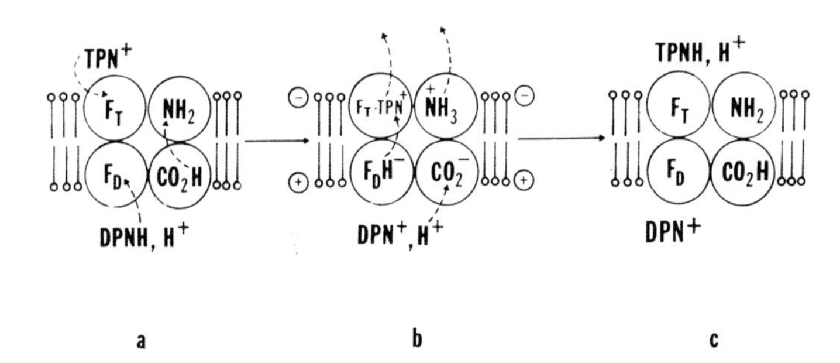

Figure 20. Postulated macromolecular structure of mitochondrial transhydrogenase.

Other Extensions of the Model

We have deliberately excluded from the present communication consideration of some mitochondrial phenomena not because the model is incapable of rationalizing these phenomena, but rather because there are limits to what can be developed satisfactorily in a single communication. Elsewhere we intend to consider pseudo-energized swelling and energized contraction. There are also a variety of observations which need explaining and which were not considered in the present paper. How the solubilization of succinic and DPNH dehydrogenases can be fitted into the framework of the octet principle; how the enzymes which carry out the citric and fatty acid cycles are integrated within the matrix system; how charged molecules like P_i, ADP and ATP penetrate the mitochondrial membranes; and finally how our postulated localization of the electron transfer chain on the matrix side of the inner membrane can be rationalized with the inaccessibility of the membrane on the matrix side to cytochrome *c*. None of these phenomena have posed insuperable problems for the model as we shall develop elsewhere.

Acknowledgements

In the development of the electromechanochemical model of energy transduction, we have drawn heavily from the insights of others. To Peter Mitchell we owe our first appreciation of the fundamental concept that the membrane is the instrument of transduction—a concept which is the foundation stone of our own approach. To V. P. Skulachev we are indebted for his incisive proof that an electric potential is generated when mitochondria are energized, and to H. T. Witt we are equally indebted for his important demonstration that the energized state of chloroplasts is characterized by an electric potential. To Lars Ernster we are indebted for his clarification of the localization of the citric cycle enzymes in the matrix space and for his discovery of energized transhydrogenation as a third coupling capability of energized mitochondria. A. L. Lehninger has for many years championed and developed the mechanochemical theme of energy transduction and his views have greatly influenced our own thinking about transduction mechanisms. The pioneer and systematic studies of E. Racker and his group on the identification of the headpiece as the site of ATPase activity have opened for us the door to the exploration of the ATPase coupling unit. The studies of W. Stein on the structure of glucose permease and of R. E. Dickerson on the x-ray crystallography of cytochrome *c* provided our introduction to the internal structure of multimeric and monomeric proteins. Several of our present and former colleagues have made independent and crucial contributions to the thinking which underlies our present model. E. F. Korman and J. McLick introduced us to the possibility of the synthesis of ATP by the direct union of ADP and P_i given the necessary conformational changes in the ATPase coupling unit. John Young was our guide to the notion of the metastable energized state and to the central role of the electric potential in active transport. Among the experimentalists in our laboratory, the work of George Blondin on mitochondrial ionophores and on the role of ion binding in active transport, the studies of Rod Capaldi and Jennie Smoly on the chemical capabilities of the matrix system, the resolution by Alan Senior of the headpiece into its component proteins, the pioneering contributions of A. Tzagoloff, D. MacLennan and K. Kopaczyk to the studies of the oligomycin-sensitive ATPase, the discovery by G. Vanderkooi of the principle of membrane construction, the superb electron microscopic work of J. Asai and T. Wakabayashi, the configurational studies of R. Harris, D. Allmann, O. Hatase and J. Penniston, the brilliant resolution of the electron transfer chain into the four complexes by Y. Hatefi, and the systematic study of the active transport of divalent metal ions and inorganic phosphate by G. Brierley, all have provided much of the experimental foundations for the structural framework of the electromechanochemical model. We acknowledge with gratitude the advice given to us over the many years by Harold Baum. It was primarily the work of his group which changed the course of our thinking about the mechanism of energy transduction.

Finally we are deeply indebted to Dr. H. Yu, Department of Chemistry and Dr. L. Shohet, Department of Electrical Engineering, both of the University of Wisconsin in Madison, for their advice in matters relating to the effect of the electric field on macromolecular systems.

The studies in our laboratory were supported in part by program project grant GM-12847 of the National Institute of General Medical Sciences of the National Institutes of Health.

References

1. J. T. Penniston, R. A. Harris, J. Asai and D. E. Green. *Proc. Nat. Acad. Sci. U.S.A.*, **59** (1968) 624.

2. R. A. Harris, J. T. Penniston, J. Asai and D. E. Green, *Proc. Nat. Acad. Sci. U.S.A.*, **59** (1968) 830.
3. D. E. Green, J. Asai, R. A. Harris and J. T. Penniston, *Arch. Biochem. Biophys.*, **125** (1968) 684.
4. D. E. Green and H. Baum, *Energy and the Mitochondrion*, Academic Press, New York, 1970, p. 77.
5. D. E. Green and D. H. MacLennan, *Metabolic Pathways*, 3rd Ed., Vol. 1, Academic Press, Inc., New York, 1967, p. 47.
6. J. D. Watson and F. H. C. Crick, *Nature*, **171** (1953) 737, 694.
7. G. Vanderkooi and D. E. Green, *Proc. Nat. Acad. Sci. U.S.A.*, **66** (1970) 615.
8. G. Vanderkooi and M. Sundaralingam, *Proc. Nat. Acad. Sci. U.S.A.*, **67** (1970) 233.
9. H. E. Huxley, *Science*, **164** (1969) 1356.
10. J. T. G. Overbeek and J. Lijklema, in: *Electrophoresis, Theory, Methods and Applications*, M. Bier (ed.), Academic Press, Inc., New York, 1959, Chap. 1.
11. H. A. Laitinen, *Chemical Analysis: An Advanced Text and Reference*, McGraw-Hill Book Company, Inc., New York, 1960, p. 66.
12. H. Fernández-Morán, T. Oda, P. V. Blair and D. E. Green, *J. Cell Biol.*, **22** (1964) 63.
13. K. Kopaczyk, J. Asai, D. W. Allmann, T. Oda and D. E. Green, *Arch. Biochem. Biophys.*, **123** (1968) 602.
14. H. S. Penefsky, M. E. Pullman, A. Datta and E. Racker, *J. Biol. Chem.*, **233** (1960) 3330.
15. J. O'M. Bockris and S. Srinivasan, *Fuel Cells: Their Electrochemistry*, McGraw-Hill Book Company, New York, 1969, p. 607.
16. A. B. Hart and G. J. Womack, *Fuel Cells: Theory and Applications*, Chapman and Hall, London, 1967.
17. V. P. Skulachev, *FEBS Letters*, **11** (1970) 301.
18. G. Vanderkooi and D. E. Green, *Bio Science*, **21** (1971) 409.
19. E. F. Korman and H. Vande Zande, *Fed. Abstracts*, **27** (1968) 526.
20. R. E. Dickerson, T. Takano, D. Eisenberg, O. B. Kallai, L. Samson, A. Cooper and E. Margoliash, *J. Biol. Chem.*, **246** (1971) 1511.
21. J. E. Falk, *Porphyrins and Metalloporphyrins*, B.B.A. Library, Vol. 2, Elsevier Publishing Company, Amsterdam, 1964, p. 53.
22. D. E. Green, D. C. Wharton, A. Tzagoloff, J. S. Rieske and G. P. Brierley, in: *Oxidases and Related Redox Systems*, T. E. King, H. S. Mason and M. Morrison (eds.), Vol. 2, John Wiley and Sons, Inc., New York, 1965, p. 1032.
23. A. Tzagoloff and P. Meagher, *J. Biol. Chem.*, **246** (1971) 7328.
24. O. Hatase, T. Wakabayashi and D. E. Green, *Configurational changes in submitochondrial particles* (manuscript in preparation).
25. A. Tzagoloff, D. H. MacLennan and K. H. Byington, *Biochemistry*, **7** (1968) 1596.
26. D. E. Green and T. Wakabayashi (unpublished observation).
27. D. W. Allmann, J. Munroe, T. Wakabayashi, R. A. Harris and D. E. Green, *Bioenergetics*, **1** (1970) 87.
28. P. Mitchell and J. Moyle, in: *Biochemistry of Mitochondria*, E. C. Slater, Z. Kaninga and L. Wojtczak (eds.), Academic Press, Inc., London, 1967, p. 53.
29. J. H. Young, G. A. Blondin and D. E. Green, *Proc. Nat. Acad. Sci. U.S.A.*, **68** (1971) 1364.
30. T. Takano, R. Swanson, O. B. Kallai and R. E. Dickerson, *Conformational Changes Upon Reduction of Cytochrome c*, Cold Spring Harbor Symposium on Quantitative Biology, June 1971.
31. D. E. Green and G. Vanderkooi, in: *Physical Principles of Biological Membranes*, G. Iverson and J. Lam (eds.), Gordon and Breach Science Publishers, Inc., New York, 1970, p. 287.
32. C. T. O'Konski and N. C. Stellwagen, *Biophysical J.*, **5** (1965) 607.
33. W. Junge, H. M. Emrich and H. T. Witt, in: *Physical Principles of Biological Membranes, Proceedings of the Coral Gables Conference*, Snell, *et al.* (eds.), Gordon and Breach Science Publishers, New York, 1970, p. 383.
34. For a discussion of the ATP synthesis mechanism via a direct union of ADP and P_i, see Korman and McLick in this issue.
35. D. Agin, *Proc. Nat. Acad. Sci. U.S.A.*, **57** (1967) 1232.
36. J. A. Donlon and A. Rothstein, *J. Membrane Biol.*, **1** (1969) 37.
37. L. Bass and W. J. Moore, *Nature*, **214** (1967) 393.
38. D. A. T. Dick, *Cell Water*, Butterworths, Inc., Washington, 1965, p. 5.
39. T. L. Hill, *Proc. Nat. Acad. Sci. U.S.A.*, **58** (1967) 111.
40. C. R. Hackenbrock, *Proc. Nat. Acad. Sci. U.S.A.*, **61** (1968) 598.
41. E. F. Korman, R. A. Harris, C. H. Williams, T. Wakabayashi, D. E. Green and E. Valdivia, *Bioenergetics*, **1** (1970) 387.
42. J. Asai, G. A. Blondin, W. J. Vail and D. E. Green, *Arch. Biochem. Biophys.*, **132** (1969) 524.

43. M. J. Lee, R. A. Harris, T. Wakabayashi and D. E. Green, *Bioenergetics*, **2** (1971) 13.
44. T. Wakabayashi, J. M. Smoly and D. E. Green, *Bioenergetics* (1972) in press.
45. G. R. Hunter and G. P. Brierley, *J. Cell Biol.*, **50** (1971) 250.
46. N. E. Weber and P. V. Blair, *Biochem. Biophys. Res. Commun.*, **36** (1969) 987.
47. N. E. Weber and P. V. Blair, *Biochem. Biophys. Res. Commun.*, **36** (1970) 821.
48. D. E. Green, *N.Y. Acad. Sci.*, Conference on Membrane structure and its Biological Applications, June 2–4, 1971 (in press).
49. G. Vanderkooi and R. A. Capaldi, *N.Y. Acad. Sci.*, *Conference on Membrane Structure and its Biological Applications*, June 2–4, 1971, in press.
50. Y. Hatefi, *Adv. Enzymol.*, F. F. Nord (ed.), **25** (1963) 275.
51. D. E. Green and R. F. Brucker, *Bio Science* (1972), **22** (1972) 13.
52. H. Muirhead and M. F. Perutz, *Cold Spring Harbor Symp. Quant. Biol.*, **28** (1963) 451.
53. J. C. Brooks, (unpublished observation).
54. E. F. Korman and J. McLick, *Proc. Nat. Acad. Sci. U.S.A.*, **67** (1970) 1130.
55. J. T. Penniston (unpublished observation).
56. P. B. Garland, R. A. Clegg, P. A. Light and C. I. Raglan, in: *Inhibitors: Tools in Cell Research*, T. Bücher and H. Sies (eds.), Springer-Verlag, New York, 1969, p. 217.
57. J. A. Berden and E. C. Slater, *Biochim. Biophys. Acta*, **216** (1970) 237.
58. E. Racker, *Mechanisms in Bioenergetics*, Academic Press, New York, 1965, p. 145.
59. G. R. Drysdale and M. Cohen, *J. Biol. Chem.*, **233** (1958) 1574.
60. P. G. Heytler, *Biochemistry*, **2** (1963) 357.
61. M. E. Pullman, H. S. Penefsky, A. Datta and E. Racker, *J. Biol. Chem.*, **235** (1960) 3322.
62. E. Racker, *J. Biol. Chem.*, **235** (1960) 148.
63. D. H. MacLennan and A. Tzagoloff, *J. Biol. Chem.*, **241** (1966) 1933.
64. H. A. Lardy and G. Feldott, *Ann. N.Y. Acad. Sci.*, **54** (1951) 636.
65. P. D. Boyer, in: *Biological Oxidations*, T. P. Singer (ed.), Interscience Publishers, New York, 1968, p. 210.
66. A. E. Senior and J. C. Brooks, *Arch. Biochem. Biophys.*, **141** (1970) 257.
67. L. Ernster, C. P. Lee and S. Janda, in: *Biochemistry of Mitochondria*, E. C. Slater, Z. Kaninga and L. Wojtczak (eds.), Academic Press, London, 1967, p. 29.
68. H. A. Lardy, J. L. Connelly and D. Johnson, *Biochemistry*, **3** (1964) 1961.
69. G. D. Greville, in: *Current Topics in Bioenergetics*, Vol. 3, Academic Press, New York, 1969, p. 1.
70. D. H. Jones and P. D. Boyer, *J. Biol. Chem.*, **244** (1969) 5767.
71. P. D. Mitchell, *Chemiosmotic Coupling and Energy Transduction*, Glynn Research Ltd., Bodmin, 1968.
72. J. H. Young, G. A. Blondin and D. E. Green, *Proc. Nat. Acad. Sci. U.S.A.*, **68** (1971) 1364.
73. L. Packer and K. Utsumi, *Arch. Biochem. Biophys.*, **131** (1969) 386.
74. R. S. Cockrell, E. J. Harris and B. C. Pressman, *Biochemistry*, **5** (1966) 2326.
75. C. Moore and B. C. Pressman, *Biochem. Biophys. Res. Commun.*, **15** (1964) 562.
76. A. Fonyo, *Biochem. Biophys. Res. Commun.*, **32** (1968) 624.
77. D. D. Tyler, *Biochem. J.*, **111**, 665 (1969).
78. H. Rasmussen, B. Chance and E. Ogata, *Proc. Nat. Acad. Sci. U.S.A.*, **53** (1965) 1069.
79. M. Montal, B. Chance and C. P. Lee, *J. Mem. Biol.*, **2** (1970) 201.
80. A. L. Lehninger, in *Biomembranes*, Vol. 2 (Ed. L. A. Manson), Plenum Publishing Corp., New York, 1971), p. 147.
81. D. H. MacLennan (unpublished observation).
82. C. P. Lee and L. Ernster, in: *Regulation of Metabolic Processes in Mitochondria*, Tager, *et al.* (eds.), Elsevier Publishing Company, Amsterdam, 1966, p. 218.
83. H. R. Mahler and E. H. Cordes, *Biological Chemistry*, Harper and Row, Publishers, New York, 1966, p. 619.
84. K. Van Dam and H. F. Ter Welle, ref. 83, p. 235.
85. D. H. MacLennan, J. M. Smoly and A. Tzagoloff, *J. Biol. Chem.*, **243** (1968) 1589.
86. C. P. Lee and L. Ernster, *Biochem. Biophys. Acta.*, **81**, (1964) 187.
87. P. Mitchell and J. Moyle, *Nature*, **208** (1965) 1205.

Part B

Membrane Structure

A selection of these articles appeared in *Journal of Bioenergetics* Volume 4

The Relationship of the ($Na^+ + K^+$)-Activated Enzyme System to Transport of Sodium and Potassium Across the Cell Membrane

J. C. Skou

Institute of Physiology, University of Aarhus, 8000 Aarhus C, Denmark

I. *Introduction*

The energy-requiring, active, transport of sodium out and potassium into the cell seems to be due to the membrane bound (Na + K)-activated enzyme system.[1–6] This system seems also to be involved in the Na:Na and K:K exchange which under certain conditions takes place across the cell membrane.[4,7] It has been named an enzyme system [1] since it not only catalyzes the hydrolysis of ATP but takes part in the reaction in the sense that the hydrolysis of ATP via this system seems to be translated into the vectorial movement of the cations against the electrochemical gradients, i.e. it seems to be identical with the transport system.

A detailed knowledge of the system and the way it functions may lead to an understanding as to how sodium and potassium are transported across the cell membrane. This would include answers to at least three questions:

(1) What is the molecular structure of the system.
(2) How is the relationship between the effect of sodium, potassium, magnesium, and ATP on the system and what is the sequence of steps in the reaction which leads to the hydrolysis of ATP.
(3) What happens on the molecular level when the system reacts with sodium, potassium, magnesium, and ATP.

The questions cannot be answered at present. There is, however, a number of observations on the system which makes it possible to discuss the sequence of some of the steps in the reaction and to discuss some of the problems to solve to get more insight in the transport mechanism.

II. *Main Characteristics of the System*

The substrate for the system is ATP, which is hydrolyzed to ADP and Pi.[1] The system requires a combined effect of sodium and potassium for activation,[8] and in the intact cell the effect of sodium is from the inside and of potassium from the outside.[9–12] The number of cations necessary for activation is unknown, but if the activator sites are identical with the carrier sites for the transport of the cations, there must be at least three sodium ions and two potassium ions necessary.[4] The system has thus two sets of sites, one set located on the inside of the membrane (the i-sites), on which sodium is necessary for activation, and another set located on the outside (the o-sites), on which potassium is necessary for activation. There is a competition between sodium and potassium both for the i- and for the o-sites.[1]

With the isolated system in a test tube it is not possible to establish an asymmetric situation with a sodium medium in contact with the i-sites, and a potassium medium with the o-sites; both sets of sites are in

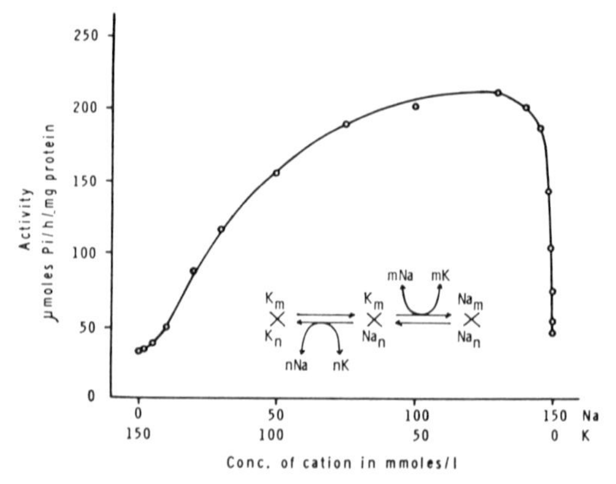

Figure 1. The effect of sodium plus potassium on the activity of the $(Na^+ + K^+)$-activated enzyme system, Mg^{2+} 6, ATP 3 mM, pH 7·4, 37°C. The enzyme was prepared from ox brain.[74]

contact with the same sodium-potassium medium. The ratio between the concentrations of sodium and potassium necessary to give half maximum saturation of the i-sites differs, however, so much from the ratio necessary to give half maximum saturation of the o-sites that it is possible with certain concentrations of sodium and potassium in the test tube to have a situation where a major part of the system is on the active K^o_m/Na^i_n form, Fig. 1. The ratios between sodium and potassium for half maximum activation read from the curve in Fig. 1 is about 1:3–4 for the ascending part of the curve (the sodium-activating sites, the i-sites), and about 100:1 for the descending part (the potassium-activating sites, the o-sites). From these values it can be calculated that the activity of the system with concentrations of sodium and potassium which give maximum activity, 130 and 20 mM, respectively, is about 80–85% of the activity which could be obtained if the i-sites were in contact with a sodium solution and the o-sites with a potassium solution. The asymmetry of the curve reflects how much the apparent sodium/potassium affinity ratio for the i-units differs from the apparent potassium/sodium affinity ratio for the o-units.

It is furthermore a characteristic property of the system that it is inhibited by cardiac glycosides.[1–6]

III. *Preparation and Purity of the Enzyme System*

The enzyme system is located in a cell membrane and preparations of the enzyme system consist of membrane pieces isolated by a differential centrifugation of a tissue homogenate. Membrane pieces which contain the system are found in all the sediments after a differential centrifugation. In the heavier sediments the (Na + K)-activated activity is manifest, while in the lighter sediments—heavy and light microsomes—part of the activity is latent. The explanation seems to be that the small membrane pieces form vesicles with no access for either MgNaATP, or potassium to the one side of the membrane.[13–15]

The latent activity can become manifest by treatment of the membrane vesicles with detergents like DOC,[16] high concentrations of salt like sodium iodide,[17] and by freezing.[14] Up to a certain concentration both salt and detergents will increase the activity while they again decrease the activity when used in higher concentrations. The effect of the detergents are highly dependent on temperature, time, concentration of protein in the solution and for the ionized detergents of pH[13]. This means that for optimum activation the treatment has to be done under control of all these parameters.

There are two effects of the activation. One is that the latent activity becomes manifest, apparently due to an opening of the vesicles;[13–15] this gives an increase in specific activity. The other is a release of inactive protein which goes into solution.[13,14] It may be protein which has been trapped inside the vesicles and is released when the vesicles are opened and/or protein released from the membrane *per se*. It amounts to 40–50% of the protein in the preparation prior to a treatment with the detergents. The membrane particles which contain the enzyme activity can be separated from the released dissolved protein by centrifugation, and this gives a further increase in the specific activity.

The membranes always contain a magnesium activated ATPase besides the (Na + K)-activated. The ratio between the two activities varies from tissue to tissue and also depends on the preparative procedure. The magnesium-activated ATPase is more labile than the (Na + K)-activated towards SH-blocking agents[18] and treatment with high concentrations of salt.[17] Most of this activity disappears in the procedures where high concentrations of salt are used to activate the system.[18] High concentrations of sodium iodide can also be used to decrease the magnesium activity in the preparations activated by detergents.[19]

Part of the magnesium-activated ATPase may stem from mitochondrial contamination of the membranes. It is unknown whether the other part of the magnesium-activated ATPase has any relation to the $(Na^+ + K^+)$-activated ATPase.

A further purification of the $(Na^+ + K^+)$-activated enzyme system can be obtained by density gradient centrifugation of the detergent activated membrane preparations,[20] by ammonium sulphate fractionation,[21] or gel filtration[22,23] of detergent solubilized preparations. The specific activity which can be obtained by the density gradient centrifugation depends on the tissue used as starting material. This may mean that membranes from different tissues has a different density of enzyme sites. The outer medulla from rabbit kidney is the tissue which has so far given preparations with the highest specific activity, namely about 1500 μM Pi/mg protein/hour.[20] In this preparation the enzyme system is still bound to membrane pieces, and is estimated to be maximally 49% pure.

The number of enzyme units/mg protein of enzyme preparations has been determined either by measuring the sodium-dependent incorporation of P^{32} from ATP^{32},[21,22,24,25] the binding of ATP^{32},[26–28] or the binding of labelled g-strophanthin.[25,28–31] In experiments with enzyme preparations from different tissues where the number of binding sites on the same enzyme preparation has been determined by P^{32} labelling and cardiac glycoside binding,[25] by P^{32} labelling and ATP binding,[27] and by ATP and cardiac glycoside binding,[28] the correlation, with one exception,[25] has been close to 1·0. Assuming one binding site per enzyme unit, the molecular activity varies from about 3000 to about 15,000 molecules Pi/enzyme unit/minute for preparations from different tissues. This may mean that the molecular activity of preparations from different tissues varies, or it may reflect that it is difficult to obtain reliable values for the site numbers; there are problems with background labelling both with labelling by P^{32}, binding of ATP^{32}, and of labelled cardiac glycosides. It must also be emphasized that the molecular activity is calculated from the maximum activity of the system which can be obtained in the test tube, and this is as discussed in the previous section not the maximum activity of the system.

Molecular weight determinations by radiation inactivation have given values of the order of 500,000[32] and 250,000[33].

On a polyacrylamide gel the partly purified preparations dissolved in sodium dodecyl sulfate-mercaptoethanol give two major bands, one with a molecular weight of 94,000[21]–90,000[22] and another with a molecular weight of about 53,000.[22] The sodium-dependent P^{32} labelling from ATP^{32} is found in the 90,000–94,000 molecular weight band,[21,34] which suggests that a polypeptide of this molecular weight is part of the system.

IV. *Lipids for Activation*

Extraction of loosely bound lipids, cholesterol and part of the phospholipids, has no effect on the activity of the $(Na^+ + K^+)$-activated enzyme system,[35] while extraction of the more firmly bound phospholipids by polar solvents inactivates the system.[35–36] The inactivation may be due to the removal of the lipids or to a denaturing effect of the solvents.

Treatment with crude preparations of phospholipase A and C[36–40] leads to an inactivation; so does treatment with a highly purified preparation of phospholipase A.[35] In the phospholipase-treated preparations which have been partly inactivated, the addition of lecithin[36] or asolectin,[40] a commercial soybean extract, gives a certain, but low reactivation, while no reactivation was seen by addition of lipids to completely inactivated preparations.

In enzyme preparations in which the latent enzyme activity has not been uncovered by treatment with detergents or high concentrations of salt, the addition of phosphotidylserine gives a certain but low increase in activity.[41,42] It seems most likely that this is due to a detergent-like effect of the phospholipids.[35]

DOC-solubilized inactivated enzyme preparations can to some extent be reactivated by addition of lipids.[23,42–47] Tanaka and Sakamuto[47] found that reactivation could be obtained by acidic phospholipids such as phosphatidic acid, phosphatidylinositol and phosphatidylserine, whereas neutral lipids as lecithin and phosphatidylethanolamine were inactive. Besides the acidic phospholipids, mono- and diacyl phosphates could activate, and they concluded that the essential structure needed for activation is a phosphate group plus one or two fatty acyl residues. Wheeler and Whittam[42] found that crude commercial samples of acidic phospholipids reactivated, but after partial purification by chromatography definite activation was only shown with the components which migrated like phosphatidylserine. It is not possible from the experiments on the reactivation of the DOC-solubilized inactivated preparations to exclude that the lipid reactivation is due to a removal of the detergents from the enzyme by the lipids and not to an effect of the lipids per se on the enzyme. On the other hand, the apparent requirement for specific lipids may suggest that it is due to an effect of the lipids.

An interesting observation concerning requirement for lipids for the transport system is that the amount of sulfatides in preparations from salt glands from duck increases parallel with the $(Na^+ + K^+)$-ATPase during a salt load,[48] while other lipids increased less.

V. *Intermediary Steps in the Reaction*

The system hydrolyzes ATP to ADP and Pi in the presence of magnesium, sodium, and potassium. ATP forms a complex with magnesium, which means that the solution contains a mixture of Mg^{2+}, MgATP, and free ATP (ATP_f). To understand the sequence of the reaction which leads to the hydrolysis of ATP it is necessary to know the relationship between the effect of all these components on the system. At present, our knowledge about this is sparse, partly because there is no way to vary Mg^{2+}, MgATP, and independently, partly because there are few ways to investigate each of the steps in the reaction independently.

A. *Without Magnesium*

At 0°C, the system binds ATP with a high affinity with no magnesium, sodium, or potassium in the medium;[26,27] the dissociation constant is about 0·2 μM. Sodium has no effect on the affinity for ATP, while potassium decreases the affinity.[26,27]

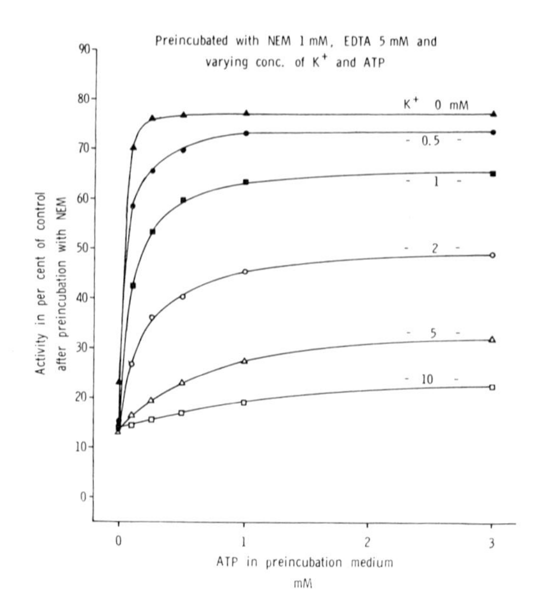

Figure 2. The effect of potassium and ATP on the inhibition of the ($Na^+ + K^+$)-activated enzyme by 1 mM n-ethylmaleimide (NEM). The enzyme was preincubated with 1 mM NEM 5mM EDTA, and the concentration of potassium and ATP shown on the figure in 30 mM Tris HCl buffer, pH 7.4 at 37°C for 30 min. After preincubation, the activity of the preparation was tested by transferring 0·1 ml of the preincubation medium to 1 ml of test solution with a final concentration of 3 mM magnesium, 3 mM ATP, 120 mM sodium, 30 mM potassium, 1 mM β-mercaptoethanol, 30 mM Tris HCl, pH 7·4, 37°C. Control was enzyme preincubated without NEM (unpublished).

The system has, as discussed above, two sets of sites, the i-sites and the o-sites with affinities for cations. The binding experiments do not tell whether the effect of potassium on the affinity for ATP is on the i- or on the o-sites. Information about this may come from experiments where the inhibitory effect of n-ethyl maleimide (NEM) has been used as a tool to test the effect of sodium and potassium on the effect of ATP on the system. ATP protects against the inhibitory effect of NEM. Potassium and sodium decreases the protective effect of ATP, but in a different way. In agreement with the results from the binding studies,

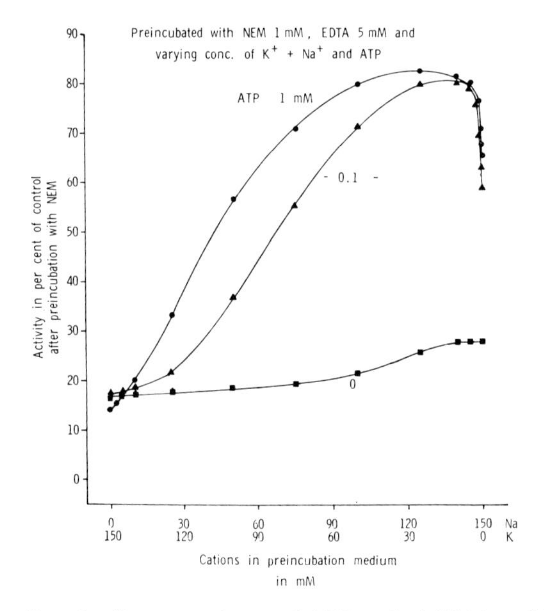

Figure 3. The effect of sodium, potassium, and ATP on the inhibition of the $(Na^+ + K^+)$-activated enzyme system by 1 mM NEM. The enzyme was preincubated with 1 mM NEM, 5 mM EDTA, and the concentration of sodium, potassium, and ATP shown on the figure in 30 mM Tris HCl buffer, pH 7·4 at 37°C for 30 min. After preincubation, the activity of the preparation was tested as described in Fig. 2 (unpublished).

potassium decreases the protection by decreasing the apparent affinity for ATP, Fig. 2, while sodium has no effect on the apparent affinity (not shown). Both potassium and sodium decreases the maximum level of protection which can be obtained by ATP, but the effect of potassium is much more pronounced than that of sodium, see Fig. 3.

Figure 3 shows that the inhibitory effect of potassium can be titrated away by sodium, and the inhibitory effect of sodium can be titrated away by potassium. A comparison between Figs. 1 and 3 (the left hand part of the curves) shows that the concentration of sodium necessary to give half maximum removal of the inhibitory effect of potassium with 1 mM ATP (Fig. 3) is of the same size as the concentration of sodium to give half maximum activation of the hydrolysis (Fig. 1). This suggests that the inhibitory effect of potassium on the protection by ATP against NEM is due to an effect of potassium on the sodium-activating site of the system, the i-sites. Figure 3 furthermore shows that ATP increases the apparent affinity of this site for sodium relative to potassium. The simplest way to explain the observations on the effect of sodium and potassium on the binding of ATP and on the protection of ATP against the effect of NEM is that the system with the i-sites on the potassium form does not react with ATP_f while without potassium, or with the i-sites on the sodium form, it does. (TS for transport system, i for the sodium-activating sites on the inside of the membrane, n is a number).

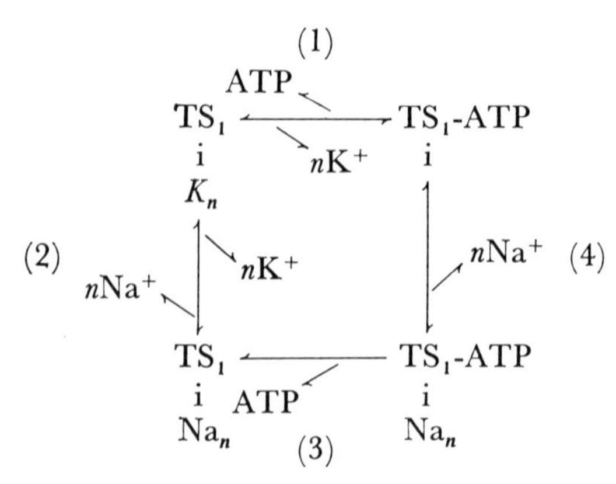

B. *With Magnesium*

The binding experiments and the experiments on the inhibition by NEM show that magnesium is not necessary for the reaction with ATP. Magnesium is, however, necessary for the hydrolysis of ATP. What is then the effect of magnesium? Which of the components, Mg^{2+}, MgATP, and ATP_f does the system react with when the medium contains magnesium plus ATP, and what is the sequence?

As a basis for the discussion of this problem it is convenient to use the following simple scheme for the overall hydrolysis of ATP by the enzyme system with magnesium, ATP, sodium, and potassium in the medium:

$$E + ATP \leftrightharpoons E\text{-}ATP \tag{5}$$

$$E\text{-}ATP \overset{Na^+,\, Mg^{2+}}{\leftrightharpoons} E \sim P + ADP \tag{6}$$

$$E \sim P + H_2O \overset{K^+}{\leftrightharpoons} E + Pi \tag{7}$$

$$E + ATP + H_2O \overset{Mg^{2+},\, Na^+,\, K^+}{\leftrightharpoons} E + ADP + Pi \tag{8}$$

It is based on the observation that the reaction with magnesium, ATP, and sodium leads to a phosphorylation-dephosphorylation of the system. The rate of phosphorylation is high, while that of dephosphorylation is low. When potassium is added to a prephosphorylated preparation, the rate of dephosphorylation is increased.[3, 49]

As seen from the scheme, the system can accomplish a magnesium–sodium-dependent ATP–ADP exchange. It was observed by Fahn *et al.*[50] that the magnesium requirement for the exchange reaction is much lower than for the ($Na^+ + K^+$)-dependent hydrolysis of ATP. This led to the suggestion that two magnesium molecules and two phospho-enzymes are involved in the reaction. One magnesium molecule for which the system has a high affinity and which is necessary for the formation of the phosphorylated intermediate which takes part in the exchange reaction, (9) and (10) in the following scheme.[50] Another magnesium molecule for which the affinity is an order of magnitude lower and which is necessary for the transformation of the phosphorylated intermediate into a form which can react with potassium and be dephosphorylated, (11), (12) and (13).

$$E + Mg^{2+} \leftrightarrows \text{Mg-E} \tag{9}$$

$$\text{Mg-E} + ATP \overset{Na^+}{\leftrightarrows} \text{Mg-E} \sim P + ADP \tag{10}$$

$$\text{Mg-E} \sim P + Mg^{2+} \leftrightarrows \text{Mg-E} \sim \text{P-Mg} \tag{11}$$

$$\text{Mg-E} \sim \text{P-Mg} \overset{(Na^+ ?)}{\leftrightarrows} \text{Mg-E-P-Mg} \tag{12}$$

$$\text{Mg-E-P-Mg} + H_2O \overset{K^+}{\leftrightarrows} \text{Mg-E} + Pi + Mg^{2+} \tag{13}$$

C. *With Magnesium, ATP, and Sodium*

Evidence for the existence of two different phosphorylated intermediates has been given by Post *et al.*[51] In experiments at 0°C they were able to show that the phosphorylated intermediates formed with a concentration of magnesium which was low and high, respectively, relative to the concentration of ATP, differed in their reactivity towards ADP, potassium, and g-strophanthin. When formed with a very low concentration of magnesium, the addition of ADP led to an increased rate of dephosphorylation, while potassium had a low or no effect. When formed with a higher magnesium concentration, the addition of potassium led to an increased rate of dephosphorylation, while ADP had a low or no effect. The ADP-sensitive phospho-enzyme did not react with g-strophanthin, while the potassium-sensitive did. This difference in sensitivity towards g-strophanthin is in disagreement with the observation that g-strophanthin inhibits the exchange reaction,[50] but apart from this, which may be due to differences in experimental conditions, the two phospho-enzymes behave as predicted from the scheme by Fahn *et al.*[50] According to this scheme, it is the native enzyme which has the high affinity for magnesium (9), and the phospho-enzyme (11), which has the low affinity for a second magnesium which is necessary for the transformation of E ~ P into E–P.

Results from experiments with g-strophanthin seem to support the view that there is a shift in the requirement for magnesium when ATP is hydrolyzed.[52] The experiments suggest, however, that the shift in the requirement for magnesium is due to a different way of interaction between sodium and magnesium with and without ATP.

Without ATP, sodium was found to decrease the apparent affinity for Mg^{2+}, while with ATP this effect disappears. Without sodium, free ATP inhibits the effect of magnesium plus ATP on the reaction by g-strophanthin. With sodium, the inhibitory effect of free ATP disappears. In other words, ATP seems to eliminate an antagonism between sodium and magnesium, while sodium eliminates an inhibitory effect of free ATP. The experiments suggest that with sodium in the medium it is the sodium-ATP form of the system which reacts with magnesium with a very high affinity. The elimination of the inhibitory effect of free ATP by sodium may either mean that the system on the sodium-ATP form has a magnesium site with a very high affinity for Mg^{2+} or that the sodium-ATP form does not require Mg^{2+}, but MgATP for the reaction with g-strophanthin.

The ADP-sensitive phospho-enzyme is formed under conditions where the concentration of magnesium is very low relative to the ATP concentration which suggests that free ATP does neither inhibit the sodium-dependent formation of the ADP-sensitive phospho-enzyme.

Considering this, it seems most likely that the sodium elimination of the inhibitory effect of free ATP means that Mg^{2+} at a magnesium site is not necessary for the reaction with g-strophanthin and for the formation of the ADP-sensitive phospho-enzyme, but MgATP. The sodium form of the system may react with MgATP with a high affinity; or, as the system binds ATP without magnesium, it may be that ATP bound to the sodium form of the system complexes Mg^{2+} with a high affinity, higher than for ATP in solution.

$$Na_n\text{-}TS_1 + MgATP \leftrightharpoons Na_n\text{-}TS_1\text{-}MgATP \tag{14}$$

or

$$Na_n\text{-}TS_1\text{-}ATP + Mg^{2+} \leftrightharpoons Na_n\text{-}TS_1\text{-}MgATP \tag{15}$$

On the Na_n-TS_1-MgATP form the system has catalytic activity, and the reaction leads to the formation of the ADP-sensitive phospho-enzyme.

$$Na_n\text{-}TS_1\text{-}MgATP \leftrightharpoons Na_n\text{-}TS_1 \sim P + Mg^{2+} + ADP \tag{16}$$

A higher magnesium concentration relative to the ATP concentration is necessary for the formation of the potassium sensitive phospho-enzyme. Experiments on the reaction of the system with pNPP as substrate suggest that there is a site on the system which reacts with

Mg^{2+} independent of substrate.[53] It seems therefore likely that the formation of the potassium-sensitive phospho-enzyme requires a reaction of the system with Mg^{2+} at a magnesium site besides the reaction with MgATP;

$$\text{Na}_n\text{-TS}_1\text{-MgATP} + \text{Mg}^{2+} \leftrightharpoons \text{Mg-Na}_n\text{-TS}_1\text{-MgATP} \qquad (17)$$

The phosphate in the potassium-sensitive phospho-enzyme seems to be bound to the same group on the system as the phosphate in the ADP-sensitive phospho-enzyme;[51] this seems to be an acyl phosphate,[54–56] which means that it is bound in a high energy bond in both phospho-enzymes. The different sensitivity towards ADP and potassium of the two phospho-enzymes can therefore not be explained by a different way of phosphorylation—they must differ in some other way. The exchange reaction, i.e. the formation of the ADP-sensitive phospho-enzyme is insensitive to oligomycin, while the formation of the potassium-sensitive phospho-enzyme is sensitive to oligomycin.[57, 58] The formation of the oligomycin-sensitive phospho-enzyme is more sensitive to a decrease in temperature than the formation of the ADP-sensitive.[57, 58] NEM which blocks the hydrolysis of ATP with sodium plus potassium in the medium increases the exchange reaction, i.e. it apparently blocks the step which leads to the formation of the potassium-sensitive but not the ADP-sensitive phospho-enzyme.[59] The effects of temperature and of NEM on the formation of the potassium-sensitive phospho-enzyme may suggest that this step involves a change in conformation of the system, TS_1 to TS_2 (cf. refs. 51, 58), and that it is this difference in conformation that gives a different sensitivity of the acyl phosphate towards ADP and towards potassium.

$$\text{Mg-Na}_n\text{-TS}_1\text{-MgATP} \leftrightharpoons \text{Mg-Na}_n\text{-TS}_2 \sim \text{P} + \text{ADP} \qquad (18)$$

In the scheme given by Fahn *et al.*,[50] and by Post *et al.*,[51] the ADP-sensitive and the potassium-sensitive phospho-enzyme represents two consecutive steps in the reaction. Another possibility is, as shown above, that either the one or the other is formed dependent on the magnesium concentration.

As mentioned above, it is apparently the same group on the system which is phosphorylated in the ADP-sensitive and in the potassium-sensitive phospho-enzyme.[51] The phosphate seems to be bound as an acyl phosphate,[54–56] which means in a bond which has normally a free energy of hydrolysis, which is of the same size as for the hydrolysis of the γ–β phosphate bond in ATP.

The phosphorylation of the system is, however, not specific for the reaction with ATP. ITP,[49, 60] AcP,[61–63] pNPP,[64] and Pi,[65, 66] can also phosphorylate the system. The phosphorylation from ITP and AcP is dependent on sodium as is the phosphorylation from ATP (see, however, refs. 67, 68). The phosphorylation from pNPP and Pi requires a reaction of the system with g-strophanthin and magnesium.

The phosphate from AcP,[62,63] pNPP,[64] and Pi[51,66] seems to be bound to the same group on the enzyme as the phosphate from ATP. (It has not been investigated for ITP). The formation of a high energy bond from a relatively low energy substrate as pNPP and from Pi shows that energy for the formation of the bond under these conditions must come from the reaction of the system with g-strophanthin and not from the substrate. The very slow rate of reaction with g-strophanthin suggests that it involves a change in conformation of the system. This may lead to a transformation of "conformational energy" into bond energy and by this to the formation of the high energy phosphate bond.[66]

The formation of the phospho-enzyme with a high energy phosphate bond is thus not specific for a reaction of the system with ATP. There is, however, a specific requirement for ATP for the transport process.[69] AcP which has a high energy phosphate bond, which gives a sodium-dependent phosphorylation of the system, and which is hydrolyzed by the system at a rate which is comparable to the rate of hydrolysis of ATP[70] cannot give a transport of sodium.[71] This shows either that formation of the phospho-enzyme is not enough for the transport process or that ATP can phosphorylate under conditions where AcP cannot. In either case it shows that there must be an effect of ATP on the system which precedes the phosphorylation (see also ref. 72).

There is a high specificity for the binding of ATP to the system, the affinity is at least 2–3 orders of magnitude higher than the affinity for CTP, GTP, and ITP.[27,73] According to the scheme (1)–(4), potassium and ATP exclude each other, which means that TS with potassium, and TS with ATP, must differ in some way (TS_1^x for the ATP form in the following). The discussed difference in the effect of sodium on the reaction of the system with magnesium with and without ATP seems to support this. The hydrolysis of AcP and pNPP is activated by potassium without sodium, which shows that these substrates must bind to the system in the presence of potassium in contrast to ATP. This suggests that AcP does not change the system in the same manner as ATP apparently does.

The change from TS_1 to TS_1^x due to ATP may be the specific effect of ATP and a necessary prerequisite for the transformation of the system from TS_1 to TS_2 when the bond between the γ and β phosphate of ATP is cleaved. The transformation may follow from the cleavage of this bond and not from the phosphorylation as such, and it may be the transformation which specifically requires an effect of ATP, sodium, and magnesium. The phosphorylation may be of importance for stabilizing the system in the TS_2 state until it can react with potassium, and by this be dephosphorylated and return to the TS_1 state.

As it seems to be the step which leads to the formation of the potassium sensitive phospho-enzyme which involves a change in configuration, it seems likely that this is the step which specifically requires the reaction with ATP. This would exclude the ADP-sensitive phospho-enzyme as an intermediate in the reaction which leads to TS_2, and suggest that it follows from an abortive reaction.

The dephosphorylation by ADP of the "low" magnesium, TS_1, phospho-form suggests that the ΔG for the formation of this phospho-enzyme is close to zero. The lack of reaction of the "high" magnesium form with ADP in spite of a high energy phosphate enzyme bond suggests that formation of the TS_2 phospho-enzyme is an energy-requiring process.

The reaction may be illustrated in a simple way by the following scheme (modified from ref. 74).

R_1, R_2, R_3 illustrate a conformation inside the system. $R_3 \sim R_2$ indicates a change in the distribution of energy inside the system.

Low Mg^{2+}

$$TS_1\text{-}R_1(R_2)(R_3) \xrightleftharpoons{\text{ATP, NA}^+,\ \text{Mg}^{2+}} Na_n\text{-MgATP-}TS_1^x\text{-}R_1(R_2)(R_3) \rightleftharpoons Na_n\text{-}TS_1\text{-}R_1(R_2)(R_3 \sim P) + Mg^{2+} + ADP$$

High Mg^{2+}

$$TS_1\text{-}R_1(R_2)(R_3) \xrightleftharpoons{\text{ATP, Na}^+,\ \text{Mg}^{2+}}$$

$$Mg\text{-}Na_n\text{-MgATP-}TS_1^x\text{-}R_1(R_2)(R_3) \longleftarrow Mg\text{-}Na_n\text{-}TS_2\text{-}R_1\text{-}R_2(R_3 \sim P) + ADP$$

$$Mg\text{-}Na_n\text{-}TS_2\text{-}R_1\text{-}R_2(R_3 \sim P) + mK^+ \rightleftharpoons K_m\text{-}TS_1\text{-}R_1(R_2)(R_3) + nNa^+ + Mg^{2+} + Pi$$

AcP as substrate

$$TS_1\text{-}R_1(R_2)(R_3) \xrightleftharpoons{\text{Mg}^{2+},\ \text{Na}^+,\ \text{AcP}} Mg\text{-}Na_n\text{-AcP-}TS_1\text{-}R_1(R_2)(R_3) \longleftarrow Na_n\text{-}TS_1\text{-}R_1(R_2)(R_3 \sim P) + Mg^{2+} + Ac$$

The formation of a high energy phosphate bond from low energy sources, Pi, when the system reacts with g-strophanthin may suggest that g-strophanthin mimics an effect of sodium plus ATP.

g-strophanthin

$$TS_1\text{-}R_1(R_2)(R_3) \xrightarrow{Mg^{2+},\ G} \text{G-Mg-}TS_1^x\text{-}R_1(R_2)(R_3) \xleftarrow{Pi} \text{G-Mg-}TS_1^x\text{-}R_1(R_3)(R_3 \sim P)$$

According to the view given above it is the reaction which leads to the phosphorylation and not the phosphorylation as such which is specific for the reaction with ATP and which is important to get the system to act as a transport system.

D. *With Magnesium, ATP, Sodium, and Potassium*

In the previous section the reaction of the system with magnesium, ATP, and sodium has been discussed. It is, however, characteristic for the system that it requires a combined effect of sodium and potassium for the overall hydrolysis of ATP, where sodium activates on the i-sites, and potassium on the o-sites. How is then the relationship between the effect of the two monovalent cations? Is it the reaction of the system with magnesium, ATP, and sodium, which brings it into a state in which it can react with potassium; or is the activation of the catalytic activity with sodium and potassium in the medium due to a combined, simultaneous effect of sodium on the i-sites and potassium on the o-sites? This problem is intimately related to another problem, namely do the sites on the system alternate between the inside and the outside of the membrane, i.e. between an i- and an o-form, a one-unit system; or, do they exist simultaneously, a two-unit system.[75]

There is at present no answer to the problem, and as the interpretation of the effect of sodium plus potassium on the enzyme system differs for a one- and a two-unit system, it may be of interest shortly to discuss both possibilities.

1. *One-unit system.* In a one-unit system the i-sites must be transformed into o-sites and back again as the reaction proceeds. As sodium activates on the i-sites and potassium on the o-sites, this means that the reaction of the system with potassium must follow that of the reaction with sodium. The reaction with sodium leads to the formation of the potassium-sensitive phospho-enzymes. Potassium added to a prephosphorylated enzyme increases the rate of dephosphorylation, which means that the reaction consists of a sodium-dependent phosphorylation followed by a potassium-dependent dephosphorylation.

The concentration of potassium necessary to dephosphorylate the potassium-sensitive enzyme is very low relative to the concentration of sodium in the medium, and it increases with the sodium concentration.[49]

This suggests that the potassium-sensitive phospho-enzyme, Mg-Na_n-$TS_2 \sim P$, is a form of the system in which the sites have a much higher apparent affinity for potassium than for sodium, i.e. the form in which the sites face the outside of the membrane, o-sites.

In the intact cell the transport system can accomplish an exchange of sodium from inside with sodium from outside, a Na:Na exchange, and this is sensitive to oligomycin.[76] Oligomycin increases the sodium-dependent labelling from ATP^{32}, and this seems to be due to a decreased rate of dephosphorylation.[49, 62, 77–79] As oligomycin has no effect or enhances the sodium-dependent ATP-ADP exchange found when the magnesium concentration is low relative to the ATP concentration,[50, 58] it seems to be the dephosphorylation of the potassium-sensitive phospho-enzyme, Mg-Na_n-$TS_2 \sim P$, which is inhibited by oligomycin, and not the dephosphorylation of Na_n-$TS_1 \sim P$. This suggests that it is the inhibition of the dephosphorylation of Mg-Na_n-$TS_2 \sim P$, which leads to an inhibition of the Na:Na exchange, and that this form takes part in the Na:Na exchange. It suggests that the system prior to the formation of the potassium-sensitive phospho-enzyme is in contact with the inside solution.

It was suggested above that the formation of the ADP-sensitive phospho-enzyme was an abortive reaction found when the magnesium concentration was too low, and that this intermediate was not part of the reaction when the magnesium concentration was high enough to give the conformational change of the system from the TS_1 to the TS_2 state, to the potassium-sensitive form. If this is correct, it must be Mg-Na_n-TS_1^x-MgATP which is the form prior to Mg-Na_n-$TS_2 \sim P$, and which is in contact with the inside solution. It leads to the scheme shown in Fig. 4a for the connection between the reaction of the system with ATP and the transport process. In the scheme it is the cleavage of the γ–β phosphate bond of ATP by the system in the form into which it has been brought due to the reaction with ATP, sodium, and magnesium that leads to the conformational change, $TS_1 \rightarrow TS_2$. This gives the transformation of i-sites into o-sites, and by this a translocation of sodium from inside to outside followed by an exchange of sodium for potassium on the o-sites. In Fig. 4b is shown the alternative possibility that the ADP-sensitive phospho-enzyme, $TS_1 \sim P$, is an intermediate in the reaction.

The shift in affinity may mean that for the TS_2 state the equilibrium between the sodium and the potassium form, $TS_2 \rightleftharpoons TS_2'$, is towards the potassium form, TS_2'. The exchange of sodium for potassium from outside is dependent on this equilibrium, and on the K/Na ratio in the external solution. On the potassium form the system is dephosphorylated, and this leads to the back-transformation of the system into the TS_1 state, $TS_2' \rightarrow TS_1'$, with a transport of potassium from outside to inside.

In the scheme shown in Fig. 4, the chemical change is intimately related to the translocation, and to the change from a sodium to a potassium affinity.[80] For a detailed discussion of the scheme, see ref. 75.

It must be emphasized that a translocation from TS_1 to TS_2 in a one-unit system not necessarily means a macroscopic (relative to the dimensions of the membrane) movement of a carrier molecule from the inside to the outside of the membrane and that phosphate is moved from inside to outside. The system may both in the TS_1 and TS_2 state be in

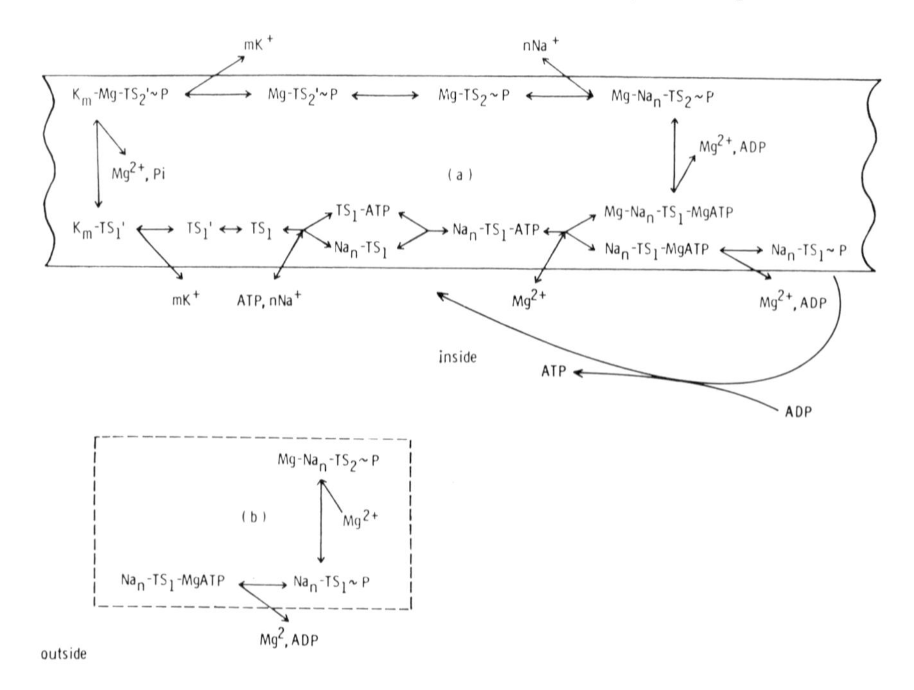

Figure 4. A one-unit model for the transport process. For explanation, see text. Modified from ref. 75.

contact with the inside of the membrane, but in the TS_1 state be able to exchange cations with the internal solution, and in the TS_2 state with the external solution, as for example in an alternating gate system.[81]

2. *Two-unit system.* In a two-unit system, see Fig. 5 there exists at the same time sites for sodium and potassium on the inside and on the outside of the membrane. The double competition between sodium and potassium suggests that each set of sites can exist on a sodium or on a potassium form: $o_S \leftrightharpoons o_P$, and $i_S \leftrightharpoons i_P$, respectively (o for outside, i for inside, S for sodium, and P for potassium). Each of the units accepts more than one cation (m and n in the figure); assuming that

each of the units accepts only sodium or potassium, i.e. that none of the units exists in a hybrid form (see, however, ref. 82), there are four combinations of the transport system, o_S/i_S, o_S/i_P, o_P/i_S, o_P/i_P, which must exist at the same time in the membrane. The ratio between these forms must depend on a number of factors. One is the built-in differences in affinities for each of the units for sodium and potassium.

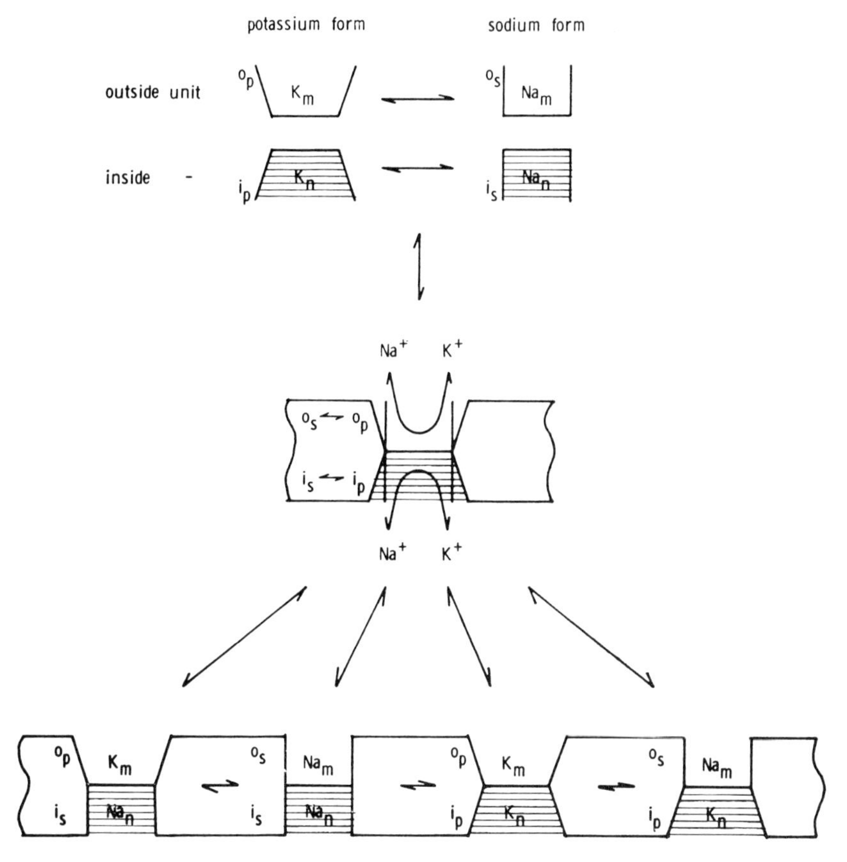

Figure 5. A two-unit model. For explanation, see text.

Another is the ratio between the concentrations of sodium and potassium in the solutions in contact with the sites. A third is the concentration of ATP; according to what has been discussed above, the i-sites on the potassium form, i_P, have a low affinity for ATP or do not react with ATP, while the sodium form, i_S, has a high affinity, which means that ATP will tend to shift the equilibrium towards the formation of is on account of i_P (see Fig. 6). The effect of the K:Na ratio on the equilibrium between the potassium and the sodium form of the o-unit seems to be independent of their effect on the equilibrium between the potassium and sodium form of the i-unit.[83]

The i-unit on the sodium-ATP form complexes magnesium with a very high affinity and reacts with Mg^{2+} with a lower affinity as has been described for a one-unit system. The difference between the reaction with MgATP and MgATP plus Mg^{2+} seems to be a change in conformation which is in some way related to the translocation of the cations (see section V, C). This may for a two-unit system mean that the reaction with Mg^{2+} besides MgATP leads to a change in the interaction between the two units from a state in which the cations on the o-unit do not

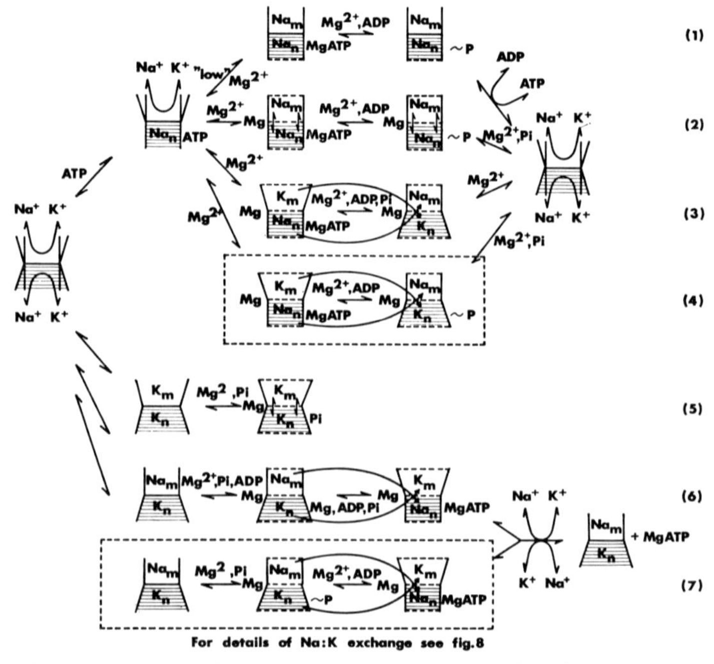

Figure 6. A two-unit model for the transport process. For explanation, see text. Modified from ref. 75.

influence the catalytic activity of the system to a state in which they do, and in which an exchange of the cations in between the two units is made possible. The change in interaction between the two units is shown on Figs. 6, 7, and 8 as a change from a situation where the two units are separated by a full-drawn line to a situation where they are separated by a dashed line, named a non-interacted and an interacted state, respectively, in the following; this indicates nothing about the molecular events—it is used to describe states which react differently. The dashed line on the surface of the units indicates a decreased exchange of the cations between the units and the surroundings.

For a one-unit system, the potassium-sensitive phospho-enzyme, $TS_2 \sim P$, formed with sodium but no potassium in the medium must be part of the reaction with sodium plus potassium, while this may or may not be the case for the ADP-sensitive phospho-enzyme, $TS_1 \sim P$.

If it is a two-unit system, the formation of $TS_1 \sim P$ and $TS_2 \sim P$ with sodium without potassium, i.e. with the system on the Na^o_m/Na^i_n form could be parallel reactions, (1) and (2) in Fig. 6, or consecutive reactions, Fig. 7, as for the one-unit system. With sodium plus potassium the hydrolysis of ATP could be due to a consecutive reaction in which the i-sites react with sodium, and this is followed by a reaction of the o-sites with potassium, Fig. 7; or, to a combined, simultaneous effect of potassium on the o-sites and of sodium on the i-sites, Fig. 6.

In the consecutive reaction shown in Fig. 7, the formation of $TS_1 \sim P$ is due to an effect of sodium on the i-sites and when $TS_1 \sim P$ reacts with Mg^{2+}, it is transformed into $TS_2 \sim P$, i.e. into the state of interaction between the two units in which the cations on the o-sites influence the rate of hydrolysis of the phospho-enzyme.

Another possibility (not shown) could be that the formation of $TS_2 \sim P$ requires a combined effect of sodium on the i- and the o-sites, and that the formation of $TS_2 \sim P$ leads to a shift of the o-unit from a sodium to a potassium form with a following exchange of sodium for

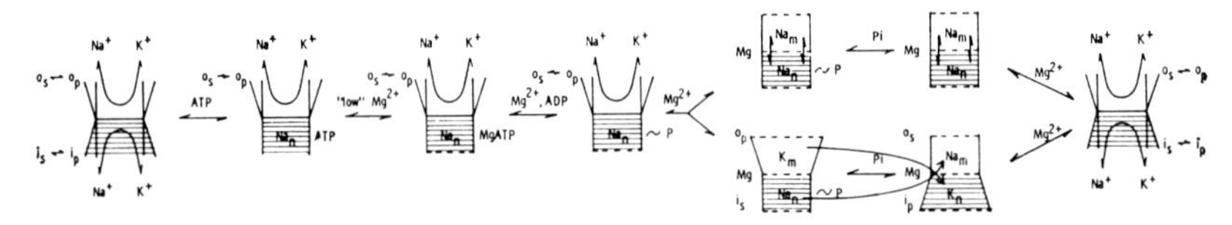

Figure 7. A two-unit model for the Na:K exchange with the ADP- and potassium-sensitive phospho-enzyme as an intermediate. For explanation, see text; see also Fig. 6.

potassium on the o-site, $Na^o_m/Na^i_n \rightarrow K^o_m/Na^i_n$, followed by a dephosphorylation and exchange of the cations in between the two units as shown in Fig. 7.

The consecutive reaction seems, however, to be ruled out by the observation that potassium in activating concentrations decreases the apparent affinity for magnesium plus ATP. In experiments where magnesium and ATP were kept at a constant 1:1 ratio, it was found that the concentration necessary for half saturation of the $(Na^+ + K^+)$-dependent hydrolysis by the enzyme system increased with the potassium concentration[84] (see also ref. 85). The effect of potassium was seen in the range in which potassium activates (the right part of the curve in Fig. 1) which suggests that it is an effect of potassium on the o-sites. It can therefore not be explained by the discussed effect of potassium on the i-sites on the affinity for ATP. A decrease in the affinity for magnesium plus ATP on the inside, on the i-unit, due to an effect of potassium on the o-sites may suggest that potassium reacts with the o-sites before the γ–β bond of ATP is cleaved. This seems to rule out the consecutive reaction described above (cf. Fig. 7). It suggests that the reaction of the system on the K^o_m/Na^i_n form follows a pathway which is different from that of the system on the Na^o_m/Na^i_n form, Fig. 6, and that the hydrolysis of ATP is due to a combined, simultaneous effect of potassium on the o-sites and sodium on the i-sites. The lower affinity for magnesium plus ATP may be due to a way of interaction between the units on the o_P/i_S form which is different from that on the i_S/i_P form.

A different pathway for the reaction on the Na^o_m/Na^i_n and the K^o_m/Na^i_n form raises the question whether the intermediate formed with the system on the Na^o_m/Na^i_n form, $TS_2 \sim P$, is also part of the reaction with the system on the K^o_m/Na^i_n form? The answer depends on the answer to another question: is the formation of $TS_2 \sim P$ with the system on the Na^o_m/Na^i_n form due to a combined effect of sodium on the o- and i-sites, or is it enough that there is sodium on the i-sites? If it requires a combined effect of sodium on the two units, $TS_2 \sim P$ cannot be part of the reaction with the system on the K^o_m/Na^i_n form. A decreased labelling found with sodium plus potassium must then be due to a decrease in the amount of the system on the Na^o_m/Na^i_n form.

If sodium on the i-sites is enough for the formation of $TS_2 \sim P$, what happens then when the system is on the K^o_m/Na^i_n form? The answer depends on the answer to another question: when the system on the Na^o_m/Na^i_n form is prephosphorylated and potassium is added, is the increased rate of dephosphorylation then due to an exchange of sodium for potassium on the o-site or on the i-site? Potassium has the effect in such low concentrations relative to the concentration of sodium that it suggests that it is an effect of potassium on the o-sites.[49] If this is the case, it is difficult to see what the result would be of an effect of sodium on the i-sites which gives a high rate of hydrolysis of the γ–β bond of ATP, and which tends to form a bond between the γ-phosphate and the enzyme and of potassium on the o-sites which tends to increase the rate of hydrolysis of the bond between the γ-phosphate and the enzyme. Will it be a phosphorylation, or will it be a hydrolysis of ATP without formation of a covalent bond between the enzyme and the phosphate ((3) in Fig. 6)?

If, on the other hand, the effect of potassium on the dephosphorylation of a prephosphorylated enzyme is due to an exchange of sodium for potassium on the i-sites, the phosphorylation found with the Na^o_m/Na^i_n form of the system could be part of the reaction with the system on the K^o_m/Na^i_n form, even if the two reactions follow a different pathway ((2) and (4) in Fig. 6).

A different pathway for the hydrolysis of ATP on the Na^o_m/Na^i_n and on the K^o_m/Na^i_n form with different affinities of the two forms for Mg^{2+} and ATP may explain why the same system seems to behave as two different systems, one found with sodium and no potassium, the other found with sodium and potassium.[86] (For a detailed discussion of the two-unit model, see ref. 75.)

It is not possible from our present knowledge to decide between the different possibilities, a consecutive reaction, Fig. 7, a simultaneous reaction without formation of a phospho-enzyme, (3) in Fig. 6, a simultaneous reaction with formation of a phospho-enzyme, (4) in Fig. 6. That means that it is neither possible to tell what it is that gives the translocation of the cations. Whether it is the hydrolysis of ATP without the phosphorylation, (3) in Fig. 6, the formation of a phosphorylated intermediate, (4) in Fig. 6, or the dephosphorylation of the prephosphorylated system, Fig. 7. Nor is it possible to tell what it is that gives the shift in the affinity of the o-sites which is necessary for the exchange of the cations.

The number of sodium ions transported per ATP hydrolyzed is about 3, while the number of potassium ions transported is lower.[4] This may mean that n is 3 and m a lower number (Figs. 5–8). Another possibility is that m and n are not identical with the number of cations transported. When ATP is hydrolyzed, and the o-unit is changed from o_P to o_S, and

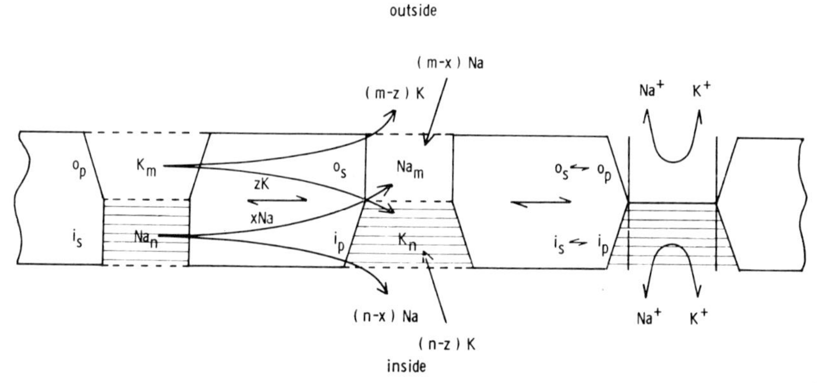

Figure 8. A model to describe the variation in the Na:K coupling ratio for a two-unit model. For explanation, see text. Modified from ref. 75.

the i-unit from i_S to i_P (see Fig. 8),[75] there may be an exchange of cations both in between the two units and between the units and the external and internal solution. Let for example m and n in Fig. 8 be equal to 4, x to 3, and z to 2. When the affinity of the two units are changed, there may be a flow of 3 sodium ions (x) from i to o, of 1 sodium ion ($n–x$) from i to the internal solution, and of 1 sodium ion ($m–x$) from the external solution to the o-unit. Simultaneously, there may be a flow of two potassium ions (z) from the o- to the i-unit, of two potassium ions ($m–z$) from the o-unit to the external solution, and of two potassium ions ($n–z$) from the internal solution to the i-unit. By this the four sites on the o-unit are filled up with sodium, and on the i-unit with potassium in spite of a transfer of only 3 sodium ions from i to o, and 2 potassium ions from o to i. When sodium on the o-unit in the following step in the non-interacted state is exchanged for potassium from the outside solution and potassium on the i-unit with sodium from the inside solution, the net result has been a transfer of 3 sodium out and 2 potassium in. This will besides a transport of the cations give an effect on the potential across the membrane, an electrogenic pump. m needs not be identical with n, and they can take any number equal to or higher than 3. The net number of cations exchanged between the two units depends on the ratio between the resistances for the flow of the cations in between the units and between the units and the external and internal solutions, respectively, and on the electrochemical gradient between the units and the external and internal solutions. The coupling ratio can vary without a variation of the number of sites for the cations on the units, m and n. For z equal to 0, the system gives a sodium efflux which is not coupled to a potassium influx, but is activated by potassium from outside.

VI. *Fluxes of Sodium and Potassium in Intact Cells in Relation to the Transport Models*

Besides the sodium efflux which is chemically coupled to a potassium influx there are a number of other carrier-mediated fluxes of sodium and potassium across the cell membrane. One is a sodium efflux which requires potassium in the external medium, but is not chemically coupled to a potassium influx, an electrogenic pump;[87,88] another is a sodium efflux which is coupled to a sodium influx, a Na:Na exchange.[12,71,89–95] A third is a sodium efflux seen without sodium and potassium in the external medium.[90,96] A fourth is a potassium influx which can either be coupled to a sodium influx, a reversal of the pump,[7,76,97] or to a potassium influx, a K:K exchange.[7]

The ratio between these fluxes depends in a complicated and only partly understood manner on the ratio between potassium and sodium in the external solution,[92,93] and in the internal solution,[71,92,94] and on the internal concentrations of ATP, ADP, and Pi,[92,98–100] and maybe also Mg^{2+}. It is common for all these fluxes that they are inhibited by cardiac glycosides. It seems therefore likely that they are all due to the same transport system in the cell membrane, and as the ($Na^+ + K^+$)-activated enzyme system is specifically inhibited by cardiac glycosides, it seems likely that this is the common transport system for the different cardiac glycoside-sensitive fluxes of sodium and potassium.

This means that the system must have a high degree of flexibility. The Na:K coupling can be switched over to a Na:Na or to a K:K coupling, and the Na:K coupling ratio may vary. How does this fit with the two discussed transport models which are mainly based on observations on the ($Na^+ + K^+$)-activated enzyme system in the test tube?

A. *Sodium-potassium Coupling*

The sodium pump can generate a potential across the cell membrane.[87,88] This effect requires potassium (or another of the cations which can activate the sodium pump) in the external solution, and it is inhibited by g-strophanthin. With a high internal sodium concentration, a hyperpolarization can be seen with a membrane potential more negative than the potassium equilibrium potential, and which is apparently not due to a depletion of potassium from the external surface by a neutral pump. It suggests that there is a transport of sodium outwards which is not chemically coupled to an inward transport of potassium, or to an outward transport of an anion. It means that the outward transport of sodium gives a transfer of net electrical charge across the membrane, and that the sodium pump can act as an electrogenic pump.

This fits with the observation on red blood cells that the number of sodium ions transported out per ATP hydrolyzed is higher than the number of potassium ions transported in, namely about 3:2.[4]

A problem is whether the 3:2 ratio is a fixed ratio found under all conditions and in all cells, and furthermore, whether this means that the pump in each cycle transports three sodium out and two potassium in, or that the pumping consists of a mixture of a neutral 3:3 pump and an electrical pump in which three sodium ions are transported out without an inward transport of potassium, but activated by potassium from outside.

Or, is the coupling ratio variable, and if so, is it a variation in the ratio between a neutral and an electrogenic pumping; or is the sodium pump more flexible and can pump sodium and potassium with a variable chemical coupling ratio, which is a function of factors like membrane potential and electrochemical potential gradients for the ions between the membrane phase and the solutions?

There is no definite answer to this problem. In experiments on the electrogenic effect of the sodium pumping in stretch receptors[101] and ganglion cells[102] under conditions where the internal sodium concentration was high, it was found that about $\frac{1}{4}$–$\frac{1}{3}$ of the sodium was extruded uncoupled in agreement with the Na:K coupling ratio of 3:2 found in red cells. In giant axons, with a high internal sodium concentration, the Na:K coupling ratio was found to be of the same order, namely 3:1[71] and 2:1.[94] If, however, the internal sodium concentration was decreased, the coupling ratio increased towards 1:1 at a low internal sodium concentration, suggesting that the coupling ratio may vary as a function of the internal sodium concentration.

A fixed coupling ratio can be explained from both the one- and the two-unit model. A variation in the coupling ratio is simpler to explain from the two-unit model (Fig. 8), and especially an efflux of sodium which is activated by potassium from outside, but which is not chemically coupled to an influx of potassium (if it exists). In the two-unit model the coupling ratio depends on the resistances for the flow of the cations inside the units and between the units and the external and internal solutions, respectively; and on the electrochemical potential gradients for the ions between the units and the solutions.

B. *Na: Na Exchange*

According to Garrahan and Glynn,[76] at least four features of the g-strophanthin-sensitive Na:Na exchange have to be taken into consideration when discussing a hypothetical model for the transport.

1. ATP is necessary for the exchange to take place in spite of low or no hydrolysis of ATP.
2. The exchange increases in rate with increasing external sodium concentration over a wide range and under conditions where the external sodium concentration is higher than the internal sodium concentration.
3. With a high internal ATP concentration relative to the internal Pi, there will be no exchange when the internal sodium is high, only with a low internal sodium and a high internal potassium concentration. With a high internal sodium, an increase in the internal Pi relative to the ATP will increase the Na:Na exchange.
4. The Na:Na exchange is oligomycin-sensitive.

And according to Glynn and Hoffman:[100]

5. ADP is necessary.

1. *One-unit system.* The requirement for ADP for the Na:Na exchange and the low hydrolysis of ATP due to this exchange seems to exclude that the exchange reaction can be due to a forward reaction of the system in which the dephosphorylation of the system in the TS_2 state is due to an effect of sodium from outside, instead of potassium, followed by a back-translocation on the sodium form.

It suggests the reaction

$$\text{Mg-Na}_n\text{-TS}_1\text{-MgATP} \leftrightharpoons \text{Mg-Na}_n\text{-TS}_2 \sim \text{P} + \text{ADP} + \text{Mg}^{2+}$$

as responsible for the exchange reaction (Fig. 4a).

Or, in the reaction scheme where the ADP-sensitive phosphoenzyme is an intermediate in the reaction (Fig. 4b), the reaction[100]

$$\text{Na}_n\text{-TS}_1 \sim \text{P} + \text{Mg}^{2+} \leftrightharpoons \text{Mg-Na}_n\text{-TS}_2 \sim \text{P}.$$

A Na:Na exchange due to these reactions requires ATP, but there is no net hydrolysis of ATP (requirement 1). They lead to a shift in affinity which makes it possible to explain requirement 2. They are oligomycin-sensitive (requirement 4). Internal sodium, magnesium, and ATP tend to shift the equilibrium towards the translocated state, and ADP towards the non-translocated state; it seems therefore possible to explain the effect on the Na:Na exchange of a decrease in internal sodium concentration, a decrease in internal ATP and an increase in internal ADP concentration from their effect on the equilibrium of the reaction (requirements 3 and 5). It is, however, difficult to explain the effect of high internal Pi.

2. *Two-unit system.* With sodium and potassium in the external medium, a decrease in the ATP/ADP $\times$ Pi concentration leads to a decrease in the potassium sensitivity of the sodium efflux and apparently to an increase in the Na:Na exchange and a decrease in the potassium influx.[98,99] This observation is simpler to explain from a two-unit model than from the one-unit model shown in Fig. 4 (for discussion of this problem for a one-unit model, see ref. 103).

The observations on the effect of potassium on the requirement for magnesium plus ATP suggested that the Na^o_m/Na^i_n form of the system has a higher affinity for magnesium and ATP than the K^o_m/Na^i_n form. A decrease in the ATP/ADP $\times$ Pi ratio at a given potassium/sodium ratio in the external solution therefore tends to shift the equilibrium from the interacted state of the Mg-K^o_m/Na^i_n-MgATP form towards the interacted state of the Mg-Na^o_m/Na^i_n-MgATP form of the system; this may explain the apparent decrease in potassium sensitivity of the sodium efflux, and an increase in Na:Na efflux (see below).

It may also explain why the addition of potassium to an external medium containing sodium decreases the g-strophanthin-sensitive sodium efflux, when the ATP/ADP $\times$ Pi ratio is low.[92] Without potassium in the external solution, a certain part of the system is on the Na^o_m/Na^i_n form, and this may as discussed below give a Na:Na exchange, when the ATP/ADP $\times$ Pi ratio is low. When potassium is added to the external solution, a part of the Na^o_m/Na^i_n form is transferred into the K^o_m/Na^i_n form. This gives a decrease in the sodium efflux coupled to a sodium influx. But due to the low ATP/ADP $\times$ Pi ratio, the equilibrium of the K^o_m/Na^i_n form is towards the non-interacted state, which means that the decrease in sodium efflux coupled to a sodium influx will not, or only to a lower extent, be replaced by a sodium efflux coupled to a potassium influx.

Removal of potassium from the external solution shifts the equilibrium towards Na^o_m/Na^i_n on account of K^o_m/Na^i_n. In the interacted state of the Na^o_m/Na^i_n form, the system may exchange sodium in between the two units (see (2) in Fig. 6), and the forward and backward reaction of Mg-Na^o_m/Na^i_n-MgATP $\leftrightharpoons$ Ma-$Na^o_m/Na^i_n \sim$ P + ADP + Mg^{2+} may give a Na:Na exchange as the parallel reaction in the one-unit system.

In the two-unit system there are, however, at least two other ways to explain the Na:Na exchange and the requirement for ADP.

Assuming that sodium in the interacted state can be exchanged not only in between the two units, but also between the units and the surroundings, the forward reaction with a low rate of turnover, because of a high concentration of ADP, may give a high rate of exchange.

A more complicated reaction would be that the catalytic reaction of Mg-Na^{o}_{m}/Na^{i}_{n}-MgATP ((2) in Fig. 6) in the interacted state tends to transform the i-unit from i_S to i_P, as is the case when the system reacts with ATP on the K^{o}_{m}/Na^{i}_{n} form. As there is no potassium on the o-unit to be exchanged for sodium on the i-unit, sodium will hinder this, and the result of the reaction is therefore not a hydrolysis of ATP, but the phosphorylation with formation of the potassium-sensitive phospho-enzyme. This phospho-enzyme may react with ADP and form ATP, but only if i_S is transformed into i_P, i.e. if sodium on the i-unit is exchanged for potassium from inside. This reaction is supported by a high internal ADP and a high internal potassium/sodium ratio. The dephosphorylation leads to a transformation into the non-interacted state. It will give a Na:Na exchange, which is increased by a high internal ADP and a high internal K/Na ratio, and which requires ATP with no net hydrolysis of ATP, and is oligomycin-sensitive. It would mean that the potassium-sensitive phospho-enzyme cannot be an intermediate in the reaction on the K^{o}_{m}/Na^{i}_{n} form. Furthermore, that the dephosphorylation of the potassium-sensitive phospho-enzyme besides potassium requires ADP, and is due to an effect of potassium on the i-unit and not on the o-unit!

There is a g-strophanthin-insensitive sodium efflux which, according to Hoffman and Kregenow,[104] is on another transport system (pump II). Brinley and Mullins[95] showed that strophanthidin increases the sodium efflux when the ATP concentration is very low, but it decreases the influx when the ATP concentration is high. The increase and decrease occur to the same level of efflux. As pointed out by the authors, this suggests that the cardiac glycoside-sensitive transport system is responsible for the cardiac glycoside-insensitive sodium efflux. It suggests that strophanthidin blocks the system in a state in which it can accomplish a low sodium efflux (Na:Na exchange?). The increase in sodium efflux by strophanthidin when the concentration of ATP is low and the decrease when the concentration of ATP is high, suggest that this state is an intermediate in the turnover of the system. It suggests that the reaction with strophanthidin with a low concentration of ATP leads to a shift in equilibrium from a state in which the system cannot give a sodium efflux, and which does not react with strophanthidin, the non-interacted state, towards a state in which it can give a sodium efflux, and which can react with strophanthidin, the interacted state. With the higher concentration of ATP, g-strophanthin blocks the system in the same state as with the low concentration of ATP. By this it blocks the turnover and decreases the sodium efflux to the same level as found with strophanthidin and the low concentration of ATP.

C. *Reversal of the Sodium Pump*

Under certain conditions, namely with external sodium, but no external potassium, and with a high internal potassium concentration and a low ATP/ADP $\times$ Pi ratio, it is possible to drive the cation pump backwards and form ATP from ADP and Pi.[7,76,97]

In the one-unit model it can be explained by reversal of the forward reaction.

In the two-unit model, the equilibrium of the o-unit is towards the potassium form, o_P, but with sodium and no potassium, the unit will be on the sodium form, o_S. When this form interacts with the i-unit on the potassium form, i_P, in the presence of magnesium, ATP, and Pi, there may be a tendency for the cations to be exchanged in between the two units with a shift in the equilibrium towards o_P/i_S, and with a formation of ATP ((6) and (7), cp. (3) and (4) respectively, in Fig. 6).

D. *K:K Exchange*

Under certain conditions, namely with a high internal concentration of potassium and Pi, a low concentration of ATP, and with potassium in the external medium, there seems to be a g-strophanthin-sensitive potassium efflux which is coupled to a potassium influx, a K:K exchange.[7]

In the one-unit system, the K:K exchange can be explained by a reversal of the last step of the transport process. This step is related to the potassium entry mechanism, which may explain why Pi is necessary.

In the two-unit system, the K:K exchange can be explained by a shuttling of the system on the K_m^o/K_n^i form between the non-interacted and interacted states ((5) in Fig. 6); this does not represent a reversal of the system in the same sense as in the one-unit system, but a reaction of the system on the K_m^o/K_n^i form independent of the reaction for the potassium-coupled sodium efflux.

VII. *Conclusion*

In the preceding section the sequence of the steps in the reaction of the $(Na^+ + K^+)$-activated enzyme system has been discussed, and an attempt has been made to relate them to the transport of cations across the cell membrane and from this to formulate a model for the transport process.

The result of this has been two principally different models, for both of which it must be emphasized that they are based on a number of assumptions, and must only be taken as suggestions. It is not possible from our present knowledge to decide whether the transport process can be described by a one-unit or by a two-unit model. The two-unit model seems to have a few advantages over the one-unit model. It is simpler to explain a variable ratio between the number of sodium and potassium ions transported, and also an external potassium activation of a sodium efflux without an influx of potassium. Furthermore, that potassium from outside influences the affinity for magnesium and ATP on the inside, and also that a decrease in the ATP/ADP × Pi ratio decreases the sensitivity of the sodium efflux towards potassium from the external solution. Each of these problems can, however, also be explained from a one-unit model.

There is, however, one set of experimental results which lends support to a two-unit system, and which only under special conditions can be explained from a one-unit system. That is the observations by Hoffman and Tosteson on sheep red cells that the apparent affinity of external potassium and of internal sodium for the g-strophanthin-sensitive sodium efflux and potassium influx is independent of the internal and external Na/K ratios, respectively.[83]

Experiments on the effect of ATP on the phosphatase activity of the system raise another problem which has not been discussed. The ($Na^+ + K^+$)-activated enzyme system can apparently hydrolyze pNPP in the presence of magnesium and potassium.[53, 62–64, 105–113] The effect of potassium on the hydrolysis of pNPP is, however, influenced by ATP and sodium.[108–112] Does that mean that the system binds ATP at a modifier site, and pNPP at a catalytic site?[111] And does that mean that the system with ATP as substrate binds two ATP molecules at two different sites, one which has a modifying effect on the hydrolysis, and another which is hydrolyzed? (See also ref. 27.)

Or, is the effect of ATP on the hydrolysis of pNPP due to a consecutive reaction of the system with ATP and pNPP, i.e. the reaction with ATP leads to a phosphorylation, and this is followed by a reaction with pNPP on the ATP-site? Or, can pNPP induce a change in the catalytic site which allows binding of both ATP and pNPP to the same site, while without pNPP only one ATP molecule is bound?

At present there is no definite answer to the problem about a one- or a two-unit system, nor to the problem about one or two sites for ATP on the system. This tells that our knowledge about the sequence of the steps in the reaction is not detailed enough. And, apart from that, we have nearly no information about the molecular structure and the molecular events related to the steps in the reaction. Where in the membrane is the system located, on the inside or on the outside or all the way through the membrane? How far are the cations transported on the system? How do the units discriminate between the cations? Do they behave as ion exchangers, or are there specialized structures of the type which has been used in bilayer studies to discriminate between sodium and potassium? This and many other questions have to be answered before we are able to understand the transport process.

References

1. J. C. Skou, *Physiol. Rev.*, **45** (1965) 596.
2. E. Heinz, *A. Rev. Physiol.*, **29** (1967) 21.
3. R. W. Albers, *Ann. Rev. Biochem.*, **36** (1967) 727.
4. I. M. Glynn, *Br. Med. Bull.*, **24** (1968) 165.
5. R. Whittam and K. P. Wheeler, *Ann. Rev. Physiol.*, **32** (1970) 21.
6. S. L. Bonting, in: *Membranes and Ion Transport*, Vol. 1, E. E. Bittar (ed.), Wiley-Interscience, New York, 1970, p. 257.
7. I. M. Glynn and V. L. Lew, *J. Gen. Physiol.*, **54** (1969) 289s.
8. J. C. Skou, *Biochim. Biophys. Acta*, **23** (1957) 394.
9. I. M. Glynn, *J. Physiol.* (Lond.), **160** (1962) 18P.
10. P. C. Laris and P. E. Letchworth, *J. Cell. Comp. Physiol.*, **60** (1962) 229.
11. R. Whittam, *Biochem. J.*, **84** (1962) 110.
12. P. J. Garrahan and I. M. Glynn, *J. Physiol.* (*Lond.*), **192** (1967) 217.
13. P. Leth Jørgensen and J. C. Skou, *Biochim. Biophys. Acta*, **233** (1971) 366.
14. O. J. Møller, *Exp. Cell Res.*, **68** (1971) 347.
15. J. Rostgaard and O. J. Møller, *Exp. Cell Res.*, **68** (1971) 356.
16. J. C. Skou, *Biochim. Biophys. Acta*, **58** (1962), 314.
17. T. Nakao, Y. Tashima, K. Nagano and M. Nakao, *Biochem. Biophys. Res. Comm.*, **19** (1965) 755.
18. J. C. Skou, *Biochem. Biophys. Res. Comm.*, **10** (1963) 1.
19. H. Matsui and A. Schwartz, *Biochim. Biophys. Acta*, **128** (1966) 380.
20. P. Leth Jørgensen, J. C. Skou and L. P. Solomonson, *Biochim. Biophys. Acta*, **233** (1971) 381.
21. S. Uesugi, N. C. Dulak, J. F. Dixon, T. D. Hexum, J. L. Dahl, J. F. Perdue and L. E. Hokin, *J. Biol. Chem.*, **246** (1971) 531.
22. J. Kyte, *J. Biol. Chem.*, **246** (1971) 4157.
23. D. W. Towle and J. H. Copenhaver, Jr., *Biochim. Biophys. Acta*, **203** (1970) 124.
24. H. Bader, R. L. Post and G. H. Bond, *Biochim. Biophys. Acta*, **150** (1968) 41.
25. R. W. Albers, G. J. Koval and G. J. Siegal, *Mol. Pharmacol.*, **4** (1968) 324.
26. J. G. Nørby and J. Jensen, *Biochim. Biophys. Acta*, **233** (1971) 104.
27. C. Hegyvary and R. L. Post, *J. Biol. Chem.*, **246** (1971) 5234.
28. O. Hansen, J. Jensen and J. G. Nørby, *Nature, New Biol.*, **234** (1971) 122.
29. J. C. Ellory and R. D. Keynes, *Nature*, **221** (1969) 776.
30. O. Hansen, *Biochim. Biophys. Acta*, **233** (1971) 122.
31. P. F. Baker and J. S. Willis, *Nature*, **226** (1970) 521.
32. M. Nakao, K. Nagano, T. Nakao, N. Mizuno, Y. Tashima, M. Fujita, H. Maeda and H. Matsudaira, *Biochem. Biophys. Res. Comm.*, **29** (1967) 588.
33. G. R. Kepner and K. J. Macey, *Biochem. Biophys. Res. Comm.*, **30** (1968), 582.
34. J. Kyte, *Biochem. Biophys. Res. Comm.*, **43** (1971) 1259.
35. B. Roelofsen, R. F. A. Zwaal and L. L. M. van Deenen, in "Membrane-Bound Enzymes", G. Porcellati and F. di Jeso (eds.), *Adv. in Exp. Med. Biol.*, Vol. 14, Plenum Press, New York, 1971, p. 209.
36. P. Emmelot and C. J. Bos, *Biochim. Biophys. Acta*, **150** (1968) 341.
37. H. J. Schatzmann, *Nature*, **196** (1962) 677.
38. T. Ohnishi and H. Kanamura, *J. Biochem.*, **56** (1964) 377.
39. J. C. Skou, in: *Transport and Metabolism*, Vol. 1, A. Kleinzeller and A. Kotyk (eds.), Academic Press, New York, 1961, p. 228.
40. G. Hegyvary and R. L. Post, in: *The Molecular Basis of Membrane Function*, D. C. Tosteson (ed.), Prentice Hall, New Jersey, 1969, p. 519.
41. Y. Israel, in: *The Molecular Basis of Membrane Function*, D. C. Tosteson (ed.), Prentice Hall, New Jersey, 1969, p. 529.
42. K. P. Wheeler and R. Whittam, *J. Physiol.* (*Lond.*), **207** (1970) 303.
43. R. Tanaka and L. G. Abood, *Arch. Biochem. Biophys.*, **108** (1964) 47.
44. R. Tanaka and K. P. Strickland, *Arch. Biochem. Biophys.*, **111** (1965) 583.
45. R. Tanaka, *J. Neurochem.*, **16** (1969) 1301.
46. L. J. Fenster an J. H. Copenhaver, Jr., *Biochim. Biophys. Acta*, **137** (1967) 406.
47. R. Tanaka and T. Sakamoto, *Biochim. Biophys. Acta*, **193** (1969) 384.
48. K. A. Karlsson, B. E. Samuelson and G. O. Steen, *J. Membrane Biol.*, **5**, (1971) 169.
49. J. C. Skou and C. Hilberg, *Biochim. Biophys. Acta*, **185** (1969) 198.
50. S. Fahn, G. J. Koval and R. W. Albers, *J. Biol. Chem.*, **241** (1966) 1882.
51. R. L. Post, S. Kume, T. Tobin, B. Orcutt and A. K. Sen, *J. Gen. Physiol.*, **54** (1969) 306s.
52. J. C. Skou, K. Butler and O. Hansen, *Biochim. Biophys. Acta*, **241** (1971) 443.
53. P. J. Garrahan, M. I. Pouchan and F. Rega, *J. Physiol.* (Lond.), **202** (1969) 305.
54. L. E. Hokin, P. S. Sastry, P. R. Galsworthy and A. Yoda, *Proc. Nat. Acad. Sci. U.S.A.*, **54** (1965) 177.
55. K. Nagano, T. Kanazawa, N. Mizuno, Y. Tashina, T. Nakao and M. Nakao, *Biochem. Biophys. Res. Comm.*, **19** (1965) 759.

56. H. Bader, A. K. Sen and R. L. Post, *Biochim. Biophys. Acta*, **118** (1966) 106.
57. S. Fahn, G. J. Koval and R. W. Albers, *J. Biol. Chem.*, **243** (1968) 1993.
58. R. Blostein, *J. Biol. Chem.*, **245** (1970) 270.
59. S. Fahn, M. R. Hurley, G. J. Koval and R. W. Albers, *J. Biol. Chem.*, **241** (1966) 1890.
60. W. Schoner, R. Beusch and R. Kramer, *European J. Biochem.*, **7** (1968) 102.
61. G. H. Bond, H. Bader and R. L. Post, *Fed. Proc.*, **25** (1966) 567.
62. Y. Israel and E. Titus, *Biochim. Biophys. Acta*, **139** (1967) 450.
63. G. H. Bond, H. Bader and R. L. Post, *Biochim. Biophys. Acta*, **241** (1971) 57.
64. C. E. Interrusi and E. Titus, *Mol. Pharmacol.*, **6** (1970) 99.
65. G. E. Lindenmayer, A. H. Langhfer and A. Schwartz, *Arch. Biochem. Biophys.* **127** (1968) 187.
66. G. F. Siegel, G. F. Koval and R. W. Albers, *J. Biol. Chem.*, **244** (1969) 3264.
67. G. Sachs, J. P. Rose and B. I. Hirschowitz, *Arch. Biochem. Biophys.*, **119** (1967) 277.
68. W. F. Dudding and C. G. Winther, *Biochim. Biophys. Acta*, **241** (1971) 650.
69. J. F. Hoffman, *Circulation*, **26** (1962) 1201.
70. H. Bader and A. K. Sen, *Biochim. Biophys. Acta*, **118** (1966) 116.
71. L. F. Mullins and F. J. Brinley, *J. Gen. Physiol.*, **53**, (1969), 704.
72. J. C. Skou, *Biochim. Biophys. Acta*, **42** (1960) 6.
73. J. Jensen and J. G. Nørby, *Biochim. Biophys. Acta*, **233** (1971) 395.
74. J. C. Skou, in: *Biomembranes*, L. A. Manson (ed.), Vol. 2, Plenum Publishing Corporation, New York, 1971, p. 165.
75. J. C. Skou, in: *Current Topics in Bioenergetics*, Vol. 4, D. R. Sanadi (ed.), Academic Press, New York, 1971, p. 357.
76. P. J. Garrahan and I. M. Glynn, *J. Physiol. (Lond.)*, **192** (1967) 237.
77. S. Fahn, G. J. Koval and R. W. Albers, *J. Biol. Chem.*, **243** (1968) 1993.
78. R. Whittam, K. P. Wheeler and A. Blake, *Nature*, **203** (1964) 720.
79. N. Gruener and Y. Avi-Dor, *Biochem. J.*, **100** (1966) 762.
80. P. Mitchell, *Advances in Enzymology*, F. F. Nord (ed.), Vol. 29, Interscience Publishers, New York, 1967, p. 33.
81. O. Jardetzky, *Nature*, **211** (1966) 969.
82. H. W. Middleton, *Arch. Biochem. Biophys.*, **136** (1970) 280.
83. P. G. Hoffman and D. C. Tosteson, *J. Gen. Physiol.*, **58** (1971) 438.
84. J. D. Robinson, *Biochemistry*, **6** (1967) 3250.
85. R. L. Post, A. K. Sen and R. S. Rosenthal, *J. Biol. Chem.*, **240** (1965) 1437.
86. A. H. Neufeld and H. M. Levy, *J. Biol. Chem.* **244** (1969) 6493.
87. R. P. Kernan, in: *Membrane and Ion Transport*, Vol. 1, E. E. Bittar (ed.), Wiley-Interscience, New York, 1970, p. 395.
88. J. M. Ritchie, in: *Current Topics in Bioenergetics*, Vol. 4, R. D. Sanadi (ed.), Academic Press, New York, 1971, p. 327.
89. P. C. Caldwell, A. L. Hodgkin, R. D. Keynes and T. I. Shaw, *J. Physiol. (Lond.)*, **152** (1960a) 561.
90. P. J. Garrahan and I. M. Glynn, *J. Physiol. (Lond.)*, **192** (1967a) 159.
91. P. J. Garrahan and I. M. Glynn, *J. Physiol. (Lond.)*, **192** (1967b) 175.
92. P. J. Garrahan and I. M. Glynn, *J. Physiol. (Lond.)*, **192** (1967c) 189.
93. P. F. Baker, M. P. Blaustein, R. D. Keynes, J. Manil, T. I. Shaw and R. A. Steinhardt, *J. Physiol. (Lond.)*, **200** (1969) 459.
94. R. A. Sjodin and L. A. Beaugé, *J. Gen. Physiol.*, **51**, No. 5, Part 2 (1968) 152.
95. F. J. Brinley and L. J. Mullins, *J. Gen. Physiol.* **52** (1968) 181.
96. P. F. Baker, *Biochim. Biophys. Acta*, **88** (1964) 458.
97. A. F. Lant, R. N. Priestland and R. Whittam, *J. Physiol. (Lond.)*, **207** (1970) 291.
98. P. C. Caldwell, A. L. Hodgkin, R. D. Keynes and T. I. Shaw, *J. Physiol. (Lond.)* **152** (1960b) 591.
99. P. de Weer, *J. Gen. Physiol.*, **56** (1970) 583.
100. I. M. Glynn and J. F. Hoffman, *J. Physiol. (Lond.)*, **218** (1971) 239.
101. S. Nahajima and K. Tabakaski, *J. Physiol. (Lond.)*, **187** (1966), 105,
102. T. Thomas, *J. Physiol. (Lond.)*, **201** (1969) 495.
103. P. C. Caldwell, *Physiol. Rev.*, **48** (1968) 1.
104. J. F. Hoffman and F. M. Kregenow, *Ann. N.Y. Acad. Sci.*, **137** (1966) 566.
105. J. D. Judah, K. Ahmed, A. E. M. Maclean, *Biochim. Biophys. Acta*, **65** (1962) 472.
106. M. Fujita, T. Nakao, N. Mizuno, K. Nagano and M. Nakao, *Biochim. Biophys. Acta*, **117** (1966) 42.
107. B. Formby and J. Clausen, *Z. Physiol. Chem.*, **349** (1968) 909.
108. H. Yoshida, K. Nagai, T. Ohashi and Y. Nakagawa, *Biochim. Biophys. Acta*, **179** (1969) 178.
109. P. J. Garrahan, M. I. Pouchan and F. Rega, *J. Membrane Biol.*, **3** (1970) 14.
110. F. Rega, P. J. Garrahan and M. I. Pouchan, *J. Membrane Biol.* **3** (1970) 26.
111. A. Askari and G. Koyal, *Biochem. Biophys. Acta*, **225** (1971) 20.
112. G. Koyal, S. N. Rao and A. Askari, *Biochim. Biophys. Acta*, **225** (1971) 11.
113. A. Askari and S. N. Rao, *Biochim. Biophys. Acta*, **241** (1971) 75.

On the Meaning of Effects of Substrate Structure on Biological Transport*

Halvor N. Christensen

The Department of Biological Chemistry, The University of Michigan
Ann Arbor, Michigan

No single approach or simple combination of approaches will bring us an understanding of membrane transport, how it occurs, how it is driven and how it drives other processes. The particularity of the minds of individual investigators usually ensures that numerous attacks on a problem are undertaken, and also that these attacks are continued, and reapplied even after other investigators may have begun to deplore them. This article was in the first instance initiated for myself, to rethink what it is that my associates and I hope to achieve by one set of approaches, namely by observing the results of modifying the structure of substrates on their transport. To a degree I am taking a precaution that we do not decide that such studies are obsolescent, that the approach is indirect, a "black box" and born of a lack of courage for direct isolative approaches, or that only superficial objectives can be pursued by the approach, simply because superficial objectives are so often pursued by it.

Perhaps it is enough to point out that even after isolation of a catalytic system or one of its components, much can be learned by looking for substrates that it handles in unusual ways. Consider how well this statement applies to chymotrypsin. No doubt it also applies to the leucine-binding protein and other binder proteins. Such isolated proteins may also be "black boxes" to direct inspection. Perhaps it is a fragmentary rather than a universal truth that the information one gets by isolation is so unique as to deserve sole attention. On the contrary, each step of isolation appears to increase the advantage with which other methods of study can be applied. Having begun my analysis, I will proceed, however, to try to illustrate what this approach can do by examples (one in some detail) of what it is now doing, a

* The experiments described here that derive from my laboratory, and the preparation of this manuscript, were supported in part by a grant (HDO1233) from the Institute of Child Health and Human Development, National Institutes, U.S. Public Health Service.

I speak also for my collaborators in these experiments, including M. E. Handlogten, A. M. Cullen, G. Ronquist and C. de Cespedes in the case of unpublished results.

procedure which can I suppose only vaguely reveal what it is capable of accomplishing.

1. *To determine whether we are observing transport in the strict sense or instead a group translocation.* Enzymologists bring to the transport field an apparently inexhaustible intuition that transport will prove to have as its basis a persisting enzymatic modification of the substrate. They also bring with them a firm understanding of how two reactions may be linked so that energy from one can be made available to drive another, but often a reluctance to accept a coupling whereby one of the two processes linked together is not a net reaction at all but a movement of a solute from one chemical potential to another—a coupling that we may call not *chemical* but *chemiosmotic*, if we allow ourselves to use that word in a broader sense than it is sometimes used. The transported sugar, the enzymologist may feel, will prove to have been mutarotated or phosphorylated, the transported amino acid will have been converted to an AMP anhydride or to an amide and so on. It is perhaps a parallel conservatism that wants to keep the coupling between electron transport and oxidative phosphorylation an organic chemical matter.

It was over three decades ago, we should remember, that three different *O*-methyl derivatives of glucose were shown to undergo typical intestinal transport, excluding the necessity of phosphorylation in either of three positions.[1] It was also years ago that anhydroglucitol ("1-deoxyglucose") was shown to be transported in an apparently typical manner both by the intestine and by the human red blood cell, and a Na^+-independent renal transport of 2-deoxyglucose has had recent extensive study. A new instance of inhibitory actions of these and other blocked sugars on the transport of galactose and 2-deoxyglucose have recently been reported for *Aspergillus nidulans*.[2]

To descend to a simpler and indeed an absurd level, it has been obvious from the first that Na^+ was transported without conversion to K^+, Ca^{2+} without conversion to Mg^{2+}, and so on. The student of inorganic-ion transport has therefore largely escaped this bias. Organic substrates of transport may also, however, be designed so they also escape modification. Amino acid analogs have been synthesized that undergo no sign of destabilization by the organisms and cells so far studied. Even distortion of the molecular structure seems largely excluded for some of these analogs, such as the one pictured in Fig. 1, and yet their transport is typical.[3]

It would be unwise at this point in history, however, to argue that metabolites are never transferred from one compartment of the cell to another by the process so lucidly described by Mitchell and Moyle a few years ago as *group translocation*,[4] i.e., in the course of an enzymatic transfer of the chemical group from a donor to an acceptor. A membrane-associated phosphotransferase action on sugars is clearly of high metabolic importance in an as yet unknown number of anaerobic bacterial species. It has been difficult to determine whether a phosphorylating enzyme actually exposes its reactive site on the external surface of the membrane to receive the sugar substrate, and then carries it through the membrane to bring it into contact with the phosphorylated *heat-stable protein*. The experiments of Gachelin indicate that it is probably another structure that binds the sugar during the translocation step, without changing the sugar into anything else, until it is brought into association with the phosphorylating system deeper in the cell.[5] Nevertheless, the phosphorylation appears to be coupled to the mediated transport step, to transfer energy to it in an unexpected way. In addition it appears inherently likely that some simpler group translocations also take place, especially in subcellular metabolism.

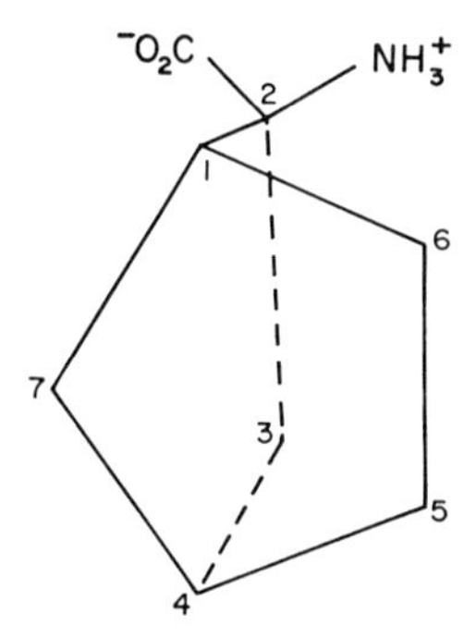

Figure 1. Structure of isomer of 2-aminobicyclo-[2,2,1]-heptane-2-carboxylic acid serving as a model substrate for transport.

If the positions of the amino and carboxyl groups are reversed one secures the *exo-endo* isomer to the structure shown. If that structural formula and the present one are viewed in a mirror, their stereoisomers will be pictured. Only the isomer shown undergoes mediated transport into *E. coli* K12 by System *LIV*, and only it stimulates release of insulin from the pancreatic islet. Furthermore this isomer corresponds more precisely than the others to the structural requirement of System *L* of various cells. Hence the orientation of the pentagon seen in the foreground is critical, as is also the general location of the space requirement of the bicyclic structure with reference to the amino and carboxyl groups.

Although on paper one can arrange almost any enzymatic sequence to present a vectorial step, i.e., so that it provides a *chemiosmotic* coupling, it is striking, perhaps even embarrassing, how much more often the chemiosmotic coupling appears not to occur at the substrate level. Perhaps nature had first to solve the problems of the transport of electrons and inorganic ions. These solutions could then be extended to other transports by way of osmotic–osmotic coupling, thereby presumably reducing the evolutionary stimulus to development through subsequent modification of membrane structure of scores of totally new chemiosmotic couples in other transport systems. Even the phosphotransferase system appears to represent an energizing reaction, coupled to a molecular transport of the usual sort.[5]

Whatever the total role of group translocation, the point I should like to make here is not merely that molecular transport really also exists everywhere we look, but also how readily specific proposals of group translocation can be tested and excluded where they happen not to apply, by examining a range of analogs of the substrate that can be transported but some of which may be immune to the proposed chemical alteration. Too often one does not see these proposals challenged as vigorously as they readily could be. To a colleague who recently proposed an enzymatic cycle of group translocations as the basis for the transport of a class of amino acids, I offered this rather rash challenge: "I'll bet that you have on your own shelves the analog that will serve to exclude your proposal." Until one has tried hard to design analogs that are typically transported but do not undergo the metabolic modification supposedly inherent to the transport, and has thus dissociated the transport event as completely as possible from other events, one has not challenged well the relevance of that modification.

Another distinct possibility that can be eliminated or established by the study of variation of substrate structure is whether a metabolite must be converted into another substance *before* it can be accepted by the transport process. According to our experience, cystine appears to enter or leave some animal cells only after conversion to cysteine. In contrast the finding that cystine and diaminopimelic acid compete for the same transport system[6] and for the same binding protein[7] in *Escherichia coli* shows that cystine is undoubtedly transported by that cell as such. It is important to know for example to what extent glutamate might be transported after conversion to glutamine, or to what extent nucleosides and nucleotides may be interconverted either as a separate event before transport or as an inherent step in transport.

2. *To determine whether we are observing transport by one, two or three or more systems.* Perhaps I have already belabored sufficiently elsewhere the advantage of knowing which properties belong to the transport system we are attempting to describe, and which properties belong to another.[8] I can make the new point that confusing relations will be observed between the properties of a binding protein and the transport activity with which it is presumed to be associated, if the latter is in reality produced by two transport systems. Furthermore, tests for the necessity for an energy input or for a co-substrate will yield confusingly complex results (e.g. indications of loose and variable coupling) if the contribution of a separate facilitated diffusion is unsuspected. We can expect to make little sense of the development of transport systems during embryonic life unless we differentiate among different systems.

In the meantime dozens of new instances of heterogeneity in the transport of a given metabolite have been reported. Hence we probably need say little more about the necessity of selecting or designing the substrate carefully if a single transport agency is to be studied in isolation, or about the confusion likely to arise when one is unknowingly studying the total activity of two or more processes at once.

3. *To describe in a complementary way the shape of a receptor site*, i.e. the spatial distribution of the several components that recognize several chemical groups (usually three or more) in a configuration presented by effective substrates; also the space available and that not available around these components for other parts of the substrate molecule. I will consider first an example where no cosubstrate is recognized, and secondly cases in which the geometric description will include the relation between the positions taken by substrate and co-substrate. I shall begin by provisionally supposing the picture reached to represent a static site, nevertheless retaining for later development important reservations on that treatment.

The transport receptor site accommodating leucine, isoleucine and valine in *E. coli* K12 ("System LIV") provides space for the bulky structure of the bicycloheptyl ring of the norbornane amino acid shown in Fig. 1, but only if it lies in one of the four possible orientations with respect to the α-amino and the α-carboxyl group.[3,9] The discrimination among these isomers could originate either from the spatial distribution of obstacles to the entry of the unsuitable substrates into the site, or from the distribution of structures providing favorable apolar bonding for the suitable substrates. Although reactivity with the amino acid transport systems usually increases with an increase in the number of methyl and methylene groups available for apolar bonding, all the isomers of the norbornane amino acid appear to provide several such groups at positions that produce reactivity for the branched-chain amino acids and their analogs. Hence it appears that the sharp rejection of three of the isomers is based on their encountering obstacles not met by the isomer shown in Fig. 1. Continuing tests of other unreactive and reactive analogs are refining the description of the receptor site. Apparently either this amino acid[3] or trifluoroleucine[10] can be used to discriminate System LIV from a leucine-specific system in *E. coli* strains.

Let me offer next an example where both a substrate and a co-substrate enter transport together. The experiments to be described were made for transport into the Ehrlich cell, the nucleated red cell of the pigeon, and the rabbit reticulocyte. The logic behind the exposition of this case began with observations for the transport system for cationic amino acids and progressed to findings for the neutral amino acid system *ASC*, for which the geometry of the receptor site can now be proposed in the greater detail. The basic amino acid System Ly^{+} reacts strongly with amino acids carrying a cationic group on the sidechain. Unless that group tightly retains its proton, so that it presents a positive charge in an apparently apolar microenvironment at the transport site, the amino acid will show a high K_m (i.e., poor reactivity) for System Ly^{+}; for example arginine and homoarginine are more reactive than lysine on that basis. Homoarginine is more reactive than arginine on another basis, presumably the added strength of the apolar bonding arising from its extra methylene group.

A number of unequivocally neutral amino acids at rather substantial concentrations are, paradoxically, able to inhibit transport of basic amino acids, or to enter countertransport with them. This activity has been found to depend strongly on the identity and concentration of the alkali–metal cation present in the suspending medium. Furthermore, the selectivity among the alkali–metal cations depends in turn on the structure of the amino acid. The neutral amino acids most effective in this reaction are also the ones whose action is most dependent on the

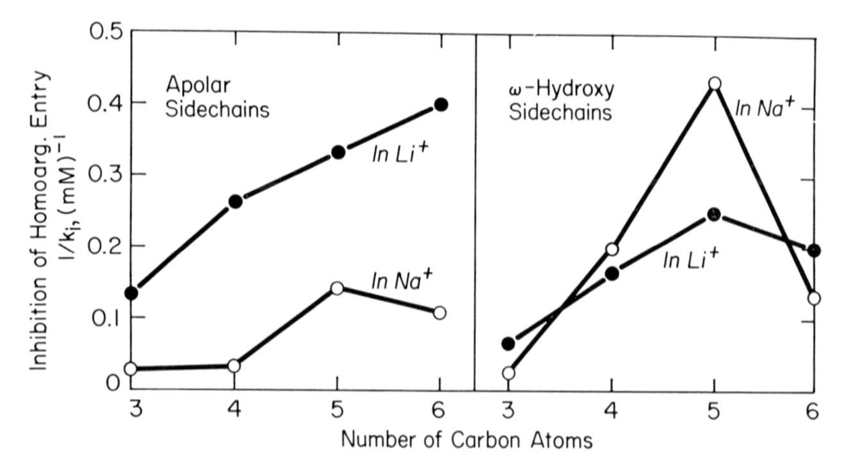

Figure 2. Effect of length of linear chains of amino acids on their inhibitory effect on homoarginine uptake by the Ehrlich cell, in the presence of Na^+ or Li^+. *Left*: For amino acids with apolar sidechains, alanine to norleucine. *Right*: For ω-hydroxy amino acids, from serine to ϵ-hydroxynorleucine. [Na^+] or [Li^+] = 105 mN. Note the sharp reversal of the effectiveness of Na^+ and Li^+ when the hydroxyl group is on carbon five. K^+, Rb^+ and Cs^+ were also tested, but the inhibitory actions were very small when they replaced Na^+ in the modified Krebs-Ringer bicarbonate medium. Corresponding but somewhat different changes in selectivity were seen for the rabbit reticulocyte and in that case when the hydroxyl group fell on carbon four rather than on carbon five. Reproduced with permission from the Journal of Biological Chemistry.[11]

presence of a specific alkali–metal cation, namely Na^+. To be at all inhibitory such amino acids must first of all meet certain structural requirements that we have come to associate with the cationic amino acid transport system (e.g. only one beta carbon; the amino group must be primary); but in addition for a strong interaction with a specific alkali–metal, a hydroxyl or mercapto group must be present on one specific carbon atom (carbon five, for the Ehrlich cell; carbon four, for the rabbit reticulocyte). Figure 2 illustrates the inversion of the preference between Li^+ and Na^+ as the hydroxyl group is shifted to carbon five for the former cell. The other three alkali–metal ions are very low in enhancing activity in that cell,[11] and hence are not represented in Fig. 2.

We suppose the neutral amino acids that interact strongly with a specific alkali–metal ion to inhibit transport by the cationic amino acid system are the ones that first of all can enter Site L_y^+ even if they do not occupy it completely. We suppose furthermore that these neutral amino acids become bonded to the inorganic ion when both substrates are bound at the receptor site, since all those identified as highly reactive at this time contain a sulfur or an oxygen atom on the sidechain and because the interaction is so sensitive to the position of that oxygen or sulfur atom. We suppose that the alkali–metal ion then occupies the position at the receptor site otherwise taken by the positively charged, distal nitrogenous group of the basic amino acid.[11–13]

We were led then to ask whether the converse inhibitory reaction also occurs: Does arginine inhibit the linked transport of neutral amino acids plus Na^+ by systems primarily reactive with neutral amino acids? An apparently ubiquitous system for straight-chain, neutral amino acids, System *ASC*, was studied in the rabbit reticulocyte and the pigeon red blood cells. Its typical substrates are alanine, serine and cysteine (hence the abbreviated designation *ASC*) and their 4- and 5-carbon homologs. In both cases arginine was inhibitory to serine uptake, although in this case no exchange of arginine for (Na^+ + serine) across the plasma membrane could be detected. Either the guanidinium, methylguanidinium or lithium ions was inhibitory to transport by competition with the co-substrate, Na^+.[14]

These inhibitory effects led us to study more closely the normal interaction between the two co-substrates, amino acid and Na^+, in System *ASC*. Both of these co-substrates must be present on both sides of the membrane for transport to take place in nucleated and reticulated red cells. The transport involves the exchange of a neutral amino acid plus Na^+ for either the same or a different neutral amino acid plus Na^+, although the number of sodium ions moving in each direction is not necessarily equal when the two amino acids are dissimilar. As the concentration of each co-substrate was raised, the apparent K_m of the other co-substrate fell. This effect ranged from very small for proline, to very large for the amino acids with a hydroxyl or sulfhydryl group on carbon three or four. Furthermore, the stoichiometry of the flux augmentations rose sharply as the hydroxyl group was placed in the optimal position. About five proline molecules entered the cell with each sodium ion, whereas about three sodium ions were taken up with each serine molecule, and about four sodium ions with each cysteine molecule. These results show that the interaction with Na^+ was again strongly influenced by the presence of an O or S atom in a particular position. Furthermore, the preferred position was distinctly different for this system than for the cationic amino acid transport system. Other evidence is also overwhelmingly against the idea that System *ASC* and the cationic amino acid system are merely two manifestations of the same entity, the K_i values being totally inconsistent with the idea.[13] Furthermore, the cationic amino acid system persists on maturation of the reticulocyte after we no longer can detect System *ASC*.[15]

Finally, the orientation of the hydroxyl group at the preferred position on carbon three or four, whether *cis* or *trans* to the carboxyl group of the amino acid, is decisive to the interaction with Na^+ in System *ASC*. If a *cis* hydroxyl group is introduced into proline, the effects on transport and on interaction with Na^+ are not favorable. If instead the hydroxyl group is *trans*, both transport and potentiation of transport by Na^+ are large, and the flux stoichiometry, $\Delta\nu_{Na^+}/\Delta\nu_{aa}$, is sharply

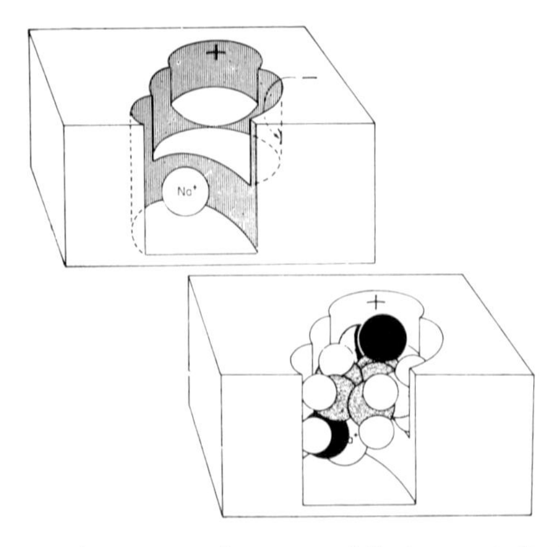

Figure 3. Diagrammatic representation proposed for transport site ASC, showing Na^+ (top) and both Na^+ and hydroxyproline (bottom) in place.

The arrow points to a structure that recognizes the protonated α-amino or α-imino group; the plus sign identifies a structure recognizing the carboxylate group. Undoubtedly the receptor structure to some degree mediates the interaction between the two cosubstrates. Structural considerations persuade us, however, to propose a direct interaction between the two, as suggested by the lower sketch. No implication is intended that Na^+ must enter the site first. See text for further discussion.

increased.[16] Because the aminoprolines are not admitted to site $L\dot{y}$ in the cells we have so far studied, we have not yet attained an unequivocal determination of the corresponding geometric feature of that site.

These results lead us to propose geometric relations in the positions taken by the amino acid and Na^+ at receptor Site *ASC* resembling those crudely pictured in Fig. 3. The Na^+ is visualized as lying nearer to carbon four than to carbon three, at a distance such that bonding can occur to an oxygen or sulfur atom attached to one of these carbon atoms. Furthermore the position of Na^+ is pictured as *trans* to the structural feature of the site that determines the position taken by the carboxyl group. We are led also to believe that apolar bonding occurs to the *cis* aspect of the methyl or methylene group represented by carbon three, because of the especially unfavorable effect of a 3-*cis* hydroxyl group.[17] The unusually high stereospecificity of System *ASC* gains added interest when we propose a different kind of bonding, in addition to apolar bonding, to the amino acid sidechain. The new kind of bonding to the site must be absolutely unavailable to the sidechain if it takes the position characteristic of D-amino acids, whereas the needed apolar bonding must be more diffusely available.

Information of this kind shows the word *co-transport* has turned out to be more precise than it might have proved: The co-substrates appear to bind in juxtaposition and to interact with each other. Until we describe this aspect of their behavior we will not be prepared to understand why each is necessary for the transport of the other.

Reference should also be made to the classical demonstration by Le Fevre and Marshall that the affinity of various monosaccharides for uptake by the human red blood cell is determined by the number of hydroxyl groups offered in an equatorial arrangement by the more stable conformer of the sugar.[18] I suspect for certain Na^+-dependent sugar transport systems that one or more of the hydroxyl groups of the sugar (possibly that on C2, because it is most decisive to Na^+-dependent transports) may bond to Na^+. Several 2-deoxy sugars are taken up by kidney cells by a Na^+-independent transport system, and not by the usual Na^+-requiring system of that tissue,[19] a contrast that may also point to the 2-hydroxyl group.

4. *To modify the selectivity among analogous co-substrates.* In the preceding section we found ourselves fortuitously already pursuing this goal: As we change the structure of the neutral amino acid, the relative effectiveness of Na^+, Li^+ and K^+ as co-substrates in the reaction with System $L\dot{y}^+$ changes (see for example Fig. 2). We are following the same approach with System *A* to see whether we can tell whether the Na^+ lies near any particular part of the amino acid molecule at that receptor site. In this case the structural features that intensify the preference of Na^+ over Li^+ are giving us very promising signals of nearness of the two co-substrates although the topography is unlike that for System *ASC*.

5. *To discriminate sequential events in transport and to determine the nature of each.*

(a) *By design of substrates not entering steps subsequent to binding.* In the preceding section we have largely assumed that our experimental results were describing for us a static receptor site. In actuality it is already possible to see that successive chemical events are occurring. For example, Li^+ plus a neutral amino acid substrate can occupy Site *ASC*; either arginine or methionine[14, 20] can occupy the site without kinetic evidence that Na^+ is binding also. In each of these three cases, however, transport does not occur. Hence both co-substrates must have the right structure; otherwise the site fills but transport does not occur. The opportunity arises to discover the nature of the event that follows

Na^+ binding, but not Li^+ binding, when these ions enter the site in juxtaposition to the amino acid substrate.

Caspary, Stevenson and Crane[21] have observed an opposite kind of case: 6-deoxy-L-galactose (L-fucose) shows a Na^+-dependent inhibition of glucose transport by the intestine. In this case even though the alkali–metal ion is the right one so that occupation of the site can presumably be completed, the monosaccharide is still structurally unsuitable for a subsequent event.

Researches on transport frequently stop short of an important opportunity when they merely show that an analog is inhibitory to the uptake of a standard substrate for a given transport system. The roles of substrate and inhibitor should be reversed, to see if the response of the receptor site to the inhibiting analog is complete transport. It becomes a more remarkable case when 1-anhydroglucitol is a *substrate* as well as an *inhibitor* of galactose transport. On the other hand a valuable investigative opportunity is uncovered if anhydroglucitol should prove instead to be only an inhibitor and not a substrate, because one could then explore the relation of the failure to the missing 1-hydroxyl group. The several opportunities mentioned above have not been well exploited yet, although we can begin to see how they can be used. One must of course think *chemically* about the differences between competing analogs. The experience described in this section counsels us, whenever we see evidence that two substrates of distinctly different charge (e.g. aspartate and fumarate) are competing for the same receptor site, to look for an inorganic ion co-substrate as the equalizing factor. Where *one* sodium ion is co-substrate to the ordinary neutral amino acids for transport, it is perhaps plausible that *two* sodium ions, even though in a different transport system, might be co-substrate to aspartate or glutamate.[22]

(b) *Design of substrates that slow a subsequent step, both for uptake and entry*. Other modifications of transport behavior by variations in substrate structure may also be informative. We have noticed, for example, that all the ordinary substrates of System *A* of the Ehrlich cell show about the same V_{max} for uptake, namely about 5 or 6 mmol per kg cell water·min. But α-(methylamino)-isobutyric acid shows a V_{max} of about one-third this value.[23] Apparently the structural features of the ordinary amino acids do not "obtrude" to change the rate of the rate-limiting event, but the *N*-methyl group does in this case obtrude. We have been able to take advantage of its slowing of the translocative event by demonstrating the phenomenon of *trans* inhibition with this amino acid (Fig. 4).[24] It had been predicted by Heinz and Durbin that *trans* stimulation would be replaced by *trans* inhibition if the rate of reorientation of the filled site could be changed from greater than that of the empty site, to a rate *lower* than that of the empty site.[25] Figure 4 shows that external AIB causes an acceleration of the exodus of either

AIB or MeAIB from the Ehrlich cell, whereas MeAIB causes a slowing of the exodus of either test amino acid. Hence the structure of the amino acid that drives the *trans* effect rather than the one that records it apparently is decisive as to whether the effect is positive or negative. *Trans* inhibition by MeAIB occurs equally well for uptake and for entry (see Fig. 6 in ref. 24); i.e. no evidence is obtained for an asymmetric action of the *N*-methyl group on translocation. Appearance of the predicted *trans* inhibition provides valuable support for the approximate correctness

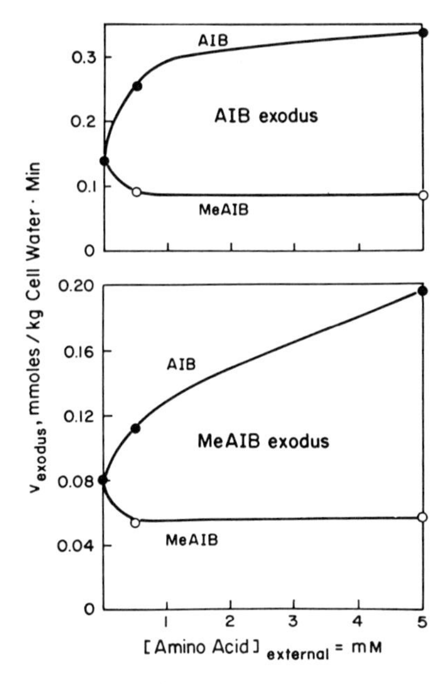

Figure 4. Contrasting production of *trans* stimulation of amino acid exodus from the Ehrlich cell by external α-aminoisobutyric acid, and *trans* inhibition of exodus by its *N*-methyl derivative.

The labeled amino acid as indicated had first been introduced into the cell at an estimated mean internal concentration of about 1·4 mmolar. The loss of the label was then observed during two min into Krebs-Ringer medium containing the indicated concentration of AIB or MeAIB. Note that the responses of labeled AIB and MeAIB are essentially the same; i.e., the results are uninfluenced whether homo- or hetero-exchange is being observed.

of aspects of the model for transport under which we are proceeding. In addition it offers some as yet unclear opportunities to discover something about the nature of the translocation event: Why does the *N*-methyl group obtrude to cause slowing when the 5-methylene group of proline does not?

This kind of *trans* inhibition lowers both of the two opposed fluxes and does not apparently modify the equilibrium position finally reached; but presumably it can slow progress of net uptake to such a degree that in the real case uptake effectively may stop far short of equilibrium, as appears to occur in *Neurospora*.[26] Such an arrangement could prevent over-charging of the amino acid pool of a cell, and for a tightly-coupled uphill transport should conserve energy. It should, however, leave the organism sensitive to an excess of any single substrate of the system concerned, because the uptake of other substrates of the same system would presumably also be interrupted.

Several years ago, Oxender and Whitmore[27] showed that a 2-min treatment of the Ehrlich cell with 0·5 mmolar 2,4-dinitrofluorobenzene minimizes or abolishes *trans* stimulation by substrates of transport System *L*, without causing a proportional slowing or diminution of their net uptake. This independent abolition of exchange seemed to indicate that exchange and net uptake occur by different components of the transport System *L*.[27] But is it not more likely that *trans* stimulation is eliminated by a specific slowing of the reorientation of the substrate-carrier complex just as it was through the methylation of AIB? In testing this hypothesis we have shown that treatment of the Ehrlich cell with DNFB slows not only uptake of 2-aminobicycloheptane-2-carboxylic acid by amino acid-depleted cells, but also its exodus into amino acid-free medium, i.e. net exodus. So far the only way we have to show that this slowing does not apply also to reorientation of the unloaded as well as the loaded carrier is the near elimination of *trans* stimulation, not unfortunately an independent proof. If we select one of the more slowly migrating substrates of System *L*, DNFB may be able convincingly to replace even the powerful *trans* stimulation of System *L* with *trans* inhibition.

A fundamental question inherent in these tests is whether one can indeed eliminate *trans* stimulation (accelerative exchange) or *trans* inhibition by changing substrate or membrane structure in any way other than by manipulation of the relative rates of migration by steps inherent to the transport process itself.

(c) *Design of substrates that accelerate a subsequent step*. An effect of this kind has presumably been encountered in the study of System *ASC* already reported. The presence of a properly orientated hydroxyl or mercapto group appears to accelerate differentially the release of Na^+ from the ternary complex, thereby sharply increasing the coupling stoichiometry of Na^+ to amino acid.[16]

Acceleration of the translocation step appears to be produced in a number of cases by the presence of a second amino group, usually on the γ-carbon atom. Oddly enough, this behavior is seen with systems for neutral amino acids, for which diamino acids would not be expected to serve as substrates. The first such case encountered was α,γ-diaminobutyric acid, which shows a remarkably high V_{max} for System *A* of the Ehrlich cell, over 60 mmol per kg cell water·min. Since $pK_2' = 8{\cdot}4$, this compound is actually cationic to the extent of about 91% in free solution at pH 7·4. A corresponding transport by System Ly^+ can easily be measured. Apparently at the transport receptor Site *A*, however, it is stabilized in the form in which the γ-amino acid group is not protonated and hence the compound also enters transport as a neutral α-amino acid by that system. Such a stabilization could well result if the γ-group falls in a highly apolar region at the transport receptor site.* We will consider later the question why the V_{max} of this amino acid should be so high.

Here we presume the hydrogen ion, rather than an alkali–metal cation, serves as the factor compensating for the difference in charge among competing substrates, a circumstance that could permit energy transfer from a movement of hydrogen ion, down a gradient. This possibility recalls the apparent service of H^+, in place of Na^+ as a co-substrate for uptake of neutral amino acids by yeast.[28]

Note that the chemical form in which we believe α,γ-diaminobutyric acid to be stabilized on its entry into transport System *A*, namely

$$H_2NCH_2CH_2CH(NH_3^+)COO^-$$

should be an excellent substrate also for System *ASC*, just as the closely analogous homoserine is. Instead the diamino acid is essentially inert with respect to System *ASC*; a 50-mM concentration shows less than a 4% effect on the uptake of L-serine by the pigeon red blood cell or the

TABLE I. Predicted route of transport of diamino acids

γ-Zwitterion	Cation *	α-Zwitterion
$CH_2(NH_3^+)CH_2CH(NH_2)$-COO^-	$CH_2(NH_3^+)$-CH_2-$CH(NH_3^+)$-COO^-	$CH_2(NH_2)$-CH_2-$CH(NH_3^+)$-COO^-
ß system?	$\underline{Ly}^+$ system	ASC and A systems
Found:		
partial inhibition by ß-alanine	partial inhibition by lysine	All criteria for powerful uptake by System A; Negligible uptake by System ASC

Conclusion: This amino acid reacts like an α-Zwitterion with System A, but with System ASC it does not!

* Note the inductive action between the two amino groups of this compound. The cationic form shown in the upper middle of this table, and also that shown in a similar position in Table III, can dissociate in either of two ways: to form an α,γ-dipolar ion as shown at the left, or to form an α,α-dipolar ion as shown at the right. For diaminobutyric acid, also for 4-aminopiperidine-4-carboxylic acid, as shown in Table III, and for several other of the basic amino acid discussed here, either of these pathways may occur rather readily and the proportion dissociating in a given way should be sharply modifiable according to the environment the two amino groups find at the receptor site. Once one of these protons has been lost, however, the other comes by an inductive effect to be more strongly retained. Conversely, stabilization of the retention of one of them causes the other to be lost more readily. If we guanylate the piperidine N (lower, central structure in Table III), the resulting guanidinium group will not readily lose its positive charge. Hence the dissociation will be stimulated to proceed along the pathway toward the lower left, to form an α,γ-dipolar ion, which is not a substrate for any of the transport systems under consideration.

rabbit reticulocyte. This finding is not paradoxical when we remember that the γ-amino group will find no apolar region at the *ASC* receptor site, if the model suggested in Fig. 3 is correct. Nevertheless the contrasting acceptance of this amino acid by System *A* and its rejection by System *ASC* tells us of an important difference in these two. Table I presents the contrasting behavior of α,γ-diaminobutyric acid just described.

Other cases in which the state of dissociation of a basic amino acid appears to be modified in the environment intrinsic to transport have been encountered for several lysine analogs and will now be discussed at considerable length. Lysine itself was found surprisingly to behave

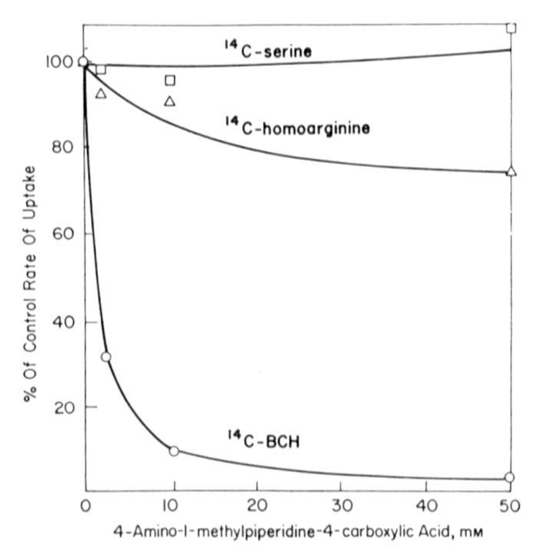

Figure 5. Inhibition by 4-amino-1-methylpiperidine-4-carboxylic acid of uptake of three labeled amino acids by the Ehrlich cell.

Uptake of the test amino acid measured during 1 min at 37° and pH 7·4 from Krebs-Ringer bicarbonate medium. The failure to inhibit serine uptake reveals negligible reaction with Systems *A* and *ASC*. The action on homoarginine uptake shows weak reaction as a cationic amino acid, whereas the inhibition of BCH uptake shows a strong reaction with System *L*.

to a considerable extent as a *neutral* amino acid, presumably through a degree of repression of the same sort of the protonation of its *epsilon* amino group.[13] For that reason it cannot be trusted as a test substrate specific to the cationic amino acid transport system. Instead arginine analogs, which are unambiguously cationic, must be recommended as the model substrates for System Ly^{+}.

Figure 5 illustrates one of the methods we used for investigating transport properties of the lysine analogs.* A piperidine amino acid shows itself ineffective in inhibiting serine uptake by the Ehrlich cell during 1 min (upper curve), effective only at high levels in inhibiting

* The results of Tables II and III and Figs. 5 to 9 have been reported in a preliminary form.[29] M. E. Handlogten and A. M. Cullen collaborated with the author in obtaining the results to be described by these tables and figures.

homoarginine uptake (middle curve), but extremely effective in inhibiting uptake of the norbornane amino acid that serves as a model substrate for System *L*, the system for neutral amino acids with bulky, apolar sidechains (lower curve). Some of the analogs tested (Table II) are either piperidine amino acids (the first three) or 4-aminocyclohexyl amino acids (the last two). Because their pK_2' values lie between pH 7 and 8, they should be present in free solution partly as cations and

TABLE II. Transport reactivity of analogs of basic amino acids in the Ehrlich cell

	K_i on homoarginine uptake (mM)	K_i on uptake of norbornane amino acid (mM)	Distr. ratio at 60 min
4-Aminopiperidine-4-carboxylic acid	50	15	12·8
1-methyl derivative	90	1	25
1-guanyl derivative	4·3	>100	12·7
Cis-1,4-Diaminocyclohexanecarboxylic acid*	30	2·7	10·2
Trans isomer*	60	8	2·4

* The *cis* isomer was separated from admixture with the *trans* isomer by conversion to the copper salts, the *cis* isomer forming a particularly insoluble salt. Structure was established by showing that diamino acid forming the soluble copper salt can be cyclized to an α-piperidone, as ornithine can, by neutralizing its methyl ester hydrochloride with excess freshly precipitated hydrous silver oxide,[30] whereas the diamino acid isolated as the insoluble copper salt cannot. Synthesis of the mixture of isomers proceeded via 4-toluene-sulfamido-cyclohexanone, converted to its aminonitrile by the Strecker synthesis, and the latter to the free amino acid by HCl hydrolysis. The Bucherer-Libe synthesis yielded mainly the *trans* isomer.

partly as dipolar ions, the latter probably largely ω-zwitterions. All of them react extremely weakly with the transport system for cationic amino acids (first column of data) unless an unambiguously cationic group is produced either by dimethylation or by guanylation of the distal N atom (third compound listed).

The simplest, unsubstituted piperidine amino acid (the first compound listed) proves to be a fairly good substrate for the transport System *L* for neutral amino acids (middle column of data). Hence here also each of the receptor sites of two transport systems appears to suppress the otherwise likely protonation of the terminal cationic group. This interpretation is strengthened by the effect of methylation of the piperidine N (second line)—the product is fifteen times as reactive with the neutral System *L*, and now it is even less reactive than before as a cationic amino acid. Presumably this methyl group lowers only moderately the pK' of the piperidine N, its main effect perhaps being to assist the accommodation of the unprotonated sidechain into the apolar microenvironment supplied for the sidechain at Site *L*.

Similar behavior is seen for what we may call the next higher homolog, 1,4-diaminocyclohexanecarboxylic acid, when the two amino groups are in the *cis* relation. It is highly reactive with System *L*, very weakly with System Ly^{+}. If, however, the two amino groups are in the *trans* orientation, the transport reactivity for System *L* is substantially weaker. It should be noted also that the several amino acids mentioned so far are concentrated 10- to 25-fold by the Ehrlich cell, except that the *trans* diamino acid (lower right) is much less strongly concentrated.

Does this *cis*, *trans* difference serve to report to us differences in the microenvironment encountered by the two sides of the amino acid sidechain? Perhaps, but not necessarily. We may instead suppose that the *cis* isomer has the advantage that its δ-amino group as it enters the apolar environment is able to donate its proton directly to the α-amino group, whereas for the *trans* isomer no such approach and donation can take place. To discriminate between these possibilities we need lysine analogs that present separately each of these two features, namely (1) a *cis* orientation of the two amino groups, and (2) approachability of the two amino groups. For the present, then, we cannot draw an unequivocal 3-dimensional picture for Site Ly^{+}.

The poor reactivity of these diamino acids as cationic substrates is unequivocably contrary to the prediction from their titration curves. In free solution they are certainly present from 25 to 50% as cations (39% in the case of the *N*-methyl derivative), but their low reactivity shows that they maintain no such structure at the Ly^{+} site. Furthermore, from the reactivity of the corresponding *N*-dimethyl and guanyl derivatives, we know there is no lack of room at the site for amino acids with the *N*-methylpiperidine or aminocyclohexane structure. Hence we suppose that the recognition site of System Ly^{+}, as well as that of *L*, presents an environment unfavorable to the retention of a proton on the distal amino group.

We will return then to the reaction of diamino acids with Site *L*. Figure 6 shows that the species of 4-amino-1-methylpiperidine-4-carboxylic acid reactive with System *L* appears as expected to be the dipolar ion MPAA±. As the pH is raised, the concentration of the transport reactive species may be seen to increase in approximate proportion to the amount of the dipolar ions generated. But that result does not tell us which dipolar ion, the α- or the γ-zwitterion. For the 1-methylpiperidine amino acid, pK_2' is 7·2 and pK_3', 9·7. The question is, which group does pK_2' represent, and which group does pK_3' represent? We can readily answer that question for the benzyl amino acid, for which $pK_2' = 6{\cdot}9$ and $pK_3' = 9{\cdot}2$ (Fig. 7). Figure 8 shows at the left the change in the UV spectrum during its titration. As the two steps pK_2' and pK_3' of the titration curve are produced, the UV absorption may be seen to

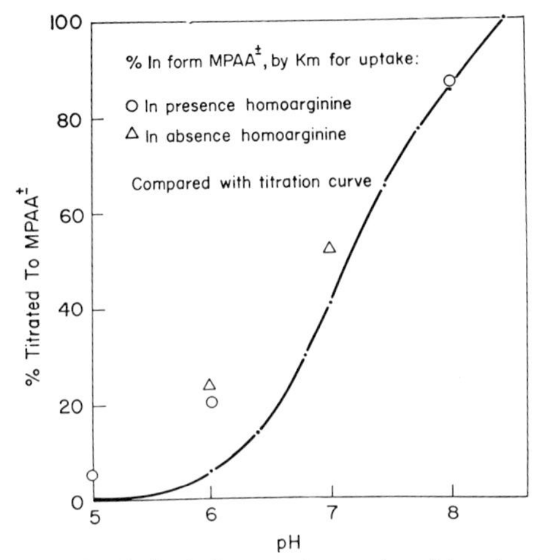

Figure 6. Identification of an isoionic form as the species of 4-amino-1-methyl-piperidine-4-carboxylic acid reactive with transport System *L*.

The solid line is a portion of the titration curve of the amino acid at 25°, $\Gamma/2 = 0{\cdot}1$, and shows the proportion present in the isoionic form $MPAA^{\pm}$ as a function of pH. The circles and triangles show the proportional amount present in the transport-reactive species as indicated by the measure of K_m for uptake at 37°, the triangles representing total uptake, and the circles, the uptake retained in the presence of 20 mM homoarginine. Krebs-Ringer phosphate medium adjusted to the indicated pH values, under an oxygen atmosphere. This result does not tell us however which isoionic form, the α- or the γ-zwitterion, is the reactive species. The demonstration is precisely valid only if System *L* is pH-independent—which as we shall see is not quite the case.

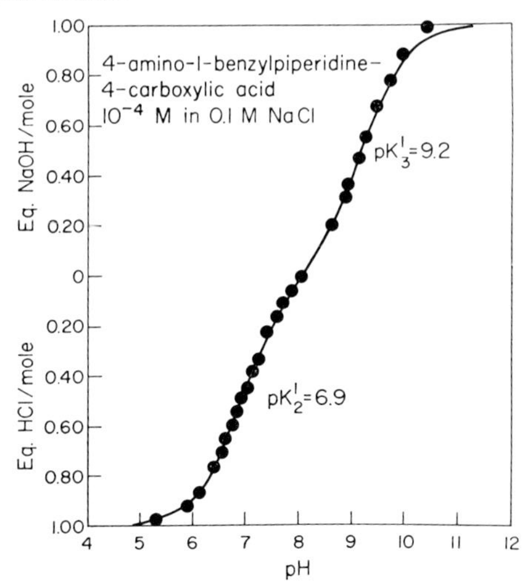

Figure 7. Titration of 4-amino-1-benzylpiperidine-4-carboxylic acid with NaOH and HCl. The amino acid was 10^{-4} M in 0·1N NaCl at 25°C. This result does not tell us of course to what degree the α- and the γ-nitrogen atoms figure in each of the two titration steps.

increase sharply at wavelengths of about 262 and 268 nanometers, producing the two peaks seen in the middle of the figure. Of the total increase in absorbancy at these peaks, 41% occurs in association with pK_2' (note separation of the lowest and the middle curve) and 59%

occurs in association with pK_3' (note separation of the middle and the highest curve).

For reference purposes we made the same titration with 1-benzyl-4-piperidone which has only a single stage of dissociation since it has no other titratable group than the piperidine N. The same development of the two peaks now takes place in the single step. This result identifies

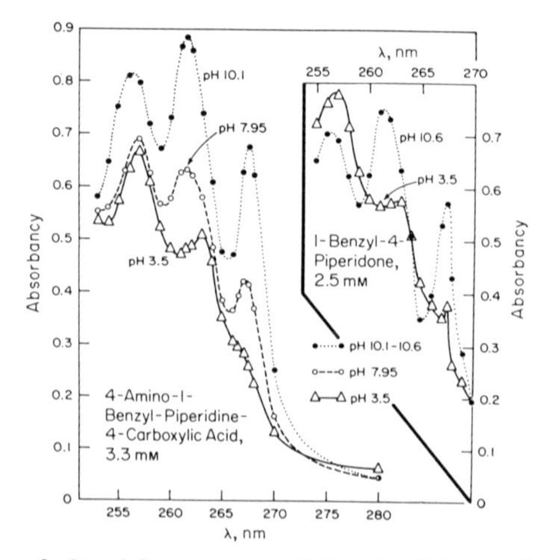

Figure 8. Change of ultraviolet spectrum of 4-amino-1-benzylpiperidine-4-carboxylic acid, 3.3 mM, and of 1-benzyl-4-piperidone, 2·5 mM, with change of pH.
NaCl = 0·1 N, 25°C. The indicated pH values were obtained by adding very small volumes of HCl or NaOH. (See text for discussion.)

the appearance of these two peaks with the formation of the unprotonated benzylpiperidine group. Accordingly, the two upper steps of the titration of the benzylpiperidine amino acid are severely hydridized, the first step involving 41% protonation of the piperidine N and 59% protonation of the α-amino group. The second step then completes both protonations.

Our use of the benzyl group, however, undoubtedly directs the initial protonation toward the piperidine N, in comparison with the 1-methyl analog. For example the pK' of 1-benzylpiperidone is 7·2, whereas that of 1-methyl-4-piperidone is 8·0. Therefore the latter amino acid will certainly form less α-zwitterion than the former in free solution. Note that this difference of 0·8 is the same we saw between the 1-benzylpiperidine and 1-methylpiperidine amino acids; but in that case the difference was divided between the two groups in the ratio 3:5. This result tells us that pK_3' of the 1-methylpiperidine amino acid is associated to a larger extent with the piperidine N than with the α-amino group, since that nitrogen atom will certainly be the more sensitive to our substitution of the benzyl group for the methyl group. From the relation of these pK' values it can be estimated that at least 5 molecules of the δ-zwitterion should be present at pH 7·4 for each molecule of the α-zwitterion. This calculation means that we should divide by 6 the K_i of the methylpiperidine amino acid describing its reactivity with System *L*, in order to obtain the value of K_i applying to the α-zwitterionic species. The value of 0·17 mM thus obtained is unrealistically low in comparison with the values shown by analogs certainly not present as the δ-zwitterion, e.g. the 1-formylpiperidine amino acid or 1-aminocyclohexane-1-carboxylic acid. Accordingly we conclude that the *L* site has stabilized the methylpiperidine amino acid in the α-zwitterionic form, thus producing an unexpected reactivity with System *L*, far beyond that to be expected from the quantity present in that form in free solution.

TABLE III

Predicted Route of Transport of Diamino Acids

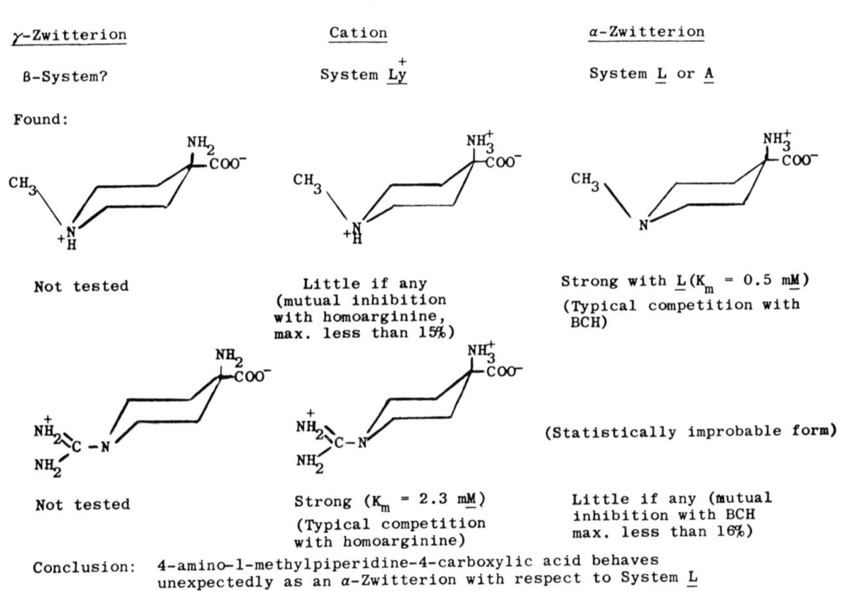

Table III summarizes the contrast between the 1-methyl and the 1-guanyl piperidine amino acid. The first behaves as a neutral amino acid with respect to System *L*, even though only a small part is present in the reactive form in free solution. In the second case, guanylation of the piperidine N has produced so high an affinity for the hydrogen ion that it reacts for transport only as a cationic amino acid.

We should emphasize that we are by no means proposing that the production of an uncharged sidechain is necessary to permit entry of these hydrophilic substrates into the bulk apolar environment of a lipid pool, e.g. that presented by a bimolecular lipid bilayer. Clearly certain parts of the amino acid molecule encounter instead polar areas at which they are bound, and only a limited and distinct part of the molecule encounters apolar bonding. I question whether there is good evidence for an indiscriminate conversion of amino acids or sugars into micromolecular, lipid-soluble derivatives during their physiological transport.

(d) *Design of substrates that intensify the asymmetry in the two-directional operation of a transport system.* In the preceding section we began by considering the design of substrates that accelerate the rate-limiting step, so that V_{max} for transport is increased. Incidental to that effort two other kinds of behavior have been encountered:

(1) Failure of cationic amino acids to be transported as though they are cations, as already noted.
(2) Intensification of the net uphill transport that can be achieved by a given system far above that obtained with any natural substrate.

The reader will note in the last column of Table II that the piperidine amino acids are concentrated much more strongly by the Ehrlich cell than are the ordinary substrates of System L at similar concentrations. Do features of their structures cause these amino acids to respond to an energy gradient across the membrane that is not sensed for example by leucine or isoleucine? By equilibrium dialysis we can detect no evidence of significant binding of 4-amino-1-methylpiperidine-4-carboxylic acid to cell contents after lysis. The difference of electrical potential across the cell membrane can scarcely be sufficient to account for the intensity of its apparent uphill transport, even if we assume that it exists largely as a cation within the cell. The question also arises for 4-amino-1-guanylpiperidine-4-carboxylic acid: Why is it concentrated so much more than arginine and lysine, to which it is an analog? Does it sense some energy-releasing flow across the membrane that arginine cannot sense or respond to? Again, no binding to cell contents could be detected after lysis. In any case we must conclude that both systems L and Ly^+ are capable of asymmetric operation, a capability that has heretofore often been equivocal, especially for System Ly^+.

A question of parallel interest emerges from comparison of maximal velocities as a function of structure. We have touched in passing on the uniquely high V_{max} of α,γ-diaminobutyric acid uptake by System A.[31] Another diamino acid, S-(2-amino-ethyl)-L-cysteine (*thialysine*) shows the same paradox of a high maximal velocity of uptake, although in this case for both systems A and L of the Ehrlich cell. This amino acid has a pK_2' of 8·39,[32] so it should be about 90% cationic in free solution. It does indeed inhibit the uptake of homoarginine (K_i 1·2 mM), and a corresponding component of its uptake is inhibitable by homoarginine. This case shows incidentally that ready entry of a diamino acid into System

L is not necessarily associated with its rejection by System Ly^{+}, a correlation that might otherwise have raised doubt as to the separateness of the two systems.

Thialysine shows a K_m of 20 mM in its MeAIB-inhibitable (System *A*) uptake, and a V_{max} of about 20 mmol per kg cell water·min, about four times the usual value for V_{max}.[23] Given that the proton on the distal amino group can be displaced in the environment of the receptor site, the observed reactivity with System *A* is not surprising because of the linear nature of the thialysine chain. Note that methionine and norleucine are highly reactive substrates for that system as well as System *L*. The K_m of thialysine for System *L* (uptake inhibitable by the norbornane amino acid) is 0·6 mM, and V_{max} about 6 mmol per kg cell water·min, the latter about three times the usual values.[33] Note also, if we correct for the proportion of thialysine present in free solution in forms other than the α-zwitterion, that the K_m of this amino acid for System *L* must be unusually low, and hence the quotient describing its rate of uptake at low levels, V_{max}/K_m, uniquely high.

Note the following parallelism: α,γ-Diaminobutyric acid $pK_2' = 8{\cdot}4$, shows a V_{max} of not less than ten times typical values for System *A*; thialysine, $pK_2' = 8{\cdot}39$,[32] shows V_{max} values for both Systems *A* and *L* three or four times the usual values. The pK_2' value of 8·4 may be almost ideally positioned to permit two effects:

(a) It is low enough so that protonation of the distal amino group may be repressed in an apolar microenvironment at the receptor site.

(b) It may be high enough to permit, subsequently but within the transport process, reprotonation of the distal amino group, which could permit the potential difference across the membrane to accelerate translocation.

This possibility we have tested for by modifying and reversing the potential difference across the plasma membrane of the Ehrlich cell, an effect sought by substituting the non-permeating anion, sulfate, for chloride in the suspending medium. The initial rate of the uptake of thialysine by the neutral amino acid systems is only moderately sensitive to this substitution. The V_{max} for uptake is, however, still exceptionally high after the chloride content of the medium has been replaced isoosmotically or isotonically with sulfate.

The transport asymmetry noted with some of these diamino acids can be observed also as an asymmetry of the accelerative exchange process, as illustrated in Table IV. Whereas internal BCH greatly stimulates the uptake of *cis*-1,4-diaminocyclohexanecarboxylic acid, and the latter when presented to the exterior of the cell stimulates correspondingly the exodus of BCH, no such responses are seen in either direction when the cyclohexane amino acid is presented within the cell

TABLE IV. Representation of sidedness in the mutual acceleration of entry and exodus produced between two amino acids

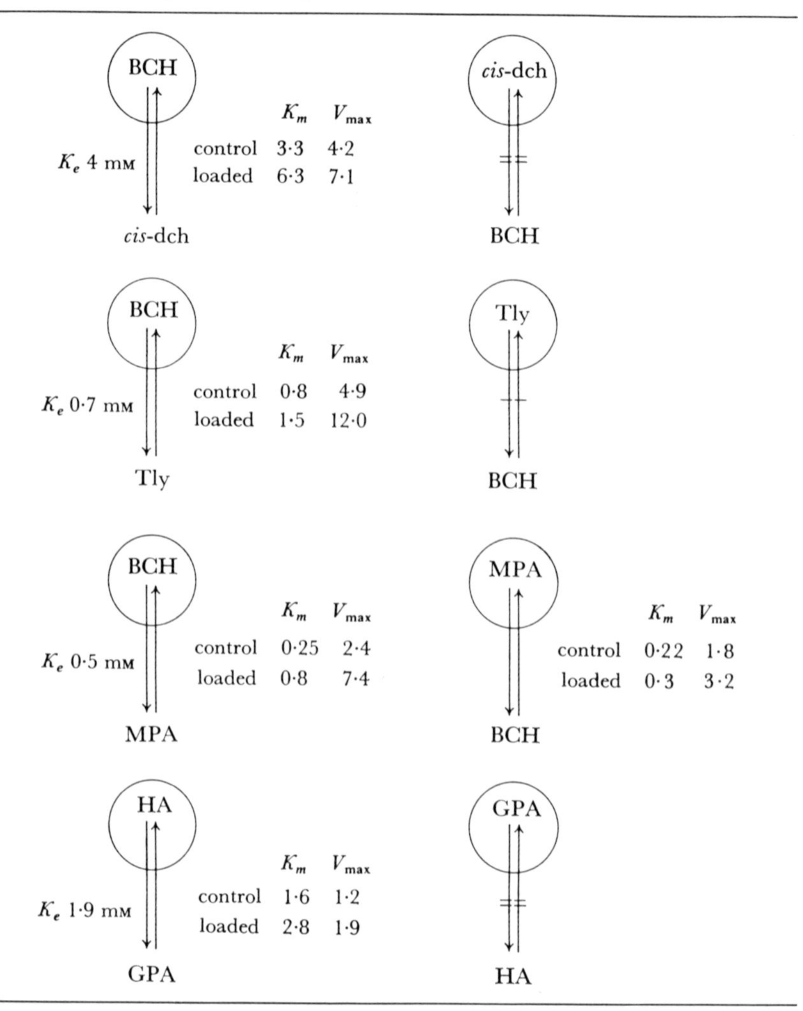

The amino acid enclosed in the circle has been introduced into the cell, in one experiment in labeled, in another in unlabeled form. In the latter case the apparent concentrations were, reading from left to right, line by line, BCH 6·2, dch 7·4, BCH 7·1, Tly 4·5, BCH 7·3, MPA 8·4, HA 5·8, GPA 3·0, all in mmoles per kg cell water. The unbarred arrows in the column at the left show that each of the two amino acids mutually accelerates the migration of the other across the membrane to produce exchange; whereas the barred arrows at the right indicate that when the two amino acids are in the opposite positions, such effects either can not be seen or in the case of thialysine are unusually small. BCH = 2-aminobicycloheptane-2-carboxylic acid; dch = *cis*-1,4 diaminocyclohexanecarboxylic acid; Tly = L-thialysine; MPA = 4-amino-1-methylpiperidine-4-carboxylic acid; GPA = 4-amino-1-guanylpiperidine-4-carboxylic acid; HA = L-homoarginine. K_e represents the concentration of the external amino acid producing a half-maximal acceleration of exodus of the internal amino acid; K_m and V_{max} describe entry of the external amino acid in the absence and in the presence inside the cell of the other amino acid at the apparent concentrations already stated. Note that the value of V_{max}/K_m was usually not increased by the prior loading; hence acceleration tends to be appreciable only at external substrate levels above K_m.

and BCH on the outside. A similar situation applies to the interaction between BCH and thialysine: Unusually weak exchange can be seen between internal thialysine and external BCH, whereas exchange is vigorous when the amino acids are presented in the reversed positions.

The behavior presented in these two cases can plausibly be attributed to a relative failure of the internally orientated receptor site to accomplish the paradoxical effect achieved by the externally orientated site for these diamino acids, which we suppose is to stimulate deprotonation of the distal amino group. Since such an effect will cause an accumulation of the diamino acid, it will have an energy cost, presumably invested in maintaining the difference in the state of the site in the two orientations. We face the question whether and how the presumed energy investment is kept smaller when the substrate of System *L* is instead leucine or phenylalanine, amino acids which are only weakly concentrated by this cell. These results may mean that an intrinsic chemiosmotic responsiveness of the plasma membrane permits transport systems to respond by stimulated energy transduction to substrates that can exist alternatively in charged and uncharged states.

Early in our studies of the heterogeneity of neutral amino acid transport, we were perhaps unduly successful in calling attention to the stronger participation of the Na^+-independent System *L* than of Na^+-dependent System *A* in amino acid exchange. Some readers apparently as a result have incorrectly equated Na^+ independence of uptake with uptake by exchange, and have therefore supposed that the heterogeneity in amino acid uptake can be eliminated by depleting the supply of endogenous amino acids available for exchange. The Na^+-dependent systems are however also exchanging systems. In reticulated and nucleated red cells System *ASC* is much more sharply restricted to exchange than is System *L*. Our present results show attainment in the absence of added Na^+ of accumulation of test amino acids in amounts exceeding the initial size of the entire endogenous amino acid pool, and with only modest depletion of that pool. We have for example obtained apparent gradients of more than 32 mM of 4-amino-1-methylpiperidine-4-carboxylic acid, a value exceeding by 30% the whole content of the endogenous pool, with only 8% depletion of that pool. The source of energy for this uptake is suggested by a heightened sensitivity of the uptake of 4-amino-1-methylpiperidine-4-carboxylic acid to the presence of 2,4-dinitrophenol or oligomycin. The latter is effective at 0.1 μg/ml in oxygen, but only at 5 or 10 μg/ml in N_2 (glucose present). These results are consistent with a utilization of mitochondrial ATP in O_2 and of ATP from glycolysis in N_2. We can find no basis, however, for any idea that the Na^+-dependent: Na^+-independent dichotomy in amino acid transport arises from the distinction between net uptake and uptake by exchange.

We are warned by the third case presented in Table IV that steep asymmetries in the operation of System L do not arise simply from a complete "sidedness" with reference to the exchangeability of two analogs across the plasma membrane. 4-Amino-1-methylpiperidine-4-carboxylic acid and BCH exchange well, no matter which is inside and which is outside; yet the first of these amino acids is concentrated twenty-five times into the Ehrlich cell. The final case tabulated provides a different warning, namely that a difference in the state of protonation of the distal cationic group is not the only way that "sidedness" in exchangeability between two amino acids arises. The exchange between homoarginine and 4-amino-1-guanylpiperidine-4-carboxylic acid is vigorous when homoarginine is the internal member of the pair, but undetectable when the other arginine analog is internal. These compounds both have unequivocally cationic sidechains, and an effective repression of the protonation of the guanidinium group of neither of these seems likely in the biological context. A much more likely explanation is a repression of the protonation of the α-amino group of the guanylpiperidine amino acid at the internal face. The positive charge on the nearby guanidine group would facilitate inductively the loss of the proton from the α-amino nitrogen. The occurrence of this event could solve another mystery: In the case of the cyclohexane *cis*-diamino acid, retention of the distal proton might be expected to cause transport via system Ly^{+} rather than System L. If an α-proton tends to be lost at the same time that one is gained by the distal amino group, however, the amino acid becomes comparatively unsuitable as a substrate for either system, as actually observed (Table III).

The presence of a distal amino group appears not to be the only structural change that serves to generate an asymmetry between entry into and exodus from the cell. While we were trying several years ago to design an amino acid obstructed to substrate action in all known transport systems, we encountered the strange case of α,α-diethylglycine, which is of course an isomer to the leucines. This substance we now recognize as an atypical substrate for System L, although at one time its unusual behavior had all but convinced us that it is accumulated instead by a unique transport system.[34] Diethylglycine is gradually accumulated into the Ehrlich cell by 10-fold or more, $K_m = 4$ mM, $V_{\max} = 2{\cdot}4$ mmol per kg cell water·min. Even though entry is accordingly quite slow, exodus is so extremely slow that high accumulation ultimately results. The two α-ethyl groups apparently are barely accommodated by the receptor site for entry, but scarcely if at all by the receptor site for exodus. Hence crowding by the two α-ethyl groups presents another structural feature, besides a distal amino group, that brings out more of the asymmetry that must be inherently possible in the bi-directional operation of System L.

As for some of the diamino acids already discussed, we see here a logical consequence of the asymmetry of the transport of this amino acid: Although external diethylglycine will accelerate the exodus of the norbornane amino acid, and internal norbornane amino acid will accelerate the uptake of diethylglycine, no *trans* stimulation is seen when the positions of the two amino acids are reversed. An even more crowded analog, α,α-dicyclopropylglycine, which shows no saturable uptake by the Ehrlich cell, is nevertheless accumulated by rat brain slices[35] and by the liver of the intact rat.[36] Amino acids that serve as characteristic substrates of System L are effective inhibitors of dicyclopropylglycine uptake by these tissues.

Perhaps the most interesting property previously reported for diethylglycine uptake is the near doubling of its initial rate when the pH is lowered from pH 7·4 to 5.[34] A slowed exodus at pH 5 makes an additional contribution to the higher accumulation at that pH. Hence the asymmetry between the fluxes is much larger at pH 5 than at 7·4.

Basis of structurally determined access to energy sources. Let me attempt a summary of the present section. We have seen that asymmetry between the two fluxes across the plasma membrane, favoring strong accumulation, can be obtained by certain modifications of substrate structure. The most effective modification is the introduction into the sidechain of a distal amino group with a rather low pK_a'. This amino group behaves as though it is not protonated when the amino acid enters a neutral amino acid transport system from outside the cell. At the interior surface, however, it fails to show this paradoxical behavior; that is, exodus of these amino acids is exceptionally slow, as though they maintain an unsuitable structure at that surface. Structures of that character should predominate in water solution.

The important feature, however, is that slow exodus is only part of the explanation of the strong accumulation of diamino acids. Their maximal rates of entry are much faster than those of the normal substrates of the neutral amino acid transport systems. Although they behave as though the sidechain were unprotonated on their first combining with the external receptor site, they behave as though the amino group had become protonated by the time they leave the membrane to enter the cytoplasm, not a strange conclusion considering that the internal receptor site apparently sees the free amino acid inside the cell in the less acceptable, ω-protonated form. Furthermore, this presumed transition from a species without net charge to a cationic species must occur *during* the translocation step, since that step is strongly accelerated by the introduction of the sidechain amino group.

The strong accumulation of these amino acids requires an energy source outside the transport system. That energy demonstrably does not come from pre-existing gradients of exchangeable amino acids, since the high initial velocities of entry are seen even at high substrate levels producing maximal rates, accompanied by relatively small losses of endogenous amino acids. The difference in electrical potential across the membrane appears not to be necessary to the high rate of accumulation.

For System *A* we have to think also of the down-gradient flows of the alkali–metal ions, particularly Na^+, as energy sources. The stoichiometry of the interdependent or linked fluxes between Na^+ and diaminobutyric acid into the Ehrlich cell we find if anything lower than the stoichiometry observed for AIB and glycine. Hence at least for that amino acid, the presence of a protonatable, sidechain amino group does not increase the proportional flow of Na^+ through System *A*. I also see no reason why the presence of this amino group should tighten coupling between the two flows. Furthermore, a deficiency in the energy available from the Na^+ flow is already a matter of concern even if tight coupling is assumed for ordinary amino acids[37] (cf. ref. 38). For System *L* the Na^+ gradient appears not to be a significant factor, and no energizing gradients except those of amino acids have so far been identified.

What then is the nature of the energy source that is tapped to an unusual extent when an amino group whose protonation is easily modified is present on the amino acid sidechain? All uphill transport must be driven, I believe in a final analysis, by either co-transport or counter-transport, given only that the species entering into linked flux with the transported substrate may be any whatever, even an electron or a proton. This statement I take to follow directly from Curie's theorem, that is, from the circumstance that a completely scalar process cannot drive a vectorial process. Co-transport and counter-transport seem to me to be the modes of linkage between two vectorial processes under ordinary carrier models.

We may then restate the question opening the preceding paragraph as follows: What flow within the membrane is sensed or stimulated to a superior degree when an amino group of suitable pK is present on the amino acid sidechain? We come readily to the response that the flow may be a flow of the hydrogen ion.

Now the H^+ gradient across the plasma membrane of the Ehrlich cell has already been examined by use of a weak acid presumed to reach a diffusion-type equilibrium for its protonated form only.[39] Although, the observed H^+ gradients vary accordingly to metabolic conditions their magnitudes appear ordinarily even less adequate than the gradients of Na^+ or K^+ to account for the accumulation of amino acids.

A provisional interpretation. Because of these observations we suppose that the selected amino acids have a superior ability to establish a vectorial linkage with gradients that are ordinarily to a large degree *confined to the membrane* somewhat as suggested by Williams,[40] and not fully expressed between the two aqueous phases separated by the membrane. It is perhaps somewhat arbitrary whether we describe this gradient as a gradient of hydrogen ion or as a gradient of polarization of molecular structure, somewhat like proposals made recently by Green and Ji[41] for the inner mitochondrial membrane.

This gradient is located we suppose in patterns of protein structure that extend here and there almost all the way through the barrier material of the membrane. This gradient may be represented as a gradient in the degree of protonation of otherwise similar proton-accepting groups of protein structure; or it may be represented as gradients of electron distribution set up by electron flows. Its effects can, however, be communicated to the composition of the two adjoining aqueous phases to variable degrees through the presence of certain substrates that fit certain corresponding receptor sites. Some of these receptor sites recognize ATP and Na^+, others recognize K^+, still others certain amino acids alone, or certain amino acids along with Na^+ or H^+. A cation presented in the apolar screening environment of an ionophore may establish a superior communication. In still other cases sugars along with Na^+ are recognized. The consequence is that the gradient within the membrane is caused to do work or to have work done on it, although the responses of the substrate do not necessarily measure the full intensity of the standing gradient, the coupling being loose, and varying according to substrate structure. The membrane responds to the stimulated energy transduction caused by transported substrates, by increasing the fundamental flow that maintains its polarization. For membranes containing an electron transport chain, the fundamental flow that maintains polarization may be one of electrons from a donor, through the structure on which the gradient is impressed, to a receptor; for others it may be the reversal of a charge-separating step produced by the catalytic hydrolysis of ATP.

Mediated transport systems may transduce variable amounts of energy via the standing gradient of the membrane of the metabolizing cell, depending among other factors on the comparative filling by the substrate of the corresponding receptor sites at the two membrane surfaces. The consequence of the presence of a gradient of polarization within the membrane is that the two receptor sites appear to a selected substrate to be dissimilar, and to respond to the substrate differently by conformational change, leading to an energy-utilizing asymmetry of fluxes. Thialysine, for example, reacts with the external receptor site for neutral amino acids as though its sidechain had entered a region in which the chemical potential of the hydrogen ion corresponds, so to speak, to a pH as much as two units above that of the external medium. At the internal surface of the membrane, no such paradox is apparent; perhaps a H^+-deficiency will be found instead.

Influence of H^+ on amino acid transport. If the above interpretation is correct, then it would be strange if the hydrogen ion distribution had no influence whatever on amino acid transport. Uptake by the Na^+-dependent systems is in fact greatly slowed by lowering the pH from 7·4 to 5, although we have attributed this effect to a competition of H^+ for the site at which the essential Na^+ binds. In contrast the uptake of such amino acids as leucine and phenylalanine is decreasing by only 10 to 20% by the same pH lowering.[33]

Our earlier observation that α,α-diethylglycine uptake is stimulated by lowering the pH[34] finally brought us to see that this response is characteristic for System *L*. We now appreciate that this amino acid serves as a marginally suitable substrate for entry (although scarcely for exodus) by that system. The norbornane amino acid also proves to have its uptake strongly accelerated and its exodus slowed when the pH of the suspending medium is lowered. If our observations are made in media in which choline replaces Na^+, the uptake of leucine, isoleucine and valine shows similar acceleration by external H^+.

Eddy and Nowacki have recently shown that yeast cells take up H^+ and lose K^+ when such amino acids as glycine, phenylalanine and methionine are taken up. These authors have proposed that these ion movements may bear the same relationship as the correspondingly opposed movements of Na^+ and K^+ in animal cells, which they show may well drive amino acid uptake by ascites cells.[28] The cation gradients across the plasma membrane may in neither case, however, be essential intermediates in the transduction between metabolic energy and osmotic energy. The actual intermediating gradients may as described above lie within the membrane, reflected only incompletely and unreliably by gradients and flows of cations between the two aqueous solutions.

Recent experiments by Gunnar Ronquist in this laboratory indicate that the accumulation of α-aminoisobutyric acid and sarcosine by the Ehrlich cell can be largely dissociated from Na^+ influx. This effect has been obtained by incubating the cells twice for 5 min each time in the presence of AIB at 10 mM, with Na^+ at 15 mM, for the purpose of extensively replacing the endogenous amino acids with AIB. The cells come as a result to contain about 15 to 18 mM AIB and about 8 to 10 mM of all other amino acids.

We observe that at $[^{22}Na^+]_0 = 120$ mN cells treated thus will take up large further quantities of AIB or sarcosine to high gradients with only a very small associated uptake of Na^+. Uptake is however small from Na^+-free, choline-containing medium. The uptake of methionine and alanine is still accompanied by approximately stoichiometric uptake of Na^+. Apparently then the structures of the amino acids present on the two sides of the membrane can determine the extent to which transport is linked with Na^+ movements in System *A* as well as in System *ASC*.

If neither Na^+ or H^+ are obligatory co-substrates for uptake of amino acids by the Ehrlich cell, then the gradients of neither of these ions between the two aqueous solutions seem likely to be obligatory mediating stages in the coupled flow of energy. On the other hand, energy from the Na^+ gradient appears to be available under normal conditions to drive the uptake of certain amino acids, and energy from the H^+

gradient may be available for other systems or in other cells for the same purpose. As we wrote in 1962, "Uphill transports may well not be placed *in series*, e.g., with alkali-metal transport driving amino acid transport; but *in parallel*, with many or all uphill transports driven by the response of the membrane to ATP cleavage [or we will now add, an electron or proton flow]. The maintenance of an activated, transport-responsive state of the membrane may then be aided by the presence of high cellular levels of potassium ion or of amino acids," i.e., of gradients already generated by transport.[42] This seems now provisionally the most plausible form of the alkali-metal gradient hypothesis.

The foregoing treatment will I hope help to show that the study of transport changes produced by modifications of substrate structure is a biochemical as well as a physiological approach to transport. The approach has undoubtedly suffered occasionally from too purely a biological interest (e.g., which of the ordinary sugars or amino acids are transported, which are not?) with perhaps insufficient attention to the meaning of the structural differences among the several analogous substrates transported by a given system.

An important feature brought out by the foregoing results is that *sensing groups* or *probes* can be introduced into transport substrates to report the location of either specific structures or special microenvironments encountered during the sequential events of biological transport. The variation permissible in substrate structure is large enough for the insertion of chemical groups that may report nearness to a specific inorganic ion or other structure, or the degree of polarity of the environments encountered by a given part of the substrate structure. Other sensing groups that might be introduced include fluorescing structures or paramagnetic structures from which signals could then be obtained. The surprising amount of space available, particularly for sidechain structures of amino acids by transport System *L*, make possible a very wide range of tests. Similar possibilities appear to exist for transport systems for other metabolites. Room is presumably also available for introduction of some types of pharmacologically active structures into these regions of transport substrates.

6. *Similar goals for recognition sites serving for the transmission of information.* Finally, I should like to point out that objectives similar to those discussed so far make important the study of substrate structure on membrane-borne recognition sites of other kinds, which likewise do not apparently call for the destabilization of the substrate. I refer particularly to receptor sites by which cells can detect the presence in their external environments of gradients of nutrients and respond by chemotaxis[43,44] or sites by which they can maintain a given degree of induction of a transport system,[45] or prevent a spontaneous intensification of transport on incubation of the embryonic chick heart *in vitro*,[46] or by which they detect elevated levels of a metabolite in their

environment, and respond for example by releasing a hormone, e.g. insulin or glucagon.[47–49] Receptor sites on neurons of the central nervous system may account for the potent stimulatory and inhibitory effects of several amino acids when released into the synaptic area.[35] A broad range of physiological functions obviously is served by receptor sites of such types. Some sites of these kinds may have an internal location, rather than on the plasma membrane; in other cases the location is unknown.

One kind of site with an internal rather than an external location has been proposed whereby an accumulated amino acid may cause a *trans* inhibition of further uptake of that amino acid, or indeed of other substrates of the same system. In *Streptomyces hydrogenans* a different specificity pattern has been shown for *trans* inhibition than for uptake *per se*.[50] Hence a separate site is believed to activate a loop for feedback inhibition of further uptake. In yeast a separate site is also believed to serve, although the evidence may not be decisive,[51] whereas in *Neurospora* the phenomenon is not believed to entail a distinct site, as discussed under Item 4 above, since the specificity for uptake and that for producing *trans* inhibition are indistinguishable.[52] It is conceivable that the structural specificity for signals to arise from substrate passage through a transport system may be narrower than the specificity for transport *per se*. One can imagine that some of the substrates of a given transport system might fail to elicit a signal arising say from the generation of the kind of new membrane asymmetry implied by *trans* stimulation simply because these substrates are translocated too slowly to cause any such new asymmetry. Hence the structural requirement for signalling an elevated metabolite level by a given transport system may be somewhat more restrictive than those for transport by it. The discrepancy noted in *S. hydrogenans* is, however, in the opposite direction; hence it is unlikely that the transport site *per se* is the source of the signal in that special case.

These categories of receptor sites show some of the characteristics of transport sites, including the absence of any necessary destabilizing of the provoking agent. In the case of chemotaxis, the same binding proteins may serve for both material and informational transport.[44] These various sites may well have common evolutionary or differentiative origins; in some instances they may serve concurrently for transport and for information transfer. The opportunity to design analogs that inhibit but do not cause the response in question may permit separate study of the filling of the recognition site on the one hand, and, on the other hand, the sending either of material or of a signal to the cell interior.

References

1. T. Z. Csáky, *Ber. ges.*, **108** (1938) 620; *Z. Physiol. Chem.*, **227** (1942) 47.
2. C. G. Mark and A. H. Romano, *Biochim. Biophys. Acta*, **249** (1971) 216.

3. H. S. Tager and H. N. Christensen, *J. Biol. Chem.*, **246** (1971) 7572.
4. P. Mitchell and J. Moyle, *Nature*, **182** (1958) 372.
5. G. Gachelin, *Eur. J. Biochem.*, **16** (1970) 342.
6. L. Lieve and B. D. Davis, *J. Biol. Chem.*, **240** (1965) 4362.
7. E. A. Berger, J. H. Weiner and L. A. Heppel, *Federation Proceedings*, **30** (1971) 1061.
8. H. N. Christensen, *Advances in Enzymology*, **32** (1969) 1.
9. H. N. Christensen, M. E. Handlogten, I. Lam, H. S. Tager and R. Zand, *J. Biol. Chem.*, **244** (1969) 1510.
10. C. E. Furlong and J. H. Weiner, *Biochem. Biophys. Res. Comm.*, **38** (1970) 1076.
11. E. L. Thomas, T.-C. Shao and H. N. Christensen, *J. Biol. Chem.*, **246** (1971) 1677.
12. H. N. Christensen and M. E. Handlogten, *FEBS Letters*, **3** (1969) 14.
13. H. N. Christensen, M. E. Handlogten and E. L. Thomas, *Proc. Nat. Acad. Sci. U.S.*, **63** (1969) 948.
14. E. L. Thomas and H. N. Christensen, *J. Biol. Chem.*, **246** (1971) 1682.
15. J. A. Antonioli and H. N. Christensen, *J. Biol. Chem.*, **244** (1969) 1505.
16. B. H. Koser and H. N. Christensen, *Biochim. Biophys. Acta*, **241** (1971) 9.
17. E. L. Thomas and H. N. Christensen, *Biochem. Biophys. Res. Comm.*, **40** (1970) 277.
18. P. G. Le Fevre and J. K. Marshall, *Am. J. Physiol.*, **194** (1959) 333.
19. A. Kleinzeller, *Biochim. Biophys. Acta*, **211** (1970) 264.
20. H. N. Christensen, E. L. Thomas and M. E. Handlogten, *Biochim. Biophys. Acta*, **193** (1969) 228.
21. W. F. Caspary, N. R. Stevenson and R. K. Crane, *Biochim. Biophys. Acta*, **193** (1969) 168.
22. P. F. Baker and S. J. Potashner, *Biochim. Biophys. Acta*, **249** (1971) 616.
23. Y. Inui and H. N. Christensen, *J. Gen. Physiol.*, **50** (1966) 203.
24. H. N. Christensen, in Roles of Membranes in Secretory Processes, Proceedings of an International Conference on Biological Membranes, Italy, 1971, North Holland Publ. Co., 1972.
25. E. Heinz and R. P. Durbin, *J. Gen. Physiol.*, **41** (1957) 101.
26. M. L. Pall, *Biochim. Biophys. Acta*, **233** (1971) 201.
27. D. L. Oxender and B. Whitmore, *Fed. Proc.*, **25** (1966) 592.
28. A. A. Eddy and J. A. Nowacki, *Biochem. J.*, **122** (1971) 701.
29. H. N. Christensen, M. E. Handlogten and A. M. Cullen, *Federation Proceedings*, **30** (1971) 1116.
30. E. Fischer and G. Zemplen, *Ber. d. chem. ges.*, **42** (1909) 4878.
31. H. N. Christensen and M. Liang, *J. Biol. Chem.*, **241** (1966) 5542.
32. L. Eldjarn, *Scan. J. Clin. Lat. Inv.*, **6**, Suppl. 13 (1966) 1.
33. D. L. Oxender and H. N. Christensen, *J. Biol. Chem.*, **238** (1963) 3686.
34. H. N. Christensen and M. Liang, *J. Biol. Chem.*, **240** (1965) 3601.
35. H. N. Christensen, in: "Role of Vitamin B_6 in Neurology," E. Ebadi (eds.), *Advances in Biochem. Pharmacol.*, **4** (1972) 39.
36. H. N. Christensen and A. M. Cullen, *Biochim. Biophys. Acta*, **150** (1968) 237.
37. J. A. Schafer and E. Heinz, *Biochim. Biophys. Acta*, **249** (1971) 15.
38. E. Heinz (ed.), Proceedings of Symposium on Coupling Between Electrolyte and Non-Electrolyte Transport in Cells, Erbach, 1971, North Holland Publishing Company, 1972.
39. D. J. Poole, T. L. Butler and W. J. Waddell, *J. Nat. Cancer Inst.*, **32** (1964) 939.
40. R. J. P. Williams, *J. Theoret. Biol.*, **1** (1961) 1.
41. D. E. Green and S. Ji, *Proc. Nat. Acad. Sci. U.S.*, **69** (1972) 726.
42. H. N. Christensen, *Biological Transport*, W. J. Benjamin, New York, 1962, pp. 86–87.
43. J. Adler, *Science*, **166** (1969) 1588.
44. J. Adler and G. L. Hazelbauer, *Nature New Biology*, **230** (1971) 101.
45. H. H. Winkler, *J. Bact.*, **101** (1971) 470: **107** (1971) 74.
46. G. D. Gazzola, R. Franchi, V. Saibene, P. Ronchi and G. G. Guidotta, *Biochim. Biophys. Acta*, **266** (1972) 407.
47. H. N. Christensen and A. M. Cullen, *J. Biol. Chem.*, **244** (1969) 1521.
48. S. S. Fajans, R. Quibrera, S. Pek, J. C. Floyd, Jr. and H. N. Christensen, *J. Clin. Endocrinol. Metab.*, **33** (1971) 35.
49. H. N. Christensen, B. Hellman, Å. Lernmark, J. Sehlin, H. S. Tager and I.-B. Täljedal, *Biochim. Biophys. Acta*, **241** (1971) 341.
50. K. Ring, W. Gross and E. Heinz, *Arch. Biochem. Biophys.*, **137** (1970) 243.
51. M. Grenson, M. Crabeel, J. M. Wiame and J. Bechet, *Biochim. Biophys. Acta*, **30** (1968) 414.
52. M. L. Pall and K. A. Kelly, *Biochem. Biophys. Res. Comm.*, **42** (1971) 940.

Performance and Conservation of Osmotic Work by Proton-Coupled Solute Porter Systems

Peter Mitchell

Glynn Research Laboratories, Bodmin, Cornwall, England

Introduction

According to the chemiosmotic coupling conception of oxidative and photosynthetic phosphorylation systems,[1–3] the hydrogen and electron carriers of the respiratory chain and photoredox chain are looped across the non-aqueous (proton-insulating) M phase of the coupling membrane in such a way that redox activity along the chain is accompanied by the translocation of protons from one side of the membrane to the other, generating a protonmotive force of some 200 to 300 mV between the aqueous (proton-conducting) phases on either side. Thus, the primary physiological function of the respiratory chain and photoredox chain systems of mitochondria, chloroplasts and microorganisms is regarded as the provision of a source of power in the form of the proton current that can be used by appropriate systems plugged through the M phase of the coupling membrane.[4–7]

Owing to the relatively widespread interest of biochemists in the mechanism of reversal of the ATPase reaction in oxidative and photosynthetic phosphorylation systems, and owing to the relative importance attached to the elucidation of the fundamental mechanism of oxidative and photosynthetic phosphorylation,[8–9] the special energetic role of the reversible proton-translocating ATPase as a user of the proton current generated by the respiratory chain and photoredox chain systems has tended to monopolize attention.

In the present paper, my object is to emphasize the physiological importance of proton-coupled solute porter systems, plugged through the coupling membrane, the function of which depends on the use of the proton current to raise or lower the concentration of specific solutes in the inner aqueous phase of the cell or organelle relative to the concentration in equilibrium with that in the outer aqueous phase (or environmental medium). Such systems may obviously be energetically important in microorganisms that grow in dilute nutrient solutions, because the atomic constituents of the media are at a much lower average concentration than that of the cell constituents. Under these circumstances, nutrient assimilation obviously requires both the chemical work involved in the bond exchanges of intermediary metabolism and the osmotic work involved in bringing the assimilated chemicals from their low concentrations in the media to their relatively high concentrations in the organism. Under other circumstances, when the distinction between the chemical work and the osmotic work involved in chemical synthesis or intermediary metabolism in this type of system is not so obvious, the conservation of osmotic work by the solute porters is nevertheless fundamentally important in the overall efficiency of energy transduction; and it is the purpose of the present article to help to explain how the diffusion reactions of solute translocation and exchange between the two aqueous compartments separated by the coupling membrane participate in the partially reversible reactions of energy transduction. The general methods and nomenclature used here follow those introduced previously.[7,10–12]

Those who are concerned with the strategy of research may find it interesting to note that although the concept of proton-coupled solute translocation in bacteria, mitochondria and chloroplasts was introduced 10 years ago,[1,4,5] and reasons were given for regarding the much-studied but little understood β-galactoside translocation system of *Escherichia coli* as a prototype[5] (see Fig. 1), this simple and unifying concept was not investigated experimentally in the established schools of bacterial membrane transport.[13–16] Meanwhile, students of energy metabolism, who found themselves in the field of membrane-transport more by accident than by design, encouraged by the pioneering work from Chappell's laboratory,[17,18] confirmed the existence of proton-linked porter systems in mitochondria[19–23] and chloroplasts[9,24,25] and began to characterize the solute specificity and stoichiometry of these systems in detail (e.g. refs. 26–30). This work recently began to encourage students of bacterial metabolism to undertake similar studies which, especially in the laboratories of Hamilton[31,32] and Harold[33,34] have yielded results that are fundamentally similar to those obtained with mitochondria. However, some microbiologists, notably Kaback,[35–39] prefer to assume that the coupling between respiratory chain activity and the transport of sugars and amino acids, observed in certain bacteria, is due to hypothetical respiratory chain components that act as specific solute carriers. A penetrating commentary was given by Greville[8] on the strategy of postulating chemical coupling intermediates to produce explanations such as those deployed by Kaback.[35–39] Greville pointed out that, compared with their chemiosmotic counterparts, "such explanations are less precise and more hypothetical, since they involve postulated but undiscovered entities ...". But he went on to suggest that the hypothetical chemical coupling intermediates might nevertheless be spared Occam's Razor because "in these days beards and dialogue are characteristics of youth and iconoclasm". I detect, however, that beards may now be going out of fashion, especially amongst young students of bioenergetics who prefer their dialogue not to be obscured by extraneous growths. Therefore I propose to sharpen the Razor, as best I can, in the light of recent work in my laboratory.[40]

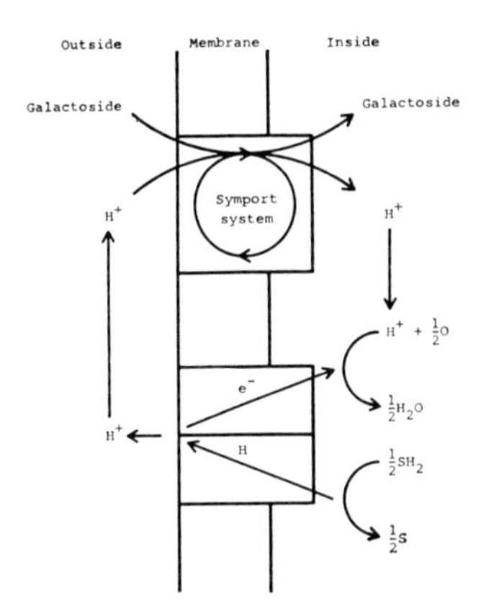

Figure 1. Diagram of cyclic coupling suggested between proton-translocating redox system and a β-galactoside-H^+ symport system in the plasma membrane of *E. coli* (from ref. 5). This diagram does not represent the stoichiometric relationships between the proton current and the redox and β-galactoside flows.

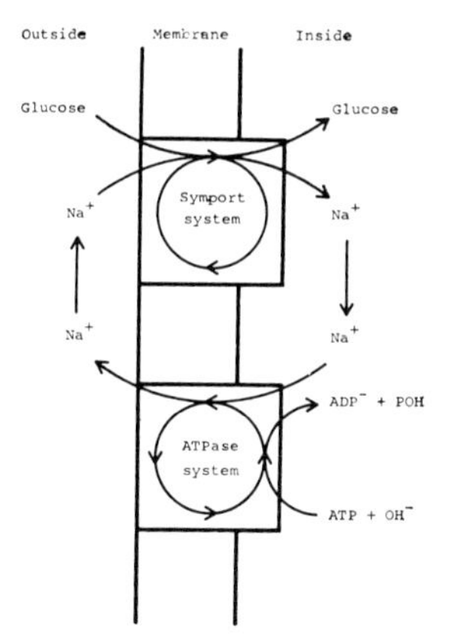

Figure 2. Diagram of cyclic coupling suggested between the cation-translocating ATPase, and the glucose-Na^+ symport system catalysing translocation of glucose across the intestinal mucosa (from ref. 5). The translocation of K^+ by the ATPase is omitted from this diagram for simplicity, and the stoichiometric relationships between the flows are not represented.

Although special attention is devoted to the proton-linked solute porter systems in the present article (because there are reasons for thinking that they may be as ubiquitous as the respiratory chain, the photoredox chain and the proton-translocating ATPase complex, with which they are associated[7]), other solute porter systems, such as that illustrated in Fig. 2 (from ref. 5), are also mentioned in order to illustrate broadly the general molecular mechanisms whereby the chemicomotive forces of vectorial metabolism are associated with the osmotic forces of solute electrochemical potential gradients. It is emphasized that, as long as certain fundamental properties of thermodynamically natural systems are not lost sight of,[4] the transmission of power through trains of solute molecules interacting by way of specific solute porters may conveniently be understood by regarding the solute porters not as abstract "transducers of energy", but as physically real miniature engines, having specifically articulated parts, designed (by natural selection) to couple the flow of one species of solute particle to that of another.[5]

The availability of several recent reviews on solute porter systems[7, 19–25, 31, 33, 34] makes it unnecessary to catalogue the numerous solute-specific porter activities that have been identified in recent years, and enables me to concentrate attention on a few selected topics.

"Osmotic Coupling" Compared with "Chemical Coupling"

The fact that Wang could recently state in an authoritative treatise on electron transport and oxidative phosphorylation[41] that the coupling mechanisms of the chemiosmotic type that I have proposed are less attractive than mechanisms involving chemical intermediates "because they are electrochemically all equivalent to the inefficient concentration cells" shows that it is still not generally appreciated that specific porter-catalysed osmotic reactions, which involve secondary and ionic bond interchanges across a coupling membrane separating two aqueous phases, may occur as reversibly as the more familiar specific enzyme-catalysed reactions, which involve covalent bond interchanges between chemical reactants and resultants. Therefore, I should like to make it particularly clear that there is no fundamental difference energetically between the reactions usually observed by chemists, which involve covalent bond exchanges, and the reactions usually observed by students of transport, which involve secondary and ionic bond exchanges—provided that appropriate catalysts are available to facilitate and channel the reactions in predetermined ways, minimizing the dissipation of available energy by the effective frictional resistance impeding the flow or interchange of chemicals in the required reactions, and minimizing the escape of chemicals (and available energy) in side-reactions.[12]

According to the rationale of vectorial metabolism and chemiosmotic coupling that I have endeavoured to foster,[4, 5] non-radiative thermodynamic energy transduction can occur only by means of the diffusion of trains or groups of material particles that move (with conserved impulse and momentum) under the influence of forces transmitted by local electrical and chemical interactions between them. Thus, a biochemically satisfactory understanding of the coupling between solute translocation reactions and the group translocation reactions of metabolism, or between one solute translocation reaction and another, can be obtained relatively simply in terms of the coupling between the flows of the chemical particles thermodynamically involved in these processes, all of which (in the form of the actual thermally mobile species) must move down their electrochemical potential gradients. This formulation of the "energetic coupling" problem is helpful to the biochemist because it explicitly focuses attention on the actual thermally mobile species involved in solute and group translocation and on the coupled flows of solutes and chemical groups channelled by the catalytic carrier and enzyme systems. It is, of course, particularly relevant that the actual thermally mobile species through which the forces and flows of energy transduction are transmitted include conformationally mobile enzymes and catalytic carriers and their substrate or passenger complexes. This was originally impressed upon me while studying a highly specific system catalysing tightly coupled exchange of phosphate for phosphate or of phosphate for arsenate across the plasma membrane of *Staphylococcus aureus*,[42–44] which has recently become of more topical interest because it appears to resemble a system catalysing the net translocation of phosphoric acid across the cristae membrane of mitochondria.[45]

Owing to the sensitivity of the staphyloccal phosphate translocation system to phenyl mercuric acetate, it was possible to estimate a maximum value for the absolute translocation rate and, by appropriate temperature studies, to obtain a minimum value for the entropy of activation, which accounted for at least 17,700 cal/mol in a total heat of activation of 37,400 cal/mol. The kinetics of this system thus resembled that of reversible protein denaturation and suggested that the translocation of phosphate involved a large conformation change in the translocater system in the membrane.[43,44]

The exchange translocation of phosphate for phosphate or of phosphate for arsenate by this system was remarkably tightly coupled, as shown by the fact that the net flux of phosphate (or of phosphate + arsenate) was not normally more than 1% of the exchange flux under non-metabolizing conditions, even when the system was operating well below its apparent dissociation coefficient for phosphate (and arsenate) on the outer side of the membrane.[43,44] It was therefore suggested[46] that, owing to the mutual satisfaction of valencies, the mobility of a translocater across a membrane may be as dependent upon its occupation by the specific passenger species as the translocation of the passenger species across the membrane may be dependent on combination with the specific translocater. I originally assumed that the tight coupling and high specificity of the translocation of phosphate by the staphylococcal system probably required the formation of a covalent compound with the translocater,[47] but it later became evident that this need not necessarily be the case.[1, 5]

The chemiosmotic theory focused attention on the components coupling the flows of particles in the biochemical systems catalysing solute, group and electron translocation.[5] Particular significance was therefore attached to the tentative suggestions by Christiansen[48, 49] and by Crane,[50] that the transport of amino acids and sugars might be coupled to the translocation of K^+ ions or Na^+ ions in the opposite or in the same direction respectively by specific carriers in the membrane. The same significance was attached to my own postulate[1] that the membranes of bacteria, mitochondria and chloroplasts contain specific systems catalysing tightly coupled exchange-diffusion of cations against H^+ ions and anions against OH^- ions. These systems evidently represented a new class of translocation catalyst corresponding to the exchange diffusion type of system originally postulated by Ussing[51] and Widdas[52] for coupling the translocation of chemically *analogous* solutes, but differing in that they were supposed to catalyse the sym- or anti-coupled translocation of chemically and sterically *unrelated* solutes, as exemplified by the suggested prototype symport systems for the coupled translocation of sugars with H^+ or Na^+ illustrated in Figs. 1 and 2 (from ref. 5).

In order to promote the precise description of coupled solute translocation reactions, it was suggested that non-coupled solute translocation should be described as uniport* and that anti- and sym-coupled solute translocations should be described as antiport and symport respectively. The general mechanism of catalysis of solute uniport, antiport and symport reactions is illustrated in terms of the operation of a cyclically mobile carrier (or carrier centre,[10] analogous to the active centre of an enzyme), represented by X, in Figs. 3, 4 and 5 (from ref. 5). These diagrams take account of the fact that the solutes involved in specifically catalysed translocation reactions through lipid membranes separating aqueous phases exist as the hydrates (written SW_L, AW_L, BW_L and SW_R, AW_R, BW_R in the left and right aqueous phases respectively) and that translocation across the lipid phase generally requires disengagement of the solute from its hydration shell by exchange of valencies with the carrier (or carrier centre) X, specific thermal mobility of the solute-X complex across the non-aqueous phase of the membrane, and re-engagement of the solute with its hydration shell so that it is released from X on the other side of the membrane.[5] The catalysis of uniport (Fig. 3) arises from the specific mobility of both unoccupied X and of the SX complex across the porter system at given concentrations of S in the aqueous phases. The catalysis of A-B symport (Fig. 5) likewise arises from the specific

* Monoport was originally suggested,[5] but was withdrawn in favour of uniport[10] for the sake of linguistic consistency.

mobility of both unoccupied X and of the inclusive complex ABX across the porter system at given concentrations of A and B in the aqueous phases; but the tightness of coupling depends on the extent to which the flux of A and B passing through the partial complexes AX

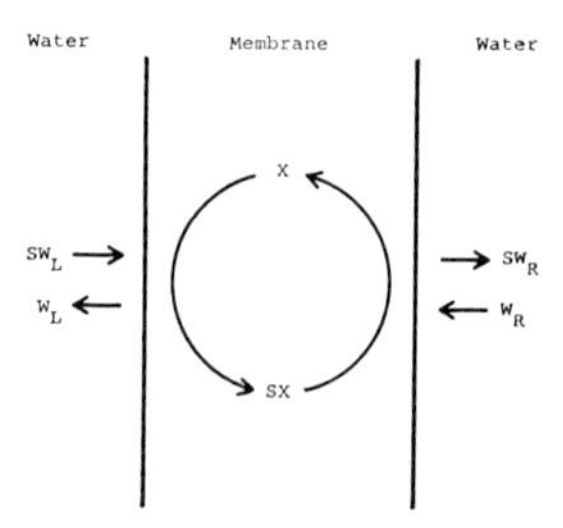

Figure 3. Diagram of uniport system for translocation of molecules or ions S combining with a carrier X in a lipoprotein membrane system between aqueous phases L and R in which S exists as the hydrate SW_L and SW_R (from ref. 5).

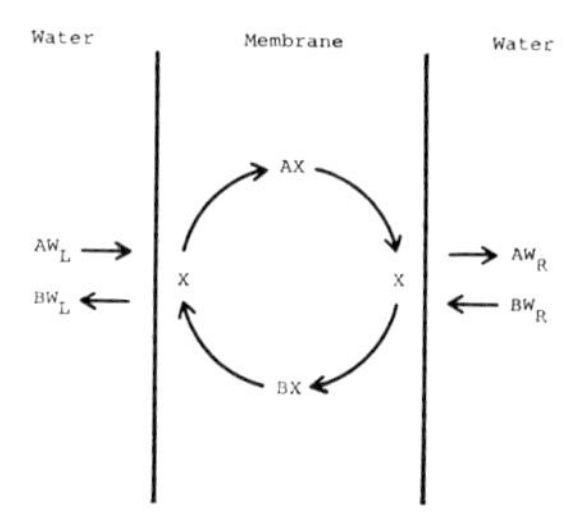

Figure 4. Diagram of antiport system for molecules or ions A and B competing for carrier X in a lipoprotein membrane system between aqueous phases in which A and B exist as hydrates WA and WB. The symbols L and R denote the aqueous phases (from ref. 5).

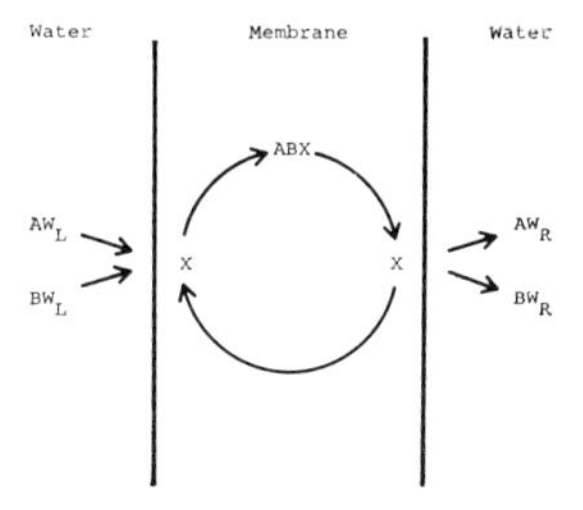

Figure 5. Diagram of symport system for molecules or ions A and B combining synergistically with carrier X. Other conventions as in Fig. 4 (from ref. 5).

and BX across the porter is minimized, compared with the flux of A and B passing through the inclusive ABX complex.[10] In the case of A/B antiport (Fig. 4) catalysis of translocation depends on the specific mobility of both AX and BX; but the tightness of coupling depends on the extent to which the flux of unoccupied X is minimized, compared with the flux of A and B passing through the AX and BX complexes.[10]

The rate of any reaction, such as that catalysed by a symporter or antiporter, along a given diffusion pathway is dependent upon the concentrations of the species actually mobile through the degrees of freedom constituting the pathway; and since, according to the Maxwell–Boltzmann law, the concentration of a given state of a system is relatively small if its potential energy is large compared with the thermal energy kT, it follows that a pathway that it may be evolutionarily desirable to close, because it corresponds to a side-reaction and causes uncoupling (such as that involving unoccupied X translocation in A/B antiport or AX and BX translocation in A-B symport), can be selected against by mutations causing it to correspond to states of the system having relatively high potential energy compared with states along the preferred, tightly-coupled, translocation pathway[12] (such as those involving translocation of AX and BX in A/B antiport or those involving translocation of XAB and X in A-B symport).

My object has been to emphasize, as stated before,[5, 12] that the meaning of ("energetic") coupling in this context is nothing more than the mutual dependence of the flows of A and B due to their passage as exclusive or inclusive complexes (AX and BX or ABX) specifically through the antiporter or symporter system, so that work can be performed by the transmission of forces and flows through the train of molecules of A and B undergoing the coupled translocation through the membrane.

It is instructive to compare the mechanism of "osmotic coupling" described for the symporter of Fig. 5 with the classical substrate-level mechanism of "chemical coupling" associated with 3-phosphoglyceraldehyde oxidation, with which we are, perhaps, more familiar. As discussed elsewhere,[12, 53] the coupling between the oxidation of 3-phosphoglyceraldehyde (PG) to 3-phosphoglycerate (PGA) and the phosphorylation of ADP through the reactions catalysed by 3-phosphoglyceraldehyde dehydrogenase (E) and 3-phosphoglycerate kinase (E^1) can conveniently be summarized by the following abbreviated equation:

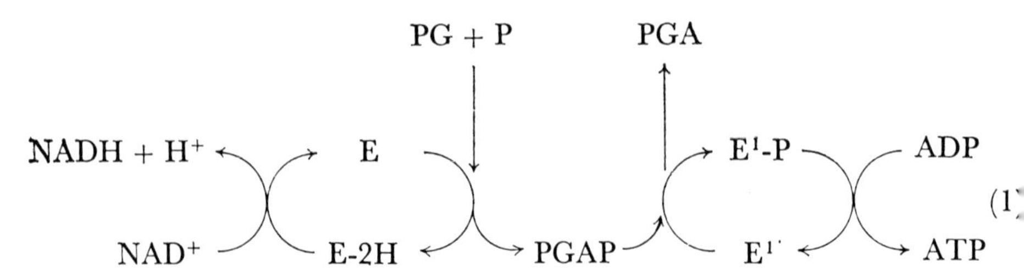

where PGAP stands for 1,3-diphosphoglycerate and E-2H and E^1-P stand for the respective enzyme-substrate-cofactor transitional complexes.

Each cycle of changes in the reactions represented by equation (1) corresponds to the flow of specific chemical groups through pathways determined by the interactions between the substrates and cofactors and the active centre regions of the enzymes E and E^1; and the tightness of coupling between the redox reaction and the phosphorylation reaction depends on the extent to which the movements of the chemical components are prescribed by the closely articulated enzyme-substrate-cofactor complexes so that the uncoupling effects of side-reactions are minimized. In particular, coupling between the production of PGA by the oxidation of PG and the production of ATP by the phosphorylation of ADP is attributable to the coupling between the flow of PGA out of the oxidation reaction and the flow of phosphoryl into the phosphorylation reaction because the PGA and phosphoryl are linked together in the inclusive intermediate PGAP.

There is a strict correspondence between the coupling function of the flow of the intermediate PGAP in the "chemically coupled" reactions of equation (1) and the coupling function of the flow of the intermediate ABX in the "osmotically coupled" reactions of Fig. 5. The fact that PGAP is a chemical compound, in which the components undergoing coupled flow are united by a strong covalent bond (with free energy of activation for uncatalysed exchange or dissociation $\gg kT$), need not produce an effectively tighter coupling in the "chemically coupled" system than can be achieved through the intermediate ABX in the "osmotically coupled" system, because strong bonding may occur between the components of ABX (with free energy of activation for uncatalysed exchange or dissociation $\gg kT$) by the co-operation of a number of secondary bonds. Indeed, components of ABX may well be linked covalently. For example, the net translocation of H^+ with $H_2PO_4^-$ catalysed by the phosphoric acid porter may involve the linkage of H^+ and $H_2PO_4^-$ in the covalent intermediate H_3PO_4. Moreover, the H_3PO_4 undergoing net translocation in the reaction catalysed by the phosphoric acid porter could conceivably be linked covalently to functional groups in the porter.[7, 47] As far as the tightness of "energy-coupling" is concerned, it is irrelevant what type of bonding is responsible for prescribing the flow of a given type of particle or what type of bonding causes the flow of one species of particle to be reciprocally dependent upon (coupled to) the flow of another species of chemical particle. The relevant factor is that the bonding prescribing the coupled group transfers or solute translocations involved in the "chemically coupled" or "osmotically coupled" systems should be such as to provide free energies of activation of exchange or dissociation that are large compared with kT, so that thermally activated dissociation or exchange is minimized, and the forces and flows can be efficiently transmitted between the two sets of particles without appreciable slip. Thus, both in enzyme-catalysed group-transfer reactions (or group-translocation reactions) and in porter-catalysed solute-translocation reactions, the uniqueness of the flow process (of chemical groups or of solutes) that is catalysed, which we customarily describe as the substrate specificity, may be taken to depend on the same type of condensed complex, with certain precisely specified conformational transitions corresponding to the chemical or translocation reaction. It would appear, therefore, that there is no reason for supposing that porter-catalysed solute-translocation reactions need be any less substrate-specific or any less well-coupled to one another than their enzymically-catalysed group-transfer counterparts.[12]

The reader will, I hope, be tolerant of the fact, reflected in this discussion, that the development of the solute porter concept originally owed much to inference from general principles, and tended to proceed somewhat ahead of the more detailed experimental knowledge needed to show that evolution actually exploited the physicochemical options thought to be available. At all events, as mentioned in the introductory section, the recent intensive studies of the highly specific and tightly-coupled proton-linked porter systems catalysing the translocation of anionic and cationic substrates in mitochondria, and studies on similar systems in bacteria and chloroplasts that are beginning to be undertaken by a few pioneers, provide very substantial experimental support for the proton-coupled solute porter concept that I have described as part of the chemiosmotic theory of energy transduction. The statement[41] that the chemiosmotic coupling mechanisms are "electrochemically all equivalent to the inefficient concentration cells" is therefore not a valid passport for the chemist, studying energy metabolism, who is anxious to provide against the accident of being landed inadvertently in the field of membrane-transport.

The Proton-Coupled β-Galactoside Porter System of Escherichia coli *as a Model for Proton-Linked Solute Translocation*

The experimental observations that West and I recently described[40] provide strong support for the proposition[5] that the specific uptake and accumulation of β-galactosides in *Escherichia coli* is primarily catalysed by a porter system, the general principle of which is illustrated in Fig. 1. In this type of system the specific solute porter catalysing the translocation of the solute is conceived as being physically and chemically separate from the proton-translocating respiratory chain system (or the proton-translocating ATPase system or pyrophosphatase system) which provides the proton current required for net solute uptake. Therefore the properties of the proposed β-galactoside-H^+ symporter can be investigated, uncomplicated by metabolic coupling, provided that the experimental conditions are arranged so that there is little or no respiratory or ATPase (or pyrophosphatase) activity. In practice, such conditions are not particularly difficult to achieve.

The β-galactosides, such as lactose, are ideal substrates for studies of proton-coupled translocation because, owing to their lack of net charge and acid/base properties near pH 7, linkage between their translocation and the translocation of H^+ ions across the coupling membrane may be recognized by the corresponding translocation of acid equivalents and positive charge across the membrane, in accordance with the following type of scheme:

gal ⟶ ⟩⟨ SYM ⟵ ⟩⟨ ⟶ gal

nH^+ ⟶ ⟩⟨ ⟶ gal-SYM-nH^+ ⟶ ⟩⟨ ⟶ nH^+ (2)

where SYM represents the translocater centre[10] of the proposed β-galactoside-proton symporter, gal stands for β-galactoside, and n is a stoichiometric coefficient. Thus, the translocation of β-galactoside would be coupled to the creation of a pH difference and a membrane potential across the membrane. According to this mechanism, the presence of specific proton conductors, such as DNP or azide, should uncouple the system by permitting the recirculation of H^+ ions back across the membrane, as illustrated by the following equation, in which 2,4-dinitrophenol is represented as the proton conductor in protonated (DNP) and deprotonated (DNP^-) forms:

gal ⟶ ⟩⟨ SYM ⟵ ⟩⟨ ⟶ gal

nH^+ ⟶ ⟩⟨ ⟶ gal-SYM-nH^+ ⟶ ⟩⟨ ⟶ nH^+

↑ ↓ (3)

nH^+ ⟵ ⟩⟨ ⟶ $nDNP^-$ ⟶ ⟩⟨ ⟵ nH^+

nDNP

Under these circumstances the β-galactoside should be able to equilibrate freely (although only by the same specific porter) across the membrane—as observed.[54] The proton-coupled mechanism of β-galactoside uptake described by equation (2) was, in fact, originally suggested[5] because it provided such a simple explanation of the observed *specific* equilibration of β-galactoside, in presence of proton-conducting reagents.

The experimental observations of the experimental paper[40] demonstrate the four main coupling relationships between the translocation of H^+ and β-galactoside described by equation (2), as follows:

1. Diffusion of β-galactoside down its concentration gradient by an NEM-sensitive reaction is coupled to the translocation of acid equivalents in the same direction.
2. Diffusion of β-galactoside down its concentration gradient across the membrane by an NEM-sensitive reaction is coupled to the translocation of positive charge in the same direction, and the number of positive charges translocated across is equal to the number of acid equivalents translocated.
3. Diffusion of H^+ ions down a pH gradient across the membrane requires the presence of β-galactoside in an NEM-sensitive reaction.
4. Diffusion of H^+ ions down an electrical potential gradient across the membrane requires the presence of β-galactoside in an NEM-sensitive reaction.

The value of n was not precisely determined, but these and previous observations by West[55] indicate that it is probably 1.

The translocation of β-galactoside through the NEM-sensitive system was found not to be directly coupled to the translocation of certain other ion species, such as Na^+ or K^+, as might have been inferred from Stock and Roseman's speculative interpretation of their experiments[56] showing that melibiose accumulation in actively metabolizing *Salmonella typhimurium* is affected by the presence of Na^+. Therefore, although there might well be a tightly-coupled Na^+/H^+ antiporter in the membrane of *E. coli*[7] as in mitochondria[57] and in *Streptococcus faecalis*,[58,59] the lack of direct coupling between Na^+ (or K^+) translocation and β-galactoside translocation showed that there was no β-galactoside-Na^+ symporter or β-galactoside-K^+ symporter and eliminated the following type of possibility:

gal SYM gal
Na^+ gal-SYM-Na^+ Na^+
(4)
Na^+ ANT-H^+ Na^+
H^+ Na^+-ANT H^+

It was still conceivable, however, that some other intermediary solute (s) could have been involved in the observed coupling between β-galactoside translocation and proton translocation according to the following general type of mechanism:

$$
\begin{array}{ccccc}
\text{gal} & & \text{ANT-}s & & \text{gal} \\
s & & \text{gal-ANT} & & s \\
\downarrow & & & & \uparrow \\
s & & \text{SYM} & & s \\
n\text{H}^+ & & s\text{-SYM-}n\text{H}^+ & & n\text{H}^+
\end{array}
\qquad (5)
$$

In this type of mechanism, β-galactoside translocation would initially involve a specific β-galactoside/s antiporter, and coupling with proton translocation would involve a specific s-$n\text{H}^+$ symporter. This possibility deserves serious consideration because it corresponds to a system known to be responsible for proton-linked dicarboxylate uptake in mitochondria, which depends upon the cyclic involvement of phosphate, according to the following type of process:

$$
\begin{array}{ccccc}
\text{dicarboxylate}^{2-} & & \text{ANT-phosphate} & & \text{dicarboxylate}^{2-} \\
\text{HPO}_4^{2-} & & \text{dicarboxylate-ANT} & & \text{HPO}_4^{2-} \\
\downarrow & & & & \uparrow \\
\text{HPO}_4^{2-} & & \text{UNI} & & \text{HPO}_4^{2-} \\
2\text{H}^+ & & \text{UNI-H}_3\text{PO}_4 & & 2\text{H}^+
\end{array}
\qquad (6)
$$

In this case, it has been shown that a specific dicarboxylate/phosphate antiporter catalyses the strictly coupled antiport represented by the upper part of equation (6), while a separate phosphoric acid uniporter with translocater centre represented by UNI (or a corresponding phosphate/hydroxyl ion antiporter) catalyses the net phosphate-proton symport represented in the lower part of the equation.[19, 26, 27, 60] It is noteworthy that the mitochondrial dicarboxylate/phosphate antiporter (in which I have represented the translocater centre as -ANT-) has separate binding sites for phosphate and dicarboxylate although there is a functional linkage that prevents both being occupied simultaneously.[60] Thus, there is an experimentally well-documented precedent for the suggestion that the observed proton-linked β-galactoside translocation could proceed via an antiport reaction with a chemically unrelated solute *s*, as represented in equation (5).

As our experiments[40] were done with suspensions of whole bacteria (which contained considerable quantities of endogenous solutes including phosphate), they were not helpful in eliminating possible anionic candidates for the role of *s* in equation (5). However, Kaback and co-workers[35–37] studied the specific β-galactoside translocation process in membrane vesicles from *E. coli* under conditions where the endogenous solutes were largely removed. Since he observed good respiration-linked β-galactoside uptake in the virtual absence of endogenous solutes, but routinely used a phosphate buffer, we can eliminate all except phosphate and hydroxyl ion from our considerations of a possible solute intermediary, *s*, in the observed proton-linked β-galactoside translocation.

As shown in the following equation:

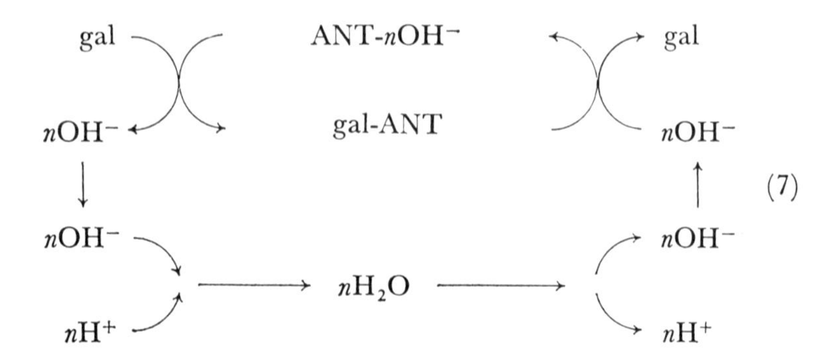

the possible involvement of the OH^- ion in the proton-coupled β-galactoside translocation through a β-galactoside/hydroxyl ion antiporter is a somewhat special case because lipid membranes have a relatively high permeability to H_2O.

In general, hydroxyl ion antiport is difficult to distinguish from proton symport because, as illustrated by equation (7), the net (proton-coupling) effect of solute/OH^- antiport is the same as that of solute-H^+ symport, because no additional solute porter is required for H_2O translocation. As discussed elsewhere,[2, 11, 12] it is conceptually convenient, for this reason, to describe all porters catalysing proton or hydroxyl ion symport or antiport as proton-coupled. But it is, nevertheless, biochemically important to distinguish between the alternative intermediary mechanisms of achieving the overall proton-coupled translocation through porters (or solutes) reacting with H^+ ions or OH^- ions.

To summarize: the experimental evidence at present available supports the proposition that specific β-galactoside translocation is coupled to net proton translocation by a proton-coupled porter system, which could be a β-galactoside-proton symporter, a β-galactoside/OH^- antiporter, or a β-galactoside/phosphate antiporter together with a phosphate-proton symporter.

The uptake of β-galactoside through the plasma membrane of *E. coli* is normally linked to respiratory activity in whole cells;[14] and the recent important experimental work of Kaback and co-workers[35–37] shows that it is also linked to the activity of the respiratory chain in membrane vesicles from which most of the soluble cytoplasmic solutes have been discharged. Therefore, there can no longer be any reasonable doubt that, as indicated, for example, by earlier observations of Kashket and Wilson[61] and of Pavlasova and Harold,[61] the specific uptake of β-galactoside is not directly coupled to the metabolism of ATP or other known metabolic intermediates, such as phosphoenolpyruvate. What, then, is the mechanism of coupling between redox activity through the respiratory chain and the translocation of β-galactosides through the specific catalytic system which is well known to catalyse β-galactoside equilibration across the membrane in presence of proton-conducting reagents,[14] and which we have characterized[40] as a catalyst of net β-galactoside-proton symport?

During respiratory chain activity in *E. coli*, as in other bacteria[4, 63, 64] and as in mitochondria,[66–68] protons are translocated outwards through the membrane. This has been observed, not only in intact *E. coli*,[40] but also in the membrane vesicle type of preparation used by Kaback and co-workers to demonstrate the respiration-linked uptake of β-galactoside and other solutes.[69] It is particularly noteworthy that Reeves[69] found that the proton translocation associated with respiration was directed outwards across the vesicle membrane, as our observations subsequently showed[40] was the case in the intact bacteria. These experimental observations directly support the proton-coupled chemiosmotic type of mechanism of β-galactoside uptake described, in principle, in Fig. 1—irrespective of the detailed biochemical mechanism of the proton-translocating respiratory chain and of the proton-coupled β-galactoside porter system.

According to Kaback and co-workers,[35–38] the transport systems in the isolated membrane vesicles of *E. coli*, which catalyse the concentrative uptake of β-galactosides, a wide variety of amino acids, galactose, arabinose, glucuronate, glucose-6-phosphate, manganese and potassium (in presence of valinomycin) are each coupled primarily to a membrane-bound flavin-linked D-lactic dehydrogenase; and the site of energy coupling is between the primary dehydrogenase and cytochrome b_1. It was further suggested[36,38] that the simplest conception that accounts for the experimental data is a model requiring that the β-galactoside-specific M protein, and the transport-specific "carriers" for the other solutes listed, are electron transfer intermediates, which undergo reversible oxidation-reduction, between D-lactic dehydrogenase and cytochrome b_1. Thus, for each transport system there should be a redox component between D-lactic dehydrogenase and cytochrome b_1, which has a binding site for that particular transport substrate. As with many similar theories of respiratory "energy-coupling" presented in the past,[8,20,22] no direct biochemical evidence for the proposed intermediates or for their special affinities and specific mechanism of participation in the energy-transduction process was provided.

One of the most significant observations in the extensive and painstaking measurements of Kaback and co-workers was that the proton-conductors DNP and carbonyl cyanide *m*-chlorophenylhydrazone caused rapid efflux of accumulated β-galactosides (and other specifically accumulated solutes) from the membrane vesicle preparations,[35–38] as from intact bacteria. But it was not explained how these reagents were supposed to affect the hypothetical redox reactions or specific carrier functions of the proposed energy-transducing intermediates in the respiratory chain system between D-lactic dehydrogenase and cytochrome b_1. Likewise, no explanation was given for the observation that the specific K^+-conductor valinomycin inhibited β-galactoside uptake by the vesicles in K^+-containing media; nor was it explained how valinomycin was supposed to participate in the hypothetical K^+-pumping units of the respiratory chain complex between D-lactic dehydrogenase and cytochrome b_1.

With regard to the interpretation of Kaback and co-workers that the translocation of β-galactoside coupled to the respiratory chain is dependent on a particular "site of energy coupling" between D-lactic dehydrogenase and cytochrome b_1, the experimental observations of Barnes and Kaback[35] showed that, with various respiratory substrates, the rates of lactose translocation into the vesicles did not run parallel to the rates of oxidation of the substrates; and it was stated that D-lactate was the more effective substrate. However, it is instructive to compare the quantitative effectiveness of the different substrates that were used in terms, not only of the rate of lactose uptake relative to that observed with D-lactate, but also in terms of the ratio (lactose uptake rate)/(oxygen uptake rate) for each substrate compared with the same ratio with D-lactate. These two indices calculated, as percentages of the values for D-lactate, from the data of Barnes and Kaback,[35] are as follows, the latter index being put in brackets:

D-lactate, 100 (100); DL-α-hydroxybutyrate, 58 (264);

succinate, 40 (24); L-lactate, 40 (146); NADH, 6·7 (8·2).

With vesicles prepared from bacteria grown on complex media, the rate of lactose uptake given by oxidation of α-glycerophosphate or formate was described as being about the same as with succinate—i.e. about 40% of the rate with D-lactate. It is noteworthy that fairly good β-galactoside uptake rates (i.e. not less than 40% of the rate with D-lactate) were given by all the substrates listed except NADH; and that when the respiratory rates with each substrate were taken into account, as shown by the bracketed numbers, DL-α-hydroxybutyrate and L-lactate were 264 and 146% as effective respectively as D-lactate. Therefore it is difficult to see why D-lactate, or the part of the respiratory chain associated with D-lactate oxidation, should be singled out as having a special relationship with the system catalysing β-galactoside translocation. The unreality of this interpretation is further emphasized by the fact that in studies on vesicles from *Staphylococcus aureus* Short, White and Kaback[39] were led to the conclusion that "amino acid transport in *S. aureus* is catalysed by mechanisms similar to those found in *E. coli* with the exception that α-glycerol phosphate, rather than D-lactate, is the primary electron donor".

The exceptionally low rate of β-galactoside uptake observed during NADH oxidation by the vesicles from *E. coli*[35] is not surprising, and would indeed be expected, because it is known that while the other "substrates" listed (see above) normally gain access to the respiratory chain system of intact bacteria from the outer medium and are normally oxidized rapidly, this is not the case with NADH. Access of NADH to the sites of the NADH dehydrogenase on the inner side of the membrane, where it is normally oxidized, presumably requires evertion, lysis or partial disorganization of the membrane. Thus, the considerable rate of NADH oxidation observed in the *E. coli* membrane vesicle preparations should be attributed to everted or disorganized membranes; and NADH would not be expected to act as a good "substrate" for proton-translocating respiratory chain activity capable of producing the outwardly-directed translocation of protons and the resultant inwardly-directed protonmotive force needed to drive β-galactoside uptake into vesicles of the same polarity as the intact cells.

Whatever may be the detailed biochemistry of the proton-translocating respiratory chain and of the proton-coupled β-galactoside porter system in *E. coli*, the observed activities of these two systems provide the basis for the simple chemiosmotic type of mechanism of β-galactoside accumulation illustrated in principle in Fig. 1, and discussed in more detail in a recent review.[7] This type of mechanism enables us to account for the specific uncoupling action of the proton-conducting reagents, such as DNP, in terms of the known specific biochemical property of these reagents. Likewise, the inhibitory effect of valinomycin (in K^+-containing media) on β-galactoside uptake can be explained by its known biochemical property of conducting K^+ ions across the membrane and collapsing the electric potential component of the protonmotive force driving β-galactoside uptake.

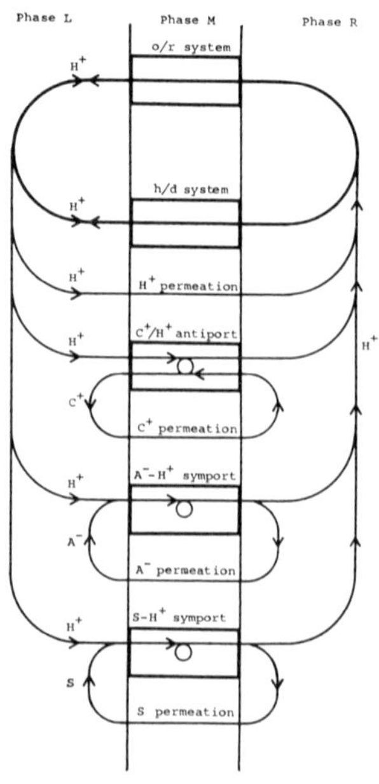

Figure 6. Composite proton circuit diagram illustrating coupling between metabolism and transport as it is thought to occur in mitochondria and in certain prokaryotic cells (after refs. 2, 5 and 7). Translocation of protons through oxido-reduction (o/r system) is shown poised against proton translocation through the reversible ATPase (h/d system). For simplicity, the oxidoreduction reactants for the o/r system and the hydrodehydration reactants (i.e. ATP, ADP, POH and H_2O) for the h/d system have been omitted from the diagram. Dissipation of part of the proton current occurs by the translocation of H^+ through substrate-specific antiporter systems for certain cations C^+ (e.g. Na^+) and symporter systems for certain anions A^- (e.g. phosphate or Krebs cycle acids) and neutral substrates S (e.g. sugars or neutral amino acids). Some of the proton-coupled porter systems may be complex, as explained in the text. The rate of dissipation of the proton current through the porter-coupled reactions in the steady state is dependent on anion, cation and neutral substrate permeation as indicated. The symbols A^- and C^+ do not denote the valency of the anions and cations and the stoichiometry of translocation is not indicated for A^-, C^+ or S.

The explanations given by Kaback and co-workers[35–39] to justify a chemical coupling mechanism of respiration-linked uptake of β-galactosides and many other solutes in *E. coli* and in other bacteria require complex *ad hoc* assumptions which have no direct biochemical foundation. The proposed chemically coupled model provides only a relatively vague account of the "energy-coupling" mechanism in the respiratory chain and does not satisfy the criterion of relative simplicity which Kaback and co-workers have themselves laid down.[36] Therefore I suggest that, since the edge of Occam's Razor has been sharpened,[40] so that it can leave a clean chemiosmotic profile,[7] we should apply Occam's Razor to the superfluous growth of hypothetical energy-transducing solute-carrying chemical intermediates attributed to the respiratory chain system.

Figure 6 shows a composite proton circuit diagram (from ref. 7) representing the cation/proton antiport (C^+/H^+ antiport) and anion-proton symport (A^--H^+ symport) systems of mitochondria[2] and the S-H^+ symport system catalysing the proton-coupled translocation of the non-ionic solute S which is based on the β-galactoside-proton symport system of *E. coli*.[5] It should be noted that phase L corresponds to the outer aqueous medium and phase R to the enclosed aqueous phase of the vesicles of the chemiosmotically coupled system. In the light of the foregoing discussion I would re-emphasize[7] the usefulness of exploring the possibility that the chemiosmotic theory, illustrated by the diagram of Fig. 6, may be found to be generally applicable to bacteria as well as to mitochondria and chloroplasts. The upper part of Fig. 6 represents the primary group-translocation systems (the respiratory chain or photoredox chain, and the ATPase or pyrophosphatase) coupling proton translocation to oxidoreductive (o/r) and hydrolytic or hydrodehydrative (h/d) metabolism. The lower part of the diagram illustrates the use of the proton current: (1) for the *extrusion* of certain cations C^+, such as Na^+ and Mg^{2+}; (2) for the *uptake* of certain anions A^-, such as phosphate, sulphate, acidic amino-acids and Krebs cycle acids; and (3) for the *uptake* of neutral substrates S, such as sugars and neutral amino-acids. The symbols A^- and C^+ in the diagram do not denote the valency of the anions and cations, and the stoichiometry of proton-linked translocation is not indicated for A^-, C^+ and S. The dependence of the poise of such proton-coupled solute porter systems on the stoichiometry and electrogenicity of the overall translocation reaction at given values of the electric component

$$\Delta\psi = \psi_L - \psi_R$$

and the chemical component

$$-Z\Delta\text{pH} = -(2{\cdot}303\,RT/F)(\text{pH}_L - \text{pH}_R)$$

of the total protonmotive force* Δp has been described in detail elsewhere,[11, 12] and is discussed briefly in the next section. Meanwhile it is noteworthy that the *non-electrogenic* classes of C^+/H^+ antiport and A^--H^+ symport reactions are particularly important both in practice and in theory, and that the *electrogenic* S-H^+ symport reactions which constitute the other most important class of translocation reactions should be taken to include, not only the translocation reactions causing uptake of non-ionic solutes, such as sugars, but also those causing uptake of solutes such as basic amino-acids that can normally exist as cations because of protonation of an electrically neutral species.

* The total protonmotive force Δp is conventionally reckoned in mV, Z then being about 60 at 25°.

For simplicity of presentation, Fig. 6 illustrates the coupling of the overall translocation reactions with the proton current and does not include the intermediary circulating solutes and their specific antiporters, such as phosphate and the specific solute/phosphate antiporters, and L-malate (see equation (6)) and the specific L-malate/tricarboxylate antiporter, known to couple the translocation of certain solutes with the translocation of protons in mitochondria.[7, 19, 26–29] As a matter of fact, circulating intermediaries, such as phosphate, were first suggested in the case of translocation reactions in bacteria,[47] but it is not yet known to what extent they may be involved in bacterial systems. These circulating intermediary solutes do, of course, affect dissipation of the proton current by non-specific permeation of the intermediaries through the M phase of the membrane, in addition to the dissipation indicated in the case of the species C^+, A^- and S in Fig. 6; but they do not affect the stoichiometry of the overall reactions under tightly coupled conditions—i.e. when the proton-current dissipation through the non-specific permeation reactions is a small fraction of the total proton current.

A recent discussion of the Na^+-linked solute translocation reactions in the small intestine by Semenza[70] implies the existence of a sodium circuit system similar to the proton circuit system described by Fig. 6.

It is, perhaps, particularly appropriate that the β-galactoside-proton symport system of *E. coli* should be recognized as a model for the further extension of knowledge of solute porter systems in bacteria because, as Kepes[14] has pointed out, the β-galactoside transport system of *E. coli* has received intensive study in the past and has come to be considered as a "classical" type of system, especially with regard to its genetic determination, which has been investigated with much ingenuity, following the initiative of the Paris school.[71]

The discussion of energetic aspects of proton-coupled solute porter systems in bacteria is made more interesting by the fact that the initial step in the metabolism of glucose (and some other sugars) in *E. coli* and other bacteria depends on a phospho-enolpyruvate phosphotransferase system, discovered by Kundig, Roseman and co-workers,[72, 73] which catalyses the accumulation of glucose-6-phosphate (and other corresponding phosphorylated sugars) in the inner aqueous phase.[16, 72–78] Evidence has been obtained in support of the proposition that the sugar-specific phosphotransferase component (called Enzyme II) of this system catalyses the translocation of the "glucose-6-" group (or other corresponding specific sugar group) inwards across the membrane where it is accepted by a phosphoryl (or phosphate) group from an intermediary phosphoryl-carrier protein (called HPr) and accumulates as glucose-6-phosphate (or other specific sugar phosphate) in the inner aqueous phase. Thus, Roseman[73] drew the conclusion "that most sugars penetrate bacterial membranes by group translocation, mediated by the respective Enzyme II for each sugar. The sugar is phosphorylated as it is transported, and the phosphate is derived from phospho-HPr". After further intensive studies in several laboratories, a detailed mechanism was suggested by Kaback[16] with the object of explaining how "a *vectorial phosphorylation* of the sugar would be accomplished, producing a group translocation type of transport system"; and arguments were advanced by Kaback[16] in favour of general mechanisms of this type.

The mechanism of sugar-group translocation suggested by Roseman[73] and by Kaback[15, 16] apparently corresponds, in general principle, to the type of system described by Fig. 7.[79] For the case of "glucose" uptake by the phospho-enolpyruvate phosphotransferase system, E would represent the glucose-specific phosphotransferase

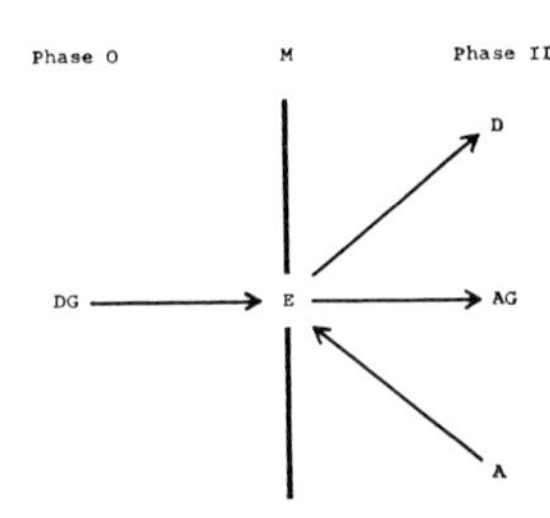

Figure 7. Diagram of group translocation system for the group G donated by the group D and accepted by the group A. The aqueous phases O and II are separated by the membrane M in which the group-transfer enzyme E is orientated so that DG can only communicate with the active centre from phase O, and D, AG, and A can only communicate from phase II (from ref. 79).

Enzyme II, DG would represent glucose (made up of a donor group D and the "glucose-6-" group undergoing transfer), and the acceptor group A would stand for the phosphoryl group (or phosphate group) of phospho-HPr which accepts the "glucose-6-" group and forms AG, representing glucose-6-phosphate. It has not so far been made clear, however, in the case of the phospho-enolpyruvate phosphotransferase system, precisely what sugar group is supposed to be translocated in the overall reaction or what other group or groups may also be translocated at the same time. Using the description "glucose-6·" to refer to the electrically neutral group remaining after removal of one OH group from the carbon atom in position 6 of glucose, the question arises, referring to Fig. 7, which of the following possible pairs of constituent groups might, for example, represent G and D: glucose-6· and OH·; glucose-6^{+} and OH^{-}; glucose-6-O· and H·; glucose-6-O^{-} and H^{+}. According to the system shown in Fig. 7, both the component groups G and D of the substrate DG would be translocated across the membrane, and the overall translocation reaction (i.e. of glucose) would not be affected by the above alternative possibilities. However, according to other group-translocation schemes that have been considered,[5] the group G might be translocated from left to right, while the group D passed back to the left. This latter type of group translocation reaction may be of practical interest, particularly if the reaction splitting the substrate DG into its constituent groups proceeds heterolytically,[5] i.e. if, for glucose translocation, G and D were, for example, to represent glucose-6$^+$ and OH^- or glucose-6-O^- and H^+.

The processes corresponding to glucose-6$^+$ translocation or glucose-6-O^- translocation can conveniently be described in the following notation, introduced for this purpose:[5]

glucose ⌒ | ⌒ phospho-HPr

OH^- ← | → glucose-6-phosphate + HPr^+ (8)

glucose ⌒ | ⌒ phospho-HPr

H^+ ← | → glucose-6-phosphate + HPr^- (9)

These possibilities, which, it will be noted, include the equivalent of (electrogenic) proton translocation across the membrane, depend on the chemical potentialities of the substrate DG (glucose), of the acceptor A (phosphoryl or phosphate) and of the carrier of A (HPr). There are additional possibilities of solute translocation, that may be associated with group translocation, depending on ligand-binding properties of E—by analogy with the binding and translocation of Na^+ and K^+ in the $3Na^+/2K^+$ antiporter ATPase during hydrolysis of ATP. The following hybrid group-translocation and solute-translocation type of reaction would not, for example, be inconceivable in view of the precedent set by the $3Na^+/2K^+$ antiporter ATPase:[5, 7, 10, 12]

H^+ + glucose ⌒ | ⌒ phospho-HPr

| → glucose-6-phosphate + HPr + H^+ (10)

The foregoing considerations are intended to illustrate two main points. The first is that the possible group-translocation process involved in the phospho-enolpyruvate phosphotransferase system requires further analysis in order to define it explicitly and unequivocally. The second is that, in view of the existence of a considerable electric potential and pH difference across the membrane of certain (if not all) bacteria, the energetics of the putative sugar group translocation reactions would be dependent upon their possible electrogenicity and protonogenicity illustrated, for example, by equations (8) to (10). In particular, it is interesting to note that the reaction shown in equations (8) or (10) would be equivalent energetically to the two-stage process involving: (i) the uptake of the sugar in a sugar-H^+ symport reaction; and (ii) the phosphorylation of the sugar by a phosphotransferase working at the same phosphate potential as that of the phospho-enolpyruvate phosphotransferase system. It is also noteworthy that if the phosphotransferase were physically positioned on the inner side of a glucose uniporter in the plasma membrane, so that glucose molecules gaining access by the uniporter would enter a closed microscopic phase leading only to the active centre of the phosphotransferase, the following reaction, equivalent to that shown in Fig. 7 would result:

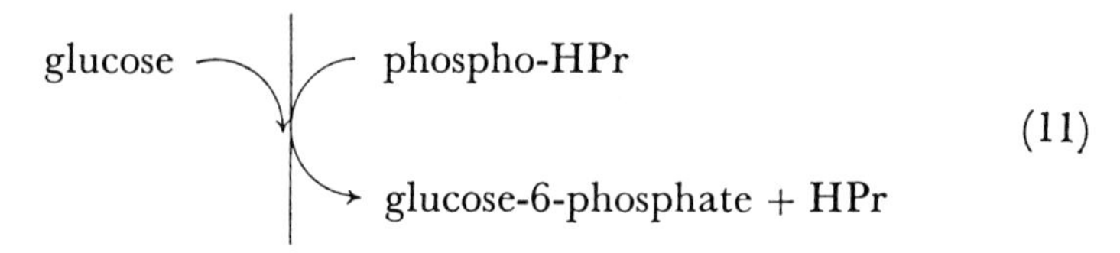

Alternatively, if there were the same type of association between a glucose-H^+ symporter and the phosphotransferase, the reaction of equation (10) would result. These possibilities illustrate the close interrelationships that are possible between enzymes catalysing group translocation and porters catalysing solute translocation,[79] especially since entry of a substrate to the active centre of an enzyme may be through a region of the polypeptide complex of the enzyme that acts like a solute porter.

Energy Transduction by Solute Porters

When I postulated the occurrence of systems catalysing anion/OH^- antiport (or anion-H^+ symport) and cation/H^+ antiport in organelles catalysing oxidative and photosynthetic phosphorylation by a chemiosmotic type of mechanism,[1] I had in mind that such "exchange diffusion" systems were required to maintain osmotic stability by using the pH component $-Z\Delta pH$ of the total protonmotive force for translocating the anions and cations back across as fast as they tended to diffuse through the relatively ion-impermeable M phase of the coupling membrane under the influence of the electric potential component $\Delta\psi$ of the total protonmotive force. Thus, these proton-coupled solute porters were conceived as catalysing electrically neutral overall translocation reactions, in which the net work done would be the balance of the work done by the hydrogen ions moving down their chemical potential gradient represented by $-Z\Delta pH$, and that done on the solute anions or cations being moved up their corresponding chemical potential gradient.[1,2,3,10–12] Using the "p" notation, as in pH, to denote $-\log_{10}$ (chemical activity) of any solute, the equilibrium distribution of anion A^{n-} or cation C^{n+} achieved only through electrically neutral A^{n-}-nH^+ symport or C^{n+}/nH^+ antiport respectively is as follows:

$$-\Delta pA = n\Delta pH \tag{12}$$

$$\Delta pC = n\Delta pH \tag{13}$$

where Δ means the value of the given variable in phase L (or outside the vesicle) minus that in phase R (or inside the vesicle). On the other hand, the equilibrium of the same anion and cation species achieved only through diffusion of the ions A^{n-} and C^{n+} across the M phase is given by the Nernst or Donnan equilibrium relationship as follows:

$$-\Delta pA = n\Delta\psi/Z \tag{14}$$

$$\Delta pC = n\Delta\psi/Z \tag{15}$$

The total protonmotive force Δp is given by

$$\Delta p = \Delta\psi - Z\Delta pH \tag{16}$$

Thus, when the total protonmotive force Δp is zero,

$$\Delta pH = \Delta\psi/Z \tag{17}$$

and it will be noted, by substituting $\Delta\psi/Z$ for ΔpH in equations (12) and (13), that under this special condition, when Δp is zero, the distribution of the anionic and cationic species A^{n-} and C^{n+} is independent of whether A^{n-} and C^{n+} equilibrate across the M phase by electrogenic ionic diffusion, or by non-electrogenic A^{n-}-nH^+ symport or C^{n+}/nH^+ antiport. However, comparing equations (12) with (14) and equations (13) with (15), it can be seen that when Δp is not zero, the species A^{n-} and C^{n+} circulate across the M phase, passing one way (as the ions) down their ionic electrochemical potential gradient and the other way down the chemical potential gradient of the A^{n-}-nH^+ or C^{n+}/nH^+ porter complexes, as illustrated in Fig. 6.

As mentioned above, this circulatory process was conceived as having a regulatory function. In mitochondria and bacteria (and possibly also in whole chloroplasts) where the pH of the inner aqueous phase containing metabolic enzymes must be regulated at a point not far from neutrality, the C^{n+}/nH^+ antiport systems deplete the $-Z\Delta pH$ component of Δp by lowering the internal concentration of cations other than H^+ and thereby maintain a large proportion of Δp in the form of $\Delta\psi$ and minimize osmotic swelling. A high Na^+/H^+ antiport activity has been observed in mitochondria[57,80] and in *Streptococcus faecalis*,[59] and a relatively low K^+/H^+ antiport activity has also been observed in mitochondria.[57] The relative abundance of K^+ compared with Na^+ in mitochondria and bacteria may well be attributable to the lower rate of K^+/H^+ antiport.[7,59,81] The circulation of the anionic solutes may also be considered to contribute to the regulatory function described for the cations because the A^{n-}-nH^+ symport systems deplete the $-Z\Delta pH$ component of Δp by raising the internal concentration of anions other than OH^- and thus help to maintain a large proportion of Δp in the form of $\Delta\psi$. The relationship shown in equation (12), which is characteristic of electrically neutral anion-proton symport, has been nicely confirmed by Palmieri and co-workers[82,83] in the case of phosphate and phosphate-linked Krebs cycle anions in mitochondria; and extensions of these important studies indicate that this relationship applies also to other anions undergoing proton-linked translocation.[21,26–30]

Regarded in the regulatory context discussed here, the anion and cation circulation would not be expected to constitute a major pathway of energy transduction, because the permeability of the M phase to the anion and cation species normally involved in these circulations is low, and only a small proportion of the total proton current available from metabolism would be dissipated.[2,3,11] In this context the proton-linked porter systems—especially the C^{n+}/nH^+ antiporters—may be regarded as being analogous to bilge pumps on a ship.

Apart from their regulatory role, the non-electrogenic A^{n-}-nH^+ symport systems fall into the class of proton-coupled substrate uptake systems of which the β-galactoside-proton symporter is the prototype.[5] Osmotic work must, of course, be done during the uptake and concentration of substrates by the proton-coupled porter systems. However, as discussed earlier in a more general context,[47] and as pointed out by Chappell[19] in the case of specific succinate/L-malate antiport accompanying succinate oxidation in mitochondria, the osmotic work required to bring in a substrate may be done by the exit of its metabolic product, thus:

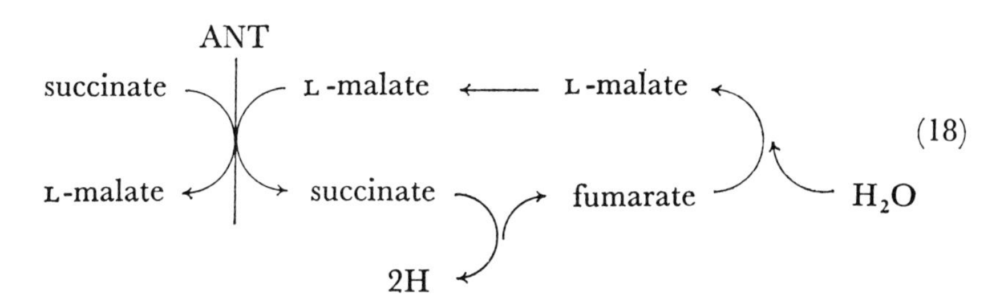

where ANT denotes the succinate/L-malate antiporter in the M phase.

Observations on the kinetics of β-galactoside translocation and hydrolysis in *E. coli* reviewed by Kepes[14] indicate that the β-galactoside porter system can catalyse β-galactoside/galactose antiport, thus:

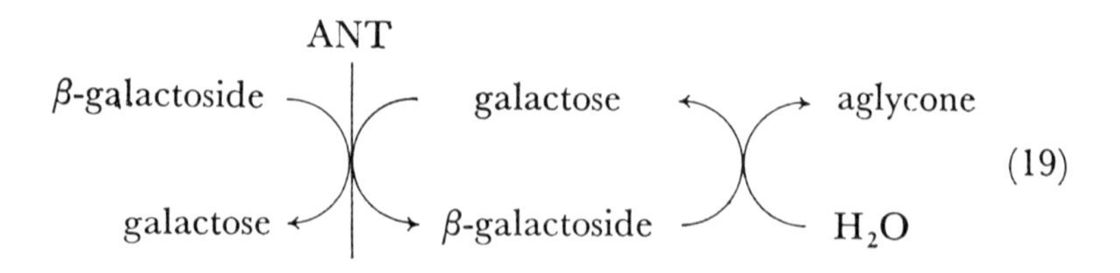

(19)

In this case, since a separate system for galactose translocation has been identified, which has similar properties to the β-galactoside-proton symport system,[13–16, 38, 71] the β-galactoside/galactose antiport reaction might be catalysed either by the β-galactoside-specific component of the β-galactoside-proton symport system, or by this component coupled by the proton current to a similar galactose-specific component of the putative galactose-proton symport system. At all events, the analogy between equations (18) and (19) is physiologically interesting because it indicates a general mechanism by which the proton-coupled porter systems can act as osmotic work transducers, enabling osmotic work to be transferred from an end-product passing down an osmotic gradient to a required metabolite passing up an osmotic gradient. This compensatory type of process would not be expected to operate in the same way, however, under conditions of rapid assimilation, as it would during steady-state (or near steady-state) metabolism; and in this respect, considerable differences may be anticipated between bacterial solute porter systems and the corresponding solute porter systems in mitochondria and chloroplasts.

The relationship between the distribution of the solute S and the components of the total protonmotive force when S equilibrates across the membrane only by S-nH$^+$ symport is as follows:

$$-\mathbf{\Delta}\text{pS} = n(\mathbf{\Delta}\text{pH} - \mathbf{\Delta}\psi/Z) \tag{20}$$

Thus, whereas the greater part of the work done by the proton current in carrying an anion inwards through an electrically neutral A^{n-}-nH$^+$ symport system is used in overcoming the electric repulsion of the anion by the membrane potential $\mathbf{\Delta}\psi$ (negative inside), and only a small fraction of the work is left to produce a higher concentration of the anion inside than outside (see equation (12)), no electric work is done in carrying the electrically neutral solute S inwards through the electrogenic S-nH$^+$ symport system, and comparatively very large concentration ratios of S may therefore occur if equilibrium is reached. For example, if $n = 1$ and $\mathbf{\Delta}p$ is 240 mV, the solute S would be about 10,000 times more concentrated in the inner aqueous phase than in the outer medium at equilibrium. Actual concentration ratios of 2000 times and 10,000 times have been recorded for DNP-sensitive lactose[84] and galactose[54] uptake respectively in respiring *E. coli*. These relatively enormous concentration ratios would be expected to cause osmotic swelling that would lead to damage or lysis of the membrane unless the external concentration of the solute being accumulated was less than 100 μM. In the light of these considerations it seems significant that the respiratory stimulation observed by Kepes[14] in suspensions of *E. coli* on adding 1 mM phenyl-β-D-thiogalactoside or 0·5 mM methyl-β-D-thiogalactoside did not last only during uptake of the β-galactoside, but continued, and was thus characteristic of an uncoupling phenomenon.

To conclude, it is interesting to note that the proton-coupled porter systems may participate significantly in the overall energy transduction process required for assimilation and organic synthesis, because, as the proton-translocating ATPase system probably translocates 2H^+ ions per ATP molecule hydrolysed,[8] the osmotic and electric work that can be done towards the assimilatory process in proton-coupled solute translocation is equivalent to half the free energy of hydrolysis of ATP per proton used in the solute translocation reaction.

Acknowledgements

I am grateful to the Cambridge University Press for permission to reproduce Figs. 1 to 6 from refs. 5 and 7, and to the Academic Press for permission to reproduce Fig. 7 from ref. 79. I thank my colleagues Drs. Jennifer Moyle and Ian West for discussion and criticism during the preparation of this paper, and I am indebted to Miss Stephanie Phillips and Mr. Robert Harper for expert assistance in preparing the manuscript and figures. I acknowledge general financial assistance from Glynn Research Ltd.

References

1. P. Mitchell, *Nature*, **191** (1961) 144.
2. P. Mitchell, *Chemiosmotic Coupling in Oxidative and Photosynthetic Phosphorylation*, Glynn Research, Bodmin, Cornwall, 1966.
3. P. Mitchell, *Biol. Rev.*, **41** (1966) 445.
4. P. Mitchell, *J. Gen. Microbiol.*, **29** (1962) 25.
5. P. Mitchell, *Biochem. Soc. Symp.*, **22** (1963) 142.
6. P. Mitchell, *Federation Proc.*, **26** (1967) 1370.
7. P. Mitchell, *Symp. Soc. Gen. Microbiol.*, **20** (1970) 121.
8. G. D. Greville, in: *Current Topics in Bioenergetics*, Vol. 3, D. R. Sanadi (ed.), Academic Press, New York, 1969, pp. 1–78.
9. D. A. Walker and A. R. Crofts, *Ann. Rev. Biochem.*, **39** (1970) 389.
10. P. Mitchell, *Advan. Enzymol.*, **29** (1967) 33.
11. P. Mitchell, *Chemiosmotic Coupling and Energy Transduction*, Glynn Research, Bodmin, Cornwall, 1968.
12. P. Mitchell, in: *Membranes and Ion Transport*, Vol. 1, E. E. Bittar (ed.), Wiley, London, 1970, pp. 192–256.
13. A. Kepes, in: *The Molecular Basis of Membrane Function*, D. C. Tosteson (ed.), Prentice-Hall, Englewood Cliffs, New Jersey, 1969, pp. 353–389.

14. A. Kepes, in: *Current Topics in Membranes and Transport*, Vol. 1, F. Bronner and A. Kleinzeller (eds.), Academic Press, New York, 1970, pp. 101–134.
15. H. R. Kaback, *Ann. Rev. Biochem.*, **39** (1970) 561.
16. H. R. Kaback, in: *Current Topics in Membranes and Transport*, Vol. 1, F. Bronner and A. Kleinzeller (eds.), Academic Press, New York, 1970, pp. 35–99.
17. J. B. Chappell and A. R. Crofts, in: *Regulation of Metabolic Processes in Mitochondria*, J. M. Tager, S. Papa, E. Quagliariello, and E. C. Slater (eds.), Elsevier, Amsterdam, 1966, pp. 293–316.
18. J. B. Chappell and K. N. Haarhoff, in: *Biochemistry of Mitochondria*, E. C. Slater, Z. Kaniuga, and L. Wojtczak (eds.), Academic Press, London, 1967, pp. 75–91.
19. J. B. Chappell, in: *Inhibitors—Tools in Cell Research*, Th. Bücher and H. Sies (eds.), Springer-Verlag, Berlin, 1969, pp. 335–350.
20. H. A. Lardy and S. M. Ferguson, *Ann. Rev. Biochem.*, **38** (1969) 991.
21. M. Klingenberg, in: *Essays in Biochemistry*, Vol. 6, P. N. Campbell and F. Dickens (eds.), Academic Press, London, 1970, pp. 119–159.
22. K. van Dam and A. J. Meyer, *Ann. Rev. Biochem.*, **40** (1971) 115.
23. J. D. McGivan and J. B. Chappell, *Biochem. J.*, **127** (1972) 54P.
24. L. Packer, S. Murakami and C. W. Mehard, *Ann. Rev. Plant Physiol.*, **21** (1970) 271.
25. R. A. Dilley, in: *Current Topics in Bioenergetics*, Vol. 4, D. R. Sanadi (ed.), Academic Press, New York, 1971, pp. 237–271.
26. J. D. McGivan and M. Klingenberg, *European J. Biochem.*, **20** (1971) 392.
27. S. Papa, N. E. Lofrumento, D. Kanduc, G. Paradies and E. Quagliariello, *European J. Biochem.*, **22** (1971) 134.
28. B. H. Robinson, G. R. Williams, M. L. Halperin, and C. C. Leznoff, *J. Biol. Chem.*, **246** (1971) 5280.
29. F. E. Sluse, A. J. Meijer and J. M. Tager, *FEBS Letters*, **18** (1971) 149.
30. H. W. Heldt and F. Sauer, *Biochim. Biophys. Acta*, **234** (1971) 83.
31. W. A. Hamilton, in: *Membranes: Structure and Function*, J. R. Villanueva and F. Ponz (eds.), Academic Press, London, 1970, pp. 71–79.
32. D. F. Niven and W. A. Hamilton, *Biochem. J.*, **127** (1972) 58P.
33. F. M. Harold, *Advan. Microbiol. Physiol.*, **4** (1970) 45.
34. F. M. Harold, *Biochem. J.*, **127** (1972) 49P.
35. E. M. Barnes and H. R. Kaback, *J. Biol. Chem.*, **246** (1971) 5518.
36. H. R. Kaback and E. M. Barnes, *J. Biol. Chem.*, **246** (1971) 5523.
37. W. N. Konings, E. M. Barnes and H. R. Kaback, *J. Biol. Chem.*, **246** (1971) 5857.
38. G. K. Kerwar, A. S. Gordon and H. R. Kaback, *J. Biol. Chem.*, **247** (1972) 291.
39. S. A. Short, D. C. White and H. R. Kaback, *J. Biol. Chem.*, **247** (1972) 298.
40. I. West and P. Mitchell, *J. Bioenergetics*, **3** (1972) 445.
41. J. H. Wang, in: *Electron and Coupled Energy Transfer in Biological Systems*, Vol. 1, T. E. King and M. Klingenberg (eds.), Marcel Dekker, New York, 1971, pp. 117–133.
42. P. Mitchell, *J. Gen. Microbiol.*, **9** (1953) 273.
43. P. Mitchell, *J. Gen. Microbiol.*, **11** (1954) 73.
44. P. Mitchell, *Symp. Soc. Exp. Biol.*, **8** (1954) 254.
45. D. D. Tyler, *Biochem. J.*, **111** (1969) 665.
46. P. Mitchell and J. Moyle, *Discuss. Faraday Soc.*, **21** (1956) 258.
47. P. Mitchell, *Symp. Biochem. Soc.*, **16** (1959) 73.
48. T. R. Riggs, L. M. Walker and H. N. Christensen, *J. Biol. Chem.*, **233** (1958) 1479.
49. D. L. Oxender and H. N. Christensen, *J. Biol. Chem.*, **234** (1959) 2321.
50. R. K. Crane, D. Miller and I. Bihler, in: *Membrane Transport and Metabolism*, A. Kleinzeller and A. Kotyk (eds.), Academic Press, New York, 1961, pp. 439–449.
51. H. H. Ussing, *Nature*, **160** (1947) 262.
52. W. F. Widdas, *J. Physiol.*, **118** (1952) 23.
53. P. Mitchell, *J. Bioenergetics*, **3** (1972) 5.
54. B. L. Horecker, M. J. Osborn, W. L. McLellan, G. Avigad and C. Asensio, in: *Membrane Transport and Metabolism*, A. Kleinzeller and A. Kotyk (eds.), Academic Press, New York, 1961, pp. 378–387.
55. I. C. West, *Biochem. Biophys. Res. Commun.*, **41** (1970) 655.
56. J. Stock and S. Roseman, *Biochem. Biophys. Res. Commun.*, **44** (1971) 132.
57. P. Mitchell and J. Moyle, *European J. Biochem.*, **9** (1969) 149.
58. F. M. Harold and D. Papineau, *J. Membrane Biol.*, **8** (1972) 27.
59. F. M. Harold and D. Papineau, *J. Membrane Biol.*, **8** (1972) 45.
60. F. Palmieri, G. Prezioso, E. Quagliariello and M. Klingenberg, *European J. Biochem.*, **22** (1971) 66.
61. E. R. Kashket and T. H. Wilson, *Biochim. biophys. Acta*, **193** (1969) 294.
62. E. Pavlasova and F. M. Harold, *J. Bacteriol.*, **98** (1969) 198.
63. P. Scholes, P. Mitchell and J. Moyle, *European J. Biochem.*, **8** (1969) 450.
64. P. Scholes and P. Mitchell, *J. Bioenergetics*, **1** (1970) 309.
65. J. W. T. Wimpenny, *J. Gen. Microbiol.*, **63** (1970) xv.

66. P. Mitchell, in: *Energy Transduction in Respiration and Photosynthesis*, E. Quagliariello, S. Papa and C. S. Rossi (eds.), Adriatica Editrice, Bari, 1971, pp. 123–149.
67. V. P. Skulachev, in: *Current Topics in Bioenergetics*, Vol. 4, D. R. Sanadi (ed.), Academic Press, New York, 1971, pp. 127–190.
68. B. Chance and M. Montal, in: *Current Topics in Membranes and Transport*, Vol. 2, F. Bronner and A. Kleinzeller (eds.), Academic Press, New York, 1972, pp. 99–156.
69. J. P. Reeves, *Biochem. Biophys. Res. Commun.*, **45** (1971) 931.
70. G. Semenza, *Biochim. Biophys. Acta*, **241** (1971) 637.
71. E. P. Kennedy, in: *The Lactose Operon*, J. R. Beckwith and D. Zipser (eds.), Cold Spring Harbor Laboratory, New York, 1970, pp. 49–92.
72. W. Kundig, S. Ghosh and S. Roseman, *Proc. Natn. Acad. Sci., U.S.*, **52** (1964) 1067.
73. S. Roseman, *J. Gen. Physiol.*, **54** (1969) 138s.
74. W. Kundig and S. Roseman, *J. Biol. Chem.*, **246** (1971) 1393.
75. W. Kundig and S. Roseman, *J. Biol. Chem.*, **246** (1971) 1407.
76. T. Nakazawa, R. D. Simoni, J. B. Hays and S. Roseman, *Biochem. Biophys. Res. Commun.*, **42** (1971) 836.
77. M. H. Saier, W. Scott Young and S. Roseman, *J. Biol. Chem.*, **246** (1971) 5838.
78. T. Korte and W. Hengstenberg, *European J. Biochem.*, **23** (1971) 295.
79. P. Mitchell, in: *Membrane Transport and Metabolism*, A. Kleinzeller and A. Koytk (eds.), Academic Press, New York, 1961, pp. 22–34.
80. G. P. Brierley, M. Jurkowitz, K. M. Scott and A. J. Merola, *J. Biol. Chem.*, **245** (1971) 5404.
81. G. P. Brierley, M. Jurkowitz, K. M. Scott, K. M. Hwang and A. J. Merola, *Biochem. Biophys. Res. Commun.*, **43** (1971) 50.
82. F. Palmieri, E. Quagliariello and M. Klingenberg, *European J. Biochem.*, **17** (1970) 230.
83. E. Quagliariello, G. Genchi and F. Palmieri, *FEBS Letters*, **13** (1971) 253.
84. A. Kepes and G. N. Cohen, in: *The Bacteria*, Vol. 4, I. C. Gunsalus and R. Y. Stanier (eds.), Academic Press, New York, 1962, pp. 179–221.

Molecular Basis for the Action of Macrocyclic Carriers on Passive Ionic Translocation Across Lipid Bilayer Membranes*

G. Eisenman, G. Szabo, S. G. A. McLaughlin and S. M. Ciani

Department of Physiology, UCLA School of Medicine, Los Angeles, California 90024

Introduction

Substantial energies in living cells are stored in ionic gradients across membranes. One of the central problems of biological energy transduction is that of understanding how these gradients arise and conversely, how the energy stored in such gradients is utilized to drive chemical reactions. These are two aspects of the reversible coupling of the energy of transmembrane ionic concentration gradients with the energy of chemical reactions—an electrochemical problem which may be thought of as involving, on the one hand, a mechanism for selective ion permeation and, on the other, a means of coupling that mechanism to the appropriate chemical reaction.

Relevant to such an understanding of the coupling between ion transport and chemical reactions should be the knowledge of the characteristics which are necessary and sufficient to produce a selective ion-translocating system, which is the principle subject of this paper. For example, schemes for metabolically-linked transport which involve conformational changes of the ion-transporter, as one step in the coupling process, from a state in which it is able to selectively bind and transport a particular ion or is rendered incapable of binding and transporting ions (cf. Skou, 1964; Whittam, 1967) require an understanding of the molecular mechanisms through which ion-selection and ion-transport arise. Similarly schemes which couple to chemical reactions the energy produced by the equilibration of an ionic (or redox) gradient (cf. Mitchell, 1961) must involve an understanding of the molecular structure of the membrane necessary to produce such selective ion equilibration. As an experimental fact, the ways in which monovalent cations exert their specific actions have been shown to follow the same

* Presented in part at the Symposium on Molecular Mechanisms of Antibiotic Action Protein Biosynthesis and Membranes, Granada, Spain, June, 1971. Supported by NSF Grant GB 16194 and USPHS Grants GM 17279 and NS 09931.

quantitative rules in metabolically dependent ionic accumulations, specific enzyme activating effects, and in purely passive membrane permeation (Eisenman, 1965).

The means by which specific ion translocation can be coupled to chemical reactions has been of considerable interest. Thus mechanisms for coupling the energy produced by electron and/or ionic transport to the production of ATP have been proposed and discussed by Lipmann (1946); Slater (1953); Lehninger (1965); Chance (1963); Boyer (1963, 1965); Lardy (1967); Green and Perdue (1966); Mitchell (1961, 1971); and others. In the direct chemical scheme the energy coupling occurs through the device of a chemical intermediate of high-energy nature generated by electron transport, which is used as a direct precursor for the formation of ATP. In electrochemical schemes, such as the "chemiosmotic hypothesis" of Mitchell (1971), coupling with the ionic gradient directly induces the formation of ATP. Detailed mechanisms by which the energy stored in ATP might be coupled to the production of ion gradients have been proposed and discussed by Patlak (1957), Skou (1964), Kepes (1964), Jardetsky (1966), Cockrell *et al.* (1967), Whittam (1967) and others. The reader interested in the reversible coupling between transport and chemical reactions is referred to Mitchell's excellent discussion (1971) as well as to other chapters in this volume. This chapter will deal with the simpler question of mechanisms by which ions can be selectively translocated across membranes under the passive conditions in which coupling to chemical reactions is not introduced. However, the ion-carrying molecules studied here can produce effects coupled to metabolism. For example, Cockrell *et al.* (1967) have shown that rotenone-treated mitochondria can synthesize ATP from ADP and inorganic phosphate when treated with valinomycin so as to enable utilization of the transmembrane K^+ gradient.

The membrane-active macrocyclic antibiotics to be considered here are particular members of a general class of amphiphilic molecules which are capable of interacting with cations through ligand oxygens, usually from the backbone of the molecule (Kilbourn *et al.*, 1967; Shemyakin *et al.*, 1969). Although typical molecules are cyclic, linear molecules can also be induced to fit around cations to form coordination complexes (Lardy *et al.*, 1967; Pressman, 1968). Complex formation comes about because the backbone oxygens of polypeptides, polyesters, or polyethers can interact with cations with energies comparable to those of hydration. Such molecules can therefore wrap around an ion and replace the hydration shell in such a way that a complex is formed in which the ion is sequestered in the polar interior of a molecule whose exterior is essentially lipophilic. The polar portion of the amphiphilic molecule decreases the electrostatic work of transferring the ion into the membrane interior; while the lipophilic portion of the molecule interacts preferentially with the membrane. When the complex so formed has the appropriate dimension and conformation, it can behave as a mobile entity and thus act as a carrier of cations (Pressman *et al.*, 1967; Finkelstein and Cass, 1968; Eisenman, Ciani and Szabo, 1968). On the other hand, when the molecule has an appropriate conformation to form a stable trans-membrane structure with a polar interior, it can provide a way to pass ions through a channel across the membrane (Urry, 1971).

These two alternatives correspond to the carrier and neutral pore mechanisms IV and V of Eisenman's (1968) summary. An important distinction between the two types of action is that the former requires a liquid-like interior for the membrane, whereas the latter does not; and recent experiments by Krasne, Eisenman and Szabo (1971) indicate that these two mechanisms of operation can be distinguished by varying the fluidity of membranes. Furthermore, neutral molecules like gramicidin, which are thought to form pores in membranes, produced discrete jumps in membrane conductance; whereas molecules like monactin, which are thought to act as carriers, do not (Hladky and Haydon, 1970). The present paper will restrict its consideration to molecules thought to be carriers, particularly to those neutral molecules whose complexes bear the charge of the complexed-ion (in which case the charged complex is the principal carrier of electrical current). The principles underlying equilibrium selectivity should, however, also apply to "electrically silent" neutral complexes between cations and negatively charged antibiotics of the Nigericin-type.

For a considerable experimental range, the chemical reactions between carriers and carried ions not only occur sufficiently rapidly in the aqueous solutions, (cf. Eigen and Winkler, 1971), but also sufficiently rapidly at the membrane-solution interface, that these reactions can be considered to be at equilibrium relative to the rate of movement of the complexes across the membrane interior. In this situation, which may usefully be called the "equilibrium domain", an understanding of equilibrium chemistry suffices to account for the observable effects of these molecules on the membranes. Thus, in this domain, comparison of the theoretically expected and experimentally observed effects of valinomycin, four macrotetralide actins, and two cyclic polyethers on the electrical conductance and potential of phospholipid bilayer membranes (as well as on the equilibrium extraction of alkali picrates into organic solvents and the formation of ion-polyether complexes in aqueous solution) demonstrates that all salient effects of these molecules can be accounted for quantitatively from appropriate thermodynamic equilibrium constants.

There is, however, an experimental range in which equilibrium considerations alone do not account for the observed membrane properties of carriers. In this latter range, which has recently been demonstrated by Läuger and Stark (1970) and Stark and Benz (1971) for strongly complexing carriers, strongly complexed ions and negatively charged lipids and by this laboratory for particular neutral lipids (Laprade *et al.*, 1972), the kinetics of the interfacial reactions become important, and thus in this range the system may be referred to as being in the "kinetic domain". A theoretical examination of the range of these domains will appear elsewhere (Ciani *et al.*, 1972), but Stark and Benz (1971) have demonstrated that both domains are observable for valinomycin, and we have also found this to be true for the macrotetralide actins (Laprade *et al.*, 1972). The present contribution confines its considerations, however, to experimental situations which fall within the equilibrium domain.

Five sets of new results are presented here and related to those from our previous studies of the macrotetralide actins.

The first section demonstrates that the effects of the depsipeptide, valinomycin, on phospholipid bilayer membranes are similar to those of the macrotetralide actins in the first order dependence of conductance on antibiotic (and cation) concentrations, and in the identity for pairs of cations (e.g. K vs. Na) of the numerical values of the permeability ratios, conductance ratios, and two-phase salt extraction equilibrium constant ratios. The range of applicability of our previous analysis for the macrotetralide actins (Ciani *et al.*, 1969, Eisenman *et al.*, 1969, Szabo *et al.*, 1969) is therefore extended to this depsipeptide.

The second section describes the considerably more complex effect on membranes of a cyclic polyether. For this molecule the conductance is found to depend on the 2nd and 3rd power of the polythere concentration. Also, the membrane conductance is only proportional to the concentration of permeant ion at low concentration (exhibiting a maximum and an inverse second order dependence at high concentrations), and the permeability ratios and conductance ratios are found to differ significantly at high concentrations of salt. Nevertheless, the present theory, when merely extended to allow for 2:1 and 3:1 polyether-cation complexes, successfully accounts for all of these observations. The range of applicability of our carrier treatment is therefore further extended to the cyclic polyethers.

The third section compares the quantitative cation selectivity patterns of valinomycin and the macrotetralides with each other and with that characteristic of biological membranes. Despite qualitative similarities in the sequences of ion effects, which are understandable in terms of the selectivity consideration of the last section, important quantitative differences are shown to exist among these molecules, and these differences may be used as "fingerprints" for certain characteristics of the molecule structure underlying cation permeation of the cell membrane. In particular, since the selectivity pattern of valinomycin for Li, Na, K, Rb, Cs and NH_4 is so close to that characteristic of cell membranes, we speculate that the molecular environment around a cation in the cell membrane may involve six polypeptide carbonyl oxygens arrayed around the cation in a manner similar to the arrangement in valinomycin.

The fourth section examines certain general features of the equilibrium energetics underlying the equilibrium selectivity of neutral ion-sequestering molecules and demonstrates how, for isosteric complexes, the ratio of selectivities for membrane properties reflects the differences in hydration energies of the cation species vs. the differences in their interaction energies within the antibiotic molecules. It also shows how the same free energy differences underlie the selectivity between cations of the overall reaction (24) by which the present carriers solubilize cations in a membrane.

Finally, the fifth section extends a previous monopolar model for the selectivity of ion exchange sites (Eisenman, 1961, 1962, 1965) to the case of neutral, dipolar sites which are more appropriate as models for the carbonyl (or ether) oxygens which are the ligands in the present molecules.

Methods and Materials

The experimental apparatus and procedures used to characterize the electrical properties of bilayer membranes formed from mixtures of n-decane and various purified phospholipids have been described previously (Szabo *et al.*, 1969; McLaughlin *et al.*, 1970), as have the methods for characterizing salt extraction equilibria (Eisenman *et al.*, 1969). However, where necessary for clarity, additional details will be given. The cyclic polyethers were generously supplied by C. Pedersen and H. K. Frensdorff of E. I. duPont de Nemours Co. The macrotetralide actins were gifts from Dr. H. Bickel of CIBA and Miss B. Stearns of Squibb. Valinomycin was a gift from Dr. G. Mallett of the Eli Lilly Co., and also was purchased from Calbiochem (the former sample had 90% the activity of the latter, presumably due to prolonged storage). Phosphatidyl ethanolamine and phosphatidyl glycerophosphate were gifts from J. Law and M. Kates.

The Effects of Valinomycin on the Electrical Properties of Bilayer Membranes and on Two-Phase Salt Extraction Equilibria

The purpose of this section is to extend and make more quantitative the previous characterizations of the effects of valinomycin on bilayer membrane potential and conductance (Mueller and Rudin, 1967; Lev and Buzhinsky, 1967; Andreoli *et al.*, 1967), as well as on two-phase salt extractions (Pressman, 1968), in order to test whether or not the theoretical considerations we have previously applied for the macrotetralide actins also hold for valinomycin. We shall show that, as with the macrotetralides, the principal effects of valinomycin on membranes are completely understandable in terms of the ability of this molecule to form a 1:1 complex with cations. Moreover, we shall verify that the ability of valinomycin to mediate the membrane permeability and conductance of the alkali metal cations is accurately accounted for by the appropriate equilibrium thermodynamic parameters (Eisenman *et al.*, 1969), as measured in two-phase salt extraction experiments. It will be helpful, before presenting experimental data to summarize briefly the salient theoretical expectations of our previously developed carrier treatment (Eisenman *et al.*, 1968; Ciani *et al.*, 1969) in order to provide a basis for interpreting the data for valinomycin and also to provide background for the subsequent extension to the cyclic polyethers in the second section.

Theoretical Considerations

For the purpose of this paper it will suffice to consider only the expectations for neutral lipids, uncomplicated by effects of charged polar head groups (Lesslauer *et al.*, 1967; McLaughlin *et al.*, 1970; Neumcke, 1970; Szabo *et al.*, 1972a), which are treated in detail elsewhere (McLaughlin *et al.*, 1970). For a membrane of the dimensions of a phospholipid bilayer (i.e., one whose thickness is less than the apparent Debye length within the hydrocarbon phase), we have deduced (Ciani *et al.*, 1969) that the cation-antibiotic complexes, IS^+, should be the major charge-carrying species within the membrane, where they are present as an excess space charge. (The concentration of these species within the membrane is sufficiently high that it determines the membrane's electrical properties, but low enough i.e. $C_{is}^* < 5 \times 10^{-5}$ M (Neumcke and Läuger, 1970) for the constant-field approximation to hold).

By integrating the Nernst-Planck flux equations for these complexes the simple equation

$$V_0 = \frac{RT}{F} \ln \frac{a_i' + \beta a_j'}{a_i'' + \beta a_j''}, \tag{1}$$

for the membrane potential, V_0, at zero current was deduced. Equation (1) expresses the potential difference between the aqueous solutions in terms of the activities, a_i', a_j', a_i'', a_j'', of the ions, I^+ and J^+, in the aqueous solutions on the two sides (′) and (″) of the membrane and a constant β, which is formally equivalent to the permeability ratio, P_j/P_i, of the Goldman-Hodgkin-Katz equation (Goldman, 1943; Hodgkin and Katz, 1949), being defined as

$$\beta = \frac{P_j}{P_i} = \frac{u_{js}^* k_{js} K_{js}^+}{u_{is}^* k_{is} K_{is}^+}, \tag{2}$$

where u_{js}^*/u_{is}^* is the ratio of the mobilities of the JS^+ and IS^+ complexes in the membrane, k_{js}/k_{is} is the ratio of the partition coefficients of the complexes and K_{js}^{+}/K_{is}^{+} is the ratio of the equilibrium constants for the formation of the complexes in aqueous solution.* No assumptions as to electroneutrality or as to profiles of potential or concentration were necessary to obtain this result, but we did assume that the equilibria at the membrane-solution interfaces were not perturbed by the flux of the complexes. Equation (1) is expected to hold for all the experimental conditions of this section.†

The membrane conductance, measured on the limit of vanishingly small applied voltage, was also deduced by evaluating the concentration profiles through integration of the Poisson–Boltzmann equation for the equilibrium situation where the aqueous solutions on both sides of the membrane have the same composition (Ciani *et al.*, 1969).‡ For the case of a single cation, J^+, the membrane conductance in the limit of zero current, $G_0(J)$, is given by:

$$G_0(\mathrm{J}) = \left[\frac{F^2}{d} u_{js}^{*} k_{js} K_{js}^{+}\right] C_s^{\mathrm{Tot}} a_j \frac{1}{1+K_{js}^{+} a_j}, \tag{3}$$

which, in the (usually encountered) limit of sufficiently low salt concentration that $K_{js}^{+} a_j \ll 1$, reduces to the simpler form

$$G_0(\mathrm{J}) = \left[\frac{F^2}{d} u_{js}^{*} k_{js} K_{js}^{+}\right] C_s^{\mathrm{Tot}} a_j. \tag{4}$$

Equations (3) and (4) indicate that the membrane conductance is expected to be proportional to the total aqueous concentration of antibiotic, C_s^{Tot}, and also to be proportional to the activity of the cation in the solution, a_j, at least at low salt concentrations. Taking the ratios of membrane conductances for J^+ vs. I^+ from equation 4,

$$\frac{G_0(\mathrm{J})}{G_0(\mathrm{I})} = \frac{u_{js}^{*} k_{js} K_{js}^{+}}{u_{is}^{*} k_{is} K_{is}^{+}}, \tag{5}$$

and comparing with the permeability ratios of equation (2), it is immediately apparent that the conductance ratios should equal the permeability ratios:

$$\frac{G_0(\mathrm{J})}{G_0(\mathrm{I})} = \frac{P_j}{P_i} \tag{6}$$

* The parameters K_{js}^{+}, k_{js} and K_j are defined in equations (24) to (28).

† Space does not permit a more extensive discussion of the range of applicability of equation (1) here other than to remark that in an analysis published elsewhere, Szabo, Eisenman and Ciani (1970) verified for the macrotetralides that equation (1) holds generally regardless of differences in antibiotic concentration in the solutions on the two sides of the membrane and the degree of complexation in the aqueous solutions, provided only that the rate of diffusion of the neutral, uncomplexed antibiotic within the membrane is rapid compared to its diffusion through the unstirred aqueous layers adjacent to the membrane and or the membrane-solution interface, as is expected for valinomycin as well as for the macrotetralides.

‡ We have assumed the diffusion of IS^+ in the membrane to be the rate-limiting step for ion transport. (This is not to be confused with the rapid diffusion mentioned in the previous footnote for the neutral, uncomplexed species, S). It should be noted that the application of the present equilibrium considerations to steady-state measurements is valid only under this condition (Läuger and Stark, 1970; Markin *et al.*, 1969a,b; and Ciani *et al.*, 1972).

This expectation was extensively tested and verified for the four macro-tetralides (Szabo *et al.*, 1969); it will be verified here for valinomycin as well.

We also showed (Eisenman, Ciani and Szabo, 1969) that the equilibrium constants, K_i, for the extraction of a salt composed of monovalent ions I^+ and X^- from the aqueous phase into a bulk solvent phase by the antibiotic molecule, S, is expressable in terms of the previous parameters as

$$K_i = \frac{k_{is} K_{is}^{+} k_x}{k_s} \tag{7}$$

where k_x and k_s are the partition coefficients of the anion and antibiotic, respectively.*

Taking the ratio of these salt extraction constants for J^+ and I^+, K_j/K_i, it is seen that

$$\frac{K_j}{K_i} = \frac{k_{js} K_{js}^{+}}{k_{is} K_{is}^{+}}, \tag{8}$$

since both k_x and k_s cancel for a given solvent in the case of a given anion and antibiotic. Comparing equation (8) with equations (2) and (5) for the permeability ratios and conductance ratios of a membrane (made of the same solvent), we see that

$$\frac{P_j}{P_i} = \frac{G_0(\mathrm{J})}{G_0(\mathrm{I})} = \frac{u_{js}^{*} k_{js} K_{js}^{+}}{u_{is}^{*} k_{is} K_{is}^{+}}, \tag{9}$$

or

$$\frac{P_j}{P_i} = \frac{G_0(\mathrm{J})}{G_0(\mathrm{I})} = \frac{u_{js}^{*} K_j}{u_{is}^{*} K_i}. \tag{10}$$

Equation (10) explicitly relates the ratio of equilibrium constants, K_j/K_i, for salt extraction into a given solvent to the electrical properties of a thin membrane made of the same solvent; while equation (9) relates the permeability and conductance ratios to the ratio of aqueous formation constants for the complex, K_{js}^{+}/K_{is}^{+}, times the ratio of partition coefficients of the complex between water and the membrane, k_{js}/k_{is}, and the ratio of the mobilities of these complexes in the membrane, u_{js}^{*}/u_{is}^{*}. Note that such characteristics as the size and shape of the complex could vary from cation to cation, and the above conclusions would still hold since no particularizing assumptions have been made as yet concerning the physical properties of the complexes. The comparison between membrane and bulk phases made in equation (10) would, however, necessitate that the membrane be made of the same solvent as the bulk phase.

* I.e., for the reactions: $I^+ + X^- + S^* \overset{K_i}{\rightleftharpoons} IS^{+*} + X^{-*}$ defined in Equation (25).

To circumvent this difficulty, we introduced the postulate of "isostericity", in which we assumed that the overall size and shape of the complex, as well as its externally-viewed electron distributions, was the same for all cations. In this case, the interaction energy with the solvent of the complex should not vary with the particular cation species bound; and the partition coefficients of all complexes are expected to be the same for any solvent so that by

$$\frac{k_{js}}{k_{is}} \simeq 1, \tag{11}$$

for all solvent (and phospholipid bilayer) compositions.

This result immediately allows us to deduce, on inserting equation 11 in equation 8, that

$$\frac{K_j}{K_i} \cong \frac{K_{js}^+}{K_{is}^+}. \tag{12}$$

Thus, the ratios of salt extraction equilibrium constants, K_j/K_i should be not only independent* of the solvent in which they are measured but, moreover, may be taken as a measure of the ratio of K_{js}^+/K_{is}^+, the ratio of stability constants for complex formation in aqueous solution.

Similarly, the mobility ratios of isosteric complexes should be independent of the composition of the membrane, being given by

$$\frac{u_{js}^*}{u_{is}^*} \simeq 1, \tag{13}$$

regardless of the particular cation species bound, since all isosteric complexes are also expected to be indistinguishable with regard to their mobilities. Inserting equation (13) in equation (10), and recalling equation (12), the remarkably simple set of relationships was found

$$\frac{P_j}{P_i} = \frac{G_0(\mathrm{J})}{G_0(\mathrm{I})} = \frac{K_j}{K_i} = \frac{K_{js}^+}{K_{is}^+}, \tag{14}$$

Equation (14) contains the principal theoretical expectations which we have previously verified for the macrotetralides and which we will now test for valinomycin. In particular, equation (14) predicts: first, that membrane conductance ratios and permeability ratios should be identical for a given pair of cations (e.g. Na^+ vs. K^+); second, that the equilibrium constant ratios measured by salt extraction should also be identical to the permeability and conductance ratios, regardless of the composition of membrane or solvent; and third, that these ratios

* This independence has been verified in model solvents for the membrane interior varying in dielectric constant from 2 (hexane) to 9 (dichloromethane). An analysis of the free energies underlying equation (12) will be presented in a further section. To avoid confusion with stability constants measured in other solvents the reader should recall that the reference state for the cations both for the 2-phase salt extraction equilibrium constant, K_j, and for the aqueous stability constant, K_{js}^+, is always the cation in water at infinite dilution.

should reflect the ratios of stability constants of the complexes in aqueous media.* From Equation (14) it is apparent why for isosteric complexes the choice of model solvent for the salt extraction studies is not critical.

The Experimentally Observed Effects of Valinomycin

There is a general agreement that the effect of valinomycin on bilayer membranes is characterized by a first power dependence of membrane conductance on antibiotic concentration (Tosteson *et al.*, 1968, McLaughlin *et al.*, 1970, Stark and Benz, 1971). This is in accord with the expectation of equations (3) and (4) and is illustrated in Fig. 1 by the slope of 1 for the logarithm of membrane conductance as a function of the logarithm of valinomycin concentration for three

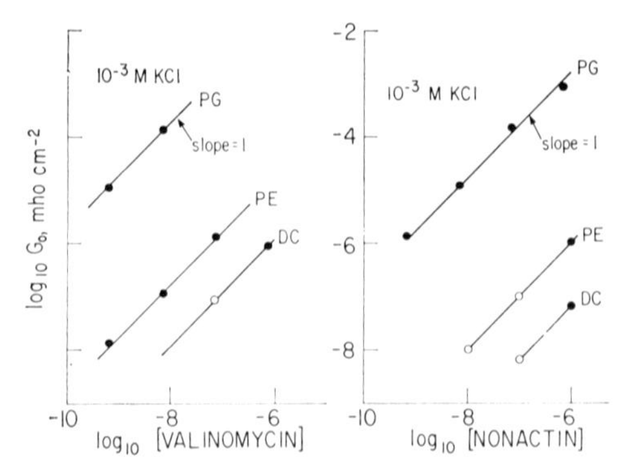

Figure 1. The first power dependence of membrane conductance on Valinomycin concentration (left) and Nonactin concentration (right). The experimentally observed conductances are indicated by data points; and the lines are drawn with a slope of 1. Open circles indicate values extrapolated from measurements made at higher KCl concentrations.

different lipids (the neutral lipid 7-dehydrocholesterol (DC), the amphoteric lipid phosphatidyl ethanolamine (PE), and the negatively charged lipid phosphatidyl glycerol (PG)). For comparison, the closely similar behavior of the typical macrotetralide, nonactin, is illustrated in the right hand portion of Fig. 1. (Note the close correspondence for valinomycin and nonactin of the dependence of effectiveness on lipid composition, as indicated by the identical displacements from lipid to lipid of the conductances due to these two antibiotics.)

The proportionality between membrane conductance and concentration of permeant cation, expected from equation (4) for uncharged

* Of course, this does not imply that any substantial number of complexes are actually present in the aqueous phase in the experimental concentration range. Indeed, to the extent that equation (4), rather than equation (3) is observed to describe the membrane conductance satisfactorily, we may deduce that a negligible fraction of antibiotic is complexed to cations in the usual aqueous media. Nevertheless, the free energies underlying the selectivity determined by the K_{js}^{+}/K_{is}^{+} ratios are the same as those underlying the selectivity determined by the K_j/K_i, P_j/P_i and $G^0(\mathrm{J})/G^0(\mathrm{J})$ ratios, discussed later.

lipids, is also verified in Fig. 2 for DC and PE by the fairly wide range of conditions over which simple proportionality holds, as judged by the lines of slope 1. Such a strict proportionality when ionic strength is allowed to vary (as in Fig. 2) is expected only for neutral lipids. For charged lipids, deviations from simple proportionality are expected (and observed) with varying ionic strength in accord with the expectations of diffuse double layer theory (cf. Lesslauer *et al.*, 1967 and McLaughlin *et al.*, 1970). These can lead to an apparent 0th power dependence for a negatively charged lipid or an apparent 2nd power

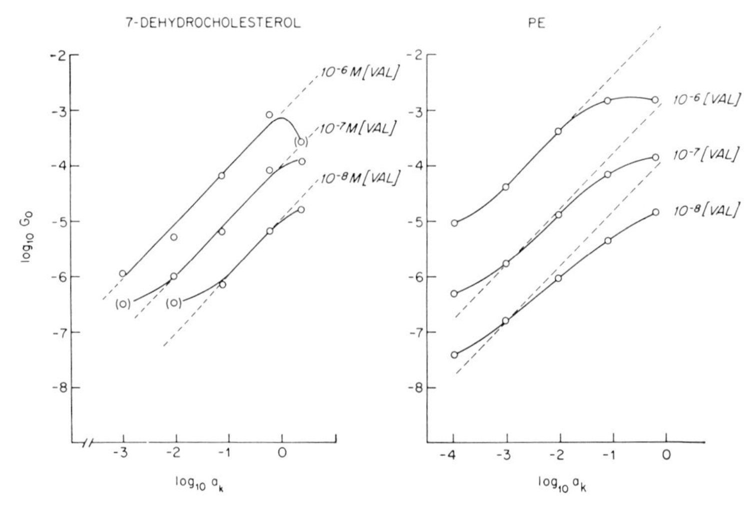

Figure 2. The proportionality between Valinomycin-mediated membrane conductance and concentration of permeant cation K^+ for the uncharged lipids 7 Dehydrocholesterol (DC) and Phosphatidyl ethanolamine (PE). Experimentally observed conductances are indicated by data points, while the lines are drawn with a slope of 1.

dependence for a positively charged lipid (Szabo *et al.*, 1970, p. 127 ff., and particularly equation (5) of McLaughlin *et al.*, 1970). This probably accounts for the departures from simple proportionality reported for valinomycin (Tosteson, 1968, Lieberman and Topaly, 1968, and Lieberman, Pronevich and Topaly, 1970). These effects can be overcome by holding ionic strength constant (Szabo *et al.*, 1969, Fig. 11). Indeed, Stark and Benz (1971) have demonstrated, in very carefully executed experiments at constant ionic strength, the strict proportionality between membrane conductance and K^+ concentration for the negatively charged lipid, phosphatidyl inositol, as well as for the neutral lipid, phosphatidyl choline.

All of the above observations are consistent with the postulate that valinomycin acts as a carrier of cations across bilayer membranes by forming 1:1 lipid soluble complexes with alkali metal cations, as do the macrotetralide actins. This similarity between valinomycin and the macrotetralide actins has recently been verified in experiments designed to test the carrier hypothesis by reversibly freezing and melting bilayer membranes (Krasne *et al.*, 1971). In these experiments it was found that the membrane effects of both valinomycin and nonactin were obliterated at the same temperature by solidifying the membranes. By contrast, the membrane effects of gramicidin A, a postulated channel former (Hladky and Haydon, 1970; Goodall, 1970), were unaltered by solidifying the membrane.*

Selectivity Among Alkali Metal Cations

The relative ability of valinomycin to mediate the effects of the alkali metal cations on membranes were first described by Mueller and Rudin (1967), who found the sequence Rb > K > Cs > Na > Li both for membrane potential and for membrane conductance. A characterization of the effects of these ions on membrane potential, in agreement with that of Mueller and Rudin but in considerably greater detail, was presented by Lev and Buzhinsky (1967). This sequence was also confirmed by Andreoli *et al.* (1967). The same sequence of cation potencies was found by Pressman (1968) for the effectiveness of valinomycin in extracting the thiocyanate salts of the alkali metals into toluene-butanol. Thus, it is seen that, at least qualitatively, the similar sequence of cation effects on permeability ratios, conductance ratios, and salt extraction potencies indicates that the expectations of Equation (14) are being fulfilled for valinomycin. To test this quantitatively we have carried out salt extraction experiments with valinomycin in exactly the same manner as previously used for the macrotetralides (Eisenman *et al.*, 1969).

Figure 3 presents the results of two sets of measurements for valinomycin for the extraction of the picrate salts of the alkali metal cations into the solvent dichloromethane.† The equilibrium constants, K_i, characteristic of these measurements are presented in the first two columns of Table I; and the last column gives the average ratios, K_i/K_K, of salt extraction equilibrium constants relative to K^+. The sequence Rb > K > Cs > Na > Li is confirmed, and the close quantitative agreement of the ratios of our salt extraction equilibrium

* Not only is valinomycin similar to the macrotetralide actins in its above mentioned effects, but also in the shape of its current-voltage characteristic, which we find to have a hyperbolic sine characteristic closely similar to that previously described for monactin by Szabo *et al.* (1969, Fig. 1), except at high salt concentrations. Such an increasing conductance with increasing voltage was first noted for valinomycin by Lev in 1966 (personal communication) and is also apparent in the published data of Andreoli *et al.* (1967, Fig. 3). By contrast, a current-limiting I–V curve has been reported as typical by Lieberman and Topaly (1968); and, indeed, we have seen this at very high KCl concentrations. Since the shape of the I–V relationship for valinomycin also depends on lipid composition (Stark and Benz, 1971), the apparent contradictions in the literature are understandable.

† This solvent has previously been established (Eisenman *et al.*, 1969) for the macrotetralides, to be satisfactory for measuring the desired equilibrium constant K_i of equation (7), without complication from ion pairing with the picrate anion.

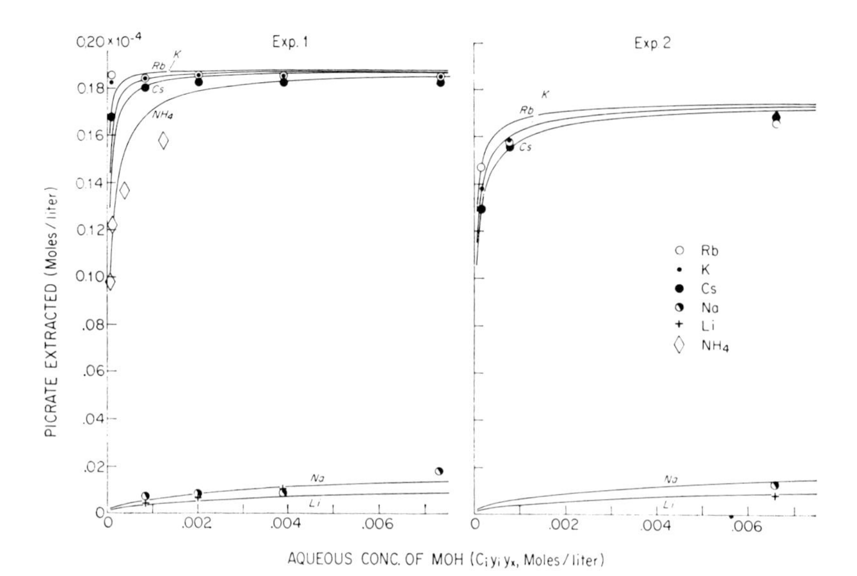

Figure 3. Extraction of alkali picrates into dichloromethane by valinomycin. The concentration of picrate extracted into the organic phase at equilibrium is plotted as a function of the aqueous concentration of MOH (corrected for activity coefficient effects). The units of the ordinate and abscissa are moles per liter. The points are experimentally observed, the curves theoretically calculated (according to Eisenman *et al.*, 1969, equation (28)) with the values of the equilibrium constants K_i given in the first two columns of Table I. The left-hand figure corresponds to experiment (1) and the right-hand figure corresponds to experiment (2) carried out under the following experimental conditions: Expt. (1), initial picrate concentrations = 10^{-4} M, initial valinomycin concentration $0{\cdot}188 \times 10^{-4}$ M, aqueous phase 10 ml, organic phase 10 ml; Expt. (2), initial picrate concentration = 10^{-4} M, initial valinomycin concentration $0{\cdot}176 \times 10^{-4}$ M, aqueous phase 2 ml, organic phase 10 ml. For methods of calculation see Eisenman *et al.*, 1969.

TABLE I. Equilibrium constants for valinomycin's ability to extract picrates into dichloromethane

	K_i	K_i	K_i/K_K
	Expt. (1)	Expt. (2)	Average
Rb	22,000	19,000	1·95
K	10,800	10,000	1·0
Cs	6,300	6,600	0·62
NH_4	2,00	—	0·19
Na	0·15	0·2	0·000017
Li	0·06	0·07	0·000007

The experimental conditions were the same as in Tables 4 and 14 of Eisenman *et al.*, 1969, using valinomycin from E. Lilly. The initial conditions for the two experiments were:

Expt. (1): $C^{\mathrm{In}}_{\mathrm{Picrate}} = 10^{-4}$ M, $C^{\mathrm{In}}_{\mathrm{Val}} = 0{\cdot}2 \times 10^{-4}$ M, $V = 10$ ml
$V^* = 10$ ml.

Expt. (2): $C^{\mathrm{In}}_{\mathrm{Picrate}} = 10^{-4}$ M, $C^{\mathrm{In}}_{\mathrm{Val}} = 0{\cdot}176 \times 10^{-4}$ M, $V = 2$ ml
$V^* = 10$ ml.

constants with literature values for the permeability ratios and conductance ratios for bilayers can be seen in Table II. Note the numerical correspondence between our K_i/K_K ratios in column (1) and the corresponding permeability and conductance ratios from the publications of Lev and Buzhinsky in column (2) and of Mueller and Rudin in columns (3) and (4). This agreement verifies that the important equality, equation (14), holds for valinomycin.*

TABLE II. Test of the identity between permeability ratios, conductance ratios, and salt extraction equilibrium constant ratios for valinomycin

	K_i/K_K*	P_i/P_K†	P_i/P_K‡	$G^0(K)/G^0(K)$†
Li	0·000007	(<0·012)	(0·0025)	(<0·005)
Na	0·000017	(<0·014)	(0·0035)	(<0·006)
K	1·0	1·0	1·0	1·0
Rb	1·95	1·9	2·3	1·5
Cs	0·62	0·44	0·53	0·25
NH_4	0·19	—	—	—

Note that the parenthesized quantities were obtained at neutral pH under conditions in which the H^+ ion can make a substantial contribution to the potential and conductance whereas the salt extraction equilibrium constants were obtained at alkaline pH, where the effects of H^+ are minimized. For this reason the data for the membranes may overestimate the relative permeabilities of the ions Li and Na, and the values listed must be considered to be maximum values for the permeabilities of these ions.

* From Table I.

† From Data of Lev and Buzhinsky, 1967, at 23°C and at 0·1 M salt (except H^+ at 0·001 M).

‡ From the data of Mueller and Rudin, 1967, at room temperature and 0·05 M salt for P_i/P_K (conductance ratios obtained at 35°C).

From all the above findings, it seems reasonable to postulate that valinomycin and the macrotetralide actins behave in the same general way and may be usefully regarded as members of the same class of carriers; they both form 1:1 complexes with cations and differ only quantitatively in the numerical values of the constants underlying their selectivity among cations, as will now be considered more extensively.

A Further Comparison of the Effects of Valinomycin with Those of the Macrotetralide Actins

With these results in hand, it is possible to compare in some detail the parameters underlying the effects of valinomycin with the corresponding parameters for the macrotetralide actins. Table III sum-

* Incidentally, the potential selectivity constants measured by Simon and his colleagues (1970) for valinomycin in thick electrodes made of diphenyl ether are also in close agreement with the data of Table II, as can be seen by comparison with his potential selectivity constants, referred to K^+, of: Li (<0.00021), Na (<0.00026), K (1), Rb (1·9), Cs (0·48), NH_4 (0·012), H (0·000056).

TABLE III. Salt extraction equilibrium constants, K_i, for the macrotetralide actins and comparison of the extraction ratios K_i/K_K with the conductance and permeability ratios for bilayer membranes

		K_i*	K_i/K_K*	P_i/P_K†	$G^0(I)/G^0(K)$†
Nonactin	Li	0·05	0·00026	<0·001	0·00042
	Na	3·2	0·017	0·0071	0·0067
	K	190	1	1	1
	Rb	90	0·47	0·58	0·58
	Cs	11·5	0·061	0·033	0·039
	NH_4	9,000	47	8‡	5‡
Monactin	Li	0·10	0·00012	<0·0005	0·00025
	Na	8	0·0096	0·0075	0·0048
	K	850	1	1	1
	Rb	290	0·34	0·5	0·34
	Cs	25	0·029	0·024	0·014
	NH_4	16,000	19	—	—
Dinactin	Li	0·15	0·000076	<0·00058	0·0002
	Na	25	0·012	0·0067	0·0081
	K	2,000	1	1	1
	Rb	800	0·40	0·42	0·48
	Cs	46	0·023	0·014	0·012
	NH_4	24,000	12	—	—
Trinactin	Li	0·23	0·000059	<0·00058	0·000033
	Na	42	0·011	0·009	0·0042
	K	4,000	1	1	1
	Rb	1,170	0·29	0·32	0·38
	Cs	75	0·019	0·015	0·013
	NH_4	46,000	11	4‡	3·5‡

* Eisenman *et al.*, 1969, Table 15.
† Szabo *et al.*, 1969, Table 5.
‡ Data of Laprade, Szabo and Eisenman for phosphatidyl ethanolamine.

marizes data for the macrotetralides from our previous studies and also includes data for NH_4^+ from work in progress in our laboratory by Raynald Laprade. The first column presents the salt extraction equilibrium constants, characterized for dichloromethane under the same experimental conditions as for valinomycin in Table II. These K_i values may therefore be directly compared to those for valinomycin. The second, third and fourth columns compare the values of the ratios K_i/K_K, P_i/P_K, and $G^0(I)/G^0(K)$, respectively. Notice the excellent agreement in the numerical values for all three of these ratios in agreement with equation (14).*

Comparison of the values of K_i for the macrotetralides in Table III with those for valinomycin in Table I indicate that valinomycin is

* Incidentally, these ratios agree closely with the potential selectivity constants measured by Simon and his colleagues (1971) for thick electrodes made from nonactin in Nujol/2-octanol, for which, the selectivities relative to K^+, were: Li (0·00056), Na (0·0067), K (1·0), Rb (0·42), Cs (0·031), NH_4 (2·5), H (0·018).

considerably more effective in its ability to extract the salts of the larger alkali metal cations than even the most effective macrotetralide, trinactin. This is illustrated in Fig. 4 where the values of log K_i are plotted for the antibiotics ranked in sequence: Nonactin, Monactin, Dinactin, Trinactin, Valinomycin.

The relative abilities of valinomycin and the macrotetralides to extract K^+ into a model solvent correlate surprisingly well with the relative effect of these molecules in the membrane's K^+ conductance. From the salt extraction data of Tables I and III, the ratio, K_j^V/K_j^M, is about 12. This correlates quite closely with the finding by Stark and Benz (1971) that the K^+ conductance produced by valinomycin is about 10 times as high as that produced by monactin under identical experimental conditions (a bilayer membrane made from dioleyl lecithin in the presence of 10^{-7} M antibiotic and 1·0 M ionic strength). This rather close correspondence may be somewhat fortuitous, as can be seen by comparison of the explicit expectations for the relationship between membrane conductance for a given antibiotic and its salt extraction equilibrium constant, which can be obtained from equation (4), on insertion of equation (7), to yield:

$$G_0(\mathrm{J}) = \left[\frac{F^2}{d}\frac{C_s^{\mathrm{Tot}}\,a_j}{k_x}\right] u_{js}^* k_s K_j. \tag{15}$$

The bracketed quantity in equation (15) will cancel when comparing ratios of membrane conductances with ratios of salt extractions carried out for two different antibiotics under the same experimental conditions. Therefore, if the ratio of conductances for valinomycin (denoted by the superscript V) and monactin (denoted by the superscript M), are compared, one obtains the expectation

$$\frac{G_0^{\mathrm{V}}(\mathrm{J})}{G_0^{\mathrm{M}}(\mathrm{J})} = \frac{u_{js}^{*\mathrm{V}}\, k_s^{\mathrm{V}}\, K_j^{\mathrm{V}}}{u_{js}^{*\mathrm{M}}\, k_s^{\mathrm{M}}\, K_j^{\mathrm{M}}}. \tag{16}$$

Since $G_0^V(\mathrm{J})/G_0^M(\mathrm{J}) = 10$, and $K_j^V/K_j^M = 12$, equation (16) implies that $u_{js}^{*V} k_s^V/u_{js}^{*M} k_s^M = 1/1{\cdot}2$. However, Stark and Benz also estimated the partition coefficients for dioleyl lecithin to be 6000 for monactin and 25,000 for valinomycin for this lipid, indicating that $k_s^V/k_s^M = 4{\cdot}2$. Taken literally this would imply that the mobility of the valinomycin complex is considerably less than that of the monactin complex if equation (16) is valid.*

It is also of some interest to try to deduce the relative values of K_{is}^+ for valinomycin vs. monactin from the relative values of K_i. This would be particularly pertinent to assessing whether effects due to complex

* However, it would not be unprecedented for the high partition coefficient for valinomycin to reflect additional interactions (possibly even a surface adsorption (Colacicco, 1969)) which might diminish its mobility. (Such opposing effects are characteristic of the movement of cations in glass, where the most strongly preferred cation is the least mobile (Eisenman, 1967).)

formation in the aqueous solution might be expected to occur at lower salt concentrations with valinomycin than with the macrotetralides. A difference in aqueous stability constants has implications for the effects of these molecules on bilayer membranes, as should be apparent from equation (3) in which it can be seen that when the term $K_{is}^{+}a_i$ is not negligible compared to unity, deviations are expected from the linear dependence of membrane conductance on salt concentration. Such an effect is a direct consequence of the fact that, when ion-carrier association in the aqueous phases becomes appreciable (i.e. $K_{is}^{+}a_i \geqslant 1$), not all of the carrier molecules are present in the free, uncomplexed form. Since the membrane conductance is proportional to the concentration of the free carrier as well as that of the salt, aqueous association results in a deviation from the proportionality seen otherwise between membrane conductance and salt concentration.

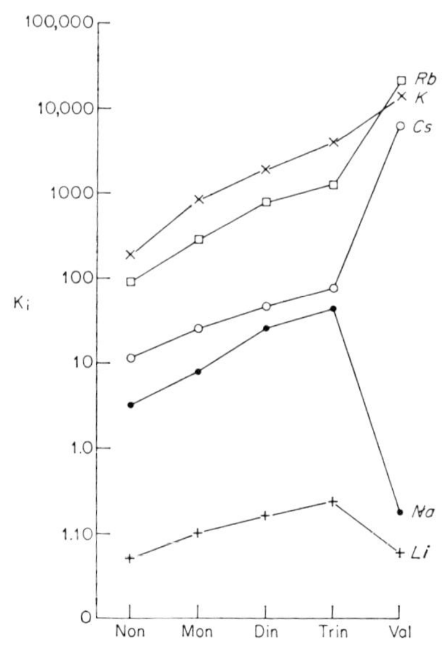

Figure 4. Summary of salt extraction equilibrium constants for the macrotetralides and valinomycin. The ordinate plots the logarithm of the values of K_i for the indicated cations as a function of the series of antibiotic molecules indicated on the abscissa.

The ratio of aqueous complex formation constants for valinomycin vs. the macrotetralides (denoted by M) can be deduced from equation (7) as:

$$\frac{K_{is}^{+\mathrm{V}}}{K_{is}^{+\mathrm{M}}} = \frac{K_i^{\mathrm{V}}\, k_s^{\mathrm{V}}\, k_{is}^{\mathrm{M}}}{K_i^{\mathrm{M}}\, k_s^{\mathrm{M}}\, k_{is}^{\mathrm{V}}}. \tag{17}$$

It seems plausible to assume that $k_s^{\mathrm{V}} k_{is}^{\mathrm{M}} / k_s^{\mathrm{M}} k_{is}^{\mathrm{V}} = 1$, in which case, equation (17) becomes:

$$\frac{K_{is}^{+\mathrm{V}}}{K_{is}^{+\mathrm{M}}} \simeq \frac{K_i^{\mathrm{V}}}{K_i^{\mathrm{M}}}, \tag{18}$$

indicating that the ratio of aqueous complex formation constants should be equal to the ratios of the salt extraction equilibrium constants. To the extent that this is true, Figure 4 can then serve to predict the relative values of K_{is}^{+}.

The Cyclic Polyethers, Bis Cyclohexyl-18-*Crown*-6 (*XXXI*) *and Bis*(*T-Butyl Cyclohexyl*)-18-*Crown*-6 (*XXXII*)

Our initial studies were concerned with the cyclic polyether bis cyclohexyl-18-Crown-6, which is diagrammed in Fig. 5 and referred to hereunder as XXXI, following Pedersen's (1967) terminology. We found that although the qualitative effects of this polyether on bilayers

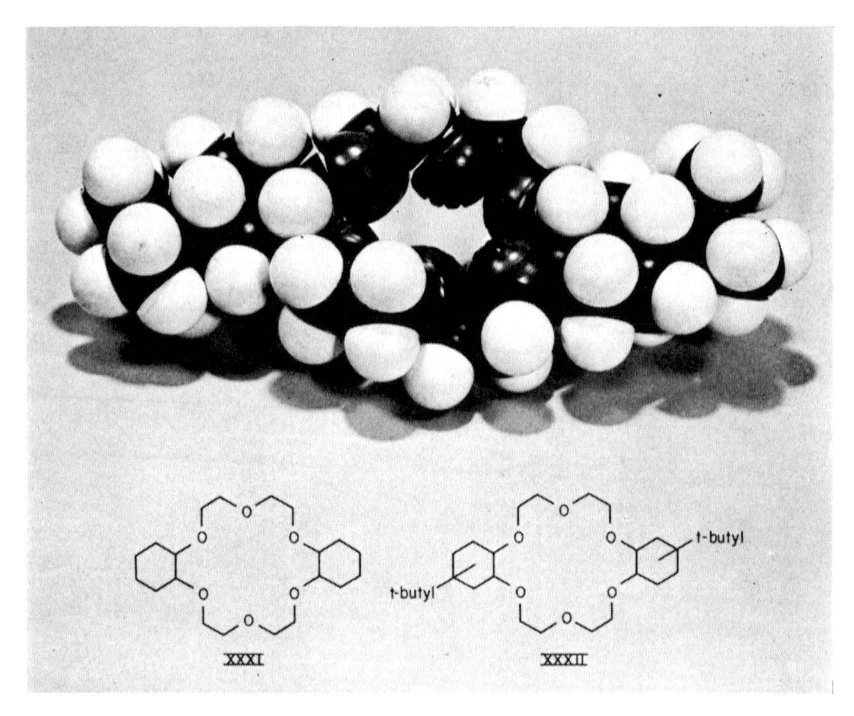

Figure 5. Chemical formulae for the cyclic polyethers XXXI (left) and XXXII (right). CPK Model of XXXII above.

and on salt extraction equilibria resembled those of the macrotetralide actins, there were serious quantitative discrepancies between the experimental observations and the theoretical expectations (Eisenman *et al.*, 1968). Specifically, for valinomycin and the macrotetralide actins, the ratios P_i/P_j, G_i/G_j and K_i/K_j were the same for a given pair of cations, whereas for the polyether the permeability and conductance ratios not only disagreed with each other, but also with the salt extraction equilibrium data.*

We have been able to resolve these apparent discrepancies by extending our studies to the more lipid soluble t-butylated analog of XXXI, bis(t-butyl cyclohexyl)-18-Crown-6, diagrammed in Fig. 5 and abbreviated hereunder as XXXII. The resolution of these apparent discrepancies involves only a direct extension of the previous theoretical

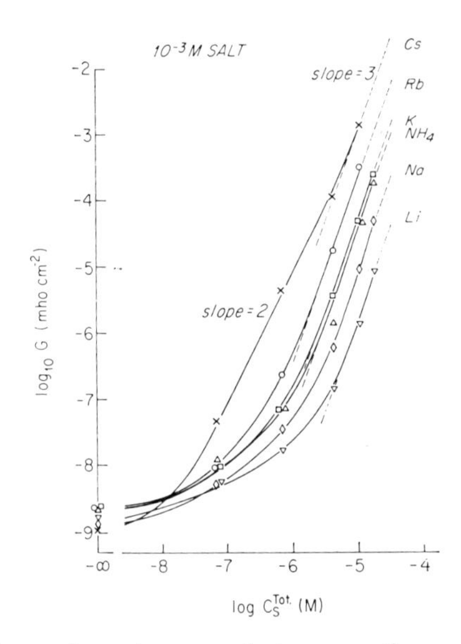

Figure 6. Dependence of membrane conductance on the aqueous concentration of polyether XXXII in the presence of 10^{-3} M chloride solutions of the indicated cations. The ordinate plots the logarithm of the observed membrane conductance; while the abscissa plots the logarithm of the total concentration of the polyether in the aqueous solution. The dashed lines are drawn with a slope of 3; while the slope of the Cs conductance between 10^{-7} and 10^{-6} M is 2. Asolectin membrane.

analysis of the carrier model to include the possibility that the polyethers form 2:1 and 3:1 carrier-cation complexes in the membrane. Such an extended theory predicts novel phenomena such as the presence of a maximum in the membrane conductance with increasing permeant ion concentration (when the ionic strength is maintained constant with an "indifferent" electrolyte) and variations in the cationic conductance sequences with salt concentrations. It enables us also to calculate the formation constants of 1:1 polyether-cation complexes in the aqueous solution for comparison with those measured directly. Since a detailed theoretical and experimental analysis has been submitted for publication elsewhere (McLaughlin, Szabo,

* The permeability ratios for XXXI were in the sequence K > Rb > Cs > Na > Li, whereas the conductance ratios were in the sequence Cs > Rb > K > Na > Li and the salt extraction sequence was K > Rb > Na > Cs > Li.

Ciani and Eisenman, 1972), only the main experimental features will be presented here.

The effects of the polyether XXXII on the membrane conductance and potential are illustrated in Figs. 6 and 7, respectively, for bilayers formed from asolectin. Fig. 6 shows the dependence of membrane conductance on the aqueous concentration of polyether in the presence of 10^{-3} M chloride solutions of the indicated cations. Comparison of the data of Fig. 6 with the previous findings for XXXI (Eisenman *et al.*,

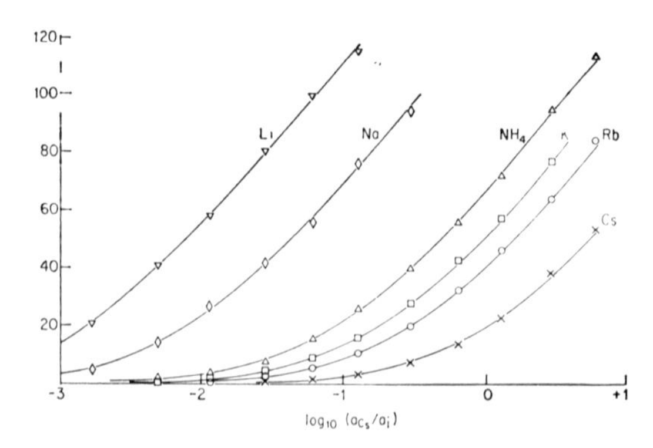

Figure 7. Membrane potentials in mixtures of cesium nitrate with the indicated alkali metal chlorides. The ordinate plots the observed values of the steady-state potential differences between the solutions on the two sides of the membrane; while the abscissa plots the logarithm of the ratio of the activity of the cesium ion to the activity of the indicated cation. The experiment was carried out in the same manner as Fig. 4 of Szabo *et al.*, 1969 by adding small volumes of cesium nitrate to one side of a membrane bathed initially on both sides in 10^{-3} M chlorides of the indicated cations. The points are experimentally measured; while the curves are drawn according to the theoretical expectations of equation 1 for the following values of permeability ratios: P_{Rb}/P_{Cs} (0·25), P_K/P_{Cs} (0·15), P_{NH_4}/P_{Cs} (0·075), P_{Na}/P_{Cs} (0·0070), and P_{Li}/P_{Cs} (0·0013). Asolectin membrane.

1968, Fig. 6c) shows that XXXII has a markedly greater ability to enhance the membrane conductance than does XXXI. This confirms John Rowell's original finding (personal communication, 1969) of the higher potency of XXXII than XXXI on bilayers. (The comparative effects of these two molecules will be discussed further in relation to Fig. 12). The conductances in a 10^{-3} M salt solution are seen to decrease in the sequence: Cs > Rb > K > NH_4 > Na > Li, identical to the sequence previously observed for XXXI (Eisenman *et al.*, 1968; Fig. 15).*

Figure 7 illustrates the corresponding effects on the membrane potential in the presence of 10^{-5} M XXXII (the points are the experi-

* The ability of XXXII to produce a very high Cs^+ conductance in bilayers has practical implications for using this molecule as a "probe" for examining cell membranes. For example, whereas it is difficult to detect an additional conductance when adding valinomycin or a macrotetralide to a squid axon (Gilbert *et al.*, 1970), presumably because of the intrinsically high K^+ conductance of this axon, it should be possible to detect the additional Cs^+ conductance for XXXII because of the low conductance characteristic of the squid axon in Cs^+ solutions (Baker *et al.*, 1962; Adelman and Fok, 1964).

mentally observed potentials; the curves are drawn from the theoretical equation (1) with the permeability ratios indicated on the figure). Note that equation (1) describes the experimental observations perfectly and that the observed Nernst slope implies the membrane is ideally permselective to cations in the presence of this polyether. It is important to note that the relative permeabilities to the various cations are in the same sequence as were the conductances, Cs > Rb > K > NH_4 > Na > Li. The close correspondence between the permeability and conductance ratios for the various cations from these two sets of data can best be compared in Table IV, where the data from Figs. 6

TABLE IV. Identity between conductance ratios and permeability ratios for the polyether XXXII at 0·001 M salt concentration

	$G^0(\mathrm{I})/G^0(\mathrm{Cs})$	P_i/P_{Cs}
Cs	1·00	1·00
Rb	0·20	0·25
K	0·044	0·15
NH_4	0·035	0·075
Na	0·0077	0·007
Li	0·0013	0·0013

and 7 are summarized. This correspondence verifies for XXXII the important identity of equation (14) and indicates that there is no discrepancy between permeability and conductance ratios for this polyether. (The reasons for the different permeability sequences observed with XXXI and XXXII will be considered at the end of this section).

Despite the general similarities between the observations so far described for XXXII and for the macrotetralides and valinomycin, certain important differences are apparent on examining Fig. 6 in greater detail. For the macrotetralides and valinomycin the membrane conductance was proportional to the first power of the antibiotic concentration, whereas the data of Fig. 6 indicate a second power dependence for cesium on polyether concentration in the region around 10^{-6} M and a third power dependence for all cations at the highest polyether concentrations. The data of Fig. 6 (see also Fig. 12), together with the observation that the membrane conductance at constant ionic strength is proportional to the first power of permeant cation concentration (at low salt concentration, as can be seen clearly for cesium in Fig. 11), strongly suggest the existence of a complex in which more than one cyclic polyether is needed to solubilize the cation in the membrane. It is seen in Fig. 6, and will become further apparent in Fig. 12, that there is a region where the 2:1 complex of Rb^+ and Cs^+ appears

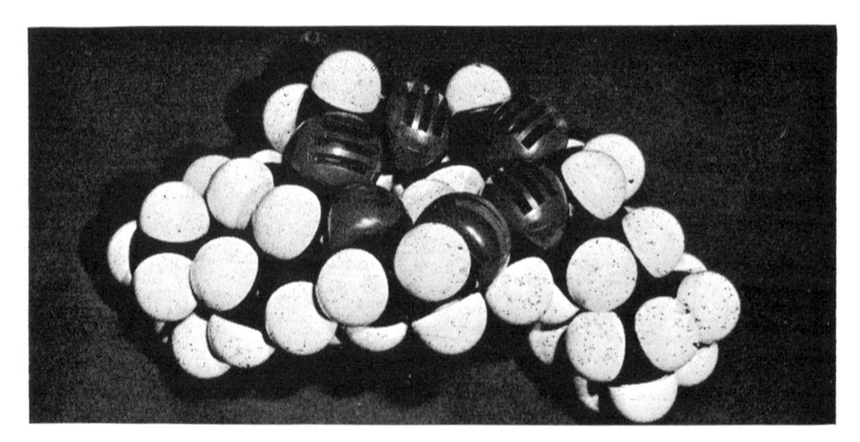

Figure 8. Two conceivable configurations for the cyclic polyether, XXXII. Top, twisted along the long axis; bottom, bent along the long axis. The 2 to 1 complex can be envisaged as sandwiching a cation between two molecules in the "planar" configurations of Fig. 5; while the 3 to 1 complex could be formed from the bottom or top configurations as a propeller-shaped, or sausage-shape complex around a cation; having an hydrophobic exterior and polar interior.

to be the permeant species; while at the highest polyether concentrations, the permeant species for all cations appears to be the 3:1 complex.* Thus, a principal difference between the cyclic polyether XXXII and the macrotetralides (as well as valinomycin) is the necessity for more than one polyether molecule to be utilized in solubilizing the

* This cubic dependence of conductance on polyether concentration is not restricted to asolectin membranes, but is also characteristic of bilayers formed from the phospholipids DC, PE and PG, as can be seen in Fig. 1 of McLaughlin *et al.*, 1970.

cation within the membrane.* The 2:1 complex is possibly a sandwich, with Cs and Rb constituting the filling, for these ions are too large to fit into the cavity of a single polyether molecule (Bright and Truter, 1970), and evidence exists for such a complex (Pedersen, 1970). The nature of the proposed 3:1 complex is unknown, but the polyether molecule can be deformed in at least two different ways, as illustrated in Fig. 8, and there are several possible ways that three such deformed molecules could arrange themselves around the cation.

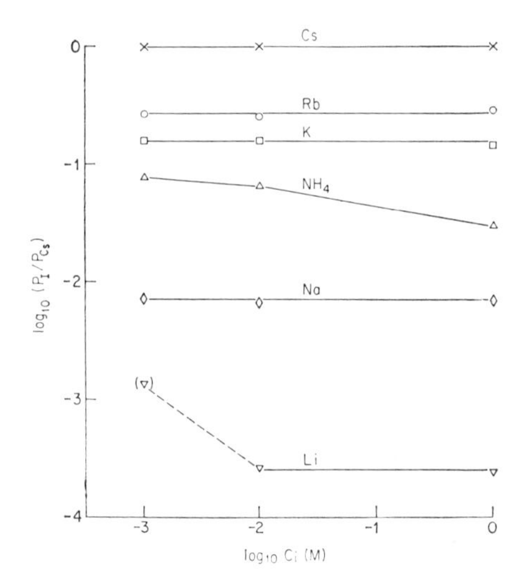

Figure 9. Constancy of permeability ratios at various salt concentrations. The permeability ratios were measured at 10^{-5} M polyether concentration as in Fig. 7 by adding $CsNO_3$ to the alkali metal chlorides at the indicated initial concentrations. Asolectin membranes.

The permeability ratios for XXXII are independent of salt concentration, as was also found to be the case with the macrotetralides and valinomycin. This is illustrated in Fig. 9, which presents the permeability ratios as a function of salt concentration (measured as in Fig. 7) for 10^{-3}, 10^{-2}, and 1 M salt concentration levels and for a 10^{-5} M concentration of polyether. The apparent variation in the lithium permeability at the lowest concentration is probably an artifact due to the presence of a low concentration of contaminants such as NH_4^+ and H^+; whereas the concentration dependence for NH_4^+ is small but real. This concentration independence of the permeability ratios is expected even when there is a significant degree (cf. Frensdorff, 1971) of complex formation in the aqueous solution.†

* Although a search for the existence of 3:1 complexes with the alkali metal cations has not been made in hydrocarbon solvents comparable to the interior of the bilayer membrane, the existence of higher complexes of polyethers with the alkali metal cations is becoming increasingly recognized (Bright and Truter, 1970; Frensdorff, 1971).

† The permeability ratios will be independent of salt concentration if the diffusion of the polyether molecules through the membrane is rapid compared to its rate of diffusing through

The conductance ratios for the polyether, however, depend markedly on the salt concentration. This can be seen clearly in the data of Fig. 10, which summarizes, for a 10^{-5} M concentration of polyether, the conductance observations from Fig. 6, together with conductances measured at higher salt concentrations. It is apparent from Fig. 10 that although the sequence of conductance ratios is the same as that of the

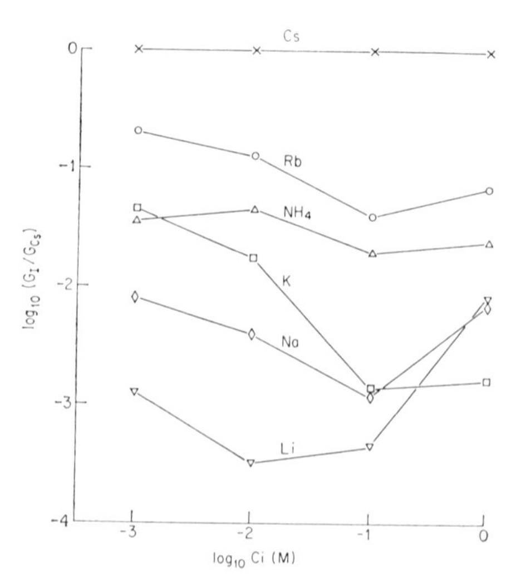

Figure 10. Dependence of conductance ratios on salt concentration. This figure summarizes the dependence of observed membrane conductances at 10^{-5} M polyether concentration from Fig. 6 and similarly obtained measurements at higher salt concentrations. Note that, in contrast to the permeability ratios, the conductance ratios depend markedly on salt concentration. Asolectin membranes.

permeability ratios at 10^{-3} M salt concentrations, the sequence differs at 10^{-2} M by inversion of the potassium and ammonium conductances. This inversion is more pronounced in the 1 M salt solution, where the conductances of lithium and sodium have both crossed potassium, so that the conductance sequence is now Cs > Rb > NH_4 > Li > Na > K. It will be shown below that these apparent complexities are a consequence of the cubic dependence of conductance on polyether concentration and the existence of significant quantities of 1:1 complexes in the aqueous solutions. Notice that the data of Figs. 9 and 10 have been obtained at 10^{-5} M polyether concentration, a concentration where the conductance depends on the cube of the carrier concentration for all cation species. This is necessary in order to simplify the treatment.

the unstirred layers or crossing the interfaces on either side of the membrane (cf. Szabo *et al.*, 1970, Case B, equations. (36)–(38)). A complete analysis of this problem has been carried out for XXXII (McLaughlin *et al.*, 1972) and it has also been verified experimentally that the rate limiting step for the movement of the neutral polyether is in fact diffusion through the aqueous unstirred layers.

We have deduced (McLaughlin, Szabo, Eisenman and Ciani, 1972) from a theoretical analysis of a carrier model that, when essentially only 3:1 complexes exist in the membrane phase and 1:1 complexes in the aqueous phase, the conductance and permeability ratios should be interrelated by:

$$\frac{G_0(\mathrm{J})}{G_0(\mathrm{I})}\cdot\frac{P_i}{P_j}=\frac{(1+K_{is}^{+}a_i)^3}{(1+K_{js}^{+}a_j)^3} \tag{19}$$

instead of the simpler identity of equation (14). Moreover, the conductance is defined by the more general equation (20) instead of equation (3).

$$G_0=\frac{F^2}{d}\,u_{is_3}^{*}\,k_{is_3}\,K_{is_3}^{+}\,\frac{(C_s^{\mathrm{Tot}})^3}{(1+K_{is}^{+}a_i)^3}\cdot a_i \tag{20}$$

In these equations, K_{is}^{+} and K_{js}^{+} are the stability constants for the formation of the 1:1 complexes in water, as before, but the subscripts "is_3" refer to the 3:1 complex. Note that the term $C_s^{\mathrm{Tot}}/(1+K_{is}^{+}a_i)$ in equation (20) is merely the free concentration of the polyether in the aqueous phase and that the product of the last three terms is, as before, the concentration of the permeant charged complex in the aqueous phase.

Equation (19) predicts that the permeability and conductance ratios should differ if significant complex formation occurs in the aqueous phase with formation constants K_{is}^{+} and K_{js}^{+}. In the limit of sufficiently low salt concentration, however (i.e. when the product $K_{is}^{+}a_i \ll 1$), the conductance and permeability ratios are expected to be identical, as indeed was observed in 10^{-3} M salt solutions (Table IV). At higher salt concentrations, it is possible to calculate values for the formation constants, K_{is}^{+}, from equation (19) by comparing the ratios of conductance and permeability at a given salt concentration. If we take species J to be lithium, and assume that its association with polyether is negligible in 10^{-2} M salt solutions (i.e. $K_{\mathrm{Li}s}\cdot 10^{-2} \ll 1$) then equation (19) reduces to the simple form

$$(1+K_{is}^{+}a_i)^3=\frac{G_0(\mathrm{Li})}{G_0(\mathrm{I})}\cdot\frac{P_i}{P_{\mathrm{Li}}} \tag{21}$$

The values of K_{is}^{+} calculated from the conductance and permeability data obtained in 10^{-2} M salt solutions by using equation (21) are given in the first column of Table V. These may be compared with values obtained for the two isomers, A and B, of XXXI (XXXII is too insoluble in water for measurements to be made on it) obtained by Frensdorff (1971) using glass electrodes and by Izatt, Nelson, Rytting, Haymore and Christensen (1971) using a calorimetric technique. The agreement in Table V is remarkably satisfactory, both as to sequence and to the magnitude of the values observed for the aqueous 1:1 complex formation constants, K_{is}^{+}.

TABLE V. Comparison of values of aqueous complex formation constants deduced from our bilayer membrane measurements for XXXII with those directly measured for XXXI

	Membrane* (potential vs. conductance)	Frensdorff† (glass electrode)		Izatt *et al.*‡ (calorimetric)	
	XXXII	XXXIA	XXXIB	XXXIA	XXXIB
K	120	150	60	100	40
Rb	34	—	—	31	4
Na	26	50	30	—	—
NH_4	19	27	6·5	22	16
Cs	12	18	7	13	—
Li	<4	5	—	—	—

All values are in liters per mole.
* Comparison between permeability ratios and conductance ratios in 0·01 M salt.
† Measurements for the two different isomers A and B by Frensdorff (1971) using cation selective glass electrodes.
‡ Calorimetric measurements by Izatt *et al.*, 1971.

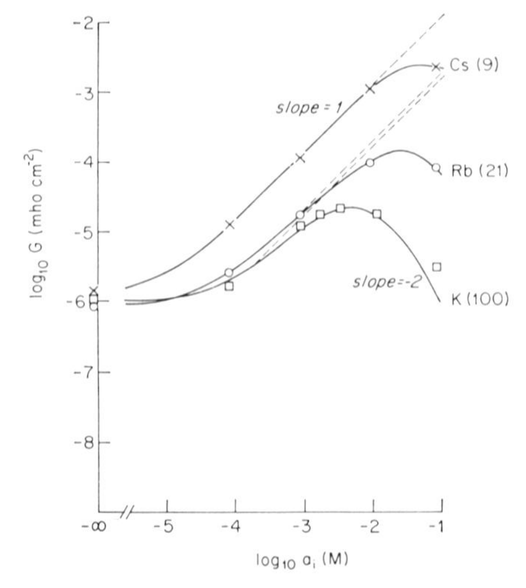

Figure 11. Dependence of membrane conductance on salt concentration at constant ionic strength. The ordinate plots the observed values of membrane conductance, while the abscissa plots the activity of indicated cation in solutions whose ionic strength was maintained constant at 0·1 M with LiCl. The points are experimental measurements; while the curves are drawn theoretically according to equation (20) for the following values of K_{is}^{+}: Cs (9), Rb (21), K (100). The dashed lines are drawn with a slope of 1. Asolectin membranes.

A further test of the present model becomes apparent on examination of equation (20). This equation predicts that the membrane conductance should first increase linearly (when $K_{is}^{+}a_i \ll 1$) with the concentration of permeant cation, a_i, should then go through a maximum (when $K_{is}^{+}a_i = 0{\cdot}5$) and then should ultimately decrease with an inverse square dependence on a_i (when $K_{is}^{+}a_i \gg 1$). This is in fact the observed behavior of the system, as is illustrated in Fig. 11. The data in Fig. 11 were obtained under conditions in which the ionic strength was held constant with the relatively impermeant or "indifferent" electrolyte, lithium chloride. The observed conductance data are plotted as points while the solid curves have been drawn according to equation (20) using the indicated values of the aqueous association constant, K_{is}^{+} (which are in a close agreement with those in Table V).

Comparison of the Properties of Polyethers XXXII and XXXI

We turn now to a comparison of the effects of XXXI and XXXII in order to reconcile the last remaining discrepancy between permeability and conductance ratios observed for XXXI, which was mentioned in the introduction to this section. The effects of the polyethers XXXI and XXXII on bilayer membranes are best compared in Fig. 12, which illustrates the dependence of membrane conductance on

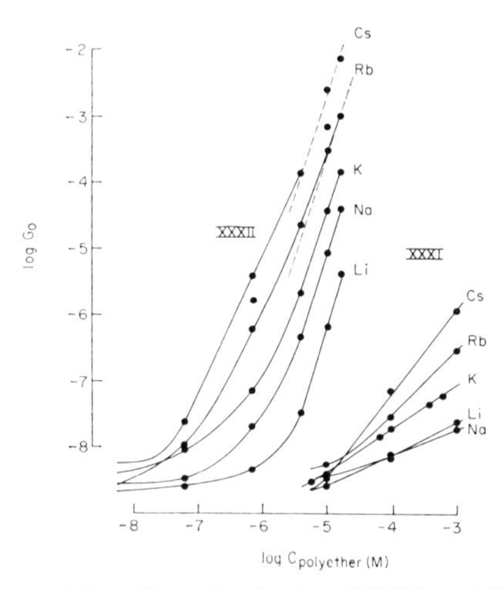

Figure 12. Comparison of the effects of polyethers XXXI and XXXII on membrane conductance. The logarithm of the membrane conductance for asolectin membranes is plotted as a function of the logarithm of the total polyether concentration at 0·1 M concentration of the indicated cations. The principal difference between the two polyethers is that XXXII is about 1000 times more effective than XXXI, in accord with their measured partition coefficients between water and hexane.

XXXI and XXXII concentrations under identical conditions. It is apparent from Fig. 12 that the effects of these polyethers on the membrane conductance are qualitatively similar. The conductance ratios for both of the polyethers are in the same lyotropic sequence. The main difference is that the tertiary butylated polyether XXXII is about a thousand times more effective than XXXI.*

* The principal effect of tertiary butylation on the polyethers is presumably to increase its membrane partition coefficient without markedly effecting its energy of interaction with the cations. Indeed, the partition coefficient of XXXII between water and hexane has been shown to be about 10^4 in favor of hexane, a factor of a thousand greater than the partition coefficient of XXXI into hexane (McLaughlin *et al.*, 1972).

Why, then, do permeability and conductance ratios previously noted for the cyclic polyether XXXI (Eisenman *et al.*, 1968) differ from one another or, alternatively, why are the permeability ratios for XXXI and XXXII different? The reason for this can be seen from the data presented in Fig. 13, where it is apparent that the Cs/K permeability ratio for XXXII is constant only when the polyether concentration in the solution exceeds 10^{-6} M. Below this concentration the

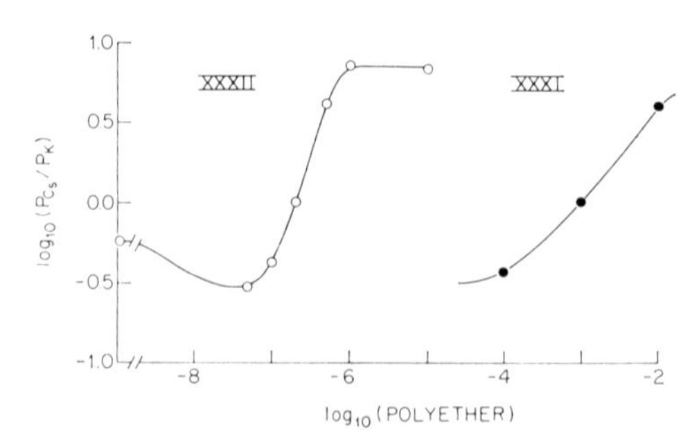

Figure 13. Dependence of permeability ratios on polyether concentration at very low (membrane) concentrations of polyether. This figure illustrates that the permeability ratio P_{Cs}/P_K is independent of XXXII concentration only above 10^{-6} M polyether; below this concentration, this permeability ratio depends on polyether concentrations. Note that all measurements of Figs. 8, 10, 11 and 13 were made at 10^{-5} M polyether. The right-hand portion of the figure illustrates a similar dependence of permeability ratio for XXXI, which naturally occurs at much higher polyether concentrations because of lower partition coefficient of XXXI than XXXII. See text for further details. Asolectin membranes.

permeability ratio, P_{Cs}/P_K, decreases and inverts, so that at low polyether concentrations the membrane is more permeable to potassium than to cesium. This phenomenon is also present for XXXI, as can be seen in the right-hand portion of Fig. 13. For the less favorably partitioned molecule, XXXI, however, we have had to increase the polyether concentration considerably above that in our previous publication in order to produce the change in the permeability ratios. Once the membrane concentrations of the two polyethers are scaled in terms of their different partition coefficients, the phenomena are seen to be remarkably comparable. Below a certain polyether concentration (i.e. about 10^{-6} M for XXXII and about 10^{-2} M for XXXI), the permeability ratios are a function of the polyether concentration (and different from the conductance ratios over a certain range), because of a combination of two effects described briefly below.*

* The phenomena illustrated in Fig. 13 presumably arise because at low polyether concentrations the most permeant species is the 1:1 complex. As the association constant for the 1:1 complex in water is greater for K than for Cs (see Table V), and therefore so should be the value of $\bar{K}_i$ for solubilizing the 1:1 K^+ complex in the membrane, it is reasonable that the membrane is more permeable to K than Cs at the lowest polyether concentrations. The inversion of the permeability ratio, P_{Cs}/P_K, occurs before the inversion of the conductance ratio, G_{Cs}/G_K, because a secondary effect of complex formation in the aqueous phase is to reduce the free polyether concentration. Permeability and conductance ratios are compared

We conclude that the behavior of XXXI is similar to that of XXXII and that the main difference between these two molecules is that XXXII has a partition coefficient about a 1000 times greater than that of XXXI. The effects that both these molecules produce on the permeability and conductance properties of bilayer membranes may be understood in terms of the present simple theoretical framework once it is recognized that a significant degree of association occurs in the aqueous phase and that at higher polyether concentrations the permeant species are no longer the simple 1:1 complexes as was the case for valinomycin and the macrotetralide antibiotics.

Quantitative Comparison of the Cation Selectivity of Valinomycin and the Macrotetralides with that of Biological Membranes

This section compares the cation selectivities characteristic of the macrotetralide actins and valinomycin with each other, as well as with the selectivities characteristic of biological membranes. It demonstrates that the quantitative pattern of selectivity among the alkali metal cations for the macrotetralide actins is distinguishable in certain important features from that characteristic of valinomycin. Moreover, the macrotetralide pattern is quite different from the pattern for cell membranes. By contrast, the quantitative selectivity of valinomycin for Li^+, Na^+, K^+, Rb^+ and Cs^+ is surprisingly close to that characteristic of the membranes of living cells. A further similarity between the data for the cell membrane and valinomycin (and a further difference between the cell membrane and the macrotetralide actins) comes from the selectivity for NH_4^+ which is similar for valinomycin and for the cell membrane, but is markedly different for the macrotetralide actins.

The rationale underlying the comparison to be presented here is that quantitative differences in selectivity are expected between molecules having different types, numbers, and orientations of the cation-binding ligands; and such differences may be useful as "fingerprints" for comparing the observed selectivity pattern in an unknown system with that characteristic of well defined molecules of known ligand type and configuration. Indeed, empirically generated selectivity patterns have been described for such diverse systems as glass electrodes (Eisenman, 1962) and cell membranes (Eisenman, 1963, 1965), which differ in important quantitative details. (Compare Figs. 15 and 16). For glass these have proven useful in systematizing the effects of a wide variety of cations on electrodes of diverse composition. In particular, it has been found possible to predict successfully the relative permeabilities of the species Li^+, Rb^+, Cs^+, NH_4^+, Ag^+, Tl^+ solely from a knowledge of

at the same total concentration of polyether, but the free concentration in the membrane for the P_K/P_{Cs} measurement is lower than that for the Cs conductance measurement and higher than that for the K conductance measurement, as is discussed in more detail elsewhere (McLaughlin *et al.*, 1972).

the P_{Na}/P_K ratio (Eisenman, 1967, cf. Figs. 4–7 of Chapter 9). For the moment, we ask the reader to bear with us in accepting the postulate that it will be instructive to compare the quantitative selectivities empirically observed for valinomycin and the macrotetralide actins to see whether or not we can identify distinguishable differences in the way these molecules select among cations for comparison with the selectivity "fingerprints" characteristic of cell membranes.

To define what we mean by an empirical selectivity pattern, let us begin with the selectivity among certain monovalent cations characteristic of glass electrodes. Figure 14 (which reproduces Fig. 9–4 of Eisenman, 1967) demonstrates the experimental evidence for the existence of a quantitative pattern of relative selectivities among the cations Li^+, Na^+, K^+, Rb^+, Cs^+, and H^+ for alumino-silicate glass electrodes, as well as air-dried collodion membranes. The ordinate plots the experimentally observed membrane potential differences, relative to K^+, for the various cations at pH 7 in 0·1 normal solutions. These values correspond to $58 \log P_i/P_K$ in present terminology. The effects of the individual cations relative to K^+ are indicated on each of the subfigures. Thus, the uppermost subfigure gives the permeability ratios of Li^+ relative to K^+. All data points below the horizontal line symbolizing K^+ represent glass compositions for which Li^+ was found to be more permeable; whereas all data points above the horizontal line represent glasses for which K^+ is more permeable. Data points at the same location on the abscissa of all subfigures represent observations for a given electrode composition. Over all of the figure except the extreme right-hand portion, the relative effects of cations for differing compositions of glasses have been ranked using the observed permeability ratio of Na^+ vs. K^+. This has been done by locating each composition from left to right by putting its point on the Na^+ "isotherm". Notice that the points must lie exactly on the straight line, since this has been chosen arbitrarily to define a left to right position for a given composition. Once this position is fixed for a given composition, the permeability ratios for the other cations are plotted according to their observed values on each of the subfigures. When this has been done, it becomes apparent that the data for each of the cations are described by the rather simple lines, henceforth called "isotherms", drawn on the figure as visual averages of the experimental points. Such simplicity need not have been the case. In this way, a set of isotherms was characterized for sodium alumino silicate glass electrodes (Eisenman, 1962) and subsequently found to apply also to collodion membranes and electrodes in which the original sodium oxide of the composition was replaced by other alkali metal oxides. The range of membrane compositions is indicated by the designations "LAS", "NAS", at the lower right of Fig. 14, which indicate lithium alumino silicate, sodium alumino silicate, etc. compositions. The isotherms from Fig. 14 could then be superposed into a set giving an overall selectivity pattern as has been done in Fig. 15. It is this sort of pattern with which we will be dealing in the remainder of this section; but before concluding this

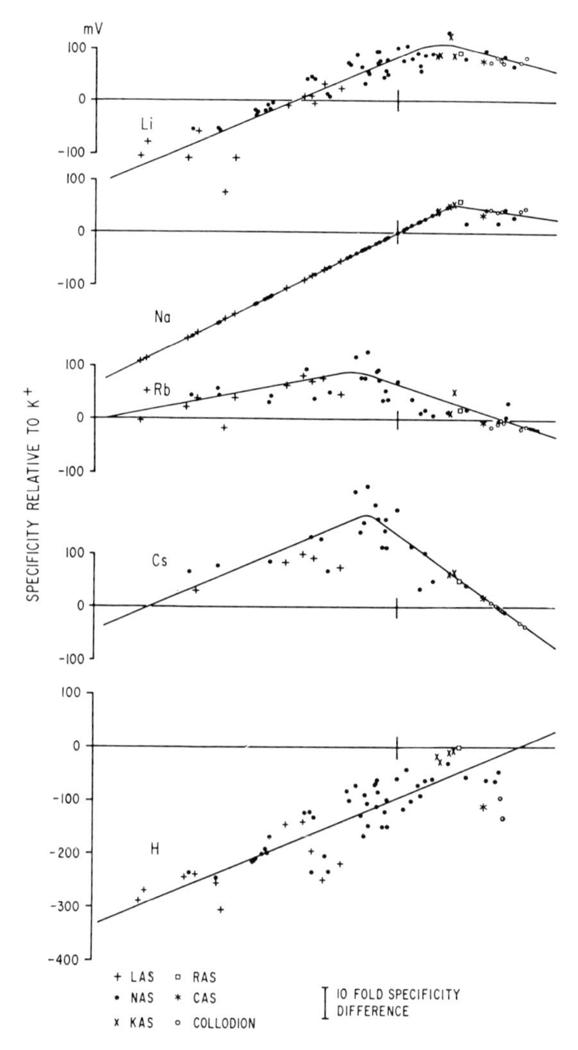

Figure 14. Experimental evidence for the existence of a quantitative pattern of cation selectivity for glass electrodes and collodion membranes. (After Fig. 9–4 of Eisenman, 1967). The membrane potential differences are indicated in millivolts (58 mV corresponds to a 10-fold selectivity difference). Data points for the indicated compositions are given by the symbols below the figure. See text for full details.

section it should be pointed out that this, entirely empirically developed, quantitative pattern has proven useful not only in describing a vast amount of data on electrodes of diverse composition but also in predicting for new compositions the relative effects of a wide variety of cations from a simple knowledge of the P_{Na}/P_K ratio. It should be apparent that *a knowledge of this ratio alone allows the permeability ratios relative to potassium for the other cations to be predicted through the selectivity pattern of Fig.* 15.*

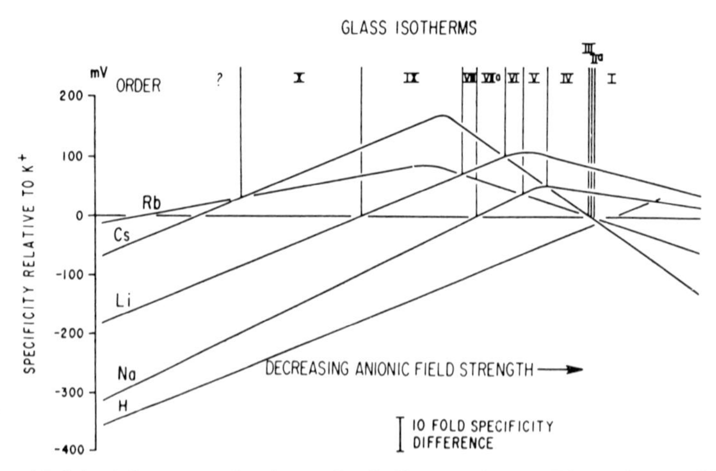

Figure 15. Selectivity pattern for glass and collodion membranes. The isotherms of Fig. 14 have been superposed to yield the selectivity pattern indicated here.

Let us now turn to comparable considerations for cell membranes. An examination of data in the literature for the permeation, accumulation, enzyme activation, and electrogenic action of Li^+, Na^+, K^+, Rb^+ and Cs^+ has indicated that these ions differ only quantitatively, and the empirical specificity pattern shown in figure 16 was proposed

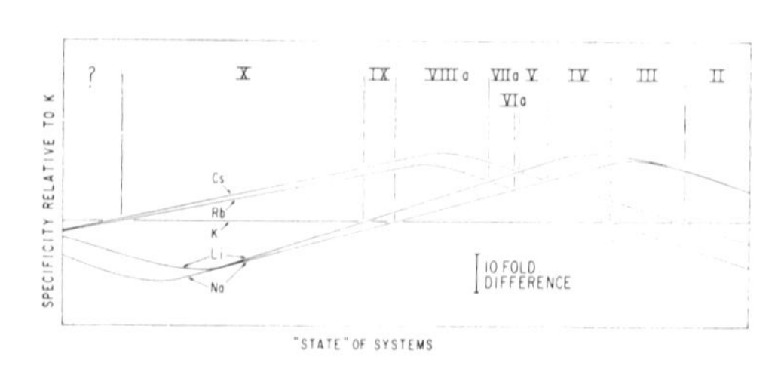

Figure 16. Biological selectivity pattern (after Figs. 8 and 9 of Eisenman, 1965). Details of the manner in which this figure was constructed are given elsewhere (Eisenman, 1965), and will be considered explicitly for cell membrane potential data in Fig. 17.

to describe these differences in biological phenomena (Eisenman, 1963). The generality of this pattern and its usefulness has received further support in more recent work (Eisenman, 1965) and from findings of analogous patterns for anions and divalent cations in biological systems

* It goes beyond the scope of the present section to discuss the fact that the Na^+ isotherm is not simply a straight line but shows a maximum at the right. In this case the data have had to be scaled by the extrapolated Cs^+ isotherm. This is a small degree of degeneracy in the data, which is discussed elsewhere (Eisenman, 1962, 1967).

by Diamond and Wright (1969).* Figure 16 presents the set of isotherms extracted in the same manner as Fig. 14 from a wide variety of biological selectivity data in the literature. Ordering the data according to the relative effects of a given pair of cations (Rb^+ and K^+ were used for convenience), it was found that the effects of the other alkali metal cations could be systematized. As for glasses, this need not have been the case. The essential feature of the "biological selectivity pattern" is that *specifying the selectivity for a given pair of cations allows one to determine quantitatively the selectivity for the other three alkali metal cations*. Even though such a correlation is empirical, and its usefulness is quite independent of any theoretical arguments, it should be noted that the existence of such a selectivity pattern is suggested from considerations of the balance of energies between ion-site vs. ion-water interactions, as examined in the last section.

The biological selectivity pattern of Fig. 16 requires that a given permeability ratio between any two cations (e.g. P_{Na}/P_K) specify a particular set of permeability ratios for all the other cations. The extent to which this is verified for cell membranes will now be critically examined in order to provide a basis for comparison with the present antibiotics. Figure 17 plots as data points the observed cationic permeability ratios (relative to K^+) for various physiological states of the nerve membrane and compares these with the solid curves which represent the postulated biological specificity isotherms of Fig. 16. (Note that these curves were deduced from a large variety of independent data.)

The ordinate plots the permeability ratio of the indicated cations (symbolized by I^+) relative to K^+. Data points for each of the cations have been located from left to right along the abscissa using the observed P_{Na}/P_K values to rank each set of observations from varying physiological "states" of the axon along the straight line segment of the Na isotherm. This uniquely locates the observed permeabilities to the other cations along the x-axis. For example, consider the set of data points labelled "18-underswing" on Fig. 17. These data represent the relative permeabilities calculated from the effects of Rb^+, Cs^+, Na^+ and K^+ on the membrane potential of the squid axon during the post-spike underswing (Baker, Hodgkin and Shaw, 1962). In their experiments the ratio P_{Na}/P_K was measured to be 0·03, the effect of 10 mM Rb^+ was found to be indistinguishable from that of 10 mM K^+, and 10 mM Cs^+ had the same effect in reducing the underswing as 3 mM K^+. From these data the appropriate permeability ratios for Rb and Cs were calculated to be: $P_{Rb}/P_K = 1$, $P_{Cs}/P_K = 0{\cdot}33$. There is only one

* It should be apparent that, although Figs. 15 and 16 exhibit sequences of cation selectivities which are qualitatively similar, the two patterns are easily distinguishable in many quantitative details. We propose to make use of this type of distinction below when we extend the comparison of empirically observed quantitative selectivity behavior to the macrotetralide actins vs. valinomycin.

place on the straight line segment of the Na^+ isotherm of Fig. 17 to locate P_{Na}/P_K, namely at the point indicated by the half filled dot. Therefore the data for Rb^+ and Cs^+, must also be located at this position along the abscissa. Notice that the Rb^+ data point corresponds precisely to the expected position of the Rb^+ isotherm here, while the Cs^+ data point corresponds quite closely to that of the Cs^+ isotherm. In the same way all of the other data points were plotted on Fig. 17.

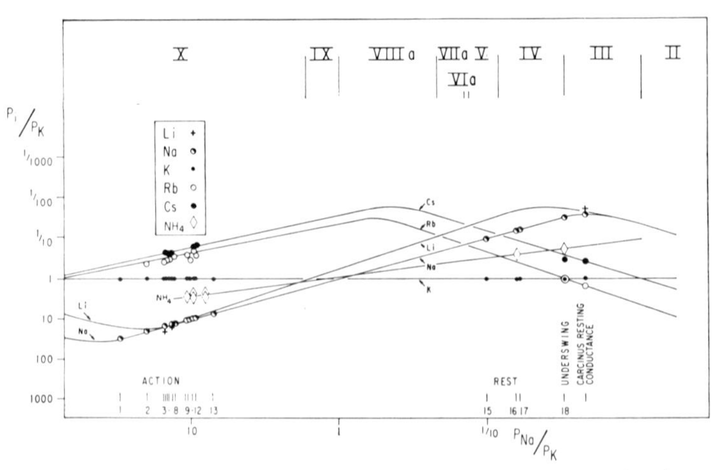

Figure 17. Selectivity pattern for cell membranes. The curves representing the empirical biological specificity isotherms of Fig. 16 are compared with the permeability ratios in various physiological states of the squid giant axon membrane, as well as the resting membrane conductance of Carcinus nerve. Data points represent observations for the permeability ratios of the indicated cations in varying physiological "states" of the axon, scaled along the abscissa by the observed P_{Na}/P_K values as described in the text using the biological Na^+ vs. K^+ selectivity isotherm (a straight line over the range of these data).

From left to right, the data come from the following sources: The points above numbers 1 and 13 represent P_{Na}/P_K measured by Baker, Hodgkin and Shaw (1962); numbers 2 through 12 are permeability measurements from individual experiments of Chandler and Meves (1964). Numbers 15 and 16 are the externally measured resting P_K/P_{Na} values of Baker, Hodgkin and Shaw (1962); while number 17 is the resting P_K/P_{Na} ratio measured by Adelman and Fok (1964). Number 18 represents the relative permeabilities in the underswing state reported by Baker, Hodgkin and Shaw (1962). In addition to these permeability ratios, Fig. 17 includes membrane conductance ratios (labelled "CRC") deduced from Hodgkin's (1947) data for *Carcinus Maenas* nerve.*

* The reason for the correspondence of conductance ratios and permeability ratios, an expectation of the independence principle, is seen intuitively from equation (14). The fact that the conductance ratios agree so well with the extrapolated expectations for the permeability ratios, albeit from different axons, is an interesting, although possibly coincidental, feature in which certain cell membranes resemble bilayer membranes treated with the present antibiotics. A further discussion of the independence principle will be found in Ciani *et al.*, 1969.

Also included in Fig. 17 are data for NH_4^+ from the recent studies of Binstock and Lecar (1969). These authors have characterized the effects of NH_4^+ on the early transient current, as well as the delayed currents, in voltage clamp studies on perfused squid giant axons. For the action potential region, Binstock and Lecar's P_{NH_4}/P_{Na} values (indicated by the diamond shaped symbol) for individual axons have been located from left to right by their observed values of P_K/P_{Na}. Although these authors did not report values for P_{Na}/P_K in the resting and underswing states of the axon, they did find P_{NH_4}/P_K to be 0·26 for the resting potential and P_{NH_4}/P_K to be 0·2 for the underswing potential. These permeability ratios for NH_4^+ have therefore been located in the resting region and underswing region, respectively by the P_K/P_{Na} ratios corresponding to these states on this figure. The NH_4^+ permeability data are seen to lie on the straight line labelled "NH_4^+" on the figure, which we will tentatively take as the ammonium isotherm for the cell membrane for comparison with the NH_4^+ data for the macrotetralide actins and valinomycin.

Note the remarkably close correspondence between the observed data points and the particular isotherms for each of the alkali metal cations, as well as for NH_4. At the very least the data of Fig. 17, therefore, indicate that a given permeability ratio observed between any two cations (e.g. P_{Na}/P_K) implies a particular set of selectivities for all the other cations. This is expected if a particular set of selectivities reflects a particular spatial array of ligand oxygens of a particular molecule.* The observed differences of selectivity from one cation to another are then a consequence of their different energies of interaction with these molecules (in competition with their energies of hydration), as examined in the last section. The empirical patterns can be likened to "fingerprints" for the detailed interactions between cations and the ion binding molecules, and comparison of the quantitative selectivity pattern for molecules with known (or at least knowable) configurations of ligand oxygens is then instructive in terms of attempting to deduce something about the intimate nature of cation-ligand interaction in the unknown biological membrane. It is to this that we now turn our attention.

Figures 18 and 19 present selectivity data for the macrotetralide actins and valinomycin for comparison with the cell membrane selectivity pattern. The observed data for the macrotetralide actins are indicated in Fig. 18 by points (K_i/K_K values on the left side, P_i/P_K values on the right side), and the biological isotherms of Fig. 17 have been indicated by fine lines for comparison. In both portions of the figure, the average selectivities for the macrotetralide actins, taken from

* A pattern could then develop from a set of selectivities of a given array by small perturbations of dipole moment of the ligand oxygens as, for example, is seen from the effect of adding the electron-withdrawing methyl groups to the members of the macrotetralide actin series of antibiotics (Eisenman *et al.*, 1969).

the more accurate salt extraction data are indicated by the heavy lines. Notice the large differences (indicated by the arrows) between the selectivities induced by these molecules and the biological selectivity isotherms. Clearly, there is *not* a quantitative correspondence between the pattern of selectivity observed for the macrotetralide actins and the pattern of selectivity for the cell membrane, the differences for the ammonium ion being the most striking.*

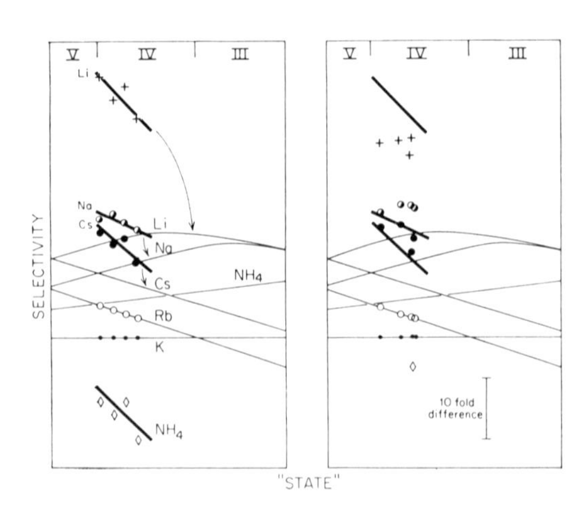

Figure 18. Comparison of the selectivity data for the macrotetralide actins with the selectivity pattern for cell membranes. Salt extraction equilibrium data (left) and permeability ratios (right) have been scaled along the abscissa according to K_{Rb}/K_K and P_{Rb}/P_K, respectively. Notice that there are marked differences between the observed cation selectivities for the macrotetralide actins (heavy lines) and the biological selectivity isotherms (light lines). So ranked, the macrotetralide actins are from left to right trinactin, monactin, dinactin, nonactin. (Dinactin is out of place by such a ranking and the discrepancy could have been removed by ranking the data by the potassium-ammonium isotherm, for example. This would not alter any of the present conclusions.) The data at the right give the corresponding permeability ratios similarly ranked (the data points from left to right are for trinactin, dinactin, nonactin, monactin; and the heavy lines trace the average selectivities from the left hand side of the figure).

In sharp contrast to this is the cation selectivity pattern manifested by valinomycin illustrated in Fig. 19 where the precisely measured permeability ratios of Lev and Buzhinsky (1967, Table I, p. 103) at 23°C and pH 6·2–6·5 are plotted in the same manner as in Fig. 18. A remarkably close correspondence is immediately apparent between the permeability ratios characteristic of valinomycin (data points) for the alkali metal cations and those characteristic of cell membranes (curves).

The individual sets of data points correspond to the selectivities observed at different ionic strengths (from left to right 0·5 M, 0·2 M,

* The data in Figs. 18 and 19 have been located from left to right using the Rb^+ isotherm (i.e., the P_{Rb}/P_K ratio) since the selectivities of the macrotetralides for Na^+ are so different from the biological values that they cannot be scaled by the biological Na^+ isotherms. This procedure alters none of the conclusions to be drawn here.

0·1 M, 0·001 M and 0·01 M). Also included on the figure are the permeability ratios measured by Mueller and Rudin (1967) at 0·05 M ionic strength.* Why the permeability ratios for valinomycin should depend upon ionic strength is not clear. Nevertheless, the close agreement between these data (particularly for K, Rb, Cs and NH_4) and the biological specificity pattern is apparent. Such a dependence of the

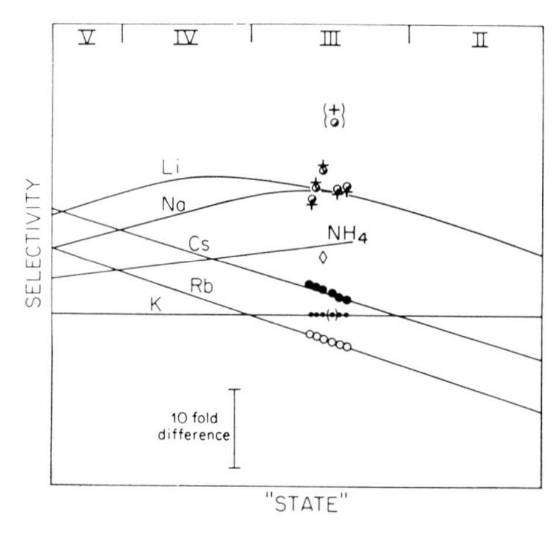

Figure 19. Comparison of the selectivity pattern for valinomycin with the cell membrane selectivity pattern. The data of Mueller and Rudin (1967) are indicated by parentheses. (See text for full details.)

"state" of the valinomycin molecule on ionic strength is not unprecedented. Similar changes in selectivity of glass electrodes, usually seen at constant ionic strength when pH is varied (Eisenman, 1962), are also observed at constant pH when salt concentration is varied.

Additional support for the similarities of ionic selectivity of valinomycin and of the cell membrane comes from close correspondence of the NH_4^+ selectivity data point for valinomycin plotted as a diamond in Fig. 19 (K_{NH_4}/K_K taken from the salt-extraction data of Table I is plotted, since membrane measurements for NH_4^+ are not presently available) and the NH_4^+ isotherm characteristic of the cell membrane. By contrast, the NH_4^+ selectivity for the macrotetralide actins of Fig. 18 are seen to be strikingly different from that of the cell membrane (and from valinomycin).

* The fact that Mueller and Rudin observed considerably lower Li and Na permeabilities (indicated by parentheses in Fig. 19) than did Lev and Buzhinsky, probably reflects the dependence of the apparent values of these permeabilities on trace components of other ions in the solution (see the discussion of Table II). The valinomycin permeability ratios at neutral pH give an overestimate of the effects of these ions on membrane potential due to the effects of H^+ among other cations. The important thing to note in Fig. 19 is that sodium and lithium permeabilities are considerably smaller than the potassium permeability. Indeed, they are probably considerably smaller than indicated both for the biological selectivity data and for the membrane selectivity data; and the apparent maximum and convergence of the Li and Na isotherms at the right of Figs. 16–18 probably reflects the effects of comparable levels of contaminating trace ions such as H^+ and K^+.

The above analysis of the quantitative differences in cation selectivity pattern between valinomycin and the macrotetralides indicate how sensitive the quantitative aspects of the cation-selectivity pattern are to the small differences in arrangements of the ligand groups in these two molecules. Indeed, we suggest that these patterns are useful as a "fingerprint" to distinguish between the sixfold carbonyl coordination of valinomycin and the four-fold coordinated carbonyl oxygens (and four-fold furane oxygens) present in the macrotetralide actins. For this reason we tentatively conclude that the cation detecting site in the resting cell membrane is not identical to that of the macrotetralide actins. It is tempting to speculate that it involves six carbonyl oxygens (possibly from the polypeptide backbone of a membrane protein) arrayed around the cation in a way essentially identical to the array characteristic of the interior of the cation-valinomycin complex.*

Irrespective of whether this speculation is correct or not, detailed comparisons of ionic specificity patterns are a powerful tool for inferring details of the likely ligand composition and configuration of the cation detecting sites in cell membranes and other biological molecules (e.g. cation-activated enzymes). The procedure is to compare the observed selectivity patterns in an unknown system with those characteristic of well defined model molecules of known ligand configuration (e.g. octahedral vs. tetrahedral) and type (e.g. carbonyl vs. ether) in the way illustrated here for cell membranes using valinomycin and the macrotetralides. A corollary is to infer the structure of the membrane's ligand groups from the selectivity of a membrane to particular "shapes" of cations. Hille (1971, 1972) has initiated such a characterization for the Na^+ and K^+ system of frog nerve.

Although the above considerations have been for passive permeation, the intracation selectivity pattern of Figs. 16 and 17 has been found to be the same as that seen in the active accumulation of ions in living cells, as well as in the cation activation of such enzymes as β-galactosidase and ATPase (Eisenman, 1965).

Energetics Underlying the Equilibrium Selectivity of Neutral Ion-Sequestering Molecules

Experiments described in the previous sections have examined the dependence for the observed membrane effects of valinomycin, the macrotetralides, and certain cyclic polyethers on the equilibrium

* It should be mentioned that, although the present selectivity considerations for the cell membrane are most easily deduced for a carrier mechanism of ion permeation, they are not at all inconsistent with the properties of a possible channel, whose selectivity could be due to either equilibrium or nonequilibrium factors. For example, if the process of loading and unloading of a channel is analogous to forming and dissociating a complex in water, the equilibrium selectivity factors, thought from the present analysis to underlie the ion interaction with molecules like valinomycin, could correlate with appropriate combinations of rate constants.

constants of certain simple reactions. In particular, identity (14), relating the ratios of membrane permeabilities and membrane conductances, salt extraction equilibrium constants, and aqueous complex formation constants has been verified experimentally:

$$\frac{P_j}{P_i} = \frac{G^0(\mathrm{J})}{G^0(\mathrm{I})} \simeq \frac{K_j}{K_i} \simeq \frac{K_{js}^+}{K_{is}^+}, \tag{14}$$

Since this result indicates that equilibrium considerations suffice to account for the principal effects of these molecules in bilayer membranes,* we will now examine the underlying equilibrium energetics in order to show how the energies contribute to the selectivity manifested in the effects of these molecules on bilayers. More detailed considerations of the energetic analysis of ionic selectivity will be found in papers by Simon and Morf (1972), by Eigen and his colleagues (Diebler *et al.*, 1969; Eigen and Winkler, 1971), and by Eisenman *et al.*, 1972, and Szabo *et al.*, 1972.

The formal relationships between the homogeneous reaction to form complexes in water and the heterogeneous reaction to solubilize cations in membranes (and salts in bulk phases).

From equilibrium considerations alone, the key to the selectivity of the present molecular carriers of cations lies in their ability to solubilize cations in media of low dielectric constant such as the interior of a phospholipid bilayer membrane. The cation solubilization can be expressed formally through equation (24) for the heterogeneous equilibrium by which a carrier molecule in the membrane, S^*_{memb}, reacts with a cation from the aqueous solution, I^+_{aq}, and solubilizes it in the membrane as the lipid soluble complex, IS^{+*}_{memb}:

$$I^+_{\mathrm{aq}} + S^*_{\mathrm{memb}} \xrightleftharpoons{\bar{K}_i} IS^{+*}_{\mathrm{memb}} \tag{24}$$

For clarity the subscripts "aq" and "memb" have been used to designate that the ions I^+ come from the aqueous phase whereas the

* It should also be noted that according to recent studies by Eigen's group (Diebler *et al.*, 1969) the ratio of the equilibrium constants for the formation of the complex in methanol have been shown for monactin to bear a direct relationship to the rate constant for loading the complex, $\vec{K}_{is}^+$, and the backward reaction of unloading the complex, $\overleftarrow{K}_{is}^+$, as defined in the reaction (written here for water):

$$I^+_{\mathrm{aq}} + S^+_{\mathrm{aq}} \underset{\overleftarrow{K}_{is}^+}{\overset{\vec{K}_{is}^+}{\rightleftharpoons}} IS^+_{\mathrm{aq}}; \qquad K_{is}^+ = \frac{\vec{K}_{is}^+}{\overleftarrow{K}_{is}^+} \tag{22}$$

where forward and backward rate constants have been indicated by arrows in the appropriate direction. The data of Diebler *et al.* in methanol suggest that the rate of formation of the complexes in water is diffusion controlled so that these rates are approximately the same for all cations. From this it is immediately possible to deduce that the ratio of formation constants must equal the ratio of the off constants for the kinetics of the reactions:

$$\frac{K_{js}^+}{K_{is}^+} = \frac{\overleftarrow{K}_{js}^+}{\vec{K}_{is}^+} \tag{23}$$

carrier molecules S and the complexes IS^+ are present in the membrane phase. (Although redundant, asterisks have been retained to designate species in the membrane for consistency with our previous publications.) The process by which this reaction enables ions to cross the membrane is indicated diagrammatically in Fig. 20. Equation (24) is an heterogeneous reaction, describing formally the process by which the IS^+ complex is formed within the membrane phase from a cation I^+ from the aqueous solution and a carrier molecule S from the membrane phase.*

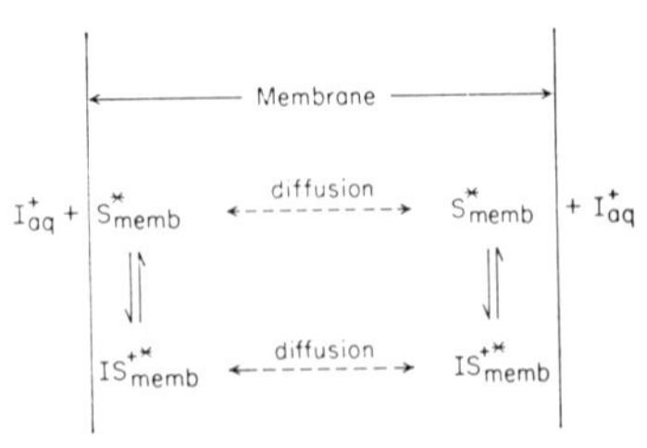

Figure 20. Diagram illustrating the process by which cations from the aqueous phase can react with carrier molecules from the membrane to form mobile complexes within the membrane.

This reaction is closely related to the reaction for extracting salt into an organic phase:

$$I^+_{aq} + X^-_{aq} + S^*_{memb} \underset{K_i}{\rightleftharpoons} IS^{+*}_{memb} + X^{-*}_{memb} \tag{25}$$

differing from it only by the partition of the anion into the organic phase to preserve electroneutrality (see Ciani *et al.*, 1969 and Eisenman *et al.*, 1969 for further details).

The similarities between the heterogeneous reactions (24) and (25) are intuitively apparent. Less apparent, but equally real, is the similarity between these reactions and the homogeneous reaction (26) for the formation of the ion-carrier complex in water:

$$I^+_{aq} + S_{aq} \overset{K^+_{is}}{\rightleftharpoons} IS^+_{aq} \tag{26}$$

This relationship is made more apparent by noting that reaction (24) can be viewed as the result of adding the subreaction (27) to, and subtracting subreaction (28) from, reaction (26).

$$IS^+_{aq} \overset{k_{is}}{\rightleftharpoons} IS^{+*}_{memb} \tag{27}$$

$$S_{aq} \overset{k_s}{\rightleftharpoons} S^*_{memb} \tag{28}$$

* Of course, this particular heterogeneous reaction cannot be distinguished in an equilibrium system from an alternative reaction path by which the IS^+ complex is formed in the aqueous phase and is then partitioned into the solvent. Stark and Benz (1971) have recently attempted to distinguish between these two possibilities, and a detailed theoretical analysis has been carried out by Ciani, Eisenman, Laprade and Szabo (1972) and is being tested experimentally in our laboratory by Raynald Laprade.

Reaction (27) represents the process of taking an IS^+ complex from an aqueous medium into the interior of the membrane; and its equilibrium constant, k_{is}, is the partition coefficient of the complex. Reaction (28) represents the process of taking a neutral molecule from an aqueous medium into the membrane, and its equilibrium constant, k_s, is the partition coefficient of the neutral carrier.*

Corresponding to these combinations of reactions, the equilibrium constant, $\bar{K}_i$, of the important heterogeneous reaction 24 is related to that, K_{is}^+, of the homogeneous equilibrium 26 through:

$$\bar{K}_i = K_{is}^+(k_{is}/k_s) \tag{29}$$

where K_{is}^+ is multiplied and divided by the equilibrium constants of reactions (27) and (28), respectively.

Taking the ratio of the equilibrium constants in equation 29 for two cations, J^+ and I^+, we get:

$$\frac{\bar{K}_j}{\bar{K}_i} = \frac{K_{js}^+ k_{js}}{K_{is}^+ k_{is}}, \tag{30}$$

which, recalling equation (8) is seen to be identical to the ratio of K_j/K_i measured from salt extraction equilibria. This shows the basic identity:

$$\frac{\bar{K}_j}{\bar{K}_i} = \frac{K_j}{K_i}, \tag{31}$$

which constitutes the rationale for using measurements of bulk phase salt extraction equilibrium constants to deduce the properties of thin membranes and allows us to generalize equation (14) as:

$$\frac{\bar{K}_j}{\bar{K}_i} \cong \frac{K_j}{K_i} \cong \frac{K_{js}^+}{K_{is}^+} = \frac{P_j}{P_i} = \frac{G^0(\mathrm{J})}{G^0(\mathrm{I})} \tag{32}$$

Because of identity (32), we can consider the salt extraction equilibrium constant ratios, K_j/K_i, to give a measure either of the quantity $\bar{K}_j/\bar{K}_i$ or of K_{js}^+/K_{is}^+. Let us therefore consider the equilibrium factors underlying the selectivity of the homogeneous complex formation reaction (26), which is the simplest case with which to begin an analysis of the origin of the selectivity of the macrocyclic molecules at equilibrium.

Free Energy and Equilibrium Selectivity for the Formation of Ion-Carrier Complexes in Aqueous Solutions

The problem of understanding the equilibrium selectivity of any reaction is always one of assessing the elementary interactions underlying the Gibbs free energy change of the reaction (cf. Eisenman, 1961, 1962, 1965). Thus, the equilibrium constant, K_{js}^+, of reaction (26) is

* Of course, each of these equilibrium constants can be expressed as the ratio of rate constants for the forward and backward reactions, but considerations of kinetics exceed the scope of this paper.

related to the standard free energy change, ΔF^0, of the reaction through:

$$\Delta F^0 = -RT \ln K_{is}^+; \tag{33}$$

and the elementary process underlying the selectivity of this reaction can be assessed by decomposing the reaction into elementary processes following a classical Born-Haber cycle (cf. Eisenman, 1961).* This cycle is illustrated in Fig. 21. Reaction 26 has been indicated at the bottom in the heaviest lettering; and process I consists of taking the cation, I^+, and the carrier molecule, S, out of the aqueous solutions into

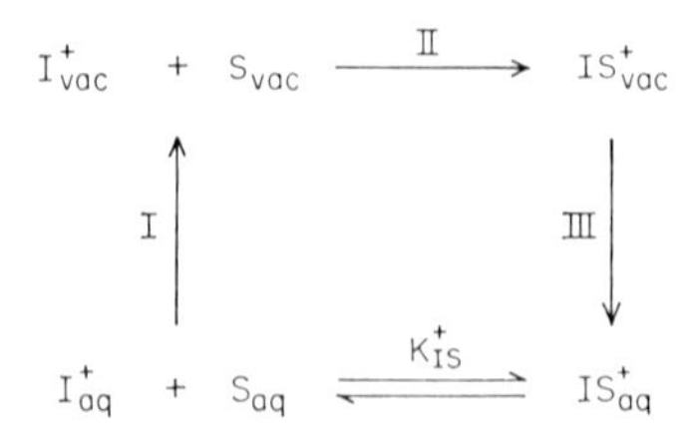

Figure 21. Born-Haber cycle decomposing the aqueous complex formation reaction into their conceptual subprocesses, as described in the text.

vacuum. Process II consists of allowing these species to combine in vacuum to form the IS^+ complex in vacuum. Process III consists of returning the IS^+ complex to the aqueous solution.

The free energy change for the overall reaction is given by the sum of these three subprocesses:

$$\Delta F_0 = -RT \ln K_{is}^+ = \Delta F_{\text{I}} + \Delta F_{\text{II}} + \Delta F_{\text{III}} \tag{34}$$

Let us now consider the separate contributions to these free energies.

The free energy of reaction I consists of the combination of the free energies of hydration of the cation, $\Delta F_{\text{hyd}}(\text{I}^+)$, and of the neutral molecule, $\Delta F_{\text{hyd}}(\text{S})$;

$$-\Delta F_{\text{I}} = \Delta F_{\text{hyd}}(\text{I}^+) + \Delta F_{\text{hyd}}(\text{S}). \tag{35}$$

$\Delta F_{\text{hyd}}(\text{I}^+)$ can be assessed from the experimentally well known hydration energies of cations plus anions, provided an assumption is made as to the relative contribution between cation and anion (usually we are concerned only with the well-defined differences in the hydration energies for pairs of cations so that this assumption is generally not a problem). $\Delta F_{\text{hyd}}(\text{S})$ is an unknown quantity.

The free energy change of process II represents the free energy change in sequestering an ion inside the carrier molecule. This contains positive attractive terms between the cation and the ligand oxygens of the molecule as well as repulsive terms between ligand oxygens and

* It is important to note that since the system is at equilibrium, the free energy difference between the initial and final states are independent of the pathway taken between those two states. This fact justifies the use of the Born-Haber cycle as well as the use of alternative pathways further on in this section.

each other. These will be lumped together and called the "electrostatic" free energy, ΔF_{el}. In addition, there are strain energy differences between the conformation of uncomplexed carrier vs. the conformation of the complexed carrier as well as entropy changes, which are difficult to assess, all of which will be included in $\Delta F_{conform}$. So that:

$$\Delta F_{II} = \Delta F_{seq}(I^+) = \Delta F_{el}(I^+) + \Delta F_{conform}(I^+) \tag{36}$$

The free energy change of process III can also be considered to consist of two parts:

$$\Delta F_{III} = \Delta F_{Born}(IS^+) + \Delta F_{Cav}(IS^+) \tag{37}$$

where, $\Delta F_{Born}(IS^+)$ represents the electrostatic energy of the Born charging process for the IS^+ complex in aqueous media vs. vacuum (since the complex is quite large, the applicability of the Born charging process for a homogeneous dielectric should be a good approximation for this energy). $\Delta F_{Cav}(IS^+)$ represents the energy of forming a cavity in water around the IS^+ complex. If one wanted to evaluate the difference in free energy for the formation of an ion-carrier complex in the aqueous phase, therefore, one would need to know the values for the differences in free energies for each of the subcomponents of the equation

$$\begin{aligned} -RT\ln K_{is}^{+} = {} & -\Delta F_{hyd}(I^+) - \Delta F_{hyd}(S) + \Delta F_{el}(I^+) + \Delta F_{conform}(I^+) \\ & + \Delta F_{Born}(IS^+) + \Delta F_{Cav}(IS^+) \end{aligned} \tag{38}$$

Although it could turn out that some of these energies are insignificant relative to others or that some of the terms cancel, not enough is known about the conformation of the molecule in these states to make the assumptions necessary to reduce this equation to any simpler form. By contrast, however, one can reduce the comparable equation for the selectivity ratio between two cations to a calculable form since, in this case, several of the terms will be the same, or sufficiently similar, that they will cancel. The following is the total equation for the difference in free energy of a carrier complexed with $[I^+]$ vs. $[J^+]$ before any terms are cancelled:

$$\begin{aligned} -RT\ln\frac{K_{js}^{+}}{K_{is}^{+}} = {} & -\overbrace{[\Delta F_{hyd}(J^+) - \Delta F_{hyd}(I^+)]}^{A} - \overbrace{[\Delta F_{hyd}(S) - \Delta F_{hyd}(S)]}^{B} \\ & + \overbrace{[\Delta F_{el}(J^+) - \Delta F_{el}(I^+)]}^{C} + \overbrace{[\Delta F_{conform}(J^+) - \Delta F_{conform}(I^+)]}^{D} \\ & + \overbrace{[\Delta F_{Born}(JS^+) - \Delta F_{Born}(IS^+)]}^{E} + \overbrace{[\Delta F_{Cav}(JS^+) - \Delta F_{Cav}(IS^+)]}^{F}. \end{aligned} \tag{39}$$

It is immediately apparent that term B equals zero and thus falls out of the equation. Furthermore, if one makes the reasonable assumption that the IS^+ and the JS^+ complexes are "isosteric" (Eisenman *et al.*, 1969), an assumption for which there is experimental evidence for several carrier-cation complexes (see the discussion of "isostericity"), then term D equals 0 because the conformations of IS^+ and JS^+ are the same, term E equals 0 because the radii of the IS^+ and JS^+ complexes are the same,* and term F equals 0 because the size of the cavities made by the IS^+ and JS^+ complexes are the same. For "isosteric" complexes, therefore, equation 39 reduces to

$$-RT\ln\frac{K_{js}^+}{K_{is}^+} = [\varDelta F_{\text{hyd}}(\text{I}^+) - \varDelta F_{\text{hyd}}(\text{J}^+)] - [\varDelta F_{\text{el}}(\text{I}^+) - \varDelta F_{\text{el}}(\text{J}^+)] \tag{41}$$

This important relation, in which the selectivity ratio reflects the differences in hydration energies of the cation species vs. the differences in their essentially electrostatic energies of sequestration within the antibiotic molecules in vacuum, underlies the analysis of the dipole model for antibiotic selectivity presented in the following section (cf. Eisenman, 1969; Eisenman *et al.*, 1972; Szabo *et al.*, 1972b).

Free Energy and Equilibrium Selectivity for Cation-binding by Membrane Carriers

If the reference state is taken as the hydrocarbon interior of a membrane rather than vacuum, a similar set of considerations can be applied for the key reaction 24 which describes the overall reaction by which the present carriers solubilize cations in a membrane.

To do this it will be helpful to refer to the physical states of the various cation complexing molecules and complexes in three different physical states schematized in Fig. 22. These states are: in water at the bottom; in the hydrocarbon region of the membrane in the middle; and in vacuum at the top. These states are interconnected by processes whose energies can be assessed by a Born-Haber cycle as in Fig. 21. Each process is designated by an arabic numeral and a verbal description. For example, the process "1. Hydration of I^+" represents the process of taking the cation I^+ from infinite dilution in vacuum to infinite dilution in water. The polar ligand oxygens have been indicated diagrammatic-

* The equation for the Born charging energy for a monovalent species is

$$\varDelta F_{\text{Born}} = -\frac{332}{2r}\left(\frac{1}{D_{\text{vac}}} - \frac{1}{D_{\text{H2O}}}\right) = -\frac{332}{2r}\left(\frac{79}{80}\right) \tag{40}$$

where r is the radius of the complex, D_{vac} is the dielectric constant of a vacuum, which is equal to 1, D_{H_2O} is the dielectric constant of water, which is equal to 80, and the energy is in kcals per moles when r is in angstroms. Note that $\varDelta F_{\text{Born}}$ will be the same for complexes with the same radii.

ally by dark ellipses, and the possibility of a conformational change in the molecule is illustrated in processes labelled "Change of conformation."*

It is possible to use the states diagrammed in Fig. 22 to analyze the process underlying the reaction (24) which is schematically indicated by the process:

$$S^*_{memb} \xrightarrow{k_s} S_{aq} + I_{aq} \xrightarrow{K^+_{is}} IS^+_{aq} \xrightarrow{k_{is}} IS^{+*}_{memb} \qquad (42)$$

The equilibrium constant $\bar{K}_i$ of this overall reaction (for processes 5, 6 and 7) can be decomposed into the sum of free energies for the four

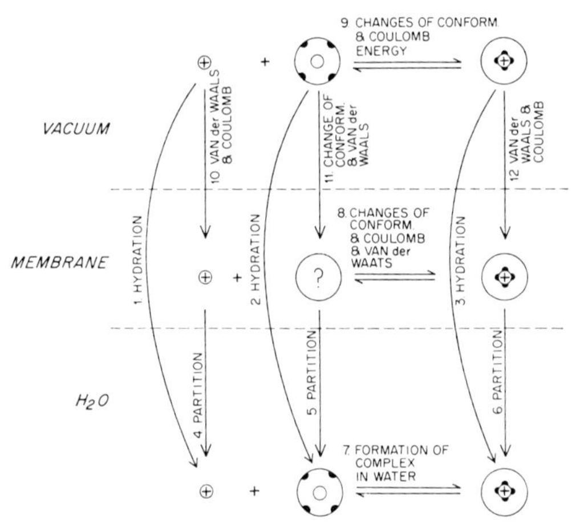

Figure 22. Diagram of the physical states of the various cation complexing molecules and their complexes.

processes (1, 11, 9 and 12 schematized in Fig. 22 such that $-RT\ln\bar{K}_i = -\Delta F_1 - \Delta F_{11} + \Delta F_9 + \Delta F_{12}$. The energies for these processes are as follows: ΔF_1 is the hydration energy of the cation. ΔF_{11} is the energy of taking the antibiotic molecule from vacuum into the membrane and contains an attractive (e.g. Van der Waals) interaction with the membrane interior (ΔF_{11a}) and a possible conformation change (ΔF_{11b}). ΔF_9 represents the ion sequestration reaction in vacuum, and has two parts: an attractive interaction (ΔF_{9a}), for which an electrostatic model will be given shortly, and a conformational energy change

* It seems reasonable to assume that the minimum energy configuration of the uncomplexed molecule will have the polar groups facing out in water. Similarly, it seems reasonable for the molecule to assume a similar configuration in vacuum since the polar ligand groups should exert repulsions on each other. It is difficult to predict, however, what configuration the uncomplexed molecule will have in the membrane since hydrophobic interaction between the hydrocarbons of the molecule and the hydrocarbons of the membrane would favor the same configuration as for the complexed molecule; whereas electrostatic repulsions between the carbonyl groups would favor the same configuration as for the uncomplexed molecule in vacuum. Hence the question mark on the figure. As none of these considerations is necessary when comparing one isosteric cation to another, since configurational effects then all cancel, we shall leave the configuration of the uncomplexed molecule in the membrane as undetermined.

(ΔF_{9b}). ΔF_{12} represents the differences of free energies of the complex in the membrane and in vacuum. This energy can be considered to consist of two parts: an electrostatic energy difference (ΔF_{12a}), and an attractive (e.g. Van der Waals) interaction energy between the complex and the membrane (ΔF_{12b}).

We can summarize these energies as follows:

$$-\Delta F_{1} = -\Delta F_{\text{hyd}}(\text{I}^+)$$

$$-\Delta F_{11} = -\Delta F_{11a} - \Delta F_{11b} = -\Delta F_{\text{memb}}^{\text{Van der Waals}}(\text{S}) - \Delta F_{\text{memb}}^{\text{conformation}}(\text{S})$$

$$\Delta F_{9} = \Delta F_{9a} + \Delta F_{9b} = \Delta F_{\text{vac}}^{\text{attractive}}(\text{IS}^+) + \Delta F_{\text{vac}}^{\text{conformation}}(\text{IS}^+)$$

$$\Delta F_{12} = \Delta F_{12a} + \Delta F_{12b} = \Delta F_{\text{memb}}^{\text{Born}}(\text{IS}^+) + \Delta F_{\text{memb}}^{\text{Van der Waals}}(\text{IS}^+).$$

Analogous with the previous calculation, there is not enough information known to reduce these equations to any simpler form; but if one looks instead at the relative selectivity between two cations, and assumes "isostericity", terms 11, 9b and 12 cancel for the two ions, and one is left with

$$-RT\ln\frac{\bar{K}_j}{\bar{K}_i} = [\Delta F_{\text{hyd}}(\text{I}^+) - \Delta F_{\text{hyd}}(\text{J}^+)] - [\Delta F_{\text{attractive}}^{\text{vac}}(\text{IS}^+) - \Delta F_{\text{attractive}}^{\text{vac}}(\text{JS}^+)]. \qquad (43)$$

Equation (43) is seen to contain the same terms as equation (41).

The problem of cation selectivity of antibiotics in membranes thus reduces to a problem of assessing the competition between the interaction energies between the antibiotic and the ions in vacuum vs. the hydration energies of the ions. Since the differences of hydration energies are well known quantities, the problem reduces to one of estimating the sequestration energy differences expected between two cations. It is to this problem that we now turn.

Electrostatic Calculations of Binding Energies

We will now extend a previous model for the selectivity of monopolar ion exchange sites (Eisenman, 1961, 1962, 1965) to neutral dipolar sites which are more appropriate as models for the carbonyl (or ether) oxygens, which are the ion-binding ligands in the macrocyclic antibiotics. The considerations here in no way imply an attempt to calculate the binding energies for particular molecules; rather we try to examine the most obvious energy terms.

Monopolar Sites

Let us begin by considering a highly simplified model (Eisenman, 1961, 1962) in which the molecular anionic ion exchange sites of typical ion exchangers are approximated by a monopolar model, as indicated diagrammatically in Fig. 23. Here we compare the selectivity expected for the competition for a given cation between such a site and a single water molecule. The energies of interaction of a monovalent cation I^+ of radius r^+ with a singly charged anionic site of radius r^-, referred to

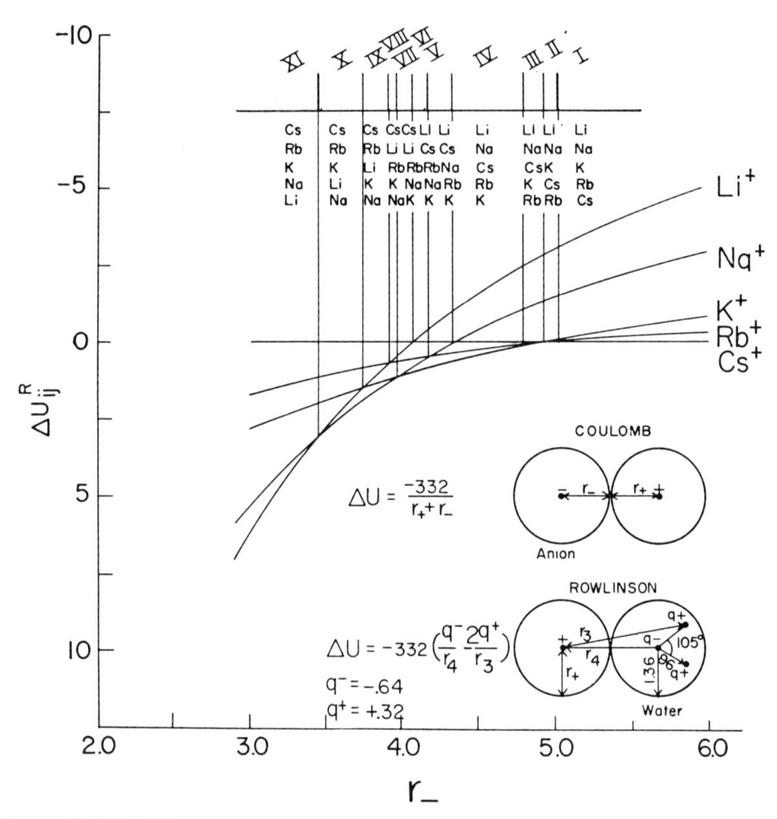

Figure 23. Selectivity pattern for the hypothetical cation exchange between a monopolar anion and a single multipolar water molecule. (Reproduced from Fig. 1 of Eisenman, 1961). The cation most strongly selected from water by the anionic "site' 'is the lowest on the chart. Above the graph are tabulated the cationic sequences (increasing specificity downwards), eleven rank order designations, as a function of decreasing site radius r_-. Units in kcal/mole and angstroms.

the ions at rest in a vacuum, are given approximately by the electrostatic attractive energy:

$$\Delta F_{\text{vac}}^{\text{attractive}} = \frac{-332}{r^+ + r^-} \tag{44}$$

While its energy of interaction with the indicated tripolar model of a water molecule is given by:

$$\Delta F^{\text{hyd}} = -332\left(\frac{q^-}{r_4} - \frac{2q^+}{r_3}\right), \tag{45}$$

representing the net attractions between the cation and the effective charge of the oxygen ($q^- = -0{\cdot}64$) at a distance r_4 from the center of the cation and the repulsions between the cation and the effective charge ($q^+ = +0{\cdot}32$) of the two protons at the greater distance r_3 from the center of the cation. (All energies are in kcal/mole for distances in angstroms, and charges are expressed as fractions of the electronic charge).

Calculating the values of these energies for the various naked (Goldschmitt) radii of the cations and inserting these in equation (41) (or 43),* one obtains the selectivity isotherms, referred to Cs^+, plotted on the ordinate in Fig. 23 (where energy in kcal/mole is labelled U^r_{ij}). This simple procedure leads to a pattern of selectivity in which, for the anion site of largest radius (i.e., lowest electrostatic field strength), the cations are preferred in the lyotropic sequence Cs > Rb > K > Na > Li while for an anionic site of sufficiently small radius the sequence of preference is reversed, being Li > Na > K > Rb > Cs. Between these extremes, the cations are seen to pass through eleven selectivity sequences, corresponding to the 11 Roman numerals designating the regions between the intercepts of the selectivity isotherms of Fig. 23.

This simple model is sufficient to account for the observed selectivities of many phenomena (as has been shown elsewhere in considerable detail (Eisenman, 1962, 1963, 1965; also see the excellent reviews by Reichenberg, 1966, and by Diamond and Wright, 1969). It can be made more realistic by replacing the interactions with the single water molecule by the differences of free energies of hydration of the cations in water, recalling that the interactions of cations with water are constant, being well known experimentally.†

Considering one limiting case, in which the sites of the exchanger are assumed to be widely separated and the water molecules are assumed to be excluded from their vicinity, the difference in free energies of interaction between cation and site ($\Delta F^{\text{attract}}_{j+} - \Delta F^{\text{attract}}_{i+}$) may be approximated by Coulomb's law (equation (44)). The selectivity expected in this case is plotted in Fig. 24.

In another limit, in which the sites are assumed to be very closely spaced (for example, with 6 sites coordinated around each cation and 6 cations around each site), the free energies are approximately given by

$$\Delta F^{\text{attract}}_{\text{vac}} = 1{\cdot}56 \frac{-332}{r_+ + r_-} \tag{46}$$

where the factor 1·56 appears as a consequence of the Madelung con-

* Equations 41 and 43 are formally identical to the equation previously used for assessing the selectivity of monopolar ion exchangers.

† In these calculations, free energies of hydration were taken from Latimer's (1952) compilation as the most accurate source. Referred to Cs^+ these are: $53{\cdot}8 \pm 1{\cdot}4$ kcal/mole for Li^+, $28{\cdot}9 \pm 0{\cdot}9$ for Na^+, $12{\cdot}7 \pm 0{\cdot}9$ for K^+, and $6{\cdot}7 \pm 1{\cdot}1$ for Rb^+.

stant (1·75) for this coordination state and a factor of 8/9 due to the Born repulsion energy.* The expected selectivities in this case are plotted in Fig. 25.

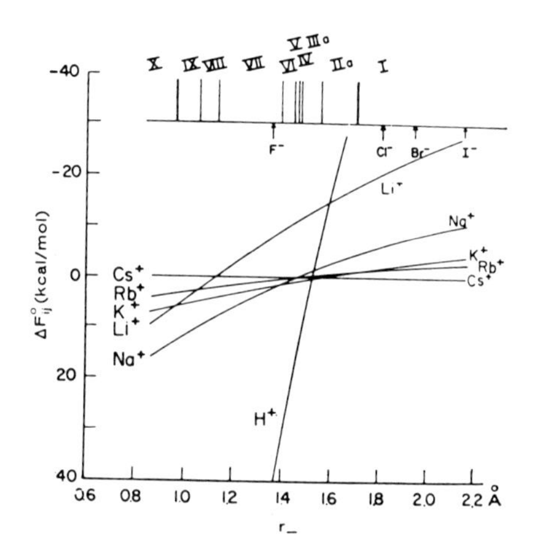

Figure 24. Selectivity isotherms for widely separated monopolar ion exchange sites. (Reproduced from Fig. 16 of Eisenman, 1962). Negative values of free energy change of reaction 30 of Eisenman, 1969 are plotted as a function of r_- for the case of widely separated sites. The Roman numerals above the figure indicate the sequences of ionic selectivity pertaining to the regions separated by the vertical lines drawn to the intersections of the various cation isotherms. The more strongly an ion is preferred by the exchanger the lower its position on the figure.

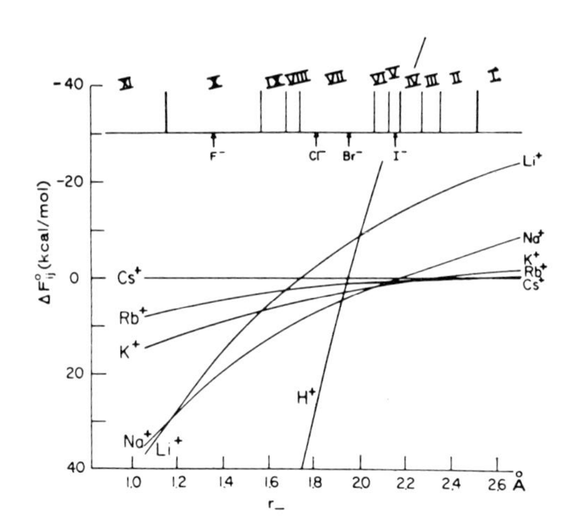

Figure 25. Selectivity isotherms for closely spaced monopolar sites. (Reproduced from Fig. 17 of Eisenman, 1962.) Plotted in the same manner as Fig. 24.

Inspection of Figs. 23–25 yields the following conclusions. First, the selectivity among cations depends in the same general way upon the

* Equation (46) is the classical Born-Lande equation for the internal energy of an alkali halide crystal lattice.

radius of the anion r_-, regardless of the differences in coordination of sites and waters. Second, a particular pattern is seen for the selectivity among group Ia cations in which only 11 sequences of cation effectiveness, indicated by the Roman numerals I to XI above the figures, are predicted out of a possible 120 permutations, with only the minor variations indicated by subscript "*a*". The underlying reason for this selectivity is the asymmetry between ion-site and ion-water interactions. Such an invariance of pattern is expected from the fact that the independent variable in these three figures is r_-, whereas the variable in the electrostatic term is $(r_+ + r_-)$. Since r_- is generally a quantity of the size of, or larger tham, r_+, any multiplication operation on the hydration energies or the electrostatic energies can always be approximately compensated for by a pure change in r_-. Indeed, if r_+ had been negligible compared to r_-, such changes in coordination number would have made no variations whatsoever in the pattern, since they could always be compensated for by merely shifting the magnitude of r_-. It is this basic asymmetry which underlies the ubiquitousness of the particular cation selectivity pattern in a wide variety of natural phenomena. It reflects the underlying fact that cation-site forces decrease more slowly with increasing distance than do cation-water forces (because water forces are essentially multipolar). Serious deviations from the above pattern are only expected when the ion interactions with the site fall off as a more rapid function of ion radius than their interactions with water. Such a situation occurs with a high atomic or molecular polarizability, as has been discussed in considerable detail elsewhere (Eisenman, 1961, 1962).

The applicability of this model has recently been critically assessed for ion exchangers by Reichenberg (1966), for zeolites by Sherry (1969), and for biological membranes by Diamond and Wright (1969).

Dipolar Sites

An understanding of the elementary factors underlying the specificity of neutral molecules which bind actions can be gained by extending the above considerations to a model illustrated in Fig. 26 for the carbonyl or ether oxygens which are the ligands in typical neutral carrier molecules.

For such a neutral dipolar site the electrostatic free energy is given by:

$$\Delta F_{\text{vac}}^{\text{attract}} = \left[\frac{-332}{r_+ + r_n} + \frac{332}{r_+ + r_p}\right](q \,.\, N), \tag{47}$$

where q is the fractional value of electronic charge, N is the coordination number of the ligands, r_+ is the (Goldschmitt) cationic radius, and r_n and r_p are the distances from the surface of the dipole of the negative and positive charges, respectively. (This calculation lumps all repul-

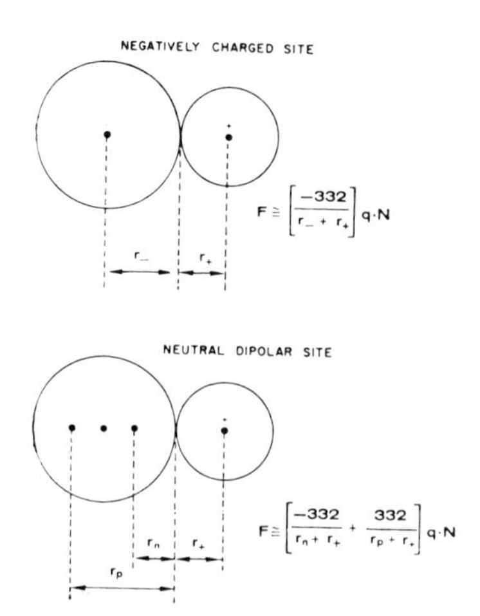

Figure 26. Models for a negatively charged ion exchange site and a neutral dipolar ligand. Described in text.

sions between ligands, as well as the Born repulsion with the cation, by diminution of effective charge.)

Calculating values for these energies and inserting in equations (41) and (43), using Latimer's (1952) experimental values for $\Delta F_{i+}^{\text{hyd}} - \Delta F_{j+}^{\text{hyd}}$, the selectivity isotherms of Fig. 27 were deduced.

Each subfigure in Fig. 27 presents a set of isotherms. The ordinate of each subfigure is the free energy change, as given at the lower left in

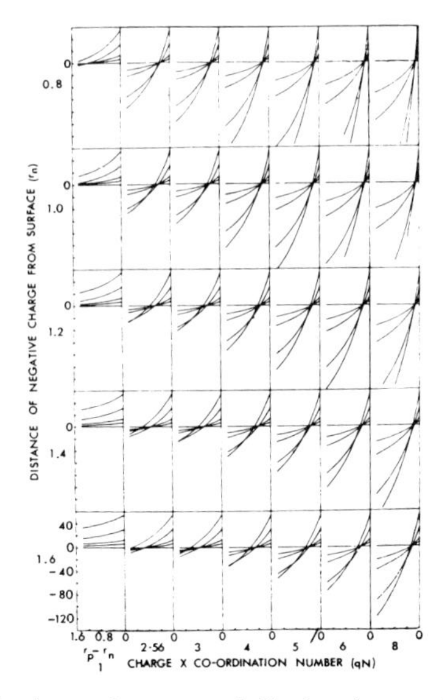

Figure 27. Selectivity isotherms for a neutral dipolar site as a function of various charge distributions and ligand coordination numbers Described in text.

kcal/mole. The lower the position of the cation, the more it is preferred. The abscissa of each subfigure represents the separation (in Å) of the positive pole from the negative pole $(r_p - r_n)$. This separation is zero at the right-hand side of each subfigure, in which case there is zero energy of interaction with the dipole; and the values of energies along the right-hand edge of each subfigure are therefore simply the differences of free energies of hydration, which are the same for all subfigures. The individual isotherms are not labelled but can be identified as follows: in all cases the horizontal line represents Cs^+, while the sequence of isotherms from bottom to top at the right-hand edge of each subfigure is Cs^+, Rb^+, K^+, Na^+, Li^+ (corresponding to the differences of free energies of hydration).

Each row of isotherms was calculated for a given distance of the negative pole from the surface (i.e., "0·8" means $r_n = 0{\cdot}8$ Å); while each column of isotherms corresponds to a particular product of charge times coordination number $(q\,.\,N)$, as indicated at the bottom of the figures. (A value $(q\,.\,N)$ equal to 3 could correspond to dipoles of unit charge in three-fold coordination or to dipoles of charge 0·5 in six-fold coordination.) These ranges of $(q\,.\,N)$, r_n, and r_p should encompass all values of dipoles likely to be encountered for carbonyl or ether oxygens in nature (the family of isotherms for the value $(q\,.\,N)$ of 2·56 represents a reasonable case of a tetrahedral coordination of ether oxygens having a net negative charge of –0·64 corresponding to the Rowlinson-type water molecule of Fig. 23). It has been assumed that the ligand oxygens of the sequestering molecules are free to coordinate around the naked cation with negligible steric restraints, at least from one cation to another.*

When the dipole separation is sufficiently large $(r_p - r_n \gg r_n)$, the selectivities must approach those for the monopolar model, since in this situation the energetic contribution due to the positive pole becomes negligible. It is not surprising to see a pattern of selectivity sequences in Fig. 26 like those of the monopole model for this charge distribution. Of more interest is the fact that over a considerable range of dipolar charge distributions, the sequences of cation selectivity seen in Fig. 27 are essentially the same as those given by the monopolar model (cf. Figs. 23–25).

This similarity is in accord with the experimentally observed selectivities for the macrocyclic molecules. Thus, sequence IV ($K^+ > Rb^+ > Cs^+ > Na^+ > Li^+$) is observed for the macrotetralide actins (see Table III and figures 8 and 9) and sequence III ($Rb^+ > K^+ > Cs^+ > Na^+ > Li^+$) for valinomcyin (see Table II). Pressman (1968) has reported sequence IIa for the solvent extraction of compound X-537, sequence V for compound X-206 and nigericin, and sequence VII for dianemycin. Sequences III, IV, and VIII are apparent in Pedersen's (1968)

* For some molecules the key variable may be differences in coordination number.

published solvent extraction data for different cyclic polyethers; Lardy and his colleagues (1967) have found similar sequences for the effects of these compounds on biological membranes. Further data for macrocyclic molecules are summarized by Diamond and Wright (1969, pp. 587–588).

In considering the selectivity of an antibiotic such as nonactin or valinomycin (or for that matter, the selectivity of a biological site) it is useful to view N in Fig. 26 (i.e. the coordination number) as a variable which is a function of the radius of the complexed cation. As a specific example, it is probably unrealistic, because of steric considerations, to imagine that eight oxygens could be close-packed around a Li^+ ion whereas it is totally reasonable to expect that eight oxygens could be close-packed around a Cs^+ ion (cf. Pauling, 1960 for a discussion of closest packing arrangements). Therefore, a molecule such as nonactin might have the four carbonyl oxygens closely packed against Li^+ and the four tetrahydrofuran oxygens at a greater distance, whereas all eight oxygens might be closely packed against Cs^+. Similarly, valinomycin might have all six of its coordinating oxygens closely packed around Rb^+ and Cs^+ whereas only four of these oxygens may be able to pack closely around K^+, Na^+, and Li^+, the other two having to be at a greater distance (or, alternatively, the minimum energy configuration might even involve a cavity larger than the coordinated cation or even involve a coordination of partially hydrated cations). Additionally, the value of q in Fig. 26 (i.e. the charge on the ligand) is likely to be different for different ligand groups within a molecule (e.g. the value of q for the ester carbonyl oxygens in nonactin is likely to be different from the q value for the tetrahydrofuran oxygens in nonactin as well as differing from that for the amide carbonyl oxygens in valinomycin). Basically then, the product $q \cdot N$ can be viewed as both a function of the complexing molecule and of the cation being complexed. A detailed calculation of such a case has not been done, but especially as the coordinations of the different cations in complexing molecules becomes known from crystallographic studies, such calculations might provide meaningful insights into the selectivity patterns for different antibiotics. Conversely, such calculations could also put constraints on the possible conformations of the ligand groups in biological cation binding sites.

It should therefore be apparent that the energies of interaction of cations with the dipolar ligand groups, in competitition with the hydration energies of the ions, can account for the salient features of the selectivity among the alkali metal cations characteristic of neutral sequestering molecules.

The realization of this basic competition between hydration energies and sequestration energies is important in interpreting selectivity data. For example, Prestegard and Chan (1970) noted a lack of selectivity in the binding manifested by nonactin in acetone solutions and therefore suggested "a rather similar dependence of the free energy of ion coordination on ion size for both nonactin and the solvent [acetone]".

It is likely that the ion size dependence of the relative energies of binding to nonactin carbonyls is sufficiently similar to that of the relative solvation energies by the acetone carbonyls that the apparent lack of selectivity is due to symmetrical interactions with the ligand groups vs. the solvent (see Szabo *et al.*, 1972). Indeed, their lack of selectivity suggests that carbonyl groups of acetone are excellent models for carbonyl groups of the macrotetralide actins.

The above realization also suffices to explain why it is that the ratios of stability constants for complex formation in a solvent such as methanol should bear a relatively simple relationship to the stability constants for the complex formation in water. That this is the case was previously discussed by Eisenman *et al.* (1970, see footnote 12, p. 340) where it was noted that this was understandable since the differences of solvation energies of cations in methanol are approximately the same as those in water and, from Fig. 22, it is apparent that the differences of Born charging energies for the complexes in methanol and water will also cancel. Therefore, the ratios of complex formation constants in methanol should indeed give a fairly accurate picture of the ratios of complex formation constants in water. This is a useful approximation since for most of the molecules of interest, the solubilities are not sufficiently high in water to measure these complex formation constants directly.

A discussion of selectivity of the present molecules would be incomplete without noting the importance of the spatial orientation assumed by the ligand groups. Although a detailed discussion of this lies beyond the scope of the present treatment, it is apparent from considerations of the unusually high ammonium ion selectivity for the macrotetralide actins (which have four carbonyl oxygens conveniently arrayed in tetrahedral coordination) as compared to the considerably lower ammonium selectivity of valinomycin (which has six carbonyl oxygens arrayed in octahedral symmetry), that the tetrahedral "shape" of the ammonium ion probably contributes an additional energy of interaction with the more suitably arrayed macrotetralide. Differences in details of the quantitative selectivity pattern among the essentially spherical alkali metal cations, and cations having definite "shape", such as ammonium, can provide empirical guide lines for attempting to decide the detailed array of the ligand groups (probably oxygens) of the binding sites in natural membranes.

Acknowledgement

It is a pleasure to acknowledge the perceptive and valuable comments and criticisms of our colleagues, Drs. Sally Krasne and Raynald Laprade.

References

Adelman, W. J. and Fok, Y. B., *J. Cell. Comp. Physiol.*, **64** (1964) 429.
Andreoli, T. E., Tieffenberg, M. and Tosteson, D. C., *J. Gen. Physiol.*, **50** (1967) 2527.

Baker, P. F., Hodgkin, A. L. and Shaw, T. I. *J. Physiol.* (*Lond.*) **164** (1962) 355.
Bean, R., in: *Membranes—A Series of Advances,* Vol. 2, G. Eisenman (ed.), Marcel Dekker, Inc., Publishers, New York.
Binstock, L. and Lecar, H. J., *Gen. Physiol.*, **53** (1969) 342.
Boyer, P. D., *Science*, **141** (1963) 1147.
Boyer, P. D., in: *Oxidases and Related Redox Systems*, Vol. 2, T. E. King, H. S. Manson and M. Morrison (eds.), Proc. Intern. Symp., Amherst, Mass., 1964, Wiley, New York, 1965.
Bright, D. and Truter, M. R., *Nature*, **225** (1970) 176.
Chance, B., in: *Energy Linked Functions of Mitochondria*, B. Chance (ed.), Academic Press, New York, 1963.
Chandler, W. K. and Meves, H., *J. Physiol.* (*Lond.*), **173** (1964) 31; also *ibid.* **180** (1965) 788.
Ciani, S., Eisenman, G. and Szabo, G., *J. Membrane Biol.*, **1** (1969) 1.
Ciani, S., Eisenman, G., Laprade, R. and Szabo, G., in: *Membranes—A Series of Advances,* Vol. 2, Chap. 2, G. Eisenman (ed.), Marcel Dekker, Inc., Publishers, New York, 1972.
Cockrell, R. S., Harris, E. J. and Pressman, B. C., *Nature* (*Lond.*), **215** (1967) 1487.
Colacicco, G., *J. Colloid and Interface Sci.*, **29** (1969) 345.
Diamond, J. M. and Wright, E. M., *Ann. Rev. Physiol.*, **31** (1969) 581.
Diebler, H., Eigen, M., Ilgenfritz, G., Maas, G. and Winkler, R., *Pure Appl. Chem.*, **20** (1969) 93.
Eigen, M. and Winkler, R., *Neurosci. Res. Program. Bull.*, **9** (1971) 330.
Eisenman, G., *Symposium on Membrane Transport and Metabolism,* A. Kleinzeller and A. Kotyk (eds.), p. 163. Academic Press, New York, 1961.
Eisenman, G., *Biophys. J.*, **2**, Part 2 (1962) 259.
Eisenman, G., *Bol. Inst. Estud. med. biol.* (*Mex.*), **21** (1963) 155.
Eisenman, G., *Proc. of the XXIIIrd International Congress of Physiological Sciences* (*Tokyo*), (1965) p. 489.
Eisenman, G., *Glass Electrodes for Hydrogen and Other Cations: Principles and Practice,* M. Dekker, New York, 1967.
Eisenman, G., *Federation Proc.*, **27** (1968) 6, 1249.
Eisenman, G., Ciani, S. M. and Szabo, G., *Federation Proc.*, **27** (1968) 6, 1289.
Eisenman, G., in: *Ion-Selective Electrodes*, R. A. Durst (ed.), National Bureau of Standards Special Publication 314, p. 1, 1969.
Eisenman, G., Ciani, S. M. and Szabo, G., *J. Membrane Biol.*, **1** (1969) 294.
Eisenman, G., McLaughlin, S. G. A. and Szabo, G., IUPAC Presymposium on "Physical Chemical Basis of Ion Transport Through Biological Membranes", Riga, U.S.S.R. (1970).
Eisenman, G., Szabo, G., Ciani, S., McLaughlin, S. and Krasne, S., *Progress in Surface and Membrane Science*, J. F. Danielli (ed.), Academic Press, 1972, in press.
Finkelstein, A., *Biochim. Biophys. Acta*, **205** (1970) 1.
Finkelstein, A. and Cass, A., *J. Gen. Physiol.*, **52** (1968) 145s.
Finkelstein, A. and Holz, R., in: *Membranes—A Series of Advances,* Vol. 2, G. Eisenman (ed.), Marcel Dekker, Inc., Publishers, New York, 1972.
Frensdorff, H. K., *J. Am. Chem. Soc.*, **93** (1971) 600.
Gilbert, D., (1970) See Stillman *et al.* (1969).
Goldman, D. E., *J. Gen. Physiol.*, **27** (1943) 37.
Goodall, M. C., *Biochim. Biophys. Acta*, **219** (1970) 28.
Green, D. E. and Perduc, J. F., *Ann. N.Y. Acad.. Sci.*, **137** (1966) 667.
Hille, B., *J. Gen. Physiol.*, **58** (1971) 599.
Hille, B., *J. Gen. Physiol.* **59** (1972) in press.
Hladky, S. B. and Haydon, D. A., *Nature*, **225** (1970) 451.
Hodgkin, A. L., *J. Physiol.*, **106** (1947) 319.
Hodgkin, A. L. and Katz, B., *J. Physiol.*, **116** (1949) 473.
Izatt, R. M., Nelson, D. P., Rytting, J. H., Haymore, B. L. and Christensen, J. J., *J. Am. Chem. Soc.* **93** (1971) 1619.
Jardetsky, O., *Nature*, **211** (1966) 969.
Kepes, A., *The Cellular Functions of Membrane Transport*, J. H. Hoffman (ed.), Prentice-Hall, New Jersey, 1964, p. 155.
Kilbourn, B. T., Dunitz, J. D., Pioda, L. A. R. and Simon, W., *J. Mol. Biol.*, **30** (1967) 559.
Krasne, S., Eisenman, G. and Szabo, G., *Science*, **174** (1971) 412.
Laprade, R., Szabo, G., Ciani, S. M. and Eisenman, G., *Abstracts*, *Biophys. Soc. Meeting*, 1972.
Lardy, H. A., Graven, S. N. and Estrada-O, S., *Fed. Proc.*, **26** (1967) 1355.
Läuger, P. and Stark, G., *Biochim. Biophys. Acta*, **211** (1970) 458.
LeBlanc, O. H., *J. Memb. Biol.*, **4** (1971) 227.
Lehninger, A. L., The Mitochondrion, Benjamin, New York, 1965.
Lesslauer, W., Richter, J. and Läuger, P., *Nature*, **213** (1967) 1224.
Lev, A. A., and Buzhinsky, E. P. *Tsitologiya*, **9** (1967) 102.
Liberman, E. A. and Topaly, V. P., *Biochim. Biophys. Acta*, **163** (1968) 125.
Liberman, E. A., Pronevich, L. A. and Topaly, V. P., *Biofizika*, **15** (1970) 612.

Ling, G. N., *A physical theory of the living state*, Blaisdell Publ. Co., Waltham, Mass., 1962.
Lipmann, F., in: *Currents in Biochemical Research*, D. E. Green (ed.), Interscience, New York, 1946, p. 137.
Markin, V. S., Krishtalik, L. I., Liberman, E. A. and Topaly, V. P., *Biofizika*, **14** (1969) 256.
Markin, V. S., Pastushenko, V. F., Krishtalik, L. I., Liberman, E. A. and Topaly, V. P., *Biofizika*, **14** (1969) 462.
McLaughlin, S. G. A., Szabo, G., Eisenman, G. and Ciani, S. M., *Proc. Natn. Acad. Sci.*, **67** (1970) 1268.
McLaughlin, S. G. A., Szabo, G., Ciani, S. and Eisenman, G., *J. Memb. Biol.* (1972). In Press.
Mitchell, P., *Nature*, **191** (1961) 144.
Mitchell, P., in: *Membranes and Ion Transport*, Vol. 1, E. E. Bittar, (ed.), Wiley, New York, 1971, p. 192.
Mueller, P. and Rudin, D. O., *Biochem. Biophys. Res. Commun.*, **26** (1967) 398.
Mueller, P. and Rudin, D. O., in: *Current Topics in Bioenergetics*, Vol. 3, Academic Press, New York, 1969, p. 157.
Neumcke, B., *Biophysik*, **6** (1970) 231.
Neumcke, B. and Läuger, P., *J. Memb. Biol.*, **3** (1970) 54.
Patlak, C. S., *Bull. Math. Biophys.*, **19** (1957) 209.
Pauling, L., *The Nature of the Chemical Bond*, Cornell Univ. Press, New York, 1960.
Pedersen, C. J., *J. Am. Chem. Soc.*, **89** (1967) 7017.
Pedersen, C. J., *Fed. Proc.*, **27** (1968) 1305.
Pedersen, C. J., *J. Am. Chem. Soc.* **92** (1970) 386.
Pressman, B. C., Harris, E. J., Jagger, W. S. and Johnson, J. H., *Proc. Natn. Acad. Sci.* **58** (1967) 1949.
Pressman, B. C., *Fed. Proc.* **27** (1968) 1283.
Prestegard, J. and Chan, S. I., *J. Amer. Chem. Soc.*, **92** (1970) 4440.
Reichenberg, D., in: *Ion Exchange*, Vol. 1, J. Marinsky (ed.), Dekker, New York, 1966, p. 277.
Shemyakin, M. M., Ovchinnikov, Y. A., Ivanov, V. I., Antonov, V. K., Vinogradova, E. I., Shkrob, A. M., Malenkov, G. G., Evstratov, A. V., Laine, I. A., Melnik, E. I. and Ryabova, I. D., *J. Memb. Biol.*, **1** (1969) 402.
Sherry, H. S., in: *Ion Exchange*, Vol. 2, J. Marinsky (ed.), Dekker, New York, 1969, p. 89.
Simon, W. and Morf, W., in: *Membranes—A Series of Advances*, Vol. 2, G. Eisenman (ed.), Marcel Dekker, Inc., Publishers, New York, 1972.
Skou, J. C., *Progress in Biophysics*, **14** (1964) 131.
Slater, E. C., *Nature*, **172** (1953) 975.
Stark, G. and Benz, R., *J. Membrane Biol.*, **5** (1971) 133.
Stillman, I., Gilbert, D. and Robbins, M., *Biophysical Society Abstracts*, A-250 (1969).
Szabo, G., Eisenman, G. and Ciani, S., *J. Memb. Biol.*, **3** (1969) 346.
Szabo, G., Eisenman, G. and Ciani, S. M., in: *Physical Principles of Biological Membranes*, F. Snell, J. Wolken, G. J. Iverson and J. Lam (eds.), Gordon and Breach Science Publishers, New York, 1970, (1970) pp. 79–133.
Szabo, G., Eisenman, G., Krasne, S., Ciani, S. and Laprade, R., in: *Membranes—A Series of Advances*, Vol. 2, Chap. 3, G. Eisenman, (ed.), Marcel Dekker, Inc., Publishers, New York, 1972.
Szabo, G., Eisenman, G., McLaughlin, S. G. A. and Krasne, S., *Ann. N.Y. Acad. Sci.*, **195** (1962a) 273.
Tosteson, D. C., *Fed. Proc.*, **27** (1968) 1269.
Urry, D. W., *Proc. Natn. Acad. Sci.*, **68** (1971) 672.
Whittam, R., *The Neurosciences*, G. C. Quarton, T. Melnechuk and F. O. Schmitt (eds.), 1967, p. 313.
Wipf, H. K., Pache, W., Jordan, P., Zähner, H., Keller-Schierlein, W. and Simon, W. *Biochem. Biophys. Res. Commun.*, **34** (1969) 387.

Mechanisms of Energy Conservation in the Mitochondrial Membrane*

Lars Ernster, Kerstin Juntti and Kouichi Asami†
Biokemiska Institutionen, Kungl. Universitetet i Stockholm, Box 6409, *S*-113 82 *Stockholm, Sweden*

Introduction

The importance of the inner mitochondrial membrane for respiratory chain-linked energy conservation is well recognized. The precise role of the membrane in the process, however, is not known. A problem of great current interest in this context concerns the mode of interaction of the mitochondrial electron-transport and ATPase systems in catalysing oxidative phosphorylation. According to the chemiosmotic hypothesis[1] this interaction is indirect and involves as an obligatory intermediate a proton gradient across the membrane. In contrast, the chemical hypothesis of oxidative phosphorylation envisages the formation of high-energy intermediates as functional links between the two systems,[2] possibly with the involvement of conformational changes of proteins[3] and localized proton gradients.[4] Since these intermediates most probably are of macromolecular nature, with only limited mobility within the membrane, such a mechanism would be likely to involve a direct interaction between the electron-transport and ATPase systems. Thus, whereas the chemiosmotic hypothesis considers the two systems as separate units, where each ATPase ought to be able to interact with any electron-transport chain within the same membrane (Fig. 1A), the chemical hypothesis is compatible with an assembly-like arrangement of the two systems, where any given ATPase may interact with only one electron-transport chain (Fig. 1B).

Indications for an assembly-like interaction of the electron-transport and ATPase systems were reported by Lee *et al.*[5] who found that addition to submitochondrial particles of oligomycin, which inhibited the ATPase activity of the particles, induced a respiratory control, the extent of which was unaffected by electron-transport inhibitors such as rotenone, antimycin, cyanide or azide. Similar indications were

* The following non-conventional abbreviations are used: ANS, 8-anilino-naphthalene-1-sulfonate; FCCP, carbonyl cyanide *p*-trifluoromethoxyphenylhydrazone; PMS, phenazine methosulfate.
† Present address: National Institute of Radiological Sciences, Chiba-shi, Japan.

obtained by Baum *et al.*[6] from measurements of the effects of rotenone and oligomycin on the rate of ATP-driven succinate-linked NAD^+ reduction. These authors concluded that their results were quite inconsistent with the chemiosmotic hypothesis. They pointed out, however, that this conclusion may be invalidated if the conserved energy were also dissipated through processes other than the reduction of NAD^+.

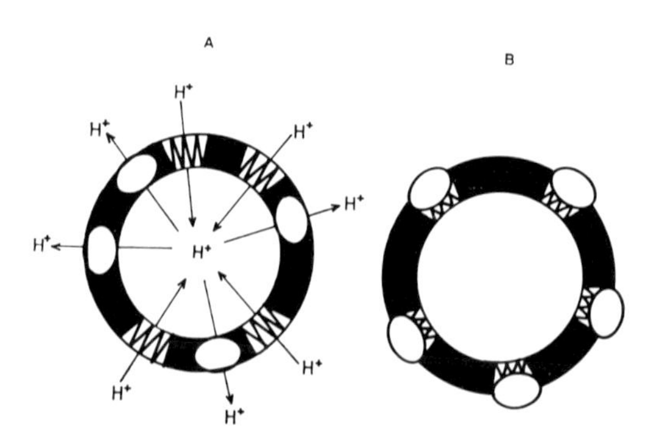

Figure 1. Functional relationship between electron-transport (zigzag line) and ATPase (ellipse) systems in submitochondrial particles. A: interaction via proton gradient; B: interaction via assemblies. (For explanation, see text.)

The purpose of this paper is to summarize the results of experiments carried out in this laboratory which appear to lend strong support to the concept of an assembly-like interaction of the mitochondrial ATPase and electron-transport systems. The evidence is based on comparative studies of the effects of an uncoupler (FCCP), an energy-transfer inhibitor (oligomycin) and the mitochondrial ATPase inhibitor described in 1963 by Pullman and Monroy[7] on various energy-linked reactions catalysed by submitochondrial particles. These results have recently been presented in a symposium lecture,[8] and a detailed account is being published elsewhere.[9]

Experimental Procedure

Submitochondrial particles were prepared from beef-heart mitochondria by sonication and isolated by differential centrifugation as described by Lee and Ernster.[10] Two types of particles were used: "nonphosphorylating" particles, obtained by sonication of mitochondria at alkaline pH in the presence of EDTA ("EDTA particles"), and "phosphorylating" particles obtained by sonication of mitochondria at neutral pH in the presence of Mg^{2+} and ATP ("Mg-ATP particles").

The reactions studied included ATP-driven succinate-linked NAD^+ reduction,[11] ATP-driven energy-linked transhydrogenase,[12] ATPase,[13] and respiration-induced ANS-fluorescence enhancement;[14] these were assayed as described in the appropriate references. Details of the assay systems are given in the figure legends.

ATPase inhibitor purified from beef-heart mitochondria was prepared and assayed as described by Horstman and Racker.[15]

Results

Effects of FCCP and Oligomycin

Figure 2 compares the effects of increasing concentrations of FCCP and oligomycin on the ATP-driven succinate-linked NAD$^+$ reduction and energy-linked transhydrogenase reaction of Mg-ATP particles. It may be seen that the half-inhibitory concentrations of both compounds

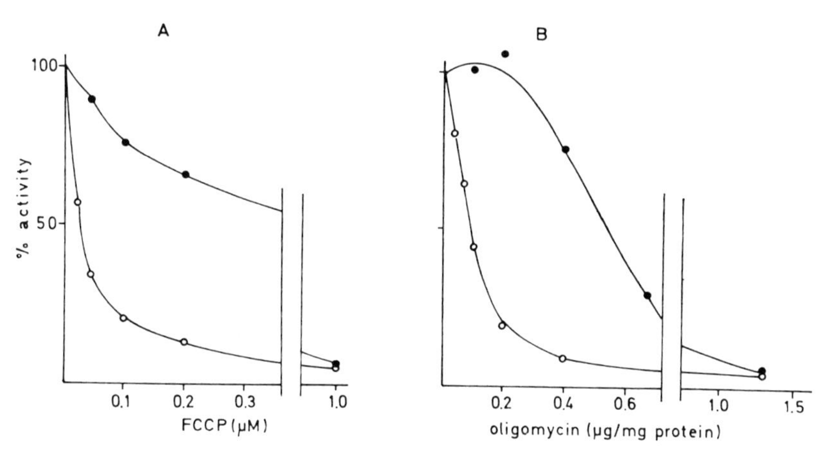

Figure 2. Effects of FCCP and oligomycin on ATP-driven succinate-linked NAD$^+$ reduction (○) and nicotinamide nucleotide transhydrogenase reaction (●) in Mg-ATP particles. The reaction mixtures contained, in a final volume of 1 ml, the following additions.

Succinate-linked NAD$^+$ reduction: 170 mM sucrose, 50 mM tris-Ac, pH 7·5, 5 mM succinate, 1 mM NAD$^+$, 3 mM $MgSO_4$, 3 mM ATP, 2 mM KCN, 0·2 mg particle protein, and FCCP or oligomycin as indicated. The activity in the absence of FCCP and oligomycin was 91 nmoles NADH formed per min and mg protein.

Transhydrogenase: 170 mM sucrose, 50 mM Tris-Ac, pH 7·5, 0·2 mM NADH, 1mM NADP$^+$, 1 mM glutathione disulfide 2 μg glutathione reductase in 1% bovine serum albumin, 1 μM rotenone, 4·5 mM $MgSO_4$, 4·5 mM ATP, 0·2 mg particle protein, and FCCP or oligomycin as indicated. The activity in the absence of FCCP and oligomycin was 80 nmoles NADPH formed per min and mg protein.

The reactions were followed spectrophotometrically at 340 nmoles. Temperature, 30°C.

were considerably lower in the case of the former than the latter reaction. The rates of the two reactions in the absence of FCCP were approximately equal.

Results similar to those described above were obtained with EDTA particles (Fig. 3), except that here, in accordance with earlier findings of Lee and Ernster,[16] low concentrations of oligomycin stimulated the ATP-driven reactions (Fig. 3B). In order to ensure maximal reaction rates, the measurements with increasing concentrations of FCCP (Fig. 3A) were performed in the presence of a maximally stimulating concentration of oligomycin. As in the case of the Mg-ATP particles, it is evident that both FCCP and oligomycin inhibited the ATP-driven succinate-linked NAD^+ reduction more efficiently than the ATP-driven transhydrogenase reaction.

The results described so far are similar to those reported in 1963 by Danielson and Ernster,[17] who found that the ATP-driven transhydrogenase reaction catalysed by rat-liver submitochondrial particles was

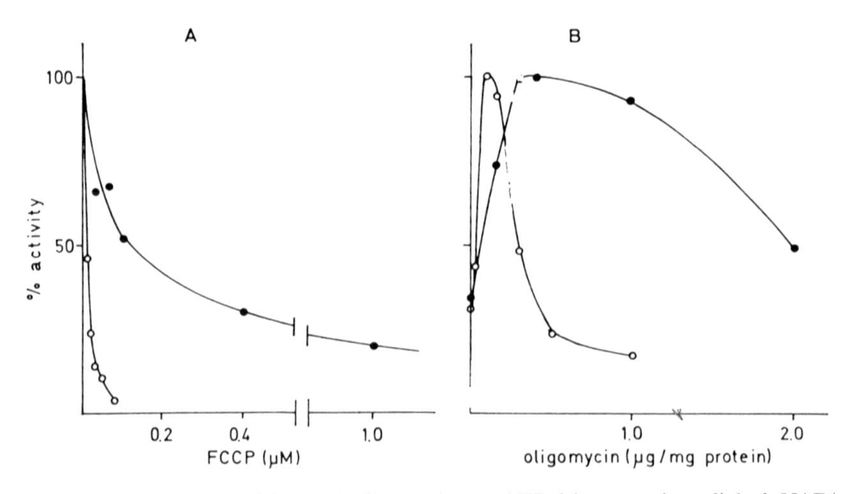

Figure 3. Effects of FCCP and oligomycin on ATP-driven succinate-linked NAD^+ reduction (○) and nicotinamide nucleotide transhydrogenase (●) in EDTA particles. The reaction mixtures were the same as in Fig. 2, except that in the series in Fig. 3A, the samples contained a maximally stimulating amount of oligomycin as deduced from the data in Fig. 3B. The maximal activities were 130 nmoles NADH formed per min and mg protein for the succinate-linked NAD^+ reduction, and 150 nmoles NADPH formed per min and mg protein for the transhydrogenase.

much less sensitive to the uncoupler 2,4-dinitrophenol than was the ATP-driven succinate-linked NAD^+ reduction. They also found that replacement of ATP by another nucleoside triphosphate, e.g. ITP, which is less efficient as substrate for ATPase than is ATP, resulted in a decrease in the rate of the energy-linked transhydrogenase reaction that was proportional to the decrease in rate of the nucleoside triphosphatase, whereas the rate of the succinate-linked NAD^+ reduction decreased much more.

Danielson and Ernster[17] interpreted these findings in terms of a difference in energy requirement between the ATP-driven succinate-linked NAD^+ reduction and transhydrogenase reactions. According to their interpretation, the thermodynamically unfavourable succinate-linked NAD^+ reduction requires a higher level of the high-energy intermediate ~X, i.e., a higher "energy pressure", in order to proceed at maximal rate, than does the thermodynamically favourable transhydrogenase reaction. Thus, when the steady-state level of ~X is decreased, either by decreasing its rate of generation from ATP (by replacing ATP with a kinetically less efficient substrate of the ATPase, such as ITP), or by increasing its rate of dissipation (by adding an uncoupler), this will affect the succinate-linked NAD^+ reduction more than the energy-linked transhydrogenase reaction. In line with this interpretation was also the finding that addition of ADP and/or P_i in combination with ATP, i.e., lowering of the phosphate potential,

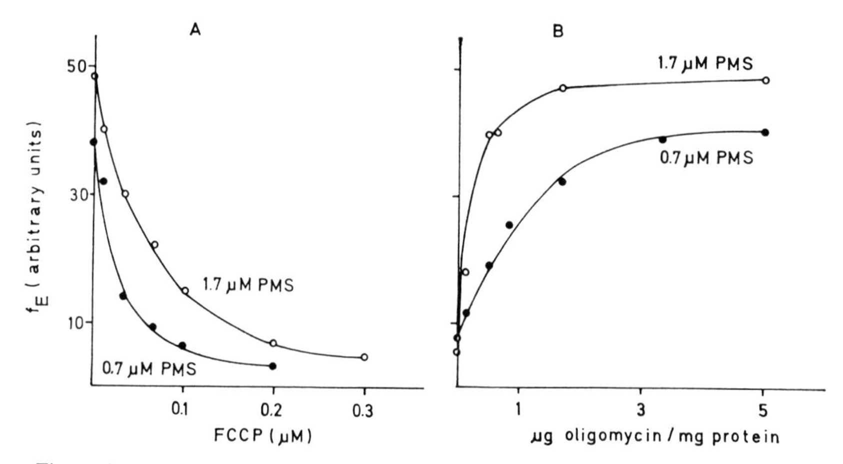

Figure 4. Effects of FCCP and oligomycin on the energy-dependent ANS-fluorescence enhancement (f_E) of EDTA particles. The reaction mixture contained, in a final volume of 3 ml, 10 mM Tris-Ac, pH 7·5, 170 mM sucrose, 1·7 μM rotenone, 17 μM ANS, 5 mM ascorbate, 0·6 mg particle protein, and 0·7 or 1·7 μM PMS. FCCP (Fig. 4A) and oligomycin (Fig. 4B) were added as indicated. In Fig. 4A, 3 μg oligomycin was added to all samples. Temperature, 24°C. (Data quoted from ref. 18.)

greatly suppressed the succinate-linked NAD^+ reduction but had virtually no effect on the energy-linked transhydrogenase reaction.

Evidently, a similar interpretation may be valid for the present data, and may thus apply also for the case of oligomycin, which, like FCCP, lowers the "energy pressure" generated from ATP. In fact, independent evidence supporting this interpretation comes from data recently published from this laboratory,[18] quoted in Fig. 4, concerning the effects of increasing concentrations of FCCP and oligomycin on the respiration-induced ANS-fluorescence enhancement in EDTA particles. Here, just as in the case of the ATP-driven reactions, uncouplers act as inhibitors, by promoting energy dissipation. Oligomycin, on the other hand, promotes energy accumulation from the respiratory chain, by tightening an "energy leak" present in these particles, and thereby enhancing respiration-supported energy-linked reactions. Using ascorbate + PMS as substrate, one can regulate the rate of respiration, and thereby the respiration-generated "energy pressure", by varying the concentration of PMS. As illustrated by the data in Fig. 4, the higher the concentration of PMS, i.e., the higher the "energy pressure", the higher is the FCCP concentration required to inhibit, and the lower is the oligomycin concentration required to stimulate the respiration-induced ANS-fluorescence enhancement.

These data seem to eliminate the possibility that the differences in FCCP and oligomycin sensitivity between the ATP-driven succinate-linked NAD^+ reduction and transhydrogenase reaction would be due to intrinsic features of the two reactions other than those concerning energy requirement.

It may thus be concluded that the ATP-driven succinate-linked NAD^+ reduction and nicotinamide nucleotide transhydrogenase reaction show different sensitivities to both uncouplers and oligomycin, and that this difference is a reflection of the different energy requirements of the two reactions. As will be shown in the following, a different situation is found when the two reactions are compared with respect to their sensitivities to the mitochondrial ATPase inhibitor.

Effects of Mitochondrial ATPase Inhibitor

In 1963 Pullman and Monroy[7] reported the purification of an ATPase inhibitor from beef-heart mitochondria. The inhibitor, a protein of an estimated molecular weight of 10,000–15,000, was shown to inhibit the ATPase activity of both submitochondrial particles and coupling factor 1 (F_1), while it had no effect on oxidative phosphorylation. The inhibitor was shown to form a complex with F_1, a finding that was subsequently confirmed by several investigators.[19–20]

In 1970, it was reported from this laboratory[21] that the ATPase inhibitor also inhibited ATP-driven energy-utilizing reactions of submitochondrial particles such as succinate-linked NAD^+ reduction, energy-linked transhydrogenase, and ATP-induced ANS-fluorescence enhancement. The same reactions when driven by the respiratory chain were unaffected.

The same year Horstman and Racker[15] described a new procedure for the purification of the ATPase inhibitor. The inhibitor so obtained was shown to require preincubation with ATP and Mg^{2+} in order to inhibit the ATPase activity of submitochondrial particles and purified F_1. Evidence was presented indicating that ATP and Mg^{2+} promote the binding of the inhibitor to F_1.

The findings of Horstman and Racker[15] were recently confirmed in this laboratory, and it was shown[22] that the inhibitor prepared according to these authors has a molecular weight 4–6 times higher than that reported for the preparation of Pullman and Monroy,[7] thus probably representing a polymer of the latter. It was also found[22] that when the inhibitor is preincubated with EDTA particles in the presence of ATP and Mg^{2+}, so as to inhibit the ATPase activity, the inhibition persisted even after dilution of the ATP and Mg^{2+} in the preincubating medium. However, the inhibition was relieved, if the diluted preincubation mixture was reincubated in the presence of succinate and a low concentration of oligomycin, i.e., under conditions which give rise to an accumulation of energy from the respiratory chain. These findings were interpreted to indicate that energy derived from ATP promotes the binding of the inhibitor to F_1, whereas energy derived from the respiratory chain promotes its release from F_1. Studies of the effects of P_i and aurovertin indicated, furthermore, that a phosphorylated high-energy intermediate probably is responsible for the respiration-induced release of the inhibitor.

Kinetic studies of the effect of the inhibitor on the ATPase reaction of EDTA particles revealed[9,23] that the inhibition is noncompetitive with respect to the substrate, leaving unaltered the K_m for ATP and ITP, and likewise the K_i for ADP. Oligomycin and aurovertin did not affect the titer of the ATPase with respect to the inhibitor, nor did the ATPase inhibitor alter the inhibition titer of the ATPase with respect to oligomycin. These data are consistent with the conclusion that the ATPase inhibitor firmly binds to a site of the ATPase which is essential for catalytic activity but is not identical with the binding site for either ATP, ADP, oligomycin, or aurovertin.

Next, the kinetics of the effect of ATPase inhibitor on the ATP-driven succinate-linked NAD^+ reduction and energy-linked transhydrogenase reaction were investigated. Mg-ATP particles, which catalyze a maximal activity of these reactions without need for the presence of a low concentration of oligomycin,[24] proved to be unsuitable for this purpose, since it was found that added ATPase inhibitor had little or no effect on their ATPase activity, in contrast to that obtained with EDTA particles (Fig. 5). Also the ATP-driven electron-transport reactions of the Mg-ATP particles showed little or no sensitivity to the ATPase inhibitor. Treatment with Sephadex G-50 resulted, in accordance with earlier findings of Racker and Horstman,[25] in a several-fold increase of the ATPase activity of Mg-ATP particles, with a simultaneous release of ATPase inhibitor. However, this treatment also caused a decrease of the ATP-driven electron-transport activities, which now required the addition of low concentrations of oligomycin to reach maximal levels, similar to EDTA particles. In view of these findings it was decided to use EDTA particles for investigating the kinetics of the effect of ATPase inhibitor on the ATP-driven electron-transport reactions.

Figure 6 compares the effects of increasing amounts of ATPase inhibitor on the ATP-driven succinate-linked NAD^+ reduction, nicotinamide transhydrogenase and ATPase activities of EDTA particles. The ATP-driven electron-transport reactions were measured

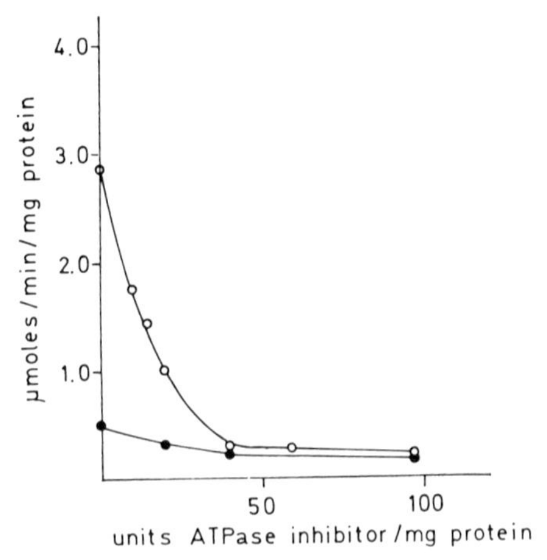

Figure 5. Effect of ATPase inhibitor on the ATPase activity of EDTA particles (○) and Mg-ATP particles (●). ATPase inhibitor (AI), prepared according to Horstman and Racker,[15] was preincubated in the particles at 30°C for 10 min in 0·5 ml of a mixture containing 0·25 M sucrose, 10 mM tris-TES, pH 6·5, 0·45 mM and 0·45 mM $MgSO_4$.

ATPase activity was assayed with an aliquot of the preincubation mixture in 1 ml of a medium containing 5 mM tris-Ac, pH 7·5, 30 mM KCl, 3 mM $MgCl_2$, 0·2 mM NADH, 0·75 mM phosphoenolpyruvate, 15 µg lactate dehydrogenase, 0·1 mg pyruvate kinase, 1·5 µM rotenone, 1 µM FCCP, and 0·02 mg particle protein.

The reaction was followed spectrophotometrically at 340 nmoles. Temperature, 30°C.

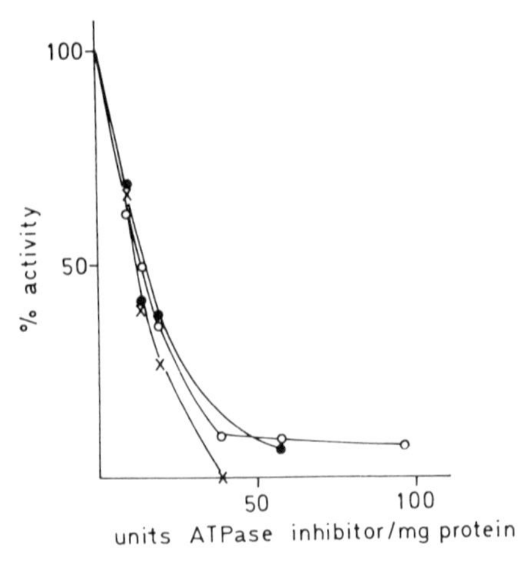

Figure 6. Effects of ATPase inhibitor on the ATPase (○) and ATP-driven succinate-linked NAD^+ reduction (×) and nicotinamide nucleotide transhydrogenase reaction (●) of EDTA particles. Preincubation with ATPase inhibitor (AI) was done as described in Fig. 5.

ATPase activity was assayed as in Fig. 5, and ATP-driven succinate-linked NAD^+ reduction and transhydrogenase were assayed as in Fig. 3A.

Activities in the absence of ATPase inhibitor were: 2·9 µmoles ADP formed per min and mg protein for ATPase; 73 nmoles NADH formed per min and mg protein for succinate-linked NAD^+ reduction; and 128 nmoles per min and mg protein for transhydrogenase.

in the presence of a maximally stimulating concentration of oligomycin (cf. Fig. 3B). It may be seen in Fig. 6 that the ATPase inhibitor inhibited with equal efficiencies the two ATP-driven reactions, in sharp contrast to FCCP and oligomycin, which inhibited the succinate-linked NAD^+ reduction much more efficiently than the transhydrogenase reaction (cf. Figs. 2 and 3). Furthermore, the ATPase inhibitor titers of the two ATP-driven reactions were the same as that of the ATPase. The ATPase activity of the non-inhibited system was about one order higher, in terms of estimated molar amounts of ATP expended, than those of the ATP-driven reactions (cf. legend of Fig. 6). These findings, which confirm preliminary observations reported from this laboratory,[21] strongly suggest that the ATPase inhibitor acts on a common rate-limiting step of the ATPase and ATP-driven electron-transport reactions.

Conclusions and Comments

The results described in this paper reveal a striking difference between the effects of the ATPase inhibitor, on one hand, and of FCCP and oligomycin, on the other, on ATP-driven electron-transport reactions of submitochondrial particles. The ATPase inhibitor inhibits the ATP-driven succinate-linked NAD^+ reduction and nicotinamide nucleotide transhydrogenase reaction with equal efficiencies, whereas FCCP and oligomycin inhibit much more efficiently the succinate-linked NAD^+ reduction than the transhydrogenase reaction. Furthermore, the inhibition of the ATP-driven electron-transport reactions by the ATPase inhibitor parallels the inhibition of the ATPase reaction, indicating a common rate-limiting step in the three reactions.

There is strong evidence in the literature[7,15] that the ATPase inhibitor acts on the terminal enzyme of the phosphorylating system, F_1. It might seem conceivable, therefore, that F_1 is the common rate-limiting step for the ATPase and ATP-driven electron-transport reactions, and that this is the reason why the ATPase inhibitor, but not FCCP and oligomycin (which act on steps intermediate between F_1 and the electron-transport system), inhibits the ATP-driven succinate-linked NAD^+ reduction and transhydrogenase reaction with equal efficiencies. This explanation, however, appears unlikely, since replacement of ATP by ITP, which most probably also limits the ATP-driven reactions at the level of F_1, has been shown to lower the rate of the succinate-linked NAD^+ reduction more than that of the transhydrogenase reaction. In fact, it is quite obvious from the data summarized in Table I that half-inhibition of the ATPase activity by either oligomycin or replacement of ATP by ITP results in a substantial decrease in the ratio of the relative activities of the two ATP-driven electron-transport reactions, whereas that by ATPase inhibitor does not.

A more probable explanation of the above findings is that the ATPase inhibitor forms, at least under the conditions here employed, a stable,

TABLE I. Relative ratio of rates of ATP-driven succinate-linked NAD^+ reduction and nicotinamide nucleotide transhydrogenase at 50% inhibition of ATPase of submitochondrial particles

Preparation	ATPase inhibited by	Relative ratio of rates of ATP-driven succ.-linked NAD^+ red. and transhydrogenase
Beef-heart Mg-ATP particles	Oligomycin	0·12
Beef-heart EDTA particles	Oligomycin	0·25
Rat-liver particles	Replacement of ATP by ITP	0·33*
Beef-heart EDTA particles	ATPase inhibitor	0·95

* Calculated from data in ref. 17.

practically undissociable complex with F_1. Thus, addition of increasing amounts of ATPase inhibitor to the particles will inactivate an equivalent amount of F_1 in an all-or-none fashion. If, as predicted by the chemiosmotic hypothesis (1), ATPases and electron-transport chains within the same membrane would interact in a random manner, by way of a common proton gradient (cf. Fig. 1A), then a partial inhibition of the ATPase activity by ATPase inhibitor would result in a greater inhibition of the thermodynamically unfavourable succinate-linked NAD^+ reduction than the thermodynamically favourable transhydrogenase reaction. The fact that the inhibition of the two ATP-driven electron-transport reactions by the ATPase inhibitor closely parallels that of the ATPase strongly suggests, therefore, that the ATPase and electron-transport systems interact in a direct, assembly-like fashion as indicated in Fig. 1B.

It should be pointed out that the above interpretation does not necessitate the postulation of the occurrence of one ATPase per electron-transport system or even a strict stoichiometry within each electron-transport chain, and it is quite possible that there is an excess of F_1 relative to the number of electron-transport systems[26] and of cytochromes relative to flavoproteins,[27] as indicated by estimates of the mitochondrial contents of these components; regarding transhydrogenase, there are no estimates available so far. What the present interpretation does imply is that those ATPases and electron-transport chains which do interact, do this in a direct, assembly-like fashion within the membrane, i.e., in a fashion incompatible with the involvement of a bulk proton gradient or membrane potential as an obligatory intermediate. The role of the membrane, though not yet fully understood, would be to promote this interaction, rather than merely to serve as an inert insulator as envisioned by the chemiosmotic hypothesis.

Acknowledgement

This work has been supported by grants from the Swedish Natural-Science Research Council and the Swedish Cancer Society.

References

1. P. Mitchell, *Nature,* **191** (1961) 144; *Biol. Rev.,* **41** (1965) 445.
2. E. C. Slater, *Nature,* **172** (1953) 975.
3. P. D. Boyer, in: *Oxidases and Related Redox Systems,* vol. 2, T. E. King, H. S. Mason and M. Morrison (eds.), Wiley, New York, 1965, p. 994.
4. R. J. P. Williams, in: *Current Topics in Bioenergetics,* Vol. 3, D. R. Sanadi (ed.), Academic Press, New York, 1969, p. 79.
5. C. P. Lee, L. Ernster and B. Chance, *European J. Biochem.,* **8** (1969) 153.
6. H. Baum, G. S. Hall, J. Nelder and R. B. Beechei, *Abstracts, Colloquium on Bioenergetics,* Pugnochiuso, 1970, p. 59; Adriatica Editrice, Bari, in press.
7. M. E. Pullman and G. C. Monroy, *J. Biol. Chem.,* **238** (1963) 3762.
8. L. Ernster, *Abstr. Commun., 7th Meet. Eur. Biochem. Soc.,* Varna, 1971, p. 53.
9. K. Juntti, K. Asami and L. Ernster, manuscript in preparation.
10. C. P. Lee and L. Ernster, *Meth. Enzymol.,* **10** (1967) 543.
11. L. Ernster and C. P. Lee, *Meth. Enzymol.,* **10** (1967) 729.
12. L. Ernster and C. P. Lee, *Meth. Enzymol.,* **10** (1967) 738.
13. M. E. Pullman, H. S. Penefsky, A. Datta and E. Racker, *J. Biol. Chem.,* **235** (1960) 3322.
14. K. Nordenbrand and L. Ernster, *European J. Biochem.,* **18** (1971) 258.
15. L. Horstman and E. Racker, *J. Biol. Chem.,* **245** (1970) 1336.
16. C. P. Lee and L. Ernster, BBA Library, Elsevier, Amsterdam, **7** (1966) 218; *European J. Biochem.,* **3** (1968) 391.
17. L. Danielson and L. Ernster, *Biochem. Z.,* **338** (1963) 188.
18. L. Ernster, K. Nordenbrand, C. P. Lee, Y. Avi-Dor and T. Hundal, in: *Colloquium on Bioenergetics,* Pugnochiuso, 1970, Adriatica Editrice, Bari, in press.
19. J. B. Warshaw, K. W. Lam, B. Nagy and D. R. Sanadi, *Arch. Biochem. Biophys.,* **123** (1968) 385.
20. A. E. Senior and J. C. Brooks, *Arch. Biochem. Biophys.,* **140** (1970) 257.
21. K. Asami, K. Juntti and L. Ernster, *Biochim. Biophys. Acta,* **205** (1970) 307.
22. K. Asami, K. Juntti and L. Ernster, manuscript in preparation.
23. K. Juntti, K. Asami and L. Ernster, *Abstr. Commun. 7th Meet. Eur. Biochem. Soc.,* Varna, 1971, p. 243.
24. C. P. Lee, G. F. Azzone and L. Ernster, *Nature,* **201** (1964) 152.
25. E. Racker and L. Horstman, *J. Biol. Chem.,* **242** (1967) 2547.
26. B. Chance, A. Azzi, I. Y. Lee, C. P. Lee and L. Mela, in: *Mitochondria—Structure and Function,* L. Ernster and Z. Drahota (eds.), Academic Press, London, 1969, p. 233.
27. M. Klingenberg, in: *Biological Oxidation,* T. P. Singer (ed.), Wiley, New York, 1968, p. 3.

Biogenesis of Mitochondria

23. The Biochemical and Genetic Characteristics of Two Different Oligomycin Resistant Mutants of *Saccharomyces Cerevisiae* Under the Influence of Cytoplasmic Genetic Modification

Carolyn H. Mitchell,* C. L. Bunn, H. B. Lukins and Anthony W. Linnane

Biochemistry Department, Monash University, Clayton, Victoria, Australia

Abstract

Two classes of *Saccharomyces cerevisiae* mutants resistant to oligomycin, an inhibitor of mitochondrial membrane bound ATPase are described. Biochemical analysis shows that *in vitro* the mitochondrial ATPase of both types of mutant are sensitive to oligomycin. *In vivo* sensitivity of the mutants to oligomycin can be demonstrated following anaerobic growth of the cells, which grossly alters the mitochondrial membrane and renders the ATPase of the mutants sensitive to oligomycin. It is concluded that the mutation to oligomycin resistance in both mutant types is phenotypically expressed as a change in the mitochondrial membrane. The intact mitochondrial membrane in the wild type cell is freely permeable to oligomycin, whereas the resistant mutant is impermeable to oligomycin; alteration of the mitochondrial membrane during isolation of the organelle or physiological modification of the membranes of the mitochondria by anaerobic growth renders the membranes permeable.

These mitochondrial membrane mutants differ in their cross-reference patterns and their genetics. One is resistant to oligomycin only, and behaves like previously reported cytoplasmic mutants. The other shows cross-resistance to inhibitors of mitochondrial protein synthesis as well as to oligomycin; although the mutant appears to arise from a single step mutation its genetic properties are complex and show part-nuclear and part-cytoplasmic characteristics. The implications of the observations are discussed.

Introduction

As an integral part of the study of the biogenesis of mitochondria in the yeast *Saccharomyces cerevisiae*, this laboratory has been studying the formation and function of mitochondrial membranes in yeast. Previous communications have described the manipulation of the composition of mitochondrial membranes by physiological and genetic methods and

* Present address: Department of Biology, California Institute of Technology, Pasadena, California, U.S.A.

the studies illustrate and emphasize the critical nature of the organization of the mitochondrial membrane in the regulation of diverse mitochondrial processes (for review[1,2]).

When grown under anaerobic conditions, *S. cerevisiae* cannot synthesize unsaturated fatty acids (UFA) or ergosterol[3,4] and such cells contain poorly developed mitochondria.[5–9] They can, however, incorporate into their membranes a wide variety of unsaturated fatty acids[10] and sterols[11] enabling the lipid composition of mitochondrial and other cell membranes to be varied by altering the lipid supplements in the growth medium.[12] The normal unsaturated fatty acid content (UFA) of yeast mitochondria is about 75% of the total organelle fatty acids; cells grown to contain less than 20% UFA contain mitochondria deficient in ribosomes[13] and the DNA dependent RNA polymerase.[14] This process is freely reversible and on restoration of the mitochondrial UFA content, the mitochondrial RNA polymerase and ribosome content returns to normal.[15]

The employment of anaerobically grown cells is of limited use in the study of the energy-linked functions of mitochondria, as the anaerobic cell lacks the normal respiratory cytochromes.[16] The biochemical analysis of genetic mutants with altered mitochondrial membranes enables investigations to be carried out on the role of the membrane in the functioning of the organelles under aerobic conditions. Using a fatty acid desaturase mutant, we have shown that under aerobic conditions limited UFA-depletion (about 30% of total fatty acids) leads to a specific loss of the ability of the mitochondria to couple oxidation to the formation of ATP and that this process is reversible on the addition of UFA.[17–19] Another class of mutant which is providing some insight into mitochondrial membrane function consists of mutants with cytoplasmically determined antibiotic resistance. Recently our laboratory has reported the isolation of cytoplasmic mutants resistant to a number of antibiotic inhibitors of mitochondrial protein synthesis, namely, mikamycin, chloramphenicol, carbomycin, lincomycin and tetracycline, and these mutants have been phenotypically characterized as probable mitochondrial membrane mutants.[1,20] The investigations have led to the suggestion that extensive mitochondrial ribosome-membrane interactions occur;[1,20] the lipid depletion studies in general support this concept.[13–15]

Oligomycin resistant mutants of yeast which show cytoplasmic inheritance have been described by Wakabayashi and Gunge,[21] Stuart,[22] and Avner and Griffiths.[23] These three groups of workers report that the ATPase of mitochondria isolated from their mutants is completely sensitive to oligomycin, although more recently Griffiths *et al.*[24] have also reported that the mitochondrial ATPase from some of their mutants shows partial *in vitro* resistance to oligomycin.

This communication is concerned with the biochemical and genetic characterization of a number of oligomycin resistant mutants. On the basis of their genetic behaviour, they fall into two classes, one classically cytoplasmic and the other a new type showing nucleo-cytoplasmic genetic interaction; they are also distinguished by their different cross-resistance patterns to a variety of chemically unrelated mitochondrial protein synthesis antibiotic inhibitors. Biochemical analysis of both genetic classes of mutant indicates that the *in vitro* and *in vivo* properties of the mitochondrial oligomycin sensitivity conferring protein (OSCP) and ATPase are unaltered, and that the mutation to oligomycin resistance is a consequence of a change in some other mitochondrial membrane component.

Materials and Methods

Isolation of Mutants

Strains described as sensitive to an antibiotic are unable to grow on a non-fermentable substrate, such as ethanol, in the presence of that drug; in particular, at concentrations per milliliter of medium, of 0·3 μg oligomycin, 0·01 mg mikamycin, 0·5 mg chloramphenicol, 0·1 mg erythromycin, 0·5 mg lincomycin, 0·3 mg carbomycin, 1·0 mg spiramycin, 0·5 mg tetracycline. Conversely, strains resistant to these antibiotics are able to grow on such media in the presence, per milliliter of medium, of 50 μg oligomycin, 0·3 mg mikamycin, 4 mg chloramphenicol, 4 mg erythromycin, 4 mg lincomycin, 0·6 mg carbomycin, 4 mg spiramycin, 2 mg tetracycline.

Cultures of the sensitive, prototrophic diploid strain, N 1300 were irradiated with ultraviolet light to give approximately a 99% kill, and the surviving cells were grown for 18 h on a yeast extract-peptone-ethanol (YEPE) medium to allow for phenotypic lag. The culture was then plated on YEPE medium supplemented with 50 μg/ml oligomycin. Resistant mutants appeared on these plates as papillae; one such diploid mutant, N 1311, was then isolated and sporulated to obtain the resistant haploid L 4000. This technique was similar to that used for the isolation from N 1300 of the mikamycin resistant mutant N 1301.[20] This mikamycin resistant diploid mutant, N 1301, and all spores from it including L 3000 examined here, were cross-resistant to oligomycin. All the mutant strains isolated were stable with respect to their antibiotic resistance.

Mating procedures, the method of tetrad analysis, and the determination of the proportion of sensitive and resistant cells that arise from single diploid zygotes (denoted by us as mixedness) have been described previously, as have the complex and synthetic media used.[25–27]

The auxotrophic haploids used in the present study are characterized by the following growth requirements, all of which derive from mutations of nuclear genes: L 3000, uracil and histidine; L 4000, adenine, uracil, and histidine; L 2200 and L 2300, adenine, tryptophan, and lysine; L 5628, adenine; L 2265, tryptophan, leucine, arginine, threonine and isoleucine.

Loss or Retention of Antibiotic Resistance in Petites

Respiratory competent, antibiotic resistant strains were grown for 18 h in yeast-extract–peptone–glucose medium containing 20 μg ethidium bromide/ml of medium. The cells were harvested, washed and crossed with an early stationary phase culture of an oligomycin-sensitive strain, and the diploid progeny from this cross analysed for resistance or sensitivity to the particular antibiotic as described.[26,27] It has been shown that such treatment with ethidium bromide yields petite (ρ^-) cultures whose cells lack mitochondrial DNA, symbolized ρ°.[28–30] Thus the loss of antibiotic resistance following this treatment, shown by the absence in the diploid progeny of the resistance characteristic of the ethidium bromide treated culture, indicates a cytoplasmic origin for the resistance mutation. Spontaneous ρ^- strains from resistant mutants were similarly examined for loss or retention of resistance characteristics by analysis of the diploid progeny of a cross with a sensitive strain.

Mitochondrial ATPase

Mitochondria were isolated from yeast cells grown on a 1% ethanol salts medium, as described previously[31] and purified on a 20–70% sucrose gradient. Mitochondrial ATPase was measured essentially as described by Somlo.[32] The reaction mixture (1 ml) contained the following: 50 mM tris-malate buffer (pH 6·2) or 50 mM tris-HCl (pH 9·4), 1 mM ATP, 1 mM $MgCl_2$, 5 mM phosphoenolpyruvate, 17 enzyme units pyruvate kinase, and, where indicated, oligomycin dissolved in methanol over the range of concentrations given in the tables. The medium was preincubated for 5 min at 30° and the reaction was started by the addition of mitochondria (0·2–0·4 mg protein). After 10 min the reaction was stopped by the addition of 0·2 ml of a solution containing 0·8 M $HClO_4$ and 0·6 M Na_2SO_4, protein was removed by centrifugation, and inorganic phosphate was measured in 0·5 ml or 1·0 ml aliquots of the supernatant.[33]

As small amounts (5–10%) of non-mitochondrial ATPase were sometimes found to be present in the mitochondrial preparations, the ATPase activity in each strain was also measured in the presence of mitochondrial F_1 inhibitor, which specifically inhibits mitochondrial ATPase from yeast or mammalian cells and has no effect on non-mitochondrial ATPase. The F_1 inhibitor was prepared from beef heart mitochondria by the method of Pullman and Monroy.[34] The ATPase assay was slightly modified in the following way: F_1 inhibitor (200 μg/mg mitochondrial protein) was preincubated with mitochondria for 10 min at 30°. A mixture, containing all other reaction components, was then added and the reaction was carried out for 10 min. The remainder of the procedure was exactly as described above. Results were corrected for any non-mitochondrial ATPase in the samples.

Specific activities are expressed as μmoles inorganic phosphate produced per min per mg mitochondrial protein.

Effect of Oligomycin on Anaerobically Grown Cells

Cells were grown anaerobically at 28°C for 22 h on a 5% glucose-1% Difco yeast extract-salts medium, in the presence or absence of the two lipid supplements, 0·5% Tween 80 and 20 mg/ml of ergosterol.[5] The cells were harvested at 0° and washed twice with sterile water.

Separate aliquots of the anaerobic cells were resuspended, at a concentration of 0·5–1·0 mg cells/ml medium in one of the following aeration media: 1% glucose–1% yeast extract-salts medium plus or minus oligomycin (10 μg/ml), (a concentration about 30 times greater than the amount required to prevent growth of sensitive strains on ethanol medium) or 1% ethanol-1% yeast extract–salts medium plus or minus oligomycin (10 μg/ml). Cells were aerated and samples were taken at 16 h to measure growth and cell viability. The respiration of washed whole cells was measured at 30° with an oxygen electrode in the presence of 50 mM potassium phosphate buffer (pH 7) and 2 mM glucose or 1% (w/v) ethanol. Whole cell cytochromes were measured on a Carey 14 recording spectrophotometer.

Protein synthesis. Assays of mitochondrial protein synthesis were carried out as previously described.[31]

Results

Mitochondrial ATPase Activity

Oligomycin is a potent inhibitor of both animal and plant mitochondrial ATPase. However while very low concentrations of oligomycin (about 1 μg oligomycin/mg mitochondrial protein) strongly inhibit ATPase in both animal and plant mitochondria, much higher levels are required to inhibit that of normal yeast mitochondria (about 20 μg oligomycin/mg mitochondrial protein). In wild type yeast mitochondrial ATPase is membrane bound and exhibits two pH maxima at 6·2 and 9·4;[35] ATPase activity at both pH's is inhibited by oligomycin. The *in vitro* effect of oligomycin on the mitochondria of two mutant strains of *S. cerevisiae* (L 3000, L 4000) resistant *in vivo* to oligomycin, and two characteristic sensitive strains (L 2200, L 410) are shown in Fig. 1. The two mutants L 3000 and L 4000 were selected for study as they differ in their genetic characteristics (see later). At both pH 6·2 and 9·4 neither of the mutant strains showed a clear pattern of *in vitro* resistance to oligomycin. Comparison of the ATPase system of the sensitive strains with L 4000 shows small differences in sensitivity at pH 9·4 and 6·2, with the resistant mutant showing some increases in resistance over the range 5–40 μg of oligomycin. The mitochondrial ATPase of strain L 3000, on the other hand, shows essentially no difference in oligomycin sensitivity from that of the wild type strains at pH 9·4; however at pH 6·2 it is even more sensitive than them.

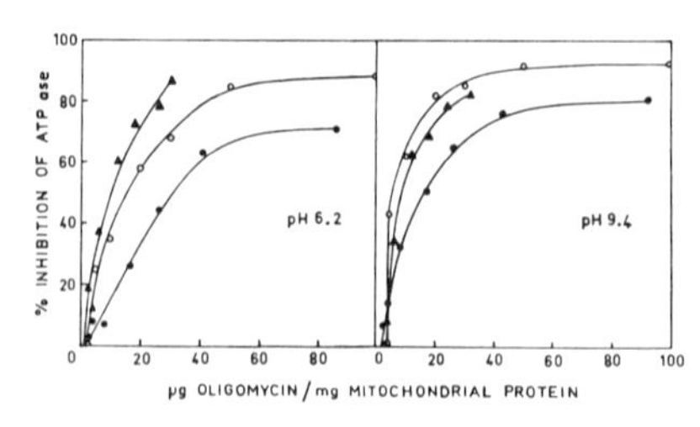

Figure 1. The effect of oligomycin on ATPase activity of isolated mitochondria. Mitochondria were isolated from ethanol grown cells and purified on a 20–70% sucrose gradient as described in Methods. Specific activities of mitochondrial ATPase, expressed as μmoles inorganic phosphate/min/mg mitochondrial protein were: strains L 410, L 2200, 1·35 (pH 6·2) and 1·96 (pH 9·4); strain L 4000, 0·58 (pH 6·2) and 1·47 (pH 9·4); strain L 3000, 1·01 (pH 6·2) and 1·12 (pH 9·4). ATPase activities in the presence of oligomycin are expressed as percentages of these controls. All values are corrected for non-mitochondrial ATPase. The ATPase of the oligomycin sensitive strains L 2200 and L 410 had almost identical *in vitro* activities in both the presence and absence of oligomycin (○—○). The results obtained with the mitochondria from the two oligomycin resistant strains are shown thus: L 3000, ▲—▲; L 4000, ●—●.

Anaerobic Growth and Aerobic Induction in the Presence of Oligomycin

The *in vivo* resistance and *in vitro* sensitivity of the oligomycin resistant mutants was reminiscent of previously studied mutants resistant to the mitochondrial protein synthesis inhibitors mikamycin and chloramphenicol; these were postulated to be mitochondrial membrane mutants.[20] Studies with these mutants showed that after anaerobic growth, during which the mitochondrial membranes are grossly altered, the resistant mutants could be rendered sensitive *in vivo* to mikamycin; they failed to develop respiration or the particulate mitochondrial cytochromes upon aeration in the presence of the drug. The experiments demonstrated that unlike some other mutants isolated in our laboratory,[25] the mitochondrial protein synthesizing system *per se* was not altered by the mutations, and that physiological alteration of membrane properties by anaerobic growth allows mikamycin to penetrate the altered membrane and inhibit mitochondrial protein synthesis, which in turn prevents cytochrome formation and the development of respiration.

The influence of anaerobic growth on oligomycin resistance was examined in a similar manner to that of mikamycin resistance. Yeast cells were grown anaerobically on lipid supplemented or lipid deficient media for 24 h and then aerated in either glucose or ethanol medium in the presence and absence of oligomycin (see Methods section). The results were the same whichever type of anaerobically grown cells were employed. Both oligomycin sensitive and resistant strains aerated for 16 h on glucose medium, with or without oligomycin, are able to grow and develop a functional respiratory system and the particulate mitochondrial cytochromes (Table I). The extent of the development of

TABLE I. The effect of oligomycin on the growth and aerobic induction of respiration in anaerobically grown cells

Yeast strain	Aeration medium	Antibiotic addition	Aerobic induction: Cell density (mg/ml) To	Cell density (mg/ml) T_{16}	Whole cell respiration T_{16}	Cytochromes a, a_3, b, c_1 T_{16}
L 2200	Glucose	None	0·57	3·8	16	Present
	Glucose	OL	0·57	2·5	23	Present
	Ethanol	None	0·43	1·53	246	Present
	Ethanol	OL	0·43	0·43	76	Present
L 3000	Glucose	None	0·57	2·4	36	Present
	Glucose	OL	0·57	2·3	21	Present
	Ethanol	None	0·76	1·54	76	Present
	Ethanol	OL	0·76	0·76	13	Present
L 4000	Glucose	None	0·57	2·4	33	Present
	Glucose	OL	0·57	2·3	22	Present
	Ethanol	None	1·07	1·34	77	Present
	Ethanol	OL	1·07	1·07	19	Present

Cells were grown anaerobically on lipid supplemented medium for 24 h, harvested, washed and resuspended in ethanol or glucose aeration media in the presence or absence of oligomycin (10 μg/ml), see Methods. At the beginning of aeration (To), cell density was measured; these cells did not respire and the mitochondrial cytochromes a, a_3, b, and c_1 were absent. Cell density, whole cell respiration and cytochromes were measured after 16 h aeration. The effect of oligomycin (OL) on the strains *in vivo* under fully aerobic conditions, and *in vitro*, on the ATPase activity of isolated mitochondria, was as follows:

L 2200 *in vivo* and *in vitro*, oligomycin sensitive.
L 3000 *in vivo*, oligomycin resistant, *in vitro*, oligomycin sensitive.
L 4000 *in vivo*, oligomycin resistant, *in vitro*, oligomycin sensitive.

Whole cell respiration: mμ-atoms O_2/min/mg dry weight cells.

respiratory capacity on glucose media, due to catabolite repression, is limited. However it is sufficient to clearly show that in both resistant and sensitive cells, oligomycin does not inhibit the elaboration of the respiratory system or, in other words, inhibit mitochondrial protein synthesis. When the cells from either sensitive or resistant strains are aerated on ethanol, the cell mass increases and mitochondrial cytochromes and respiration develop. The extent of the growth of the organisms on glucose medium is considerably greater than on ethanol medium as the anaerobically grown cells can immediately extensively utilize glucose for growth by non-oxidative pathways. On the other hand cells aerated on ethanol initially have no capacity to utilize ethanol. However both sensitive and resistant cells aerated on ethanol in the presence of oligomycin fail to grow but develop mitochondrial cytochromes and respiration.

It is well established that the mitochondrial membrane structure differs depending on whether the cells are grown anaerobically in the presence of an excess of unsaturated fatty acids and ergosterol or they are grown in the presence of growth limiting amounts of these lipids.[5,8,9] However, the alteration of the mitochondrial membrane properties as a consequence of anaerobic growth, irrespective of the presence of an added source of lipids, is extensive enough in both instances to allow oligomycin access to the oxidative phosphorylation system, and thus prevent derivation of energy from the respiratory chain and growth on ethanol.

Cross Resistance to Antibiotics

Previous work in this laboratory has shown that a number of cytoplasmically determined mutants are resistant to various combinations of inhibitors of mitochondrial protein synthesis, such as mikamycin, erythromycin, chloramphenicol, carbomycin, lincomycin, tetracycline and spiramycin and appear to be mitochondrial membrane mutants.[20] It was therefore of interest to examine possible cross-resistance patterns of oligomycin mutants to protein synthesis inhibitors. The two oligomycin resistance mutations can be readily distinguished one from the

TABLE II. Cross resistance in oligomycin resistant mutants

	Phenotype: *In vivo* / *In vitro*						
Yeast strain	Mikamycin	Chloramphenicol	Lincomycin	Carbomycin	Spiramycin	Erythromycin	Oligomycin
L 3000	R/S	R/S	R/S	R/S	S/S	S/S	R/S
L 4000	S/S	S/S	S/S	S/S	S/S	S/S	R/S

In vivo phenotypes, R (resistance) and S (sensitivity) denote the ability and inability, respectively, of cells to grow on ethanol medium containing the particular antibiotic, at concentrations as described in Methods.

In vitro phenotypes R and S denote resistance, defined as less than 25% inhibition of ^{14}C leucine incorporated into protein by isolated mitochondria, and sensitivity, defined as greater than 60% inhibition by 0·1 mM of the antibiotics mikamycin, chloramphenicol, lincomycin, carbomycin, spiramycin or erythromycin (see Methods).

In vitro resistance to oligomycin is defined as the lack of inhibition of ATPase from isolated mitochondria in the presence of the drug; sensitivity denotes 50% or more inhibition of mitochondrial ATPase by oligomycin (30 μg/mg mitochondrial protein).

other on the basis of their cross-resistance patterns so that the gene products involved are apparently not identical. The results are shown in Table II. The mutation to oligomycin resistance in strain L 4000 does not confer any cross-resistance to the antibiotics tested. On the contrary oligomycin resistance in strain L 3000 was simultaneously accompanied by cross-resistance *in vivo* to mikamycin, chloramphenicol, lincomycin, and carbomycin but not to spiramycin and erythromycin. However *in vitro* amino acid incorporation into protein by the isolated mitochondria from both L 3000 and L 4000 was inhibited by all the protein synthesis inhibiting antibiotics (Table II).

Inheritance Characteristics of the Oligomycin-resistant Mutants

The phenomenology of cytoplasmic genetics in yeast is still in an early stage of development. However some criteria have been suggested as descriptive of the process of cytoplasmic genetics.[1,20,25,36,37] The criteria are as follows: (a) on crossing a haploid wild type cell and a haploid mutant cell the diploid zygote formed contains both alleles of the cytoplasmic gene and these segregate during vegetative growth, to eventually give rise to a stable population of diploid progeny, some with the wild type characteristic and others with the mutant characteristic. This characteristic has been referred to as "mixedness" whereby a single zygote gives rise to a genetically mixed clone. The final ratio of stable resistant to sensitive diploid cells is a characteristic of each cross, but varies among individual zygotes. (b) On sporulation of purified diploid cells the cytoplasmic determinant segregates all or none into the four spores of each ascus. Thus in the case of an antibiotic resistant diploid cell sporulation gives rise to 4 resistant: O sensitive spores, and the converse is true for sensitive diploid cells. (c) The cytoplasmic gene may be deleted with ethidium bromide, a specific cytoplasmic mutagen. Indeed the drug has been shown to eliminate mitochondrial DNA from the cells without affecting cell viability.[29,30] The two classes of mutant isolated were examined for each of these criteria.

Characteristics of Diploid Progeny

Table III shows that when the oligomycin resistant strains L 4000 and L 3000 are crossed with several oligomycin sensitive strains L 2200, L 5628, L 2300 or L 2265, both stable oligomycin resistant and oligomycin sensitive diploids appeared in the diploid progeny from these crosses. Hence the cytoplasmic inheritance characteristic of mixedness of diploid progeny is observed with both these mutants.

Loss of Resistance with Petite Mutation

The elimination of the oligomycin resistance determinant from L 4000 by ethidium bromide treatment was demonstrated by growing

Table III. Mixedness characteristic of cytoplasmic inheritance

Cross			Diploid Progeny Oligomycin	
Oligomycin-R		Oligomycin-S	%R	%S
L 4000	×	L 2200	58	42
L 4000	×	L 5268	53	47
L 3000	×	L 2300	25	75
L 3000	×	L 2265	19	81

Cultures of each strain were crossed for 4 h, and diploids selected by prototrophic selection. Diploid progeny were assayed for oligomycin resistance (R) and sensitivity (S) as described in Methods: the degree of mixedness is reported as the percentage of oligomycin resistant and sensitive diploids in the total diploid progeny population examined, about 1000 cells per cross.

cultures overnight in the presence of the drug (20 μg/ml of medium), and subsequent crossing with untreated oligomycin sensitive strain L 2200. As a control, L 2200 was also treated similarly with ethidium bromide, and then crossed with untreated L 4000. The other oligomycin resistant strain L 3000, after treatment with ethidium bromide, was crossed with the untreated oligomycin sensitive strain L 2300; the control experiment was carried out similarly with ethidium bromide treated L 2300. The results are presented in Table IV and analysis of

TABLE IV. Loss of oligomycin resistance and sensitivity alleles on treatment with ethidium bromide

Ethidium bromide treated cells	Consequent cross	Diploid progeny
L 4000	L 4000 (ρ°) × L 2200 (ρ^{+}S)	All ρ^{+}S
L 2200	L 2200 (ρ°) × L 4000 (ρ^{+}R)	All ρ^{+}R
L 3000	L 3000 (ρ°) × L 2300 (ρ^{+}S)	All ρ^{+}S
L 2300	L 2300 (ρ°) × L 3000 (ρ^{+}R)	All ρ^{+}R

Cultures of oligomycin resistant (R) and sensitive (S) strains were treated with ethidium bromide (EtBr) at 20 μg drug/ml of medium for 18 h as described in Methods. Respiratory competent (ρ^{+}) cells are made respiratory deficient and lack mitochondrial DNA; such cells are designated ρ°.

Cultures of each haploid strain were crossed, and diploids selected and assayed for oligomycin resistance or sensitivity as described in Methods.

the zygotic progeny for resistance and sensitivity to oligomycin showed that only the characteristic of the untreated parent is expressed in the diploids from these crosses. The elimination of the determinant by ethidium bromide is thus evident.

Spontaneously arising petites from strains L 4000 and L 3000 were also examined for retention or loss of the oligomycin resistance determinant by crossing with oligomycin sensitive strains L 2200 and L 2300 respectively, and analysing diploid progeny for oligomycin resistance as before. A number of these spontaneously occurring petites of both resistant strains retained the resistance gene in contradistinction to the ethidium bromide treated cells. Eight out of 24 randomly selected petites of strain L 4000 retained the resistance gene, while 25 out of 48 petites of strain L 3000 similarly retained the resistance gene. Thus both these mutations to oligomycin resistance, in L 4000 and L 3000, are clearly linked to the classical cytoplasmic petite mutation.

Tetrad Analysis: Two Classes of Oligomycin Resistant Mutants

The foregoing data suggests that both classes of mutant are apparently cytoplasmic. However on examination of the segregation of oligomycin resistance in tetrads following meiosis, a different pattern of inheritance for each mutant becomes apparent; these are shown in Table V.

TABLE V. Tetrad analysis of oligomycin resistant mutants

Diploid/Cross	Resistance of diploid colony selected	No. of tetrads analysed	Segregation in spores Oligomycin Resistant:Sensitive
* N 1311	R	4	All 4:0
L 4000 × L 2200	R	5	All 4:0
	S	4	All 0:4
* N 1301	R	3	All 4:0
L 3000 × L 2300	R	13	All 2:2
	S	14	All 2:2

Strains N 1311 and N 1301 are oligomycin-resistant diploid mutants derived from the oligomycin-sensitive diploid N 1300 (see Methods). L 4000 is an oligomycin-resistant haploid strain derived from the sporulation of N 1311; L 3000 is similarly derived from N 1301. L 2300 and L 2200 aer oligomycin-sensitive strains. Cultures of each strain were crossed, and diploids selected as described in Methods. R and S denote oligomycin resistant and sensitive clones selected, sporulated and dissected as described in Methods. Nuclear alleles segregated 2:2 in all tetrads.

* Spore viability was poor in strains N 1311 and N 1301. However, random spore dissection yielded 72 spores from N 1311 and 66 spores from N 1301, all of which were oligomycin resistant.

When L 4000, an oligomycin resistant haploid derived from the oligomycin resistant mutant N 1311, was crossed with the oligomycin sensitive strain L 2200, mixed diploid progeny resulted. Sporulation of resistant diploids gave cytoplasmic 4:0 segregation in tetrads for resistance:sensitivity, and sensitive diploids gave the cytoplasmic 0:4 segregation. The diploid mutant N 1311 itself, when sporulated, showed 4:0 segregation in tetrads for oligomycin resistance to sensitivity. These results clearly indicate the cytoplasmic character of the mutation to oligomycin resistance in strains N 1311 and L 4000.

On the other hand, the cross of strain L 3000 with L 2300 yielded mixed diploid progeny as described, but on sporulation, all tetrads examined showed nuclear 2:2 patterns of inheritance for oligomycin resistance to sensitivity whether from sporulated resistant or sensitive diploids (Table V). These tetrad data are inconsistent with the cytoplasmic inheritance characteristics of mixedness and ethidium bromide elimination of antibiotic resistance exhibited by strain L 3000. However it should be emphasized that direct sporulation of the oligomycin resistant mutant N 1301 from which L 3000 was derived shows 4:0 segregation in tetrads, indicative of cytoplasmic inheritance.

The oligomycin resistant strain L 3000 shows cross-resistance to the antibiotics mikamycin, chloramphenicol and carbomycin. Simultaneous examination of the segregation of these cross-resistances in diploids, and in all haploid spores from these diploids showed that they segregated with oligomycin resistance in all cases; also oligomycin sensitive diploids and spores were sensitive to all these antibiotics. Thus no separation of cross-resistances during vegetative growth, or following meiosis, was obtained. Similarly, all cross-resistances were lost simultaneously on ethidium bromide treatment. Spontaneously arising petites from L 3000 had retained or lost all drug resistances concomitantly. It appears that a single mutation in L 3000 had conferred simultaneously resistance to all these antibiotics.

A series of other mutants independently selected for oligomycin resistance have also been genetically analyzed. Four of these mutants behave identically to N 1311 and L 4000, and three behave in the same way as N 1301 and L 3000.

Discussion

The mitochondria of cells of the two representative mutants L 3000 and L 4000 are resistant to oligomycin; in addition the mitochondria of L 3000 are cross resistant *in vivo* to the protein synthesis inhibiting antibiotics mikamycin, chloramphenicol, lincomycin and carbomycin whereas L 4000 is sensitive to these drugs. These results establish a clear biochemical difference between the two mutants. However the resistance characteristics of the mitochondria have not to date been demonstrated *in vitro*; mitochondria isolated from strains L 3000 and L 4000 are sensitive to oligomycin and all the other antibiotics. The ATPase and OSCP do not therefore appear to be the site of either of the two mutations. Indeed, that these two protein complexes are not directly affected by the mutations is further shown by the demonstration that growth of the two oligomycin resistant mutants under anaerobic conditions renders both types of cells sensitive to oligomycin inhibition. Clearly the alteration of the mitochondrial membrane properties produced by anaerobic growth results in the basically sensitive ATPase complex being accessible to oligomycin.

The results indicate that the mitochondrial membrane during isolation of the organelles is physically modified and hence the mitochondrial ATPase and protein synthesizing system are rendered accessible to inhibition by oligomycin and the protein synthesis inhibiting antibiotics. Similarly, physiological modification of the mitochondrial membranes by anaerobic growth renders the organelle membranes permeable to each of the antibiotics. The data are interpreted to indicate that the two mutations affect a protein (*s*) of the mitochondrial membrane.

The mutation to oligomycin resistance in L 4000 indicates the exclusion of this drug from access to the mitochondrial ATPase; it might therefore be described as a simple mitochondrial membrane mutation specifically affecting the permeability of the organelle to oligomycin. One possible candidate for the likely altered membrane protein of the mitochondria of strain L 4000 may be that recently described by Subik *et al.*[38] and Tzagaloff[39] which appears to be required for the binding of mitochondrial ATPase and OSCP to the electron transport system. However explanations of this kind are probably too simplistic when the properties of L 3000 are considered.

The observation that in strain L 3000 both the mitochondrial protein synthesizing system and the ATPase complex are shielded *in vivo* from oligomycin and some chemically unrelated antibiotics (mikamycin, chloramphenicol, lincomycin, carbomycin) but not others (erythromycin, spiramycin) suggests a more complicated situation may be being observed. It may be considered that the membrane alteration in L 3000 is such that conformational changes are induced which simultaneously shield the ATPase complex from oligomycin and partly shield the mitochondrial ribosome (which is attached to the mitochondrial membrane[2,40]) from some antibiotics, but not others.[20] This interpretation is supported by the isolation in our laboratory of other mutants with a wide variety of cross resistance patterns to antibiotics herein under examination; these mutants are then envisaged as membrane mutations inducing conformational membrane changes which in turn lead to different extents of ATPase and ribosome shielding.[1,2,40,41] Thus the attachment of the mitochondrial ribosome to the membrane provides one rationale for different single mutations producing such complex cross-resistance patterns among the protein synthesis inhibitors and the correlation that apparently a single mutation can in addition confer oligomycin resistance. It must be recalled that many of the antibiotics are chemically unrelated and those drugs which are chemically related, do not necessarily group in cross-resistance patterns. Further, chemically related compounds such as carbomycin, spiramycin and erythromycin react at different ribosomal sites,[42–44] so a membrane conformational change could conceivably screen one or more of the reaction sites, leaving the others exposed. Thus all antibiotics could *in vivo* penetrate the mitochondrial organelle, but dependent upon the conformational screening of the ribosome only some antibiotics could react with the ribosomes.

The inheritance characteristics of strain L 4000 are entirely consistent with the phenomena associated with cytoplasmic inheritance outlined in Results. Hence, this cytoplasmically determined mitochondrial mutant resembles, in its genetic properties, the previously described cytoplasmic mutants, resistant to erythromycin,[25] paromomycin[45] and chloramphenicol.[37] These mutants are still biochemically poorly characterized. The erythromycin mutation has been shown to involve a change in the mitochondrial protein synthesizing system,[46] apparently in the mitochondrial ribosome.[47] The mutation to paromomycin resistance also appears to be expressed as a change in the mitochondrial protein synthesizing system[45] while the chloramphenicol mutants have not been biochemically investigated.

The other mitochondrial membrane mutant L 3000 simultaneously shows characteristics of both nuclear and cytoplasmic inheritance. In crosses between resistant and sensitive strains, it shows the cytoplasmic inheritance pattern of the production of mixed diploid progeny arising from single primary zygotes. A cytoplasmic association of oligomycin resistance is also shown by both its retention or loss in cytoplasmic petites. Only following crosses between oligomycin resistant and sensitive cells and subsequent meiosis is a nuclear inheritance characteristic observed, namely as a nuclear 2:2 segregation ratio of 2 OL^r:2OL^s in tetrads. During the processes of cytoplasmic segregation, meiosis and petite formation the antibiotic cross-resistance pattern (oligomycin, mikamycin, chloramphenicol, carbomycin, lincomycin) shows no separation, all the antibiotic resistances stay together. The mutation to multiple antibiotic resistance in L 3000 appears to be a single mutation.

In a study of a tetracycline resistant mutant, Wilkie[45] has reported that on sporulation of sensitive and resistant diploids, 4 sensitive:0 resistance spores and 2 sensitive:2 resistant spores respectively are obtained. Also he has reported that there is no association between the tetracycline mutation and the rho factor.[45] These observations are clearly different to those reported herein but also involve some type of nucleo-cytoplasmic genetic interaction.

Both cytoplasmic and nuclear patterns of inheritance have been separately reported for oligomycin resistant mutants.[1, 21–24, 48] However the present data should not be confused with those previously reported, the results clearly establish both nuclear and cytoplasmic genetic behaviour interacting in the one mutant. It is not established as yet whether the primary mutation is cytoplasmic or nuclear.

However it is clear that the rho factor determines the phenotype of the cell. Thus in a cross involving one cell which has been converted to ρ° (mit-DNA absent) it is the other ρ^{+} cell in the cross which determines whether the diploids formed will be resistant or sensitive to the antibiotic. Yet on sporulation of the resultant ρ^{+} diploids whether sensitive or resistant, 2 sensitive and 2 resistant haploid spores are obtained, which in subsequent crosses can vegetatively transmit the apparent cytoplasmic characteristic of sensitivity or resistance. Any simple concept of dominant or recessive nuclear alleles does not appear to account for the observations. Little is known of the fate of the mitochondrial DNA during sporulation; such knowledge now appears to be crucial to a full understanding of cytoplasmic genetic phenomena in yeast. It may be that possibly we are observing part of a global phenomenon, whereby all cytoplasmic genomes be they virus determining, bacterial sex or R factors, mitochondrial DNA, chloroplast DNA and as yet uncharacterized DNA's of the endoplasmic reticulum and cell surface etc. are of a nature that they can all on occasion, be integrated into the nuclear genome and thus all have some properties similar to the classical bacterial episome.

An alternative and more attractive hypotheses can be advanced. Consider that the primary mutation to resistance is nuclear and that the gene product (a mitochondrial membrane component) requires an appropriate mitochondrial membrane conformation for its acceptance into the membrane; the phenotypic expression of the gene product then becomes possible. The apparent cytoplasmic inheritance would then be envisaged as being determined by the transmission and segregation of preformed resistant or sensitive mitochondrial membranes among progeny cells. In heterozygotes formed in crosses between sensitive and resistant ρ^{+} cells the pre-existing membrane conformations would then play their essential roles in the progeny by determining whether it is the nuclear gene product of the sensitive or resistant allele which is inserted into developing mitochondrial membrane. It is envisaged in this hypothesis that the mutation to petite leads to loss or retention of resistance as determined by the extent of the membrane deformation caused by the petite mutation rather than a direct effect of the ρ^{-} mutation on a cytoplasmic oligomycin resistance gene. It follows that resistance retention in the petite mutants requires the mitochondrial membrane structure to be still capable of accepting the nuclear resistance gene product; loss of resistance means that the particular petite mutation has altered the mitochondrial membrane to such an extent that it cannot accept the nuclear resistance gene product. The membranes have, in this hypothesis, a degree of self determination. These interpretations owe much to the insight and concepts developed by Sonneborn to explain the heredity of the cortical patterns of *Paramecium*.[49]

A preliminary study of some of the genetic properties of strain L 3000 in relation to mikamycin resistance has been reported earlier.[20] At the time it appeared that the mutations to antibiotic resistance in strain L 3000 and several other strains (L 3200, L 3300) were classical cytoplasmic mutations analogous to the erythromycin resistance mutation.[25] However it is clear from the results reported herein that the situation is not so simple and that nucleo-cytoplasmic interactions are occurring in strain L 3000. The genetic characteristics of strain L 3000 are probably not unique but are representative of a considerable number of mutants recently isolated in our laboratory which are phenotypically characterized by a variety of cross resistance patterns different to that of L 3000. A full account of this work is being prepared for publication elsewhere.[41]

References

1. A. W. Linnane and J. M. Haslam, in: *Current Topics in Cellular Regulation*, B. L. Horecker and B. R. Stadtman (eds.), Vol. 2, Academic Press, New York, 1970, p. 101.
2. A. W. Linnane, J. M. Haslam and I. T. Forrester, *Biochemistry and Biophysics of Mitochondrial Membranes*, G. F. Azzone, E. Carafoli, A. L. Lehninger, E. Quagliariello and N. Siliprandi (eds.), Bressanone, Italy, 1971, Academic Press, New York, in press.
3. A. A. Andreson and T. J. B. Stier, *J. Cell. Comp. Physiol.*, **41** (1953) 23.
4. D. K. Bloomfield and K. Bloch, *J. Biol. Chem.*, **235** (1960) 337.
5. P. G. Wallace, M. Huang and A. W. Linnane, *J. Cell. Biol.*, **37** (1968) 207.
6. R. S. Criddle and G. Schatz, *Biochemistry*, **8** (1969) 322.
7. H. Plattner and G. Schatz, *Biochemistry*, **8** (1969) 339.
8. K. Watson, J. M. Haslam and A. W. Linnane, *J. Cell Biol.*, **46** (1970) 88.
9. K. Watson, J. M. Haslam, B. Veitch and A. W. Linnane, in: *Autonomy and Biogenesis of Mitochondria and Chloroplasts*, N. K. Boardman, A. W. Linnane and R. M. Smillie (eds.), North Holland Press, Amsterdam, 1971, p. 162.
10. K. Bloch, B. Baronowsky, H. Goldfine, W. J. Lennarz, R. Light, A. T. Norris and G. Scheuerbrant, *Fed. Proc.*, **20** (1961) 921.
11. J. W. Proudlock, L. W. Wheeldon, D. J. Jollow and A. W. Linnane, *Biochim. Biophys. Acta*, **152** (1968) 434.
12. D. J. Jollow, G. M. Kellerman and A. W. Linnane, *J. Cell Biol.*, **37** (1968) 221.
13. I. T. Forrester, K. Watson and A. W. Linnane, *Biochem. Biophys. Res. Commun.*, **43** (1971) 409.
14. K. A. Ward and A. W. Linnane, "Cell Differentiation", submitted.
15. I. T. Forrester, K. A. Ward and A. W. Linnane, in preparation.
16. B. Ephrussi, *Harvey Lectures*, **46** (1950) 45.
17. J. W. Proudlock, J. M. Haslam and A. W. Linnane, *Biochem. Biophys. Res. Commun.*, **37** (1969) 847.
18. J. W. Proudlock, J. M. Haslam and A. W. Linnane, *J. Bioenergetics*, **2** (1971) 327.
19. J. M. Haslam, J. W. Proudlock and A. W. Linnane, *J. Bioenergetics*, **2** (1971) 351.
20. C. L. Bunn, C. H. Mitchell, H. B. Lukins and A. W. Linnane, *Proc. Natl. Acad. Sci. (U.S.)* **67** (1970) 1233.
21. K. Wakabayashi and N. Gunge, *FEBS Letters*, **6** (1970) 302.
22. K. D. Stuart, *Biochem. Biophys. Res. Commun.*, **39** (1970) 1045.
23. P. R. Avner and D. E. Griffiths, *FEBS Letters*, **10** (1970) 202.
24. D. E. Griffiths, P. R. Avner, W. Lancashire and J. R. Turner, *Biochemistry and Biophysics of Mitochondrial Membranes*, G. F. Azzone, E. Carafoli, A. L. Lehninger, E. Quagliariello, and N. Siliprandi (eds.), Bressanone, Italy, 1971, Academic Press, New York, in press.
25. A. W. Linnane, G. W. Saunders, E. B. Gingold and H. B. Lukins, *Proc. Natl. Acad. Sci. (U.S.)*, **59** (1968) 903.
26. G. W. Saunders, E. B. Gingold, M. K. Trembath, H. B. Lukins and A. W. Linnane, in: *Autonomy and Biogenesis of Mitochondria and Chloroplasts*, N. K. Boardman, A. W. Linnane and R. M. Smillie (eds.), Amsterdam, North Holland Press, 1971, p. 185.
27. E. B. Gingold, G. W. Saunders, H. B. Lukins and A. W. Linnane, *Genetics*, **62** (1969) 735.
28. P. Nagley and A. W. Linnane, *Biochem. Biophys. Res. Commun.*, **39** (1970) 989.
29. E. S. Goldring, L. I. Grossman, D. Krupnick, D. R. Cryer and J. Marmur, *J. Mol. Biol.*, **52** (1970) 323.

30. G. Michaelis, S. Douglass, M. Tsai and R. S. Criddle, *Biochem. Genet.*, **5** (1971) 487.
31. A. J. Lamb, G. D. Clark-Walker and A. W. Linnane, *Biochem. Biophys. Acta*, **161** (1968) 415.
32. M. Somlo, *European J. Biochem.*, **5** (1968) 276.
33. G. Gomori, *J. Lab. Clin. Med.*, **27** (1942) 955.
34. G. C. Monroy and M. E. Pullman, in: *Methods in Enzymology*, Vol. 10, R. W. Estabrook and M. E. Pullman (eds.), Academic Press, New York, 1967, p. 510.
35. L. Kovac, H. Bednarova, and M. Greksak, *Biochim. Biophys. Acta*, **153** (1968) 32.
36. B. Ephrussi, H. Jakob and S. Grandchamp, *Genetics*, **54** (1966) 1.
37. D. Coen, J. Deutsch, P. Netter, E. Petrochilo and P. P. Slonimski, in: *The Control of Organelle Development*, P. L. Miller (ed.), p. 449, Cambridge University Press, 1970.
38. J. Subik, S. Kuzela, J. Kolorov, L. Kovac and T. M. Lachowicz, *Biochim. Biophys. Acta*, **205** (1970) 513.
39. A. Tzagaloff, *J. Biol. Chem.*, **246** (1971) 3050.
40. D. Plummer, A. A. Green, K. Watson, K. A. Ward and A. W. Linnane, in preparation.
41. C. L. Bunn, H. B. Lukins and A. W. Linnane, in preparation.
42. R. E. Monro and D. Vazquez, *J. Mol. Biol.*, **28** (1967) 161.
43. H. Teraoka, M. Tamaki and K. Tanaka, *Biochem. Biophys. Res. Commun.*, **38** (1970) 328.
44. J. C. H. Mao and E. R. Robishaw, *Biochemistry*, **10** (1971) 2054.
45. D. Wilkie, in *The Control of Organelle Development*, P. L. Miller (ed.), Cambridge University Press, 1970, p. 71.
46. A. W. Linnane, A. J. Lamb, C. Christodoulou and H. B. Lukins, *Proc. Natn. Acad. Sci. (U.S.)*, **59** (1968) 1288.
47. L. A. Grivell, L. Reijanders and H. de Vries, *FEBS Letters*, **16** (1971) 159.
48. J. H. Parker, I. R. Trimble and J. R. Mattoon, *Biochem. Biophys. Res. Commun.*, **33** (1968) 590.
49. T. M. Sonneborn, in: *The Nature of Biological Diversity*, J. M. Allen (ed.), McGraw-Hill, New York, 1963, p. 165.

A Physico-chemical Basis for Anion, Cation and Proton Distributions between Rat-liver Mitochondria and the Suspending Medium

E. J. Harris

Biophysics Department, University College, London, W.C.1

The distribution of ions, including protons, between the mitochondrial interior and the medium can be treated, most simply, as a physico-chemical problem akin to that met with in cell suspensions, particularly when sufficient is known about the system. More is now known about mitochondrial properties, and the aim of the present review is to summarize the evidence, some new and much derived from earlier reports, that the distributions of permeant anions and of protons between the mitochondrial interior and the exterior provide an example of the Gibbs–Donnan law applied to a situation with an ionized internal buffer. This can only hold of course when the metabolic fluxes are low. Some anions behave as non-penetrants and so can take on any distribution ratio. The cations Ca^{2+} and K^{+} are subject to energy-linked inward transport.

The Basis of the Physico-chemical Interpretation

The classical Gibbs–Donnan system has 2 compartments ("in" and "out"), and one at least carries a complement of charged particles which cannot pass to the other. The requirement for equalities of electrochemical potential of the salts of permeable species lead to the equations:

$$r = H^{+}_{\text{out}}/H^{+}_{\text{in}} = \text{Anion}^{1-}_{\text{in}}/\text{Anion}^{1-}_{\text{out}} = (\text{Anion}^{2-}_{\text{in}}/\text{Anion}^{2-}_{\text{out}})^{1/2} \qquad (1)$$

for protons and permeant anions bearing the number of charges shown as superscript. Across the membrane there is potential difference $E = RT/F \ln r$.

When, as in the present case, cations are moved by energy-linked process, the charge balance leads to:

$$H^{+} + K^{+} + 2Ca^{2+} = [X^{-}] + rA^{-}_{\text{out}} + r^{2} A^{2-}_{\text{out}} + r^{3} A^{3-}_{\text{out}} + \sum nB^{n-}_{\text{in}} \qquad (2)$$

In this A^{-} denotes the sum of singly charged penetrant anions and so on.

and B_{in}^{n-} is an internally generated n-charged anion which remains trapped inside.

The values of r and the potential difference will rise or fall as the energy-linked cation accumulation process provides more, or fewer internal cations. In addition the parameters will respond to the content of non-buffer anions denoted by A^- and B^- in equation (2). A further factor related to r is the internal pH (pH_i) because it determines the degree of ionization of the buffer (X^- in equation (2)). As pH_i falls the buffer becomes protonated and to maintain internal electroneutrality more anions A^- and/or B^- can be accommodated.

Between pH 7 and 8, in presence of uncoupler to preclude energized ion movements, Harris *et al.* (1966) found that mitochondria have a buffer power of about 50 nequiv per mg dry weight. The value corresponds to about 55 nequiv per mg protein because the dry weight is about 1·1 × the biuret protein. The result means that when internal pH falls from 8 to 7 electroneutrality requires entry of 55 nequiv anions, provided no cations are lost. This explanation of why mitochondrial anion capacity increases with acidity, a phenomenon described by Palmieri *et al.* (1970) does not involve assumptions about the mechanism of proton movement.

To relate the ionization of the internal buffer to the anion content the transmembrane pH difference (ΔpH) can be calculated from the concentration ratio holding for a selected permeant anion using the relation:

$$\Delta pH = \frac{1}{n} \log \frac{A_{in}^{n-}}{A_{out}^{n-}}$$

TABLE I. Comparison between transmembrane pH differences deduced from distributions either of dimethyloxazolidinedione (DMO) or phosphate using the data of Hoek *et al.* (1971) and the curve relating phosphate concentration ratio to Donnan ratio given by Harris (1970)

Phosphate concn. in medium	μM	200	560	700
ΔpH deduced from DMO ratio		1·10	0·93	0·87
ΔpH deduced from phosphate ratio		1·17	0·99	0·96

Hoek *et al.* (1971) have shown that the ΔpH deduced from the distribution of the weak acid dimethyloxazolidinedione agrees fairly well with that deduced from the phosphate distribution. Calculated values derived from their data are given in Table I. A slight excess phosphate concentration (giving an apparently higher pH) can be ascribed to binding, probably to Ca^{2+}. Palmieri *et al.* (1970) demonstrated an inverse relation between the ΔpH (deduced from the dmo distribution) and the concentration ratio for acetate and the square root of the ratio for the doubly charged malate anion. Although these authors do not accept the applicability of the Donnan concept with its associated internal positivity their data are in precise accord with the behaviour predicted by the theory.

In the light of equation (2), let us consider how the situation is changed when a penetrant anion is added. As more anion enters the mitochondria there is a trend to saturation because the maximum capacity is limited to equality with the cation equivalents. Hence, the concentration ratio between inside and out falls and ΔpH falls,

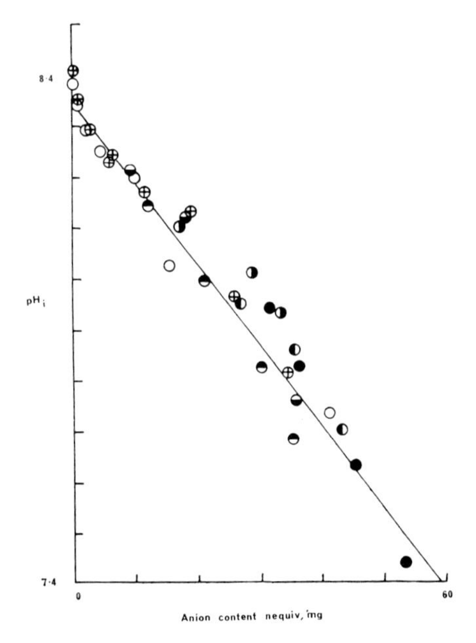

Figure 1. The relation between the content of internal anions in equivalents and the internal pH. The latter was deduced by adding:

$$\Delta\text{pH} = \frac{1}{n}\log\frac{A_{\text{in}}^{n-}}{A_{\text{out}}^{n-}}$$

to the external pH. The differently marked points refer to different experiments in which various anions were used. Experimental details of the experiments are given elsewhere (Harris and Bangham, 1972).

with the interior becoming less alkaline. The internal buffer is protonated ($X^- \rightarrow HX$) equivalent to the anion gain so, in effect, the internal buffer is titrated by the anions entering. This fact is used by plotting the ΔpH deduced from the ratio of a permeant anion against the total content of anions other than the buffer. In this way a curve is obtained corresponding to a buffering power of about 60 nequiv per pH per mg protein (Fig. 1).

This explanation of the nature of the response of internal pH to anions requires a means by which protons can be supplied to combine with X^-. As originally suggested by Chappell and Crofts (1966) a cycling of the two forms of the phosphate ion present at physiological pH can act as proton transporter. That phosphate distributes according to the ratio holding for other permeant anions has been shown by Harris (1970). Most mitochondrial preparations carry sufficient phosphate to obscure the requirement when relatively small anion movements are studied. However, Chappell and Crofts (1966) did find a phosphate requirement for penetration in quantity of dicarboxylates.

A demonstration of the applicability of another feature of equation 1; namely, the dependence of anion content on cation content is perhaps most directly obtained by comparing anion and cation contents before and after induction of K^+ uptake by valinomycin. Permeant anions are found to enter in amount almost equivalent to the K^+ gain (Harris, 1969). Less extensive movements of anions with Ca^{2+} have also been described (Rasmussen *et al.*, 1965; Rossi *et al.*, 1967; Kimmich and Rasmussen, 1968; Harris and Berent, 1970). Uptake of phosphate along with Ca, even when the prevalent concentration of phosphate is only that provided by leakage from the mitochondria, will provide a quantitative explanation of the acidification of the medium accompanying Ca^{2+} uptake. In absence of penetrant anions, about 1 H^+ appears in the medium per Ca^{2+} removed from it. If one calculates the acidity left in the medium when trebly charged phosphate is removed to form either tricalcium phosphate or hydroxyapatite, the results (Table II) depend on external pH as do the observed values. Best agreement is obtained on the assumption of formation of a compound having the composition of hydroxyapatite, though evidently this is not formed in the crystalline state (see review by Lehninger, 1970).

Another consequence of Ca^{2+} uptake which has attracted considerable interest is the transient internal alkalinization which accompanies it (Chance and Mela, 1966; Reynafarge *et al.*, 1967; Rossi *et al.*, 1966;

TABLE II. Calculation of ratio of protons in medium to Ca^{2+} uptake for different pH media. It is assumed that the Ca^{2+} is accompanied by phosphate to form internal hydroxyapatite: $3.Ca_3(PO_4)_2.Ca(OH)_2$

pH of medium	7·62	7·15	6·48
Net charge on P_i	−1·85	−1·60	−1·30
Protons freed when 6 P_i converted to P_i^{3-}	6·9	8·4	10·2
Protons freed by dissocn. of water to provide $2OH^-$	2	2	2
Protons per hydroxyapatite (10 Ca)	8·9	10·2	12·2
Protons/Ca calc.	0·89	1·02	1·22
Protons/Ca obs.	0·89	1·00	1·13

The experimental determination (last line) was made in a medium having 150 mM KCl, 20 mM tris-chloride, 10 mM tris-hydroxybutyrate, 60 mM mannitol.

Addanki *et al.*, 1968; Ghosh and Chance, 1970). This effect is predictable from equations (1) and (2) because cation gain leads to gain of anions and so to a higher inside/outside concentration ratio with, in turn, fewer internal protons. The fact that the entering Ca^{2+} combines with phosphate means that there will be a tendency for any other available anions to accumulate inside, but in any event the system should adjust so that the ratio of ionized internal phosphate to external phosphate is higher than before the Ca^{2+} addition, because of removal of phosphate from the medium. One corollary of the proposal made is that one would expect that substances forming highly insoluble phosphates (such as the rare earths) should inhibit uptake of Ca^{2+}, as indeed they do (Mela and Chance, 1968; Ghosh and Chance, 1970).

Having discussed the consequences of changes in anion concentration, cation content and internal pH, one may turn to the effect of external pH. This factor has two opposing results. As noted by Harris and Berent (1970), acidification can lead to discharge from the mitochondria of Ca^{2+} with citrate. Loss of cation and anion corresponds to a diminished accumulation of anions. On the other hand, to the extent that the interior moves towards acidity in response to acidification of the medium there will be a protonation of the buffer and an increase of non-buffer anions. What actually happens experimentally depends on the content of Ca^{2+} at the time the acidification is carried out. When the mitochondrial Ca^{2+} is more than about 8 nmole/mg protein, acidification leads to loss of Ca^{2+} and hence of anions. In Table 3, Expts. 1 and 2, Ca^{2+} losses of 41 and 21 nequiv/mg occur to offset the protonation of the buffers so smaller anion changes are needed than in Expts. 3 and 4 where the Ca^{2+} contents are little changed. Examples of concomitant changes of citrate and Ca^{2+} are given by Harris and Bangham (1972).

TABLE III. Effect of acidification on Ca^{2+} content of rat liver mitochondria

Expt.	pH of medium	Ca^{2+} nmole/mg	Expt.	pH of medium	Ca^{2+} nmole/mg
1	7·4	26	3	7·2	6·9
	7·0	21·5		6·8	6·0
	6·75	13·2		6·2	5·2
	6·2	5·5			
2	7·2	16	4	7·4	1·8
	6·8	7·5		6·8	2·2
	6·2	5·5		6·2	2·7

The medium contained 150 mM KCl, 30 mM sucrose and 29 mM tris chloride.

Other Evidence for, and Consequences of Internal Positivity

The most direct evidence that the interior is positive and that the potential behaves like that of a Donnan system towards anions has been given by Tupper and Tedeschi (1969), who used microelectrodes. Protagonists of the view that the interior is negative have rejected the result on the grounds that it is an artifact. However, the behaviour of the cations also points to internal positivity. Ca^{2+} uptake is associated with energy consumption, as is also the case when K^+ uptake is induced by valinomycin (Harris *et al.*, 1966). The electrochemical gradients against which the cations move will be higher if the interior is positive than if it is negative. Indeed, with sufficient internal negativity K^+ uptake would not demand energy expenditure. It has been shown that discharge of the internal K^+ to an initially K-free medium can be used to phosphorylate ADP, either endogenous (Cockrell *et al.*, 1967) or exogenous (Rossi and Azzone, 1970). About 1 ATP is obtained per 4 K^+ ions discharged. If one assumes an internal K concentration of 100 mM and an average external concentration during the discharge of 1 mM and an internal positivity amounting to 45 mv (so that unibasic anions are accumulated by a factor of 6) then the energy obtainable per K lost is 3700 cal per g-ion. Loss of 4 K g-ions furnishes 14,800 cal which is close to the energy required to phosphorylate ADP under comparable conditions of concentration. This stoichiometry has been reported by Rossi and Azzone (1970). If the interior had been negative many more K ions would have had to be lost. Cl^- is normally an impermeant anion. Its distribution will however reach the same ratio as that applying to the permeant pyruvate after induction of Cl permeability by alkyl tin salts (Harris, Bangham and Zukovic, 1972).

Further Discussion

Mitchell (1968) has used as evidence for an energy consuming proton ejection process the observation that restoration of energy to previously stored particles leads to an acidification of the medium. Most reports concerning phosphate movements are couched in terms of necessary phosphate-OH^- exchanges. It has been shown that the acidification of the medium is linked to entry of leaked Ca^{2+} (Thomas *et al.*, 1969) which in the light of the calculations of Table 2 has its effect on pH because of co-entry of phosphate. It is not necessary to invoke an obligate OH^--for-phosphate exchange if it is accepted that both protons and phosphate ions are at electro-chemical equilibrium.

The interaction between Ca^{2+} and phosphate to form a non-crystalline compound having the composition of hydroxyapatite could provide a sink for OH^-. Such a process, associated with protonation of X^- to remove H^+ would allow removal of water from phosphate or ADP ions at an enclosed locus so as to promote formation of ATP.

Acknowledgements

Thanks for financial support are due to the Medical Research Council, The Muscular Dystrophy Associations of America Inc., and the Wellcome Trust.

References

Addanki, S., Cahill, F. D. and Sotos, J. F., *J. biol. Chem.*, **243** (1968) 2337.
Chappell, J. B. and Crofts, A. R., in: *Regulation of Metabolic Processes in Mitochondria*, J. M. Tager *et al.* (eds.), Elsevier, Amsterdam, 1966, p. 293.
Chance, B. and Mela, L., *J. biol. Chem.*, **241** (1966) 4588.
Cockrell, R. C., Harris, E. J. and Pressman, B. C., *Nature, Lond.*, **215** (1967) 1487.
Ghosh, A. K. and Chance, B., *Arch. Biochem. Biophys.*, **138** (1970) 483.
Harris, E. J., in: *Energy Level and Metabolic Control in Mitochondria*, S. Papa *et al.* (eds.), Adriatica, Bari, 1969, p. 31.
Harris, E. J., *F.E.B.S. Letters*, **11** (1970) 225.
Harris, E. J., Judah, J. D. and Ahmed, K., in: *Current Topics in Bioenergetics*, Vol. 1, D. Sanadi (ed.), Academic Press, New York, 1966, p. 255.
Harris, E. J. and Bangham, J. A., *J. Membrane Biol.*, (1972) in press.
Harris, E. J., Bangham, J. A. and Zukovic, B. (1972) in preparation.
Harris, E. J. and Berent, C., *F.E.B.S. Letters*, **10** (1970) 6.
Harris, E. J., Cockrell, R. C. and Pressman, B. C., *Biochem. J.* **215** (1967) 1487.
Hoek, J. B., Lofrumento, L. E., Meijer, A. J. and Tager, J. M., *Biochim. Biophys. Acta*, **226** (1971) 297.
Kimmich, G. and Rasmussen, H., *Fed. Proc.*, **27** (1968) 528.
Lehninger, A. L., *Biochem. J.*, **119** (1970) 129.
Mela, L. and Chance, B., *Biochemistry*, **7** (1968) 4059.
Mitchell, P., *Chemiosmotic Coupling and Energy Transduction*, Glynn Research Ltd., Bodmin, Cornwall, 1968.
Palmieri, F., Quagliariello, E. and Klingenberg, M., *Euro. J. Biochem.*, **17** (1970) 230.
Rossi, C., Azzone, G. F. and Azzi, A., *Euro. J. Biochem.*, **1** (1967) 141.
Rossi, E. and Azzone, G. F., *Euro. J. Biochem.*, **12** (1970) 319.
Thomas, R. C., Manger, J. R. and Harris, E. J., *Euro. J. Biochem.*, **11** (1969) 413.
Tupper, J. T. and Tedeschi, H., *Science, Wash.*, **166** (1969) 1539.

Microwave Hall Mobility Measurements on Heavy Beef Heart Mitochondria

D. D. Eley, R. J. Mayer* and R. Pethig†

Department of Chemistry, University of Nottingham

* *Department of Biochemistry, The Medical School, University of Nottingham*

Abstract

The observed initial microwave Hall mobility values at 1·21 tesla of heavy beef heart mitochondria is at least six times greater than that observed for bovine serum albumin at similar resistivity values. The respiratory inhibitor cyanide significantly reduces the initial Hall mobility values for HBHM and for a preparation of HBHM cytochrome oxidase.

The four enzymic complexes of the respiratory chain were partially or completely separated. Of these complexes cytochrome oxidase exhibits the largest microwave Hall mobility.

The maximum hydration content of loosely bound water for freeze-dried preparations of cytochrome oxidase is 5% by weight; 60% of this hydration content is driven off by microwave power. Since the effective ac resistivity of the samples of cytochrome oxidase did not appreciably vary with changes in hydration content, the true resistivity of cytochrome oxidase has a value of the order 5×10^3 ohm cm and possibly much lower.

The electron transport pathway (as measured by Hall signal) of cytochrome oxidase is irreversibly damaged by prolonged exposure to microwave irradiation at 9·2 GHz. This is accompanied by the complete loss of capacity to oxidise ferrocytochrome *c*. Such changes do not occur with HBHM or with the other respiratory complexes.

There appears to be a direct relationship between observed Hall signals and the capacity of cytochrome oxidase to oxidize ferrocytochrome *c*. There is a "background" signal which is not directly related to electron transport but which is dependent on the conformation of the cytochrome oxidase.

The observed electronic parameters of cytochrome oxidase do not depend appreciably on its redox state.

Acid denaturation of cytochrome oxidase drastically reduces the Hall signal, to include almost complete removal of the "background" signal. It also more than doubles ac resistivity.

An electron tunnelling model is outlined.

Introduction

The mechanisms of charge transfer in the respiratory chain are not fully understood. The fact that charge transfer is greatly inhibited and

† Present address: School of Engineering Science, University College of North Wales, Bangor, Caerns.

possibly completely stopped at liquid nitrogen temperatures has been interpreted in favour of transport of electrons via mobile carriers (e.g. coenzyme Q, cytochrome *c*) rather than mechanisms based on resonance energy transfer or long range electron transport through conduction bands.[1,2]

The observed temperature variation does not, in fact, preclude the possibility that charge transfer occurs via conduction bands within individual molecular complexes of the cytochrome system. Transfer between complexes could involve a temperature activated tunnelling or hopping mechanism between the individual conduction band systems. Potential energy barriers could also exist within a particular cytochrome complex, the barrier shape and hence temperature variation of charge transfer being dependent on the molecular conformation.

In previous work[3] it was shown that freeze dried preparations of rat liver mitochondria gave an N-type Hall signal, much greater than that obtained from bovine serum albumin at the same resistivity values. This signal was significantly reduced by cyanide, but not by rotenone or Antimycin A.

Materials and Methods

The microwave Hall mobility measurements described here have been obtained using a microwave system based on that described by Trukhan.[4] A detailed account of the pertinent theory and experimental procedure has been given elsewhere.[5]

Preparations of heavy beef heart mitochondria (HBHM) were obtained as described by Smith.[6] The enzymic activities of the electron transport chain were partially or completely resolved to give Complex I + III (NADH-cyt. *c* reductase[7]), Complex II (succinate-coenzyme Q reductase[8]) and Complex IV (cytochrome oxidase[9]).

NADH-cytochrome *c* reductase was assayed as described by Halefi and Rieske,[7] in the absence of added phospholipid. Cytochrome oxidase was assayed as described by Wharton and Tzagoloff.[10] Succinate-cyt. *c* reductase (Complex II + III) activity was assayed as described by King.[11] Succinate-coenzyme Q reductase was not assayed.

No contamination of cytochrome oxidase by NADH-cyt. *c* reductase or succinate cyt. *c* reductase could be demonstrated. The NADH-cyt. *c* reductase was contaminated to less than 5% with cytochrome oxidase and less than 0·01% with succinate-cyt. *c* reductase.

Protein was determined by the method of Lowry, Rosebrough, Farr and Randall.[12]

Preparations of cytochrome oxidase were treated as follows: to 1 ml samples were added (a) 1 ml of aqueous KCN (final conc. 10^{-4} M); 0·2 ml of hydrogen peroxide (10 vols); sodium dithionite (1 mg); 1 ml of ferrocytochrome *c* (1%, produced as described by Wharton and

Tzagoloff;[10] 1 ml of ferrocytochrome *c* followed by 1 ml of KCN (final conc. 10^{-4} M) or sodium azide (final conc. 10^{-4} M).

A sample of cytochrome oxidase was denatured by titration with 6 N-HCl to pH 1 in the presence of Triton X-100 (final conc. 0·5%).

Preparations were freeze-dried for 18 hr prior to making discs for the Hall signal measurements.

Results

The results presented in Table I show that samples of HBHM give Hall signals which, as in preparations of mitochondria from rat liver,[3] are significantly affected by cyanide, but not by antimycin-A and rotenone. The Hall mobility is proportional to the square root of the intensity of the Hall signal.[4]

TABLE I. Initial permittivity, resistivity and Hall mobility measurements on freeze-dried samples of HBHM

	ϵ'	Resistivity (ohm cm)	Hall Mobility* (cm^2/V.sec)
HBHM	2·27	$4{\cdot}15 \times 10^3$	12·2
HBHM + potassium cyanide	2·32	$2{\cdot}84 \times 10^3$	8·15
HBHM + antimycin-A	2·24	$4{\cdot}60 \times 10^3$	11·2
HBHM + rotenone	2·39	$5{\cdot}80 \times 10^3$	11·7

* Calculated on basis that disc is 100% protein (see text).

All the Hall effect results are N-type and the applied field was 1·21 tesla. 1 ml of a preparation of HBHM or 1 ml of a preparation of HBHM treated with aqueous KCN (final conc. 10^{-4} M), ethanolic antimycin-A (0·125 μg/mg protein) or ethanolic rotenone (0·25 μg/mg protein) were freeze dried. Samples of these preparations were used for the measurements.

The Hall mobility values are expressed on the basis that all the sample placed in the bimodal cavity of the Hall apparatus is protein. This approach is used since previous work has shown that mitochondrial lipid gives a small Hall signal.[3]

The marked effect of cyanide in reducing the observed Hall signal suggests that the signal obtained from untreated HBHM may originate from the cytochrome oxidase part of the respiratory chain, since the site of action of cyanide is with the cyt.-a_3 moiety of cyt. oxidase[13] and the associated copper.[14]

To test this hypothesis, the various respiratory complexes were completely or partially purified from HBHM. Hall signal measurements were made on samples of the complexes. Typical results are given in Table II.

The greatest Hall mobility is given by cytochrome oxidase. Smaller Hall signals were given by NAD-cyt. *c* reductase and succinate-coenzyme Q reductase.

TABLE II. Initial permittivity, resistivity and Hall mobility measurements on freeze-dried samples of respiratory complexes

	ϵ'	Resistivity (ohm cm)	Hall mobility* (cm²/V.sec)
HBHM	2·27	$4{\cdot}15 \times 10^3$	12·2
Complex I + III (NADH-cyto. *c* reductase)	2·46	$2{\cdot}54 \times 10^3$	4·32
Complex II (Succinate-coenzyme Q reductase)	2·38	$5{\cdot}13 \times 10^3$	7·5
Complex IV (cytochrome oxidase)	2·56	$3{\cdot}82 \times 10^3$	26·4

* Calculated on basis that disc is 100% protein.
All the Hall effect results were N-type and the applied field was 1·21 tesla. Complexes prepared as given in *Materials and Methods*.

Apart from exhibiting the largest Hall signal, cytochrome oxidase differed from HBHM and the other respiratory complexes in that the observed Hall signal decayed appreciably with time. The Hall signal given by cytochrome oxidase decayed from 26·4 to 15·75 cm²/V.sec in approx. 24 hr whereas the signal from HBHM decayed from 12·2 to 11·5 cm²/V.sec in 6 hr and appeared to reach a constant value.

The decay of observed Hall signal given by cytochrome oxidase is shown in Fig. 1.

Not only did the Hall signal from the cytochrome oxidase preparation decay with time, but the overall shape of the Hall signal changed in that the normal Hall signal shape around 0·2–0·5 tesla, where paramagnetic absorption effects are expected to be observed, decreased in magnitude and extent with time. This is also shown in Fig. 1.

The decay in Hall signal from cytochrome oxidase was paralleled by a decrease in the ability of cytochrome oxidase (when reconstituted in an aqueous medium) to oxidize ferrocytochrome *c*. This result, on which a preliminary note has been published,[15] is for convenience again presented in Fig. 2.

This result strongly suggests that there exists a close relationship between the observed Hall signal and the capacity of cytochrome oxidase to oxidize ferrocytochrome *c*, and that the prolonged exposure to microwave radiation (60 mW) produces deleterious effects on the cytochrome oxidase complex. A large residual Hall signal is given by the cytochrome oxidase preparation when extrapolated to zero enzymic activity. This may indicate charge transfer pathways not connected with electron transport in the respiratory chain.

In the microwave cavity, the samples lose weight. A 3% weight loss occurs after 4 hr in the cavity while a specimen left in the cavity for

24 hr (in a vacuum of 0·133 N/m^2) shows a loss of weight of 4·93%. The loss of weight is interpreted to be a loss of water from the sample.

The observed reduction in the Hall signal of cytochrome oxidase

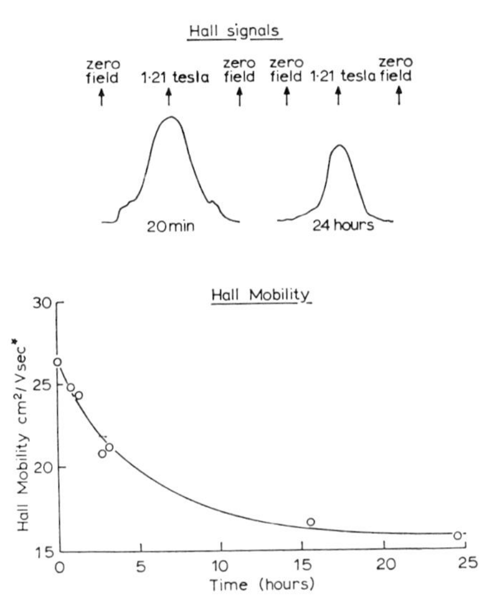

Figure 1. Variation of Hall signal, and Hall mobility of cytochrome oxidase with time spent in the bimodal cavity. * Hall mobility corrected to 100% protein.

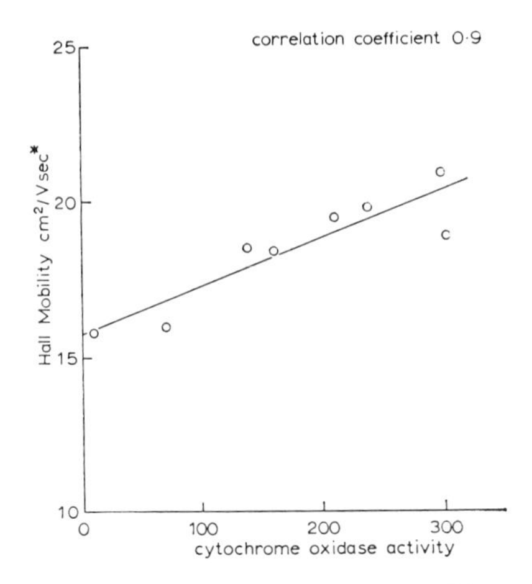

Figure 2. Variation of Hall mobility with cytochrome oxidase activity (from ref. 15). Cytochrome oxidase was assayed as indicated in ref. 10. * Hall mobility corrected to 100% protein.

with time in the bimodal cavity could possibly result from the oxidation of cytochrome oxidase (although this effect must be minimal since the bimodal cavity is flushed with N_2 during the course of each experiment) or the denaturation of the sample by microwave power.

To examine these possibilities, experiments were performed with fully reduced and oxidized cytochrome oxidase. The effect of supplying the electron transport chain with excess free electrons by adding ferro-cytochrome *c* to cytochrome oxidase was also tested, together with the effects of cyanide and azide in the presence of these excess free electrons (Table III).

The addition of a drop of H_2O_2 to 1 ml of the cytochrome oxidase preparation caused an immediate precipitation of the cytochrome oxidase in a similar manner to the acid denaturation of cytochrome oxidase. Hence, the change in Hall signal probably reflects denaturation rather than oxidation of cytochrome oxidase.

Both cyanide and azide, known inhibitors of cytochrome oxidase, drastically reduce the Hall signal compared with an untreated cyto-chrome oxidase preparation. The addition of ferrocytochrome *c* as a free electron source did not appreciably alter the observed electronic parameters of cytochrome oxidase or the way in which the Hall signals decayed with time spent in the microwave cavity, although the Hall signals did exhibit larger paramagnetic absorption effects in the field region around 0·2–0·5 tesla than those for the untreated cytochrome oxidase preparation. Addition of cyanide or azide to the ferrocyto-chrome *c* reduced preparations produce the same fall in Hall signal as in the absence of a free electron source. Addition of dithionite to reduce the cytochrome oxidase did not change the observed electronic parameters or make an appreciable difference in the way the Hall signals decayed with time spent in the microwave cavity.

TABLE III. Initial permittivity, resistivity and Hall mobility measurements for cytochrome oxidase preparations

	ϵ'	Resistivity (ohm cm)	Hall mobility* ($cm^2/V.sec$)
1. Cytochrome oxidase	2·56	$3{\cdot}82 \times 10^3$	26·4
2. Cytochrome oxidase + cyanide	1·99	$2{\cdot}18 \times 10^3$	9·92
3. Cytochrome oxidase + H_2O_2	2·49	$3{\cdot}21 \times 10^3$	4·94
4. Cytochrome oxidase + dithionite	2·12	$4{\cdot}91 \times 10^3$	16·92
5. Cytochrome oxidase + ferrocyto-chrome *c*	2·61	$3{\cdot}10 \times 10^3$	13·4
6. Cytochrome oxidase + ferrocyto-chrome *c* + cyanide	2·44	$3{\cdot}19 \times 10^3$	5·74
7. Cytochrome oxidase + ferrocyto-chrome *c* + azide	2·29	$4{\cdot}13 \times 10^3$	8·33
8. Acid denatured cytochrome oxidase	2·02	$1{\cdot}07 \times 10^4$	4·08

* Calculated on basis that disc is 100% protein.

Details of treatments applied to samples of cytochrome oxidase are given in Materials section.

Acid denaturation of cytochrome oxidase causes a drastic alteration in electronic properties. Besides the reduction in Hall signal, the resistivity of the sample changed significantly. Although a drastic fall in Hall signal occurred on H_2O_2 treatment, no change in resistivity was observed. This indicates a major difference in the action of the two sets of treatments.

Discussion

The observed reduction of the Hall signals with time for cytochrome oxidase was attributed to irreversible denaturation rather than oxidizing effects. Supporting evidence for such a conclusion comes from the observation of Muijers *et al.*[16] that, whereas at least three, possibly four different conformations of oxidized cytochrome oxidase can be distinguished depending upon the presence of molecular oxygen, a denaturation of a small amount of this enzyme during oxygenation might have a more profound effect on the observed oxidation rate than conformational charges induced by oxygenation. Also, conformation changes of cytochrome *c* upon oxidation have apparently no influence on the oxidation rate constants.[17]

From measurements on whole mitochondria from rat liver, it was concluded that the observed Hall signals corresponded to a true microscopic Hall mobility of 50–80 $cm^2/V.sec$, measurements on cytochrome oxidase prepared from HBHM gave Hall mobility values of the order 20 $cm^2/V.sec$ when calculated on a protein content basis. The fact that addition of cyanide and also azide, reduced the observed Hall effect for cytochrome oxidase strongly suggests that the observed signals are associated with electron transport in cyt. a_3 and/or the associated copper atoms. Since the molecular weight of cytochrome oxidase is probably some four times greater than that of cyt. a_3, the true microscopic Hall mobility for cyt. a_3 can be estimated to be of the order of 80 $cm^2/V.sec$.

Dickerson *et al.*[18] have recently reported their results for the 2·8 Å structure determination of horse ferricytochrome *c*. One of the reasons for obtaining this structure was to aid the understanding of how electron transfer occurs into or out of cyt. *c*. The essential feature of the cyt. *c* structure is that the haem is tightly enveloped in hydrophobic groups, with two "channels" filled with hydrophobic side chains leading to the right and left from the haem to the surface of the molecule. It was observed that the aromatic groups have a remarkable tendency to occur in approximately parallel pairs and these authors speculated on the transfer of electrons via the overlap of aromatic π-electron bonds. Another possible electron pathway could involve residues 92 to 104 which form an α-helix. Three of the residues forming the α-helix (Leucine 94 and 98, Iso-leucine 95) have side chains which pack against the haem, whilst the two lysine[99,100] residues lie at the edge of the right "channel", which contains two parallel aromatic groups. It would be of interest to see if theoretical calculations applied to this combination of α-helix and overlapping π-electron bonds could indicate the existence of a well defined energy band for rapid electron transfer. If electron transport was confined to a similar pathway in cyt. a_3 (i.e. of the order 15% of the total molecule) then the true microscopic Hall mobility for cyt. a_3 would be one of the order 500 $cm^2/V.sec$.

It can be readily shown that the magnitude (S) of the oscillatory path described by the electrons under the influence of the microwave electric field is given by

$$S = \mu\hat{E}/2\pi f$$

where μ is the electron conductivity mobility, $\hat{E}$ the maximum electric field strength value in the sample and f is the microwave frequency. For our cavity and Klystron power source and assuming a specimen relative permittivity value of 2·5, then the value for $\hat{E}$ is of the order 11 V/cm. For a true microscopic Hall mobility of 500 $cm^2/V.sec$ (and assuming a 1:1 relationship between the Hall and conductivity mobilities), then for a frequence of 9 GHz, the electron pathway is of the order 9·7 Å, not an unreasonable value.

If the complete electron pathway within the cyt. a_3 molecule does involve a combination of α-helix and overlapping π-electron bonds, then studies of the variation of the Hall mobility with conformation changes should be of interest. Temperature variation measurements may reveal the activation energy required for the electron to traverse the "gap" between the π-bonds and the α-helix.

A Semiconductor Model for the Mitochondria

In order to construct a working hypothesis, we shall take the biological macromolecule as a potential energy box containing a full valence band of electrons and an empty conduction band separated by an energy gap $\Delta\epsilon$.[19,20] An electron in the conduction band is regarded as tunnelling through the potential energy barrier between one macromolecule and the next, to give the observed conductivity of films and crystals of proteins and other organic semiconductors, see Fig. 3. This model has been advanced on general physical grounds over many years to account qualitatively for a large volume of experimental data, and for details reference should be made to a review.[21] In the absence of information on the state of hydration of proteins in the mitochondrial membrane, we assume they are held in their lipoidal environment in an essentially dry state. We know that a very large number of proteins, including cyt. c in the very dry state have $\Delta\epsilon \simeq 2{\cdot}6$ eV and we accept the idea of energy bands, width about 0·2 eV, based on mobile π electrons in the

$C = O \ldots HN$ hydrogen bond system,[22] realizing that modern quantum calculations give $\Delta\epsilon$ values of around 5 eV[23] which, however, may be brought down to the experimental value if polarization terms be included.[24] It is possible in favourable circumstances that these hydrogen-bonds combine with the mobile π electrons of phenyl side chains to give a specially favourable pathway, as we have just considered for cyt. *c*. The above interpretation regards the dry protein as an essentially intrinsic semiconductor with a very high room temperature resistivity of around 10^{17} ohm cm. By contrast, cytochrome oxidase shows a higher conductivity, with $\Delta\epsilon = 2{\cdot}0$ eV at >50°C and

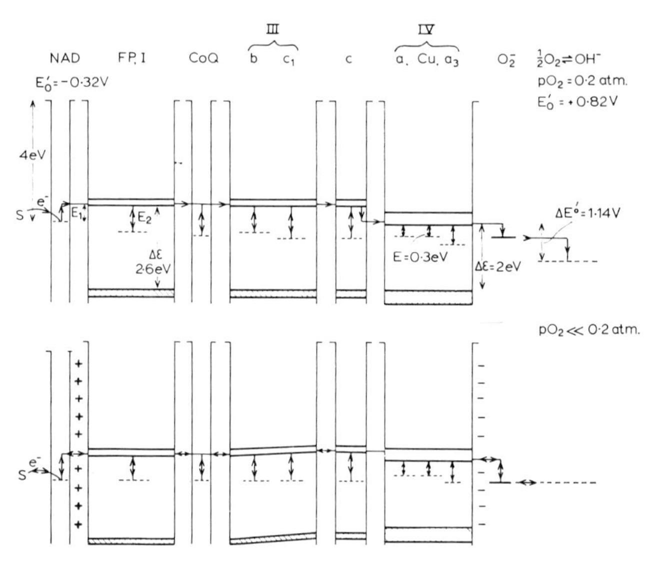

Figure 3. An electron potential energy box model for electron transport through the protein complexes of the mitochondrion. NAD nicotine adenine dinucleotide; FP I Flavoprotein Complex I; CoQ Coenzyme Q; III *b* c_1, Complex III, cytochromes *b* and c_1; C cytochrome *c*; IV *a*, Cu, a_3, Complex IV, cytochrome *a*, copper, cytochrome a_3. The zero of energy is the electron at infinity and the model is explained in the text.

$\Delta\epsilon = 0{\cdot}6$ eV ($E = 0{\cdot}3$ eV) below 50°C.[25] While the first value may be the intrinsic energy gap, the second value is almost certainly an impurity semiconduction based on the copper atoms in the molecule. To date, it is interesting that we have no d.c. dark conduction evidence for the haems acting as impurity levels, although we believe that infrared photoconduction we have detected in haemprotein systems may very well arise from excitation of trapped electrons on these prosthetic groups, so that dark conduction data may be forthcoming in future.[26]

For the present purpose, we shall neglect the succinate-Fraction II pathway to coenzyme Q and consider the electron transfer series below, the components following in physical juxtaposition the order of their redox potentials, values from Mahler and Cordes.[27] The abbreviations have their usual significance.

$$\begin{array}{lcccccccccc} & & & \text{I} & & & & \text{III} & & & \\ & \text{NAD} & \rightarrow & \text{FPN} & \rightarrow & \text{CoQ} & \rightarrow & (\text{cyt. } b)_2 \ldots \text{cyt. } c_1 & \rightarrow & \text{cyt. } c & \rightarrow \\ E^{0'}\ (\text{V}) & -0{\cdot}32 & & 0{\cdot}00 & & 0{\cdot}10 & & 0{\cdot}12 \qquad\quad 0{\cdot}21 & & 0{\cdot}23 & \end{array}$$

$$\begin{array}{ccccc} & & \text{IV} & & \\ \rightarrow & (\text{cyt. } a)_n\ \text{Cu}_m\ (\text{cyt. } a_3)_n\ \text{Cu}_n & & \rightarrow & \text{O}_2 \\ & 0{\cdot}29 \qquad\qquad\qquad\quad 0{\cdot}35 & & & 0{\cdot}82 \end{array}$$

More recent[28] relevant values of $E^{0'}$ are b 0·038, $c + c_1$ 0·227, a 0·205 and a_3 0·365 V. Erecinska *et al.*[29] give $E^{0'} = 0{\cdot}24$ V for the 830 nm copper signal in intact pigeon heart mitochondria, which is less than the 0·35 V quoted. In Fig. 3, we have arranged single-electron potential energy boxes representative of the different molecules from left to right, with dotted lines denoting the $E^{0'}$ levels, which represent the standard free energy of the electron on the prosthetic group, regarded as an impurity level below the conduction band of the protein concerned. The redox potential level may be regarded on this picture as a Fermi level between the actual impurity level and the bottom of the conduction band. The energy zero for the diagram is the electron at infinity, and the NAD redox level is fixed with respect to this by regarding it as equivalent to an electron affinity for the solvated molecule of 4·0 eV (cf. crystalline adenine 5·3 eV[30]). A value of 3·5 eV is ascribed to the electron affinity i.e., the bottom of the conduction band for the proteins, with the exception of cytochrome oxidase with its lower $\Delta\epsilon$ (2·0 as against 2·6 eV), where the valence band levels are matched. The band width for cytochrome oxidase is put at 0·4 eV, wider than the other proteins, on mobility grounds (see later). These small electron-transfer units may be regarded as fixed to the inner matrix side of the mitochondrial membrane, each unit held rigidly, apart from the mobile units coenzyme Q and cyt. c.[31] The linear dimension of the potential box is drawn to bear a rough relationship to the molecular weight of the molecule or molecular aggregate concerned. Each transfer unit will occupy about 50,000 Å^2 of membrane surface[32] and this gives about 3×10^4 transfer units on the cristae membrane of one liver cell mitochondrion, total area $16\mu^2$.[33] A later more detailed view[32] doubles the chain back across the *thickness* of the membrane, with cyt. a_3 on the matrix side leading to c on the "outer" or c side and back to the flavoprotein on the matrix side. This model is concerned with proton transfer across the membrane and is probably the best for further consideration from our viewpoint. However, at present we omit any attempt to link our model with the established models of proton transfer and phosphorylation[34] and concentrate on the electron path.

We suggest that the electrons travel mainly via the conduction bands, the redox levels acting as trapping states, as indicated by the arrows. An electron from substrate S is transferred to nicotine adenine dinucleotide (NAD) and hence via an activation energy of E_1 to the conduction band of the flavoprotein oxido-reductase, Complex I. Whether the measured activation energy of 14 Kcal mole^{-1} (0·6 eV)[35] refers to injection from substrate E_1, or excitation from the redox level E_2, is left open. The electron then descends down the redox level gradient, probably via conduction bands as shown, although tunnelling between impurity levels may need consideration (depending on the distance between neighbouring redox groups). Microwave Hall mobility values of 50 cm^2/Vs for the mitochondrion[36] and 25 cm^2/Vs for cytochrome oxidase (this paper) both reduceable by inhibitor concentration of KCN, identify cytochrome oxidase, complex IV as the particle of maximum electron mobility. This is emphasized by the correlation between the oxidase activity of cytochrome oxidase and its microwave Hall electron mobility.[14] For this reason we have ascribed it bands of 0·4 eV width, wider than the other proteins. However, even so, at the cyt. *c*–cytochrome oxidase interface it appears necessary for phonon emission to accompany electron tunnelling between conduction bands, a process known in solid state physics as inelastic phonon-assisted tunnelling.[37] This interface is one of the three mitochondrial sites for undergoing phosphorylation and it may be speculated that the emitted phonon (an energy quantum of lattice vibration) would provide the energy trigger for this process (cf. considerations by Cope and Straub[25]). It has been suggested that dark conduction in cytochrome oxidase and its enzymic activation energy for substrates, are equal to a Fermi level depth of 0·3 eV (energy level depth of 0·6 eV) below the conduction band.[38] There is very good evidence that a possibly similar reaction, the cyt. *c*-bacterio chlorophyll photoactivated reaction occurs by tunnelling with zero or very low activation energy below 50°K.[39] The O_2^- radical is a logical chemisorbed intermediate and this is not ruled out by biochemical evidence.[40] We have placed the O_2^- energy level at about 4·5 eV, by comparison with its energy level adsorbed on inorganic oxides like ZnO.[41] The final state, of course, is OH^-, the total drop in standard redox potential from NAD to OH^- being 1·14 V.

The dc drift mobility of 2×10^{-2} cm^2/Vs[42] in mitochondria on this view corresponds to the tunnelling probability between protein molecules.[20] The microwave Hall mobility of 25–50 cm^2/Vs already mentioned, on the other hand, refers to the to-and-fro motion of the electron over about 10 Å at 9·2 GHz, along the path of highest mobility, identified in this and the preceding paper[15] as cytochrome oxidase.

If we de-energize the mitochondria by reduction of the oxygen pressure, the equilibrium redox potential will rise (i.e. tend from positive to zero values) as shown in the bottom part of Fig. 3. In other words, as electrons pass down the chain, they will accumulate on the active centre of cytochrome oxidase, giving rise to a *change* in the electrostatic potential difference across the transfer unit and causing all the energy levels to rise. Where the redox levels throughout are equal we shall have equal numbers of electrons passing in both directions and an electrostatic potential of 1·14 V across the transfer unit. However, it might be more reasonable to allow for a change of 0·1 V in redox potential. Allowing an overall length per unit of 500 Å, so there will be a voltage drop of 2×10^4 V/cm, which should exert a marked piezoelectric strain (contraction) on a "soft" biopolymer system. The changes which result will depend on the geometrical relationship between transfer unit and membrane, but these could be quite large for certain geometries which might favour the "bimetallic strip" type of behaviour. Large conformational changes are, of course, observed in the membrane in its different states.[43]

We have made no attempt to speculate on the states of hydration of the proteins, although these can profoundly increase conductivity by introducing donor impurity levels on one view,[44] or changing the microscopic dielectric constant and contracting the energy gap between conduction and valence bands on another.[24] It is possible also that we have neglected a charge transfer role of coenzyme Q, which has been found to induce conductivity in cyt. *c* by a charge-transfer interaction (quinones act as electron acceptors with proteins[44]). In an early publication, Cardew and Eley[45] pointed out that a redox driving potential of 1 V could only drive a current through a dry protein fibre 1 micron long × 50 Å diameter (a primitive mitochondrial model) 10^{16} times smaller than the electron transfer rate due to respiration. Furthermore, the necessary activation energy of $\frac{1}{2}\Delta\epsilon$ i.e. 1·3 eV was much greater than the 0·48 eV usually observed for the overall process. Rosenberg and Postow[24] more recently reconciled the magnitudes by allowing for hydration of the protein. In the present model we are concerned essentially with electrons injected from the substrate over a barrier of 0·5 eV into the conduction band, so there should be no problem even with the dry protein, although we have not attempted as yet detailed calculations. The misleading concept in our past calculation was to base our calculations on electrons excited over $\Delta\epsilon = 2{\cdot}6$ eV from the valence band. We also note that Cope and Straub[25,38] have reconciled the oxygen kinetics on the basis of a slow process of electron transfer from substrate through cytochrome oxidase.

In conclusion, we claim no more than that there is a logical case for considering membrane bound electron transfer enzymes from the viewpoint of organic semiconductors, and that the present model serves to reconcile the scanty evidence at present available. We hope that it will serve to stimulate and guide further experimental work.

Acknowledgements

We are glad to acknowledge the award of an I.C.I. Fellowship to R.P., the financial support from the Science Research Council, and the valuable work of Messrs W. E. Porter, R. E. Parsons and A. G. Hands on the microwave Hall equipment.

References

1. B. Chance and E. L. Spencer, *Disc. Faraday Soc.*, **27** (1959) 200.
2. M. Klingenberg, in: *Biological Oxidations*, by T. Singer (ed.), Wiley, New York, 1968, p. 19.
3. D. D. Eley and R. Pethig, *J. Bioenergetics*, **2** (1971) 39.
4. E. M. Trukhan, *Pribory tekhm. eskper*, **4** (1965) 198.
5. D. D. Eley and R. Pethig, *Disc. Faraday Soc.*, **51** (1971) 164.
6. A. L. Smith, *Methods in Enzymology*, Vol. X, R. W. Estabrook, and M. E. Pullman (eds.), Academic Press, New York and London, 1967, p. 81.
7. Y. Hatefi and J. S. Rieske, *Methods in Enzymology*, Vol. X, R. W. Estabrook and M. E. Pullman (eds.), Academic Press, New York and London, 1967, p. 225.
8. D. Ziegler and J. S. Rieske, *Methods in Enzymology*, Vol. X, R. W. Estabrook and M. E. Pullman (eds.), Academic Press, New York and London 1967, p. 231.
9. D. E. Griffiths and D. C. Wharton, *J. Biol. Chem.*, **236** (1961) 1850.
10. D. C. Wharton and A. Tzagoloff, *Methods in Enzymology*, Vol. X, R. W. Estabrook and M. E. Pullman (eds.), Academic Press, New York and London (1967) p. 245.
11. T. E. King, *Methods in Enzymology*, Vol. X, R. W. Estabrook and M. E. Pullman (eds.), Academic Press, New York and London, 1967, p. 216.
12. O. H. Lowry, N. J. Rosebrough, A. L. Farr and R. J. Randall, *J. Biol. Chem.*, **193** (1951) 265.
13. D. Keilin and E. F. Hartree, *Proc. Roy. Soc., London, B*, **127** (1939) 167.
14. B. F. Van Gelder and A. O. Muijers, *Biochim. Biophys. Acta*, **81** (1964) 405.
15. D. D. Eley, R. J. Mayer and R. Pethig, *J. Bioenergetics*, **3** (1972) 271.
16. A. O. Muijers, R. H. Tiesjema and B. F. Van Gelder, *Biochem. Biophys. Acta*, **234** (1971) 468.
17. D. W. Urry and P. Daty, *Amer. Chem. Soc.*, **87** (1965) 2756.
18. R. E. Dickerson, T. Takano, D. Eisenberg, O. B. Kallai, L. Samson, A. Cooper and E. Margoliash, *J. Biol. Chem.*, **246** (1971) 1511.
19. D. D. Eley and D. I. Spivey, *Trans. Faraday Soc.*, **56** (1960) 1432.
20. D. D. Eley and M. R. Willis, in: *Symposium on Electrical Conductivity in Organic Solids*, H. Kallman and M. Silver (eds.), Wiley (Interscience), New York, 1961, p. 257.
21. D. D. Eley, in: *Organic Semiconducting Polymers*, J. E. Katon (ed.), Arnold, London, Marcel Dekker, Inc., New York, 1968, p. 259.
22. M. G. Evans and J. Gergely, *Biochim. Biophys. Acta*, **3** (1949) 188.
23. M. Suard, G. Berthier and B. Pullman, *Biochim. Biophys. Acta*, **52** (1961) 254.
24. B. Rosenberg and E. Postow, *Ann. N.Y. Acad. Sci.*, **158** (1969) 161.
25. F. W. Cope and K. D. Straub, *Bull. Math. Biophys.*, **31** (1969) 761.
26. D. D. Eley and E. Metcalfe, *Nature*, in press.
27. H. R. Mahler and E. H. Cordes, *Biological Chemistry*, Harper and Row, New York, 1966, pp. 568, 600.
28. P. L. Dutton, D. F. Wilson and Chuan-Pu Lee, *Biochemistry*, **9** (1970) 5077.
29. M. Erecinska, B. Chance and D. F. Wilson, *FEBS Letters*, **16** (1971) 284.
30. F. Gutmann and L. E. Lyons, *Organic Semiconductors*, Wiley, New York, 1967, p. 704.
31. M. Klingenberg, in *Biological Oxidations*, T. P. Singer (ed.), Wiley (Interscience), New York, 1968, p. 16.
32. A. L. Lehninger, *The Mitochondrion*, W. A. Benjamin, New York, 1965, p. 30.
33. E. Racker, *Essays in Biochemistry*, **6** (1970) 1.
34. E. C. Slater, *Quarterly Rev. of Biophysics*, **4** (1971) 35.
35. A. Gierer, *Biochem. Biophys. Acta*, **17** (1955) 111.
36. D. D. Eley and R. Pethig, *J. Bioenergetics*, **2** (1971) 39.
37. C. B. Duke, *Tunnelling in Solids*, Solid State Physics Supplement 10, Academic Press, New York, 1969, p. 115.
38. F. W. Cope, *Bull. Math. Biophysics*, **33** (1967) 642.

39. D. De Vault, J. H. Parkes and B. Chance, *Nature*, **215** (1967) 642.
40. Q. H. Gibson, in: *Biological Oxidations*, T. P. Singer (ed.), Wiley (Interscience), New York, 1968, p. 403.
41. A. Terenin and I. Akimov, *Zeit. Physikal Chem.*, **217** (1961) 307.
42. A. V. Vannikov and L. I. Boguslavskii, *Biofizika*, **14** (1969) 421.
43. E. F. Korman, A. D. F. Addink, T. Wakabayashi and D. E. Green, *J. Bioenergetics*, **1**, (1970) 9.
44. K. M. C. Davis, D. D. Eley and R. S. Snart, *Nature*, **188** (1960) 724.
45. M. H. Cardew and D. D. Eley, *Disc. Faraday Soc.*, **27** (1959) 115.

Possible Mechanisms for the Linkage of Membrane Potentials to Metabolism by Electrogenic Transport Processes with Special Reference to *Ascaris* Muscle

P. C. Caldwell

Department of Zoology, University of Bristol, Bristol, U.K.

Introduction

Although electrogenic transport processes in cell membranes are being increasingly invoked to account for certain membrane phenomena (see for example Kernan, 1962; Cross *et al.*, 1965; Kerkut and Thomas, 1965; Moreton, 1969; Brading and Caldwell, 1971; Marmor, 1971) there has been comparatively little discussion of likely mechanisms and of the ways in which these might be linked to metabolism and affect the membrane potential.

A recent example for which electrogenic processes have been postulated is the highly abnormal resting membrane potential of the muscle cells of the pig roundworm *Ascaris lumbricoides* (Brading and Caldwell, 1971). Under normal conditions this potential is remarkably insensitive to changes in the concentrations of extracellular ions (Brading and Caldwell, 1964, 1971; del Castillo *et al.*, 1964) and in many situations the membrane potential, V, is given by the following form of the constant field equation

$$V = \frac{RT}{F} \ln \frac{P_K K_o + P_{Na} Na_o + P_{Cl} Cl_i + x}{P_K K_i + P_{Na} Na_i + P_{Cl} Cl_o + y} \qquad (1)$$

where $x = 290$ and $y = 1300$ if the permeability constant for potassium, $P_K = 1$ and if the ionic concentrations K_o etc. are expressed in mM. The sign of V is that observed on entering the cell. The terms x and y in (1) are large compared with the other terms and if they arise from electrogenic processes it would seem that the *Ascaris* muscle membrane potential is predominantly determined by these processes.

Moreton (1969) and Marmor (1971) have considered the situation in which an electrogenic sodium pump contributes a constant current I to the membrane currents, this current being independent of membrane potential. This leads to the following form of equation (1)

$$V = \frac{RT}{F} \ln \frac{P_K K_o + P_{Na} Na_o + P_{Cl} Cl_i + IRT/VF^2}{P_K K_i + P_{Na} Na_i + P_{Cl} Cl_o + IRT/VF^2} \qquad (2)$$

Although Marmor (1971) has obtained evidence that (2) holds for *Anisodoris* giant neurones, with I tending to remain constant and to be independent of membrane potential, (2) does not appear to hold for *Ascaris* muscle fibres. If (2) held for *Ascaris* muscle fibres, then x and y in (1) should be the same, which they are not, and they should be dependent on V.

In an attempt to account for equation (1) Brading and Caldwell (1971) took into consideration the electrogenic sodium pump mechanism discussed by Cross *et al.* (1965) shown in Fig. 1, and suggested that

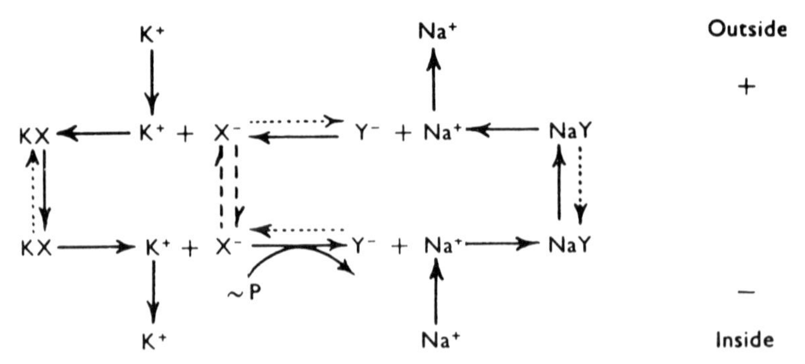

Figure 1. The scheme for an electrogenic sodium pump mechanism proposed by Cross, Keynes and Rybová (1965). Y^- is the sodium carrier, and it is only able to cross the membrane in the form of the complex NaY. X^- is the potassium carrier and it can cross the membrane either on its own or else as the complex KX.

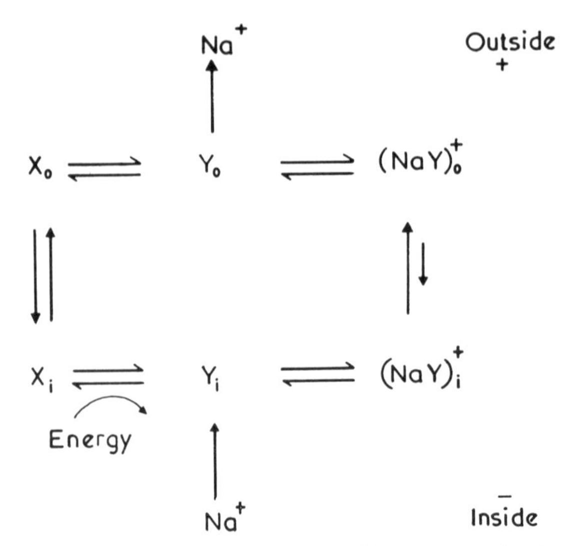

Figure 2. The scheme for the simple electrogenic sodium pump mechanism discussed in the text. Y is the sodium carrier which is uncharged and can only cross the membrane as the complex $(NaY)^+$. X is another form of the carrier which is uncharged, can cross the membrane by itself and does not form a complex with sodium or other ions. The subscripts refer to the amounts of X, Y and $(NaY)^+$ on the inside and outside of the membrane.

the movements of the negatively charged ion-free carrier X^- would contribute a current to the membrane currents to give the following form of the constant field equation

$$V = \frac{RT}{F} \ln \frac{P_K K_o + P_{Na} Na_o + P_{Cl} Cl_i + P_x X_i}{P_K K_i + P_{Na} Na_i + P_{Cl} Cl_o + P_x X_o} \quad (3)$$

where P_x is a permeability constant for X^- and X_i and X_o are its concentrations at the inner and outer surfaces of the membrane. If $P_x X_i$ and $P_x X_o$ were larger than the other terms and X_i and X_o were near to electrochemical equilibrium, then the values of x and y in equation (1) could be explained.

The electrogenic sodium pump model shown in Fig. 1 is not the simplest available and attempts to express X_i and X_o in terms of other quantities lead to quite complex expressions, even if considerable simplifications are introduced. The simpler model shown in Fig. 2 is easier to deal with and it will therefore form the basis of the discussion which follows.

Membrane with Potential Determined Solely by a Simple Electrogenic Sodium Pump with a Neutral Carrier

Figure 2 illustrates one of the simplest possible mechanisms for an electrogenic sodium pump. X and Y are neutral, X being able to cross the membrane on its own while Y can only cross in company with sodium. The sodium transport and transfer of charge are accomplished by the outward movement of the sodium:carrier complex $(NaY)^+$ across the membrane. The system can be kept in operation either by displacement of the concentrations of X and Y from equilibrium on the inside of the membrane as illustrated or by displacement on the outside. The displacement from equilibrium is brought about by a suitable coupling with energy yielding metabolic processes such as the splitting of ATP.

It is likely that all but one of the stages shown in Fig. 2 will operate near thermodynamic equilibrium (c.p. Caldwell, 1969), the exception being the movement of the complex $(NaY)^+$ across the membrane. X will move very rapidly compared with $(NaY)^+$ so its internal and external concentrations in the membrane will be almost equal.

Therefore

$$X_o = X_i = \alpha Y_o = \beta Y_i \quad (4)$$

where X_o, X_i, Y_o, Y_i are external and internal concentrations of free X and Y in the membrane, and α and β are constants relating these concentrations. If the amount of $(NaY)^+$ in the membrane is small in relation to the amount of free carrier, i.e., if Y only reacts weakly with Na, then the total carrier ϕ is given approximately by

$$\phi = (X_i + X_o + Y_i + Y_o) \quad (5)$$

Then

$$Y_i = \frac{\alpha\phi}{(2\alpha\beta + \alpha + \beta)} \tag{6}$$

$$Y_o = \frac{\beta\phi}{(2\alpha\beta + \alpha + \beta)} \tag{7}$$

Since only a small proportion of the carrier Y is assumed to be in combination with sodium the concentrations of $(NaY)^+$ at the inner and outer surfaces of the membrane will be given by the equations

$$(NaY^+)_i = \gamma Na_i Y_i \tag{8}$$

$$(NaY)^+_o = \gamma Na_o Y_o \tag{9}$$

where γ is the product of the partition coefficient of sodium between the aqueous and membrane phases and the association constant for the interaction of the free membrane sodium and Y.

The current carried into the cell by $(NaY)^+$, $I_{(NaY)}$, is given by the following equation (see for example Goldman, 1943; Hodgkin and Katz, 1949) if the sign of the membrane potential V is that observed on entering the cell.

$$I_{(NaY)} = \frac{u_{(NaY)}FV}{a}\left[\frac{(NaY)^+_o - (NaY)^+_i e^{FV/RT}}{e^{FV/RT} - 1}\right]$$

$$= P_{(NaY)}\frac{F^2 V}{RT}\left[\frac{Na_o Y_o - Na_i Y_i e^{FV/RT}}{e^{FV/RT} - 1}\right] \tag{10}$$

where $u_{(NaY)}$ is the mobility of $(NaY)^+$ in the membrane, a is the membrane thickness. $P_{(NaY)}$ is a constant which is defined as being equal to $u_{(NaY)}\gamma RT/aF$. It is not strictly comparable with the permeability constants P_K, P_{Na} and P_{Cl} which are introduced later.

If a membrane contained the electrogenic mechanism shown in Fig. 2 but had no passive permeability to ions, then charge would only be carried by $(NaY)^+$. In the absence of an applied potential the membrane would reach a stable potential when $I_{(NaY)} = 0$. When $I_{(NaY)} = 0$, $Na_o Y_o - Na_i Y_i e^{FV/RT} = 0$, and, from equations (6) and (7), the stable potential, V, is given by

$$V = \frac{RT}{F}\ln\frac{Na_o Y_o}{Na_i Y_i} = \frac{RT}{F}\ln\frac{\beta Na_o}{\alpha Na_i} \tag{11}$$

Equation (11) shows the way in which the membrane potential would be determined if the simple type of electrogenic mechanism shown in Fig. 2 was the only factor involved.

The potential is determined by the concentrations of sodium ion on the two sides of the membrane and by the values of α and β. If no energy is supplied to the mechanism then $\alpha = \beta$ and (11) reduces to the simple

Nernst equation for a membrane permeable to sodium. In these circumstances the electrogenic mechanism becomes equivalent to a channel permeable to sodium. If on the other hand energy is fed into the mechanism so that either α is changed relative to β or β relative to α, then (11) shows that the membrane potential depends on β/α as well as on Na_o/Na_i. In principle any value of the membrane potential can be obtained providing a suitable coupling of metabolism to the mechanism can produce the appropriate value of β/α.

Membrane with Potential Determined by a Simple Electrogenic Sodium Pump with Neutral Carriers and by the Passive Movement of Potassium, Sodium and Chloride

The mechanism to be considered is the same as that in Fig. 2 with the addition of passive movements of potassium, sodium and chloride. The inward currents carried by potassium, sodium and chloride, if the sign of V is that observed on entering the cell, are given by the following equations (see for example Hodgkin and Katz, 1949)

$$I_K = P_K \frac{F^2 V}{RT}\left[\frac{K_o - K_i e^{FV/RT}}{e^{FV/RT} - 1}\right] \tag{12}$$

$$I_{Na} = P_{Na} \frac{F^2 V}{RT}\left[\frac{Na_o - Na_i e^{FV/RT}}{e^{FV/RT} - 1}\right] \tag{13}$$

$$I_{Cl} = P_{Cl} \frac{F^2 V}{RT}\left[\frac{Cl_i - Cl_o e^{FV/RT}}{e^{FV/RT} - 1}\right] \tag{14}$$

A stable potential is reached when the total current across the membrane, $I_{NaY} + I_K + I_{Na} + I_{Cl}$, is zero so that, from 10, 12, 13 and 14

$$V = \frac{RT}{F} \ln \frac{P_K K_o + P_{Na} Na_o + P_{Cl} Cl_i + P_{NaY} Na_o Y_o}{P_K K_i + P_{Na} Na_i + P_{Cl} Cl_o + P_{NaY} Na_i Y_i} \tag{15}$$

From 6, 7 and 15

$$V = \frac{RT}{F} \ln \frac{P_K K_o + P_{Na} Na_o + P_{Cl} Cl_i + P_{NaY} Na_o \beta\phi/(2\alpha\beta + \alpha + \beta)}{P_K K_i + P_{Na} Na_i + P_{Cl} Cl_o + P_{NaY} Na_i \alpha\phi/(2\alpha\beta + \alpha + \beta)} \tag{16}$$

The effect of the electrogenic component in equation (16) depends on the size of the terms $P_{NaY} \beta\phi/(2\alpha\beta + \alpha + \beta)$ and $P_{NaY} \alpha\phi/(2\alpha\beta + \alpha + \beta)$ in relation to P_K, P_{Na} and P_{Cl}. If these terms are a good deal larger than P_K, P_{Na} and P_{Cl} then (16) reduces to (11). If they are a good deal smaller then (16) reduces to the conventional constant field equation.

The term $P_{NaY} Na_i \alpha\phi/(2\alpha\beta + \alpha + \beta)$ will tend to remain fairly constant and, if $P_{NaY} \alpha\phi/(2\alpha\beta + \alpha + \beta)$ is large, it could contribute the large constant term y needed in equation (1) to account for the resting membrane potential of *Ascaris* muscle. The term $P_{NaY}Na_o\beta\phi/(2\alpha\beta + \alpha + \beta)$ will vary with Na_o, although it can be a good deal smaller than $P_{NaY}Na_i\alpha\phi/(2\alpha\beta + \alpha + \beta)$ if $\beta \ll \alpha$. If β is very small then $P_{NaY}Na_o\beta\phi/(2\alpha\beta + \alpha + \beta)$ will be much smaller than $P_{Na}Na_o$ and can be neglected. If it is larger then the effective permeability constant for external sodium ions will be increased to $(P_{Na} + P_{NaY}\beta\phi/(2\alpha\beta + \alpha + \beta))$. It seems possible that in *Ascaris* muscle $P_{NaY}\beta\phi/(2\alpha\beta + \alpha + \beta) \approx 10P_{Na}$ since treatment with γ amino-butyric acid, which seems to inactivate the electrogenic mechanisms, drastically reduces the values of x and y needed in equation (1) and simultaneously reduces the value needed for P_{Na} from 1 to 0·1.

Equation (16) can therefore account for the large constant y in equation (1) and the variable value of P_{Na}. It does not account however for the smaller constant x. For this the independent electrogenic transport of an anion, possibly chloride, must be postulated.

Membrane with Potential Determined by Simple Electrogenic Sodium and Chloride Pumps with Neutral Carriers and Passive Diffusion of Potassium, Sodium and Chloride

Suppose that chloride is transported as well as sodium but by an independent electrogenic mechanism, analogous to that in Fig. 2, with chloride replacing sodium. If carriers X^1 and Y^1 are involved, if the constant total amount of carrier is θ and if $X_o^1 = X_i^1 = AY_o^1 = BY_i^1$, then

$$Y_o^1 = B\theta/(2AB + A + B) \tag{17}$$

$$Y_i^1 = A\theta/(2AB + A + B) \tag{18}$$

The inward current $I_{(ClY^1)}$ carried by the mechanism will be given by

$$I_{(ClY^1)} = P_{(ClY^1)}\frac{F^2 V}{RT}\frac{Cl_i Y_i^1 - Cl_o Y_o^1 e^{FV/RT}}{e^{FV/RT} - 1} \tag{19}$$

where $P_{(ClY^1)}$ is a constant, the definition of which is similar to that for $P_{(NaY)}$. If the current given by (19) is added to the currents given by (10), (12), (13) and (14) the following expression is obtained

$$V = \frac{RT}{F}\ln\frac{P_K K_o + P_{Na}Na_o + P_{Cl}Cl_i + P_{NaY}Na_o\beta\phi/(2\alpha\beta + \alpha + \beta) + P_{ClY^1}Cl_i A\theta/(2AB + A + B)}{P_K K_i + P_{Na}Na_i + P_{Cl}Cl_o + P_{NaY}Na_i\alpha\phi/(2\alpha\beta + \alpha + \beta) + P_{ClY^1}Cl_o B\theta/(2AB + A + B)} \tag{20}$$

If $P_{ClY^1}B\theta/(2AB + A + B)$ is small compared with P_{Cl} so that $P_{ClY^1}Cl_oB\theta/(2AB + A + B)$ can be neglected, (20) can be used to explain many of the features of the normal resting membrane potential of *Ascaris* muscle fibres.

$P_{\mathrm{Na\,Y}}\mathrm{Na_i}\alpha\phi/(2\alpha\beta+\alpha+\beta)$ and $P_{\mathrm{Cl\,Y^1}}\mathrm{Cl_i}\mathrm{A}\theta/(2\mathrm{AB}+\mathrm{A}+\mathrm{B})$ will tend to remain fairly constant and since they can have different values they can be made equivalent to y and x in equation (1). Further, the term $P_{\mathrm{Na\,Y}}\mathrm{Na_o}\beta\phi/(2\alpha\beta+\alpha+\beta)$ can contribute to the effects of external sodium and its reduction, due to inactivation of the electrogenic mechanisms, can be postulated to explain the reduction in the effective value of P_{Na} observed in the presence of γ amino-butyric acid.

Membrane with an Electrogenic Sodium Pump with Negatively Charged Carriers and Passive Diffusion of Potassium, Sodium and Chloride

The electrogenic transport models which have been discussed so far can account quite well for the behaviour of the membrane potential of *Ascaris* muscle fibres under various conditions. They do not however lead to a condition under which the current contributed by the electrogenic sodium pump is constant and independent of membrane potential. As was mentioned earlier this situation has been discussed in connection with *Helix* neurones by Moreton (1969) and experimental evidence for it has been obtained for *Anisodoris* neurones by Marmor (1971).

An alternative approach is to use the model in Fig. 2 but to assume that X and Y carry single negative charges as they do in Fig. 1. If the movement of negatively charged X is assumed to be very rapid so that it can carry a current equivalent to the net outward movement of sodium but remain at approximately equilibrium concentrations so that $X_i = X_o e^{FV/RT}$ then equation (4) becomes

$$X_i = X_o e^{FV/RT} = \beta Y_i = \alpha Y_o e^{FV/RT} \tag{21}$$

From (5) and (21) it can be shown that

$$Y_i = \frac{\alpha\phi\, e^{FV/RT}}{(\beta+\alpha\beta+\alpha\beta\, e^{FV/RT}+\alpha\, e^{FV/RT})} \tag{22}$$

$$Y_o = \frac{\beta\phi}{(\beta+\alpha\beta+\alpha\beta\, e^{FV/RT}+\alpha\, e^{FV/RT})} \tag{23}$$

Since the current, I_x, carried by X is equivalent to the net movement of sodium

$$I_x = \mathrm{KNa_i}\, Y_i - \mathrm{KNa_o}\, Y_o \tag{24}$$

where K is a constant and Y is far from saturation.

I_x as defined by (24) is normally an outward current. If the value of I_x given by (24) is subtracted from the potassium, sodium and chloride currents given by (12), (13) and (14), then

$$V = \frac{RT}{F}\ln\frac{P_{\mathrm{K}}\mathrm{K_o}+P_{\mathrm{Na}}\mathrm{Na_o}+P_{\mathrm{Cl}}\mathrm{Cl_i}+\mathrm{K}(\mathrm{Na_i}\,Y_i-\mathrm{Na_o}\,Y_o)\,RT/VF^2}{P_{\mathrm{K}}\mathrm{K_i}+P_{\mathrm{Na}}\mathrm{Na_i}+P_{\mathrm{Cl}}\mathrm{Cl_o}+\mathrm{K}(\mathrm{Na_i}\,Y_i-\mathrm{Na_o}\,Y_o)\,RT/VF^2} \tag{25}$$

The values of Y_i and Y_o given by equations (22) and (23) can be substituted in (25) to give a rather complex expression for V.

A simplification can be introduced if both $e^{FV/RT}/\beta$ and α are very large and β is small so that $Y_i \approx \phi$.

(25) then becomes

$$V = \frac{RT}{F} \ln \frac{P_K K_o + P_{Na} Na_o + P_{Cl} Cl_i + KNa_i \phi RT/VF^2}{P_K K_i + P_{Na} Na_i + P_{Cl} Cl_o + KNa_i \phi RT/VF^2} \quad (26)$$

Since ϕ is constant, $KNa_i\phi$ will tend to remain constant and equal to the roughly constant outward current, independent of membrane potential, which the electrogenic mechanism will now generate. (26) is in fact equivalent to (2), which is Moreton's equation, and implies the condition of a constant outward current independent of membrane potential found by Marmor (1971) in *Anisodoris* neurones.

Conclusion

The foregoing discussion shows that certain of the features of membrane systems in which electrogenic processes may operate can be described by quite simple models if a few reasonable assumptions are made. Since the model based on neutral carriers seems to explain the behaviour of *Ascaris* muscle fibre membranes, whereas the model based on negative carriers may be more appropriate for *Helix* and *Anisodoris* neurones, it seems that more than one type of electrogenic transport mechanism may exist. The need to postulate an electrogenic chloride pump in the case of *Ascaris* muscle also raises the possibility that ions other than sodium may be subject to electrogenic transport.

From the point of view of bioenergetics, one of the most interesting points which has been raised is the possibility that the membrane potential could be linked directly and rapidly to the free energy made available from metabolism through an electrogenic sodium pump. Equation (11) shows that if the membrane potential is generated almost exclusively by the type of mechanism shown in Fig. 2, then it is in principle possible to obtain a membrane potential of any value. This could be accomplished by a suitable gearing of the electrogenic pump mechanism to free energy yielded by metabolism to produce the required value of β/α. A case of particular interest would arise if $Na_o/Na_i \approx 10$ and $\beta/\alpha \approx 0\ 01$. Equation (11) shows that under these circumstances the resting membrane potential would be about -58 mV even though it arose from a mechanism based on the movements of sodium rather than potassium. If the supply of free energy to the mechanism was momentarily cut off, the value of β/α would change to 1·0 and the membrane potential would change to that given by the Nernst equation for sodium ions, namely about $+58$ mV. The electrogenic sodium pump mechanism shown in Fig. 2 could therefore give rise to membrane potentials corresponding to normal resting and action potentials without any contribution by other ions to the membrane currents.

References

Brading, A. F. and Caldwell, P. C., *J. Physiol.*, **173** (1964) 36P.
Brading, A. F. and Caldwell, P. C., *J. Physiol.*, **217** (1971) 605.
Caldwell, P. C., *Current Topics in Bioenergetics*, **3** (1969) 251.
Castillo, J. del, de Mello, W. C. and Morales, T., *J. gen. Physiol.* **48** (1964) 129.
Cross, S. B., Keynes, R. D. and Rybová, R. (1965). *J. Physiol.*, **181** (1965) 865.
Goldman, D. E., *J. gen. Physiol.*, **27** (1943) 37.
Hodgkin, A. L. and Katz, B., *J. Physiol.*, **108** (1949) 37.
Kerkut, G. A. and Thomas, R. C., *Comp. Biochem. Physiol.*, **14** (1965) 167.
Kernan, R. P., *Nature*, **193** (1962) 986.
Marmor, M. F., *J. Physiol.*, **218** (1971) 599.
Moreton, R. B. *J. exp. Biol.*, **51** (1969) 181.

The Effect of Redox Potential on the Coupling Between Rapid Hydrogen-Ion Binding and Electron Transport in Chromatophores from *Rhodopseudomonas Spheroides*

Richard J. Cogdell, J. Barry Jackson and Antony R. Crofts

Department of Biochemistry, The Medical School, University of Bristol, Bristol, BS8 1TD, England

Introduction

The light-driven hydrogen-ion pump of chromatophores from photosynthetic bacteria is related to the processes of electron transport and the phosphorylation of ADP.[1,2,3] Using pH indicator techniques[4,5,6] a rapid component of H^+-binding, whose kinetic parameters lie within the time range of the chromatophore electron transport reactions[7] has been detected. A study of the effects of inhibitors, uncoupling agents and ion-transporting antibiotics on this phenomenon has led to the suggestion that hydrogen-ions bind at two distinct sites in the electron transport chain.[6] Two models have been proposed to explain rapid H^+-binding.

(i) The H^+ disappearing is directly involved in the reduction of a H-accepting redox component; the H^+ carrier is envisaged as one of a series of alternating electron-carriers and hydrogen carriers, which are arranged anisotropically within the membrane such that the transport of reducing equivalents is coupled to H^+-binding on the outside and H^+-release on the inside of the chromatophore vesicle.[1,6,8]

(ii) Initiation of coupled electron transport converts the chromatophore membrane to an activated state that is necessary but not sufficient for energy coupling and which is characterized by enhanced proton binding; the "membrane Bohr effect".[5]

Recently, the thermodynamic properties of the cytochromes of *Rhodopseudomonas spheroides* chromatophores have been characterized by redox titration,[7] and the dependence of flash-induced electron flow and carotenoid spectral shift on the ambient redox potential of the chromatophore suspension have been investigated.[9] It was therefore

Non-Standard Abbreviations: FCCP—carbonyl cyanide *p*-trifluoromethoxyphenylhydrazone; BChl—bacteriochlorophyll; BCP—bromocresol purple.

of some importance to examine the redox potential dependence of the rapid H^+-binding reactions. The results of this investigation, which are presented in this communication, give further support to the chemiosmotic mechanism of proton translocation but are not easily reconciled with a "membrane Bohr effect."

Methods

Batch cultures of *Rps. spheroides*, strain Ga (the "green" mutant) were grown anaerobically in the light as previously described.[1] Chromatophores were prepared with a French Press. The chromatophores were washed and resuspended in a medium containing 100 mM choline chloride and 1 mM N-morpholino ethanesulphonic acid at pH 6·5. This low buffer medium was used throughout the preparation which was otherwise similar to the standard procedures employed in this laboratory.[1]

The colour changes of the pH indicator were measured with a sensitive and rapidly responding single beam spectrophotometer which has been described elsewhere.[10] Actinic illumination was provided by either a 20 ns Q-switched ruby laser pulse or a 200 μs xenon flash. A redox titration vessel, similar to that designed by Dutton[11] was positioned in the spectrophotometer in a cell housing modified to permit stirring by a magnetic flea. The following dyes were employed to facilitate redox equilibration between the electron transport carriers of the chromatophores and the platinum electrode; potassium ferri/ferrocyanide ($E_{m\,7\cdot0} = +430$ mV), diaminodurol ($E_{m\,7\cdot0} = +240$ mV), phenazine methosulphate ($E_{m\,7\cdot0} = +80$ mV), phenazine ethosulphate ($E_{m\,7\cdot0} = +55$ mV), pyocyanine ($E_{m\,7\cdot0} = -34$ mV). The oxidation/reduction potential of the suspension was controlled by injecting small volumes of freshly prepared potassium ferricyanide and sodium dithionite solutions through the rubber septum of the side arm of the vessel.

Small changes in pH of the suspension resulting from additions, were corrected with dilute HCl or KOH, thus keeping the total transmittance of the sample, at 583 nm, constant. In this way it was possible to maintain the pH of the chromatophore suspension within ±0·02 units of the desired value. Control experiments showed that transmittance changes at 583 nm resulting from the direct effects of pH or redox potential on the chromatophores and redox dyes were negligible. Repeated additions of sodium dithionite and KOH solutions produced a gradual increase in the buffering capacity of the chromatophore suspension. For this reason we divided each titration, as shown in Figs. 3A and 3B, into three parts and, by making use of the chromatophore "endogenous reductant",[7] avoided as far as possible the use of dithionite. Calibrations of the phenol red change with dilute HCl and

KOH before and after each experiment ensured that the buffering capacity did not change significantly during the course of the titration.

Choice of pH Indicator

In contrast to bromothymol blue[12] bromocresol purple remains largely unbound by the chromatophore membrane and accurately reflects the pH of the extra-vesicular medium, at least during the first second of illumination.[4] In preliminary experiments (with P. L. Dutton and B. Chance) it was found that bromocresol purple was irreversibly destroyed by exposure to low redox potentials. We have therefore screened a number of other indicators for their accuracy and reliability under a variety of experimental conditions. The screening procedures have been described in detail elsewhere[4] for bromothymol blue and bromocresol purple. On the basis of the experiments summarized below, phenol red (phenol-sulphonphthalein) was selected as the most suitable pH indicator for the present series of experiments.

When a chromatophore suspension was treated with between 1–100 μM phenol red, 95–97% of the indicator was recovered in the supernatant after complete centrifugation of the particles. Lack of appreciable binding of the indicator was also suggested by the failure of the chromatophores to shift the pK_a from its usual value of 7·8. Illumination of an unbuffered suspension of *Rps. spheroides* chromatophores containing 30–100 μM phenol red produced a colour change at 587–625 nm (isobestic in the absence of phenol red) in the direction of increasing alkalinity. The kinetics of the change on a slow time-scale were indistinguishable from those measured simultaneously with a glass electrode, and the extents of the changes estimated after calibration with dilute HCl were similar in each case. Addition of buffer to the suspension suppressed the phenol red and glass electrode changes in parallel. Treatment of the chromatophores with valinomycin, nigericin or FCCP had quantitatively the same effect on the light-induced phenol red change and the glass electrode response.[1,4]

In the rapidly responding single beam spectrophotometer, flash excitation elicited a rapid absorption change of phenol red at 583 nm. At this wavelength there was no light induced absorbancy change in the absence of phenol red, or at high buffer concentrations when phenol red was present. The spectrum of the flash-induced change, computed from either the difference between the flash-induced change in the presence and absence of indicator or the difference between the flash-induced change at high and low buffer concentrations, closely resembled the absorption spectrum of the alkaline form of phenol red.

Phenol red was not chemically changed by exposure to redox potentials down to at least –200 mV, and is therefore an ideal indicator for studying the dependence of rapid hydrogen-ion binding upon the oxidation-reduction potential of a chromatophore suspension.

Results

Rapid H^+-binding Indicated by Phenol Red

The kinetics and effects of inhibitors and ionophores on the rapid H^+-binding of aerobic chromatophore suspensions at pH 7·6 indicated by phenol red were essentially the same as those previously reported for changes indicated by bromocresol purple (Fig. 1). Some of these parameters are summarized in Table I.

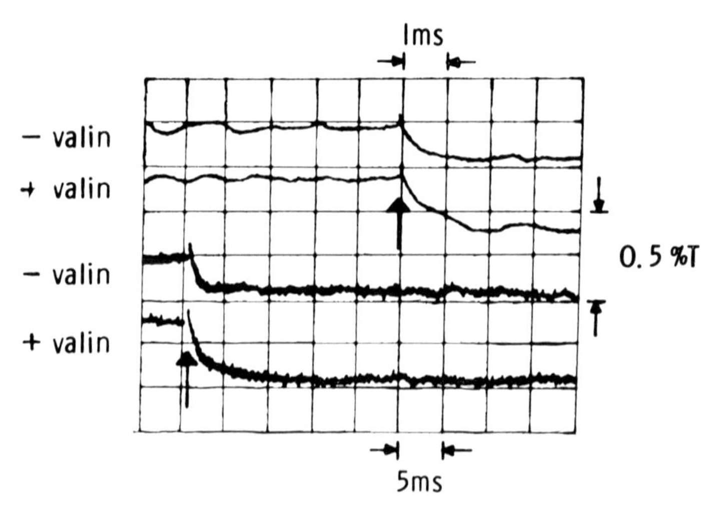

Figure 1. Laser-induced phenol red changes in an aerobic suspension of chromatophores from *Rps. spheroides*. Chromatophores, containing 0·125 mg BChl were suspended in 2·5 ml of 100 mM KCl and 50 μM phenol red at pH 7·6 and 25°C in an aerobic cuvette. The upper two oscilloscope traces were recorded at 1 ms/cm, the lower two at 5 ms/cm. The amplifier RC = 100 μs. The suspension was exposed to a 20 ns Q-switched ruby laser pulse at the time indicated by the arrows. An upward change of 0·5% transmission was equivalent to the addition of 1·34 ng ion H^+. Valinomycin, at a concentration of 4 μM was present where indicated.

As with the rapid H^+-binding indicated by bromocresol purple, a relatively high concentration of valinomycin (0·4 μM for a half maximal effect) was required for stimulation of the phenol red change. The stimulation of the laser-induced response was seen as a slower phase of H^+-binding following the 300 μs rapid phase, giving a ratio of $1H^+/125$ BChl as compared with $1H^+/180$ BChl in the untreated chromatophores (Fig. 1). Antimycin A was without effect on the laser-induced phenol red change in the absence of valinomycin but completely inhibited the slow phase resulting from treatment with the ionophore. These data are completely in accordance with the results previously reported for bromocresol purple (see Table I).

The pH Dependence of Rapid H^+-binding

The finding of two indicators of differing pK_a values, which accurately reflect the pH of a chromatophore suspension has enabled us to

TABLE I. Characteristics of rapid H^+-binding by chromatophores of *Rps. spheroides* in aerobic suspension

Preparation	Flash	Indicator	Additions	Phase I $t_{\frac{1}{2}}$	Phase I extent	Phase II $t_{\frac{1}{2}}$	Phase II extent	Total H^+/BChl
A	Laser	BCP	None	300 μs	1·0			1/170
B	Laser	Phenol red	None	300 μs	1·0			1/180
A	Xenon	BCP	None	360 μs	1·4			1/120
B	Xenon	Phenol red	None	470 μs	1·85			1/100
A	Laser	BCP	Valinomycin	250 μs	1·6			1/100
B	Laser	Phenol red	Valinomycin	600 μs	1·55			1/125
A	Xenon	BCP	Valinomycin	500 μs	2·3	2·5 ms	1·1	1/50
B	Xenon	Phenol red	Valinomycin	560 μs	2·3	2·5 ms	1·45	1/50
A	Xenon	BCP	Val. + antimycin	250 μs	1·4			1/120
B	Xenon	Phenol red	Val. + antimycin	250 μs	1·9			1/100

(A) *Rps. spheroides* chromatophores containing 0·163 mg BChl were suspended in 2·5 ml of 100 mM Kcl, 40 μM BCP at pH 6·3 and 25°C.
(B) As in figure 1. When present valinomycin and antimycin A were at a concentration of 4 μM. All the extents have been normalized so that the laser induced extent with no additions was 1·0.

investigate the pH dependence of the rapid H^+-binding reactions. Figure 2 shows good agreement between the extent of the change measured with either bromocresol purple or phenol red in the overlap region between pH 6·5 to 7·0. Between pH 6·0 and 7·8 both the extent and half risetime of the reaction were constant. Above pH 7·8 the extent of the change was attenuated but measurement of the rate of change at these pH values was unreliable owing to the high background

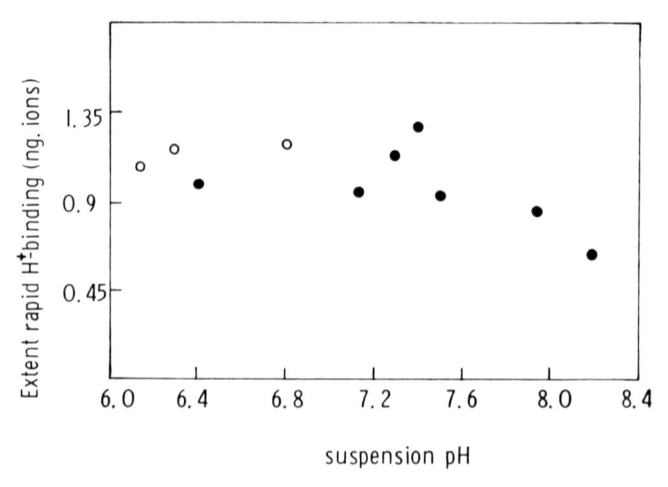

Figure 2. The pH dependence of chromatophore laser-induced rapid H^+-binding. The experimental conditions were similar to those for Fig. 1, except that phenol red was present at 465 μM (●) or BCP at 160 μM (○) and the chromatophores contained 0·13 mg BChl. The pH of the suspension was adjusted with HCl or KOH by comparison with solutions of indicator dye in buffer and was measured after each experiment.

absorption of the alkaline form of phenol red. Chance *et al.*[5] found a twofold difference in the risetime of the bromocresol purple change between pH 5·1 and 7·1 in chromatophores prepared from *Chromatium*.

Rapid H^+-binding as a Function of the Redox Potential of the Chromatophore Suspension

The extent of the rapid H^+-binding reactions of *Rps. spheroides* chromatophores elicted by a 20 ns ruby laser pulse or a 200 μs xenon flash depended on the ambient redox potential of the suspension as shown in Fig. 3A. The laser-induced change, in the absence of valinomycin was attenuated below a midpoint value of $E_{m\,7\cdot5} = +5$ mV and above approximately $E_{m\,7\cdot5} = +430$ mV. These values compare with midpoint potentials of -40 mV for the primary acceptor and $+450$ mV for the reaction centre bacteriochlorophyll at the same pH.[7, 13] The low potential midpoint of attenuation of the phenol red change was consistently 40 mV more positive than the midpoint potential of the primary acceptor,[7, 14, 15, 16, 17] but within the limits of experimental error, the high potential midpoint was similar to that of the reaction centre bacteriochlorophyll.[7, 13] A small but reproducible (25–40%) stimulation of the extent of H^+-binding was observed as the potential of the suspension was lowered through the 350–300 mV range (see

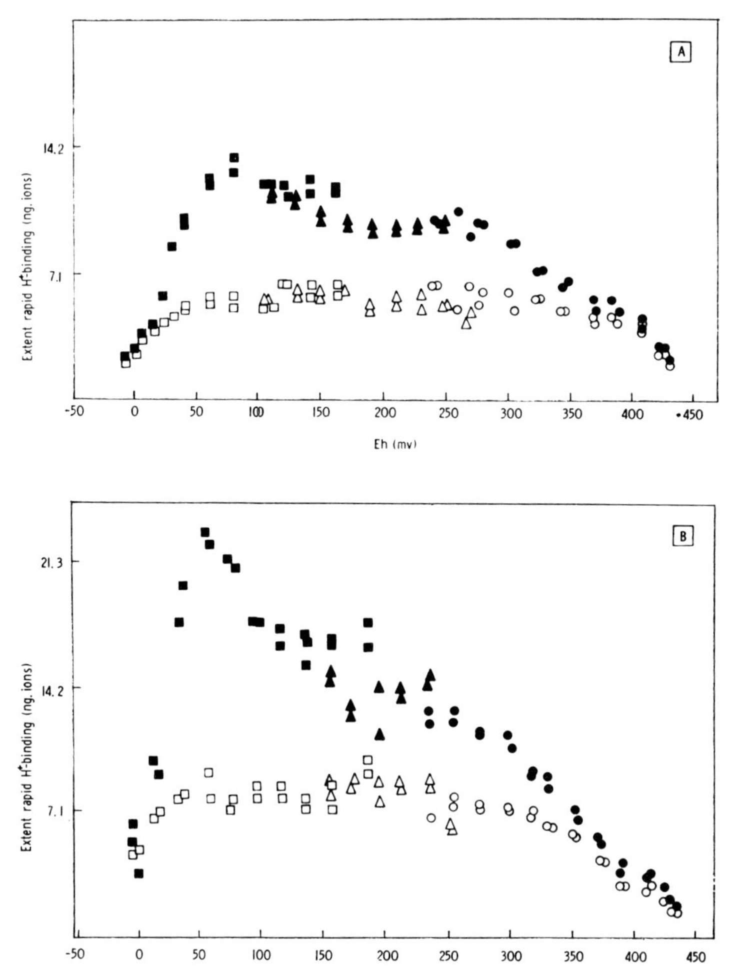

Figure 3. The dependence of rapid H^+-binding upon the redox potential of the chromatophore suspension. *Rps. spheroides* chromatophores (0·72 mg BChl) were suspended in 7·5 ml of a medium containing 100 mM KCl and 465 μM phenol red, final pH 7·5, in the anaerobic redox titration vessel. Each titration was performed in three parts with the following mediating oxidation/reduction dyes: □, ■ 10 μM phenazine methosulphate, 10 μM phenazine methosulphate, 10 μM diaminodurol, 7 μM pyocyanine; ▲, △ 10 μM phenazine methosulphate, 10 μM phenazine ethosulphate, 10 μM diaminodurol; ○, ● 250 μM potassium ferricyanide and, below +300 mV, 10 μM diaminodurol. The open symbols show the extent of H^+-binding after a laser flash, the closed symbols H^+-binding after a 200 μs xenon flash. Following addition of dithionite or ferricyanide, and compensating KOH addition the system was left to equilibrate for at least 3 min. The titrations were mainly carried out in the direction of reduction, making use of the slow endogenous reductant. Occasional calibrations with standard HCl and KOH were performed to ensure that the suspension buffering capacity remained constant. A, no valinomycin; B, in the presence of 2 μM valinomycin.

Fig. 3A and Table II), when the chromatophore cytochrome c ($E_{m\,7\cdot0}$ = +293 mV) was becoming chemically reduced before the flash.

After a 200 μs flash, H^+-binding was most extensive when the redox potential of the suspension was poised between +70 and +100 mV. With further lowering of the oxidation/reduction poise, the H^+-binding reactions were suppressed in two steps; first around a midpoint value of +40 mV and finally across the span that resulted in attenuation of the laser-induced change (at $E_{m\,7\cdot5}$ = +5 mV). To the high potential side of the +90 mV maximum, the binding reactions were attenuated in three steps, $E_{m\,7\cdot5}$ = +130 mV, +320 mV and +430 mV. These features may be correlated with certain chromatophore electron transport carriers the "+160 mV component" or "Z" ($E_{m\,7\cdot0}$ = +160 mV), cytochrome c ($E_{m\,7\cdot0}$ = +293 mV) and reaction centre bacteriochlorophyll ($E_{m\,7\cdot0}$ = +450 mV).[7] There was an interesting similarity between this profile of rapid H^+-binding (Fig. 3A) and that of the flash-induced carotenoid shift as a function of redox potential.[9]

Preliminary observations (J. B. Jackson and R. J. Cogdell, unpublished work) have shown that in chromatophores prepared from *Chromatium* the rapid H^+-binding reaction, indicated by phenol red is attenuated below $E_{m\,6\cdot9} \simeq -70$ mV. This compares with the data of Case and Parson for the midpoint potentials of the primary acceptor X, $E_{m\,7\cdot7} \simeq -130$ mV and the secondary acceptor Y, $E_{m\,7\cdot7} \simeq -90$ mV.

The pH Dependence of the Low Potential Attenuation Midpoint of the Laser-induced Rapid H^+-binding

The indicator-dye technique for measuring rapid hydrogen-ion changes is only satisfactory within a narrow pH range on either side of the indicator pK. With phenol red we have been able to perform suitably accurate redox titrations between pH 6·7 and 7·7 (see Fig. 4A). Figure 4B shows that the E_m of the component associated with attenuation of the rapid H^+-binding reaction at low potentials increases with decreasing pH with a slope of approximately 60 mV/pH unit. We have attempted to fit theoretical curves to the data as shown in Fig. 4A. The points shown for pH = 6·75, 7·0 and 7·2 approximate to a theoretical one electron transfer ($n = 1$) but at pH = 7·4 and 7·6, an $n = 2$ curve is a better fit. This discrepancy reflects the degree of scatter in the data, especially at pH values approaching the indicator pK_a.

When the chromatophore suspension was poised in the redox potential range +60 to +100 mV and below approximately pH 7·0, the decay of the laser or flash induced phenol red change was slightly increased. The increased decay may be a reflection of a higher intrinsic permeability of the chromatophore to ions at low pH,[1] or possibly of interaction with the oxidation reduction dyes. The experimental approach pre-supposes that the oxidation/reduction dyes do not react so rapidly with the chromatophore electron transport systems as to.

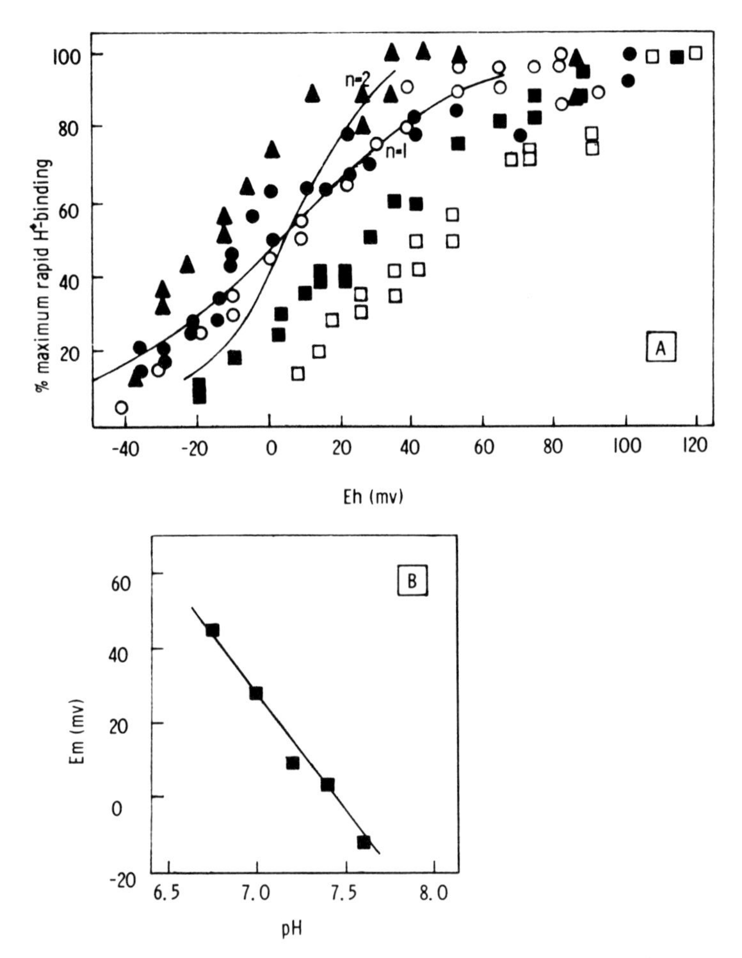

Figure 4. The pH dependence of the potential of the oxidation/reduction component responsible for rapid H^+-binding. The conditions were as described for Fig. 3 except that the chromatophores contained 0·70 mg BChl and 10 μM phenazine methosulphate, 10 μM phenazine ethosulphate and 7 μM pyocanine were used as mediators. A; □ pH 6·75, maximum extent of H^+-binding, 4·3 ng ion H^+; ■ pH 7·0, maximum extent of H^+-binding, 4·2 ng ion H^+; ○ pH 7·2, maximum extent of H^+-binding, 3·3 ng ion H^+; ● pH 7·4, maximum extent of H^+-binding, 3·2 ng ion H^+; ▲ pH 7·6, maximum extent of H^+-binding, 3·6 ng ion H^+. The solid lines are theoretical $n = 1$ and $n = 2$ plots for the data at pH 7·2 ($E_{m7\cdot2} = +6$ mV). B; the midpoint potentials from A, plotted as a function of pH.

modify the response immediately after the flash. To avoid ambiguity we have confined our data to measurements within 20 ms of flash excitation, during which time dye interference is minimal.[7, 9]

The Redox Potential Dependence of the Valinomycin Stimulation of Rapid H^+-binding

A comparison of Figs. 3A and 3B shows that valinomycin stimulation of the rapid H^+-binding occurs only when the redox poise of the suspension is such that cytochrome *c* is partially reduced before the

TABLE II. The effects of valinomycin, antimycin A and *ortho*-phenanthroline on the kinetics of laser-induced rapid H^+-binding

Preparation	Additions	o/r potential (mv)	H^+-binding (ng. ion/mg BChl)	o/r potential (mv)	H^+-binding (ng. ion/mg BChl)	o/r potential (mv)	H^+-binding (ng. ion/mg BChl)	o/r potential (mv)	H^+-binding (ng. ion/mg BChl)
A	None	385	8·9	241	12·4	166	13·0	84	13·9
A	Valinomycin	384	10·0	241	14·3	166	19·5	85	26·0
A	Valinomycin plus Antimycin A	383	10·0	242	12·8	166	14·5	85	17·4
B	None	380	8·1	248	11·7	160	12·4	99	10·0
B	Antimycin A	378	6·5	248	11·6	160	10·5	99	8·4
B	None	378	7·9	235	13·4	159	12·9	95	12·7
B	*o*-phenanthroline	378	3·9	238	8·9	159	8·6	93	8·3

The experimental conditions for this table are described in Fig. 5. Preparation A was that used in frames A → D and preparation B was used in frames E → J.

flash. Between +50 and +300 mV the ratio of hydrogen ions bound following a laser pulse to the number of chromatophore bacteriochlorophyll molecules was increased from 1/130 to 1/95 by treatment with valinomycin. Table II shows that in some preparations the extent

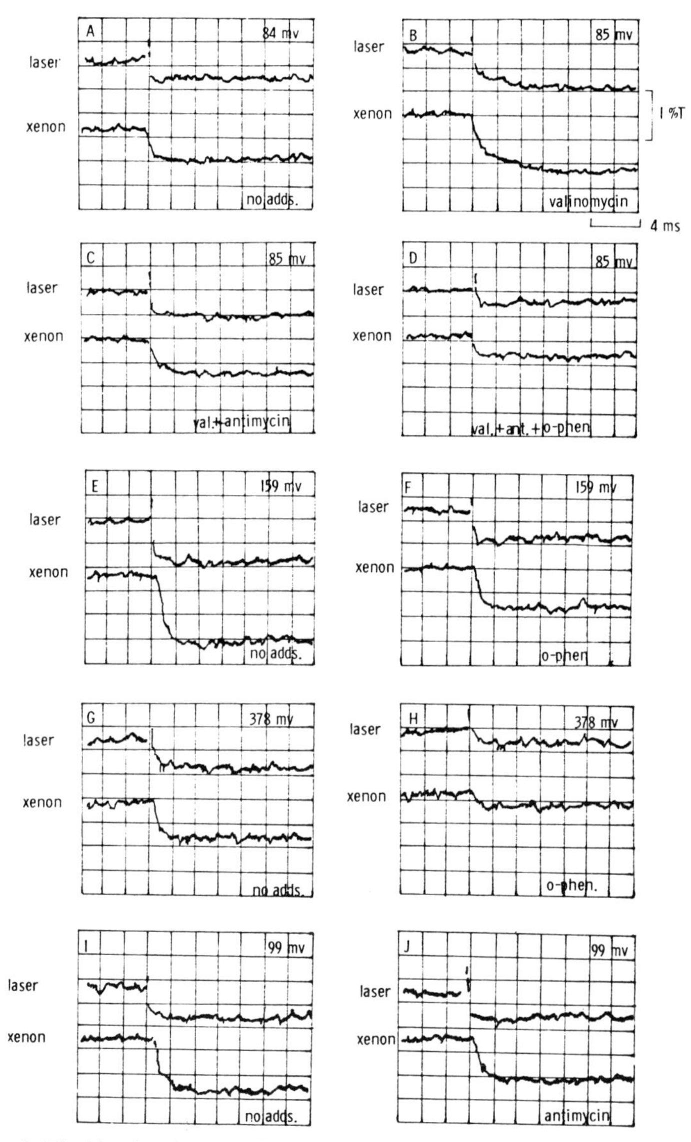

Figure 5. The kinetics of rapid H^+-binding; the effect of valinomycin, antimycin A and *o*-phenanthroline. Conditions as for Fig. 3 except that chromatophores contained 0·46 mg BChl in experiments A–D and 0·42 mg BChl in experiments E–J. Calibration with dilute HCl showed that 1% ΔT was equivalent to 8·3 ng ion H^+ in experiments A–D and 6·1 ng ion H^+ in experiments E–J. Where shown valinomycin was present at 2 μM, antimycin at 2 μM and *o*-phenanthroline at 4 mM. The redox potential of the suspension is shown in the top right-hand corner of each frame.

of valinomycin stimulation of the laser-induced phenol red change was greater between +90 and +160 mV than at higher potentials. This was always the case for H^+-binding following a xenon flash (Figs. 3A and 3B). At +70 mV the H^+-BChl ratio reached 1/30 following a xenon flash in the presence of valinomycin.

The laser induced H^+-binding reaction of chromatophores untreated with valinomycin, was complete is less than 1 ms (Fig. 5). The half-risetime of the change was about 300 μs but was not a simple first-order change. Further kinetic resolution will not be possible without the application of repetitive pulse techniques. The extra H^+-binding following valinomycin treatment was slower, $t\frac{1}{2} \simeq 2$ ms (Fig. 5) in agreement with earlier work on aerobic chromatophore suspensions.

The Inhibitory Effects of Antimycin A and ortho-Phenanthroline on the rapid H^+-binding reactions of Rps. spheroides chromatophores

In the absence of valinomycin, antimycin A was without significant effect on the H^+-binding reactions elicited by a laser pulse, regardless of the redox poise of the chromatophore suspension (Table II). However, low concentrations of antimycin A inhibited the valinomycin-stimulated phenol red change, and this was especially apparent at low redox potentials. Approximately 75% of the extra H^+-binding resulting from valinomycin treatment was inhibited by the antimycin A. Figure 5 shows that the inhibitor removed the slow phase of the H^+-binding, leaving the faster phase unaffected.

Inhibition of the rapid H^+-binding by ortho-phenanthroline was also dependent on the ambient redox potential of the chromatophore suspension. The change was 50% inhibited by 4 mM *ortho*-phenanthroline when the potential was poised such that cytochrome *c* was chemically oxidized before the flash (above +350 mV) but only 30–35% inhibited by the same concentration at redox potentials between 40 and 300 mV (Fig. 5 and Table II). Other effects of *o*-phenanthroline at lower potentials are discussed separately elsewhere.[17]

Discussion

Rapid H^+-binding at the Level of a Secondary Electron Acceptor

Rapid H^+-binding was elicited by laser pulse or xenon flash activation of a *Rps. spheroides* chromatophore suspension whenever the reaction centre bacteriochlorophyll ($E_{m\,7\cdot5} = +450$ mV) was reduced and a component of $E_{m\,7\cdot5} = +5$ mV was oxidized, before the flash. The dependence of the midpoint potential on pH indicates that this component accepts $1H^+$/electron on reduction.

The potential of the component ($E_{m\,7\cdot5} = +5$ mV) was 40 mV more positive than the midpoint potential of the primary acceptor and 20 mV more negative than cytochrome *b*, so that it cannot be identified with any of the previously characterized electron transport carriers of *Rps. spheroides* chromatophores.

There is kinetic as well as thermodynamic justification for placing the component responsible for rapid H^+-binding (henceforth called Y^1) at the level of the secondary acceptor pools. The reduction of Y^1 after a flash must have a half risetime of less than 300 μs. This is intermediate between the very rapid reaction between the primary donor and acceptor (10^{-11} s)[18] and the oxidation/reduction reactions of the *b*-type cytochromes (1–2 ms).[7]

The double pulse technique of Parson and Case [19] has shown that the half rise-time of the reaction between the primary acceptor (X) and secondary acceptor (Y) in *Chromatium* chromatophores is between 60 and 80 μs. Unfortunately species differences between *Chromatium* and *Rps. spheroides* forbid comparison of the double pulse data with the phenol red experiments reported here. However, the similarity between the rates of the two processes is encouraging and may point to the chemical identity of Y and Y^1. Our preliminary data for the midpoint potential of the hydrogen carrier of *Chromatium* chromatophores, $E_{m\,6\cdot9} \simeq -70$ mV are in good agreement with the value for Y. $E_{m\,7\cdot7} \simeq -90$ mV reported by Case and Parson,[20] since the value changes to the extent of 30 mV per pH unit.

H^+-binding and H^+-translocation

No rapid appearance of hydrogen ion was ever observed under the circumstances of these experiments (Figs. 1–5). This suggests that if any H^+-releasing reactions accompany electron transport, as might occur for example upon reduction of a cytochrome by a flavoprotein or quinone, then either the resulting protons must be released on the inside of the chromatophore vesicle, or the release of H^+ is fortuitously balanced by uptake of H^+ from the medium. This latter possibility seems unlikely. Further evidence for a proton translocation mechanism is shown in Fig. 6, where the dependence on uncoupler concentration of the decay the phenol red change was the same as that for the decay of the field-indicating carotenoid shift. Uncoupling agents of the FCCP type have been shown to act as proton conductors through natural and artificial membranes.[21,22,23] It appears that dissipation of the membrane potential generated by the light pulse is achieved by outward, electrophoretic H^+ translocation mediated by FCCP. Since the decay of the phenol red change is associated with a *trans-membrane* H^+-efflux, we may assume that the light-driven rapid H^+-binding is associated with an inward translocation of H^+.

A scheme which accounts for many of our observations on the H^+-binding reaction is shown in Fig. 7. Since the photochemical reaction involves the transfer of electrons across the chromatophore membrane, giving rise to a membrane potential,[10,24] it is possible that the H^+-binding reaction at the level of a secondary electron acceptor Y^1 serves to stabilize the electric field. These reactions may constitute the first energy conserving site or loop[8] in the chromatophore electron transport chain.

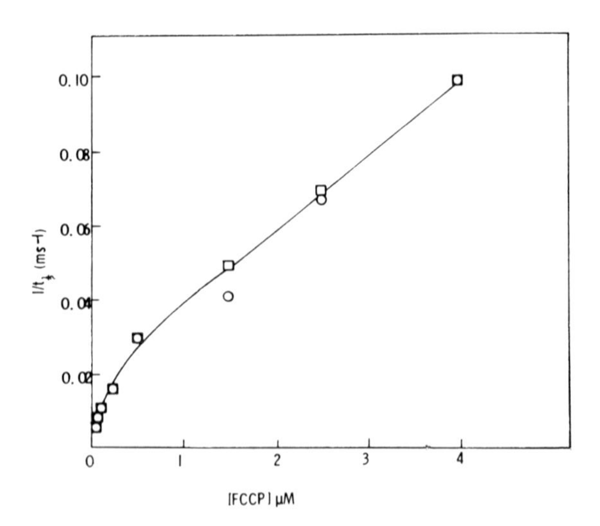

Figure 6. The effect of FCCP on the decay of the rapid H^+-binding and the carotenoid shift. ○, The reciprocal of the half decay time of phenol red change following a xenon flash measured under conditions similar to those described in Fig. 1 but on a longer time scale. The chromatophores contained 0·062 mg BChl and the phenol red concentration was 305 μM. □, The reciprocal of the half decay time of the xenon flash induced carotenoid shift, measured on a single beam spectrophotometer at 490 nm.[10] The same amount of chromatophore preparation was used as in the phenol red experiment. The suspending medium contained 100 mM KCl, 20 mM tricine, pH 7·6.

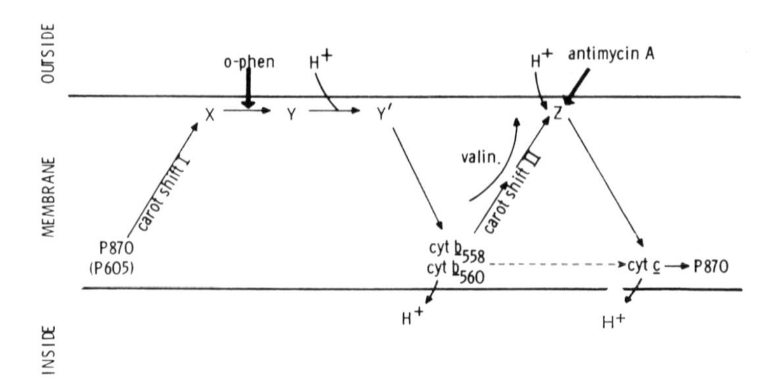

Figure 7. Model for rapid H^+-binding to *Rps. spheroides* chromatophores. Based on the model proposed in ref. 6 with additional data from this report and from refs. 7, 9 and 17. (See text.)

The Second "Site" of Photosynthetic Energy Conservation in Rps. spheroides Chromatophores

There is evidence that electrical work is done at two sites in the electron transport chain of chromatophores from *Rps. spheroides*,[9] and that coupling occurs at site II only if cytochrome c, $E_{m\,7\cdot0} = +293$ mV[7] is chemically reduced before the flash.[9] It is likely that the extra H^+-binding following a laser pulse at potentials below +300 to +350 mV (see Fig. 3 and Table II) is associated with the second site of energy conservation. Complete expression of site II H^+-binding however, was only observed in the presence of valinomycin. If we assume that valinomycin acts simply as a trans-membrane K^+ transporter,[25] then it could only directly stimulate those oxidation/reduction reactions which involve the transfer of electrical charge across the membrane. Previous studies[9] have shown that the re-reduction of cytochrome c after a laser pulse was accelerated by valinomycin. Cytochrome c oxidation and oxidation/reduction reactions of the b cyotchromes and reaction centre bacteriochlorophyll were unaffected. It is probable therefore that an electrogenic reaction and an H^+-binding reaction are located at the second site of energy conservation to the low potential side of cytochrome c.

The failure to detect significant site II H^+-binding after a laser pulse in the absence of valinomycin is puzzling since redox titration of the carotenoid shift[9] suggests that, at a redox poise between +40 and +120 mV, at pH 7·0, one electron per reaction centre bacteriochlorophyll traverses a second electrogenic site. It is conceivable however, that two electron equivalents per hydrogen carrier participate in the H^+-binding reaction. For example, if the hypothetical site II hydrogen carrier (see Fig. 7) is a quinone or a flavoprotein

$$FP + 2e^- + 2H^+ \rightleftharpoons FPH_2$$

valinomycin would facilitate transfer of the second electron to Z by lowering the electrical potential against which the reaction must do work.

The Action of Antimycin A and o-Phenanthroline on Rapid H^+-binding

The model of Fig. 7, discussed above, is consistent with the effects of antimycin shown in Table II and Fig. 5. This inhibitor has been shown to act on the electron transport reactions at the level of the second site of energy conservation.[7,9] In the absence of valinomycin or in the presence of valinomycin when the suspension was poised above about +300 mV (when site II H^+-binding is inoperative) antimycin A had little effect on the phenol red change following a laser pulse. In the presence of valinomycin, when the redox potential was below 300 mV, antimycin A inhibited by about 75% that portion of the H^+-binding associated with site II.

It is significant that *o*-phenanthroline, an inhibitor of the reaction between the primary and secondary electron acceptor pools[17,19,26] has a more pronounced inhibitory effect at redox potentials above +300 to +350 mV where site I H^+-binding alone was observed. At lower redox potentials, where site II H^+-binding was operating, the inhibitory action was less marked.

The Membrane Bohr Model for Rapid H^+-binding

In the terms of the membrane Bohr effect two changes in membrane conformation must be postulated, one for each energy conserving site, directly or indirectly driven by electron transport at either site.

H^+ uptake by chromatophores measured by a glass electrode under conditions of continuous illumination is generally considered to be a trans-membrane, electrogenic process[1,27] and is too extensive to be stoichiometrically related to a membrane conformation change of the type discussed by Chance *et al.*[5] Rapid H^+-binding and H^+-uptake would therefore be unrelated phenomena despite their similar response to treatment with uncoupling agent unless the H^+-binding is assumed to be a cyclic process reflecting an electrogenic H^+-pump.

The light-induced red shift of carotenoid absorption of the photosynthetic bacteria has been shown to be a response to the formation of a membrane potential[10,24,28] and Fig. 6 suggests that it is closely related to the rapid H^+-binding reactions. The nature of this relationship must account for the following observations: (i) the kinetics of the two phenomena are very different (compare Figs. 1 and 5 with ref. 10); (ii) valinomycin increases the extent of the rapid H^+-binding but stimulates the decay of the carotenoid shift; (iii) nigericin stimulates the decay of the rapid H^+-binding and increases the extent of the carotenoid shift.[10]

While these observations follow naturally from the chemiosmotic model proposed here, the postulated membrane Bohr effect can only account for them by additional hypotheses.

Acknowledgements

We are grateful to Miss Tricia Edwards for expert technical assistance, and to the Science Research Council for support and equipment.

References

1. J. B. Jackson, A. R. Crofts and L.-V. von Stedingk, *European J. Biochem.*, **6** (1968) 41.
2. P. Scholes, P. Mitchell and J. Moyle, *European J. Biochem.*, **8** (1969) 450.
3. A. Thore, D. L. Keister, A. Shavit and A. San Pietro, *Biochemistry*, **7** (1968) 3499.
4. J. B. Jackson and A.R. Crofts, *European J. Biochem.*, **10** (1969) 226.
5. B. Chance, A. R. Crofts, M. Nishimura and B. Price, *European J. Biochem.*, **13** (1970) 364.
6. A. R. Crofts, R. J. Cogdell and J. B. Jackson, in: *Energy Transduction in Respiration and Photosynthesis*, E. Quagliariello, S. Papa, and C. S. Rossi (eds.), pp. 883–901, Adriatica Editrice, Bari, 1971.

7. P. L. Dutton and J. B. Jackson, *European J. Biochem.*, submitted.
8. P. Mitchell, in: *Chemiosmotic Coupling in Oxidative and Photophosphorylation*, Glynn Research Laboratories, Bodmin, Cornwall, 1966.
9. J. B. Jackson and P. L. Dutton, *Biochim. Biophys. Acta*, submitted.
10. J. B. Jackson and A. R. Crofts, *European J. Biochem.*, **18** (1971) 120.
11. P. L. Dutton, *Biochim. Biophys. Acta*, **226** (1971) 63.
12. B. Chance, M. Nishimura, M. Avron and M. Baltscheffsky, *Arch. Biochem. Biophys.*, **117** (1966) 158.
13. I. D. Kunz, P. A. Loach and M. Calvin, *Biophys. J.*, **4** (1964) 227.
14. P. A. Loach, *Biochemistry*, **5**, (1962) 592.
15. D. W. Reed, K. L. Zankel and R. K. Clayton, *Proc. Natn. Acad. Sci. U.S.*, **63** (1969) 42.
16. W. A. Cramer, *Biochim. Biophys. Acta*, **189** (1969) 54.
17. J. B. Jackson, R. J. Cogdell and A. R. Crofts, in preparation.
18. K. L. Zankel, D. W. Reed and R. K. Clayton, *Proc. Natn. Acad. Sci. U.S.*, **61** (1968) 1243.
19. W. W. Parson and G. D. Case, *Biochim. Biophys. Acta*, **205** (1970) 232.
20. G. D. Case and W. W. Parson, *Biochim. Biophys. Acta*, **253** (1971) 187.
21. J. Bielawsky, T. E. Thompson and A. L. Lehninger, *Biochem. Biophys. Res. Commun.*, **24** (1966) 948.
22. V. P. Skulachev, A. A. Sharof and E. A. Liberman, *Nature*, **216** (1967) 718.
23. P. J. F. Henderson, J. D. McGivan and J. B. Chappell, *Biochem. J.*, **111** (1969) 521.
24. J. B. Jackson and A. R. Crofts, *FEBS Letters*, **4** (1969) 185.
25. P. Mueller and D. O. Rudin, *Biochem. Biophys. Res. Commun.*, **26** (1967) 398.
26. B. Chance, J. McCray and P. Thornber, *Biophys. Soc. Abstr.*, 124(a) (1970).
27. M. Nishimura and B. C. Pressman, *Biochemistry*, **8** (1969) 1360.
28. S. Schmidt, R. Reich and H. T. Witt, in: *Abstracts for the Second International Congress on Photosynthesis Research*, Stresa, Italy, 1971.

A Note on Some Old and Some Possible New Redox Indicators

Robert Hill

Department of Biochemistry, University of Cambridge

Summary

Starting with the frequently used dye 2,6-dichloroindophenol some of the chemical properties of the quinonimide class of dyes are described. Consideration of the effects of completing a hetero-six-membered ring, as in the azine, thiazine and oxazine classes, on the properties of the compounds is suggested as a lead towards the development of some redox indicators perhaps more desirable than the indophenols. This approach developed from a study of the Liebermann nitroso-reaction for phenols. Some properties of. the redox indicators which resulted from that work are described in relation to energy transduction in the chloroplast.

In studies on energy transduction, especially that involving preparations of chloroplasts much use has been made of dichlorophenol indophenol (I). This was one of the indophenol dyes developed by Mansfield Clark and colleagues[1] and was first used to determine photochemical activity with chloroplast preparations by Holt and French[2] in 1948.

The use of indophenols in biology was described by Ehrlich[3] in 1885. The now obsolete term "Indophenol Oxidase" which referred in part to the enzyme cytochrome *c* oxidase (EC 1.9.3.1.) and in part to polyphenol oxidases[4] (*para*, EC 1.10.3.2. and *ortho*, EC 1.10.3.1.) originated in the following way. The "Indophenol reagent" used contained a mixture of unsymmetrical dimethyl paraphenylene diamine and α-naphthol.[5] These components by the loss of four molecular equivalents of hydrogen formed the insoluble blue vat dye originally known as indophenol (II). This dye was formerly used as an adjunct to indigo or to replace it. But it is destroyed by acids, a property shared by all indophenols and the other quinonimide dyes of a similar type. The dye, when produced from the reagent could be reduced by hydrogen donors produced in living cells. The leuco compound was easily oxidized in presence of air by the oxidase. Three processes would have to be considered when using the reagent, synthesis of the dye (partially an irreversible process), reduction of the dye to the leuco compound and reoxidation to the dye. The second and third processes only involve two hydrogen equivalents. The use of the reagent could result in a confused interpretation. Lack of colour given could either indicate no activity at all or a very high activity of reducing systems.

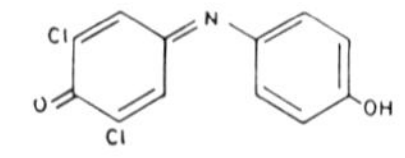

(I)

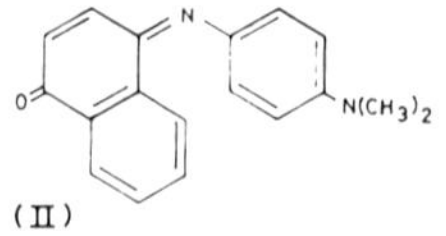

(II)

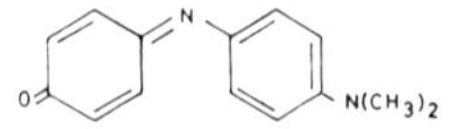

(III)

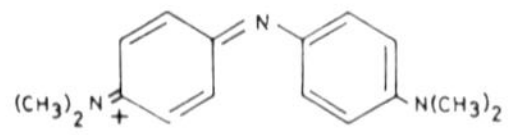

(IV)

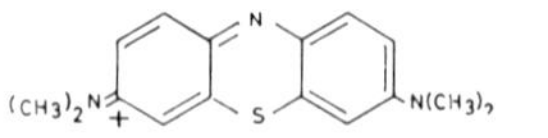

(V)

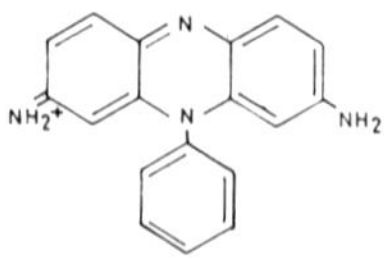

(VI)

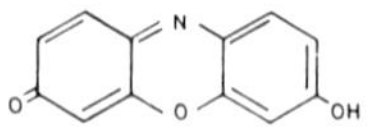

(VII)

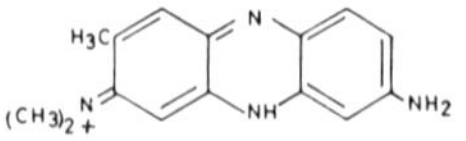

(VIII)

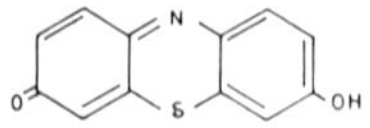

(IX)

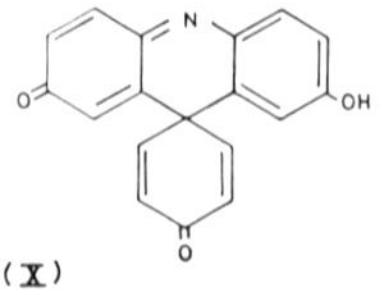

(X)

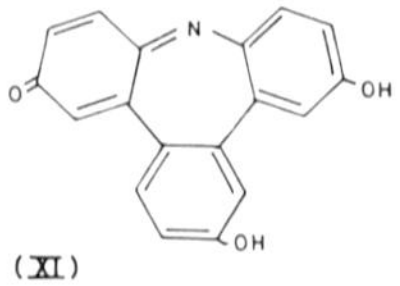

(XI)

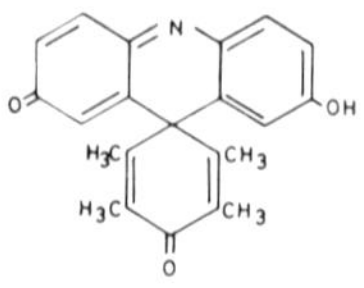

(XII)

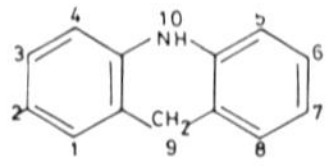

(XIII)

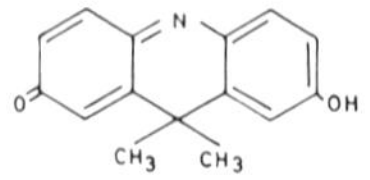

(XIV)

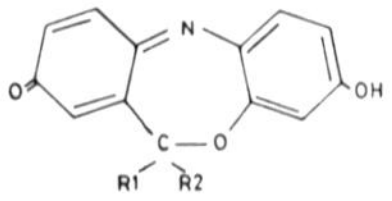

(XV)

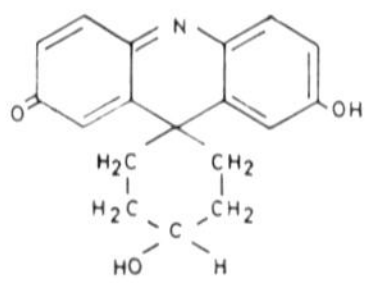

(XVI)

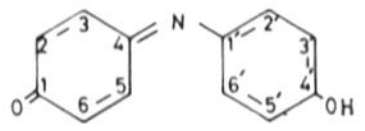

(XVII)

In 1929 it was shown by David Keilin[6] that a cold suspension of living yeast would oxidize the reagent when sufficiently aerated. At ordinary temperatures there was no reaction and the oxidase had formerly been considered absent from yeast cells. Keilin explained the result in terms of the difference in rates of oxidation and reduction processes, the latter having a higher temperature coefficient. This was an important experiment in the development of knowledge of hydrogen or electron transport in cellular respiration. Now it happens that the dye produced from the "indophenol reagent" should not properly be called an indophenol. It belongs, according to Mansfield Clark[7] to a class known as *indoanilines* represented by the so called phenol blue (III). When both oxygen atoms in the indophenol are replaced by an amine or imine group the compounds were termed *Indamines* as for example Bindschedler's green (IV). This dye corresponds to methylene blue (V) without the sulphur atom. The sulphur completes the heterocyclic ring connecting two phenyl residues. Methylene blue is comparatively resistant to acids. The other types of quinonimide compounds which do not have the heterocyclic ring are easily decomposed. Phenolindophenol for example gives initially in acid solution quinone, ammonia and probably *p*-aminophenol. It follows that it is not usually practicable to isolate the free indophenols for storing. Mansfield Clark stated that the neutral or alkaline solutions of the indophenols tend to decompose and that these dyes are best stored as their dry alkali salts.

For the detection of biological oxidizing systems the original indophenol reagent was some times used without the α-naphthol. The coloured product known as "Wursters red" resulted from a partial oxidation of the asymmetrical dimethyl-*p*-phenylene diamine. This gives a quinhydrone or semiquinone type of compound. There is, however, no stable link between two phenylene residues and the system is liable to undergo further reactions which are irreversible in terms of hydrogen transport, just as in the formation of the blue dye in presence of α-naphthol. In addition the simple quinone imides easily break down in aqueous solution to quinones and ammonia.

Indophenols have the advantage that in general they are soluble near the neutral point and that the leuco compounds are not rapidly oxidized by atmospheric oxygen[8] so that they can be used as indicators of hydrogen transfer in open vessels. They also react rapidly with a variety of hydrogen donors. This property depends on the fact that the oxidation-reduction (O.R.) potentials are in a more oxidizing range than is the case with the other quinonimide dyes such as methylene blue (thiazine), saffranine (VI) (azine) or resorufin (VII) (oxazine). With these dyes, where a quinonimide nitrogen forms part of the heterocyclic six-membered ring, the name of the class is taken from the atom which completes the ring. Thus the parent substance for methylene blue is phenothiazine. The effect on the O.R. potential of the quinonimide dye when the heterocyclic structure is completed is to change the characteristic potentials at pH 7 (E_m) towards a more reducing potential. Bindschedler's green (IV) E_m 0·225 V, pH 7, and methylene blue (V) E_m 0·015 V show the effect of the sulphur linkage, and also on the position of the absorption bands which have been displaced towards the violet end of the spectrum. In the azine series the effect of the nitrogen atom is much greater, as neutral red (VIII),

$E_m - 0{\cdot}252$ V. Resorufin $E_m - 0{\cdot}06$ V shows that the effect of the oxygen atom on the O.R. potential is similar to that of sulphur.

With the two symmetrically placed substituents the change of a phenolic to an amino group does not in general cause a marked effect. Thionol (IX) derived from dihydroxy phenothiazine has E_m recorded = 0.037 V, slightly more positive than methylene blue. Phenol indophenol with E_m 0.228 is very near to Bindschedler's green at pH 7.

The effect of different single additional substituents in the case of indophenols, quinones and indigo sulphonates was studied by Clark and by Fieser and others and it will not be reviewed here. The previous discussion was intended to show the influence of the bivalent substitutions in positions 5 and 6′ (using Clark's numbering) of a quinonimide dye which produce the heterocyclic ring.

Suggestions for possible new redox indicators resulted from a study of the Liebermann nitroso-reaction with phenol[9] (Hill *et al.*, 1970). It was found that the dye described by Liebermann[10] could not wholly be accounted for in terms of an indophenol. The interpretation given throughout the literature originated from an early publication by Baeyer and Caro[11] who prepared nitrosophenol and obtained indophenol from it, thereby concluding that the dye obtained by Liebermann must also be indophenol. The indophenols in weakly alkaline solutions do not show visible fluorescence; the product produced from ordinary phenol as described by Liebermann shows a marked red fluorescence of the blue alkaline solution. By following the fluorescence together with the characteristic absorption spectrum a stable crystalline substance was obtained. This was found, in agreement with the original work of Liebermann to have 6 carbon atoms more than the indophenol. A bivalent group in positions 5 and 6′, giving a more stable molecule and accounting for the fluorescence, was indicated. Nietzki (1888) had already put Liebermann's C_{18} dye into the oxazine class with resorufin, which is strongly fluorescent, when he wrote one of the first text books on chemistry of dyestuffs.[12] It was concluded that this group was 4 cyclohexanone 2,5-diene (1,1); in fact the compound (X) was produced in small yield when indophenol itself was treated with phenol in sulphuric acid. In this reaction as 4 equivalents of hydrogen have to be removed this results in reduction of the indophenol to the leuco compound. In Liebermann's original method there was a virtual excess of nitrite which acted as the oxidizing agent. If the leuco compound of (X), which is a spirocyclohexane derivative, as treated with aqueous mineral acid reoxidation gives an amorphous product soluble in weak alkali with a violet colour showing no fluorescence. It is concluded that an intermolecular reaction has resulted in the cyclohexane ring becoming aromatic and forming a phenylene residue which then gives a seven-membered heterocyclic ring (XI). The structure (XI) must at present be regarded as provisional until further evidence has been obtained. That this substance was also present in the original preparations of Liebermann seems certain as he described the colour as being "pure cobalt blue". With the substance giving the red fluorescence the colour of the alkaline solution would be described as pure blue without a reddish tinge apart from the fluorescence. Both these products have more oxidizing O.R. potentials than the corresponding phenolic derivatives in the series thiazines or oxazines. At a neutral pH "Liebermann's blue" is a little more reducing than indophenol itself while "Liebermann's violet" (XI) is a little more oxidizing. When phenol indophenol is treated with durenol (2,3,5,6-tetra methyl phenol) in sulphuric acid a Liebermann's blue (XII) is formed. The methyl groups in the cyclohexane residue would seem to account for the fact that the leuco compound remains unchanged when treated with acid.

The substance (X) is to be regarded as an acridan (XIII) derivative and would therefore belong to a series of dyes discovered by Kehrmann and colleagues[13] (1919) which they called carbazines. This name was used because a carbon atom completed a six-membered heterocyclic ring as the sulphur does in the thiazines. But the parent substance of Kehrmann's carbazines is known as 9,10 dehydroacridine or acridan. These dyes, as phenolic compounds, may be prepared from an indophenol (see (XVII) for numbering) with disubstituted methanol group in position 5 or 6′. The leuco compound of this tertiary alcohol readily forms the acridan ring on warming with mineral acid by the loss of a molecule of water. The 9,9-dimethyl acridan derivative (XIV) was prepared by Goldstein and Kopp (1928)[4] and was termed by them 2-hydroxy-C-dimethyl carbazone (7). The secondary and primary alcohols do not react in this way to give the acridan derivatives.

It was observed, however, that indophenols prepared from 3-hydroxyphenol carbinols by action of quinone chlorimide in dilute alkali, showed a colour change from blue to purple on keeping the reaction mixture for 3–4 days. This change required the presence of air. The products could be isolated as stable crystalline substances, dissolving in weak alkalies to give brilliant purple solutions. Dr. B. R. Webster and Dr. G. R. Bedford[15] very kindly examined 2 samples of these compounds. The mass spectra suggested molecular ions 16 mass units higher than the originally proposed structures which represented the compounds as a dihydroxy acridine and a dihydroxy methyl acridine. The mass spectrum of the desmethyl compound gave a molecular ion at *m/e* 227 which measured 227·0584, $C_{13}H_9NO_3$ requires 227·0582. This was consistent with the structure (XV) proposed from the n.m.r. spectrum. The n.m.r. spectra were interpreted by Dr. Bedford as shown in Table I. Two patterns were shown for the aromatic rings suggesting they were not identical and a CH_2 group in the desmethyl compound and a $CHCH_3$ group in the methyl compound

TABLE I. The n.m.r. spectra of "desmethyl purple" (D.M.P.) and "methyl purple" (M.P.) in D_6 dimethylsulphoxide (d_6 d.m.s.o.) solution can be described as an equilibrium mixture of forms I and II

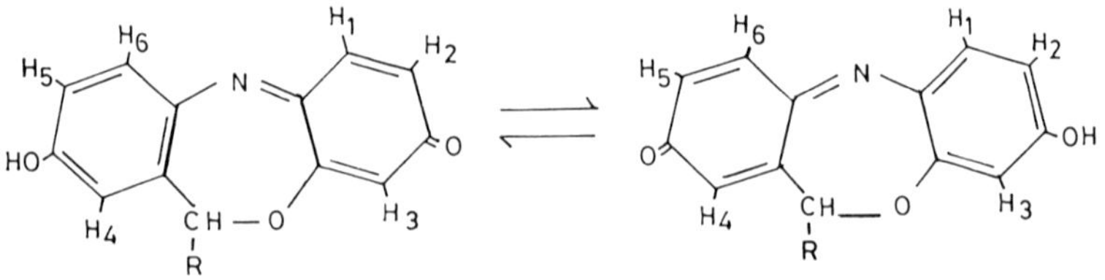

The n.m.r. spectra were interpreted as follows:

Compound R =	H	CH_3
H_1	7·28 doublet J = 9 Hz	7·29 doublet J = 9 Hz
H_2	6·55 doublet of doublet J = 9, 2 Hz	6·55 doublet of doublet J = 9, 2 Hz
H_3	6·05 doublet J = 2 Hz	6·00 doublet J = 2 Hz
H_4	6·75 doublet J = 2 Hz	6·70 doublet J = 2 Hz
H_5	6·85 doublet of doublet J = 8, 2 Hz	6·84 doublets of doublets J = 8, 2 Hz
H_6	7·47 doublet J = 8 Hz	7·46 doublet J = 8 Hz
CH	4·97 singlet	5·13 quartet J = 7 Hz
R	4·97 singlet	1·52 doublet J = 7 Hz

whereas a =CH and a =CCH_3 would be expected on the originally proposed structures. These chemical details are given here because the structure (XV) represents a possibly new and easily accessible type of quinonimide dye. Corbett[6] (1970) in a study of structure and spectra of the indoaniline dyes had shown the effect of the asymmetry as regards the two aromatic rings, as shown in (III). While solubility properties of these methylene oxy compounds are influenced by the nature of the substituents R_1R_2 on the methylene carbon the visible absorption spectra are very little affected. The compound $R_1 = H$ and $R_2 = CH_3$ has been found by me to be very useful as an indicator of hydrogen transport with chloroplast preparations. The region of absorption is sufficiently removed from the red end of the spectrum at pH 7 and above so that it does not, in contrast to (I) obscure the red absorption band of chlorophyll.

The compound (XVI) was prepared from (X) by reduction of the cyclohexanonediene residue to the 4′-hydroxycyclohexane with AlNi alloy in alkaline solution. (XVI) has a very reduced fluorescence compared with (X). The absorption spectrum is similar to (XIV) but this has a strong red fluorescence in its alkaline solution. It is thought that the lack of fluorescence shown by XVI may be due to "boat" and "chair" forms in the cyclohexane part. Dr. Norman Good[17] observed a ten-fold increase in fluorescence when the alkaline salt was present in glycerol. The compound (XII) in spite of the 4 methyl groups has a more marked fluorescence than (X). Both the compounds (XII) and (XVI) are reduced by chloroplast preparations in light. They do not obscure the red absorption band of chlorophyll to an appreciable extent when used in sufficient concentration for direct measurement of oxygen. The reduction of (XVI) is found closely to resemble that with an equivalent amount of ferricyanide.[18] This is shown in Fig. 1. The effect of adding NH_3 as an uncoupler also is

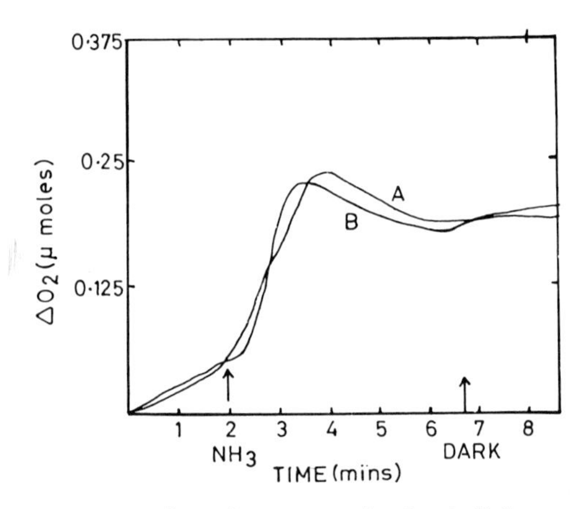

Figure 1. Simultaneous recording of oxygen production in light comparing equivalent amounts of dye and of ferricyanide.[17]

Pea chloroplasts prepared in 0·33 M sorbitol and 50 mM orthophosphate pH 6·5, broken in electrode vessel, assayed in 0·33 M sorbitol and 50 mM Hepes pH 7·6. In 3 ml: 100 μg chlorophyll, equivalent dye (0·5 μmoles) and ferricyanide (1 μmole). Trace A, ferricyanide, B, dye (XVI), NH_4Cl added to final concentration of 6 mM (Stokes and Walker 1970).

practically identical when (XVI) is compared with ferricyanide. This is in contrast with the uncoupling effect which is shown with 2,6-dichloroindophenol. It was found that there was an appreciable rate of reoxidation of the leuco compound in the dark as there is with the indophenol. This would be of little consequence under the appropriate conditions of chloroplast concentration and light intensity.

The accepted chemical nomenclature for (X) which here was referred to as Liebermann's blue is 7-hydroxy spiro [acridine-9, 1′-cyclohexa-2′5′-diene]-2(9*H*),4-dione. This does not suggest either a convenient trivial name, nor a convenient initial letter abbreviation. The trivial name "dichroin" was given in 1888 by Brunner and Chuit.[9] This term was used by Nietzki in his textbook in 1888. In 1923 the name appeared in Beilstein[20] as *verbindung* $C_{18}H_{15}O_3N$,α-phenyl dichoin but as it had not been actually characterized as a pure compound later editions omit it. Mansfield Clark (1923) refers to the name phenoldichroin as an alternative one for phenol indophenol. There is now a basis for reviving this trivial name, which obviously refers to the simultaneous appearance of blue and red tints in the alkaline solution. If the term phenol now is used to indicate the origin of the cyclohexane residue then compound (XII) would be "durenol dichroin" and compound (XVI) "reduced dichroin". These can be abbreviated as DC, DDC and RDC. These three compounds have the plane of the cyclohexane residue at right angles to the plane of the quinonimide part. The substance here referred to as "Liebermann's violet" could be termed "dichroin violet" or DCV. If the numbering of the indophenol residue (XVII) according to Clark is considered then 2,6 dichloro dichroin DCDC would be intelligible as would be 2,6 dichloro dichroin sulphonate (3'). It was most convenient to name the purple compounds after the substituents R_1R_2 as methyl purple, $R_1 = H$, $R_2 = CH_3$ desmethyl purple, $R_1 = R_2 = H$, and dimethyl purple, $R_1 = R_2 = CH_3$. But the name methyl purple is already in use for a pH indicator. So the purple compounds might be called simply methyleneoxy indophenols (5,6'), keeping the carbon atom on the left as (XV). They could be indicated by initial letters with the nature of the substituent groups R_1 and R_2 indicated.

From the point of view of the use of the quinonimide dyes as indicators of oxidation-reduction it is a matter for regret that they cannot be described on a uniform basis. For example the numbering of acridine nucleus according to some authors begins next to the carbon atom, while in the oxazines the numbering begins next to the nitrogen atom. The asymmetrical numbering that Mansfield Clark used for the indophenols bears no relation to the other quinonimide dyes. It is the structural relationship in the series that we emphasize here as being of importance. The "bridging group" for indophenol (5,6') has an important influence on the oxidation reduction properties and on the absorption spectra. This effect is apparently minimal when the bridging group, forming the six-membered ring, is of a saturated character; it would then be regarded as potentially incapable of being involved in any tautomeric changes. This is perhaps the simplest generalization concerning the range of properties exhibited by the compounds cited here. The *practical* advantage of a bridging group would be to give a structure more stable than the original indophenol.

Acknowledgements

The writer wishes to thank Professor D. A. Walker, Dr. Norman Good, Dr. G. R. Bedford and Dr. B. R. Webster for their help. He is grateful to Professor F. G. Young, F.R.S. for facilities in the Department.

References

1. H. G. Gibbs, W. L. Hall and W. Mansfield Clark, Supplement No. 69 to the Public Health reports, Washington DC (1928).
2. A. S. Holt and C. S. French, *Arch. Biochem.*, **19** (1948) 368.

3. P. Ehrlich, *Das Sauerstoff-Bedürfniss des Organismus*, Hirschwald, Berlin, 1885, see also "The Requirement of the Organism for Oxygen" tr. editors, in *The Collected Papers of Paul Ehrlich*, Vol. 1, H. H. Dale, F. Himmelweit and M. Marquardt (eds.), Pergamon Press, 1956, p. 433.
4. International Union of Biochemistry: Report of the Commission on Enzymes (1964). (Reprinted in *Comparative Biochemistry*, Vol. 13, M. Florkin and E. H. Stotz (eds.), 1965, p. 84.
5. F. Rohmann and W. Spitzer, *Chem. Ber.*, **28** (1895) 567.
6. D. Keilin, *Proc. R. Soc. London* B **104** (1929) 206.
7. W. Mansfield Clark, *Oxidation-Reduction Potentials of Organic Systems*, Baillière, Tindall and Cox, London, 1960.
8. E. S. G. Barron, *J. Biol. Chem.* **97** (1932) 287.
9. R. Hill, G. R. Bedford and B. R. Webster, *J. Chem. Soc.* C **1970**, 478.
10. C. Liebermann, *Chem. Ber.*, **7** (1874) 1098.
11. A. Baeyer and H. Caro, *Chem. Ber.*, **7** (1874) 963.
12. R. Nietzki, *Chemistry of the Organic Dyestuffs*, tr. A. Collin and W. Richardson, Gurney and Jackson, London, 1892, p. 165.
13. F. Kehrmann, H. Goldstein and P. Tschudi, *Helv. Chim. Acta*, **2** (1919) 2315.
14. H. Goldstein and W. Kopp, *Helv. Chim. Acta*, **11** (1928) 478.
15. B. R. Webster and G. R. Bedford, personal communication, 1970.
16. J. F. Corbett, *J. Chem. Soc.* B **1970**, 1418.
17. N. E. Good, personal communication, 1971.
18. D. M. Stokes and D. A. Walker, personal communication, 1970.
19. H. Brunner and P. Chuit, *Chem. Ber.*, **21** (1888) 249.
20. Beilstein's Handbuch, 4 Aufl., **6** (1923) 137.

The Role of Lipid-Linked Activated Sugars in Glycosylation Reactions

W. J. Lennarz and Malka G. Scher

Department of Physiological Chemistry, The Johns Hopkins University, School of Medicine, Baltimore, Maryland

Introduction

In the prokaryotic eubacteria the single cytoplasmic membrane appears to serve many of the functions performed by the numerous subcellular organelles found in eukaryotic organisms. Thus, the cytoplasmic membrane of the bacteria is the site of oxidative phosphorylation; it is the site of active transport; and it is the site of biosynthesis of many of the macromolecular components of the cell surface. In this chapter attention will be focused on biosynthesis of the sugar-containing macromolecules that ultimately reside within or outside the membrane surface. More specifically it will deal with the mechanism of energy utilization in the biosynthetic processes involving glycosyl transfers from high energy, lipid-linked sugar intermediates first found in the eubacteria. It should be noted, however, that although it is in the prokaryotic bacteria where such systems have been most definitely established, preliminary reports suggest that analogous reactions may occur in eukaryotic cells.[1,2,3]

Relationship of High Energy Hydrophobic Sugar Intermediates to ATP Production

It is worthwhile to consider first the overall sequence of events involved in synthesis of the glycans of the cell envelope. Studies indicate that in most bacteria the origin of ATP is the cytoplasmic membrane, wherein the enzymes of oxidative phosphorylation reside. ATP thus formed presumably diffuses from the membrane to other locations for utilization in a great variety of energy-dependent biosynthetic processes. One class of compounds whose formation involves ATP either directly or indirectly is the other nucleoside triphosphates. Cytoplasmic enzymes catalyze the formation of these compounds. Also present in the cytoplasm are enzymes that catalyze the synthesis of the nucleoside diphospho sugars from the nucleoside triphosphates. These sugar nucleotides are, of course, the classical donors of glycosyl groups in formation of a great variety of polysaccharides, as first demonstrated by Leloir and coworkers for the biosynthesis of glycogen.[4] However, in the bacteria some of the glycans that are found within or beyond the cytoplasmic membrane are formed in part not by direct glycosylation reactions via sugar nucleotides, but rather by glycosylations after the hexose is transferred to a carrier lipid. Since the membrane constitutes a hydrophobic permeability barrier, a lipid is particularly well suited to serve as carrier of activated glycosyl groups involved in reactions within the membrane.

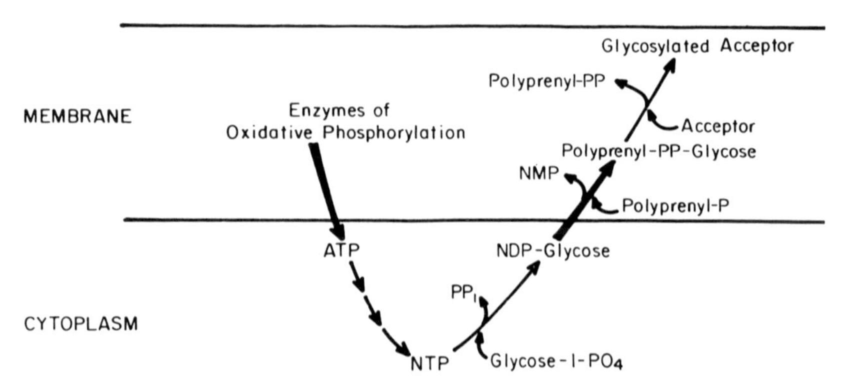

Figure 1. Production and utilization of high energy phosphate compounds for polysaccharide synthesis. NTP, NDP, and NMP represent nucleoside triphosphate, diphosphate and monophosphate, respectively.

Thus, as summarized in Fig. 1, following ATP synthesis at the membrane, a number of reactions in the cytoplasm lead to formation of sugar nucleotides. These sugar nucleotides can react with the carrier lipid in the membrane, thereby being converted into hydrophobic, high-energy glycosyl compounds that can participate in biosynthetic processes in the membrane. Thus, by this rather indirect route the energy of ATP synthesized in the membrane is utilized for biosynthesis of glycan components in the membrane.

Discovery of the Involvement of a Carrier Lipid in Glycan Synthesis

Cell wall peptidoglycan synthesis has been studied most extensively in *Staphylococcus aureus* and *Micrococcus lysodeikticus*. During investigations on peptidoglycan synthesis in these organisms, Strominger and his coworkers discovered that their particulate enzyme preparations catalyzed the incorporation of sugar residues supplied as UDP-*N*-acetyl-D-glucosamine and UDP-*N*-acetyl-D-muramyl pentapeptide into a lipophilic compound as well as into polysaccharide.[5] Synthesis of the lipid measured as a function of time indicated that it was formed prior to formation of the polysaccharide. It was also necessary to demonstrate that the lipophilic compound could function independently as a substrate for peptidoglycan synthesis. Therefore, the lipid intermediate was generated in an initial incubation of particulate enzyme with the sugar nucleotides. The particular preparation was washed free of sugar nucleotide precursors and was supplied as the only substrate in a further incubation. Peptidoglycan was, in fact, produced in this second reaction. Thus, a disaccharide-containing lipophilic compound was proposed as an intermediate in synthesis of peptidoglycan from sugar nucleotides (Fig. 2, reactions 1 and 2). Subsequent investigations established the structure of this novel lipid to be undecaprenyl phosphate and elucidated the cyclical nature of its involvement in the

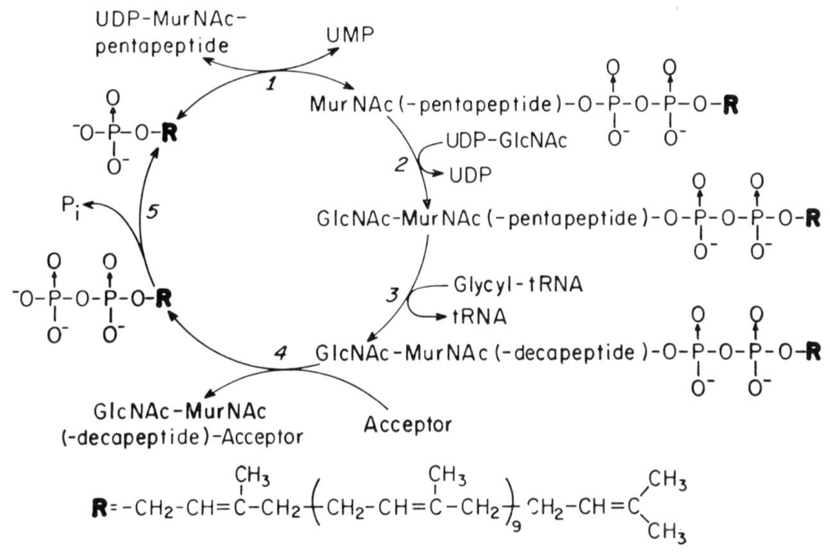

Figure 2. Sequence for the biosynthesis of the cell wall peptidoglycan. The abbreviations used here are: GlcNAc (*N*-acetyl-D-glucosamine); MurNAc (*N*-acetyl-D-muramic acid).

biosynthetic pathway. As indicated, the initial step in the biosynthetic pathway is reversible.[6] From the high energy bond of UDP-*N*-acetylmuramyl pentapeptide, initially a cytoplasmic component, a high energy glycosyl-1-phosphoryl bond is synthesized in the membrane.[7] This high energy bond remains intact until the final transfer of sugar residues from the lipid intermediate to form a glycosidic bond with acceptor polysaccharide. The second step in the reaction sequence produces the disaccharide containing lipid, and at this stage modifications of the peptide chain occur, such as the sequential transfer of glycine units from t-RNA to the lipophilic intermediate to form a pentaglycine side chain (reaction 3).[8,9] Reaction 4 is the transfer of the disaccharide to endogenous polysaccharide acceptor and concomitant release of pyrophosphoryl undecaprenol. It is not yet known if the disaccharide is further polymerized on the lipid before transfer to polysaccharide. Phosphoryl undecaprenol must be enzymatically regenerated (reaction 5) before participating again as acceptor lipid. This reaction is the site of inhibition of peptidoglycan synthesis by bacitracin.[10] Stone and Strominger have only recently presented data which suggests that the mechanism of inhibition is the formation of a complex between bacitracin, C_{55}-polyisoprenyl pyrophosphate and divalent cations in which the cation serves as a bridge between the lipid and the antibiotic.[11] The synthesis of peptidoglycan is completed by transpeptidation resulting in cross-linking of the peptide chains via the pentaglycine bridge.[12] The biosynthesis of peptidoglycan, as well as other polysaccharides discussed in this review, has been described in depth in recent review articles.[13,14,15]

Similar pathways involving undecaprenyl phosphate intermediates operate in the synthesis of the *O*-antigen portion of the lipopolysaccharide associated with the cell envelope of *Salmonella typhimurium*[16] and *S. newington*.[17] The *O*-antigen chain of *S. typhimurium* consists of branched tetrasaccharide repeating units. The repeating trisaccharide is mannosylrhamnosylgalactose, and abequose linked to mannose forms the branch. *S. newington* contains the same repeating trisaccharide but the configurations are different and there is no branch sugar. It was initially reported that *S. typhimurium* cell envelope preparations catalyzed incorporation of radioactive sugars from the appropriate sugar nucleotides into a lipophilic product which contained galactose-1-phosphate as the reducing terminal sugar residue.[16] The radioactivity of the intermediate in the lipid phase was transferred to the macromolecular product by raising the temperature of the reaction from 10° to 37°. At the same time Wright *et al.*[17] reported synthesis of a butanol-extractable radioactive compound with an *S. newington* preparation in the presence of the labeled sugar nucleotides UDP-galactose, TDP-rhamnose and GDP-mannose. The addition of GDP-mannose subsequent to incorporation of sugar residues from UDP-galactose and TDP-rhamnose into lipid caused a decrease in the level of radioactivity in the lipid, but an increase in radioactivity in the non-dialyzable polymer.

The initial step in the reaction sequence (Fig. 3) is the reversible transfer of galactose-1-phosphate from UDP-galactose to acceptor phosphate, with release of UMP.[16,17,18] Following formation of galactosyl-pyrophosphoryl-undecaprenol, rhamnose and mannose are transferred sequentially to the monosaccharide lipid intermediate. Galactosylpyrophosphoryl-antigen carrier lipid and rhamnosylgalactosylpyrophosphoryl-antigen carrier lipid have been isolated and used as acceptors for the addition of rhamnose and mannose, respectively.[19] At this stage polymerization occurs in the *S. newington* system. Addition of abequose catalyzed by *S. typhimurium* preparations requires the trisaccharide-lipid intermediate as a substrate, but the polymerized trisaccharide-lipid will not accept abequose (from CDP-abequose).

A dimer of the tetrasaccharide attached to lipid has been isolated and characterized, indicating that polymerization for the *O*-antigen system does occur at the level of lipid intermediates.[20,21] Polymerized oligosaccharide is then transferred to core lipopolysaccharide and undecaprenyl pyrophosphate is released. Undecaprenyl phosphate is regenerated by dephosphorylation of lipid pyrophosphate in a bacitracin sensitive reaction.[22]

Robbins and his coworkers also discovered that *O*-antigen synthesis proceeds by addition of trisaccharide repeating units at the reducing

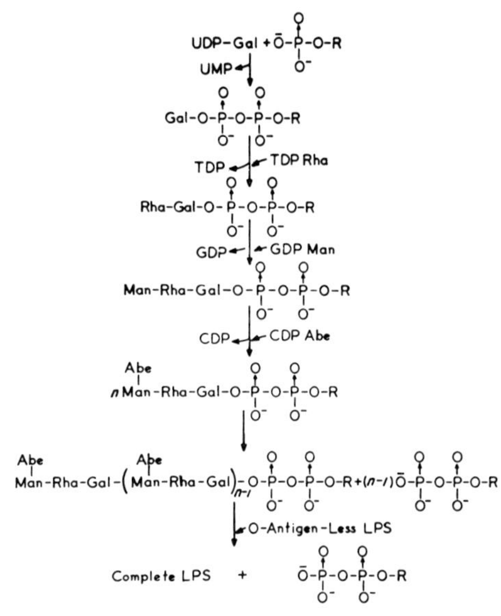

Figure 3. Sequence for the biosynthesis of the *O*-antigen chain of the lipopolysaccharide of *S. typhimurium*. The abbreviations used here are: Gal (D-galactose); Rha (L-rhamnose); Man (D-mannose); Abe [abequose (3,6-dideoxy-D-galactose)]; LPS (lipopolysaccharide); and R (undecaprenol).

end of growing chains.[23] Their study involved pulse-labeling of growing cultures with ^{14}C-glucose or pulse labeling a cell free system with UDP-galactose-^{14}C followed by borohydride treatment of *O*-antigen chains which converted the reducing terminal galactose unit to galactitol, thus permitting one to distinguish the reducing from the non-reducing end of the chain.

The lipid intermediates from the peptidoglycan and *O*-antigen synthesizing systems have been isolated and purified. In independent studies from the laboratories of Strominger and Robbins the presence of a pyrophosphate linkage between sugars and the lipid was established and the lipid moiety of each intermediate released upon acid hydrolysis was analyzed by mass spectroscopy.[24,25] The structure thereby established was a polyisoprenoid alcohol containing 11 isoprene units, each unit having a double bond (cf. Fig. 2). Whether or not identical isomers of undecaprenol are involved in both polysaccharide synthesizing systems has not been determined, but the significance of this mechanism for synthesis of polysaccharides which are located external to the cytoplasmic membrane increased when it was discovered that undecaprenyl phosphates also mediate the synthesis of other steps in biosynthesis of *O*-antigen as well as the synthesis of a variety of other complex glycans.

The Role of Carrier Lipids in Other Bacterial Systems

Recently two enzymatic systems which utilize polyisoprenol intermediates to modify the *O*-antigen chain during lipopolysaccharide biosynthesis have been reported, and both of these systems involve transfer of a single hexose residue to a preformed oligosaccharide or polysaccharide. In addition the sugar residue in both intermediates is linked to the lipid via a phosphodiester bridge. Wright has studied the biosynthesis of the lipopolysaccharide in E group *Salmonella*[26,27] and has reported upon the mechanism of the modification reaction in which glycosyl groups are added to the *O*-antigen in lysogenic cells. The *O*-antigen of E group *Salmonella* is a heteropolysaccharide composed of mannosylrhamnosylgalactosyl units. In cells lysogenic for either bacteriophage ϵ^{15} or ϵ^{34} the galactosyl units are of the β-configuration. In cells doubly lysogenic for ϵ^{34} and ϵ^{15} the galactosyl units are also substituted with α-D-glucosyl groups. Incubation of particulate enzyme from doubly lysogenic cells with ^{32}P-UDP-glucose-^{14}C results in transfer of the ^{14}C but not ^{32}P to the particulate preparation. Lipid-linked ^{14}C-glucose comprises 54% of the radioactivity transferred and the remainder was characterized as ^{14}C-lipopolysaccharide. The lipid intermediate that upon further incubation can transfer ^{14}C-glucose to lipopolysaccharide has a ratio of glucose to phosphate of 0·9:1. The sugar moiety is β-glucose-1-phosphate. The free lipid produced by acid hydrolysis was analyzed by mass spectrometry and the results indicated that the lipid is a mixture of C_{55} and C_{50} polyisoprenoid alcohols, with each isoprene unit having one double bond. Thus, without knowledge of the geometric configuration, the lipid moiety is identical in structure to the lipid moieties involved in peptidoglycan, *O*-antigen, capsular polysaccharide, and mannan synthesis. The lipid is distinguished from peptidoglycan and *O*-antigen intermediates by the presence of a phosphodiester bridge rather than a pyrophosphoryl bridge but it is similar in this respect to mannosylphosphoryl-undecaprenol (see below).

The acceptor for C^{14}-glucose in the glucose transfer reaction is proposed to be the growing *O*-antigen chain attached to antigen carrier lipid. Experimentally, this hypothesis was tested in a system which contained incomplete core lipopolysaccharide that was incapable of accepting O-antigen chains. In this system it was still possible to observe transfer of glucose from the carrier to O-antigen attached to lipid. Thus, it seems likely that glucosyl units are transferred from glucosyl lipid stepwise to the growing O-antigen chain still linked to lipid. A hexasaccharide ($x = 2$ in Fig. 4) would be the shortest possible sequence of repeating units that act as a glucose acceptor, since glucosylation requires β-galactosyl units and these are formed during polymerization of the repeating units catalyzed by the ϵ^{15}-specific O-antigen polymerase. Further experiments will be necessary to substantiate this proposal.

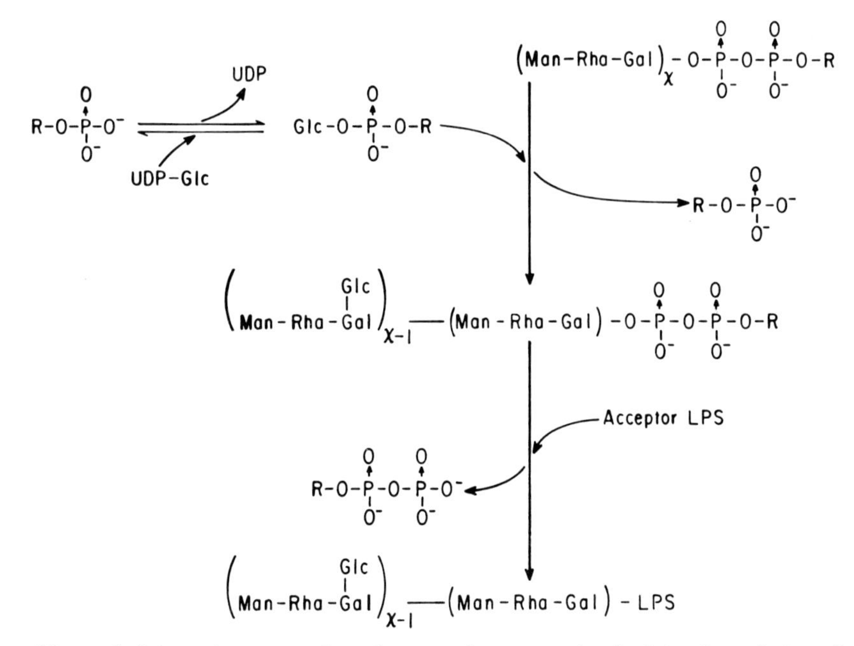

Figure 4. Schematic presentation of proposed sequence involved in glycosylation of *Salmonella* ϵ^{15}, ϵ^{34} O-antigen. The abbreviations used here are the same as those identified in Fig. 3 and Glc (D-glucose).

O-antigen factor 12_2 present in *Salmonella* groups A, B, and D, is determined by branches of the O-antigen which are nonreducing glucose residues linked to galactose units of the O-antigen side chain. Nikaido *et al.*[28] discovered that a cell envelope preparation from *Salmonella* strains possessing O-antigen 12_2 catalyzes the transfer of glucose from UDP-glucose to an endogenous acceptor. Results of chromatography of the radioactive product indicate that it is most probably lipopolysaccharide and not merely the O-antigen side chains. Moreover, the lipopolysaccharide is characteristic of strains which bear a modified O-antigen because radioactivity is present in the product as glucose and the oligosaccharide glucosylgalactosylmannosylrhamnose was obtained after partial acid hydrolysis. The time course of incorporation of radioactivity from ^{32}P-UDP-glucose-^{3}H into lipid and lipopolysaccharide suggested that the glucosyl-lipid was an intermediate in the transfer of glucose, and glucosyl-lipid isolated from an incubation mixture served as a direct donor of glucose for lipopolysaccharide synthesis. The acid lability and chromatographic behavior after catalytic hydrogenation of this lipid intermediate and rhamnosyl-galactosyl antigen-carrier lipid were similar. However, phosphate is not transferred to the lipid along with glucose and the ratio of glucose to phosphate in the intermediate is approximately 1. Although the

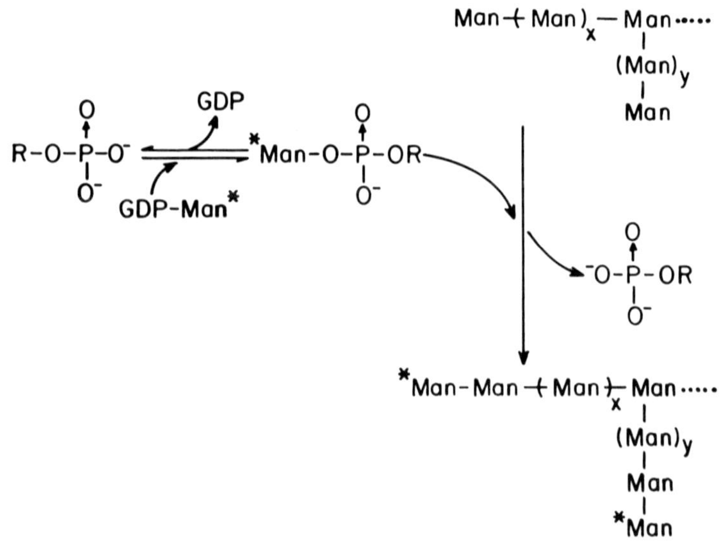

Figure 5. Role of mannosyl-1-phosphoryl undecaprenol in mannan synthesis in *M. lysodeikticus*. R is undecaprenol.

degree of unsaturation was not determined, the glucosyl-lipid carrier appeared to be the same as antigen carrier lipid and the structure of the intermediate was therefore proposed to be glucosyl-1-phosphoryl-polyisoprenol.[29] Using a *Salmonella* mutant which could not transfer completed *O*-antigen chains from antigen carrier lipid to the defective core of the lipopolysaccharide produced by this organism it was demonstrated that ^{14}C-glucose was transferred from UDP-glucose-^{14}C to the *O*-antigen side chains attached to antigen carrier lipid[30] in agreement with the hypothesis of Wright discussed above.

A phosphodiester linked lipophilic sugar intermediate is also involved in mannan synthesis catalyzed by a *Micrococcus lysodeikticus* particulate preparation.[31] The system utilizes GDP-mannose as the hexose source and the acceptor lipid required for the initial transfer of mannose has been isolated, purified and characterized as undecaprenyl phosphate[32] (Fig. 5). The lipid intermediate produced in this reaction has also been purified, and analysis established the structure to be mannosyl-1-phosphoryl-undecaprenol.[33] The isoprenoid moiety of the intermediate appears to be identical to the intermediates involved in peptidoglycan and *O*-antigen synthesis, although the sugar is linked to the lipid via a phosphodiester bridge rather than a pyrophosphoryl bond. Clearly, more than one sugar residue is attached to the lipid intermediates of peptidoglycan and *O*-antigen synthesis, but it has not been possible to detect a lipid intermediate containing mannose oligosaccharides in the mannan biosynthetic system. Moreover, ^{14}C-mannosyl units are transferred only to the non-reducing termini of endogenous mannan. Mannosyl-1-phosphoryl-undecaprenol thus seems to function only to complete the synthesis of endogenous mannan, in a manner similar to the glucosyl-1-phosphoryl-undecaprenol intermediates which serve in the formation of branches in lysogenic *Salmonella O*-antigen. Addition of mannosyl units to the non-reducing termini is analogous to the classical method of synthesizing glycogen and starch by addition of glucosyl units to the non-reducing ends of growing chains[34] and is in contrast to the mechanism of *O*-antigen synthesis.

Biosynthesis of capsular polysaccharide with the participation of lipid intermediates in *Klebsiella* (*Aerobacter*) *aerogenes* has been reported by Troy *et al.*[35] The polysaccharide is composed of the repeating trisaccharide galactosylmannosylgalactose with glucuronate branches on each mannose residue (Fig. 6). Initially it was observed that ^{14}C-galactose accumulates in the lipid phase as a result of incubation of a cell envelope fraction with UDP-galactose-^{14}C at 12°. The first step of the reaction cycle is reversible and results in production of galactosyl pyrophosphoryl-undecaprenol. The acceptor lipid has been definitively characterized primarily by means of mass spectrometry as undecaprenyl phosphate. This lipid restores enzymatic activity to a lipid depleted particulate preparation and it is active as acceptor lipid for mannan synthesis.[36] The functional equivalence of the acceptor lipids from *M. lysodeikticus* and *A. aerogenes* does not necessarily indicate identical structures since the enzymes may not have absolute specificity for the acceptor. Mannosyl and glucuronyl units are sequentially added to the galactosyl lipid intermediate. Glucuronic acid must be incorporated into the lipid intermediate before the fourth sugar, another galactose, can be added to the repeating units. An octasaccharide which consists of 2 tetrasaccharide repeating units was obtained after release from the carrier phospholipid indicating that polymerization occurs at the level of lipid intermediate. Ultimately the radioactive oligosaccharide is transferred to an acceptor to form polysaccharide product. The identification of endogenous acceptor for the polymerized oligosaccharide attached to lipid remains unknown. The polysaccharide formed by the system was characterized as capsular material with the aid of specific phage induced capsular polysaccharide depolymerase which has endogalactosidase activity. Participation of a lipid intermediate, the structure of which is not clearly defined, has been described in synthesis of capsular polysaccharide from another strain of *Klebsiella*.[37]

Douglas *et al.*, have detected accumulation of a lipid intermediate containing *N*-acetylglucosamine-^{14}C during biosynthesis of *Staphylococcus lactis* I3 teichoic acid which consists of *N*-acetylglucosamine and glycerol linked via phosphodiester bridges.[38] In the absence of CDP-glycerol, *N*-acetylglucosaminyl lipid accumulation was stimulated and the sugar was transferred to teichoic acid upon further incubation of the

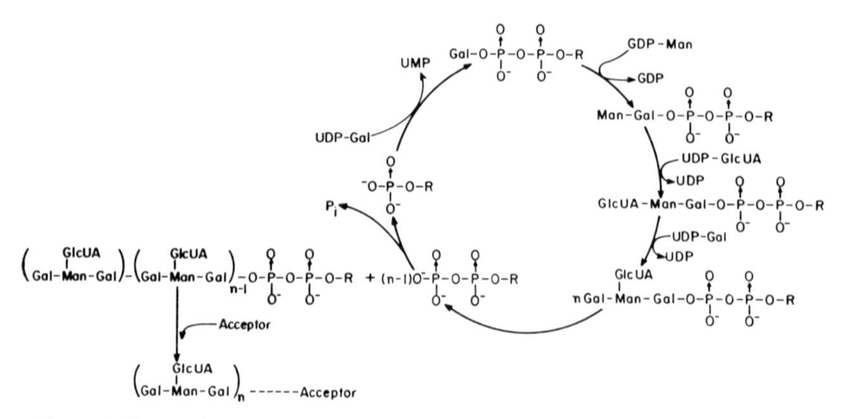

Figure 6. Proposed sequence for capsular polysaccharide synthesis. The abbreviations used here are those in Fig. 3. GlcUA designates a glucuronic acid residue.

washed particulate preparation with CDP-glycerol. Neither the structure of this lipid nor the structure of a similar *N*-acetylglucosaminyl lipid which has been detected during synthesis of the *S. lactis* 2102 wall polymer[39,40] has been determined. The acid lability of these pyrophosphate containing intermediates suggests the presence of a double bond β to a phosphate group and, therefore, the authors suggest that the lipid may be of the polyisoprenoid type.

If the intermediates in teichoic acid biosynthesis were found to have the same structures as the other isoprenoid lipid carriers, this system and the peptidoglycan synthesizing system would compete for the isoprenoid acceptor lipid. Watkinson *et al.* have investigated the competition between the two systems in *S. lactis* I3 by measuring teichoic acid synthesis alone and teichoic acid synthesis in the presence of peptidoglycan synthesis.[41] Teichoic acid synthesis was measured in the presence of CDP-glycerol-^{14}C, UDP-*N*-acetylglucosamine, particulate enzyme preparation and appropriate cofactors. Upon the addition of UDP-*N*-acetylmuramyl pentapeptide, which is required to form the lipid intermediate for peptidoglycan synthesis, the percentage of substrate incorporated into teichoic acid decreased from 83 to 65%, suggesting a possible competition of the teichoic acid and peptidoglycan system for the lipid. Upon the addition of bacitracin or vancomycin as well as UDP-*N*-acetylmuramyl peptapeptide to the reaction mixture synthesizing teichoic acid, the percentage of substrate incorporated into teichoic acid was further reduced. These antibiotics do not modify the extent of substrate incorporation into teichoic acid in the absence of UDP-*N*-acetylmuramyl pentapeptide. However, they are known to inhibit peptidoglycan synthesis by preventing the dephosphorylation of undecaprenyl pyrophosphate (bacitracin) and by inhibiting the sugar transfer from lipid intermediates to acceptor (vancomycin).[7] The net effect of either antibiotic, therefore, is a depletion of the supply of acceptor lipid, undecaprenyl phosphate, necessary to maintain the cyclical nature of peptidoglycan synthesis. The observed inhibition in teichoic acid synthesis in the presence of these antibiotics and UDP-*N*-acetylmuramyl pentapeptide thus is consistent with the theory of competition between the two systems, since the inhibition of teichoic acid synthesis may be viewed as a consequence of the reduced.quantity of undecaprenyl phosphate in the presence of the antibiotics.

In another report it was demonstrated that chloramphenicol can exert inhibition directly on synthesis of teichoic acids, and this effect was independent of any effect on protein synthesis.[42] Chloramphenicol effectively inhibited only those enzymatic systems which produced polymers containing glucose in the main chain and which had been shown in unpublished experiments to involve lipid intermediates. The authors concluded that the site of inhibition in these cases, therefore, was at the stage of transfer of glucose from the nucleotide precursor to the lipid. However, it is clear that further studies must be made before the mechanism of the chloramphenicol inhibition is understood.

Concluding Comments

The concept which unifies these independent investigations on undecaprenol intermediates is the participation of a lipid, effectively serving as a coenzyme, that is associated with the cell membrane. The location of the lipid in the membrane permits it to mediate the transfer of low molecular weight, hydrophilic compounds that serve as the building blocks of macromolecules localized within or beyond the hydrophobic cell membrane.

The most distinct difference among the lipid intermediates formed in the different systems described here is the two types of linkages formed between the sugar moieties and the lipid moiety. A pyrophosphate group forms the bridge between sugars and lipid in the intermediates for peptidoglycan, *O*-antigen, and capsular polysaccharide synthesis,[24,25,35] but a phosphodiester linkage is present between sugar and lipid in the intermediate for mannan synthesis and glucosylation of *O*-antigen (Fig. 7).[27,28,33] Nevertheless, the first step in all of the cycles of polysaccharide biosynthesis is the same: that is, a freely reversible reaction resulting in the synthesis of a new glycosyl-1-phosphoryl bond that serves to link the sugar to the hydrophobic lipid. The phosphodiester and pyrophosphate linked intermediates thus

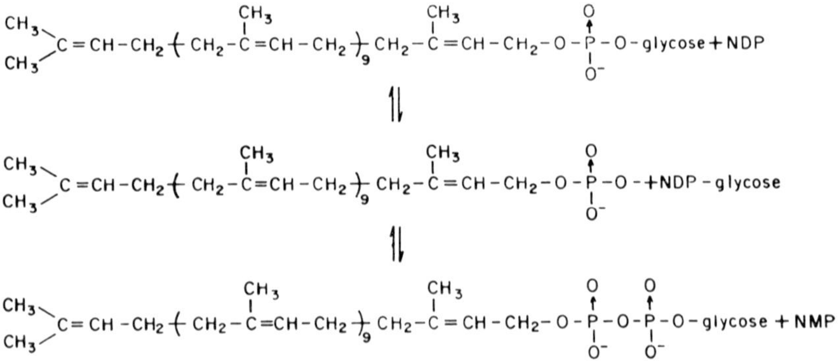

$$(CH_3)_2C{=}CH{-}CH_2{-}(CH_2{-}\overset{CH_3}{\overset{|}{C}}{=}CH{-}CH_2)_9{-}CH_2{-}\overset{CH_3}{\overset{|}{C}}{=}CH{-}CH_2{-}O{-}\underset{O^-}{\underset{|}{\overset{O}{\overset{\uparrow}{P}}}}{-}O{-}\text{glycose}+\text{NDP}$$

$$\rightleftharpoons$$

$$(CH_3)_2C{=}CH{-}CH_2{-}(CH_2{-}\overset{CH_3}{\overset{|}{C}}{=}CH{-}CH_2)_9{-}CH_2{-}\overset{CH_3}{\overset{|}{C}}{=}CH{-}CH_2{-}O{-}\underset{O^-}{\underset{|}{\overset{O}{\overset{\uparrow}{P}}}}{-}O^-+\text{NDP-glycose}$$

$$\rightleftharpoons$$

$$(CH_3)_2C{=}CH{-}CH_2{-}(CH_2{-}\overset{CH_3}{\overset{|}{C}}{=}CH{-}CH_2)_9{-}CH_2{-}\overset{CH_3}{\overset{|}{C}}{=}CH{-}CH_2{-}O{-}\underset{O^-}{\underset{|}{\overset{O}{\overset{\uparrow}{P}}}}{-}O{-}\underset{O^-}{\underset{|}{\overset{O}{\overset{\uparrow}{P}}}}{-}O{-}\text{glycose}+\text{NMP}$$

Figure 7. Structure of phosphate and pyrophosphate containing intermediates and the transfer reactions leading to their synthesis. NDP-glycose, NDP and NMP represent a sugar nucleotide, nucleoside diphosphate, and nucleoside monophosphate, respectively.

retain essentially the same reactivity as the sugar nucleotides and therefore, with regard to energetics, are just as effective as the sugar nucleotides in the biosynthesis of glycans.

References

1. C. L. Villimez, *Biochem. Biophys. Res. Commun.*, **40** (1970) 636.
2. J. F. Caccam, J. J. Jackson and C. H. Eylar, *Biochem. Biophys. Res. Commun.*, **35** (1969) 505.
3. N. H. Behrens, A. J. Parodi, L. F. Leloir and C. R. Krisman, *Arch. Biochem. Biophys.*, **143** (1971) 375.
4. L. F. Leloir, J. M. Olavarria, S. H. Goldenberg and H. Carminatti, *Arch. Biochem. Biophys.*, **81** (1959) 508.
5. J. S. Anderson, M. Matsuhashi, M. A. Haskin, and J. L. Strominger, *Proc. Natn. Acad. Sci.*, **53** (1965) 881.
6. M. G. Heydanek, Jr., W. G. Struve and F. C. Neuhaus, *Biochem.*, **8** (1969) 1214.
7. J. S. Anderson, M. Matsuhashi, M. A. Haskin and J. L. Strominger, *J. Biol. Chem.*, **242** (1967) 3180.
8. M. Matsuhashi, C. P. Dietrick and J. L. Strominger, *Proc. Natn. Acad. Sci.*, **54** (1965) 587.
9. T. Kamiryo and M. Matsuhashi, *Biochem. Biophys. Res. Commun.*, **36** (1969) 215.
10. G. Siewert and J. L. Strominger, *Proc. Natn. Acad. Sci.*, **57** (1967) 767.
11. K. J. Stone and J. L. Strominger, *Proc. Natn. Acad. Sci.*, **68** (1971) 3223.
12. J. Ghuysen, *Bacteriol. Rev.*, **32** (1968) 425.
13. M. J. Osborn, *Ann. Rev. Biochem.*, **38** (1969) 501.
14. E. C. Heath, *Anm. Rev. Biochem.*, **40** (1971) 29.
15. A. Wright and S. Kanegasaki, *Physiol. Rev.*, **51** (1971) 748.
16. I. M. Weiner, T. Higuchi, L. Rothfield, M. Saltmarsh-Andrew, M. J. Osborn and B. L. Horecker, *Proc. Natn. Acad. Sci.*, **54** (1965) 228.
17. A. Wright, M. Dankert and P. W. Robbins, *Proc. Natn. Acad. Sci.*, **54** (1965) 235.
18. M. Dankert, A. Wright, W. S. Kelley and P. W. Robbins, *Arch. Biochem. Biophys.*, **116** (1966) 425.

19. S. Kanegasaki and A. Wright, *Proc. Natn. Acad. Sci.*, **67** (1970) 951.
20. M. J. Osborn and I. M. Weiner, *J. Biol. Chem.*, **243** (1968) 2631
21. J. L. Kent and M. J. Osborn, *Biochem.*, **7** (1968) 4419.
22. P. W. Robbins and M. J. Osborn, unpublished observations, cited in ref. 13.
23. P. W. Robbins, D. Bray, M. Dankert, and A. Wright, *Science*, **158** (1967) 1536.
24. A. Wright, M. Dankert, P. Fennessey, P. W. Robbins, *Proc. Natn. Acad. Sci.*, **57** (1967) 1798.
25. Y. Higashi, J. L. Strominger and C. C. Sweeley, *Proc. Natn. Acad. Sci.*, **57** (1967) 1878.
26. A. Wright, *Federation Proc.*, **28** (1969) 658.
27. A. Wright, *J. Bacteriol.*, **105** (1971) 927.
28. H. Nikaido, K. Nikaido, T. Nakae and P. H. Mäkelä, *J. Biol. Chem.*, **246** (1971) 3902.
29. K. Nikaido and H. Nikaido, *J. Biol. Chem.*, **246** (1971) 3912.
30. M. Takeshita and P. H. Mäkelä, *J. Biol. Chem.*, **246** (1971) 3920.
31. M. Scher and W. J. Lennarz, *J. Biol. Chem.*, **244** (1969) 2777.
32. M. Lahav, T. H. Chiu and W. J. Lennarz, *J. Biol. Chem.*, **244** (1969) 5890.
33. M. Scher, W. J. Lennarz and C. C. Sweeley, *Proc. Natn. Acad. Sci.*, **59** (1968) 1313.
34. L. F. Leloir, M. A. R. de Fekete and C. E. Cardini, *J. Biol. Chem.*, **236** (1961) 636.
35. F. A. Troy, F. E. Frerman and E. C. Heath, *J. Biol. Chem.*, **246** (1971) 118.
36. F. E. Frerman, E. C. Heath, M. Lahav, T. H. Chiu and W. J. Lennarz, unpublished observations cited in ref. 35.
37. I. W. Sutherland and M. Norval, *Biochem. J.*, **120** (1970) 567.
38. L. J. Douglas and J. Baddiley, *FEBS Letters*, **1** (1968) 114.
39. D. Brooks and J. Baddiley, *Biochem. J.*, **113** (1969) 635.
40. D. Brooks and J. Baddiley, *Biochem. J.*, **115** (1969) 307.
41. R. J. Watkinson, H. Hussey and J. Baddiley, *Nature New Biology*, **229** (1971) 57.
42. M. Stow, B. J. Starkey, S. C. Hancock and J. Baddiley, *Nature New Biology*, **229** (1971) 56.

Lipid–Protein Interactions in the Structure of Biological Membranes

Giorgio Lenaz
Istituto di Chimica Biologica, Università di Bologna, Italy

I. *Introduction*

In 1971 the general ideas on the structure of biological membranes are perhaps less clear than before 1960; the reason is that, while at that time there was but one model generally accepted for membrane structure, now we have a variety of different models.

The Danielli–Davson model[82] has satisfied the requirements for the properties of natural membranes for three decades. In the most recent version of the Danielli hypothesis,[386] lipids form a bilayer with the polar heads turned toward the outside; proteins of two kinds cover the

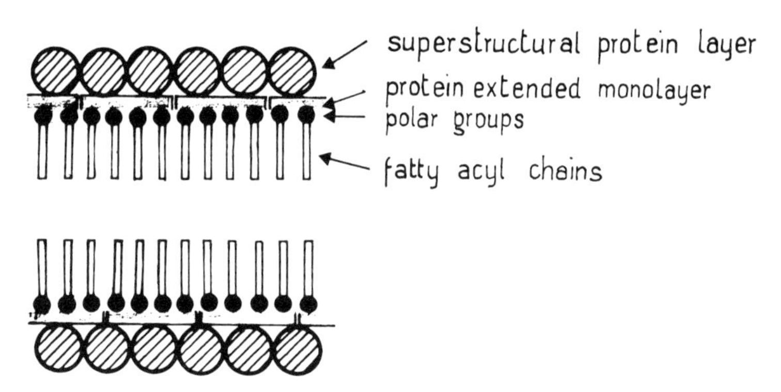

Figure 1. The Danielli model of membrane structure.

outer surfaces: a monomolecular layer, consisting of juxtaposed molecular protein units, and "superstructural" globular protein units (Fig. 1).

Recently some completely new hypotheses have been presented which reject the classical model;[18, 144, 199, 395] they have been proposed to account for hydrophobic bonding of membrane proteins with the nonpolar lipid chains. Lipoprotein subunits are envisaged to determine the membrane continuum in some of these models.

Experimental work directed to prove or disprove the various models has resulted in refinements of the previous versions or even in the proposal of new models. Recently the conflicting experimental data have led to the tendency for a moderate compromise, so that some allowances have been made by both sides.

There is little doubt that a lipid bilayer is a common feature of all membranes; what must be clarified is whether there is protein penetration into the bilayer, and, if there is penetration, to what extent does it contribute to membrane structure and how much it subverts the classical model. One party still supports the classical Danielli model,[159,386] but the possibility that proteins penetrate in part into the bilayer is not absolutely rejected; other investigators believe that protein subunits are immersed into a discontinuous lipid bilayer.[131,342,387] The most recent models are not equivalent and may be differentiated in the regularity of protein disposition in the membrane.

The clue for the final solution of membrane structure is the relation between the membrane molecular components, i.e. fundamentally lipids and proteins. Clearly researches may be articulated toward two main directions. What is the nature of the interactions responsible for the bonds holding together the variety of lipid and protein molecules? And what is the relative spatial disposition of the different molecules? A well-defined nature of the lipid-protein bonds is implicit in several models: while in the Danielli model the interaction of the lipids with the (first) protein layer must be fundamentally polar (ionic and hydrogen bonds, etc.), in most of the other models the proteins are in contact with the hydrocarbon chains of the phospholipids, and the binding has been largely assumed to be hydrophobic. Experimental evidence bearing on this point is one of the main reasons for preferring one or another model.

This review is concerned with an examination of the available evidence on lipid-protein interactions in model systems and in natural membranes in relation to membrane structure. Also steric orientation, physical state, environment and specificity of the lipids and conformation of the protein are examined; the final aim is a critical evaluation of the role of lipids in membrane structure and functions.*

II. *Lipid–Protein Interactions in Model Systems*

A. *Types of Interactions*

Several types of interactionshave been considered for lipid-protein bonds in natural membranes.[56,141,320,384]

Ionic interactions occur between molecules bearing net opposite charges: complexes are formed when anionic lipids like phosphatidyl serine interact with basic proteins;[64,65,83] zwitterionic compounds do not interact at their isoelectric points. Salts prevent ionic interactions

* Analysis of the literature generally covers a period up to the summer of 1971.

and dissociate the formed ionic complexes by neutralization of the surface charges.

Hydrophobic interactions[180] are the result of the repulsion for water of nonpolar groups or molecules; hydrocarbon chains have little affinity for water and hence preferentially assume a conformation in which they are removed from the aqueous medium. Hydrophobic bonds form because of favourable thermodynamics; since water molecules have lower affinity for nonpolar groups than for other water molecules, their quasi-crystalline order increases in proximity of nonpolar groups; if these groups leave the aqueous environment, a large gain in entropy occurs in the overall system and the order of water decreases. The actual strength of the bonds depends on formation of effective van der Waals attractions between nonpolar groups. Lipids in water assume particular configurations where the nonpolar fatty acyl chains "squeeze out" of water and the polar heads remain in contact with water.[61] The actual stability of the "bonds" depends on chain length; unsaturation or branching decrease the force of the binding since stable van der Waals attractions cannot become operative.

Increasing the ionic strength of the medium enhances hydrophobic interactions; however certain anions such as SCN^-, ClO_4^-, I^-, Br^-, are thought to weaken hydrophobic bonds.[155] These "chaotropic ions" induce "melting" or breakage of water structure;[391] water around these ions becomes more lipophilic, and the transfer of apolar groups to water is favored. Hydrophobic bonds, whose intrinsic nature is the repulsion of lipophilic molecules for water, are therefore weakened.

Hydrogen bonds have also been considered important;[386] many groups in lipids and proteins have hydrogen donor or acceptor characteristics; lipids capable of hydrogen bonding could be protein-bound through water bridges reinforced by extensive van der Waals interactions.[386]

Bonds mediated by divalent cations may also be present in membrane structure: divalent cations could bridge two phospholipid molecules between one another[386] or phospholipid with protein molecules.[172,246]

B. *Lipid–Lipid and Lipid–Water Interactions*

The physical behaviour of lipid molecules is a subject of extensive investigations in relation to the more complex lipid–water–protein systems like model or natural membranes.

(a) *Physical Properties of Amphipathic Lipids*

Membrane lipids are essentially amphipathic, i.e. they have a polar hydrophilic end and a long hydrophobic moiety formed by the fatty acyl chains. For classification and distribution of amphipathic lipids in relation to membranes several reviews are available (cf. e.g. refs. 12, 319, 379, 402).

When a pure anhydrous amphipathic compound like a phospholipid is heated, numerous phase transitions occur between the collapse of the tridimensional crystalline structure and complete melting;[56,222,402] differential calorimetric analysis has shown that when a phospholipid is heated, it reaches a transition point where a marked endothermic

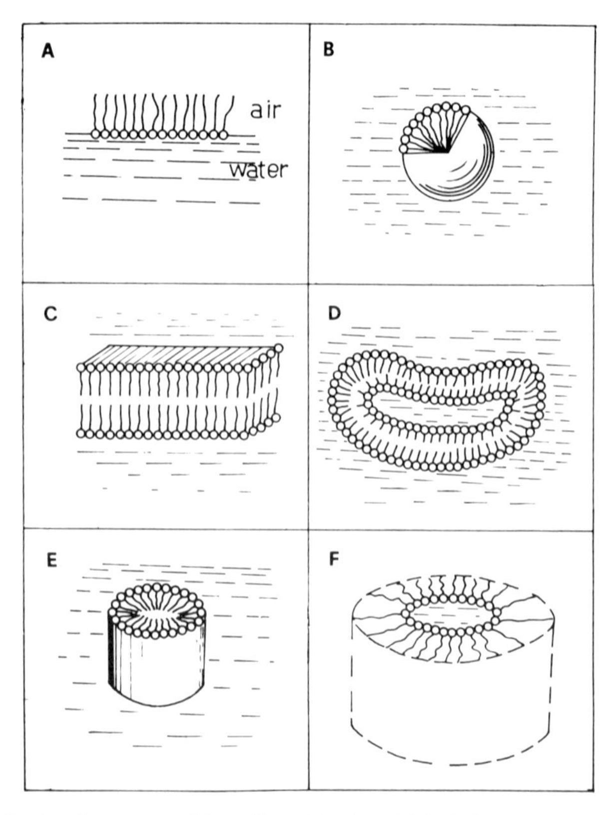

Figure 2. Lyotropic mesomorphism of amphipathic lipids. A. Monolayer at the air–water interface. B. Micelle in water. C Lamellar (smectic) phase in water. D. Lamellar phase: section through a sonicated liposome. E. Hexagonal phase I (oil-in-water). F. Hexagonal phase II (water-in-oil). The circles represent the polar heads of the lipid molecules.

change occurs and the aliphatic chains "melt" while the polar groups remain unchanged. The *mesomorphic* phases in which the paraffin chains take up disordered conformation are called *liquid crystalline phases*.

When phospholipids are hydrated, they exhibit both *lyotropic* and *thermotropic mesomorphism*, in other words they assume a variety of liquid crystalline phases which are a function of water content and temperature (Fig. 2). At low concentration in water amphipathic lipids are disposed in monolayers at the water–air interface. At higher concentrations the lipids dispose themselves in the aqueous medium;[11] the lamellar (smectic) phases are those predominant for lecithin and phospholipid mixtures at physiological temperatures. Several other phases besides the lamellar are possible.[222] Only at low concentration, but above the critical micelle concentration and above the so-called Kraft temperature, true micellar solutions are formed for certain amphipaths like lysolecithin. The lyotropic phases exhibit thermotropic mesomorphism: when increasing amounts of water are added, the transition temperatures are progressively lowered till a constant value is reached.[56]

The mesomorphic behaviour of phospholipids depends on their chemical nature. Besides the nature of the polar groups and the total arrangement of the molecule, also the nature of the hydrocarbon chains affects the phase transitions. The more unsaturated the phospholipids, the lower the temperature at which liquid crystalline phases are found. The reason is in the lower cohesion by van der Waals attraction between hydrocarbon chains having *cis* double bonds; at the normal temperatures in biological systems the natural phospholipids are above the transition temperatures, and therefore liquid-crystalline. Different molecules such as cholesterol[56] or steroid hormones[157] are found to affect the phase diagrams.

(b) *Phospholipids as Experimental Models for Membranes*

(i) *Monolayers.* Although monolayers of lipid molecules at an air–water interface are used as very simple systems to evaluate the chemical and physical interactions between molecules, many authors believe that such films are not completely satisfactory as structural models for biological membranes[219] since a monolayer does not separate two aqueous compartments as most membranes do. Experimentally the properties measured are surface pressure, potential, and viscosity as a function of surface area. Widely used are pressure-area curves: the area of the monomolecular film can be accurately determined under varying pressures. Different phospholipids exhibit different pressure-area curves: at room temperature, saturated and long chain phospholipids are more condensed than unsaturated and short chain phospholipids. In other words the area occupied by one molecule of phospholipid (Å/molecule) is lower for the saturated and long-chain compounds. Also the polar head will influence the packing of the monolayer.[48] Cholesterol induces a condensation effect in lecithin monolayers when unsaturated fatty acids are present in the lecithin; this interesting effect is still debated and not very clear.[47]

A great deal of studies concern the effects of subphase changes on the pressure-area curves in relation to excitability and transport properties of membranes as functions of phospholipids.[47,285] Particular importance in relation to the subject of this review is the interaction of protein added in the subphase with the monolayer (Section II. C).

(ii) *Bilayers*. When an amphipath above the transition temperature is smeared across a small hole joining two compartments in a plastic cup, a bimolecular film is formed, which now separates the two compartments. Formation of the film can be followed visually by appearance of a black spot which in a variable time extends to the whole film.

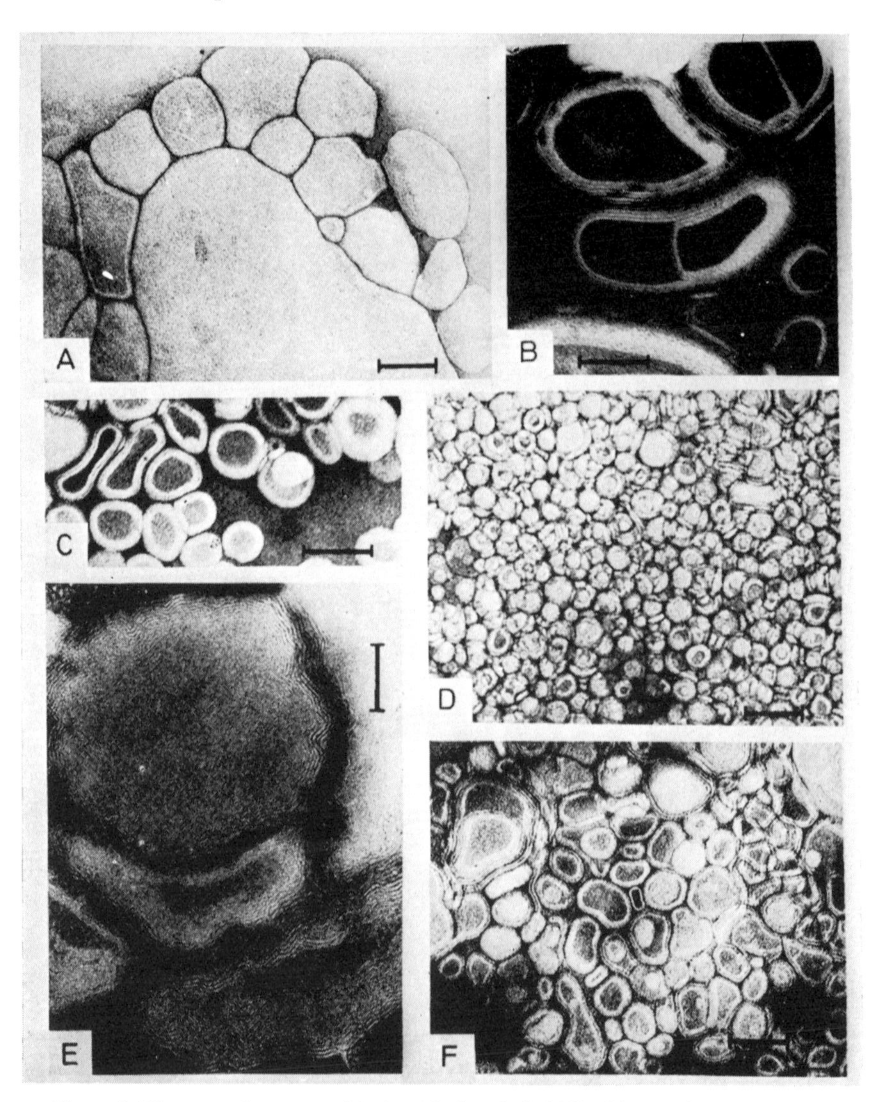

Figure 3. Electron microscopy of hydrated phospholipid liquid crystals. A, D, E, F dispersed in 145 mM ammonium acetate, subsequently stained with ammonium molybdate. B, C: dispersed in 2% ammonium molybdate. A, B: phosphatidylcholine mechanically shaken; C: phosphatidylcholine sonicated for 20 min; D: phosphatidylinositol sonicated for 50 min; E: phosphatidylethanolamine mechanically shaken; F: phosphatidylcholine-phosphatidylethanolamine (1:1) mechanically shaken in the presence of cytochrome *c* (0·03 μmole/μmole of lipid). Magnification: marker = 0·1 μ (from Papahadjopoulos and Miller (278), reproduced with kind permission of Dr. Papahadjopoulos).

These "black" lipid membranes are widely used as model membranes.[132,360] Addition of molecules or ions to the two aqueous compartments is used to modify the physical properties of the bilayer in relation to membrane physiological functions.

(iii) *Liposomes.* Myelin figures and liposomes are formed when phospholipids are dispersed in excess water. Prolonged ultrasonic irradiation of amphipathic lipids in water gives rise to homogeneous dispersions of vesicles of fairly constant diameter which retain the lamellar structure[12,360] (Fig. 3). Liposomes are very popular as model membrane systems.

The interaction of natural proteins or synthetic polypeptides has been studied, besides with synthetic long chain cationic or anionic detergents,[113,308] also with lipid monolayers, bilayers, or liposomes. The results obtained may be relevant to membrane structure and will be summarized here.

C. *Protein Interactions with Monolayers*

Lipid-protein interactions at the air–water interface have been conveniently studied with the monolayer approach.[285]

After the initial studies of Schulman,[96,245] Eley and Hedge[98,99] proposed that extended polypeptide chains of denatured protein

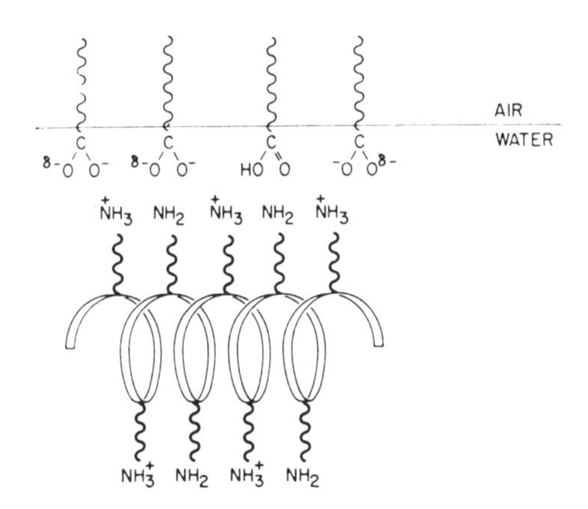

Figure 4. Schematic representation of interaction between stearic acid and poly-L-lysine in monolayers at pH 11. (D. O. Shah,[335] reproduced with kind permission of Dr. Shah and of Plenum Press).

penetrate the lipid monolayer by intercalating the nonpolar side chains of the amino acids between the hydrophobic chains of the lipid. It is now considered that at least secondary structure may be retained at the interface;[157,235] maximal (ionic) interaction of polylysine with stearic acid monolayers occurs at pH 11, where polylysine is α-helical[335] (Fig. 4). If the protein covered entirely the polar groups of the lipids, it would not be accessible to lipases: since lecithin is readily available to phospholipase A after interaction with proteins, and since the proteins are also available to proteolytic enzymes, Colacicco[69] has proposed a mosaic model where packages of lipids alternate with packages of proteins. Penetration of protein in the monolayers has been measured by increase in the film pressure and by actual measure of

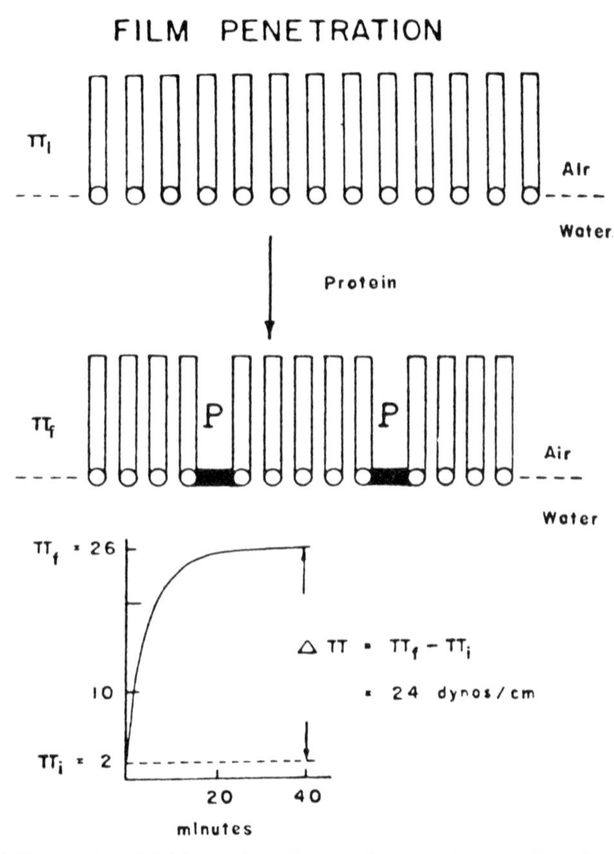

Figure 5. Penetration of lipid monolayer by protein at the air–water interface. The letter P, in central panel, designates the protein, which is filling some spaces between the lipid (Colacicco,[69] reproduced with kind permission of the author).

protein which has left the aqueous subphase (Fig. 5). In a first mechanism, *free penetration*, protein accommodated in the hydrophobic intermicellar lipid spaces accounts for the increase in surface pressure; the excess protein which does not contribute to the increase in surface pressure can be accommodated in a thick layer extending in the aqueous subphase. In *binding-mediated penetration* protein is first adsorbed under the lipid (as shown by an initial rapid loss of film pressure) and then penetrates the film. A third mechanism, *binding-inhibited penetration*, is characterized by a small $\Delta\pi$ value: the protein bound to the lipid hydrophilic groups blocks further penetration. The different mechanisms depend on the nature of the lipids.

Dawson and Quinn[86] have studied the interaction of proteins with different lipid monolayers at increasing surface pressures: at low pressure the protein molecules enter gaps in expanded films of lecithin; as the surface pressure increases, penetration (shown as interfacial radioactivity of ^{14}C-carboxymethylated proteins) is inhibited; at high pressures there is complete desorption of previously penetrated protein. A different situation occurs for the adsorption of cytochrome *c* with acidic phospholipids: interaction occurs up to 24 dynes/cm but without any penetration. On compressing a film of phosphatidyl ethanolamine (PE)* containing penetrated cytochrome *c* to 40 dynes/cm, only a portion of the protein is ejected from the monolayer; however if 1 M NaCl is added to the subphase, all the protein is ejected. This shows the role of ionic forces in protein interaction with phospholipid films.[294]

Kimelberg and Papahadjopoulos[185] studied the interaction of negatively charged monolayers of PS with different basic proteins at differing initial pressures. Lysozyme, cytochrome *c* and albumin at acid pH fall into one group which shows appreciable positive $\Delta\pi$ up to highly initial pressures of approximately 40 dynes/cm. Ribonuclease and albumin at neutral pH however only penetrate up to an initial monolayer pressure of approx. 23 dynes/cm. An increase in π *above* the collapse pressure of the protein (~15 dynes/cm) is considered to involve penetration of some part of the protein into the hydrocarbon portion of the monolayer, although at low initial pressures simple nonspecific adsorption of the protein itself may occur at the interface. The likely explanations of the π increase are in *initial electrostatic interactions followed by hydrophobic interactions*; such hydrophobic interactions involve conformational changes in the protein with exposure of hydrophobic sites which either penetrate into the monolayer or interact with the liquid highly mobile tails of the fatty acyl chains (with "deformation" of the monolayer).

It is considered that studies with monolayers are subjected to uncertainties in their application to natural membranes: first of all, membranes are not usually equivalent to monolayers; secondly the variables of low or high initial surface pressures may not be easily related to bilayer membranes. A value of 23 dynes/cm has been however considered significant by Kimelberg and Papahadjopoulos[185] since it was calculated that each phospholipid molecule in an aqueous

* Abbreviations used are; PE (phosphatidyl ethanolamine); PS (phosphatidylserine); PI (phosphatidyl inositol); CD (circular dichroism); ORD (optical rotatory dispersion); EM (electron microscopy); NMR (nuclear magnetic resonance); IR (infrared); SDS (sodium dodecyl sulphate); LDM (lipid depleted mitochondria); SP (structural protein); FCCP (p-tri-fluoro methoxy-carbonyl cyanide phenylhydrazone); ANS (1-anilinonaphthalene-8-sulphonate).

dispersion occupies an area of 68 Å corresponding to a pressure of 20–25 dynes/cm for a monolayer at an air–water interface.[277] Another difficulty concerns the use of water-soluble proteins, often unrelated to membranes. Cytochrome *c* is a soluble membrane protein, but many systematic studies with insoluble membrane proteins are not available. Studies of surface properties of isolated proteins[108] have related differences in tertiary structure or other levels of organization of native proteins to their characteristics at the air–water interface. Specific proteins having high surface activity would be likely to be found at an interface together with lipids.[49] Total erythrocyte proteins solubilized in 2-chloro-ethanol and purified from the lipids[410] yield reasonably stable membranes at a benzene–water interface;[289] these are formed by two distinct layers resembling those seen in lipid-depleted natural membranes, but only in the lipid–protein membranes are the electrical properties restored. In studies of Colacicco,[70] the large increase of surface tension induced by the so-called mitochondrial "structural protein" (SP)[76] in comparison with soluble proteins strengthens the idea of *amphipathic proteins*[139] which are naturally suited for penetration in membrane lipids in contrast with the water-soluble "extrinsic proteins" (see Section III. A. b); intrinsic membrane proteins orient much more rapidly at an air–water interface than do soluble proteins.[68]

D. *Protein Interactions with Bilayers*

Maddy *et al.*[232] studied the interaction of membrane proteins solubilized from beef erythrocytes with "black" lipid membranes and showed that it induces a decrease of the surface tension of the bilayer and a marked increase in intensity of the light reflected from the bilayer surface. Binding of cytochrome *c* to acidic phosphoinositide bilayers[351] is a function of ionic strength; binding to lecithin, which has zero net charge, is at least 20 times smaller as compared with phosphoinositide.

Electrical properties of bilayers have been studied as a function of protein interactions[153,367] or as a consequence of an enzyme-substrate or antigen–antibody reaction.[91] The most remarkable modifications are those occurring through interaction of a lipid bilayer with a proteinaceous material (EIM: excitability inducing material[259]) or with the cyclic peptide alamethycin.[258]

An ATPase from *S. faecalis*, involved in active transport of cations, interacting with bilayer membranes in presence of Mg^{2+} and at optimal Na^{+} and K^{+} concentrations, induces a 10^{2}–10^{4} increase in the electric conductance of the bilayer.[302]

Addition of sialic acid-free erythrocyte apoproteins increases the conductance and reduces the stability of lipid bilayers; reflectance and infrared measurements (showing changes to unordered conformation) indicate that the protein is extended over the membrane surface.[66]

E. *Protein Interactions with Phospholipid Dispersions*

Electrostatic interactions are usually considered to play a major role in binding of basic proteins with acidic phospholipids.[141] In the pioneering studies of Chargaff[64, 65, 276] it was found that by interaction with basic proteins the aqueous layers separating crude cephalin lamellae were reduced in thickness. The most studied interaction concerns cytochrome *c* with acidic phospholipids. Two general types of complexes have been described: in one type[304] the hydrated complex is recovered by centrifugation; in the other type[83] the complex is extracted into a nonpolar solvent like isooctane. The molar ratio phospholipid: cytochrome *c* increases from 9–10 for acidic phospholipids like cardiolipin to 24 for PE, whereas purified lecithin (no net charge) does not react;[84] a mixture of lecithin and acidic phospholipids however reacts and the complex becomes isooctane soluble. Under appropriate conditions also insulin is capable of interaction with phosphatidic acid and phosphatidylcholine, but not with sphingomyelin and cholesterol to form chloroform-soluble complexes;[281] the interaction is likely to be ionic and probably such that the macromolecular oganization of the phospholipid micelles is altered in such way to affect the distribution of small molecules like glucose between aqueous and organic phase.

The solubility in hydrocarbons is best explained by formation of a complex where the nonpolar residues face the solvent and the protein is buried inside the complex; the interaction is ionic, and this is confirmed by the effect of salts.[83] X-ray techniques have suggested that the lipid-cytochrome *c* complexes *in isooctane* contain many protein units (30–40) with a ratio of phospholipid to protein of 20:1 and 30:1 respectively in two forms of complexes.[338] The *aqueous complexes* of phospholipids with cytochrome *c* consist basically of lamellar units[150, 278, 339] (cf. Fig. 3). As shown by electron microscopy of osmium-fixed thin sections,[183] cytochrome *c* increases the order of lecithin–cardiolipin lamellae; the thickness of the lamellae is increased, due to increase in width of the dense lines; this is further evidence that cytochrome *c* binds electrostatically to the surface polar groups of the lamellae (Fig. 6). About 50% of the bound cytochrome *c* cannot be removed by 0·15 M KCl, although the interaction is competitively inhibited by KCl. The cytochrome *c* resistant to salt washing appears to be trapped inside the lamellae, and is separated from the bulk solution by closed membrane barriers. This compartmentalized cytochrome *c* is analogous to the endogenous cytochrome *c* of sonicated submitochondrial particles[206] (Fig. 7).

In contrast with the pure ionic character assumed for the interaction of cytochrome *c* with PI or cardiolipin, in the system lysozyme-cardiolipins two principal lamellar phases were observed by Gulik-Krzywicki *et al.*:[150] in the phase called III, the thickness of the lipid leaflet is the same as in lipid–water, whereas in phase IV a shrinkage of the lipid is interpreted to involve hydrophobic contacts of protein with the lipid molecules. The CD spectral changes observed after the interaction could be ascribed to arise mainly from orientation and interactions of the protein molecules. Also Ulmer *et al.*[373] found no changes

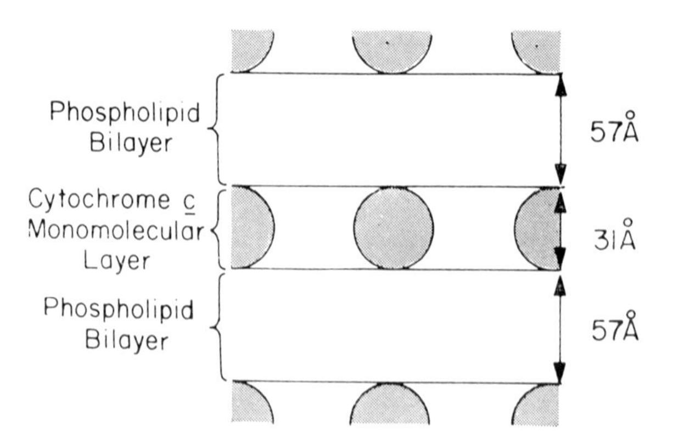

Figure 6. Scheme of cytochrome *c*–phospholipid interactions. Shaded circles represent the approximately spherical cytochrome *c* molecules; open areas between lines represent bimolecular phospholipid membranes, with the polar head groups on the surfaces (from Kimelberg, Lee, Claude and Mrena,[183] reproduced with kind permission of Dr. Kimelberg).

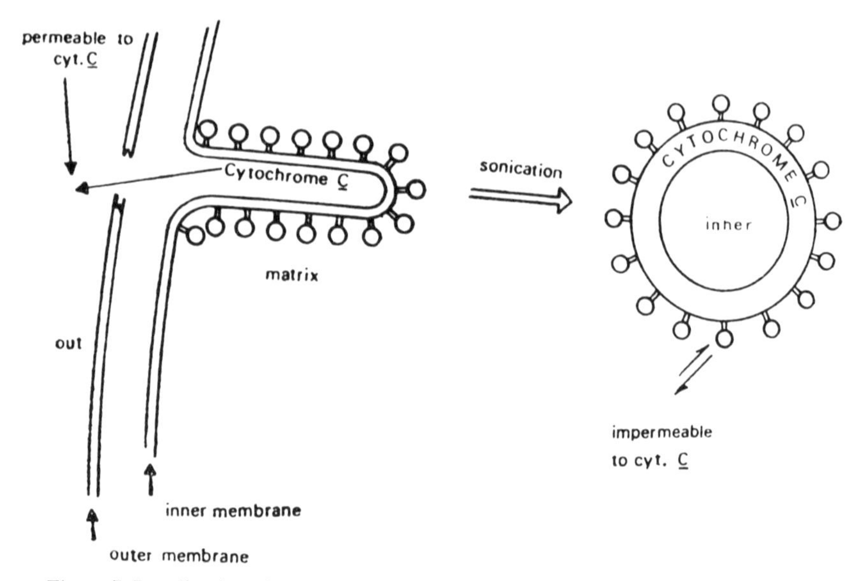

Figure 7. Localization of cytochrome *c* in the inner mitochondrial membrane (Lenaz and MacLennan[206]).

in protein conformation by interaction of cytochrome *c* with phospholipids. The hydrophobic type and electrostatic type phases (Fig. 8) could be differentiated also by use of fluorescent probes (which are sensitive to the hydrophobicity of their local environment[397] and whose location has also been studied in lipid bilayers;[214] either free probes like anilino–naphthalene sulphonate (ANS) were used[149] or probes covalently bound to protein like dansyl-Cl:[331] in the hydrophobic phases the microenvironment of the probe is of lower polarity and ANS appears in closer contact with the protein.

Kimelberg and Papahadjopoulos[184,185] have suggested that also cytochrome *c*, after initial attractive electrostatic binding, can penetrate into the interior of negatively charged phospholipid bilayers with

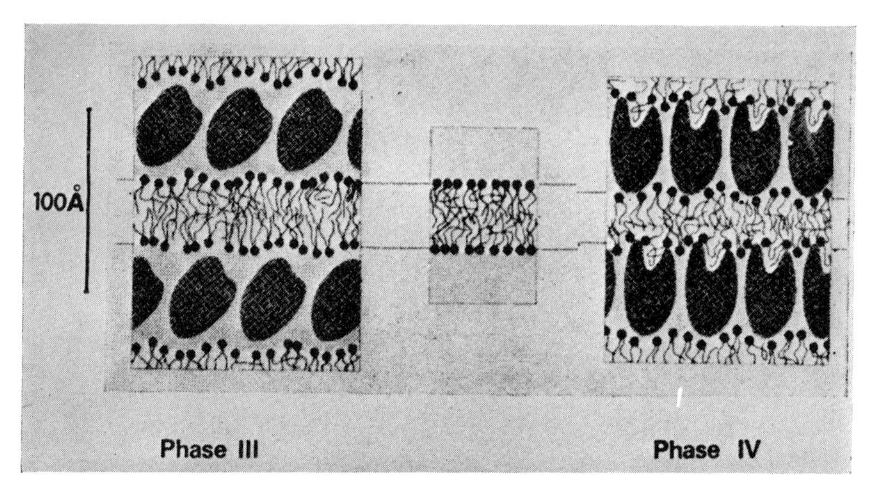

Figure 8. Lysozyme–lipid–water system: structure of the lamellar phases. The sections are perpendicular to the plane of the lamellae. The densely hatched areas represent the protein molecules. Middle frame, lamellar lipid–water phase; left hand frame, lamellar phase of the electrostatic type; right hand frame, lamellar phase of the hydrophobic type (from Gulik-Krzywicki, Schechter, Luzzati and Faure;[10] reproduced with kind permission of Dr. Luzzati).

subsequent hydrophobic interaction. This suggestion is supported by the very sensitive relation of the type of interaction with ion permeability: lysozyme, cytochrome *c*, and albumin at acidic pH interact with phospholipid vesicles and increase $^{22}Na^{+}$ permeability, although to different extents; native ribonuclease and albumin at neutral pH interact without inducing large increases in $^{22}Na^{+}$ permeability. The former proteins were likewise found to penetrate PS monolayers, whereas the latter did not (Section II. C); proteins which penetrate monolayers and enhance $^{22}Na^{+}$ permeability are assumed to be initially attracted electrostatically and then hydrophobically bound in the bilayer.

Likewise, the erythrocyte protein fraction spectrin interacts with phospholipid vesicles and penetrates phospholipid monolayers, but only at low pH[172]; however Ca^{2+} promotes spectrin-vesicle interaction at physiological pH, causing profound changes in $^{22}Na^{+}$ permeability.

One explanation suggested for the effect of Ca^{2+} is formation of a tridentate complex in which anionic sites on the protein and on the vesicles are the ligands for Ca^{2+}.

In an attempt to relate lipid–protein interactions with molecular conformation of the protein, Hammes and Schullery[152] have investigated the interactions of aqueous phospholipid dispersions with a variety of water-soluble polypeptides. CD, ORD, EM, NMR and stopped-flow techniques showed major changes in polypeptide conformation. An important requirement for interaction of lipid and peptide is electrostatic attraction; the interaction is partially stabilized by hydrophobic bonding, but this probably does not involve extensive penetration of the bilayer.

X-ray diffraction studies of precipitates at pH 3·33 between (cationic) serum albumin and (anionic) PC-cardiolipin vesicles[300] show both *interlamellar* and *intralamellar* interactions. The former interactions (which underlie the problem of cell contact) are increased in protein–lipid lamellae in comparison with pure lipid lamellae. Intralamellar molecular packing is interpreted as evidence that after initial ionic attraction there must be hydrophobic interaction, either by spreading the lipid molecules on hydrophobic protein sites, or by protein penetration into the bilayer, mediated by large conformational changes of the protein.

Similar conclusions were reached by Dawson[85] in a review of his studies on the reaction of phospholipases at the phospholipid–water interface. The demonstration of enzyme activity implies that the enzyme has approached the interface and has become stereochemically locked with the phospholipid, forming the enzyme-substrate complex. The reaction of the phospholipases tested (B, C, D, A, and triphosphoinositide phosphomono- and di-esterase) requires definite electrostatic conditions (although differing for the various reactions), at the lipid–water interface; for example, a positive z-potential, induced e.g. by Ca^{2+}, is required for activity of phospholipase C; also in the case of phospholipase A, Ca^{2+} is necessary for lecithin hydrolysis in presence of ether, but the concentration required has virtually no effect on the z-potential, and Ca^{2+} cannot be replaced by Mg^{2+}; a negative z-potential is required for hydrolysis of PE in absence of ether. Figure 9 shows the electrical "double layer" in proximity of an anionic lipid lamella.

If the initial approach of the protein molecule through the electrical double layer is likely to be electrostatic, a reinforcement can be induced by hydrogen and hydrophobic bonding of the side chains penetrating into spaces between the lipid molecules. As also remarked by Dawson, van der Waals attractions with limited or no penetration could still occur between methylene groups of ethanolamine and choline in the phospholipids and those in the protein side chains. Also the association of lecithin or cholesterol–lecithin mixtures with the enzyme lecithin–cholesterol acyltransferase was assumed to be partially ionic in character.[161]

Evidence for hydrophobic bonding between lipid and protein was first derived from the studies of Richardson *et al.*[310,311] on the interaction of mitochondrial structural protein (SP)[76] with sonicated phospholipid vesicles. Polymeric SP binds mitochondrial phospholipids to form a complex containing 20–25% phospholipid by weight; unlike the complex with cytochrome *c*, that with SP is not dissociated by salt. The extent of binding depends on the length of the hydrocarbon chains.[141] Vandenheuvel[386] has explained these results by greater stability derived

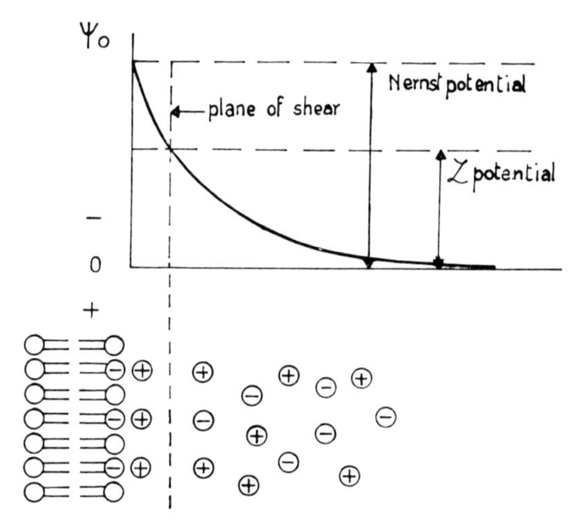

Figure 9. Representation of charged layers on the surface of a phospholipid bilayer.

from stronger nonpolar binding of longer acyl chains between each other in the lipid bilayer, and has suggested that hydrogen bonding of lipid polar heads to the protein or to an intermediate water layer would make complex formation independent to changes in ionic strength, without any need to assume lipid–protein hydrophobic bonding. Stoeckenius[352] has presented EM evidence that the SP-phospholipid complex is morphologically identical to the ionic cytochrome *c*-phospholipid complex; however the reliability of the electron microscopic trilaminar pattern as an indication of the unit membrane is highly questionable.[190] Moreover binding of SP to the surface of the bilayer appears difficult to reconcile with the fact that the SP-phospholipid complex has 90% of the ionic groups of the phospholipids still available for further ionic interaction with cytochrome *c*,[141] and that the amount of SP bound to phospholipid vesicles is doubled when previous (ionic) interaction has occurred between lysozyme and the phospholipids.[315]

III. *Lipid–Protein Interactions in Biological Membranes*

The ideal method to study lipid–protein interactions in natural membranes has not been found; combination of different direct or indirect methods will probably give satisfactory answers. I review here some of the methods which have been used, classified according to the following order:

(a) Selective and controlled dissociation and solubilization of membrane components (lipids, proteins, lipoproteins).
(b) Recombination of lipids with lipid-free membranes.
(c) Physical studies of intact or dissociated membrane systems, in relation to spatial organization, environment and conformation of the lipids and proteins.
(d) Chemical and enzymatic studies of intact or dissociated membrane systems directed to the same objectives.

A. *Dissociation of Membrane Components*

(a) *Lipid Extraction from Membranes*

(i) *Nonpolar solvents and ether.* It is well known that nonpolar solvents do not extract membrane lipids, although extracted lipids are freely soluble in hydrocarbons, ether, or chloroform: this fact is usually referred to the presence of protein. But, as a matter of fact, even phospholipid dispersions in water cannot be solubilized by nonpolar solvents: the main reason must be in the surface charge of the bilayer, because, when the charge is neutralized by salt, the phospholipids are transferred into the organic phase[37,208] (Fig. 10). Lecithin, a zwitterion, can be extracted by ether in the presence of cholesterol,[37] perhaps as a result of both internal charge neutralization of the choline and phosphate groups and decrease in the charge density induced by the interposition of the sterol molecules.

Ionic interaction of phospholipids with cytochrome *c* drastically modifies the extraction characteristics: phospholipids may now be extracted in absence of salt, bound to cytochrome *c* by saline linkages; on the other hand interaction of phospholipids with lipid-depleted mitochondria does not change the ether extraction characteristics, and phospholipids (without any protein) are extracted only in presence of salt (Parenti-Castelli, G. and Lenaz, G., unpublished observations) (Table I). These simple models may be kept in mind before applying nonpolar solvent extraction to natural membranes. When aqueous suspensions of mitochondria or submitochondrial particles are shaken with ether, either in absence or in presence of salt, very little extraction takes place;[208] if however the membranes have been previously depleted of soluble or detachable proteins (extrinsic proteins) a larger amount of phospholipids is available for extraction (Table II). An interpretation

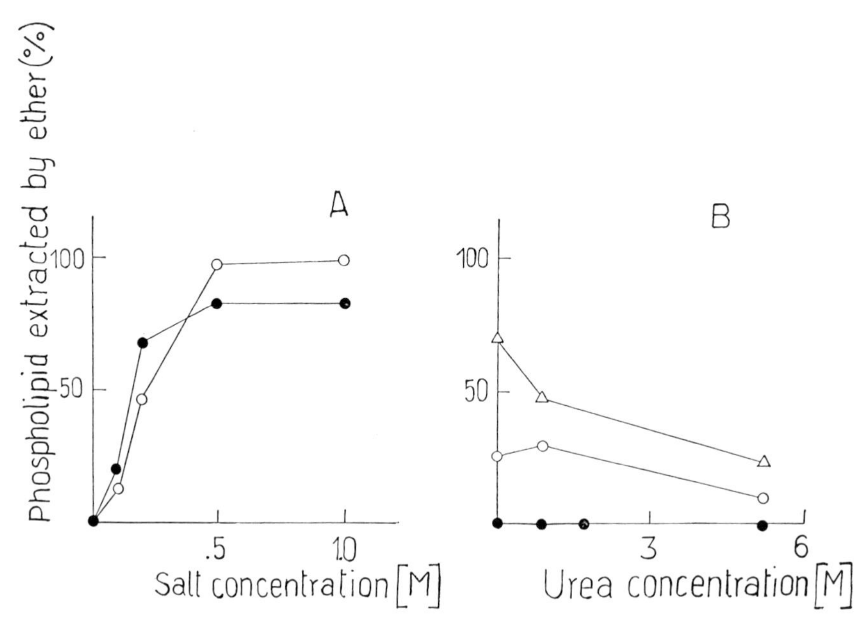

Figure 10. Effect of salt and urea on phospholipid extraction by diethyl ether from aqueous phospholipid dispersions. Values are expressed as % of P in the organic phase. Left: ●——●, NaCl; ○——○, NaSCN. Right: ●——●, urea; ○——○, urea in presence of 0·1 M NaCl; △——△, urea in presence of 0·2 M NaCl (from Lenaz, Parenti-Castelli, Sechi and Masotti[208]).

TABLE I. Conditions affecting diethyl ether extraction of sonicated phospholipid dispersions (Lenaz[202])

Exp.	Protein added	Salt added	P extracted (%)
1	—	—	1
	—	NaCl 0·1 M	52
	—	NaCl 1 M	70
	Cytochrome *c* (4 mg)	—	54*
2	—	—	3
	—	NaCl 1 M	51
	LDM (4·7 mg)	—	3
	LDM (4·7 mg)	NaCl 1 M	60

* Cytochrome *c* was extracted by ether together with the phospholipids.

of these findings is that, in presence of salt, bonds of phospholipids with extrinsic proteins are not broken by ether, whereas bonds with intrinsic proteins are broken by the solvent. The latter are presumably hydrophobic bonds, whereas the nature of the bonds that are not broken by ether is less clear: they cannot just be ionic interactions, otherwise they would be affected by salt. Indirect reasons suggest that Mg^{2+} could be involved;[203] but other bonds such as hydrogen bonds etc. may be present.

TABLE II. Diethyl ether extraction of phospholipids from submitochondrial particles (ETP) under different conditions

Exp.	Treatment previous to extraction	%P extracted*
1	—	15·0
	10 mM HCl	54·0
2	—	21·0
	3·5 M NaBr	55·5
3	—	8·3
	Sephadex, urea	25·5

* In presence of 1 M NaCl.

Cunningham *et al.*[78] studied the effect of solvent treatment on structure and activities of the inner mitochondrial membrane: iso-octane produces little change; with ether there are significant membrane alterations, shown by negative staining, with loss of the "head-pieces".

Following the pioneering studies of Parpart and Ballentine,[279] Roelofsen *et al.*[313] confirmed that ether removes only a specific portion of the lipids from lyophilized erythrocyte membranes; since the differential extraction is not due to specific solubility of different lipid classes in the organic solvent, they concluded that different types of interactions have to be present with the membrane proteins; all classes of phospholipids are present in both "loosely bound" and "strongly bound" lipids, but quantitatively the ethanolamine containing lipids prevail in the loosely bound fraction. Differential extractability was also found by Sauner and Lévy[236] by stepwise treatment of lyophilized mitochondrial and microsomal membranes with ether, ether–ethanol (96 %) and chloroform–methanol 2:1.

(ii) *Alcohols*. Lenaz *et al.*[208] have observed that alcohol–water mixtures extract phospholipids from mitochondria and submitochondrial particles; the efficacy of extraction is a function of alcohol chain length (Fig. 11): also several functions are affected with an efficacy that depends on the hydrophobicity of the alcohol.[207] The conclusion was reached that the organization of the inner mitochondrial membrane is largely based on hydrophobic interactions.

Treatment with n-butanol is effective in separating lipids from proteins in erythrocyte membranes;[230, 231, 303, 415] in contrast with n-butanol, n-pentanol induces solubilization of lipoproteins in the aqueous phase;[414] 3-pentanol removes over 90% of the lipids from *M. lysodeikticus*; the remaining lipids are enriched in cardiolipin.[263]

A number of organic solvents mixed with water under isotonic conditions causes lysis of the erythrocytes at a sharply defined concentration for each solvent: the haemolytic action correlates with the dielectric constant of the medium;[314] it was suggested that at a critical value of the constant, essential bonds between lipids and proteins are broken. It is conceivable that these are hydrophobic bonds, since ionic bonds would be strengthened by decreasing the dielectric constant of the medium.

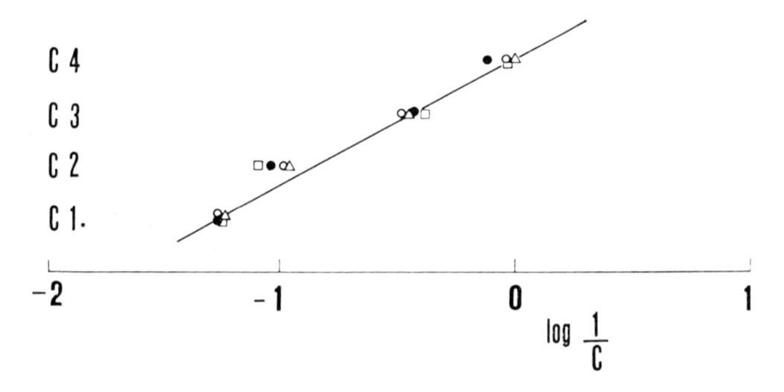

Figure 11. Effect of monohydric alcohols on phospholipid extraction from beef heart mitochondrial preparations. The logarithm of the reciprocal of the concentration of alcohol required to extract 50% of the phospholipids is plotted against alcohol chain length. ●——●, beef heart mitochondria; ○——○, mitochondria extracted with HCl; □——□, sonicated submitochondrial particles ETP; △——△, ETP extracted with HCl (from Lenaz, Parenti-Castelli, Sechi and Masotti[208]).

(iii) *Other solvents.* The most effective procedures for extracting membrane lipids are mixtures of polar and apolar solvents, like chloroform-methanol (2:1)[128] the fact that solvents remove membrane lipids may be an indication that hydrophobic forces are effective in membrane organization, although it is reasonable that hydrogen and other bonds be also broken by the solvent mixtures.

Acetone–water mixtures are capable of removing lipids from membranes: 96% acetone extracts only neutral lipids like Coenzyme Q;[215] complete removal of phospholipids is obtained by 90% acetone made alkaline with ammonia; when 90% acetone is used in absence of ammonia, cardiolipin is not extracted.[124] Indeed tightly bound cardiolipin, which can be extracted with alkaline chloroform–methanol, has been found in cytochrome oxidase.[7a] The gross microscopic appearance of the inner mitochondrial membrane is not changed after acetone extraction of the lipids,[125] raising the question how much the lipids are structural components of the membrane; in the quoted studies of Cunningham *et al.*[78] acetone extraction caused loss of the "knobs" and other structural changes, visible by negative staining but not in membrane sections.

Different types of membranes may differ in their lipid–protein interactions, although the conformation of their proteins is apparently alike;[396] after acetone extraction the contribution of the lipid to the IR spectrum of endoplasmic reticulum is almost completely abolished, but not in the case of the plasma membrane: chemical analysis has shown that only 50–60% of phospholipid is extracted from the plasma membrane, whereas over 90% is extracted from endoplasmic reticulum. Similar differences were found between mitochondrial and erythrocyte membranes.[332]

(b) *Protein Extraction*

(i) *Extrinsic proteins*. A useful distinction of membrane proteins has been suggested by Green;[139] certain proteins, considered amphipathic in nature, are firmly bound to the membrane (*intrinsic proteins*) whereas others may be removed easily and when detached are water-soluble (*extrinsic proteins*). Water solubility is an indication that these proteins are not in contact with the nonpolar interior of the membrane; weak hydrophobic bonds and different polar bonds may be involved[263] (also see D. E. Green and R. F. Brucker, personal communication). Cytochrome *c* is removed with 0·15 M KCl from mitochondria;[166,228] since a comparable salt concentration breaks the cytochrome *c*–phospholipid complex *in vitro*,[83] a similar complex may exist *in vivo*.

Mitochondrial ATPase is a key enzyme in the terminal sequence of oxidative phosphorylation.[296] An oligomycin insensitive ATPase, F_1, is a soluble coupling factor of oxidative phosphorylation and has been identified with the headpiece visible in negative staining of the inner mitochondrial membrane[178,293] (cf. Fig. 12); when bound back to the membrane, F_1 becomes oligomycin sensitive (Fig. 13). A protein factor conferring oligomycin sensitivity to F_1 (OSCP) has been isolated and identified with the cylindrical stalk connecting the headpiece with the membrane.[227] The ATPase components and other extrinsic proteins are extracted from mitochondria by dilute HCl.[142,201,204] Alcohols make the ATPase oligomycin insensitive, and the efficacy of this transformation depends on alcohol chain length, suggesting hydrophobic interaction of ATPase with the membrane;[207] on the other hand the observation that phospholipids are extracted by ether only when F_1 and other extrinsic proteins have been removed[208] suggests a concomitance of other types of binding. Abrams and Baron[3,4] have studied the effect of Mg^{2+} on binding of a bacterial ATPase to the membrane, and demonstrated the requirement for another protein called nectin.[13] Mg^{2+} is not required for F_1 binding to the mitochondrial membrane[368] but appears to enhance the binding.[213] Bulos and Racker[41] have shown that Mg^{2+} (0·8 mM) and other cations at higher concentrations are necessary for binding of F_1 to TUA particles. Kopaczyck *et al.*[188] found that Mg^{2+} is required for binding of an isolated oligomycin sensitive ATPase to solubilized membranes. Of the mitochondrial Mg^{2+} 5% is bound in the inner membrane; in heart mitochondria this corresponds to 4·6 nmoles/mg of fraction protein[25a] which is the order of magnitude of components like coenzyme Q and is sufficiently in excess over the amount of mitochondrial ATPase.

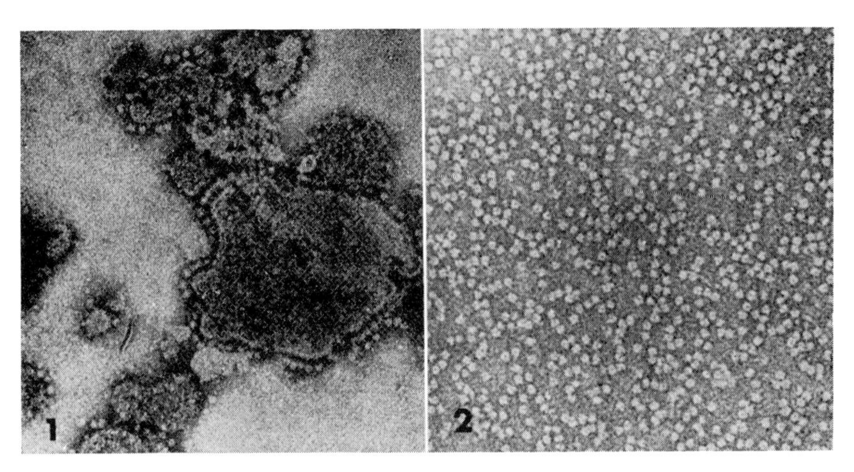

Figure 12. Electron micrograph of the inner mitochondrial membrane and of isolated oligomycin insensitive ATPase F_1. (1) A-particles (submitochondrial particles obtained by sonication in presence of ammonia). (2) isolated F_1. Negative staining with phosphotungstate. Final magnification ×140,000 (from Racker,[295] reproduced with kind permission of the author).

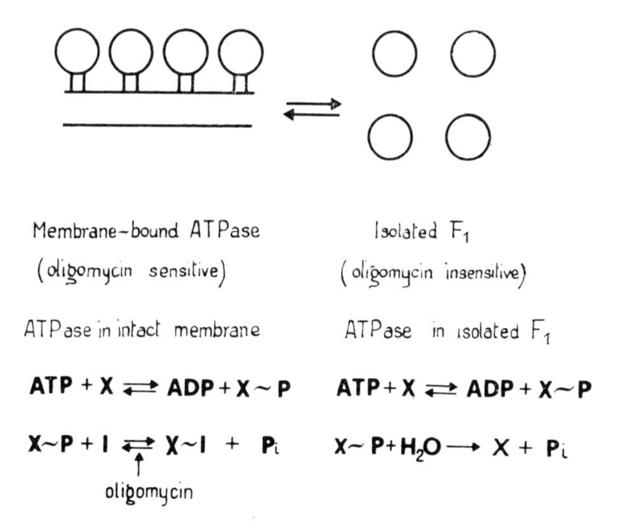

Figure 13. Schematic representation of the properties of the mitochondrial ATPase.

Attempts to detach as well as quantitate and characterize water-soluble proteins have been particularly concerned with erythrocyte ghost membranes. Thirty to 50% of erythrocyte membrane proteins can be solubilized by manipulations of pH, ionic strength, and chelating agents;[154, 163, 173, 233, 237, 238, 267] these proteins are presumably bound by ionic forces.[173] Protein solubilization may occur under the usual haemolytic procedures employed for preparing erythrocyte ghosts; the extent of solubilization varies in different preparations, mainly in relation to the osmolarities used for haemolysis; conditions under which the amount of haemoglobin and other intracellular components are retained by ghosts can be however controlled.[28,253] A loss of non-haemoglobin protein below 20 osm might be responsible for changes in

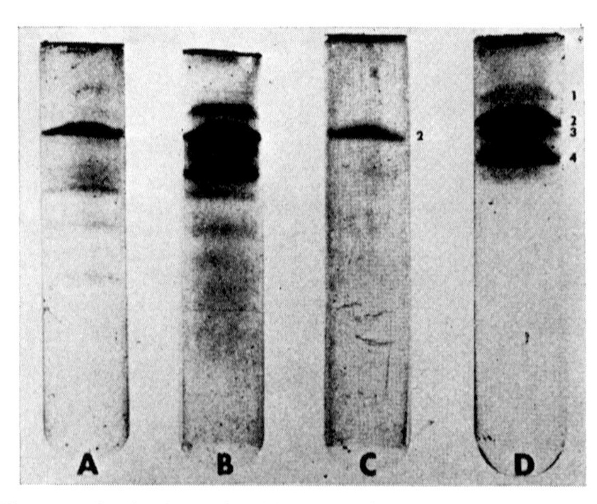

Figure 14. Photograph of polyacrylamide gel profiles of structural protein preparations. (A) prior to purification; (B) prior to purification (oxidized by performic acid); (C) after purification with 8 M urea; (D) after purification (oxidized by performic acid) (from Green, Haard, Lenaz, and Silman[142]).

permeability, enzymatic properties, or fragility phenomena described for certain ghost preparations. These proteins, released at low ionic strength, and possibly present at the inner surface of the erythrocyte membrane, must have pronounced effects upon the membrane properties.[28]

To this purpose, Green *et al.*[143] have suggested that the entire glycolytic system of erythrocytes may be bound to the plasma membrane. The idea that many enzymes generally considered soluble may be permamently or occasionally membrane-bound deserves consideration.

(ii) *Intrinsic proteins*. Several investigations report membrane fractionation by use of detergents with isolation and separation of proteins or lipoproteins.

The so-called "structural protein"[76, 311] was claimed to represent a common basis for membrane structure; its heterogeneity[142,151,204,205] (Fig. 14) and the fact that at least one component is also found in mitochondrial ATPase[330, 333] suggest that the original postulations were not correct. Other authors claim that SP contains aggregates of very hydrophobic proteins like the cytoplasmic protein described by Shannon and Mill[337] or "miniproteins" of low molecular weight.[196,261] No "miniproteins" have been found in erythrocyte membranes by Trayner *et al.*[363] and Fairbanks *et al.*;[110] however heating in low SDS and high salt produces diffuse bands of low molecular weight in gel electrophoresis, due to proteinase degradation.[110]

Membrane solubilization with detergents leads to confusing results in the interpretation of the nature (protein or lipoprotein) of the subunits and on the size and homogeneity of the polypeptide units, even in studies on the same membrane, like e.g. sarcoplasmic reticulum.[240,242,248,409]

The results of differential protein extraction from the membranes of *M. lysodeikticus* by a variety of means[263,321] indicate an onrandom distribution of proteins and lipids in the membrane: the strengths and types of cohesive forces cover a wide range from the relatively weak association of the ATPase to the firmly bound components of the respiratory system, requiring a powerful dissociating agent like SDS.

The proteins of the plasma membranes of the Ehrlich ascite carcinoma cells have been solubilized in 2-chloroethanol and purified from the lipids by gel filtration;[410] the proteins precipitate when dialyzed against different salt solutions.[411] Mitochondrial membranes are soluble in acidified chloroform–methanol[142] and can be separated by gel filtration in this solvent mixture.[79,80] Maddy[231] has solubilized 95% of erythrocyte membrane proteins in aqueous butanol.

In spite of the often harsh procedures employed in separating erythrocyte ghost proteins,[386] it has been reported that *all* proteins in ghosts become water-soluble when cations are removed.[307,309,363] Although ghost proteins may be made water-soluble to large extents, they exhibit unique structural properties leading to easy specific aggregation.[25] These unique properties well agree with the postulation of *amphipathic proteins* as a basis for membrane structure.[387] According to D. E. Green and R. F. Brucker (personal communication) several lines of evidence support the concept that intrinsic membrane proteins are amphipathic: their lower total polarity, their rapid orientation at air–water interfaces, their ready solubilization by reagents destabilizing hydrophobic associations, and the tendency to polymerization together with the capacity of membrane formation.

(c) *Membrane Fractionation and Lipoprotein Subunits*

Four lipoprotein complexes have been isolated from beef heart mitochondria by use of bile salts and several purification procedures[145] (Table III); respiratory complexes have also been isolated from mitochondria by other investigators.[186] The complexes easily reform membranes when bile salts are removed[247] or the ionic strength is

TABLE III. The four complexes of the respiratory chain (Green and Silman[145])

Reactions

Complex I	NADH – Coenzyme Q reductase
	$NADH + CoQ \rightarrow NAD + CoQH_2$
Complex II	Succinate – Coenzyme Q reductase
	$Succinate + CoQ \rightarrow Fumarate + CoQH_2$
Complex III	Coenzyme Q – cytochrome *c* reductase
	$CoQH_2 + 2\ cyt.\ c\ Fe^{3+} \rightarrow CoQ + cyt.\ c\ Fe^{2+} + 2\ H^+$
Complex IV	Cytochrome *c* oxidase
	$2\ cyt.\ c\ Fe^{2+} + \frac{1}{2}O_2 \rightarrow 2\ cyt.\ c\ Fe^{3+} + O^{2-}$

Components of the complexes

	I	II	III	IV
Cytochrome a	—	—	—	1
Cytochrome a_3	—	—	—	1
Copper	—	—	—	2
Cytochrome b	—	1	2	—
Cytochrome c_1	—	—	1	—
Core protein	—	—	3–4	—
Nonheme iron	8	8	2	—
Flavin (acid extractable)	1	—	—	—
Flavin (extractable after acid digestion)	—	1	—	—

decreased;[353] they form mixed membranes when mixed together before removal of bile salts (Fig. 15): reconstituted membranes carry out electron transfer from NADH or succinate to O_2.[140] Stoeckenius[352] has

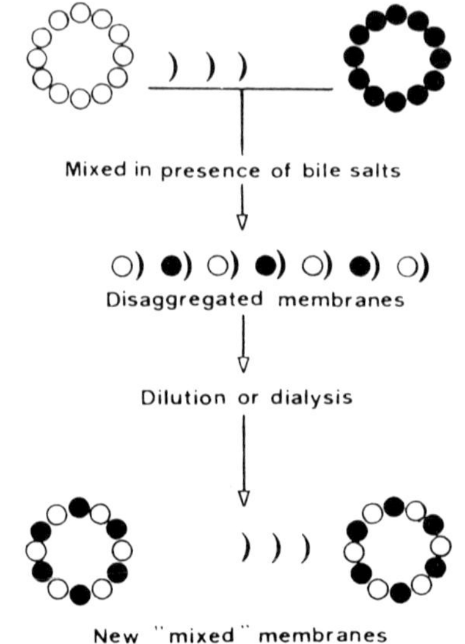

Figure 15. Reconstitution of mixed membranes from disaggregated complexes (Green and Tzagoloff[146]).

repeated some of the studies with Complex IV (cytochromoxidase) and concluded that it is not monodispersed in bile salts, but consists of small pieces of inner membrane that carry 55 Å particles on their surface. As pointed out previously, no conclusive evidence can rest on electron microscopic data alone since the preparative techniques may have induced aggregation. Particle weight determination for Complexes III and IV are in accord with the existence of monodispersed particles in presence of bile salts: by light scattering in 0·33% taurodeoxycholate, Complex III had a molecular weight of 300,000 and Complex IV of 290,000 (230,000 when corrected for lipid content).[372]

Detailed studies have been reported on disaggregation and aggregation of the membranes of *Mycoplasma laidlawii*. Solubilization of the membranes with SDS releases structures first thought to be true lipoprotein subunits[301] and then found to consist of separate lipid-SDS and protein-SDS complexes.[103] Removal of detergent leads to formation of lipoprotein material, which has membranous appearance under certain conditions in presence of Mg^{2+}.[101,102,357] Bakerman and Wasemiller[10] separated SDS-solubilized erythrocyte membrane into two principal molecular weight classes differing in chemical composition and ultracentrifugal behaviour. Pollak *et al.*[291] extracted soluble lipoproteins from endoplasmic reticulum; such complexes apparently also exist *in vivo* and may be relevant to the biogenesis of that membrane system.

Besides detergents, also other agents disaggregate membranes: alcohols, like pentanol,[415] sonication,[187,371] high pH[282,283] mechanical disintegration,[194] are effective at least partially and under certain conditions. True lipoprotein subunits (or fragments) may be obtained by use of bile salts or sonication: SDS[103] and Triton-X-100[26,217] apparently separate lipid from protein. NMR studies[61] of erythrocyte membranes are in accord with the different effects of bile salts and SDS.

Binding forces between subunits. A detailed analysis of the conditions leading to membrane disruption may be a tool to understand the lipid–lipid, protein–protein and lipid–protein binding forces, and whether the membrane is ruptured at certain specific positions, and in such a case why is it so.

Disaggregation by detergents or solvents indicates the role of apolar interactions in subunit aggregation; the detergent molecules intercalcate with the lipids or substitute the lipids between proteins. On the other hand, the role of Mg^{2+} in reaggregation of SDS-solubilized membranes from *M. laidlawii*[357] and the observation of Sun *et al.*[353] that membrane formation can be accomplished in presence of high concentrations of nonionic detergent at low ionic strength suggest a role of ionic interactions in intersubunit cohesion.

According to Bonsall and Hunt[26] however widespread ionic binding between protein and lipid is inconsistent with the finding that erythrocyte ghosts and lecithin dispersions are similarly solubilized by *non-ionic* surfactants: the protein components offer no protection to the lipids against solubilization. The mechanism of ghost solubilization has been analyzed and considered to involve initial formation of lipid-surfactant micelles:[26] a transient micelle at low surfactant concentration may be responsible for haemolysis, a more stable micelle at higher surfactant concentration will induce solubilization.

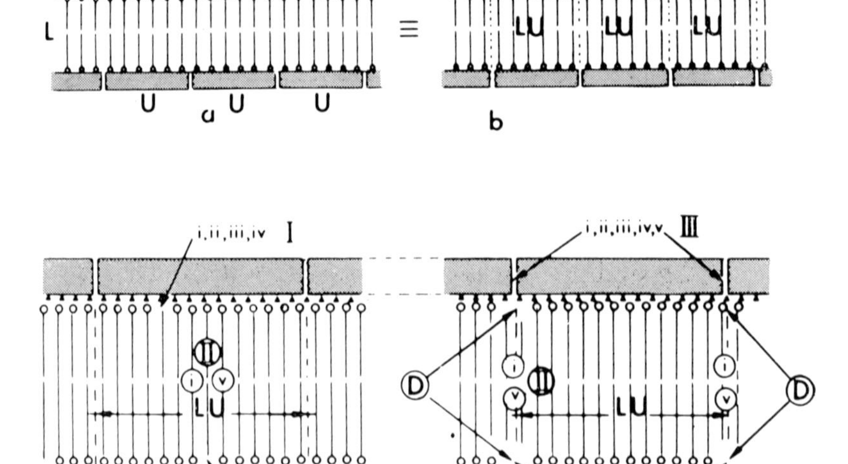

Figure 16. Schematic representation how lipoprotein subunits may be derived from a Danielli membrane. Superstructural elements have not been represented in surface protein units U. L is lipid component. Model (a) is equivalent to (b) showing subunits LU. (c) Intrasubunit, and (d) intersubunit cohesional forces. Forces i, ii, iii, iv and v are localized as shown. D indicates position of divalent cation bridges (Vandenheuvel,[386] reproduced with kind permission of Dr. Vandenheuvel and of Academic Press, New York).

The possibility to isolate specific complexes having fixed characteristics suggests that these subunits may be intrinsically present in the membrane, as also indicated by electron microscopy, although keeping in mind the severe limitations of this technique at the molecular level (Section III. C. a). The role of hydrophobic interactions in the isolated lipoprotein complexes is usually considered implicit; Vandenheuvel[384, 385, 386] has however analyzed the possibility that lipoprotein subunits may be derived from a Danielli membrane, only assuming that the forces holding the individual proteins together are weaker than those holding lipid to protein (Fig. 16).

Crane's group has carefully investigated the conditions leading to fragmentation of mitochondria[75] and with the jointed use of chemical and microscopic analysis they have concluded that the membrane is preferentially ruptured along its axis, suggesting the presence of two apposed protein (lipoprotein) monolayers (Fig. 17); this possibility appears confirmed by freeze-etching microscopy.[31]

The oligomycin sensitive ATPase may be isolated from mitochondria in absence of respiratory components;[177, 189, 369] in such preparations a headpiece attached to a stalk is evident and a membrane-like continuum supports the stalk. The ATPase complex has been characterized in many of its component proteins;[226] the knowledge of the interactions and spatial organization of the different parts is much less advanced.

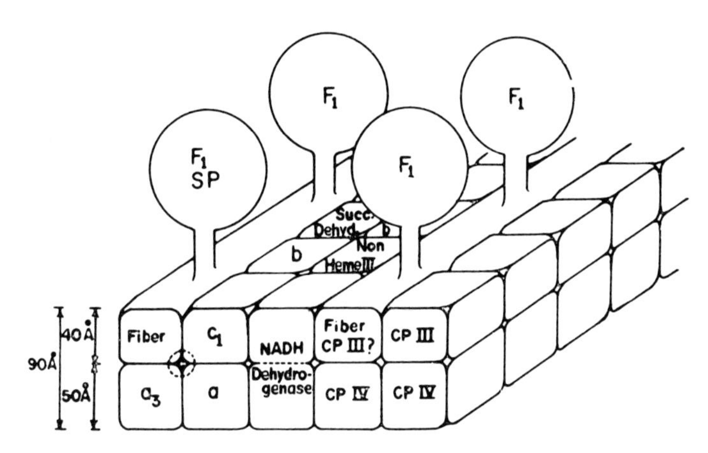

Figure 17. Diagram of binary structure in mitochondrial cristae membranes (from Crane, Arntzen, Hall, Ruzicka and Dilley,[75] reproduced with kind permission of Dr. Crane and of North-Holland Publishing Company, Amsterdam).

(d) *Proteolipids*

Folch and Lees[127] have found that, when brain tissue is homogenized in chloroform–methanol (2:1), substantial amounts of protein are soluble in the extracting medium and remain in the organic phase after partition with water; they coined the term "*proteolipids*" to indicate lipid–protein complexes soluble in organic solvents. Subsequently a number of other methods have been employed to separate these substances.[197] Isolation of lipid-free proteins from proteolipids has been achieved by dialysis against a succession of solvents of increasing polarity; these proteins retain their solubility in chloroform–methanol, but become also soluble in water. Purified proteolipids contain 1/3 lipids; the principal constituents are the acidic phosphatides; the lipids can be classified into three groups in relation to binding;[126] the strongly bound very acidic triphosphoinositides can only be removed by acidified organic solvents.[356] The binding with protein must be electrostatic in nature: Folch[126] has suggested that the polyanionic phosphoinositides bind several protein units together.

Other lipid–protein complexes have been extracted from nervous tissue with relatively nonpolar solvents at acid pH.[406] McClare[246] has studied proteolipid complexes in bacterial membranes and has proposed an interesting model where acidic lipids (phosphatidyl glycerol) are linked to protein by Mg^{2+} bridges (Fig. 18). The presence of Mg^{2+}

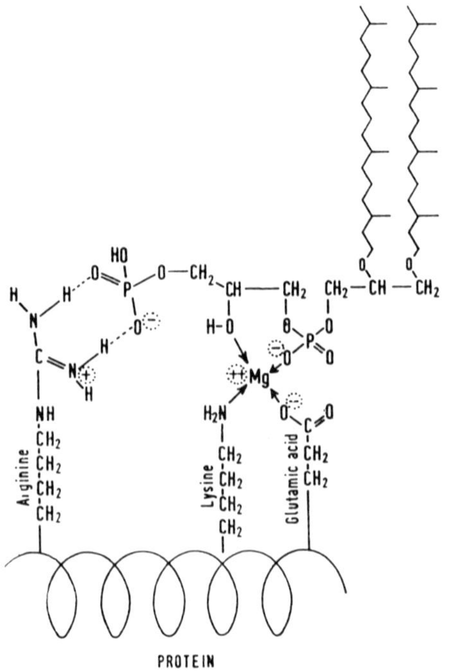

Figure 18. A possible binding mechanism between lipid and protein in the envelope of "Halobacterium halobium" (from McClare,[246] reproduced with kind permission of the author).

in membranes, its importance in membrane stability, the isolation of natural phospholipid-Mg^{2+} complexes,[182] and the well known capacity of certain phospholipids to bind divalent cations[5, 156, 277, 285, 336] are well in favour of this possibility.

B. *Lipid–Protein Interactions in Recombination Studies*

The possibility to remove lipids from membranes by solvent extraction without inducing gross morphological changes[125] has been used for recombination studies of lipids to lipid-depleted membranes.[141] Lipids can be added back to lipid-depleted mitochondria (LDM) with recovery of succinoxidase activity[122] showing that the membrane retains at least in part its native configuration.

The nonionic character of the binding of phospholipids to LDM was suggested by De Pury and Collins[92, 93] because lecithin, which bears no net charge at a wide pH range, can be bound to LDM. Studies in this laboratory have demonstrated that salts do not affect the binding even at concentrations higher than those required to break ionic bonds;[209] high concentrations of chaotropic anions, which break hydrophobic bonds by increasing affinity of nonpolar residues for water,[155] inhibit the binding.[210, 211] The characteristics of the interaction suggest its hydrophobic nature. The phospholipids in the reconstituted membranes can be readily extracted by ether in presence of salt in the same manner as in protein-free phospholipid dispersions (Section III. A. a); since the lipid-protein linkages are not broken by salt,[209] they must be broken by the nonpolar solvent; this is further indication that the bonds should be hydrophobic in nature.[208]

Hendler[159] raises the possibility that in the recombination experiments high salt may stimulate nonpolar binding at the same time that it depresses ionic interactions. The biphasic curve of the binding vs. ionic strength (Fig. 19) may corroborate this hypothesis but the fact that inhibition of binding at high ionic strength depends upon the *nature* of the anions and not upon their *charges*, is an indication of a chaotropic and not a ionic effect.[210–212]

Phospholipid vesicles and phospholipid-LDM complexes bind ionically to the basic protein lysozyme with the same stoichiometry (Table IV); in order for the polar groups of the phospholipids to be available for lysozyme cationic groups, the LDM protein must be hydrophobically buried into the phospholipid bilayer.

TABLE IV. Binding of phospholipid vesicles to lysozyme*†

Addition	P added (nmoles)	Lysozyme bound (nmoles)	P/Lysozyme	P/Lysozyme (corrected)‡
Asolectin	3500	133	26	
RM	2580	157	16	23
LDM	395	58	7	

* G. Lenaz, E. Bertoli, G. Parenti-Castelli and P. Pasquali (unpublished data).

† The ionic complexes formed by phospholipids (Asolectin) or membranes reconstituted from LDM plus Asolectin (called RM) with excess lysozyme were centrifuged and the amount of lysozyme bound was measured by difference from the O.D. at 280 nm in the supernatant.

‡ Corrected for the amount of lysozyme bound to the same amount of original LDM protein.

Zwaal and Van Deenen[179,416,417] have studied the recombination of butanol-solubilized erythrocyte membrane proteins with membrane lipids; in contrast with the studies in the lipid-free mitochondrial membranes, recombination of lipids does not occur at physiological pH or in presence of 0·2 M NaCl; however, once the complex is formed, changes in pH or increased ionic strength do not split the complex. The association of proteins and lipids involves electrostatic attractions

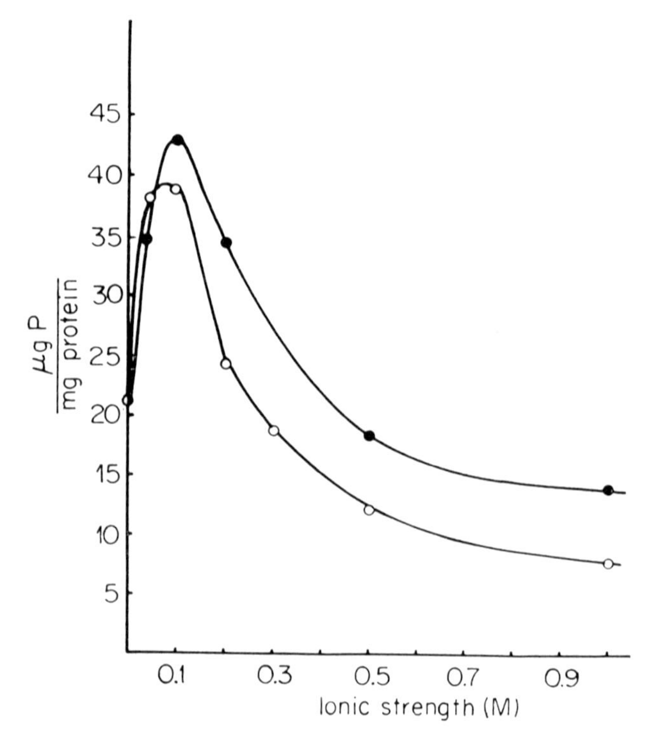

Figure 19. Phospholipid binding to LDM at increasing concentrations of NaCl and NaSCN. ●——●, NaCl; o——o, NaSCN (from Lenaz, Sechi, Masotti, and Parenti-Castelli[211]).

that are followed by apolar interactions. It would be of interest to correlate the need for an initial ionic attraction (see also Section II. E) with the water-solubility of the protein used; a conformational change may result from the initial attraction, followed by extensive hydrophobic binding. On the other hand LDM, which are initially water-insoluble with their hydrophobic groups already directed outwards, may be directly bound to phospholipids in their final configuration.

The hydrophobic nature of the lipid–protein linkages in membranes does not rule out the possibility that phospholipids in the membrane are in a bilayer arrangement. Indeed this possibility is favoured not only from enzymic digestion data (Section III. D) and physical studies of intact membranes (Section III. C), but also from certain characteristics of phospholipid interaction with LDM. When phospholipids are bound to LDM, respiratory activity is restored;[122] the restoration of activity requires unsaturation of the phospholipids;[46,202] saturated phospholipids (dipalmitoyl–lecithin and –PE) are effective only at higher temperatures (Section IV. B); the data are consistent with the idea that only phospholipids in a liquid crystalline state can restore respiratory activity. The fact that the *paraffin chains* affect the activity of the enzymic proteins is in accord with the idea that proteins and lipid paraffin chains are in contact; the fact that *the physical state* of the chains affects activity demonstrates that phospholipids must also be in contact with each other and not individually linked to the protein hydrophobic groups.

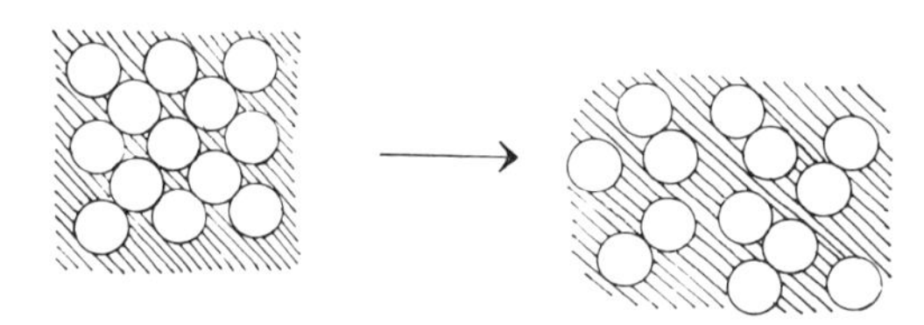

Figure 20. Possible modifications of intersubunit cohesion in the inner mitochondrial membrane. Surface view of a membrane according to the Vanderkooi–Green model.[388] The circles represent proteins; the lines between the proteins represent the polar surface of the lipid bilayer.

The presence of lipids in the bilayer form in the inner mitochondrial membranes raises the question about the restriction of the amount of lipid which can be bound in a membrane. Beef heart mitochondria contain 16–18 μg of lipid P per mg of protein;[122] can the amount of lipid be changed without affecting membrane structure? LDM usually bind a physiological lipid amount; at low salt concentration (0·05–0·2 M) however the amount of lipid which can be accommodated in the membrane may be doubled[211] (cf. Fig. 19). Rat brain mitochondria and myelin[95] and rat heart mitochondria[94] take up polar lipids and cholesterol; the binding appears nonpolar in nature since the lipids used are nonacidic. Mitochondrial membranes (HCl-treated and free of extrinsic proteins)[204] incorporate very large amounts of phospholipids.[211] Liver reticulosomes can take up lipids from a crude phospholipid emulsion.[292] All of these observations are in accord with a mosaic of protein units which are in contact with each other perhaps partially through electrostatic interactions; the crevices between the protein units are filled by bilayer lipids and their amount is usually just enough for this purpose. Under certain conditions electrostatic constrictions between the protein units may be removed and higher amounts of lipids may fill the gaps (Fig. 20). In certain membranes the subunits may be normally separated by a more continuous lipid bilayer: the outer mitochondrial membrane has a lower protein/lipid ratio than the inner membrane (1:0·829 against 1:0·275) (W. Thompson and D. F. Parsons, quoted by Chapman and Leslie).[59] Acetone extraction, which leaves apparently intact the protein skeleton of the inner membrane, completely dissolves the outer membrane.[325]

Also the exchange of lipids between lipoprotein molecules and erythrocyte membranes can bear directly on membrane structure. The exchange of free cholesterol between rat erythrocyte ghosts and human low-density lipoproteins is greatly increased in presence of compounds affecting local water structure: the effect of acetone and dimethylsulphoxide is higher at 37° than at 8–10°:[38] it is concluded that hydrophobic bonding must be of primary importance in the organization of the membrane. Vandenheuvel[384, 386] explains lipid exchange as a result of Brownian motion: certain lipid molecules acquire sufficient kinetic energy, as a result of bombardment by water molecules, to escape from the membrane or to be recaptured. Bruckdorfer and Green[38] observe that if this were the case, the process should be markedly temperature dependent, and this is not so in absence of water breakers.

C. *Physical Studies*

(a) *Electron microscopy*

A survey of all electron microscopic studies related to membrane structure is out of the purpose of this review, since electron microscopy (EM) gives only indirect evidence on lipid–protein interactions in membranes. More than any other technique, EM has been criticized about its possibility to give insights into the molecular architecture of membranes.

(i) *Fixation and thin sections.* The trilaminar "*unit membrane*" picture (Fig. 21) observed by Robertson[312] in several membranes is *qualitatively* in accord with the X-ray data and is also found in protein–free lipid lamellae.[352] Vandenheuvel[386] has described how the trilaminar picture can arise from a Danielli membrane by effect of osmium fixation alone or in combination with glutaraldehyde and/or permanganate: further detail observable by magnification of this simple pattern may be meaningless. Notwithstanding the limitations of the interpretation of EM of fixed sections, the unit membrane concept includes such an interpretation at the molecular level. The dense lines of the trilaminar picture are equated with the proteins and polar groups of the lipids while the light layer is equated with the nonpolar groups. Korn[190] has critically discussed the location of the fixatives and concludes that the dense lines in osmium-fixed membranes reveal nothing about the molecular orientation of the phospholipids in the original membrane. It may well be that osmium stains the surfaces of any lamellar structure: hollow spherules formed from proteinoid material stained with OsO_4 show a trilaminar unit membrane even though they contain no lipid.[129] Solvent extraction of nearly all of the lipid from membranes does not destroy the membranous appearance of the

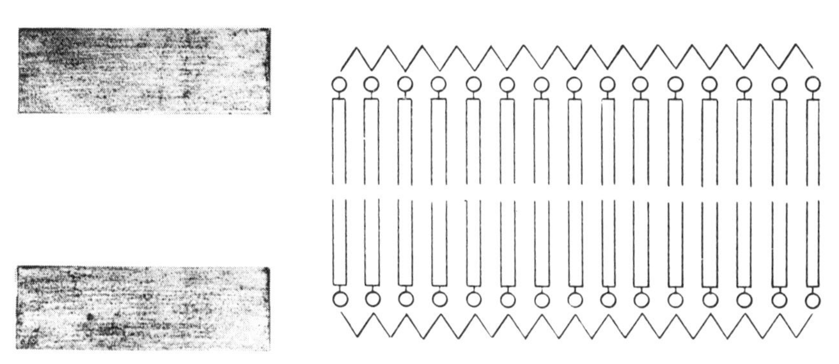

Figure 21. Schematic representation of the trilaminar structure visible by electron microscopy and the "unit membrane" representation.

material; the lipid-free membranes are the same thickness as they were before extraction.[78,125,256,266] The lack of collapse of membrane structure has been interpreted as an indication that protein is the backbone of the membrane,[387,388] but also that (reversible) denaturation has occurred in a Danielli membrane.[159]

Sjöstrand[341–343] by applying different preparative conditions has presented evidence for membrane subunits of different sizes, and proposed a two-dimensional array of globular protein units and lipoprotein complexes as the basic membrane structure (Fig. 22). Although

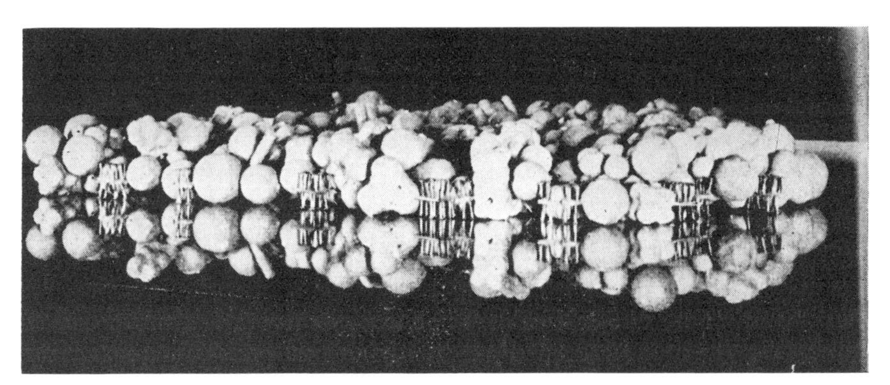

Figure 22. Mitochondrial membrane structure according to Sjöstrand.[342] In this model the molecules are arranged in three-dimensional assemblies and not in a two-dimensional monolayer. The holes between the protein molecules are assumed to contain lipid molecules in, for instance, bilayers. For details, see Sjöstrand[342] (reproduced with kind permission of Dr. F. Sjöstrand and of Academic Press, London).

the EM evidence for this model has been severely criticized;[386] evidence for membrane subunits and for importance of protein rests also on other observations.

(ii) *Negative staining.* By this technique a negative image of the unsectioned material appears against a background of electron-dense stain like Na phosphotungstate or uranyl acetate. Many membranes which appear uniform after fixation and sectioning, after negative staining show a granularity which has been interpreted as evidence for subunits.[112] Green and Perdue[144] took as a specific example of membrane subunits the tripartite structure (headpiece–stalk–basepiece) of the inner mitochondrial membrane (Fig. 23). Membranes result from

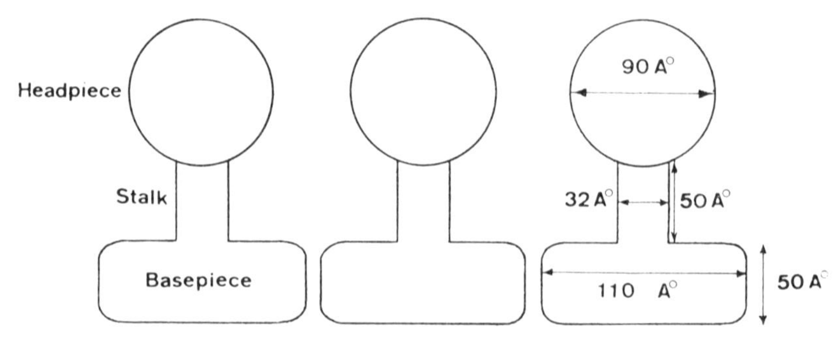

Figure 23. The tripartite structure of the repeating units of the inner mitochondrial membrane according to Green and Perdue.[144] The "headpieces" (knobs, inner membrane spheres) have been identified with the ATPase complex;[178] the "basepieces" have been identified with the complexes of the respiratory chain.[146]

nesting together of repeating units, each composed of a fixed sector (basepiece) and a detachable sector (headpiece–stalk in the inner mitochondrial membrane). The reality of the basepieces, identified with the respiratory chain complexes[146] has been questioned;[191] they are larger than Sjöstrand's granular substructures. Also the morphological appearance of the stalked particles has been considered an artifact.[342]

Negatively stained preparations of phospholipid dispersions[219] and of plasma membranes[16] have shown a pattern suggesting a micellar substructure; a micellar substructure is the basis of the model of Lucy[218] for biological membranes (Fig. 24).

Caution must be exerted in too much relying on negative staining, because its limited resolution allows to estimate only the general shape and size of small objects.[386] Alterations and structural changes may occur with the usual stains but apparently not with ammonium molybdate.[262]

(iii) *Freeze-etching.* Freeze-etched preparations are obtained by water sublimation from fracture faces of deep-frozen tissue. Branton[29, 30] suggested that the surfaces seen in freeze-etching are fracture faces along the axis of the membrane between the lipid chains; this view is strengthened by the observation that in osmium-fixed liposomes there is a decrease in fracture faces concomitant with an increase of unsaturation of membrane fatty acids (because of the formation of polymers across the bilayer).[168] The identification of the fracture faces is controversial[260, 344] and a clarification is highly desirable, since several investigations are reported with this technique.

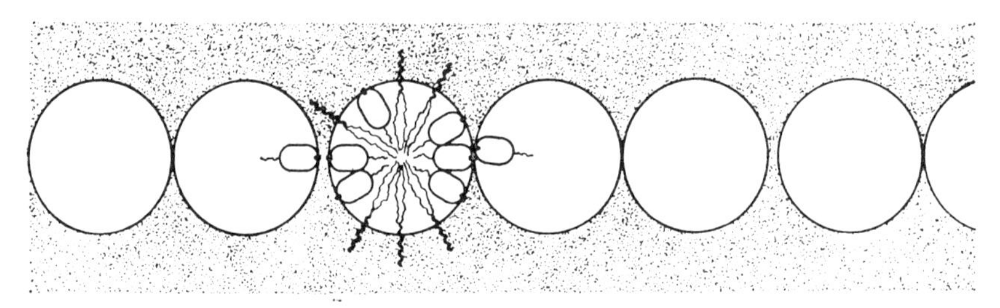

Figure 24. A cross sectional view of a model for a biological membrane in which the lipids are in small globular micelles. The micelles are arranged in a plane and a layer of protein and/or glycoprotein is shown on each side of this plane of micelles. The organization of lipid molecules (phospholipid and cholesterol) within the micelles is illustrated diagrammatically, but not stoichiometrically, for one of the micelles in the row (from Lucy,[219] with kind permission of Dr. Lucy and of Academic Press, London).

(b) *X-Ray Diffraction*

This technique has given invaluable information on the tridimensional structure of lipid–water systems[222] and also on natural membranes. The initial studies of the myelin sheath[114, 119] were qualitatively in agreement with the unit membrane; the electron density maps[408] were interpreted in terms of a bimolecular leaflet in which the phospholipid molecules are oriented outwards. A detailed model of how the lipid and protein constituents of myelin may be assembled to give a structure compatible with EM and X-ray data has been developed by Vandenheuvel[382, 383] on the lines of the Danielli model.

Studies in Finean's group with other membranes[71, 115–117, 358, 359] were also in accord with the unit membrane, but dehydration was necessary to provide suitable specimens. Recent X-ray studies with several types of membranes with techniques which do not require layered membranes nor dehydration[100, 216, 401] confirm that the main structural feature is the lipid bilayer; the location of the protein is not easy to assess at least above the phase transition temperature when the hydrocarbon tails are in a mobile condition.[100]

At least in the case of the outer segments of the retinal receptors, direct evidence for subunits has been obtained by X-ray techniques;[21–23] Vanderkooi and Sundaralingam[389] have calculated electron density distribution across a model they have proposed for the retinal rod disk membrane, and found a good agreement with the experimental values of Blaurock and Wilkins[24] which were compatible with a bilayer model (Fig. 25). In this *protein–liquid crystal model* of the membrane[388] (Fig. 26) the rhodopsin molecules are embedded

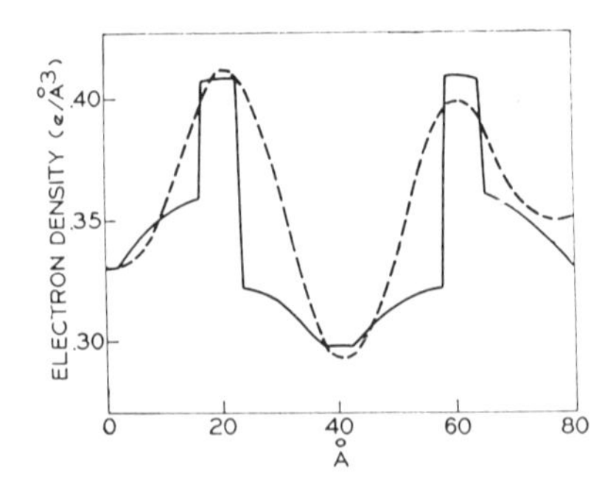

Figure 25. The calculated electron density distribution across the membrane model shown in Fig. 26 (solid line) together with the Fourier synthesis map of Blaurock and Wilkins (dashed line) (from Vanderkooi and Sundaralingam,[389] reproduced with kind permission).

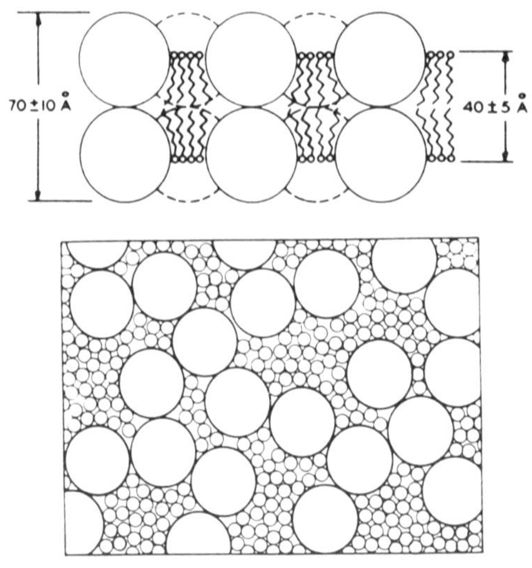

Figure 26. The protein-crystal model for membranes. A. Diagrammatic cross section of a membrane, showing a double layer of protein molecules with lipid bilayer regions filling the pores between them. The proteins drawn with dashed circles are understood to be in contact with the solid circle proteins behind the plane of the section. B. Surface view of the retinal rod disk membrane, showing the random arrangement of the proteins and lipids. The large circles represent the proteins, and the small circles the lipids (from Vanderkooi and Green,[388] reproduced with kind permission).

irregularly as individual disjointed islands in a more or less discontinuous lipid bilayer; the dynamic character of the membrane makes the protein fluctuating and assuming new protein–protein bonds each time. The proteins and lipids are thus in constant thermal motion in the plane of the membrane, rather than in a rigid crystal lattice. In this model which is a variation of another model[387] (Fig. 26) in order to account for the dynamic character of most membranes, the surface of the membrane contains both the proteins and the polar heads of the phospholipids, in accord with the X-ray data of Engelmann;[100] under certain conditions, membrane proteins assume a more regular distribution as shown by Blaisie and coworkers[22,23] at low temperatures. It seems plausible to suppose that fluidity is a general feature of membrane systems, as discussed in the next section.

(c) *The Fluidity of Biological Membranes*

Thermotropic mesomorphism of amphipathic lipids (Section II. B) may be relevant to membrane structure, since the fluidity of membranes may be related to the transition temperatures of their phospholipids. Steim *et al.*[348] observed that *M. laidlawii* do not grow well if the ambient temperature is below the transition temperature of their phospholipids; differential thermal analysis of *Mycoplasma* membranes[305,348] has shown that the phase transition of the membrane lipids occurs at

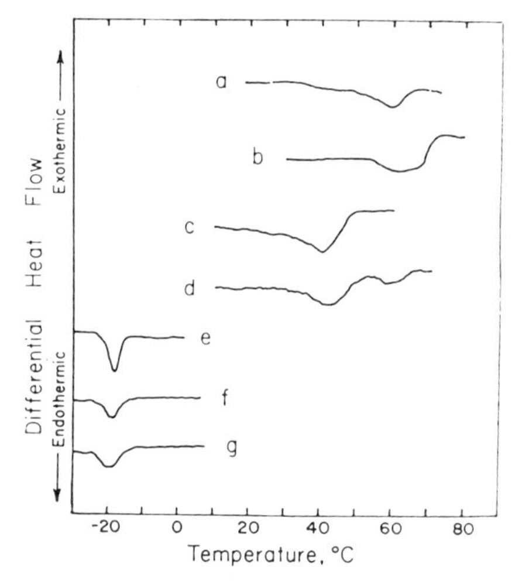

Figure 27. Calorimeter scans of "M. laidlawii" lipids, membranes and whole cells. (a) Total membrane lipids from cells grown in tryptose with added stearate; (b) membranes from stearate-supplemented tryptose; (c) total membrane lipids from unsupplemented tryptose; (d) membranes from unsupplemented tryptose; (e) total membrane lipids from cells grown in tryptose with added oleate; (f) membranes from oleate-supplemented tryptose; (g) whole cells from oleate-supplemented tryptose. The first four preparations were suspended in water; for the latter three scans, the solvent was 50% ethylene glycol containing 0·15 M NaCl (from Steim, Tourtellotte, Reinert, McElhaney and Rader,[348] reproduced with kind permission of Dr. Steim).

the same temperature in viable organisms, membranes isolated from the organisms, and isolated membrane lipids (Fig. 27). Similar results have been obtained with membranes from *Micrococcus lysodeikticus*[6a] and *E. coli.*[346a] The enthalpy of the transition agrees within 90% in the membrane and in membrane lipids, suggesting that 90% of the lipid in the membrane is in the bilayer arrangement. The data have been interpreted as giving little support to hydrophobic interaction between lipid and protein, but Vanderkooi and Sundaralingam[389] have suggested that the lipid–lipid and lipid–protein interaction energies are about the same, so that the lipid phase transition of a lipoprotein membrane should not greatly differ from that of a pure lipid bilayer. On the other hand *electrostatic* interaction of cytochrome *c* with *Mycoplasma* lipids *lowers* by 10° the transition temperature,[60] indicating that the interpretation previously given to the calorimetric studies of Steim[305,348] must be taken with much reservation. In a recent X-ray diffraction study of *E. coli* membranes,[106] the transition temperatures were *higher* in the membrane than in lipid extracts, indicating an immobilizing effect of the nonlipid constituents.

Thermal transitions are not detectable in myelin[195] but are detectable in cholesterol-free myelin lipids; in the absence of cholesterol, part of myelin is crystalline at body temperature; the organization of cholesterol in the membrane appears to prevent the lipids from crystallizing.

It is plausible that fluidity in the plane of the membrane is a general feature of membrane systems: the lipid composition of a membrane will affect its fluidity. The necessity for a well defined degree of fluidity may be the reason why the lipid composition of poichilothermic animals varies with the environmental temperature, so that the unsaturation of their constituent fatty acids is proportional to the temperature decrease.[169] Mitochondria from a chilling sensitive plant and a homeothermic animal display transition temperatures at 12° and 23° respectively; these temperatures coincide with those below which the organisms are injured upon exposure[298] and below which a break in the temperature dependency of O_2 uptake is observed.[223,224] No such transition is observed in mitochondria from a chilling-resistant plant and a poichilothermic animal. Disruption of the mitochondrial membranes with detergent changes the discontinuous Arrhenius plot exhibited by mitochondrial enzymes in the homeothermic animal and chilling sensitive plant into a straight-line Arrhenius plot[299] (Fig. 28). The discontinuity in the Arrhenius plots in succinoxidase is evident in liver mitochondria of active ground squirrels but disappears in hibernating squirrels.[297]

An *E. coli* mutant unable to synthetize unsaturated fatty acids has been grown in presence of different fatty acids at different temperatures; increasing amounts of *cis*-unsaturated acids are incorporated at decreasing temperatures, suggesting the presence of a regulatory mechanism controlling the composition of saturated and unsaturated acids in order to maintain the physical properties of phospholipids within narrow limits.[105] A liquid-like state of the lipid phase (controlled with the monolayer technique) is required for growth, respiration, and efflux of thiomethylgalactoside.[274] Arrhenius plots for β-galactoside and β-glucoside transport in the organism grown on oleic or linoleic acid are biphasic, with changes in slope at 13° and 7° respectively.[405] The membrane may therefore exist in two physical states which are only determined by the composition of the lipid phase.[404, 405] In contrast with other studies[223, 224, 298] in the *E. coli* mutant the temperatures at which physical changes in membrane lipids were detected by X-ray diffraction do not coincide with temperatures at which breaks in the

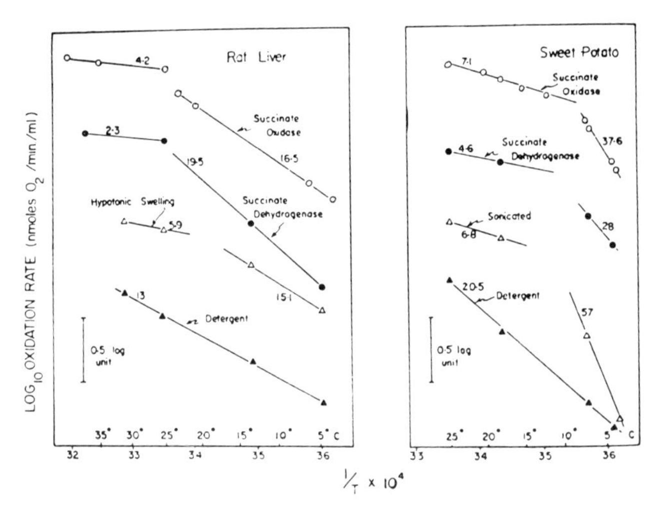

Figure 28. Arrhenius plots of the activity of the succinate oxidase system and of succinate dehydrogenase for mitochondria from rat liver and sweet potato tissue. The number beside each line shows the activation energy in kcal/mole calculated from the slope of the line. ○——○, succinate oxidase of intact mitochondria; ●——●, succinate dehydrogenase of intact mitochondria; △——△, succinate oxidase of mitochondria after swelling in 0·02 M phosphate buffer pH 7·6, containing 0·02% w/v of bovine serum albumin; ▲——▲, succinate oxidase of mitochondria in the presence of 0·05% v/v of Nonidet P_{40} (from Raison, Lyons and Thomson,[299] reproduced with kind permission of Dr. Lyons).

Arrhenius plots of proline uptake and succinic dehydrogenase activity were observed,[106] suggesting that the distribution of lipids within the membrane is heterogeneous, so that the lipids associated with these activities are in a more fluid state (cf. also ref. 404).

Unpublished studies in our laboratory have also shown that different mitochondrial activities undergo breaks in the Arrhenius plots at different temperatures: caution must be exerted in ascribing breaks in the temperature dependencies to changes in enzyme environment only.

The fatty acid composition of different membranes may explain their metabolic and functional characteristics. According to O'Brien[268] myelin owes its stability to its particular lipid composition; myelin contains largely less unsaturated fatty acids than any other subfraction of calf brain.[236] The introduction of unsaturated fatty acids into a bimolecular leaflet leads to a more loosely packed, less stable structure, which would fit better highly functional membranes like mitochondria.[270]

(d) *The Environment of the Protein*

The calorimetric and X-ray studies cited in the previous sections indicate that the phospholipids in membranes are in a bilayer configuration: however only 75% of the lipids of intact *M. laidlawii* membranes appear to "melt" at the transition temperature;[348] the remaining 25% could perhaps be involved in a more tightly coupled interaction with the membrane proteins. The controversial effect of the protein on the transition temperatures has been also described in the previous section. Other studies point out that part of the lipids are more tightly bound to membrane proteins[379] (cf. Section III. A. a for extraction studies and Section III. D. b for enzymic studies). It has been suggested that different lipid classes may be distributed in a heterogeneous way in lipid–lipid and lipid–protein interactions in the native membrane structure.[60, 106, 313, 404]

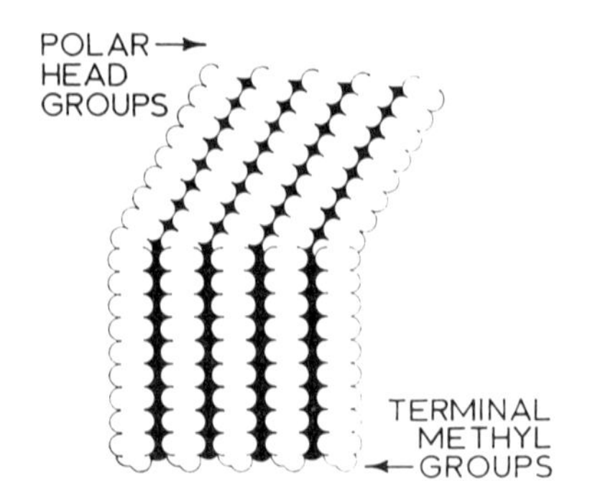

Figure 29. Bending of the fatty acyl chains in lecithin (from McFarland and McConnell,[249] reproduced with kind permission of Dr. McConnell).

Other studies indicate physical differences of lipids in intact membranes and after isolation. IR spectroscopy shows that there is little all-*trans* character of the CH_2 groups when they are organized in the erythrocyte membrane,[58] perhaps indicating a less compact packing of the lipids. On the other hand NMR has shown that erythrocyte membranes dispersed in D_2O exhibit regular signals of the protons in the choline groups, but the expected signal from the protons in the CH_2 groups is not apparent; since the signal is visible in the extracted lipids, Chapman and Kamat[57] suggest that the damping effect on acyl chain vibration is due to hydrophobic binding to protein. There are not enough protein hydrophobic groups to account for individual immobilization of the fatty acyl chains; although the damping effect has been considered the result of bonding of lipid heads to rigid protein units with reduction of free energy and increased lipid–lipid interaction,[386] lipid immobilization could also derive from hydrophobic bonds with protein and cooperation of lipid chains along the plane of the bilayer. The likelihood of the latter interpretation is strengthened by the fact that binding of cytochrome *c* to the lipid polar heads results in a lowering of the transition temperature.[60]

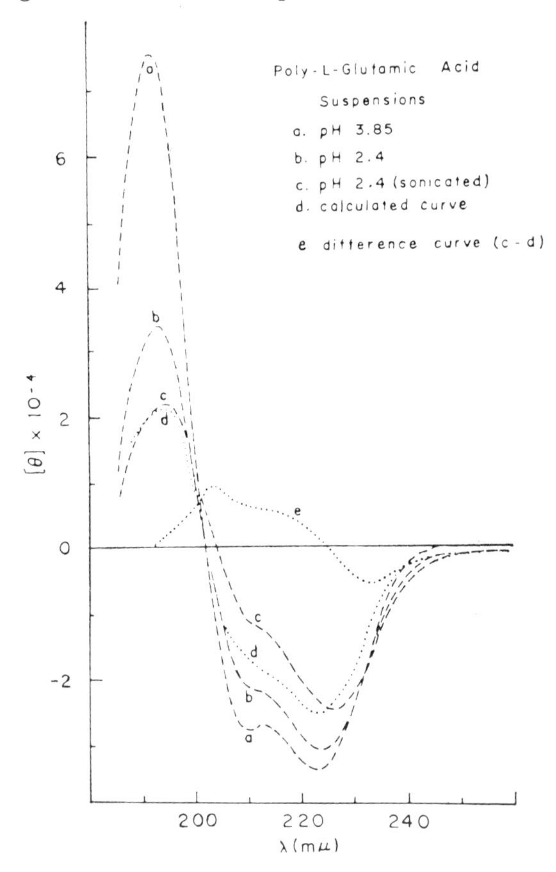

Figure 30. Circular dichroism curves for poly-L-glutamic acid in the reference state, pH 3·85, and as suspensions. The concentration of PGA is constant at 2 mg/ml, only the particle size is varied (from Urry, Hinners and Masotti,[376] reproduced with kind permission of Dr. Urry).

Fluorescent dyes ANS and dansyl-chloride are strongly immobilized by their association to excitable membrane fragments; solubilization by Triton-X-100 is accompanied by a dramatic increase of motion.[392] ESR of spin labelled lecithin bilayers[249] shows bending of the fatty acyl chains (Fig. 29) with possibility of hydrophobic pockets in the bilayer, which are speculated to serve for binding of hydrophobic molecules such as proteins. Spin labelling of *M. laidlawii* lipids[318, 362] shows a significant increase in mobility of the lipids when extracted from the membranes. Spin labelling of sarcoplasmic vesicles[332a] shows that the first 7 carbon–carbon bonds adjacent to the carboxyl group of the spin labelled fatty acid experience a highly ordered environment, whereas for $n > 7$ a pronounced increase in the flexibility of the hydrocarbon chain is observed although the system is still more ordered than

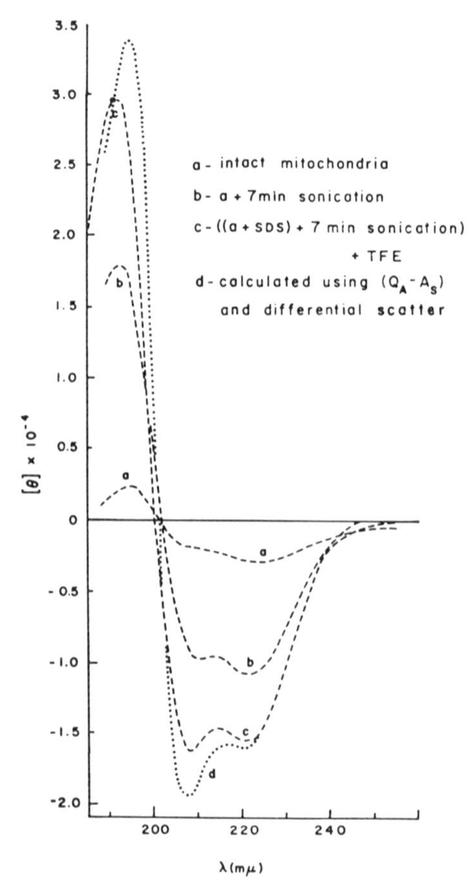

Figure 31. Circular dichroism curves of mitochondria. Curve (a) is for the intact mitochondria and curve (b) is the spectrum obtained after 7 min of sonication. The concentration of protein was 2·24 mg/ml and the path length used was 0·10 mm. Curve (c) is a solubilized state of the mitochondrial membrane in which the $[\theta]^{224}$ is closely that initially approximated for the corrected mitochondrial membrane. The state in the curve (c) was obtained by addition of SDS followed by sonication and then addition of 4 parts of trifluoroethanol. Curve (d) is the calculated curve (Urry, Masotti and Krivacic,[377] reproduced with kind permission of Dr. Urry).

pure lipid dispersions. Hydrophobic interaction with membrane protein subunits is supported by the fact that high temperatures or glutaraldehyde treatment do not alter the protein. Although phospholipase C has little effect on the temperature sensitivity of protein conformation in erythrocyte ghosts[131] there is an indication that phospholipids confer to lipoproteins high resistance to denaturation.[88, 135, 327]

The conformation of membrane proteins is a debated subject and may be relevant to membrane structure. The IR spectra of plasma membranes show evidence for α-helix and no β-conformation;[395] the spectra are unaffected by lipid extraction. Optical rotatory dispersion (ORD) and circular dichroism (CD) have shown a characteristic "red shift" which has been variably interpreted.[134, 199, 347, 378, 393] Urry and coworkers[375, 376] have clearly indicated that the red shift and other spectral anomalies are due to absorption flattening and dispersion distortion arising from the particulate nature of the samples (Fig. 30). Corrections applied to the spectra[376] suggest that protein in mitochondria, red cell membranes and other plasma membranes can well be in about 50% α-helical configuration (Fig. 31) whereas the values of absorbancies and ellipticities of sarcotubular vesicles are lower, suggesting different conformations. [243, 377] Artifacts in the CD spectra have been recently recognized by others.[130, 133] Other authors give different interpretations to CD spectra; evidence for 50% β-structure has been presented for *Mycoplasma* membranes.[67] Studies of the red cell ghosts also suggest that at least 15% of the protein is α-helix.[199, 234]

Differences in conformation in relation to energy changes have been described by IR analysis,[136, 137, 394] CD[244] and fluorescent probes.[9]

The presence of proteins in α-helical configuration in myelin[191] would leave large amounts of the bilayer uncovered by protein. Proteins in α-helical conformation do not exclude the unit membrane model *per se*, since disruption of secondary structures at interfaces may not occur.[167]

D. *Spatial Organization of Proteins and Lipids: Chemical Studies*

(a) *Proteins*

Brown[36] after solubilization of halophilic bacteria at low ionic strength found that in the solubilized membranes all the α- and ε-amino groups of proteins were available for titration without loss of lipid, but some of the carboxyl groups became available only after removal of the lipid. In addition, the amino groups could be succinylated without loss of lipid. This suggests that there can be no ionic bonding between anionic phospholipids and protein.

Studies of lipid–protein organization by use of proteolytic and lipolytic enzymes or combination of enzymes and solvents have been directed to understanding soluble lipoprotein structure.[109, 239, 346]

As for membranes, a number of studies point out that at least some membrane proteins may be available to the action of proteolytic enzymes and protein reagents. Coleman *et al.*[73] have investigated the effects of increasing trypsin concentrations upon rat muscle microsomes; at low concentrations modification of lipoprotein structure is accompanied by loss of ability to accumulate Ca^{2+} without loss of ATPase activity; higher concentrations produce further alterations and loss of ATPase. Maddy[231] found that the bulk of erythrocyte membrane bears sialic acid; since all the sialic acid in erythrocytes is accessible to neuraminidase, which does not pass through the membrane, Maddy suggested, among other possibilities, that the sialoprotein is continuous throughout the thickness of the membrane.

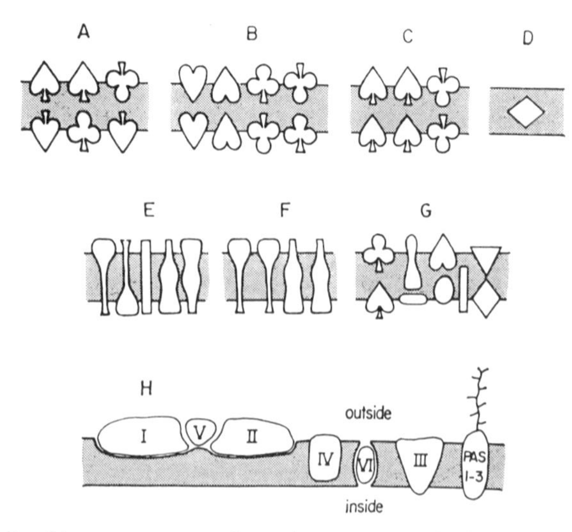

Figure 32. Possible arrangements of protein components in the erythrocyte membrane cross-section. Each unshaded form denotes a protein species in a particular distribution and orientation with respect to a hypothetical barrier to penetration of the enzyme probes (shaded zone). Models A–C all represent a double distribution of protein species which have symmetrical (A), random or mobile (B), or asymmetrical (C) orientation. In model D, the protein is buried within the barrier and is therefore inaccessible. Model E shows penetrating proteins of symmetrical, random, or mobile orientation. Asymmetrically oriented, penetrating proteins are depicted in model F. Model G shows an asymmetrical distribution of components across the barrier which could have any orientation. Model H illustrates a possible disposition for each of the major protein and glycoprotein components of erythrocyte membrane. Based on their behaviour in elution studies[110] components I, II, and V have been pictured as being rather superficial in location and physically associated *in situ*, while component VI has been assigned an intermediate position (from Steck, Fairbanks and Wallach,[345] with kind permission of Dr. Steck and of the American Chemical Society).

Only a limited portion of the protein in human erythrocytes is accessible to pronase.[15] Specific labelling of superficial proteins in intact erythrocytes showed that only a single protein[286, 287] or few proteins[20, 32, 173] are labelled (depending perhaps on the size of the reagent); all membrane proteins are labelled by formylmethionyl sulphone methyl phosphate when erythrocyte ghosts rather than intact cells are exposed to the reagent;[32] fingerprint maps show that additional peptides of a major protein component become labelled in ghosts in comparison with intact erythrocytes, indicating that this component spans the membrane;[33] on the other hand some proteins can be iodinated only when free from membranes, indicating that they are not exposed on either side.[287] Steck *et al.*[345] separated vesicles from ghosts into two fractions of normal and inside-out orientation; by treating each species with proteolytic enzymes the two faces were selectively digested, showing asymmetric localization of proteins; one protein is intrinsically resistant, while some are exposed on both sides, suggesting that they span the entire thickness of the membrane (Fig. 32). On the other hand most proteins could be attacked by pronase in purified plasma membranes from *M. laidlawii*.[256]

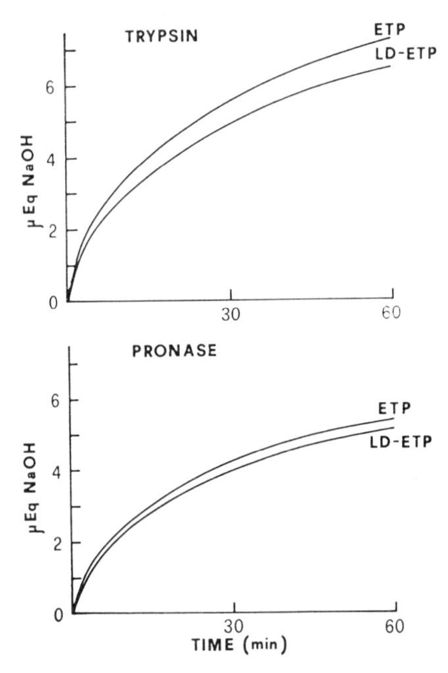

Figure 33. Action of trypsin and pronase on intact and lipid-depleted submitochondrial particles ETP (P. Pasquali, M. Monsigni and G. Lenaz, unpublished data). Almost identical curves were obtained for SU-particles (which are devoid of ATPase) and lipid-depleted SU-particles.

The problem of investigating the topology of the proteins in erythrocyte and other membranes may be however much more complex; T. L. Steck (personal communication) considers that permeability and composition of the membrane fractions can be of extreme importance for the susceptibility to proteolytic digestion; also the labelling of proteins by nonpenetrating reagents may depend on such conditions as structural intrinsic reactivity of a given protein and organizational changes following lysis and not related to membrane sidedness. (See also Section III. A. b.)

More general indications on the distribution of proteins and lipids can be however obtained by proteolytic digestion. In experiments in our laboratory using submitochondrial particles, proteolytic enzymes pronase and trypsin were able to hydrolyse protein in intact membranes at the same rate and to the same extent as in lipid-depleted membranes (Fig. 33). This finding, together with the fact that probably none of the proteins of *M. laidlawii* are completely buried and inaccessible to pronase,[256] suggests that lipids do not protect protein from enzymic digestion; a strict interdigitation of lipid and protein in the membrane as in Benson's model[18] (Fig. 34) would be expected to lead to impaired protein digestion.

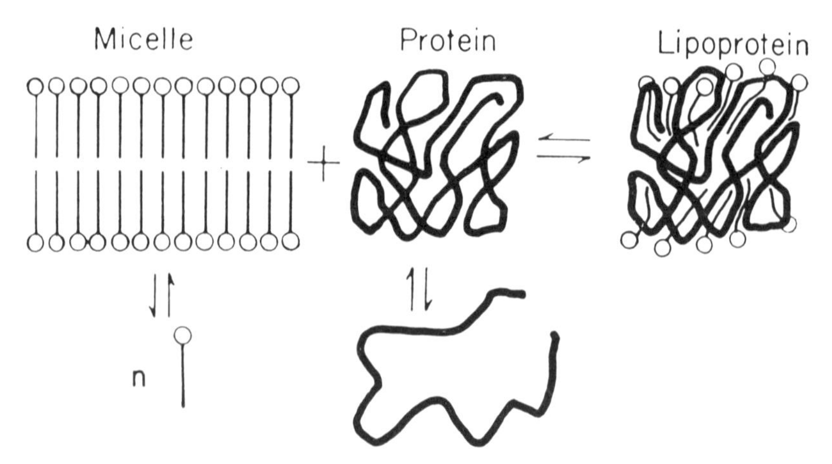

Figure 34. Benson's model for membrane structure[18] (reproduced with kind permission of Dr. Benson).

(b) *Lipids*

On the other hand it has been known for some time that membrane phospholipids are hydrolyzed by exogenous phospholipases and that a number of functions are altered by the treatment (Section IV. B). Finean and Martonosi[118] found that phospholipase C hydrolyzes 60–70% of lecithin in microsomes. The vesicles are reduced in size but the membrane still shows the identical trilaminar structure. All the choline from erythrocyte membranes is released by phospholipase D.[37] In chloroplasts, galactose is rapidly released from galactolipids.[325] Lenard and Singer[200] showed that crude phospholipase C from *Cl. perfringens* hydrolyzes about 70% of phospholipids from red cell membranes, but has no significant effect on the CD spectrum of the protein. Purified phospholipase C from *B. cereus* releases 62%[131] and 61–68%[273] of membrane P. After phospholipase C treatment 3/4 of the fatty acyl chains become more mobile; no alteration of protein conformation is shown by CD. According to Glaser *et al.*,[131] the polar heads of the phospholipids in the intact membrane must be accessible to the active site of the enzyme. Since the phospholipase C products (diglycerides) accumulate as lipid droplets in the membrane[72] most lipids must indeed be present as lipid–lipid (and not lipid–protein) complexes.[173] Spin-labelling suggested that the lipids proximal to the proteins are affected by phospholipase A_2 from *Naja naja* venom and in turn affect protein structure,[340] in contrast with phospholipase C treatment. Similar findings upon phospholipase A digestion have been reported by using the fluorescent probe ANS.[398]

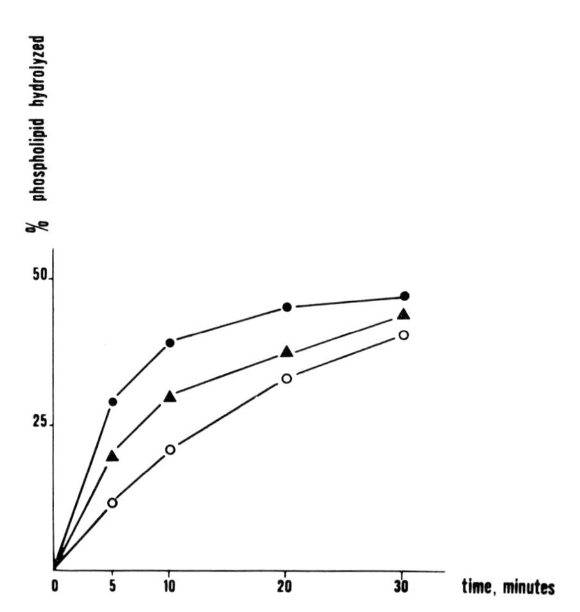

Figure 35. Phospholipase C hydrolysis of submitochondrial membranes in comparison with mitochondrial phospholipid dispersions. ▲——▲, mitochondrial phospholipids; ●——●, HCl-extracted mitochondria; o——o, sonic submitochondrial particles ETP (Lenaz[203]).

In other studies of erythrocyte membrane lipids[51, 52, 264, 265] it was concluded, by use of phospholipases and immunological techniques, that sphingomyelin must be located at the outside of the erythrocyte membrane, whereas PE and PS are located inside.

Both Brownian motion and lipid exchange (and therefore extrusion from the membrane) have been considered to account for phospholipase action in biological membranes, without any need to assume that lipids are fully exposed at the surface.[30, 159, 386]

Studies in our laboratory[201, 202, 280] have directly compared the effect of *Cl. welchii* phospholipase C on aqueous phospholipid dispersions and in submitochondrial membranes. The enzyme hydrolyzes the dispersions and the membranes to similar extents (50%) and at similar rates (Fig. 35). The phospholipid polar heads must be available at the membrane surface to the enzyme action; if a protein layer extensively covered the bilayer surface, phospholipids would not be easily hydrolyzed in the membrane, or at least (even allowing for Brownian motion of the subunits) they would be hydrolyzed at significantly lower rates. To this purpose we have shown (unpublished data) that a lysozyme–phospholipid ionic complex is not hydrolyzed by phospholipase C to any significant extent. It was also stated[30] that in many investigations phospholipases hydrolyze much more than 50% of the phospholipids, suggesting that penetration of the enzyme through the membrane may occur to hydrolyze the phospholipids of the unexposed side. Penetration of an enzyme protein into a bilayer would however be an indication (a) that certain bilayer areas must be exposed, and (b) that protein, namely the phospholipase enzyme, can interact hydrophobically with bilayer lipids. In any case when hydrolysis of 50% of the phospholipids has taken place, it is not unreasonable that a rearrangement of the remaining phospholipids to reform a bilayer allows more of the ester bonds to be hydrolyzed. Danielli[81] has given evidence that bilayers in water are energetically favoured over other structures. Moreover, breakage of the vesicular structures may ensue to the enzymatic action, leaving both sides of membranes available to digestion.

E. *Current Models of Membrane Structure*

Models of membrane structure may be divided into at least four main groups although several variations have been proposed. I shall discuss the various models on the basis of available evidence discussed in the previous sections.

(a) *Proteins are Adsorbed on the Surface of a Lipid Bilayer*

The original Danielli model and variations thereof assume that the interior of the membrane is occupied only by the nonpolar portions of the lipid molecules. There is no continuity of the protein from one side to the other side of the membrane. In order to account for the permeability of the plasma membrane, the unit membrane model has been however modified to include polar pores in the membrane structure;[350] the pores are formed by protein connecting the two membrane surfaces and bound to the lipid chains; the binding should be hydrophobic, however there is no need for *extensive* penetration of the proteins into the bilayer. If subunits are inherent in the Danielli model[386] (Section III. A. c), specialized transport could occur between subunits, and the specific reactive sites for transported molecules would be at their periphery in small clusters of superficial amino acid residues; the movement of the solute to the other side would be a consequence of a sufficient displacement of two subunits, without any need for carriers or permeases.

Among the many observations in favour of the classical model is the fact that only bilayers fulfil the thermodynamic requirements for lipid arrangement in water;[81] on the other hand bilayers in water share several properties with biological membranes.[132] Data in favour of the bilayer arrangement in membranes have been collected in the previous sections. The demonstration of hydrophobic interactions between proteins and lipids is not a conclusive argument *per se* for protein being in the inner structure of the bilayer (cf. Fig. 6, and ref. 87 for a recent model of erythrocyte membrane which is based on similar assumptions).

Two kinds of proteins are envisaged in the Danielli model (Fig. 1); a monomolecular layer consisting of juxtaposed protein molecules adsorbed on the polar heads of the lipids, and "superstructural" globular protein units. Two kinds of proteins are also differentiated by solubility characteristics: the identification of the two protein layers with intrinsic or extrinsic proteins in Green's terminology is not clear.

(b) *The Membrane is a Lipoprotein Monolayer*

Benson[18, 19] has suggested the following model. The membrane is a lipoprotein monolayer; amphipathic membrane lipids (and chlorophylls in the lamellar membranes of the chloroplasts) are associated hydrophobically with globular membrane proteins. Lipid–protein binding involves individual contact of the fatty acyl chains of the amphipathic lipids with hydrophobic side chains of protein (Fig. 34). The result is that adjacent lipoprotein subunits, where the bilayer structure is not any more apparent, make up the membrane continuum. Benson[17] has suggested that there are two classes of membranes which are both structurally and functionally dissimilar. The first, which includes myelin, has essentially the Danielli structure; the second class, which includes mitochondria and chloroplasts, consists of lipoprotein subunits. The lipid composition of these membranes (prevalence of saturated and long chain fatty acids in the first class) has led O'Brien[269] to support this suggestion.

This model accounts for hydrophobic binding of amphipathic lipids to membrane proteins and for the evidence that lipid-depleted mitochondria do not collapse. The model however does not account for the following facts: (a) X-ray and calorimetric evidence that lipids in membranes are arranged in bilayer structure; (b) the insufficient amount of hydrophobic residues in intrinsic membrane proteins to account for extensive hydrophobic binding; (c) the ready availability of membrane (intrinsic) proteins to proteolytic enzymes, and (d) most of the lipids in membranes appear to behave as separate units from the proteins: proteins and lipids are independently affected by a number of treatments.

(c) *The Membrane is a Mosaic of Protein Subunits*

The advantages of the Benson model are maintained, and the objections may be overcome in a model where globular protein subunits are the main backbone of the membrane. The lipids can be accommodated in the spaces and interstices between the subunits. In the model proposed by Glaser *et al.*[131] proteins constitute a mosaic of subunits immersed in a lipid bilayer; some units transverse completely the membrane continuum; if present in large amounts they may not be compatible with X-ray data showing presence of a low-density region in the middle of a membrane transversal section (Fig. 25). The same objection can be made to the model of Sjöstrand[342] (Fig. 22) as it is

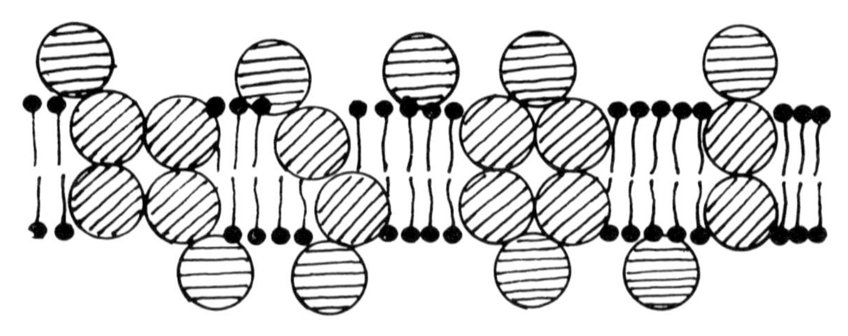

Figure 36. Binding of intrinsic and extrinsic proteins to a membrane. The intrinsic proteins are represented in clusters of variable number of units along the general lines of the model of Vanderkooi and Green (Fig. 26). Extrinsic proteins might be bound, at both sides of the membrane, either to the lipid polar heads or to the exterior of intrinsic proteins or to both components.

formulated, since there is too little account for a lipid bilayer. The model of Vanderkooi and Green[388] (Fig. 26) is compatible with the presence of a low-density region as evidenced by X-ray analysis (Fig. 25), since a double layer of globular protein molecules is responsible for the membrane continuum; the lipids fill the spaces between the proteins and are in the bilayer arrangement. A portion of the lipids may be more strongly bound in the membrane, since it is directly apposed to the protein units; evidence for tightly bound lipids has been discussed in previous Sections (III. A. a, III. C. d, III. D. b). The relationships between this model and a previous model of Green and Perdue (Fig. 23)[144] are not completely defined, although the large repeating units (basepieces) of the previous model may be complexes of the individual protein subunits of the recent model.

Although the model is compatible with available experimental data, its full acceptance waits for the following demonstrations. (a) Unambiguous evidence for the amphipathic nature of intrinsic membrane proteins under native conditions. Although there is circumstantial evidence, part of which has been described in the previous sections (amino acid composition, water insolubility, monolayer studies), the structural characterization of most membrane proteins is still lacking. (b) Demonstration that there is significantly high penetration of protein into the lipid continuum in a *bilayer* membrane. (c) Unambiguous evidence for hydrophobic binding of proteins and lipids in the natural membrane. Such evidence does not automatically demonstrate point (b) (cf. Fig. 6). Although there is evidence for hydrophobic bonding, many properties of the lipid–protein interactions might be explained by hydrogen bonding. The extraction and recombination studies described in sections III. A. a, and III. B and the properties of structural protein–phospholipid–cytochrome *c* complexes (Section II. E) and of LDM–phospholipid–lysozyme complexes (Section III. B), are in favour of hydrophobic binding with penetration.

This model leaves room for *binding of extrinsic proteins* to the membrane in terms of the classical Danielli model (ionic bonds, hydrogen bonds, divalent cation-mediated interactions)[203,212] (Fig. 36).

(d) *Presence of Lipid Micelles*

This model is due to Lucy[218] (Fig. 24). Biological structures that might be regarded as bimolecular leaflets may in certain cases be formed by the association of globular micelles of lipids as building blocks. Globular lipid micelles are not energetically favoured, but they can be in equilibrium with large areas where the lipids are arranged in a bilayer form. Adjacent micelles may be held together by hydrogen bonds, while ionic interactions would play a more or less important part according to the nature of the phospholipid molecules.[219] Protein subunits might be intercalated with lipid micelles or could penetrate into them. Phase transitions from lamellar to micellar structures may be possible, since they occur under conditions approaching those found in biological systems.[222] Such transitions, also suggested in a flexible membrane model by Kavanau,[181] might explain dynamic properties like movement, endocytosis, exocytosis, transport, fusion, division etc.[221,407] The micellar model has the advantage of explaining the details of dynamic transitions in molecular terms. However most of the membrane must exist in the bilayer form, not only for energetic considerations, but to explain physical data (Section III. C). It must also be borne in mind that most physiological lipids do not form true micelles.[11,222] However the inclusion of lysolecithin in a phospholipid bilayer will reduce its stability and favour the formation of globular lipid micelles;[157] the presence in lysosomes of phospholipases which produce lysolecithin is consistent with a high proportion of lysosomal lipids in micellar configuration in comparison with other membranes: membrane fusion is indeed involved in lysosomal physiology.[220] A mosaic model allowing for micellar transitions *in limited areas* could best represent membrane structure and functions at the state of our available present knowledge.

IV. *The Role of Lipids in Membranes*

The knowledge of how lipids and proteins interact together is the premise for the understanding of membrane functions. Why certain functions are associated to the lipid-containing membranes and not just to proteins or protein complexes? This question brings us directly to the problem of the role of lipids in biological membranes, and why lipids do interact with proteins to form a membrane.

The bilayer arrangement of the lipids (in the Danielli model) and of both lipids and proteins (in the model of Vanderkooi and Green) to form structural barriers will explain at least the permeability, transport, and control properties of membranes. Lipids are then structural components necessary for membrane stability and formation.[147] The notion that several enzyme activities of membranes are lipid-dependent and that phospholipids may be involved in protein metabolism and amino acid transport[364] raises further questions on the role of lipids in biological membranes. Is a structural function sufficient to explain all of the known facts about lipid requirements?

A. *Lipids and Membrane Formation*

Subunit structures of several membranes may be isolated by means of detergents or solvents (Section III. A. c). The depolymerized units may contain phospholipids and the detergent used for the disaggregation process. The phenomenon is reversible, and removal of the solubilizing agent in presence of the lipids allows the lipoprotein subunits to realign spontaneously and reform vesicular membranes.[140,146,179,247,252,353,416,417] Reservations have been advanced on the reality of the depolymerized state obtained by bile salts[352] which have been discussed in Section III. A. c.

Membranes of *M. laidlawii* have been disaggregated in SDS;[101–103, 357] removal of detergent in presence of Mg^{2+} leads to a reaggregated structure with membrane-like properties. Although the reaggregates are indistinguishable from native membranes by composition, density and E.M., NMR analysis and fluorescent probes suggest that at least some proteins must be incorrectly reassembled.[250,251] Membranes reaggregated from SDS-solubilized subunits of *M. lysodeikticus* are not susceptible to the same disrupting agents as the original membranes[45] suggesting changes in the nature of the bonds stabilizing the membrane.

When solubilized cytochromoxidase is depleted of the lipids, the preparation aggregates after removal of residual bile salts but does not form membranes;[247] on the other hand membrane formation in the lipid-depleted preparation can be induced by adding back mitochondrial phospholipids. Green and Tzagoloff[147] have suggested that the process of membrane assembly consists of the bidimensional association of lipoprotein "subunits" along a plane, and the function of phospholipids is to provide a hydrophilic surface preventing polymerization in the third dimension (Fig. 37). In the terminology of Green *et al.*,[140] *molecularization* rather than *aggregation* is induced.

The role of lipids for membrane formation *in vitro* can perhaps be transposed *in vivo* to the process of membrane assembly during the biogenesis of various membrane systems.

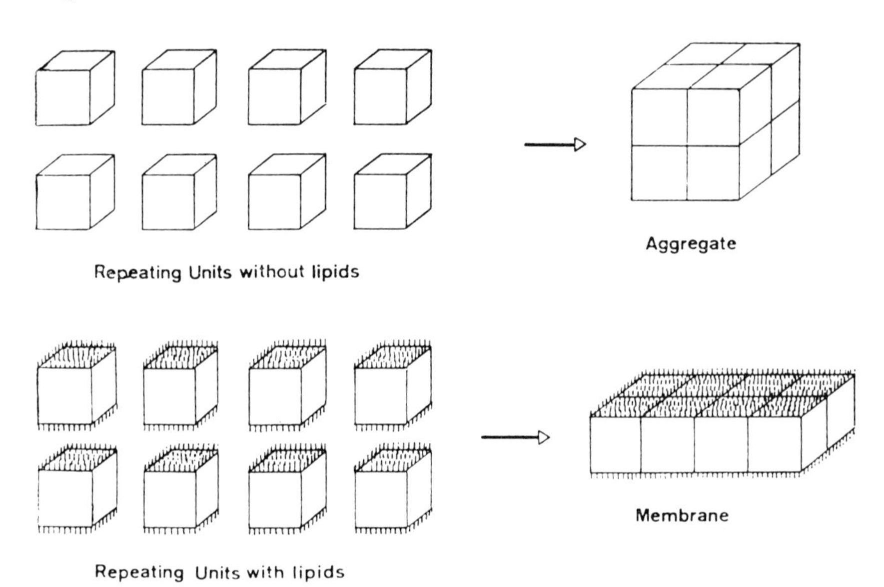

Figure 37. Role of lipids on membrane formation from disaggregated repeating units (modified from Green and Tzagoloff[147]).

B. *Lipid-dependent Enzyme Systems*

The involvement of phospholipids in the function of mitochondrial enzymes was first demonstrated by showing that rebinding of phospholipids to solvent-extracted mitochondria leads to reactivation of respiratory activity.[122] Subsequently it was demonstrated that phospholipids are required for each segment of the respiratory chain, both in intact mitochondria and in purified systems;[34, 35, 40, 370] solvent treatment destroys NADH-oxidase[78] but a lipid requirement for this enzyme system has also been shown by adding back lipids after digesting the endogenous lipids with phospholipase A.[123, 124]

Phospholipids are also essential in the activation of the particulate mitochondrial ATPase and in restoration of oligomycin sensitivity.[40–42, 176, 177, 189] This rather complex phenomenon has been investigated by Racker: in particular, the isolated oligomycin insensitive ATPase F_1[293] does not require phospholipids for activity. A colourless membrane fraction (CFo) has been obtained[177] which is capable of binding F_1: an almost complete suppression of ATPase activity results from complex formation; the inhibition is released by addition of phospholipids and the activated ATPase is oligomycin sensitive. CFo made from resolved particles (TUA) lacks the oligomycin sensitivity conferring factor (Fc or OSCP): this deficient CFo binds F_1 even in the absence of added Fc and phospholipid, but the bound ATPase is only partially sensitive to oligomycin or rutamycin; addition of Fc to the complex results in inhibition of ATPase activity, which is recovered by phospholipids.[42]

The function of the phospholipids may be relevant also to the effect of inhibitory compounds like oligomycin. Simple addition of a crude phospholipid preparation to the incubation medium prevents the oligomycin induced inhibition of ATPase.[275] The protective effect can be reproduced using individual purified phospholipids. Phospholipids may provide a suitable medium for the interaction with lipid-soluble molecules: exogenous phospholipids effectively compete for oligomycin with the phospholipids of the enzyme preparation. The competition can be so effective to induce a redistribution of oligomycin even when the antibiotic is already bound to the enzyme complex.[39] According to Bruni[39] this effect can be considered relevant from a pharmacological point of view; the possibility of a specific interaction between drugs and phospholipids indicates that the lipid can be considered a mean to channel the drug to its final target.

Other enzymic activities have been found to require phospholipids. The list includes, among other activities: pyruvate oxidase from *E. coli*;[77] pyruvate and α-ketoglutarate dehydrogenases from mitochondria;[140] solubilized β-hydroxybutyric dehydrogenase;[174,175] solubilized succinate dehydrogenase;[53–55] Na^+-activated transport ATPase;[111,271,354,355,400] microsomal ATPase;[241] glucose-6-phosphatase;[97,413] UDP-glucuronyl transferase;[138] acyl-CoA-L-glycerol-3-phosphoacyl transferase;[2] stearoyl CoA desaturase;[170] microsomal NADH-cytochrome *c* reductase;[171] adrenal steroid 11-β-hydroxylase;[403] phospatidic acid phosphatase;[74] phosphoryl choline cytidyl transferase;[120] NAD(P) transhydrogenase;[284] glucagon-stimulated adenyl cyclase;[290] plasma membrane adenyl cyclase;[306] UDP-galactose-lipopolysaccharide-α, 3-galactosyl transferase of *Salmonella typhimurium*;[316,317] C-35 isoprenoid alcohol phosphokinase from *S. aureus*;[160,322] pyridoxal-5-phosphate kinase of mouse brain (G. Toffano and P. Vecchia, personal communication). An involvement of phospholipids in mitochondrial oxidative phosphorylation has been often proposed[141] and more directly demonstrated by Burstein *et al.*[43,44]

In many of the cited cases the requirement for phospholipids was demonstrated in particulate membrane preparations, but at least in a few instances phospholipids were necessary to activate soluble enzymes.

Phospholipids and membrane binding. Phospholipids have also been found to determine the binding of certain enzymes to the membrane. Soluble monoamine oxidase binds phospholipids in solution forming soluble complexes, and phospholipids recombined with LDM forming insoluble complexes.[272] Membrane phospholipid digestion with *Naja naja* phospholipase A has been employed to extract NADH-dehydrogenase[8] and β-hydroxybutyric dehydrogenase[121] from mitochondrial membranes; PE, lecithin and cardiolipin are digested in the order, and release of NADH dehydrogenase is correlated with hydrolysis of cardiolipin;[8] inability of *Crotalus adamanteus* venom to release NADH-dehydrogenase was attributed to its inert nature toward cardiolipin. The release of glutamic-aspartic transaminase and glutamic dehydrogenase from rat liver mitochondria incubated at 30° is inhibited by purified phospholipids, which protect the membrane from the effect of activation of endogenous phospholipases.[107] These observations suggest that the released enzymes are bound to phospholipids in the intact membrane. The activating effect of phospholipids on enzymic activities is a separate phenomenon.

Difficulties in assessing lipid requirement. Caution should be exerted on suggesting a lipid requirement for any enzymic activities which are destroyed by solvent treatment or lipase digestion, even when the activities are restored by exogenous phospholipids. A first reason for caution concerns the products of phospholipase action. Augustyn *et al.*[7] found that there is a remarkable similarity among the overall effects produced in mitochondria by phospholipase A, the classical uncouplers, and oleic acid; these observations suggest that the effect of phospholipase is not due only to phospholipid splitting but also to the uncoupling action of released free fatty acids and the surface action of lysophosphatides. Only in studies where free fatty acids and lysophosphatides are removed, e.g. by extensive washing with albumin, a clear requirement for phospholipids may be demonstrated.[124]

Inactivation of enzyme activities by treatment with phospholipases (specially when crude enzymes are used) or with solvents is no proof for lipid involvement, since enzymic protein alteration and denaturation may have occurred. Roelofsen *et al.*[314] observed that reactivation of $(Na^+ + K^+)$-stimulated ATPase of erythrocytes is possible only when phospholipid removal has been incomplete; highly purified phospholipase A completely hydrolyzes ethanolamine-, choline- and serine-containing phosphoglycerides, while the Mg^{2+}-dependent as well as the $(Na^+ + K^+)$-ATPase activities are completely lost, and cannot be reactivated by any phospholipid. Similar results were found by Heggivary and Post.[158] The ATPase can be activated by phospholipids *in untreated membranes*,[165, 314, 400] and this effect is not due to loss of "loosely bound" phospholipids;[314] the same activation in the case of the $(Na^+ + K^+)$-activated ATPase is also brought about by detergents. In view of these characteristics, Roelofsen *et al.*[314] conclude that one has to be very careful with the interpretation of experiments concerning reactivation of membrane-bound ATPase by lipids. As long as there is some remaining activity, the observed increase by phospholipid addition may be caused rather by an activation of the remaining activity than by a reactivation of the activity which has been lost. The same word of caution may be applied to other reconstitution studies; however in several instances genuine reactivation must have taken place.

Specificity of phospholipids. Are phospholipids specific for restoration of membrane enzymic activities? Morrison *et al.*[257] reported a preparation of cytochrome oxidase which contains little lipid, yet retains significant amounts of activity which depends on the presence of detergent. Green *et al.*[140] interpreted this finding as a result of "molecularization": as long as the enzyme is monodispersed, it is irrelevant whether amphipathic phospholipids or amphipathic detergents produce the effect. It was subsequently found that the membrane-bound oxidase has an absolute requirement for phospholipids: detergent does not substitute for phospholipids, but instead inhibits lipid reactivation.[412] The most likely explanation according to Zahler and Fleischer[412] is that solubilization and isolation of cytochrome oxidase results in alteration of the enzyme with loss of phospholipid requirement.

Specificity of the types of phospholipids has also been studied in many of the investigations on enzyme reactivation. Certain enzymes appear to have an absolute requirement or at least a preference for certain phospholipid classes. The $(Na^+ + K^+)$-activated ATPase apparently requires PS or other negatively charged phospholipids: since mono- and diacyl phosphates are also effective,[355] the essential structures needed for activation are a phosphate group plus one or two fatty acyl residues. In view of the cation binding properties of negatively charged phospholipids (Section III. A), this can be a specialized type of specificity in the case of transport ATPase. PS also displays unusual high activity in the restoration of ATPase in phospholipid-depleted submitochondrial particles.[39a]

The ATPase in a mitochondrial fraction (TUA-particles) is activated by phospholipids: some purified phospholipids such as PS and cardiolipin give full restoration of the masked ATPase activity but low rutamycin sensitivity; other phospholipids like PE give only partial reactivation but high rutamycin sensitivity.[42]

A purified soluble β-hydroxybutyric dehydrogenase from beef heart mitochondria has an absolute requirement for lecithin:[174] the most reactive species of lecithin, as well as the species with highest affinity for the apoenzyme, is one in which the fatty acid residues show a high degree of unsaturation. A further requirement for maximal activity depends on the physical state of the lecithin added (sonicated vesicles are active, undispersed suspensions are essentially inactive). Lecithin forms an active complex with the apodehydrogenase in presence of a sulphydryl compound:[175] when the enzyme is released from the membrane, it readily forms at pH 8·5 a dimer, stabilized by one or more disulphide links, which is water-soluble and inactive. When the dimer is monomerized by sulphydryl reagents, lecithin stabilizes the monomer in the form of the enzyme–phospholipid complex. Acidic phospholipids also depolymerize the dimer but form enzymically inactive complexes. According to Green *et al.*[140] this is a particular example of "molecularization", although molecularization alone does not explain the absolute requirement for lecithin.

Pyruvate oxidase from *E. coli* is reactivated by all *E. coli* phosphatides,[77] and the K_m for various phosphatides ranges from 0·9 to 2·2 × 10^{-6} M (6·5 × 10^{-6} M for PS); the diacyl phospholipids exhibit normal Michaelis–Menten kinetics, whereas lysophosphatides diverge from normal kinetics. The hydrophobic moieties of lecithin activate pyruvate oxidase whereas the hydrophilic portions have no stimulatory effect; also fatty acids (maximally palmitoleic and oleic) stimulate the oxidase.

There is not much specificity of the phospholipid classes in restoration of respiratory activity: different mitochondrial phospholipid fractions are equally active providing that binding is achieved; cardiolipin is effective at lower concentrations than lecithin.[122,370] Also the specificity of the phospholipid nonpolar moieties (the fatty acyl chains) has been investigated in restoration of respiratory activity. De Pury and Collins[92,93] have restored succinate-cytochrome *c* reductase activity of acetone-extracted rat liver mitochondria with phospholipids prepared from livers of normal and essential fatty acid (EFA)-deficient rats. These latter phospholipids are devoid of fatty acids of the linoleic series, and enriched in fatty acids of the oleic and palmitoleic acid series. Restoration of activity is proportional to the amount of phospholipids bound by mitochondrial protein but, once bound, phospholipids from normal and deficient rats are equally effective. However the rate of binding can be influenced by the fatty acid composition of the phospholipids: arachidonic acid binds more slowly, but is more firmly retained. A review of the role of EFA in membrane structure and function has been recently published by Guarnieri and Johnson;[148] although a role of EFA appears evident, it remains to be demonstrated that membrane alterations are on the basis of the gross manifestations of EFA deficiency. Although no requirement of the respiratory chain could be demonstrated for phospholipids containing essential fatty acids, some degree of unsaturation of the phospholipids appears to be required for succinoxidase activity in beef heart mitochondria.[46,202] Aqueous dispersions of synthetic dipalmitoyl lecithin and PE do not restore succinoxidase activity of acetone extracted mitochondria at 30° but are effective at 43° (Fig. 38). In this case the specificity is not a chemical one but is determined by the physical state of the phospholipids which are in a liquid crystalline state only above 40°.[59] Any phospholipid above the transition temperature appears effective for mitochondrial respiratory activity. These data agree with the general knowledge that membrane phospholipids *in vivo* are in the liquid crystalline state (Section III. C. c) and that membranes are functional only above the transition temperatures (cf. e.g. ref. 332a for Ca^{2+}-transport ATPase).

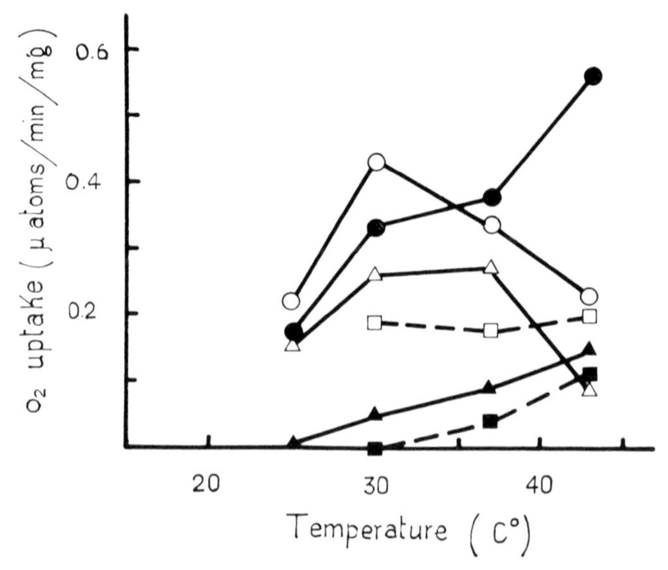

Figure 38. Effect of temperature on restoration of succinoxidase activity of LDM by different lipids (A. M. Sechi, G. Parenti-Castelli and G. Lenaz, unpublished data). ○——○, Asolectin; ●——●, myelin phospholipids; △——△, egg PE; ▲——▲, dipalmitoyl PE; □----□, egg lecithin, ■----■, dipalmitoyl lecithin. The values are reported after subtraction of residual activity (0·01–0·05 μatoms O_2/min/mg).

Role of phospholipids in enzyme activities. What are the possible functions of the phospholipids in the reactivation of enzyme systems? No unique role has to be necessarily ascribed to the phospholipids, and their functions could vary and be more or less specific in different enzymic systems. The problem of the potential functions of lipids in lipid-dependent enzymes has been widely discussed.[141, 147, 364, 365, 381]

Green and Tzagoloff[147] have considered that the structural role of the lipids may also explain the lipid-dependence of several enzymic systems. Aggregation, as opposed to "molecularization"[140] will not allow a functional contact of reactive groups and substrates. Soluble enzymes are spontaneously "molecularized" in water, and are not usually lipid-dependent; in the case of water-insoluble enzymes, molecularization represents restriction of polymerization along a plane, in other words it consists in membrane formation. Such a method of membrane construction could provide a satisfactory explanation of membrane biogenesis and structural incorporation of permeases, receptors, translocases, etc.[365]

The lipid dependence of several enzymic systems cannot be *exclusively* explained by the molecularization process of membrane formation; the observation that lipid-depleted mitochondria maintain a bidimensional appearance by E.M.[125] but are devoid of respiratory activity[122] is not fully compatible with that interpretation.

Another general function of the lipids could be to provide a hydrophobic environment where lipid-soluble cofactors or substrates are

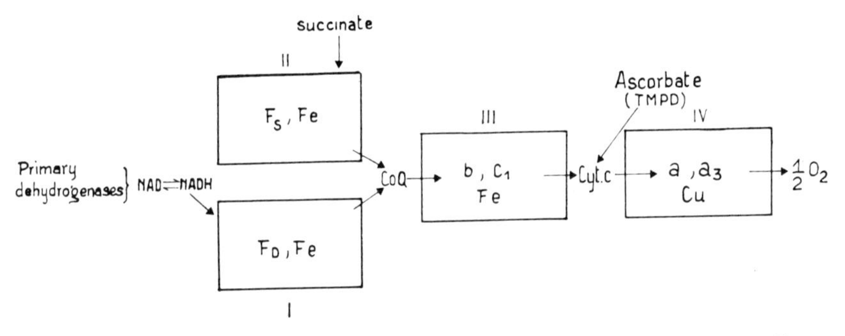

Figure 39. The mitochondrial respiratory chain. The chain is represented according to Green[145] with the four electron transfer complexes. f_D and f_S are the flavoproteins of NADH- and succinic-dehydrogenases respectively; Fe is nonheme iron; b, c_1, a, a_3 are the corresponding cytochromes.

dissolved in order to allow a proper catalytic function: the lipid-soluble coenzyme Q appears to be involved in the respiratory chain between flavoproteins and cytochrome *b*[145] (Fig. 39). Similar may also be the case of the lipid requirement of enzymes involved in lipid metabolism, in order to provide an appropriate binding area for substrates and an apparent micellar surface.[365] Also the involvement of lipids in mitochondrial oxidative phosphorylation may be a generical requirement for a low dielectric constant medium where water may not be accessible to hydrolyze high-energy intermediates (in the chemical coupling theory)[104] or which must be impermeable to protons (in the chemiosmotic coupling hypothesis).[254] Up to 70% of phospholipids can be cleaved by phospholipase C without impairing the capacity for oxidative phosphorylation, but diglycerides which remain in the membrane after digestion cannot be removed without causing complete loss of activity: only the hydrophobic moiety of the phospholipids appears to be necessary for coupling.[43,44] For all of these generical functions, lipids must be in a liquid–crystalline state, probably to provide a fluid medium in order for the interactions and transfer reactions to occur by random collision of the involved molecules. The physiological implications have already been discussed in Section III. C.

Certain studies suggest that lipids are required for the stability and subunit organization of the proteins involved in a given function. The lipid-free enzymic proteins may not be in the proper environmental conditions, and loss of activity may occur; in other cases activity may be enhanced, not only by a gross "opening phenomenon" for loss of compartmentation,[225] but also by a change in the protein environment and organization. Such may be the case for membrane-bound enzymes which are detached when lipids are removed: membrane localization may represent a control feature for the activity of several enzymes. In relation to the phenomenon of enzyme latency, it was found that the activity of lysozyme associated with liposomes[334] or with phospholipid–SP complexes[315] is latent; little or no activity is demonstrated unless the spherules are disrupted with Triton-X-100. Vessey and Zakim[390] found that treatment with phospholipase A and C *activates* UDP-glucuronyl transferase. Other agents altering microsomal lipids activate the enzyme, by increasing V_{max} and decreasing the binding affinity. In this case activation is not due to loss of compartmentation, since the enzymic properties change upon solubilization, but probably to reversal of phospholipid-induced modifications of enzyme organization. The phospholipids constrain the potential activity of the enzyme and could be physiologically involved as regulators of enzymic activities. In lipid-depleted cytochrome oxidase the apparent V_{max} decreases more than the apparent K_m for cytochrome *c*:[412] thus catalysis rather than substrate binding is most affected by lipid removal, suggesting that lipid is directly involved in the catalytic function of the enzyme.

Several other findings suggest the possibility that lipids are involved in the catalytic function of membrane enzymes, of course not directly as coenzymes or activators, although a few claims have been made in this direction: a direct role of phospholipids has been suggested in succinic dehydrogenase;[1] it was claimed that in rhodopsin retinaldehyde is linked to PE[288] but this possibility was ruled out by extensive extraction of the phospholipids; the Schiff base of *trans*-retinal with PE is indeed present, but at very low concentrations, so that it could be a storage form of highly reactive aldehyde.[6] Phospholipids could rather act through changing the conformation of enzyme proteins. Fluid lipids, needed for the restoration of mitochondrial respiration, would assure the proper alignment of the interacting sites which may transfer electrons through conformational changes of the enzymic proteins.[9, 14] Conformational changes and conformational stability may be induced directly by lipids in membrane proteins. There are indirect indications that conformational changes are induced by lipids on membrane enzymes. It has been suggested that a widening of the microsomal membrane after phospholipase C hydrolysis is the result of a change in protein conformation.[366] A cell-bound penicillinase and the secreted soluble enzyme from *Bacillus licheniformis* differ in their size behaviour and binding to deoxycholate [323, 324] suggesting a change in conformation from hydrophobic to hydrophilic monomers. It cannot be excluded that erythrocyte proteins when water-soluble[307] are in a different conformational state in comparison with the same proteins in the membrane environment.

Lipids in some instances could be envisaged in a broad sense as allosteric activators which modify membrane conformation allowing optimal activity (high or low according to the specific necessities) to be reached. In a way, this is a refinement of the molecularization theory, allowing a more dynamic representation of the membrane-bound enzymes. Of course this speculation waits for direct demonstration of conformational changes induced by lipids on membrane enzymes. CD is a promising technique in this direction once corrections are applied to artifacts arising from the particulate nature of the membranes.[376]*

Specificity of the lipids is apparent for some enzymes, but not for others. The conditions of lipid requirement may not be identical in different cases but vary according to the specific functions.

C. *Lipids and the Transport of Ions and Molecules*

It is not yet clear how much transport across membranes depends on the lipids and how much on the proteins of a membrane. The transport properties of several lipid membranes have been the object of various studies.[12, 132, 360, 399] The hydrophobic interior of a lipid bilayer is a barrier to the passage of ions and polar molecules. The permeability of a lipid membrane is increased by interaction with several molecules, including ionophore peptides or proteins. Particularly important for the mechanism of oxidative phosphorylation is the observation that uncouplers increase the proton permeability of phospholipid membranes.[63, 164, 361] Potassium ion transport through natural and artificial membranes is facilitated by valinomycin[62, 255] which is considered an example of a lipid-soluble carrier of natural membranes. On the other hand, polar pores could be established across membranes by lipophilic transmembrane structures like the π (L, D) helix proposed by Urry[374] for gramicidin A.

The differences in permeability for certain small molecules in erythrocytes of different animal species have been related[379] to differences

*We have recently shown that the *corrected* CD spectra of acetone-extracted mitochondria greatly differ from those of intact mitochondria, but reincorporation of mitochondrial phospholipids or cardiolipin brings about a restoration of the original spectra (L. Masotti, D. W. Urry and G. Lenaz, unpublished results). In contrast with these results, it was found that digestion of submitochondrial particles with phospholipase A does not modify the CD spectrum of the membrane (Zahler, W. L., Puett, D., and Fleischer S., *Biochim. Biophys. Acta*, **255** (1972)365).

in lipid and fatty acid composition of the membrane. Concerning ions, the net charge on the membrane[12] but also the nature of the fatty acids and the phospholipid-sterol ratio[380] have a large effect; the valinomycin-mediated exchange of $^{86}Rb^{+}$[89] and the permeability of liposomes to K^{+}[90, 329] in presence of FCCP (to make the K^{+} (inside): H^{+} (outside) exchange independent of H^{+} penetration) and valinomycin depend on the unsaturation of the phospholipids: the rate of K^{+} equilibration through unsaturated bilayers is more than 8 times higher than that through bilayers made of relatively saturated phospholipids. The increase in permeability with unsaturation is less for Na^{+} than for K^{+}; in absence of FCCP liposomes are impermeable to H^{+} irrespective of the degree of unsaturation.

Phospholipids themselves have been considered as possible carriers for cations and anions across the membrane;[379] this suggestion is supported by the well-known observation that phospholipids bind cations forming lipid-soluble undissociated salts (Section III. A. d). It is however difficult to visualize phospholipids in a bilayer to "flip-flop" between the two composing monolayers; it has been shown by spin-labelling techniques that this event in lecithin lamellae occurs less than 2×10^{-5}/sec[192] while lateral diffusion of the phospholipid molecules is at least eight orders of magnitude more frequent.[193] A "flip-flop" mechanism could however involve some protein or lipoprotein with a series of conformational changes leading to the transfer of ions or other molecules from one side to the other side of a membrane.[162, 349]

No need for carriers or pores, but only intersubunit weakness and subunit sliding due to Brownian motion has been envisaged by Vandenheuvel[386] for the transport mechanism; specificity only concerns the peripheral binding sites. (See Section III. E. a.)

Irrespective to the mechanism, lipids appear essential for transport; the first obvious reason is that they constitute the barrier which has to be crossed: either the barrier itself or the ion or molecule which is transported must be modified in this process. Another reason is that lipids are required for enzymic systems which are needed for transport, like the ($Na^{+} + K^{+}$) activated ATPase in the Na^{+}–^{+}K pump, or the respiration-dependent ion movement in mitochondria.

According to Lehninger,[198] Ca^{2+} transport in mitochondria involves a specific carrier; mitochondria possess specific high-affinity sites and low-affinity binding sites; these last sites have been assumed[328] to involve phospholipids, and more precisely PE; LDM lose their osmotic properties, but when the phospholipid content is restored, the osmotic behaviour reappears; however the reconstituted mitochondria cannot be rendered ion permeable by treatment with valinomycin or gramicidin;[328] on the other hand blowfly mitochondria possess negligible low affinity Ca^{2+} binding, yet possess large amounts of PE.[50]

The Ca^{2+} transport of sarcotubular membranes is very little sensitive to large variations in lipid structure induced by phospholipases.[242] The importance of proteins in this transport system is emphasized by the isolation of a Ca^{2+} binding acidic protein, calsequestin, from sarcoplasmic reticulum.[229]

V. *Conclusions*

The knowledge of membrane structure is the necessary premise for an understanding of membrane functions. The disposition of lipids and proteins and other components to build up the molecular architecture of biological membranes is the starting point for any model of the complex membrane functions. It has been shown in this review that enzyme activities are modified by the mere fact of being compartimentalized or membrane-bound (Section IV. B). It must be added that many functions appear specifically and indissolubly linked to membrane systems. Such functions, the mechanisms of which are largely unknown, must depend upon specific properties of the membrane structure. Any theory directed to unveil the mechanism of such functions as transport or energy transductions, must first take into account the molecular structure of the membrane. These functions appear utterly dependent on the intactness of the structure where they belong. An example is sufficient to show this point: the mechanism of mitochondrial oxidative phosphorylation is still an open question. The recognition that an intact vesicular structure of the inner mitochondrial membrane is required for energy coupling has stimulated the search for a mechanism where this property is accounted for. The chemiosmotic hypothesis of Peter Mitchell[254] has the merit of considering a proton gradient across a proton-impermeable membrane as the *primum movens* of the coupling process: the properties of phospholipid membranes and the mechanism of action of uncouplers (Section IV. C) are actively investigated for testing the chemiosmotic hypothesis.

Progress in the study of membrane functions is moving relatively slowly, notwithstanding the large efforts of experimentation, because it is conditioned by the slow progress made in the solution of membrane structure. The application of new analytical techniques of protein and lipid chemistry to the field of membrane chemistry is beginning to give new important answers; it is not too optimistic to believe that the basic problems of membrane structure are not far from a final solution. On the other hand it has become evident that investigations carried out on different membranes are not completely comparable; each membrane has a peculiar composition and peculiar properties and functions, thus each membrane must have a specific molecular organization.

Knowledge of the structure of biological membranes is also a premise for the study of their biogenesis. Although the biosynthesis of the individual components (lipids and proteins) is known, it is still obscure how the same components are assembled together. Studies on membrane formation *in vitro* from disaggregated units can be relevant to membrane biogenesis *in vivo*; on the other hand the understanding of the assembly mechanisms for membranes can also help to study membrane structure.

The subject of this review, lipid–protein interactions, may bear also on experimental and human pathology. A better knowledge of membrane structure will undoubtedly reveal that many diseases find their pathogenesis in membrane alterations. Although changes in membrane *composition* are already known for many pathological alterations, the final aim will be to connect these changes with changes in membrane structure and molecular interactions. Membrane phenomena and alterations are involved (primarily or secondarily) in a number of pathological disorders, like in degenerative diseases with changes in permeability and excitability (demyelinization processes, liver steatosis and cyrrhosis, muscular dystrophy, atherosclerosis), in autoimmunologic and allergic alterations, in changes in cell adhesion (like in malignant neoplasms), in the action of drugs and venoms, and in many others.

The field of the molecular interactions in natural membranes will certainly be a major area of interest and investigations also in the coming years, and it is believed that new chapters of biology and medicine will find application and answers in an improved understanding of membrane structure and functions.

References

1. Abdullah, Y. H. and Davison, A. N., *Biochem. J.*, **96** (1965) 106.
2. Abou-Issa, H. M. and Cleland, W. W., *Biochim. Biophys. Acta*, **176** (1969) 692.
3. Abrams, A. and Baron, C., *Biochemistry*, **6** (1967) 225.
4. Abrams, A. and Baron, C., *Biochemistry*, **7** (1968) 501.
5. Abramson, M. B., Katzman, R. and Gregor, H. P., *J. Biol. Chem.*, **239** (1964) 70.
6. Anderson, R. E. and Maude, M. B., *Biochemistry*, **9** (1970) 3625.
6a. Ashe, G. B. and Steim, J. M., *Biochim. Biophys. Acta*, **233** (1971) 810.
7. Augustyn, J. M., Parsa, B. and Elliott, W. B., *Biochim. Biophys. Acta*, **197** (1970) 185.
7a. Awashti, Y. C., Chuang, T. F., Keenan, T. W. and Crane, F. L. *Biochim. Biophys. Acta*, **233** (1971) 810.
8. Awashti, Y. C., Ruzicka, F. J. and Crane, F. L., *Biochim. Biophys. Acta*, **203** (1970) 233.
9. Azzi, A., Chance, B., Radda, G. K. and Lee, C. P., *Proc. Natn. Acad. Sci. USA*, **62** (1969) 612.
10. Bakerman, S. and Wasemiller, G., *Biochemistry*, **6** (1967) 1100.
11. Bangham, A. D., *Advan. Lipid Res.*, **1** (1963) 65.
12. Bangham, A. D., *Progr. Biophys. Mol. Biol.*, **18** (1968) 29.
13. Baron, C. and Abrams, A., *J. Biol. Chem.*, **246** (1971) 1542.
14. Baum, H., Rieske, J. S., Silman, H. I. and Lipton, S. H., *Proc. Natn. Acad. Sci. USA*, **57** (1967) 798.
15. Bender, W. W., Garan, H. and Berg, H. C., *J. Mol. Biol.*, **58** (1971) 783.
16. Benedetti, E. C. and Emmelot, P., *J. Cell. Biol.*, **26** (1965) 299.
17. Benson, A. A., *Ann. Rev. Plant Physiol.*, **15** (1964) 1.
18. Benson, A. A., *J. Am. Oil Chem. Soc.*, **43** (1966) 265.
19. Benson, A. A., Gee, R. W., Ji, T. H. and Bowes, G. W., in: *Autonomy and Biogenesis of Mitochondria and Chloroplasts*, N. K. Boardman, A. W. Linnane and R. M. Smillie (eds.), North Holland, Amsterdam, 1970. p. 18.

20. Berg, H. C., *Biochim. Biophys. Acta*, **183** (1969) 65.
21. Blaisie, J. K., Dewey, M. M., Blaurock, A. E. and Worthington, C. R., *J. Mol. Biol.*, **14** (1965) 143.
22. Blaisie, J. K. and Worthington, C. R., *J. Mol. Biol.*, **39** (1969) 417.
23. Blaisie, J. K., Worthington, C. R. and Dewey, M. M., *J. Mol. Biol.*, **39** (1969) 407.
24. Blaurock, A. E. and Wilkins, M. H. F., *Nature*, **223** (1969) 906.
25. Blumenfeld, O. O., Callop, P. M., Howe, C. and Lee, L. T., *Biochim. Biophys. Acta*, **211** (1970) 109.
25a. Bogucka, K. and Wojtczak, L., *Biochem. Biophys. Res. Commun.*, **44** (1971) 1330.
26. Bonsall, R. W. and Hunt, S., *Biochim. Biophys. Acta*, **249** (1871) 266.
27. Borggreven, J. M. P. M., Rotmans, J. P., Bonting, S. L. and Daemen, F. J. M., *Arch. Biochem. Biophys.*, **145** (1971) 290.
28. Bramley, T. A., Coleman, R. and Finean, J. B., *Biochim. Biophys. Acta*, **241** (1971) 752.
29. Branton, D., *Proc. Natn. Acad. Sci. USA*, **55** (1966). 1048.
30. Branton, D., *Ann. Rev. Plant Physiol.*, **20** (1969) 209.
31. Branton, D. and Park, R. B., *J. Ultrastruct. Res.* **19** (1967) 283.
32. Bretscher, M. S., *J. Mol. Biol.*, **58** (1971a) 775.
33. Bretscher, M. S., *J. Mol. Biol.*, **59** (1971b) 351.
34. Brierley, G. P. and Merola, A., *Biochim. Biophys. Acta*, **64** (1962) 205.
35. Brierley, G. P., Merola, A., and Fleischer, S., *Biochim. Biophys. Acta*, **64** (1962) 218.
36. Brown, A. D., *J. Mol. Biol.*, **12** (1965) 491.
37. Bruckdorfer, R. R., Edwards, P. A. and Green, C. *Europ. J. Biochem.*, **4** (1968) 506.
38. Bruckdorfer, R. R. and Green, C., *Biochem. J.*, **104** (1967) 270.
39. Bruni, A., Contessa, A. R. and Palatini, P., In: *Membrane-bound Enzymes*, G. Porcellati and F. di Jeso (eds.), Plenum Press, New York, 1971, p. 195.
39a. Bruni, A., Pitotti, A., Contessa, A. R. and Palatini, P., *Biochem. Biophys. Res. Commun.*, **44** (1971) 268.
40. Bruni, A. and Racker, E., *J. Biol. Chem.*, **243** (1968) 962.
41. Bulos, B. and Racker, E., *J. Biol. Chem.*, **243** (1968a) 3891.
42. Bulos, B. and Racker, E., *J. Biol. Chem.*, **243** (1968b) 3901.
43. Burstein, C., Kandrach, A. and Racker, E., *J. Biol. Chem.*, **246** (1971a) 4083.
44. Burstein, C., Loyter, A. and Racker, E., *J. Biol. Chem.*, **246** (1971b) 4075.
45. Butler, T., Smith, G. L. and Grula, E., *Can. J. Microbiol.*, **13** (1967) 1471.
46. Cabo-Soler, J., Sechi, A. M., Parenti-Castelli, G. and Lenaz, G., *J. Bioenergetics*, **2** (1971) 129.
47. Cadenhead, D. A., *Progr. Surface Science*, **3** (1970) 169.
48. Cadenhead, D. A., Demchak, R. J. and Phillips, M. C., *Kolloid Z. u. Z. Polymere*, 220 (1967) 59.
49. Camejo, G., Colacicco, G. and Rapport, M., *J. Lipid Res.*, **9** (1968) 562.
50. Carafoli, E., Hansford, R. G., Sacktor, B. and Lehninger, A. L., *J. Biol. Chem.*, **246** (1971) 964.
51. Casu, A., Nanni, G., Marinari, U. M., Pala, V. and Monacelli, R., *Ital. J. Biochem.*, **18** (1969) 154.
52. Casu, A., Nanni, G., and Pala, V., *Ital. J. Biochem.*, **17** (1968) 301.
53. Cerletti, P., Caiafa, P., Giordano, M. G. and Giovenco, M. A., *Biochim. Biophys. Acta*, **191** (1969) 502.
54. Cerletti, P., Caiafa, P., Giordano, M. G. and Testolin, G., *Lipids*, **5** (1970) 953.
55. Cerletti, P., Giovenco, M. A., Giordano, M. G., Giovenco, S. and Strom, R., *Biochim. Biophys. Acta*, **146** (1967) 380.
56. Chapman, D., *Lipids*, **4** (1969) 251.
57. Chapman, D. and Kamat, V. B., in *Regulatory Functions of Biological Membranes*, J. Järnefelt (ed.), Elsevier, Amsterdam, 1968, p. 99.
58. Chapman, D., Kamat, V. B. and Levene, R. J., *Science*, **160** (1968) 314.
59. Chapman, D. and Leslie, R. B., in: *Membranes of Mitochondria and Chloroplasts*, E. Racker (ed.), Van Nostrand Reinhold Co., New York, 1970, p. 91.
60. Chapman, D. and Urbina, J., *FEBS Letters*, **12** (1971) 169.
61. Chapman, D., and Wallach, D. F. H., in: *Biological Membranes: Physical Fact and Function*, D. Chapman (ed.), Academic Press, New York, 1968, p. 125.
62. Chappell, J. B. and Crofts, A. R., in: *Regulation of Metabolic Processes in Mitochondria*, Vol. 7, J. M. Tager, S. Papa, E. Quagliariello and E. C. Slater (eds.), BBA Library, Elsevier, Amsterdam, 1966, p. 293.
63. Chappell, J. B. and Haarhoff, K. H., in: *Biochemistry of Mitochondria*, E. C. Slater, Z. Kaniuga and L. Wotzjak, (eds.), Academic Press, New York, 1967, p. 75.
64. Chargaff, E., *J. Biol. Chem.*, **125** (1938) 661.
65. Chargaff, E. and Ziff, M., *J. Biol. Chem.*, **131** (1939) 25.
66. Cherry, R. J., Berger, K. U. and Chapman, D., *Biochem. Biophys. Res. Commun.*, **44** (1971) 644.
67. Choules, G. L. and Bjorklund, R. F., *Biochemistry*, **9** (1970) 4759.

68. Colacicco, G., *J. Colloid Interface Sci.*, **29** (1969) 345.
69. Colacicco, G., *Lipids*, **5** (1970) 636.
70. Colacicco, G., in: Conference on Membrane Structure and its Biological Applications, *Ann. N.Y. Acad. Sci.* (1972) in press.
71. Coleman, R. and Finean, J. B., *Biochim. Biophys. Acta*, **125** (1966) 197.
72. Coleman, R., Finean, J. B., Knitton, S. and Limbrick A. K., *Biochim. Biophys. Acta*, **219** (1970) 81.
73. Coleman, R., Finean, J. B. and Thompson, J. E., *Biochim. Biophys. Acta*, **173** (1969) 51.
74. Coleman, R. and Hübscher, G., *Biochim. Biophys. Acta*, **73** (1963) 257.
75. Crane, F. L., Arntzen, C. J., Hall, J. D., Ruzicka, F. J. and Dilley, R. A., in: *Autonomy and Biogenesis of Mitochondria and Chloroplasts*, N. K. Boardman, A. W. Linnane, R. M. Smillie (eds.), North-Holland, Amsterdam, 1970, p. 53.
76. Criddle, R. S., Bock, R. M., Green, E. D. and Tisdale, H., *Biochemistry*, 1 (1962) 827.
77. Cunningham, C. C. and Hager, L. P., *J. Biol. Chem.*, **246** (1971) 1575.
78. Cunningham, W. P., Prezbindowsky, K. and Crane, F. L., *Biochim. Biophys. Acta*, **135** (1967) 614.
79. Curtis, P. J., *Biochim. Biophys. Acta*, **183** (1969a) 239.
80. Curtis, P. J., *Biochim. Biophys. Acta*, **194** (1969b) 513.
81. Danielli, J. F., in: *Formation and Fate of Cell Organelles*, K. B. Warren (ed.), Academic Press, New York, 1967, p. 239.
82. Danielli, J. F. and Davson, H., *J. Cell. Comp. Physiol.*, **5** (1935) 495.
83. Das, M. L. and Crane, F. L., *Biochemistry*, **4** (1964) 859.
84. Das, M. L., Haak, E. D. and Crane, F. L., *Biochemistry*, **4** (1965) 859.
85. Dawson, R. M. C., in: *Biological Membranes. Physical Fact and Functions*, D. Chapman (ed.), Academic Press, London, 1968, p. 203.
86. Dawson, R. M. C. and Quinn, P. J., in: *Membrane-bound Enzymes*, G. Porcellati and F. di Jeso (eds.)., Plenum Press, New York, 1971, p. 1.
87. Deamer, D. W., *Biophys. J. Abstracts*, **11** (1971) 114a.
88. Dearborn, D. G. and Wetlaufer, D. B., *Proc. Natn. Acad. Sci. USA*, **61** (1969) 179.
89. De Gier, J., Haest, C. W. M., Mandersloot, J. G. and Van Deenen, L. L. M., *Biochim. Biophys. Acta*, **211** (1970) 373.
90. De Gier, J. and Scarpa, A., 7th FEBS Meetings, Varna, Abstract 520 (1971).
91. Del Castillo, J., Rodriguez, A., Romero, C. A. and Sanchez, V., *Science*, **153** (1966) 185.
92. De Pury, G. G. and Collins, F. D., *Chem. Phys. Lipids*, **1** (1966a) 1.
93. De Pury, G. G. and Collins, F. D., *Chem. Phys. Lipids*, **1** (1966b) 20.
94. Dobiasova, M. and Linhart, J., *Lipids*, **5** (1970) 445.
95. Dobiasova, M. and Radin, N. S., *Lipids*, **3** (1968) 439.
96. Doty, P. and Schulman, J. H., *Discuss. Faraday Soc.*, **6** (1949) 21.
97. Duttera, S. M., Byrne, W. L. and Ganoza, M. C., *J. Biol. Chem.*, **243** (1968) 2216.
98. Eley, D. D. and Hedge, D. G., *J. Colloid Sci.*, **11** (1956) 445.
99. Eley, D. D. and Hedge, D. G., *J. Colloid Sci.*, **12** (1957) 419.
100. Engelman, D. M., *J. Mol. Biol.*, **58** (1971) 153.
101. Engelman, D. M. and Morowitz, H. J., *Biochim. Biophys. Acta*, **150** (1968) 376.
102. Engelman, D. M. and Morowitz, H. J., *Biochim. Biophys. Acta*, **150** (1968) 385.
103. Engelman, D. M., Terry, T. M. and Morowitz, H. J., *Biochim. Biophys. Acta*, **135** (1967) 381.
104. Ernster, L. and Lee, C. P., *Ann. Rev. Biochem.*, **33** (1964) 729.
105. Esfahani, M., Barnes, E. M. and Wakil, S. J., *Proc. Natn. Acad. Sci. USA*, **64** (1969) 1057.
106. Esfahani, M., Limbrick, A. R., Knutton, S., Oka, T. and Wakil, S. J., *Proc. Natn. Acad. Sci. USA*, **68** (1971) 3180.
107. Estrada, O. S., Carabez, A. T. and Cabeza, A. G., *Biochemistry*, **5** (1966) 3422.
108. Evans, M. T. A., Mitchell, J., Musselwhite, P. R. and Irons, L., in: *Surface Chemistry of Biological Systems*, M. Blank (ed.), Plenum Press, New York, 1970, p. 1.
109. Evans, R. J., Bandemer, S. L., Heinlein, K. and Davidson, J. A., *Biochemistry*, **7** (1968) 3095.
110. Fairbanks, G., Steck, T. L. and Wallach, D. F. H., *Biochemistry*, **10** (1971) 2606.
111. Fenster, L. J. and Copenhaver, J. H., *Biochim. Biophys. Acta*, **137** (1967) 406.
112. Fernandez-Moràn, H., Oda, T., Blair, P. V. and Green, D. E., *J. Cell Biol.*, **22** (1964) 63.
113. Finean, J. B., *Biological Ultrastructure*, Academic Press, London, 1967, p. 274.
114. Finean, J. B. and Burge, R. E., *J. Mol. Biol.*, **7** (1963) 672.
115. Finean, J. B., Coleman, R. and Green, W. G., *Ann. N.Y. Acad. Sci.*, **137** (1966a) 414.
116. Finean, J. B., Coleman, R., Green, W. G. and Limbrick, A. R., *J. Cell Sci.*, **1** (1966b) 284.
117. Finean, J. B., Coleman, R., Knutton, S., Limbrick, A. R. and Thompson, J. E., *J. gen. Physiol.*, **51** (1968) 19 S.
118. Finean, J. B. and Martonosi, A., *Biochim. Biophys. Acta*, **98** (1965) 547.
119. Finean, J. B. and Robertson, J. D., *Brit. Med. Bull.*, **14** (1968) 267.
120. Fiscus, N. G. and Schneider, W. C., *J. Biol. Chem.*, **241** (1966) 3324.
121. Fleischer, B., Casu, A. and Fleischer, S., *Biochem. Biophys. Res. Commun.*, **24** (1966) 189

122. Fleischer, S., Brierley, G. P., Klouwen, H. and Slautterback, D. G., *J. Biol. Chem.*, **237** (1962) 3264.
123. Fleischer, S., Casu, A. and Fleischer, B., *Fed. Proc.*, **23** (1964) 476.
124. Fleischer, S. and Fleischer, B., *Methods Enzymol.*, **10** (1967) 406.
125. Fleischer, S., Fleischer, B. and Stoeckenius, W., *J. Cell Biol.*, **32** (1967) 193.
126. Folch, J., in: *Protides of the Biological Fluids*, H. Peeters (ed.), Elsevier, Amsterdam, 1966, p. 21.
127. Folch, J. and Lees, M., *J. Biol. Chem.*, **191** (1951) 807.
128. Folch, J., Lees, M. and Sloane-Stanley, G. H., *J. Biol. Chem.*, **226** (1957) 497.
129. Fox, S. W., Harada, K. and Kendrick, J., *Science*, **129** (1969) 1221.
130. Glaser, M. H. and Singer, S. J., *Biochemistry*, **10** (1971) 1780.
131. Glaser, M. H., Simpkins, H., Singer, S. J., Sheetz, M. and Chan, S. I., *Proc. Natn. Acad. Sci. USA*, **65** (1970) 721.
132. Goldup, A., Ohki, S. and Danielli, J. F., *Progr. Surface Science*, **3** (1970) 193.
133. Gordon, D. J. and Holzwarth, G., *Proc. Natn. Acad. Sci. USA*, **68** (1971) 2365.
134. Gordon, A. S., Wallach, D. F. H. and Straus, J. H., *Biochim. Biophys. Acta*, **183** (1969) 405.
135. Gotto, A. M., Levy, R. I. and Fredrickson, D. S., *Proc. Natn. Acad. Sci. USA*, **60** (1968) 1436.
136. Graham, J. M. and Wallach, D. F. H., *Biochim. Biophys. Acta*, **193** (1969) 225.
137. Graham, J. M. and Wallach, D. F. H., *Biochim. Biophys. Acta*, **241** (1971) 180.
138. Graham, A. B. and Wood, G. C., *Biochim. Biophys. Res. Commun.*, **37** (1969) 567.
139. Green, D. E., in: Conference on Membrane Structure and its Biological Applications, *Ann. N.Y. Acad. Sci.* (1972) in press.
140. Green, D. E., Allmann, D. W., Bachmann, E., Baum, H., Kopaczyk, K., Korman, E. F., Lipton, S., MacLennan, D. H., McConnell, D. G., Perdue, J. F., Rieske, J. S. and Tzagoloff, A., *Arch. Biochem. Biophys.*, **119** (1967) 312.
141. Green, D. E. and Fleischer, S., *Biochim. Biophys. Acta*, **70** (1963) 554.
142. Green, D. E., Haard, N. F., Lenaz, G. and Silman, H. I., *Proc. Natn. Acad. Sci. USA*, **60** (1968) 277.
143. Green, D. E., Murer, E., Hultin, H. O., Richardson, S. H., Salmon, P., Brierley, G. P. and Baum, H., *Arch. Biochem. Biophys.*, **112** (1965) 635.
144. Green, D. E. and Perdue, J. F., *Proc. Natn. Acad. Sci. USA*, **55** (1966) 1295.
145. Green, D. E. and Silman, H. I., *Ann. Rev. Plant Physiol.*, **18** (1967) 147.
146. Green, D. E. and Tzagoloff, A., *Arch. Biochem. Biophys.*, **116** (1966a) 293.
147. Green, D. E. and Tzagoloff, A., *J. Lipid Res.*, **7** (1966b) 587.
148. Guarnieri, M. and Johnson, R. M., *Adv. Lipid Res.* **8** (1970) 115.
149. Gulik-Krzywicki, T., Schechter, E., Iwatsubo, M., Ranck, J. L. and Luzzati, V., *Biochim. Biophys. Acta*, **219** (1970) 1.
150. Gulik-Krzywicki, T., Schechter, E., Luzzati, V. and Faure, M., *Nature*, **223** (1969) 1116.
151. Haldar, D., Freeman, K. and Work, T. S., *Nature*, **211** (1966) 9.
152. Hammes, G. G. and Schullery, S. E., *Biochemistry*, **9** (1970) 2555.
153. Hanai, T., Haydon, D. A. and Taylor, J., *J. theor. Biol.*, **9** (1965) 433.
154. Harris, J. R., *J. Mol. Biol.* **46** (1969) 329.
155. Hatefi, Y. and Hanstein, W. G., *Proc. Natn. Acad. Sci. USA*, **62** (1969) 1129.
156. Hauser, H. and Dawson, R. M. C., *Europ. J. Biochem.*, **1** (1967) 61.
157. Haydon, D. A. and Taylor, J., *J. theor. Biol.*, **4** (1963) 281.
158. Hegyvary, G. and Post, R. L., in: *The Molecular Basis of Membrane Function*, D. C. Tosteson (ed.), Prentice Hall, New Jersey, 1969, p. 519.
159. Hendler, R. A., *Physiol. Rev.*, **51** (1971) 66.
160. Higashi, Y. and Strominger, J. L., *J. Biol. Chem.*, **245** (1970) 3691.
161. Ho, W. K. K. and Nichols, A. W., *Biochim. Biophys. Acta*, **231** (1971) 185.
162. Hokin, L. E. and Hokin, M. R., *Fed. Proc.* **22** (1963) 8.
163. Hoogeveen, J. T., Juliano, R., Colemann, J. and Rothstein, A., *J. Membrane Biol.*, **3** (1970) 156.
164. Hopfer, U., Lehninger, A. L. and Thompson, J. E., *Proc. Natn. Acad. Sci. USA*, **59** (1968) 487.
165. Israel, Y., in: *The Molecular Basis of Membrane Function*, D. C. Tosteson (ed.), Prentice Hall, New Jersey, 1959, p. 529.
166. Jacobs, E. E. and Sanadi, D. R., *J. Biol. Chem.*, **235** (1960) 531.
167. James, L. K. and Augenstein, L. G., *Adv. Enzymol.*, **28** (1966) 1.
168. James, R. and Branton, D., *Biochim. Biophys. Acta*, **233** (1971) 504.
169. Johnston, P. V. and Roots, B. I., *Comp. Biochem. Physiol.*, **11** (1964) 303.
170. Jones, P. D., Holloway, P. W., Peluffo, R. O. and Wakil, S. J., *J. Biol. Chem.* **244** (1969) 744.
171. Jones, P. D. and Wakil, S. J., *J. Biol. Chem.*, **242** (1967) 5267.
172. Juliano, R. L., Kimelberg, H. K. and Papahadjopoulos, D., *Biochim. Biophys. Acta*, **241** (1971) 894.

173. Juliano, R. L. and Rothstein, A., *Biochim. Biophys. Acta*, **249** (1971) 227.
174. Jurtshuk, P., Sekuzu, I. and Green, D. E., *Biochem. Biophys. Res. Commun.*, **6** (1961) 76.
175. Jurtshuk, P., Sekuzu, I. and Green, D. E., *J. Biol. Chem.*, **238** (1963) 3595.
176. Kagawa, Y. and Racker, E., *J. Biol. Chem.*, **241** (1966a) 2461.
177. Kagawa, Y. and Racker, E., *J. Biol. Chem.*, **241** (1966b) 2467.
178. Kagawa, Y. and Racker, E., *J. Biol. Chem.*, **241** (1966c) 2475.
179. Kamat, V. B., Chapman, D., Zwaal, R. F. A. and Van Deenen, L. L. M., *Chem. Phys. Lipids*, **4** (1970) 322.
180. Kauzmann, W., *Adv. Protein Chem.*, **14** (1959) 1.
181. Kavanau, J. L., *Nature*, **198** (1963) 525.
182. Khalil Rayman, M., Gordon, R. C. and MacLeod, R. A., *J. Bacteriol.*, **93** (1967) 1465.
183. Kimelberg, H. K., Lee, C. P., Claude, A. and Mrena, E. *J. Membrane Biol.*, **2** (1970) 235.
184. Kimelberg, H. K. and Papahadjopoulos, D., *J. Bioι. Chem.*, **246** (1971a) 1142.
185. Kimelberg, H. K. and Papahadjopoulos, D., *Biochim. Biophys. Acta*, **233** (1971b) 805.
186. King, T. E., *Adv. Enzymol.*, **28** (1966) 115.
187. Kirk, R. G., *Proc. Natn. Acad. Sci. USA*, **60** (1968) 614.
188. Kopaczyk, K., Asai, J., Allmann, D. W., Oda, T. and Green, D. E., *Arch. Biochem. Biophys.*, **123** (1968a) 602.
189. Kopaczyk, K., Asai, J. and Green, D. E., *Arch. Biochem. Biophys.*, **126** (1968b) 358.
190. Korn, E. D., *Science*, **153** (1966) 1491.
191. Korn, E. D., in: *Theoretical and Experimental Biophysics*, Vol. 2, A .Cole (ed.), M. Dekker, New York, 1969, p. 1.
192. Kornberg, R. D. and McConnell, H. M., *Biochemistry*, **10** (1971a) 1111.
193. Kornberg, R. D. and McConnell, H. M., *Proc. Natn. Acad. Sci. USA*, **68** (1971b) 2564.
194. Kuehn, G. D., McFadden, B. A., Johnson, R. A., Hill, J. M. and Shumway, L. K., *Proc. Natn. Acad. Sci. USA*, **62** (1969) 407.
195. Ladbrooke, B. D., Jenkinson, T. J., Kamat, N. B. and Chapman, D., *Biochim. Biophys. Acta*, **164** (1968) 101.
196. Laico, M. T., Ruoslahti, E. I., Papermaster, D. S. and Dreyer, W. J., *Proc. Natn. Acad. Sci. USA*, **67** (1970) 120.
197. Le Baron, F. N., in: *Structural and Functional Aspects of Lipoproteins in Living Systems* E. Tria and A. M. Scanu (eds.), Academic Press, London, 1969, p. 201.
198. Lehninger, A. L., *Biochem. J.*, **119** (1970) 129.
199. Lenard, J. and Singer, S. J., *Proc. Natn. Acad. Sci. USA*, **86** (1966) 1828.
200. Lenard, J. and Singer, S. J., *Science*, **159** (1968) 738.
201. Lenaz, G., *Ital. J. Biochem.*, **19** (1970) 54.
202. Lenaz, G. in: *Biochemistry and Biophysics of Mitochondrial Membranes*, G. F. Azzone, E. Carafoli, A. L. Lehninger, E. Quagliariello and N. Siliprandi (eds.), Academic Press, New York, 1972a, p. 417.
203. Lenaz, G., Conference on membrane structure and its biological applications, *Ann. N.Y. Acad. Sci.* (1972b) in press.
204. Lenaz, G., Haard, N. F., Lauwers, A., Allmann, D. W. and Green, D. E., *Arch. Biochem. Biophys.*, **126** (1968a) 746.
205. Lenaz, G., Haard, N. F., Silman, H. I. and Green, D. E., *Arch. Biochem. Biophys.*, **128** (1968b) 203.
206. Lenaz, G., and MacLennan, D. H., *J. Biol. Chem.*, **241** (1966) 5260.
207. Lenaz, G., Parenti-Castelli, G., Monsigni, N. and Silvestrini, R. G., *J. Bioenergetics*, **2** (1971a) 119.
208. Lenaz, G., Parenti-Castelli, G., Sechi, A. M. and Masotti, L., *Arch. Biochem. Biophys.* **148**, (1972) 391.
209. Lenaz, G., Sechi, A. M., Masotti, L. and Parenti-Castelli, G., *Biochem. Biophys. Res. Comm.*, **34** (1969) 392.
210. Lenaz, G., Sechi, A. M., Parenti-Castelli, G. and Masotti, L., *Arch. Biochem. Biophys.*, **141** (1970a) 79.
211. Lenaz, G., Sechi, A. M., Masotti, L. and Parenti-Castelli, G., *Arch. Biochem. Biophys.*, **141** (1970b) 89.
212. Lenaz, G., Sechi, A. M., Masotti, L., Parenti-Castelli, G., Castelli, A., Littarru, G. P. and Bertoli, E., in: *Autonomy and Biogenesis of Mitochondria and Chloroplasts*, N. K. Boardman, A. W. Linnane and R. M. Smillie (eds.), North Holland, Amsterdam, 1970c p. 119.
213. Lenaz, G., Sechi, A. M., Parenti-Castelli, G. and Silvestrini, M. G., 7th FEBS Meetings, Varna, Abstr. 511, 1971b.
214. Lesslauer, W., Cain, J. and Blaisie, J. K., *Biochim. Biophys. Acta*, **241** (1971) 547.
215. Lester, R. L. and Fleischer, S., *Biochim. Biophys. Acta*, **47** (1961) 358.
216. Lévine, Y. K. and Wilkins, M. H. F., *Nature*, **230** (1971) 69.
217. Loach, P. A., Hadsell, R. M., Sekura, D. L. and Stemer, T. A., *Biochemistry*, **9** (1970) 3127.
218. Lucy, J. A., *J. theor. Biol.*, **7** (1964) 360.

219. Lucy, J. A., in: *Biological Membranes, Physical Fact and Function*, D. Chapman (ed.), Academic Press, London, 1968, p. 233.
220. Lucy, J. A., in: *Lysosomes in Biology and Pathology*, Vol. 2, J. T. Dingle and H. B. Fell (eds.), North-Holland, Amsterdam, 1969, p. 313.
221. Lucy, J. A. *Nature*, **227** (1970) 815.
222. Luzzati, V., in: *Biological Membranes. Physical Fact and Function*, D. Chapman (ed.), Academic Press, London, 1968, p. 71.
223. Lyons, J. M. and Raison, J. K., *Comp. Biochem. Physiol.*, **37** (1970a) 405.
224. Lyons, J. M. and Raison, J. K., *Plant. Physiol.*, **45** (1970b) (386.
225. Mackler, B. and Green, D. E., *Biochim. Biophys. Acta*, **21** (1956) 1.
226. MacLennan, D. H., in: *Current Topics in Membranes and Transport*, Vol. I, Academic Press, London, 1970, p. 177.
227. MacLennan, D. H. and Asai, J., *Biochem. Biophys. Res. Commun.*, **33** (1968) 441.
228. MacLennan, D. H., Lenaz, G. and Szarkowska, L., *J. Biol. Chem.*, **241** (1966) 5251.
229. MacLennan, D. H. and Wong, P. T. S., *Proc. Natn. Acad. Sci. USA*, **68** (1971) 1231.
230. Maddy, A. H., *Biochim. Biophys. Acta*, **88** (1964) 448.
231. Maddy, A. H., *Biochim. Biophys. Acta*, **117** (1966) 193.
232. Maddy, A. H., Huang, C. and Thompson, T. E., *Fed. Proc.*, **25** (1966) 933.
233. Maddy, A. H. and Kelly, P. G., *Biochim. Biophys. Acta*, **241** (1971) 290.
234. Maddy, A. H. and Malcolm, B. R., *Science*, **150** (1960) 1616.
235. Malcolm, B. R., *Nature*, **195** (1962) 901.
236. Manzoli, F. A., Stefoni, S., Manzoli-Guidotti, L. and Barbieri, M. *FEBS Letters*, **10** (1970) 317.
237. Marchesi, S. L., Steers, E., Marchesi, V. T. and Tillack, T. W., *Biochemistry*, **9** (1970) 50.
238. Marchesi, V. T. and Steers, E., *Science*, **159** (1968) 203.
239. Margolis, S. and Langdon, R. G., *J. Biol. Chem.*, **241** (1966) 485.
240. Martonosi, A., *J. Biol. Chem.*, **243** (1968) 71.
241. Martonosi, A., Donley, J. and Halpin, R. A., *J. Biol. Chem.*, **243** (1968) 61.
242. Masoro, E. J. and Yu, B. P., *Lipids*, **6** (1971) 357.
243. Masotti, L., Urry, D. W. and Krivacic, J. R., *Biochim. Biophys. Acta*, **266** (1972) 7.
244. Masotti, L., Long, M. M., Sachs, G. and Urry, D.W. *Biochim. Biophys. Acta*, **255** (1972) 420.
245. Matalon, R. and Schulman, J. H., *Disc. Faraday Soc.*, **6** (1949) 27.
246. McClare, C. W. F., *Nature*, **216** (1967) 766.
247. McConnell, D. G., Tzagoloff, A., MacLennan, D. H. and Green, E. D., *J. Biol. Chem.*, **241** (1966) 2373.
248. McFarland, B. H. and Inesi, G., *Arch. Biochem. Biophys.*, **145** (1971) 456.
249. McFarland, B. G. and McConnell, H. M., *Proc. Natn. Acad. Sci. USA*, **68** (1971) 1274.
250. Metcalfe, J. C., Metcalfe, S. M. and Engelman, D. M., *Biochim. Biophys. Acta*, **241** (1971a) 412.
251. Metcalfe, S. M., Metcalfe, J. C. and Engelman, D. M., *Biochim. Biophys. Acta*, **241** (1971b) 422.
252. Miller, D. M., *Biochem. Biophys. Res. Commun.*, **40** (1970) 716.
253. Mitchell, C. D., Mitchell, W. B. and Hanahan, D. J., *Biochim. Biophys. Acta*, **104** (1965) 348.
254. Mitchell, P., *Fed. Proc.*, **26** (1967) 1370.
255. Moore, C. and Pressman, B. C., *Biochem. Biophys. Res. Commun.*, **15** (1964) 562.
256. Morowitz, H. J. and Terry, T. M., *Biochim. Biophys. Acta*, **183** (1969) 276.
257. Morrison, M., Bright, J. and Rouser, G., *Arch. Biochem. Biophys.*, **114** (1966) 50.
258. Mueller, P. and Rudin, D. O., *Nature*, **217** (1968) 713.
259. Mueller, P., Rudin, D. O., Tien, H. T. and Westcott, W. C., *Circulation*, **26** (1962) 1167.
260. Mühlethaler, K., in: *Biochemistry of Chloroplasts*, Vol. V, T. W. Goodwin (ed.), Academic Press, London, 1966, p. 49.
261. Munkres, K. D., Swank, R. T. and Sheir, G. I., in: *Autonomy and Biogenesis of Mitochondria and Chloroplasts*, N. K. Boardman, A. W. Linnane and R. M. Smillie (eds.), North-Holland, Amsterdam, 1970, p. 152.
262. Muscatello, U. and Guarriero-Bobyleva, V., *J. Ultrastruct. Res.*, **31** (1970) 337.
263. Nachbar, M. S. and Salton, M. R. J., in: *Surface Chemistry of Boilogical Systems*, M. Blank (ed.), Plenum Press, New York, 1970, p. 175.
264. Nanni, G., Casu, A., Marinari, U. M. and Baldini, I., *Ital. J. Biochem.*, **18** (1969a) 25.
265. Nanni, G., Marinari, U. M., Baldini, I., Ferro, M. and Casu, A., *Ital. J. Biochem.*, **18** (1969b) 123.
266. Napolitano, L., Le Baron, F. and Scaletti, J., *J. Cell Biol.*, **34** (1967) 817.
267. Neville, D. M., *Biochim. Biophys. Acta*, **133** (1967) 168.
268. O'Brien, J. S., *Science*, **147** (1966) 1099.
269. O'Brien, J. S., *J. theor. Biol.*, **15** (1967) 307.
270. O'Brien, J. S., 6th FEBS Meetings, Madrid, Abstr. 45, 1969.
271. Ohnishi, T. and Kawamura, H., *J. Biochem. (Tokyo)*, **56** (1964) 377.
272. Olivecrona, T. and Oreland, L., *Biochemistry*, **10** (1971) 332.

273. Ottolenghi, A. C. and Bowman, M. H., *J. Membrane Biol.*, **2** (1970) 180.
274. Overath, P., Schairer, H. U. and Stoffel, W., *Proc. Natn. Acad. Sci. USA*, **67** (1970) 606.
275. Palatini, P. and Bruni, A., *Biochem. Biophys. Res. Commun.*, **40** (1970) 186.
276. Palmer, K. J., Schmitt, F. O. and Chargaff, E., *J. Cell Comp. Physiol.*, **18** (1914) 43.
277. Papahadjopoulos, D., *Biochim. Biophys. Acta*, **163** (1968) 240.
278. Papahadjopoulos, D. and Miller, N. *Biochim. Biophys. Acta*, **135** (1967) 624.
279. Parpart, A. K. and Ballentine, R., in: *Modern Trends of Physiology and Biochemistry*, E. S. G. Barron (ed.), Academic Press, New York, 1952, p. 135.
280. Pasquali, P., Monsigni, N. and Lenaz, G., 7th FEBS Meetings, Varna, Abstr. 512, 1971.
281. Perry, M. C., Tampion, W., and Lucy, J. A., *Biochem. J.*, **125** (1971) 179.
282. Person, P., Felton, J. H., O'Connell, D. J., Zipper, H. and Philpott, D. E., *Arch. Biochem. Biophys.*, **131** (1969a) 470.
283. Person, P., Zipper, H. and Felton, J. H., *Arch. Biochem. Biophys.*, **131** (1969b) 457.
284. Pesch, L. A. and Peterson, J., *Biochim. Biophys. Acta*, **96** (1965) 390.
285. Pethica, B. A., in: *Structural and Functional Aspects of Lipoproteins in Living Systems*, E. Tria and A. M. Scanu (eds.), Academic Press, London, 1969, p. 37.
286. Phillips, D. R. and Morrison, M., *Biochem. Biophys. Res. Commun.*, **40** (1970) 284.
287. Phillips, D. R. and Morrison, M., *Biochemistry*, **10** (1971) 1766.
288. Poincelot, R. P., Glenn Millar, P., Kimbel, R. L. and Abrahamson, E. W., *Biochemistry*, **9** (1970) 1809.
289. Pohl, G. W., *Biophysik*, **7** (1971) 236.
290. Pohl, S. L., Kraus, H. M. J., Kozyreff, V., Birnbaumer, L. and Rodbell, M., *J. Biol. Chem.*, **246** (1971) 4447.
291. Pollak, J. K., Malor, R., Morton, M. and Ward, K. A., in: *Autonomy and Biogenesis of Mitochondria and Chloroplasts*, N. K. Boardman, A. W. Linnane and R. M. Smillie (eds.), North-Holland, Amsterdam, 1970, p. 27.
292. Pollak, J. K., Ward, K. A. and Shorey, C. D., *J. Mol. Biol.*, **16** (1966) 564.
293. Pullman, M. E., Penefsky, H. S., Datta, A. and Racker, E., *J. Biol. Chem.*, **235** (1960) 2322.
294. Quinn, P. J. and Dawson, R. M. C., *Biochem. J.*, **116** (1970) 671.
295. Racker, E., *Fed. Proc.*, **26** (1967) 1335.
296. Racker, E., in: *Membranes of Mitochondria and Chloroplasts*, E. Racker (ed.), Van Nostrand Reinhold Co., New York, 1970, p. 127.
297. Raison, J. K. and Lyons, J. M., *Proc. Natn. Acad. Sci. USA*, **68** (1971) 2092.
298. Raison, J. K., Lyons, J. M., Mehlhorn, R. J. and Keith, A. D., *J. Biol. Chem.*, **246** (1971a) 4036.
299. Raison, J. K., Lyons, J. M. and Thomson, W. W., *Arch. Biochem. Biophys.*, **142** (1971b) 83.
300. Rand, R. P., *Biochim. Biophys. Acta*, **241** (1971) 823.
301. Razin, S., Morowitz, H. J. and Terry, T. T., *Proc. Natn., Acad. Sci. USA*, **54** (1965) 219.
302. Redwood, W. R., Müldner, H. and Thompson, T. E., *Proc. Natn. Acad. Sci. USA*, **64** (1969) 989.
303. Rega, A. F., Wud, R. J., Reed, C. F., Berg, G. G. and Rothstein, A., *Biochim. Biophys. Acta*, **147** (1967) 297.
304. Reich, M. and Wainio, W. W., *J. Biol. Chem.*, **236** (1961) 3058.
305. Reinert, J. C. and Steim, J. M., *Science*, **168** (1970) 1580.
306. Réthy, A., Tomasi, V. and Trevisani, A., *Arch. Biochem. Biophys.*, **147** (1971) 36.
307. Reynolds, J. A., Conference on Membrane Structure and its Biological Application, *Ann. N.Y. Acad. Sci.* (1972) in press.
308. Reynolds, J. A. and Tanford, C., *Proc. Natn. Acad. Sci. USA*, **66** (1970) 1002.
309. Reynolds, J. A. and Tanford, C., *Biophys. J. Abstract*, **11** (1971) 289a.
310. Richardson, S. H., Hultin, H. O. and Fleischer, S., *Arch. Biochem. Biophys.*, **105** (1964) 254.
311. Richardson, S. H., Hultin, H. O. and Green, D. E., *Proc. Natn. Acad. Sci. USA*, **5** (1963) 821.
312. Robertson, J. D., *Symp. Biochem. Soc.*, **16** (1959) 3.
313. Roelofsen, B., de Gier, J. and Van Deenen, L. L. M., *J. Cell Comp. Physiol.*, **63** (1964) 233.
314. Roelofsen, B., Zwaal, R. F. A. and Van Deenen, L. L. M., in: *Membrane-bound Enzymes*, G. Porcellati and F. di Jeso (eds.), Plenum Press, New York, 1971, p. 209.
315. Romeo, D. and De Bernard, B., *Nature*, **212** (1966) 1441.
316. Romeo, D., Girard, A. and Rothfield, L. *J. Mol. Biol.*, **53** (1970a) 475.
317. Romeo, D., Kinckley, A. and Rothfield, L. *J. Mol. Biol.*, **53** (1970b) 491.
318. Rottem, S., Hubbell, W. L., Hayflick, L. and McConnell, H. M., *Biochim. Biophys. Acta*, **219** (1970) 107.
319. Rouser, G., Nelson, G. J., Fleischer, S. and Simon, G., in: *Biological Membranes. Physical Fact and Functions*, D. Chapman (ed.), Academic Press, London. 1968, p. 5.
320. Salem, L., *Can. J. Biochem.*, **40** (1962) 1287.

321. Salton, M. R. J. and Nachbar, M. S., in: *Autonomy and Biogenesis of Mitochondria and Chloroplasts*, N. K. Boardman, A. W. Linnane and R. M. Smillie (eds.), North-Holland Amsterdam, 1970, p. 42.
322. Sandemann, H. and Strominger, J. L., *Proc. Natn. Acad. Sci., USA*, **68** (1971) 2441.
323. Sargent, M. G. and Lampen, J. O., *Arch. Biochem. Biophys.*, **136** (1970a) 167.
324. Sargent, M. G. and Lampen, J. D., *Proc. Natn. Acad. Sci. USA*, **65** (1970b) 962.
325. Sastry, P. S. and Kates, M., *Biochemistry*, **3** (1964) 1280.
326. Sauner, M. T. and Lévy, M., *Biochim. Biophys. Acta*, **241** (1971) 97.
327. Scanu, A. M., Pollard, H., Hirz, R. and Kothary, K., *Proc. Natn. Acad. Sci. USA*, **61** (1969) 171.
328. Scarpa, A. and Azzone, G. F., *Biochim. Biophys. Acta*, **173** (1969) 78.
329. Scarpa, A. and De Gier, J., *Biochim. Biophys. Acta*, **241** (1971) 789.
330. Schatz, G. and Saltzgaber, J., *Biochim. Biophys. Acta*, **180** (1969) 186.
331. Schechter, E., Gulik-Krzywicki, T., Azerad, R. and Gros, C., *Biochim. Biophys. Acta*, **241** (1971) 431.
332. Sechi, A. M., Masotti, L., Parenti-Castelli, G. and Lenaz, G., *Boll. Soc. ital. Biol. sper.*, **45** (1969) 286.
332a. Seelig, J. and Hasselbach, W., *Europ. J. Biochem.*, **21** (1971) 17.
333. Senior, A. E. and MacLennan, D. H., *J. Biol. Chem.*, **245** (1970) 5086.
334. Sessa, G. and Weissman, G., *J. Biol. Chem.*, **245** (1970) 3295.
335. Shah, D. O., in: *Surface Chemistry of Biological Systems*, M. Blank (ed.), Plenum Press, New York, 1970, p. 101.
336. Shah, D. O. and Schulman, J. H., *J. Lipid Res.*, **6** (1965) 341.
337. Shannon, C. F. and Mill, J. M., *Biochemistry*, **10** (1971) 3021.
338. Shipley, G. G., Leslie, R. B. and Chapman, D., *Biochim. Biophys. Acta*, **173** (1969a) 1.
339. Shipley, G. G., Leslie, R. B. and Chapman, D., *Nature*, **222** (1969b) 561.
340. Simpkins, H., Tay, S. and Panko, E., *Biochemistry*, **10** (1971) 3579.
341. Sjöstrand, F., *J. Ultrastruct. Res.*, **9** (1963) 561.
342. Sjöstrand, F., in *Structural and Functional Aspects of Lipoproteins in Living Systems*, E. Tria and A. M. Scanu (eds.), Academic Press, London, 1969, p. 73.
343. Sjöstrand, F. and Bárajas, L., *J. Ultrastruct. Res.*, **32** (1970) 306.
344. Staehelin, L. A., *J. Ultrastruct. Res.*, **22** (1968) 326.
345. Steck, T. L., Fairbanks, G. and Wallach, D. F. H., *Biochemistry*, **10** (1971) 2617.
346. Steer, D. C., Martin, W. G. and Cook, W. H., *Biochemistry*, **7** (1968) 3309.
346a. Steim, J. M., in: *Liquid Crystals and Ordered Fluids*, Plenum Press, New York, 1970, p. 1.
347. Steim, J. M. and Fleischer, S., *Proc. Natn. Acad. Sci. USA*, **58** (1967) 1292.
348. Steim, J. M., Tourtellotte, M. E., Reinert, J. C., McElhaney, R. N. and Rader, R. L., *Proc. Natn. Acad. Sci. USA*, **63** (1969) 104.
349. Stein, W. D., Conference on membrane structure and its biological applications. *Ann. N.Y. Acad. Sci.* (1972) in press.
350. Stein, W. D. and Danielli, J. F., *Disc. Faraday Soc.*, **21** (1952) 239.
351. Steinemann, A. and Läuger, P., *J. Membrane Biol.*, **4** (1971) 74.
352. Stoeckenius, W., in: *Membranes of Mitochondria and Chloroplasts* E. Racker (ed.), Van Nostrand Reinhold Co., New York, 1970, p. 53.
353. Sun, F. F., Prezbindowski, K. S., Crane, F. L. and Jacobs, E. E., *Biochim. Biophys. Acta*, **153** (1968) 804.
354. Tanaka, R., *J. Neurochem.*, **16** (1969) 1301.
355. Tanaka, R. and Sakamoto, T., *Biochim. Biophys. Acta*, **193** (1969) 384.
356. Tenenbaum, D. and Folch, J., *Biochim. Biophys. Acta*, **115** (1966) 141.
357. Terry, T. M., Engelman, D. M. and Morowitz, H. J., *Biochim. Biophys. Acta*, **135** (1967) 391.
358. Thompson, J. E., Coleman, R. and Finean, J. B., *Biochim. Biophys. Acta*, **135** (1967) 1074.
359. Thompson, J. E., Coleman, R. and Finean, J. B., *Biochim. Biophys. Acta*, **150** (1968) 405.
360. Thompson, T. E. and Henn, F. A., in: *Membranes of Mitochondria and Chloroplasts*, E. Racker (ed.), Van Nostrand Reinhold Co., New York, 1970, p. 1.
361. Trien, H. J. and Diana, A. L., *Nature*, **215** (1967) 1199.
362. Tourtellotte, M. E., Branton, D. and Keith, A., *Proc. Natn. Acad. Sci., USA*, **66** (1970) 909.
363. Trayer, H. R., Nozaki, Y., Reynolds, J. A. and Tanford, C., *J. Biol. Chem.*, **246** (1971) 4485.
364. Tria, E. and Barnabei, O., in: *Structural and Functional Aspects of Lipoproteins in Living Systems*, E. Tria and A. M. Scanu (eds.), Academic Press, New York, 1969, p. 143.
365. Triggle, D. J., *Progr. Surface Sci.*, **3** (1970) 273.
366. Trump, B. F., Duttera, S. M., Byrne, W. L. and Arstila, A. U., *Proc. Natn. Acad. Sci. USA*, **66** (1970) 433.
367. Tsofina, L. M., Liberman, E. A. and Babakov, A. V., *Nature*, **212** (1966) 681.
368. Tzagoloff, A., *J. Biol. Chem.*, **245** (1970) 1545.
369. Tzagoloff, A., Byington, K. H. and MacLennan, D. H., *J. Biol. Chem.*, **243** (1968) 2405.

370. Tzagoloff, A. and MacLennan, D. H., *Biochim. Biophys. Acta*, **92** (1965) 476.
371. Tzagoloff, A., McConnell, D. G. and MacLennan, D. H., *J. Biol. Chem.*, **243** (1968) 4117.
372. Tzagoloff, A., Yang, P. C., Wharton, D. C. and Rieske, J. S., *Biochim. Biophys. Acta*, **96** (1965) 1.
373. Ulmer, D. D., Vallee, B. L., Gorchein, A. and Neuberger, A., *Nature*, **206** (1965) 825.
374. Urry, D. W., *Proc. Natn. Acad. Sci., USA*, **68** (1971) 672.
375. Urry, D. W. and Ji, T. H., *Arch. Biochem. Biophys.*, **128** (1968) 802.
376. Urry, D. W., Hinners, T. A. and Masotti, L., *Arch. Biochem. Biophys.*, **137** (1970) 214.
377. Urry, D. W., Masotti, L. and Krivacic, I. R., *Biochim. Biophys. Acta*,, **241** (1961) 600.
378. Urry, D. W., Mednieks, M., and Bejnardvicz, E., *Proc. Natn. Acad. Sci., USA*, **57** (1967) 1043.
379. Van Deenen, L. L. M., *Progr. Chem. Fats and Other Lipids*, Vol. VII, Part 1, Pergamon Press, Oxford, 1965, p. 1.
380. Van Deenen, L. L. M., *Fed. Proc.*, **30** (1971) 1032.
381. Van Deenen, L. L. M. and de Haas, G. H., *Ann. Rev. Biochem.*, **35** (1966) 157.
382. Vandenheuvel, F. A., *J. Am. Oil Chem. Soc.*, **40** (1963) 455.
383. Vandenheuvel, F. A., *Ann. N.Y. Acad. Sci.*, **122** (1965) 57.
384. Vandenheuvel, F. A., *J. Am. Oil Chem. Soc.*, **43** (1966) 258.
385. Vandenheuvel, F. A., *Protoplasma*, **63** (1967) 188.
386. Vandenheuvel, F. A., *Advan. Lipid Res.*, **9** (1971) 161.
387. Vanderkooi, G. and Green, D. E., *Proc. Natn. Acad. Sci. USA*, **66**, (1970) 615.
388. Vanderkooi, G. and Green, D. E., *Bioscience*, **21** (1971) 409.
389. Vanderkooi, G. and Sundaralingam, M., *Proc. Natn. Acad. Sci. USA*, **67** (1970) 233.
390. Vessey, D. A. and Zakim, D., *J. Biol. Chem.*, **246** (1971) 4649.
391. Van Hippel, P. H. and Schleich, T., in: *Structure and Stability of Biological Macromolecules*, S. N. Timasheff and G. D. Fasman (eds.), Marcel Dekker, New York, 1969, p. 417.
392. Wahl, P., Kasai, M. and Changeux, J. P., *Europ. J. Biochem.*, **18** (1971) 332.
393. Wallach, D. F. H. and Gordon, A. S., in: *Protides of Biological Fluids*, H Peeters (ed.), Elsevier, Amsterdam, 1968, p. 47.
394. Wallach, D. F. H., Graham, J. M. and Fernbach, V. R., *Arch. Biochem. Biophys.*, **131** (1969) 322.
395. Wallach, D. F. H. and Zahler, P. H., *Proc. Natn. Acad. Sci. USA*, **56** (1966) 1552.
396. Wallach, D. F. H. and Zahler, P. H., *Biochim. Biophys. Acta*, **150** (1968) 180.
397. Weber, G., *Biochem. J.*, **51** (1952) 155.
398. Weidekamm, E., Wallach, D. F. H. and Fischer, H., *Biochim. Biophys. Acta*, **241** (1971) 770.
399. Wenner, C. E. and Dougherty, T. J., *Progress in Surface and Membrane Science*, J. F. Danielli, M. D. Rosenberg and D. A. Cadenhead (eds.), Academic Press, New York, 1971, p. 351.
400. Wheeler, K. P. and Whittam, R., *Nature*, **225** (1970) 449.
401. Wilkins, M. H. F., Blaurock, A. E. and Engelman, D. M., *Nature*, **230** (1971) 72.
402. Williams, R. M. and Chapman, D., *Progr. Chem. Fats and Other Lipids*, **11** (1970) 3.
403. Williamson, D. G. and O'Donnell, V. J., *Biochemistry*, **8** (1969) 1289.
404. Wilson, G. and Fox, C. F., *J. Mol. Biol.*, **55** (1971) 49.
405. Wilson, G., Rose, S. P. and Fox, C. F., *Biochem. Biophys. Res. Commun.*, **38** (1970) 617.
406. Wolfgram, F., *J. Neurochem.*, **13** (1966) 461.
407. Wolman, M., *Progr. Surface Science*, **3** (1970) 261.
408. Worthington, C. R., *Proc. Natn. Acad. Sci. USA*, **63** (1969) 604.
409. Yu, B. P. and Masoro, E. J., *Biochemistry*, **9** (1970) 2909.
410. Zahler, P. H., and Wallach, D. F. H., *Biochim. Biophys. Acta*, **135**(1967) 371.
411. Zahler, P. H. and Weibel, E. R., *Biochim. Biophys. Acta*, **219** (1970) 320.
412. Zahler, W. L. and Fleischer, S., *J. Bioenergetics*, **2** (1971) 209.
413. Zakin, D., *J. Biol. Chem.*, **245** (1970) 4953.
414. Zwaal, R. F. A. and Van Deenen, L. L. M., *Biochim. Biophys. Acta*, **150** (1968a) 323.
415. Zwaal, R. F. A. and Van Deenen, L. L. M., *Biochim. Biophys. Acta*, **163** (1968b) 44.
416. Zwaal, R. F. A. and Van Deenen, L. L. M., *Chem. Phys. Lipids*, **4** (1970c) 311.
417. Zwaal, R. F. A. and Van Deenen, L. L. M., *Biochem. J.*, **122** (1971) 62.

Structure-Function Unitization Model of Biological Membranes

D. E. Green, S. Ji and R. F. Brucker

Institute for Enzyme Research, University of Wisconsin, 1710 University Avenue, Madison, Wisconsin 53706

In 1966 Green and Perdue[1,2] proposed the repeating unit model of membrane structure, the central concept of which is the postulate that membranes are built up of lipoprotein repeating units. Each such unit was assumed to be a set of proteins associated hydrophobically with phospholipid. At that time many molecular features of the membrane were unresolved. Until these molecular details were clarified further progress in developing the repeating unit model was hampered. During the past several years many of the basic molecular parameters have been satisfactorily defined both experimentally and theoretically. The predominantly lamellar or bilayer character of phospholipid in membranes has been established firmly.[3-5] The characteristic features of intrinsic (integral) membrane proteins and the distinction between intrinsic and extrinsic (peripheral) membrane proteins have been sharply drawn.[6-11] The manner in which intrinsic proteins can associate hydrophobically with bilayer phospholipid has been more realistically evaluated.[12] The models of Vanderkooi and Green[12] and of Singer and Nicolson[8] rationalize in a satisfactory fashion the above mentioned molecular parameters of membrane structure. The way was thus cleared for a re-examination of the supramolecular features of membrane structure and for a more rigorous development of the lipoprotein repeating unit concept of Green and Perdue.[1,2]

The objective of the present communication is to present a general model of biological membranes which relates structure not only to function but also to membrane biogenesis. We shall initially consider the postulates which underlie this new model as well as some of the relevant evidence for these postulates; then describe the salient features of the model; and finally demonstrate the capability of the model for rationalizing a wide variety of membrane phenomena.

I. Basic Postulates Underlying the Model

1. *Intrinsic Versus Extrinsic Membrane Proteins*

There are two kinds of proteins in membrane systems—the intrinsic proteins which are part of the membrane continuum and hydrophobically associated with phospholipid, and the extrinsic proteins which are associated with (usually electrostatically) but are not part of the membrane continuum. The differences between the extrinsic and intrinsic classes of proteins have been considered in detail in several reviews.[6, 8, 13] In the present communication our concern will be exclusively with intrinsic membrane proteins. To avoid possible confusion we stress that "protein" implies one polypeptide chain.

2. *Bimodality of Intrinsic Membrane Proteins*

Proteins which are intrinsic to the membrane must be capable of orientation at a water-hydrocarbon interphase[7, 14] and this capability requires that the surface groups of such proteins should be predominantly polar in one sector and predominantly nonpolar in an adjoining sector. We shall refer to proteins possessing this type of distribution of surface groups as bimodal proteins. The simplest case of a bimodal protein would be a globular protein with the surface groups of one hemisphere predominantly polar and the surface groups of the other hemisphere predominantly nonpolar. The actual ratio of polar:nonpolar surface area in membrane proteins is probably variable.[7] In describing a membrane protein as bimodal, we are merely inferring that the protein is capable of stable orientation at the water–lipid interphase of the membrane continuum and that the distribution of surface groups is compatible with such orientation. Elsewhere we have considered in some detail the various molecular tactics by which bimodal distribution of surface groups can be achieved as well as the optional surface geometries which bimodal protein molecules may assume.[6, 7] The mounting experimental evidence for the bimodality of intrinsic membrane proteins has been reviewed in several recent articles.[6, 7, 9, 10, 11]

A biological membrane presents two water–lipid interphases—one at each of the two surfaces. A simple bimodal protein with polar and nonpolar hemispheres (the P-N type) can orient at only one of the interphases. A more complex bimodal protein with two polar extremities separated by a nonpolar band (the P-N-P type) can orient at the two interphases. The simple bimodal protein penetrates half way into the membrane continuum whereas the complex bimodal protein spans the membrane continuum. We shall refer to the proteins with the P-N-P type of bimodality as "through" membrane bimodal proteins.

Both phospholipid and intrinsic membrane proteins are bimodal molecules in the sense defined above. Since membranes are built up from arrays of bimodal molecules oriented at right angles to the plane of the membrane,[12] it would be expected that membranes would have a bimodal character. Indeed, the pattern for all membranes is that of a thin sheet (60–100 Å thick) with two polar surfaces separated by a nonpolar interior.

3. *The pairing of Bimodal Molecules in Membranes*

The essence of a biological membrane is the concept of paired bimodal molecules. The bilayer pattern of phospholipid arrays is an expression of this pairing principle. The pairing has its roots in the thermodynamic necessity for bimodal molecules to orient in a fashion which minimizes exposure of hydrophobic surfaces to water. When arrays of bimodal molecules are paired by apposition of hydrophobic surfaces, the resulting "membrane" with polar groups on the surface and hydrophobic groups in the interior represents the minimum free energy state and hence the most stable configurations for such arrays. The paired molecules are always oriented at right angles to the plane of the membrane. In the continuum of biological membranes, we have paired arrays both of proteins and phospholipids and as we shall discuss later, these arrays are not randomly distributed. The combination of paired arrays of protein and phospholipid leads to a more stable membrane than that composed of protein or lipid alone as evidenced by the fact that phospholipid avidly combines with lipid-free membrane proteins.[15]

Although it might appear that bimodal proteins could equally well pair with phospholipids as with other bimodal proteins, we are postulating that pairing of like with like (protein with protein and phospholipid with phospholipid) is the universal pattern in biological membranes. There are two reasons that have led us to this postulate of like pairing—first the electron microscopic evidence that the double tier structure of the membrane remains even after extraction of phospholipid[16]—and second, the evidence to be developed later that protein and phospholipids appear to form separate domains in biological membranes. We suspect that there is a much more compelling theoretical basis for like with like pairing but this has yet to be recognized. The possibility of unlike pairing of protein and phospholipid at the present stage of our knowledge cannot be excluded, but it may not be a viable possibility for reasons of stability.

The through membrane bimodal protein may be looked upon as a fusion of two paired bimodal proteins of the P-N type and hence as an extension of the pairing principle. As we shall discuss later on, the through membrane bimodal protein appears to be specialized for immunochemical processes and may be an exception rather than the rule for intrinsic membrane proteins.

High resolution electron microscopy has established that membranes in cross section consistently show two tiers of staining centers[17] and these two tiers can be equated with the pairing of protein and lipid bimodal molecules in the membrane continuum. Electron microscopic examination of freeze fractured membranes has led to the now widely accepted interpretation[18–20] that the 60–100 Å thick membrane sheet can be split down the hydrophobic middle into two sheets of half the thickness.

4. *Lipoprotein Repeating Structures as the Units of Membrane Construction and Function*

Biological membranes generally can be depolymerized to lipoprotein repeating units which spontaneously can coalesce to generate vesicular membranes when the depolymerizing reagent is removed[21–25] (see Fig. 1). These lipoprotein repeating units have been found to correspond to multimeric sets of proteins (complexes) associated hydrophobically with a complement of phospholipid—the units being stabil-

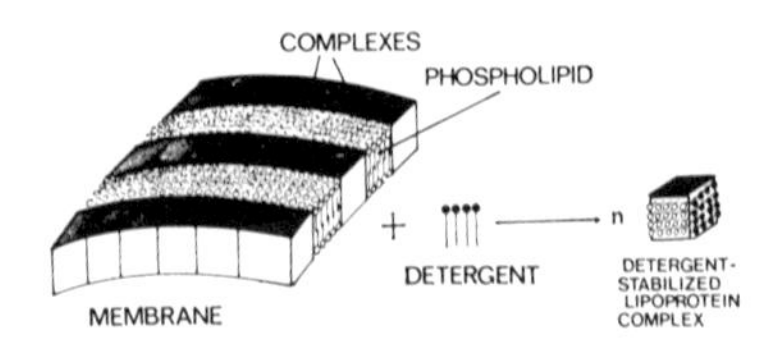

Figure 1. Depolymerization of membranes into lipoprotein repeating units.

ized by the detergents used for depolymerization of the membrane. We may consider the lipoprotein repeating units as unitized elements of the membrane stabilized by detergents and capable of generating *de novo* vesicular membranes. The ultimate structural unit of a membrane is thus a multimeric set of proteins (complex) with its complement of associated phospholipid. When the multimeric complex is further depolymerized into its component proteins, then at that point the structural identity of the complex is lost perhaps irreversibly.

A large number of functionally defined complexes have been isolated from a wide variety of membranes as lipoprotein particles stabilized by detergent and whenever tested, they have been shown to be capable of generating membranes *de novo* upon removal of the stabilizing detergent. We may conclude that the lipoprotein repeating units which generate membranes *de novo* are in fact functionally defined complexes with their associated complement of phospholipid. That is to say, the units of membrane construction are also the units of membrane function. Included among the complexes thus identified are the complexes of the mitochondrial electron transfer chain,[22,23] the electron transfer complexes of the chloroplast thylakoid membranes,[27,28] the Na^+-K^+ ATPase of the plasma membrane,[29,30] the Ca^{2+}-ATPase of the sarcoplasmic reticulum,[31] the electron transfer complexes of the bacterial electron transfer chain,[32–34] and the two complexes which collectively catalyze TPNH-dependent hydroxylations.[35,36] Since all membranes must be built up of complexes as judged by the capacity of membranes to undergo depolymerization to membrane-forming lipoprotein units, then complexes clearly must be the units for all categories of membrane function and not only for the category of enzymic catalysis. Such functions would include active transport, transprotonation, facilitated transport, photoreception, nerve transmission, etc. While at present we are still unaware of the full scope of membrane functions, our postulate is that whatever the function, the unit of its expression will be a complex (see item 1 of addendum).

In view of the wide assortment of functions which membranes can subserve additional to that of catalysis, statements frequently made that certain membranes like myelin are inert and devoid of function can only be described as inaccurate.[37] Every membrane fulfills some category of function, be it catalysis of transport, electrical insulation, transmission of a perturbation, or response to hormonal triggering, etc. Myelin, it must be remembered, arises from a loop of the plasma membrane of a nerve cell and it would hardly be expected that a plasma membrane would be devoid of some function.

We are proposing that membranes generally are constructed from lipoprotein complexes and this proposal is based on the above mentioned evidence that all membranes that have been tested have been found without exception to have the capability for depolymerization to lipoprotein repeating units. In turn the depolymerized lipoprotein repeating units have been found to have the capability for *de novo* membrane formation. There is a strong impression in the literature that the rod outer segment membrane would not conform to the pattern of lipoprotein repeating units and that in this membrane the unit of both structure and function is a single molecule of rhodopsin. In a later section of the present article, we shall be considering the problem of rhodopsin in some detail. For now it is sufficient to point out that the available experimental data are insufficient to rule out the possibility that rhodopsin exists in membranes as part of a complex as required by the present model.

5. *Complexes as Informational Sets of Protein Molecules*

A fundamental distinction has to be made between the individual bimodal proteins and multimeric sets of bimodal proteins (complexes). The distinction is not simply the difference between the parts and the whole. The complex has an ordered three dimensional structure[38] and it is this unique order not achievable by self-assembly that is the essence of the membrane dilemma. The corollary of the unique position of the complex is that the component proteins of the complex are never the unit of structure, function or biogenesis.

The link between membrane and hereditary process may well be the complex (see Fig. 2). Since the complex is most likely assembled on the ribosome presumably by a polycistronic message,[39] and since only the lipoprotein particles derived from complexes can give rise to membranes *de novo*, it may be stated that DNA is the ultimate determinant of membrane structure. How the complex is assembled and when and where the complement of phospholipid is added to the complex are still unanswered questions. We are aware from the work of Pollak[40] that the reticulosomes are paracrystalline structures which generate membranes when supplemented with phospholipid—a token that reticulosomes are arrays of complexes still unassociated with phospholipid.

A complex represents a set of intrinsic membrane proteins in a defined and invariant sequence with stable noncovalent links that maintain the sequential order. Given the order imposed by the

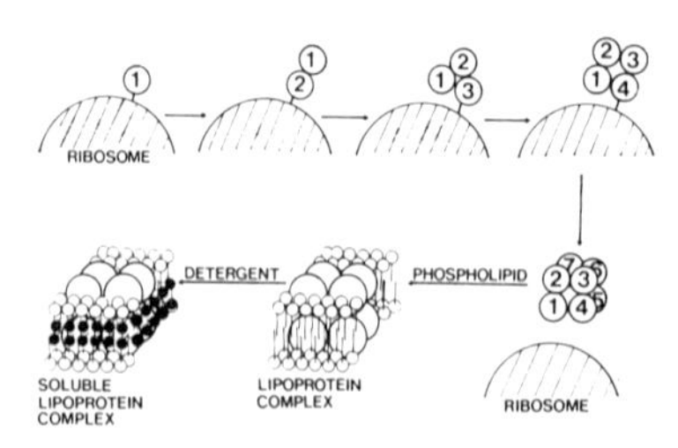

Figure 2. Biogenesis of complexes.

hereditary synthetic process, the complex will spontaneously rearrange when exposed to phospholipid to generate an element of a membrane which can associate with other such elements to generate a vesicular membrane *de novo*. The sequence of the intrinsic proteins in a complex may stand in the same relation to the three dimensional form of the complex as does the sequence of amino acids in a polypeptide chain to the final conformation of folded protein. It may be the order of associated proteins in a complex that eventually determines the way in which this set of proteins in presence of phospholipid will fold and rearrange to form an element of a membrane.

Implicit in the notion of the ribosome-dependent assembly of complexes is the postulated incapacity of the individual intrinsic proteins to form the complex by self-assembly. There have been many attempts to demonstrate self-assembly of membrane complexes from the proteins of the depolymerized complex. However, all but one have been unsuccessful.[41–43] Racker and his associates attempted to reconstitute Complex III from its component proteins and reported some limited success.[44] However other workers have raised serious doubts whether any reconstitution of Complex III was in fact achieved.[45] Green and Hechter[38] have already considered the reasons why self-assembly of membrane complexes from the component proteins would be an unlikely process.

Although the functionally unique membrane complex cannot be spontaneously self-assembled, nonetheless intrinsic membrane proteins can generate spontaneously what may be described as nonsense complexes, and these like the physiological complexes can interact with phospholipid to form membranes *de novo*. The capacity for forming nonsense complexes appears to be inherent in all intrinsic membrane proteins other than glycoproteins. The nonsense complex may consist of one polypeptide species or a mixture thereof.

6. *Cooperativity in Membranes and Protein Domains*

A considerable set of membrane phenomena can be rationalized satisfactorily only in terms of the cooperativity of membranes. Cooperativity appears to involve protein–protein interactions, the effects of which radiate throughout a membrane. It has been calculated that when a red blood cell membrane is exposed to 80 molecules of growth hormone, the change in fluorescence of tryptophane residues in the membrane is of a magnitude that would require perturbation of all the proteins in the membrane.[48] The absorption of one quantum of light by rhodopsin can trigger a nerve impulse.[49] This means that the excitation of one molecule of rhodopsin by a single photon can trigger a perturbation of the rod outer segment membrane that can radiate from the point of excitation to the point of junction of the rod outer segment membrane with a sensory membrane where an impulse travelling to the optic center of the brain can be triggered. A few molecules of acetyl choline can trigger precisely that kind of radiating perturbation in a nerve membrane (a wave of discharge of the nerve potential and a wave of ion movements).[50] When mitochondria are energized by electron transfer, the cristae can undergo a major change in configuration which bespeaks a significant conformational change in each of the repeating structures in the membrane.[51–53] All these clear examples of membrane cooperativity argue for a structure of the membrane that will allow of rapid transmission of perturbations. The principle of protein domains provides a basis for the rationalization of cooperativity in membranes.

From the wide spread occurrence of crystallinity in membranes,[6,54,55,56] it has been deduced that proteins and phospholipids form separate domains.[7] In other words, both protein and lipid are in separate continua within the membrane and there is alternation of the respective domains (see Fig. 3). We may define the protein domain as the domain in which complexes are lined up one behind the other in a continuum with noncovalent links for keeping the complexes in tight associations. Each such protein domain would be one complex wide and would extend through the thickness of the membrane. If we accept this simplistic interpretation of membrane structure, it would follow that in a plane normal to the surface of a membrane, each complex is surrounded on two sides by phospholipid (the phospholipid domains) and on the other two sides by complexes (the protein domain). The depolymerization of membranes to lipoprotein repeating units induced by detergents would be satisfactorily rationalized in terms of this domain hypothesis. The detergent would weaken the interactions between complexes and the membrane would fall apart into individual complexes with their complement of phospholipid. In order for such a lipoprotein unit to be stable in an aqueous medium, the phospholipid arrays would have to reorient through an angle of 90°, and the two hydrophobic faces of the complex exposed by depolymerization, would have to be covered by molecules of the bimodal detergent.

In formulating the hypothesis that protein domains provide structural bases for membrane cooperativity, we have assumed that com-

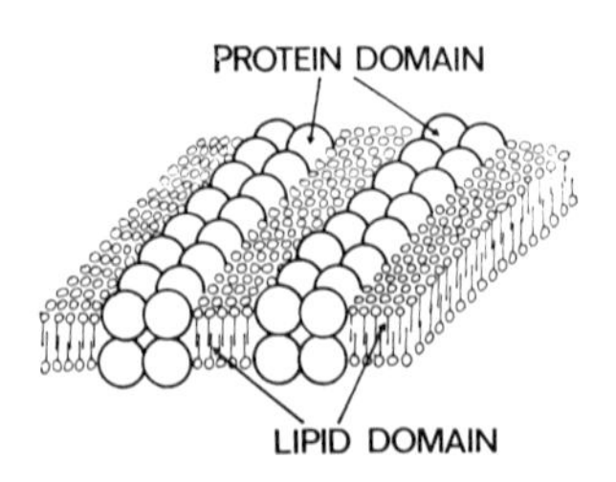

Figure 3. Domains of protein and phospholipid in membranes. The geometry of protein domains and the relation of protein to lipid domains are highly variable from membrane to membrane. The simplistic pattern shown in this figure merely illustrates the domain principle.

plexes penetrate the thickness of a membrane. In other words, each complex has a double tier structure like that of the membrane continuum.

7. *The Cavity-Channel Principle in the Construction of Complexes*

Thus far we have been vague about the macromolecular structure of a complex. How many proteins are there in a complex and how are these arranged? Are all complexes built up of the same number of proteins? We are in no position to provide final answers to any of these questions. There is sufficient analytical data to specify the probable subunit compositions of only two complexes of the mitochondrial electron transfer chain. Complex III has a molecular weight of about 200,000 and it contains 7–8 proteins.[57, 58] Complex IV has a molecular weight of about 100,000[59] and the average molecular weight of the monomers (12,000)[60] would argue for 8 protein molecules in the complex. At least for these two complexes, an octet pattern would appear to be the closest approximation to the probable number of protein molecules. Assuming 8 molecules of protein per complex, the dimensions of the two respective complexes would accommodate a double tier cuboidal pattern with four bimodal proteins on one tier (each set of four being in a square pattern). The hydrophobic faces of the paired bimodal proteins would be apposed to maximize thermodynamic stability.

Elsewhere Green and Brucker[7] have considered the concept of a central cavity in the interior of a complex with channels leading to this cavity from the exterior phase. The cavity as well as the channels could be either polar, nonpolar or some blend thereof. It is to be noted that polar cavities and polar channels are not incompatible with the bimodality of the intrinsic proteins but for such a cavity and such channels to be stable, the association of polar (or nonpolar) patches in the nesting proteins of the complex has to be precise. Otherwise the energy price would be excessive.

The central cavity and the associated channels of complexes provide a device whereby polar molecules can move from one side of the membrane to the other without "seeing" the hydrocarbon phase or the phospholipids (see Fig. 4), and a device whereby the substrates of a chemical reaction catalyzed by a complex can be insulated from the

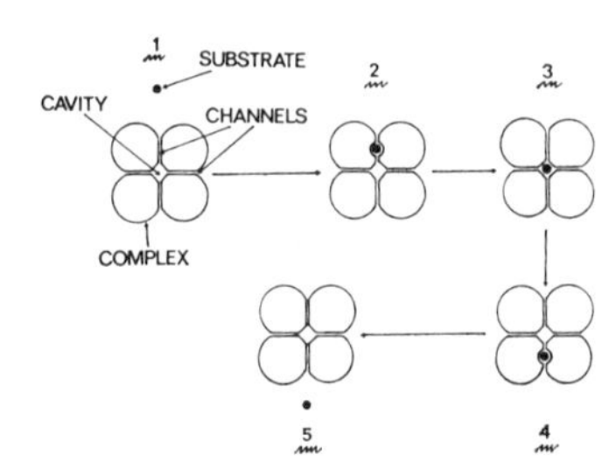

Figure 4. Cavity-channel principle of membrane complexes.

aqueous phase. Entry into the central cavity has to be conformationally mediated, and a high degree of selectivity has to be exercised in respect to the molecules capable of such conformationally controlled entry into and exit out of the central cavity.

The concept of internal cavities and channels opens the door to a new way of looking at complexes. A complex is no longer viewed merely as a structureless collection of proteins sitting in the water-lipid interphase but is regarded as an elaborate system of internal cavities and channels endowed with the capability of carrying out specific enzymic and transport functions. The geometric shape of internal cavities and channels need not be static but may be highly dynamic due to the precisely programmed conformational maneuvers accompanying the catalytic and transport functions of a complex. The polarity of the walls of the internal cavities and channels may also vary with time in such a way that the electrostatic microenvironments created within cavities and channels are regulated by conformational changes of the component proteins. In view of the fact that complexes are the most probable species to harbor active sites of membrane catalysis and transport, it is logical to anticipate that complexes possess highly ordered internal structures that are determined genetically to fulfill specific functions.

The notion of internal cavities or clefts within single proteins such as cytochrome *c*,[61] myoglobin,[62] and lysozyme[63] is well established. M. Perutz was one of the first to emphasize the fundamental significance of internal cavities in enzymes in general.[64] There is thus a continuity between individual proteins and complexes of proteins with respect to the cavity concept.

II. Structure-Function Unitization Model

We are proposing a model of biological membrane based on the principle that the unit of structure, function and biogenesis is one and the same and that this unit (the complex and its associated complement of phospholipid) has universal validity for biological membranes. It was E. Korn who first recognized the inevitability of this unitary principle.[86] But until the complex as a membrane entity was systematically explored, the identification of the complex as the unit of structure, function and biogenesis was not immediately obvious.

It has not been easy to find a convenient name for designating the model. We are suggesting that the model be described as the structure-function unitization model (or the unitization model for short). We have picked only one of the central ideas of the model as the basis of the name but a more complete name would of course be too unwieldy for general use.

It may be appropriate at this point to recapitulate the basic concepts which are crucial for the structure-function unitization model, although these concepts have already been considered in the previous section (1) Intrinsic proteins are bimodal; (2) sets of intrinsic proteins form complexes which are the units of both structure and function; (3) membranes are composites of interdigitating protein and lipid domains; (4) protein domains are arrays of associated complexes; (5) complexes with associated phospholipid are the units of membrane biogenesis; (6) the protein domain provides the structural basis for membrane cooperativity; (7) membrane complexes are ribosome-assembled informational supramolecules which cannot be self-assembled from the component bimodal proteins; and (8) complexes possess precisely structured internal cavities and channels that provide protected microenvironments for most membrane-centered physical and chemical processes.

One could schematize the purely structural aspects of the model by means of a simple geometric diagram such as is shown in Fig. 3. But it should be emphasized that no one diagram can cover adequately the structural nuances of the model no less the functional and biogenetic. There are many crystalline states[6,54–56,65] which membranes can exhibit—an indication of the variable way in which the protein and phospholipid domains can be arranged in different membranes. If we take into consideration the possible variation in the subunit structure of complexes (e.g. octet versus sextet) and the possible variation in the geometric pattern of the protein and lipid domains, then the inadequacy of a single geometric representation of the structural features of the model becomes obvious.

In specifying that complexes are the repeating units of the protein domain, we are not excluding the possibility that some intrinsic proteins may occur singly and that such proteins may be mobile in the phospholipid domain (see Fig. 5). Marchesi[66] has shown by sequence analysis and other evidence that an intrinsic glycoprotein of the red blood corpuscle plasma membrane is a through membrane bimodal protein. Capaldi[67] has further demonstrated that this glycoprotein unlike the other intrinsic proteins of the red blood cell membrane cannot be cross

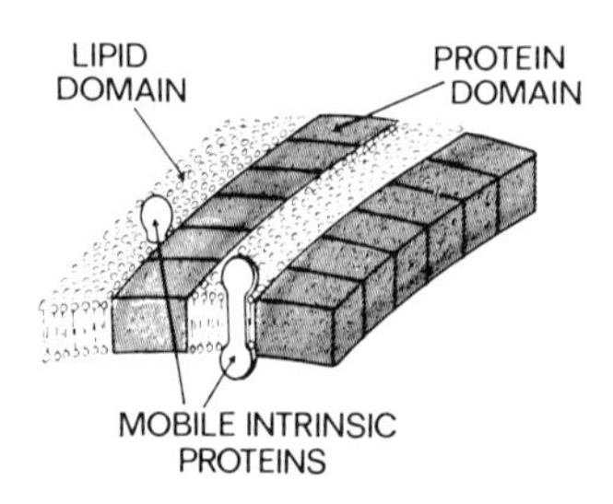

Figure 5. Glycoproteins as mobile species in membranes.

linked to other proteins by glutaraldehyde—a token that this protein is not part of a fixed set of proteins. In view of an unusual feature of the glycoprotein, namely a helical hydrophobic sector some 10 Å thick which connects the two polar regions of the molecule (see Fig. 5), it could be inferred that the glycoprotein would be capable of translational freedom in the phospholipid phase of the membrane. The narrow helical section would make it possible for the glycoprotein to serve as a mobile intrinsic protein in the membrane, and as a mobile component, the glycoprotein would not be subject to the translational restraints of the intrinsic proteins which are components of membrane complexes.

Cooperativity within the membrane may be operative at two levels—within the proteins of a complex and between complexes, i.e., at the intra- and inter-complex levels. The degrees of freedom which a bimodal protein can manifest in a membrane are highly reduced. In fact, we may say that one of the biological functions of a membrane is to diminish the translational degree of freedom of intrinsic proteins in the membrane continuum so that the protein–protein interactions can be more effectively controlled. This is not to say that protein and phospholipid domains may not fluctuate translationally in limited local regions but this fluctuation may operate within the restraint that the relative positions of proteins and phospholipids do not change appreciably. It is the domains which can fluctuate translationally, not the individual proteins nor phospholipids separately. Enzymology in the protein domain of a membrane is the new frame of reference that has to be considered. The proteins in a catalytically active membrane complex are subjected to a highly specialized and controlled environment. As alluded to above, each complex is exposed to three environments simultaneously, namely other complexes, bilayer phospholipid and a water phase.

Given the high degree of variability among membranes, is the unitization model sufficiently flexible to account satisfactorily for the full range of structural variation? Can one structural module (i.e., the complex) rationalize the properties of membranes so different as the inner mitochondrial membrane and myelin? There are multiple devices by which membrane properties can be modulated without changing the basic constructional pattern. Cholesterol and glycolipids may respectively affect both the permeability of membranes and the ratio of protein to lipid required for maximum stability of the membrane.[68,69] The sialic acid groups on intrinsic glycoproteins may augment by a large factor the charge density of the membrane surface[70] and this modulation of the charge density may have profound effects on various membrane parameters. In a large number of membranes, headpiece-stalk projections are attached at very regular intervals to one side of the membrane[71] and thus contribute to the asymmetry of biological membranes. These are only a few of the chemical devices for modulation of membrane properties. Therefore, we see no difficulty, in principle, of rationalizing the full gamut of membrane variation within the framework of the structure-function unitization model.

It is important to recognize that the structure-function unitization model of the membrane has an analogy in the unitization of structure and function among polymeric soluble enzymes; a polymeric enzyme is the unit of structure, function and biogenesis. The unitization model of the membrane thus has deep roots in biological precedents.

The thesis has been developed by Green and Goldberger that all living cells operate within the framework of a universal set of principles and that all the fundamental processes and structures fall within the framework of these universals.[72] The universals include the principles of heredity and energy transduction, and the fundamental metabolic processes. The principles of membrane structure and function undoubtedly would fall within the framework of biological universals if for no other reason than that the membrane is the very essence of living systems.

This principle of universality has enormous tactical advantages in the sense that it provides assurance that underneath the bewildering variety of specific membrane properties there is always a set of properties which is independent of the source or specialization of the membrane. It is the description of this set of universal attributes of membrane systems to which the present model is addressed. Moreover, in reaching decisions in a field with an abundance of soft and often times ambiguous data, the principle of universality may be invaluable as a guide to the separation of the wheat from the chaff.

The structure-function unitization model which we are proposing has many similarities to Changeaux's model of membrane structure.[73] The approach of Changeaux has been prophetic of the new direction which the membrane field is now pursuing. The unitization model of the membrane can readily accommodate the essential elements of the Changeaux model.

Finally, a few comments are in order about the tactical approach which we have followed in the development of the unitization model. While the model is indeed steeped in an extensive substratum of experimental evidence it goes beyond available experiment. The *a priori* approach has thus played a key role in the genesis of the model. The ultimate test of a model is neither the logic of nor the experimental justification for its genesis but rather the extent to which it rationalizes the major membrane phenomena, introduces order and clarity in the field, and predicts new relationships. The crucial test of the model, therefore, lies in this fitting of theory and experiment.

III. Applications of the Model

In this section, we shall be considering a selected set of membrane phenomena which demonstrate the versatility and scope of the structure-function unitization model.

Crystalline cytochrome oxidase. Purified preparations of cytochrome oxidase show a highly characteristic herring bone crystalline character when examined by negative staining with uranium acetate, in positively stained thin sections, or in thin section without any stain.[65] There are two quite different aspects of this crystallinity that are worthy of note.

The crystalline pattern of cytochrome oxidase is observed when the oxidase is in the oxidized state (nonenergized). When the oxidase is reduced and thereby energized, the crystalline pattern disappears. Vanderkooi *et al.*[65] have shown that in negatively stained preparations the individual complexes are visualized, and each complex is oriented in a precise herring bone pattern with respect to its neighboring complexes. When the complex is reduced, this regular orientation of complexes disappears and the orientation appears to be randomized. Several important conclusions may be drawn from these observations. The geometry of the protein-lipid domains is not constant. The domains can fluctuate from a highly regular pattern to a random pattern. This fluctuation of pattern must mean that the angle of packing of complexes in the protein domain is variable. Note that the dimensions of the individual complex do not change appreciably during the oxidized to reduced transition.[54] Reduction of the complex among other things may lead to charge separation (separation of electrons and protons) and conformational strain.[74] As a result of one or both of these perturbing influences, the individual complexes in their reduced form may assume new orientations relative to their neighbor complexes.

Hayashi *et al.*[65a] have found that a variety of conditions can abolish crystallinity in the oxidized form of cytochrome oxidase, namely high pH (>8·3) and high ionic strength (e.g. 1·5 M NaCl and 0·7 M NaBr). These observations suggest that the orientation of complexes within the protein domain is highly sensitive to environmental factors and only in a narrow range of conditions will complexes orient in the characteristic crystalline pattern.

The crystallinity of cytochrome oxidase is three dimensional as shown by the identical herring bone pattern in thin section.[65] In negatively stained preparations (surface view of dried preparations), the crystallinity is a reflection of the orderly orientation of complexes and the orderly alternation of protein and lipid domains. In positively stained, thin sectioned preparations (cross sectional view of dehydrated preparations), the same interpretation could be made regarding the arrangement of proteins and phospholipids in the crystal lattice.

Based on these observations, we may conclude that the double tier of globular particles is clearly visualizable in thin sectioned crystalline preparations and that these particles are seen even without external staining. Hence, the problem of staining is eliminated. Presumably by virtue of the perfect alignment of particles in a crystalline array, the individual protein particles have been visualized directly. The dimensions of the particles are consistent with the known mass of the proteins in cytochrome oxidase. Thus in the case of the crystalline cytochrome oxidase membrane, the component proteins in the two tiers of the membrane appear to be arranged in a highly regular and precise herring bone pattern.

Viral membrane formation. What is known about the genesis of viral membranes may be summarized by two observations, first that the membrane is pinched off from the host cell membrane, and second, that the protein composition of this pinched off membrane is different from the protein composition of the host membrane.[75–78] This means that viral proteins are introduced into the cell membrane and then pinched off by the developing virus. The point we would like to make is that the final structure of viral membranes suggests that repeating structures with projecting headpiece-stalks are introduced into the cell membrane.

These structures are induced by the interaction of viral nucleic acid with the hereditary apparatus of the cell. The conclusion that the viral membranes contain repeat structures follows from the fact that the viral membrane shows highly regular center to center distance between the projecting elements.[79] We infer therefore that viral membrane formation involves the introduction of sets of complexes with associated headpiece-stalk projections into an already existing cell membrane and the subsequent pinching off of this packaged set of complexes by the viral nucleic acid as it leaves the cell with formation of a viral membrane. Since viruses are formed in practically every type of cell, it would follow that the formation of the host cell membrane in all these different types of virus-susceptible cells depends upon the principle of complexes as the building blocks of membranes. This argument is based on the postulate that viral membrane formation is an exact counterpart of cell membrane formation. That is to say, if the viral membrane arises from complexes containing virus-induced intrinsic proteins, then the cell membrane which is used as the vehicle for viral membrane formation must arise in the same way from preformed complexes.

Viral coat formation. Virus particles such as the tobacco mosaic virus (TMV) do not have a lipoprotein membrane but rather a protein envelope which is one molecule thick.[80] In the biogenesis of this protein envelope, Klug has shown that the starting point for assembly is a set of double discs.[81] Each disc corresponds to a multimeric ring of proteins and the discs always come in pairs. The paired proteins in the apposed discs may stand in the same relation to one another as the pairing of bimodal proteins in a membrane. We might draw an analogy between the biogenesis of the viral coat and the biogenesis of membranes. The double disc of viral proteins may be considered as the equivalent of a membrane complex. The paired discs associate with one another to form the protein continuum of the viral coat. The double tier structure thus applies both to the viral discs and to the complexes. We are predicting that the pairing of viral proteins in the paired disc is a reflection of the bimodality of these proteins although this bimodality may not be as clear-cut as that of intrinsic membrane proteins as judged by the water solubility of the viral coat proteins. However, the tendency of viral coat proteins to form polymeric associations is an indication of partial hydrophobic character.[81] Perhaps the most outstanding difference between the paired viral discs and the intrinsic membrane complex is that the former but not the latter can be self-assembled.[82] But this difference may reflect the greater water solubility of the viral coat proteins—a property which allows for the extensive experimentation required for self-assembly.

Induction of intrinsic enzymes in membranes. Reagents such as barbiturates can induce enzymes in microsomal membranes which are intrinsic to the membrane.[83] It is important to emphasize the point that the enzymes which we are considering in the present context are not membrane-associated (extrinsic) but integral to the membrane (intrinsic). The mixed function oxidases are among the intrinsic membrane systems that have been found to be inducible and these oxidases are unambiguously membrane-forming complexes.[84] The theory of membranes which underlies our model requires that the induction of enzymes involve the induction of sets of protein (complexes) which are at the same time the building blocks of membranes. Insofar as these induced intrinsic enzymes have been studied from the standpoint of their structure, it would appear that the results to date are fully consistent with this prediction. Intrinsic induced enzymes are invariably enzyme complexes with the capability for *de novo* membrane formation.[85]

Crosslinking of intrinsic proteins by glutaraldehyde. Since intrinsic membrane proteins come in sets or complexes, it would be predictable that reagents such as glutaraldehyde which can cross link tightly associated proteins should completely alter the molecular weight pattern of the monomeric protein components of membrane complexes. All the monomeric species should disappear and a new set of dimeric, trimeric, etc. species should appear. Capaldi[87] has indeed verified this prediction for the behavior of the proteins in a membrane generated by purified cytochrome oxidase. The monomeric protein species largely disappeared upon glutaraldehyde treatment and a new set of polymeric protein species appeared. This dramatic demonstration of the cross linking of all proteins within the cytochrome oxidase membrane establishes the capability of the glutaraldehyde technique for "measuring" the distances separating intrinsic proteins in a membrane. Cross linking of two proteins by glutaraldehyde would require that the two proteins should be no more than 5 Å apart and probably considerably less than 5 Å. The same technique applied to the proteins of a submitochondrial particle (ETP) which is essentially an inner membrane preparation also yielded the same results as for the cytochrome oxidase membrane, namely disappearance of the monomeric protein species. Thus, in a membrane which is known to contain only complexes, the glutaraldehyde technique verifies the tight association of intrinsic proteins within the membrane implicit in the concept of complexes.

The plasma membrane of the red blood corpuscle contains both extrinsic (40%) and intrinsic proteins (60%). The mass ratio of extrinsic to intrinsic proteins (1:1·5) has been determined by Capaldi.[87] A vesicular membrane can be generated by the intrinsic protein fraction of the red blood corpuscle membrane. In such a membrane stripped of extrinsic proteins, the glycoproteins account for 5% of the total mass. It would be anticipated that all the intrinsic proteins except for the glycoproteins should show crosslinking with glutaraldehyde by virtue of being components of complexes whereas the glycoproteins would be unaffected since these are mobile species which can move freely through the phospholipid domain. Capaldi has fully confirmed this prediction.[87] The glycoproteins but none of the other intrinsic proteins were unaffected by glutaraldehyde in respect to molecular weight. In other words, components of complexes were crosslinked whereas mobile intrinsic protein species were not crosslinked by glutaraldehyde.

Paracrystalline structures in spaces between membranes. In mitochondria paracrystalline arrays (parallel bars) have been observed in the intracristal space linking outer and inner membranes or linking two neighboring and apposed cristal membranes.[88] These arrays have a periodicity of about 100–115 Å. Similar paracrystalline structures with periodicities in the same range have been reported in the synaptic gap between nerve membranes and in the spaces between apposed cell membranes.[89,90] The periodicity of the paracrystalline structures is identical with the periodicity of protein domains in membranes (90–120 Å between the centers of two domains). We, therefore, interpret the intermembrane paracrystalline structures in terms of soluble proteins in the aqueous space separating two membranes undergoing polymerization in periodic bands. These bands of polymerized protein may connect protein domains in one membrane to protein domains in the apposed membrane. It is to be noted that the membranes must be very close before these cross bridges are formed. This very proximity may trigger the polymerization required for the proteins to form cross bridges between the two membranes. The paracrystalline appearance of these cross bridges may be no more than a reflection of the exact periodicity of protein domains in a membrane—a periodicity established by the evidence from crystalline membranes.

Cell walls of bacteria. Bacterial cells of the gram negative class have basically two enveloping membranes—the so-called cell wall and the conventional cell membrane.[91] The two membranes are separated but are possibly linked through a rigid network system known as the peptidoglycan layer.[104] Let us consider how the cell wall membrane may be looked upon as a variation on the theme of biological membranes. At high resolution the cell wall membrane shows a "double tier" character (paired stain centers) as does the cell membrane.[104] Moreover, the cell wall membrane can be depolymerized with appropriate detergents into lipoprotein repeating structures which can reform the cell wall membrane when the detergent has been removed by suitable means.[92] These two observations suggest that the cell wall membrane is built up on the same constructional pattern as the cell membrane, and that complexes and associated lipid are the units of membrane construction.

The cell wall membrane differs from the cell membrane in two important respects—first in respect to its mechanical rigidity and second in respect to the presence in the membrane continuum of a third bimodal molecule, namely lipopolysaccharide. Let us consider each of these two novel features in turn. The rigidity of the cell wall membrane is in large measure referable to the peptidoglycan support structure.[91a] Protein projections extend periodically from the peptidoglycan network to appropriate sites on the cell wall membrane and these cross connections apparently increase to a very high degree the mechanical stability of the cell wall membrane. Procedures which readily rupture ordinary membranes such as ultrasonic irradiation usually have no effect on the cell wall membrane. But when the peptidoglycan support structure is disrupted by appropriate means, the cell wall membrane loses its mechanical stability and behaves like ordinary membranes such as the cell membrane.

Cell wall membranes contain about equal parts of two lipids, namely phospholipid and lipopolysaccharide. The ratio of protein:total lipid (phospholipid plus lipopolysaccharide) is about 1:1. Lipopolysaccharides generally can generate vesicular membranes *de novo*.[94] The thickness of such membranes (about 100 Å) would argue that the membranes arise by the pairing of lipopolysaccharide monomers in the same fashion as phospholipids formed paired arrays in the generation of liposomal membranes. In view of the very different dimensions of phospholipid versus lipopolysaccharide molecules (25 Å versus about 50 Å) it would be a reasonable prediction that there are three domains in cell wall membranes—protein, lipopolysaccharide and phospholipid—probably arranged in the order protein–lipid–protein–lipopolysaccharide.

Capitalize Endotoxins of gram negative bacterial cells have all been identified with the lipopolysaccharides of the cell wall membrane.[95] These endotoxins are all capable of *de novo* membrane formation and of fusing with cell membranes in the sense that the endotoxin becomes incorporated into the membrane.[96] The interaction of certain cell membranes with endotoxin can lead in many instances to the complete loss of membrane function.[97]

The stability of membranes as a function of interactions between complexes. Implicit in the unitization model of the membrane is the notion that the protein domains are composed of associated complexes in such a way that each complex in the continuum is linked to two other complexes in one direction and to phospholipid on both sides in the other direction (see Fig. 3). The stability of the links between complexes obviously must affect the stability of the membrane and weakening of the links must lead to the fragmentation of membranes into smaller and smaller vesicles and eventually into lipoprotein particles. There are several instances of precisely this kind of transition. Tzagoloff *et al.*[98] showed that inner membrane submitochondrial particles (ETP) became depolymerized to quasi-soluble lipoproteins by adjusting the pH of the particle suspension to about 9·0 and then sonicating in absence of added salts. A substantial proportion of the particles were no longer sedimentable after this treatment. Electron microscopic examination showed that the membranes had been depolymerized to lipoprotein particles of the dimensions of single complexes. When salt was added back to the lipoprotein dispersion or the pH was readjusted to neutrality, the dispersed lipoprotein units regenerated vesicular membranes *de novo*. The experiments of Tzagoloff *et al.* thus demonstrated that when complex–complex interactions were weakened by increasing the charge repulsion between complexes (alkalinization) and by decreasing the screening effect of salt (low ionic strength), then membranous vesicles of the inner mitochondrial membrane underwent fragmentation to quasi-soluble lipoprotein units.

The plasma membrane of the red blood corpuscle can become completely solubilized when exposed to EDTA for extended periods.[99] This observation of J. Reynolds was followed up by R. Capaldi[100] who showed that membranous vesicles of the ghost membrane can be depolymerized to quasi-soluble lipoprotein particles during such exposure and that these lipoprotein particles will generate vesicular membranes *de novo* when Mg^{2+} is added back to the medium and EDTA is eliminated. Here again EDTA by tying up Mg^{2+} or Ca^{2+} exerts two effects: (a) increases the repulsion between complexes by virtue of unscreening the charge repulsive action of sialic acid (in the complexes of the ghost membrane, glycoproteins with high sialic acid content play a dominant role and divalent metals are required to neutralize the charge of the sialic acid); and (b) decreases the interaction between complexes possibly by eliminating metallochelate links. In the *halobacteria*, Brown[101] has shown that membranes which are stable in 5 M salt media depolymerize to quasi-soluble lipoproteins when the molarity of the medium is decreased below 1–2 M. The intrinsic proteins of the complexes of the *halobacteria* have the properties of polycarboxylic acids.[101] High salt is required to reduce the charge repulsion between neighboring complexes; dilution of salt then weakens the interaction between complexes to the point that the membrane depolymerizes reversibly to quasi-soluble lipoprotein particles.

Membrane fluidity. The thermal transition points of the hydrocarbon chains of phospholipids in membranes nearly match those of the hydrocarbon chains of phospholipids in micellar dispersions.[5] Is this near identity of transition points compatible with the interdigitation of protein and lipid domains? There is no reason why the phospholipids in the lipid domain of membranes should behave any differently than phospholipids in dispersions. In both cases the phospholipids are in bilayer arrays, and the interactions between the nonpolar groups of proteins and the fatty chains of phospholipids are no different from the nonpolar interactions between one phospholipid and another.[116]

A second question is why the nature of the fatty acid chains in the phospholipid of membranes should have a profound influence on processes such as active transport or bacterial growth.[102] This correlation poses no special difficulty for a model of membrane structure based on the principle of complexes, and of protein and lipid domains. Integral

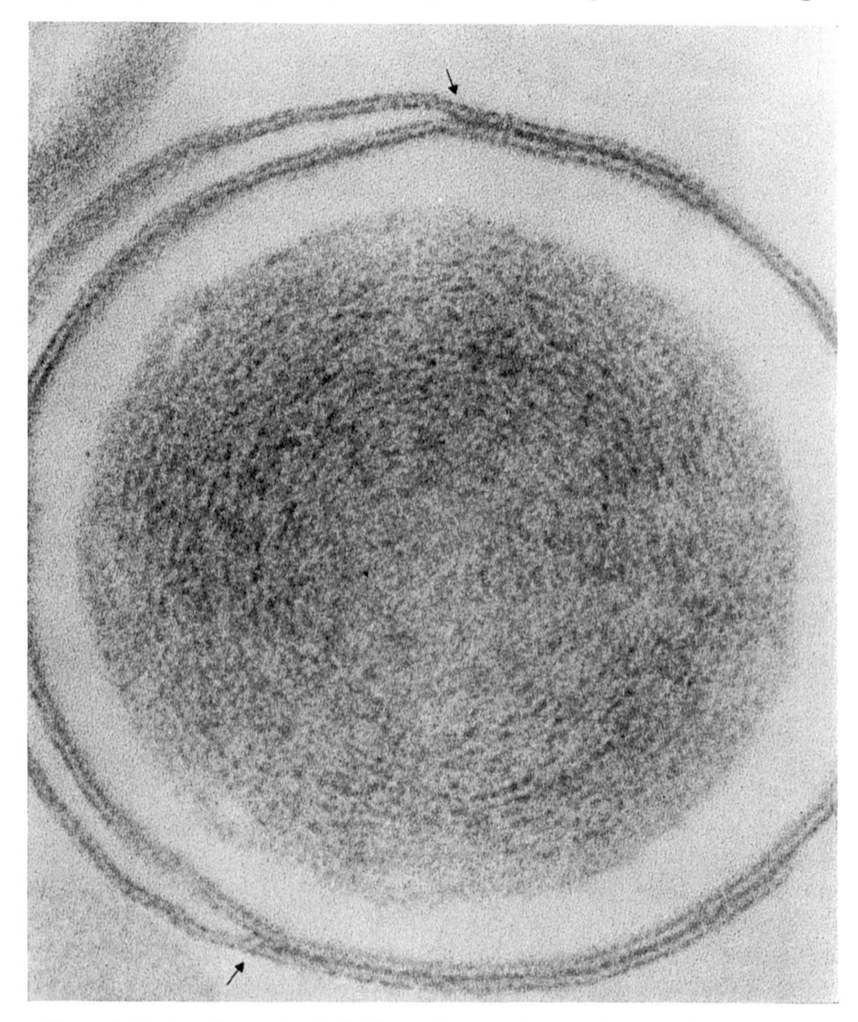

Figure 6. Fusion of two phospholipid membranous sheets to form a hybrid triple tiered structure. Osmium stained thin section. The electron micrograph was kindly supplied by Dr. Hideo Hayashi.

to the function of a complex is the conformational cycle by which substrate molecules enter and leave the central cavity of the complex. This conformational cycle will clearly be influenced by the lipid environment which bathes the complex on both sides. Thermal energies would play an important role in such conformational cycles and the fluidity of the hydrocarbon chain will perforce affect the thermally-activated conformational transitions required for transport or other functions.

Fusion of membranes. Green *et al.*[103] have shown that two membranous sheets of bilayer phospholipid can fuse to form a characteristic three tiered fusion "membrane". Divalent metals such as Mg^{2+} or basic proteins such as cytochrome *c* and protamine are required to initiate the fusion process. Electron microscopically, each of the two membranous sheets prior to fusion show double tiered structures in osmium-stained thin sections, whereas the fusion "membrane" shows three tiers of staining centers (see Fig. 6) with the middle tier much more electron dense than the two outer tiers. Brucker *et al.*[103a] have interpreted this fusion of phospholipid membranous sheets in terms of a phase transition in the state of the phospholipid on one side of each of the two apposed bilayer sheets (see Fig. 7) with formation of a hybrid "membrane"

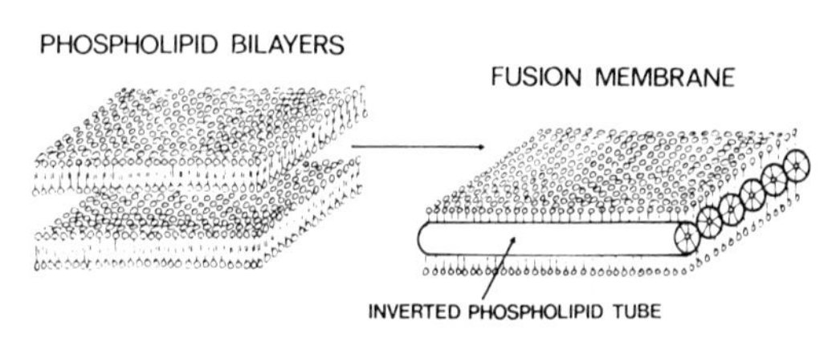

Figure 7. Mechanism of fusion of phospholipid membranous sheets and formulation of the structure of the hybrid structure formed by fusion (see item 2 of addendum).

arising from the two membranous sheets participating in fusion. In the center tier the phospholipid is assumed to be arranged in a concentric and tubular fashion with the polar heads interiorly directed and the aliphatic fatty chains exteriorly directed. The tubular cylinder of phospholipid presents a continuous hydrophobic surface which is apposed to the monolayer of phospholipid molecules on both sides.

Exactly the same kind of fusion may take place between two apposed biological membranes (either identical or nonidentical membranes), since the same reagents are required for fusion of membranes as for fusion of phospholipid sheets, namely Mg^{2+} or basic proteins. Moreover, the electron microscopic appearance of the hybrid membrane formed by fusion of two apposed membranes is indistinguishable from the hybrid structure formed by fusion of two phospholipid sheets. A triple tiered hybrid membrane is formed with an electron dense central tier. According to Brucker *et al.*,[103a] the structure of the fusion membrane can be formulated as shown in Fig. 8. Implicit in this formulation is the postulate of a phase transition in the state of phospholipid triggered by Mg^{2+} or basic proteins (only half the total phospholipid is involved in this transition), and then a rearrangement both of proteins and phospholipids to accommodate to this phase transition.

There are two fundamental processes involved in membrane fusion which merit close attention: (a) the rearrangement of complexes from a double tier to a single tier pattern; and (b) the rearrangement of a membrane from a double tier to a triple tier pattern. We shall consider each of these two processes separately.

If we consider a complex as a set of eight linked bimodal proteins, this set can assume a linear arrangement (in the fusion membrane) or a double tier arrangement (in each of the two membranes prior to fusion). Figure 9 shows diagrammatically this transition of a complex from the linear to the double tier arrangement. To achieve this type of arrangement, the links between the bimodal proteins must be relatively stable. Since membrane fusion is a reversible process[6, 103] it would appear that complexes can readily undergo this type of reversible rearrangement.

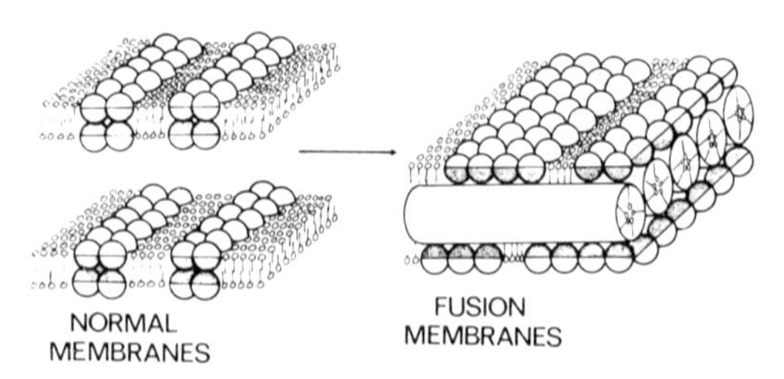

Figure 8. Mechanism of fusion of apposed membranes and formulation of the structure of the hybrid membrane formed by fusion.

The maneuver which a complex must undergo during membrane fusion is highly informative about the possible mechanism of the biogenesis of complexes. Given a linear set of eight linked bimodal proteins, this set will automatically assume a double tier pattern when exposed to bilayer phospholipid and will revert to a linear pattern when exposed to inverted, concentric phospholipid. The precise order of bimodal molecules in an octet set may be crucial for assembly of a membrane-forming complex. Unless this precise order is achievable, the resulting set may not have the properties of a membrane-forming complex.

The inner mitochondrial membrane has 90 Å headpieces projecting periodically from its matrix side.[71] In fact, the headpieces are so closely clustered that we may consider the membrane covered with a mosaic wall of projecting 90 Å headpieces. Yet two inner membranes in presence of Mg^{2+}, protamine, or cytochrome *c* can approach one another on the headpiece side and achieve fusion. During fusion the 90 Å headpieces disappear and the proteins of the headpiece appear to be incorporated into the fusion membrane. At first glance this may seem an improbable maneuver. But the rationale of the maneuver becomes more obvious when the headpiece is considered to be a complex (probably an octet). During fusion this complex appears to undergo a transition from a double tier to a single tier arrangement in precisely the same fashion as the intrinsic complex. There is one basic difference between the extrinsic and intrinsic complexes. The plane dividing each bimodal protein into polar and nonpolar halves is rotated 90° in the intrinsic as compared to the extrinsic complex as shown in Fig. 10.

The second of the fundamental features of fusion membranes is that membranes can exist in two states—the classical, bilayer state and the fusion triple tier state. In the classical state paired bimodal proteins pack together in the same continuum with paired phospholipid molecules. The orientation of the paired molecules, whether protein or phospholipid, is at right angles to the plane of the membrane. In the fusion membrane the pairing is eliminated because half the phospholipid is in the concentric inverted state and half in the orientation of the bilayer. The bimodal proteins originally paired are now lined up in linear array on one or the other side of the central hydrophobic tier of inverted phospholipid.

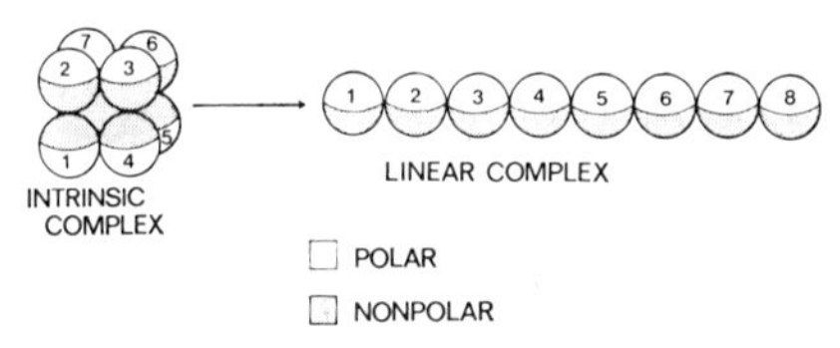

Figure 9. Transition of intrinsic membrane complexes from the double-tiered to the single tier pattern.

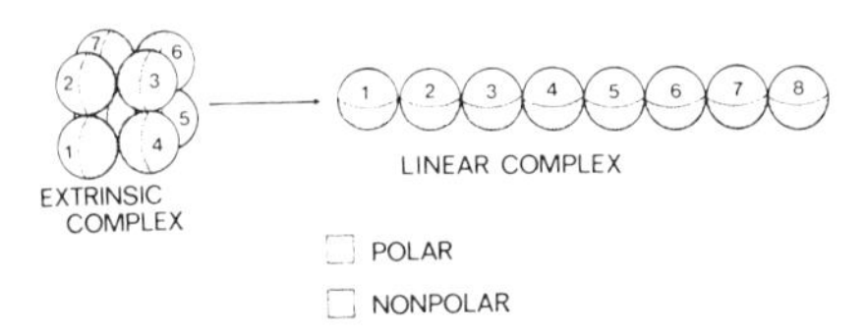

Figure 10. Transition of the ATPase complex of the projecting headpiece of the mitochondrial inner membrane from the double-tiered to the single-tier pattern.

The plasma lipoproteins appear to be constructed in the same fashion as fusion membranes.[6] These lipoprotein particles have a central core of inverted phospholipid covered by a thin shell of bimodal protein.[106] The proteins appear to be flattened discs (10 Å thick) with one side polar and the other side nonpolar. The polar face is directed exteriorly and the nonpolar face is directed interiorly to the inverted phospholipid core. The bimodality of the apolipoproteins appears to be a consequence of a helical arrangement in which the side chains on one side of the helix are polar and the side chains on the other side are non-polar.[105]

The structure of myelin. Myelin is a membrane which is generally used as a yardstick in assessing any membrane model. This has its amusing side in that myelin is about as "far out" a membrane as can be found and for myelin of all membranes to be used as a yardstick without appreciation of its idiosyncracies appears to be unwarranted to say the least. Nonetheless, in view of the widely claimed importance which attaches to myelin, it is all the more necessary to show how our unitization model which allegedly has universal applicability can rationalize the properties of this membrane.

Agreement is general that C.N.S. myelin arises by the fusion of two loops of the plasma membrane of the Schwann cell.[107] The fusion membrane then is wrapped concentrically around the axon with the ribbons of fused membranes being tightly apposed one to another in laminar concentric rings. How unique is this capability of the myelin membrane for undergoing fusion, and how unique is the capability of fused membranes to form laminar concentric rings? The capacity for undergoing fusion appears to be a general property of membranes and the capacity of fused membranes to form laminar sets depends entirely on providing the necessary ingredients in the medium—either divalent metals such as Mg^{2+} or basic proteins such as protamine.[103] Myelin contains such a basic protein which accounts for about 25% of the total protein content.[108] It is of interest to note that in demyelinating diseases where the ribbons of fused membranes tend to separate, the fundamental lesion in many cases is a deficiency of the basic protein.[109] The cristae of the inner mitochondrial membrane can be induced to form laminar sets of fused membranes (by addition of Mg^{2+} or protamine) which are indistinguishable from the laminar sets of myelin.[6, 103] At least with regard to the lamination of myelin, we are dealing with a property which is inducible under appropriate conditions in other membranes.

We have mentioned in a previous section that fusion is a reversible process. Indeed sonication of myelin apparently leads to the defusion of the concentric ribbons of fused membranes to the two partner membranes.[6, 103] This defusion process is recognizable by the transition from a triple-tiered to a pair of double-tiered membranes.

Myelin, like all other biological membranes, can be depolymerized to lipoprotein repeating units given the proper detergent, and these units will generate vesicular membranes *de novo* when the detergent is removed by appropriate means.[110] As we have emphasized many times in this article, the capability for *de novo* membrane formation is the essence of biological membranes because it implicates complexes as the units of membrane structure, and the complex is the heart of the structure-function unitization model.

The myelin membrane has one feature which to many investigators is so overriding that all evidence bearing on the normalcy of myelin is swept aside. This feature concerns the ratio of phospholipid to intrinsic protein. In most membranes this ratio is about 1:1;[86] in myelin the ratio is about 2:1 when the value for intrinsic protein is corrected for the content of basic protein.[112] It is this apparent surplus of phospholipid which is taken to be a token of a new constructional principle for myelin. There are, however, other compositional features of myelin which have to be taken into account in evaluating the phospholipid–protein ratio. Firstly, myelin has almost as much cholesterol as phospholipid.[113] Secondly, myelin has an unusually high level of galactolipids equivalent to about half the level of phospholipid.[114] The effect of both the cholesterol and the galactolipid with its complement of C_{24} fatty acid residues is to decrease fluidity and increase rigidity of the membrane.[115] Myelin is less fluid a membrane than other membranes despite the high phospholipid to protein ratio and this appears to be due to multiple causes—laminar fusion and a high content of cholesterol and galactolipid.

The only atypical feature that needs rationalization is the high phospholipid to protein ratio in myelin membranes. How can such a low protein content be reconciled with the lipoprotein repeating unit postulate? Experimentally the lipoprotein repeating units are demonstrable in myelin[110] and this must mean that the protein domains in myelin have to be stabilized by considerably larger lipid domains (domains of phospholipid and galactolipids) than are necessary for other membranes. But this apparent discrepancy may have to be viewed in somewhat different light. If as would be predictable, the galactolipids are not randomly dispersed in the phospholipid domain but rather are clustered and intimately associated with complexes[115a] then the orderly association of protein and lipid domains could be maintained. The protein domains in effect become expanded by this postulated tight association with galactolipids. Such an association would be reasonable since the galactolipids by virtue of the C_{24} chain length of the hydrocarbon tails would be expected to pack more readily with proteins than with phospholipids.

The role of internal cavities and channels in intrinsic membrane complexes. The cavity-channel postulate invoked for the construction of intrinsic complexes is consistent with the experimental observations suggesting that membranes contain selective macromolecular channels for the facilitated or energy-linked transmembrane transport of polar molecules,[117] and also provides a rationale for the selective entry of substrate molecules into the interior of complexes where controlled chemical interactions between substrate and enzymic active sites can take place. Phenomena are known in mitochondria which hitherto have been without explanation and these phenomena can be readily rationalized in terms of the cavity-channel concept.

Antimycin inhibits electron transfer of Complex III at exceedingly low concentrations—equivalent to one or two molecules of antimycin per complex.[118] The various effects of antimycin A and the characteristics of this inhibition are summarized in Table I. The simple notion that antimycin can selectively enter the hypothetical cavity of Complex III provides a sufficient basis for rationalizing all the effects of antimycin summarized in the table. The rationalizations are also listed in the same table. Here is a case in which one of the most carefully studied phenomena of mitochondriology defied explanation until rationalized in terms of the cavity-channel concept.

TABLE I. The interaction of antimycin A with Complex III and rationalization of the inhibition phenomena in terms of the cavity–channel principle

Phenomenon	Rationalization
1. Cleavage of Complex III into particulate cytochrome *b* and soluble cytochrome c_1 by exposure to taurodeoxycholate in a temperature-dependent reaction.[129]	1. Taurodeoxycholate must enter the internal cavity of Complex III in order to achieve cleavage. At low temperatures entry is interdicted.
2. Antimycin prevents cleavage of Complex III by taurodeoxycholate.[129]	2. Antimycin once in the cavity prevents entry of taurodeoxycholate into the cavity.
3. Antimycin prevents electron transfer from cytochrome *b* to cytochrome c_1 but not the reduction of cytochrome *b*.[118]	3. Antimycin prevents the conformational change by which cytochrome *b* is brought close enough to cytochrome c_1 for electron transfer to take place.
4. Antimycin can completely inhibit electron transfer in Complex III when the molar ratio of antimycin: Complex III is 1. No covalent link is formed between antimycin and Complex III.[129]	4. Antimycin is not removable from the cavity once it has entered, except by disruption of the complex.
5. Reduction of the complex prevents inhibition by antimycin.[129]	5. Reduction of the complex leads to a closure of the cavity and the cavity becomes inaccessible to antimycin.
6. Reduction of the complex prevents cleavage by taurodeoxycholate.[129]	6. The cavity in the reduced complex becomes inaccessible to taurodeoxycholate.
7. The potential of cytochrome *b* in Complex III is altered when antimycin is present in the complex.[130]	7. Antimycin changes the environment of the cavity and thereby modifies the oxidation-reduction potential of cytochrome *b*.

The 90 Å headpieces of the inner membrane are linked to the membrane via a cylindrical stalk (30 Å diameter and 50 Å long).[71] The stalk is capable of extension from the membrane and collapse into the membrane.[119] This must mean that the stalk can move in and out of a sleeve in the membrane. The headpiece-stalk projections are known to be linked to a membrane-forming complex.[22,120] Thus the three parts, the intrinsic complex in the membrane, the stalk, and the extrinsic complex of the headpiece, form one integrated unit, called the oligomycin sensitive ATPase.[120] It would appear, therefore, that the stalk fits into a cavity in the interior of the intrinsic complex in the membrane and that there is some control device which regulates the extent to which the stalk penetrates into the cavity. When the stalk is pushed deep into the cavity, the headpiece collapses on the membrane; when the stalk is extruded from the cavity, the headpiece projects away from the membrane.

Oligomycin inhibits ATPase activity only when the headpiece is attached to the membrane via the stalk.[121,120a] It has no effect on the ATPase activity of the headpiece itself (F_1). It has also been known that the presence of the stalk is essential for oligomycin-sensitivity and that oligomycin sensitivity is controlled by some protein which is neither in the headpiece nor in the stalk.[122] This protein, according to Beechey,[122] appears to be in the membrane. A viable explanation for this phenomenon is that oligomycin can selectively enter the channel of the complex which serves as the sleeve for the stalk. When oligomycin is in the cavity, the stalk is perforce extruded and the headpiece assumes a conformation incompatible with ATPase activity. When the headpiece is detached from the membrane and the stalk, then oligomycin can exert no effect on ATPase activity. The Beechey protein which appears to control oligomycin sensitivity may be one of the proteins in the intrinsic complex to which the stalk is anchored.

The rod outer segment membrane and the unit of photoreception. The rod outer segment membrane would appear to be one of the most highly organized membranes as judged by several criteria—electron microscopy, cooperativity, and crystallinity. It is completely classical in respect to some fundamental properties—depolymerizability to lipoprotein structures which can form vesicular membranes *de novo*,[123] the capacity for assuming a crystalline pattern,[124] and a normal ratio of protein to lipid (about 1:1).[125] These properties would suggest that the rod outer segment membrane conforms in basic respects to the structure-function unitization model, namely that the membrane is built up of complexes and associated bilayer phospholipid, and that there are alternating domains of protein and phospholipid.

This simple picture is in conflict with a body of evidence which suggests that the rod outer segment membrane is highly fluid,[126] that rhodopsin molecules are randomly arranged in the membrane,[127] that there is an asymmetric distribution of rhodopsin molecules on the two sides of the membrane (hence like pairing would be excluded),[128] that rhodopsin molecules not only have freedom of translation but they can spin at high speed during photoexcitation.[128a,128b] It is our present view that these observations will eventually be rationalized without compromise of the structural principles we consider to be universal for biomembranes. Some fundamental studies on the structure of the rod outer segment membrane have yet to be carried out. Until then a final decision will have to be deferred.

IV. A Critique of the Singer–Nicolson Fluid Mosaic Model of Biological Membranes

Singer and Nicolson[8] have recently proposed a model of membrane structure which in one important respect is opposed to the structure-function unitization model which we have presented in the present communication. In respect to certain molecular features such as hydrophobic bonding of protein and phospholipid, the bimodal character of intrinsic membrane proteins, the bilayer character of the phospholipid, and the distinction between intrinsic and extrinsic membrane proteins, the two models are in agreement. But the models diverge sharply in respect to the organization of the intrinsic membrane proteins. According to the fluid mosaic model, there is only one continuum, namely bilayer phospholipid. The intrinsic proteins are randomly interspersed in the bilayer lipid. The proteins, like the lipids, are capable of moving freely through the membrane so that there is no fixed position in the membrane for any molecule. The only restriction imposed by the fluid mosaic model on the movement of proteins is the flipping of a protein molecule from one side of the membrane to the other. It is to be noted that the notions of complexes and protein domains are essentially rejected for membranes generally although Singer and Nicolson admit the possibility that certain specialized membranes such as the inner membrane of the mitochondrion may have sets of intrinsic proteins rather than individual intrinsic proteins.

The phenomena on which we have based the unitization model provide the most powerful refutation of the basic thesis of the fluid mosaic model—phenomena such as lipoprotein repeating units, *de novo* membrane formation, crystalline membranes, the ultrastructural evidence of repeat structure in membranes, viral membrane formation and induction of intrinsic membrane enzymes. As we see it, an acceptable model of the membrane must deal with and provide a satisfactory rationale for all membrane phenomena. This neglect of critical membrane phenomena appears to be a major deficiency of the fluid mosaic model.

What were the compelling lines of evidence which led Singer and

Nicolson to the formulation of the fluid mosaic model? Basically there were three lines of evidence: first, the x-ray data of Blasie *et al.*[127] on the circular symmetry of the rod outer segment membrane and the inference of randomness in the arrangement of rhodopsin in the membrane; second, the immunochemical evidence that the proteins in the red blood corpuscle membrane are randomly distributed[130]; and third, a miscellaneous collection of observations which point to the conclusions that proteins in membranes have high mobility and can move around freely.[8]

With respect to the assumption of a high degree of mobility for intrinsic proteins in the membrane, we have pointed out that only a special group of intrinsic proteins, namely the glycoproteins, may show this mobility. The fluid mosaic model may be restricted in application to the intrinsic proteins concerned in immunochemical reactions.

Acknowledgements

The present investigations were supported in part by program project grant GM-12847 of the National Institute of General Medical Sciences of the National Institutes of Health in Bethesda, Maryland. In the preparation of the manuscript we drew heavily on the expertise and advice of Drs. Julius Adler, Paul Kaesberg, Carl Schnaitman, Jordi Folch-Pi, Derek Bownds and Lawrence Rothfield. We also wish to pay tribute to the monumental contributions of Dr. Y. Hatefi to the development of our knowledge of membrane complexes.

References

1. D. E. Green and J. F. Perdue, *Proc. Natn. Acad. Sci. USA*, **55** (1966) 1295.
2. D. E. Green and J. F. Perdue, *Ann. N.Y. Acad. Sci.*, **137** (1966) 667.
3. H. M. McConnell and B. G. McFarland, *Quant. Rev. Biophys.*, **3** (1970) 91.
4. M. E. Tourtellotte, D. Branton and A. Keith, *Proc. Natn. Acad. Sci. USA*, **66** (1970) 909.
5. J. M. Steim, M. E. Tourtellotte, J. C. Reinelt, T. McElhaney and R. L. Rader, *Proc. Natn. Acad. Sci. USA*, **63** (1969) 104.
6. D. E. Green, *Ann. N.Y. Acad. Sci.*, **195** (1972) 150.
7. D. E. Green and R. F. Brucker, *BioScience*, **22** (1972) 13.
8. S. J. Singer and G. L. Nicolson, *Science*, **175** (1972) 720.
9. R. A. Capaldi and G. Vanderkooi, *Proc. Natn. Acad. Sci. USA*, **69** (1972) 930.
10. G. Vanderkooi and D. E. Green, *BioScience*, **21** (1971) 409.
11. G. Vanderkooi and R. A. Capaldi, *Ann. N.Y. Acad. Sci.*, **195** (1972) 135.
12. G. Vanderkooi and D. E. Green, *Proc. Natn. Acad. Sci.*, **66** (1970) 615.
13. G. Vanderkooi, *Ann. N.Y. Acad. Sci.*, **195** (1972) 6.
14. G. Colacicco, *Ann. N.Y. Acad. Sci.*, **195** (1972) 224.
15. S. Fleischer, G. Brierley, H. Klouwen and D. B. Slautterback, *J. Biol. Chem.*, **237** (1962) 3264.
16. S. Fleischer, B. Fleischer and W. Stoeckenius, *J. Cell Biol.*, **32** (1962) 193.
17. T. Wakabayashi, E. F. Korman and D. E. Green, *Bioenergetics* **2** (1971) 233.
18. D. Branton, *Proc. Natn. Acad. Sci. USA*, **55** (1966) 1048.
19. P. Pinto da Silva and D. Branton, *J. Cell Biol.*, **45** (1970) 598.
20. T. W. Tillack and V. T. Marchesi, *J. Cell Biol.*, **45** (1970) 649.
21. D. E. Green, D. W. Allmann, E. Bachmann, H. Baum, K. Kopaczyk, E. F. Korman, S. Lipton, D. MacLennan, D. G. McConnell, J. F. Perdue, J.S. Rieske and A. Tzagoloff, *Arch. Biochem. Biophys.*, **119** (1967) 312.
22. K. Kopaczyk, J. Asai and D. E. Green, *Arch. Biochem. Biophys.*, **126** (1968) 358.
23. A. Tzagoloff, D. H. MacLennan, D. G. McConnell and D. E. Green, *J. Biol. Chem.*, **242** (1967) 2051.
24. D. H. MacLennan, P. Seeman, G. H. Iles and C. C. Yip, *J. Biol. Chem.*, **246** (1971) 2702.

25. S. Razin, Z. Ne'eman and I. Ohad, *Biochim. Biophys. Acta*, **193** (1969) 277.
26. J. S. Rieske, S. H. Lipton, H. Baum and H. I. Silman, *J. Biol. Chem.*, **242** (1967) 4888.
27. J. C. Thornber, M. Stewart, M. W. C. Halton and J. L. Bailey, *Biochemistry*, **6** (1967) 2006.
28. I. Shibuya, H. Honda and B. Maruo, *J. Biochem.*, **64** (1968) 371.
29. A. Atkinson, A. D. Gatenby and A. G. Lowe, *Nature New Biol.*, **233** (1971) 145.
30. A. Kahlenberg, N. C. Dulak, J. F. Dixon, P. R. Golsworthy and L. E. Hokin, *Arch. Biochem. Biophys.*, **131** (1969) 253.
31. A. Martonosi and R. A. Halprin, *Arch. Biochem. Biophys.*, **144** (1971) 66.
32. B. Revsin, E. D. Marquez and A. F. Brodie, *Arch. Biochem. Biophys.*, **139** (1970) 114.
33. S. Taniguchi and E. Itagaki, *Biochim. Biophys. Acta*, **44** (1960) 263.
34. J. H. Bruemmer, P. W. Wilson, J. L. Glenn and F. L. Crane, *J. Bact.*, **73** (1957) 113.
35. A. Y. H. Lu and M. J. Coon, *J. Biol. Chem.*, **243** (1968) 1331.
36. D. H. MacLennan, A. Tzagoloff and D. G. McConnell, *Biochim. Biophys. Acta*, **131** (1967) 59.
37. J. S. O'Brien, *Science*, **147** (1965) 1099.
38. D. E. Green and O. Hechter, *Proc. Natn. Acad. Sci. USA*, **53** (1965) 318.
39. M. Singer and P. Leder, *Ann. Rev. Biochem.*, **35** (1966) 195.
40. J. K. Pollak, R. Malor, M. Morton and K. A. Ward, in: *Autonomy and Biogenesis of Mitochondria and Chloroplasts*, North-Holland Publishing Co., 1970, p. 27.
41. H. Baum and J. Rieske, unpublished studies.
42. E. F. Korman, unpublished studies.
43. J. C. Brooks and A. E. Senior, unpublished studies.
44. S. Yamashita and E. Racker, *J. Biol. Chem.*, **244** (1969) 1220.
45. J. A. Berden and E. C. Slater, *Biochim. Biophys. Acta*, **216** (1970) 237.
46. D. E. Green and O. Hechter, *Proc. Natn. Acad. Sci. USA*, **53** (1965) 318.
47. W. Weidel and H. Pelzer, *Advan. Enzymol.*, **26** (1964) 193.
48. M. Sonenberg, *Proc. Natn. Acad. Sci. USA*, **68** (1971) 1051.
49. S. Hecht, S. Shlaer and M. H. Pirenne, *J. Gen. Physiol.*, **25** (1942) 819.
50. D. Nachmansohn, *Proc. Natn. Acad. Sci. USA*, **61** (1968) 1034.
51. D. E. Green, J. Asai, R. A. Harris and J. T. Penniston, *Arch. Biochem. Biophys.*, **125** (1968) 684.
52. C. H. Williams, W. J. Vail, R. A. Harris, M. Caldwell, D. E. Green and E. Valdivia, *Bioenergetics*, **1** (1970) 147.
53. D. E. Green and R. A. Harris, *FEBS Letters*, **5** (1969) 241.
54. T. Wakabayashi, A. E. Senior and D. E. Green, *Bioenergetics*, (1972) in press.
55. T. Oda, in: *Profiles of Japanese Science and Scientists*, Kodansha Ltd., Tokyo, 1970, p. 107.
56. C. C. Remsen, F. W. Valois and S. W. Watson, *J. Bact.*, **94** (1967) 422.
57. H. Baum, H. I. Silman, J. S. Rieske and S. H. Lipton, *J. Biol. Chem.*, **242** (1967) 4876.
58. H. I. Silman, J. S. Rieske, S. H. Lipton and H. Baum, *J. Biol. Chem.*, **242** (1967) 4867.
59. B. Love, S. H. P. Chan and E. Stotz, *J. Biol. Chem.*, **245** (1970) 6664.
60. R. A. Capaldi and H. Hayashi, *FEBS Letters* (1972) in press.
61. T. Takano, R. Swanson, O. B. Kallai and R. E. Dickerson, *Cold Spring Symposia on Quantitative Biology*, **36** (1971) 397.
62. R. E. Dickerson and I. Geis, *The Structure and Action of Proteins*, Harper & Row, Publishers, New York, 1969, p. 47.
63. C. C. Blake, L. N. Johnson, G. A. Mair, A. C. T. North, D. C. Phillips and V. R. Sarma, *Proc. Royal Soc.*, **B167** (1967) 378.
64. M. Perutz, *Proc. Royal Soc.*, **B167** (1967) 448.
65. G. Vanderkooi, A. E. Senior, R. A. Capaldi and H. Hayashi, *Biochim. Biophys. Acta*, **274** (1972) 38.
65a. H. Hayashi, G. Vanderkooi and R. A. Capaldi, *Biochem. Biophys. Res. Commun.* (1972) in press.
66. J. P. Segrest, R. L. Jackson, W. Terry and V. T. Marchesi, *Fed. Proc.*, **31** (1972) 736.
67. R. A. Capaldi, *Eur. J. Biochem.* (1972) in press.
68. J. De Gier, J. G. Mandershool, L. L. M. Van Deenen, *Biochim. Biophys. Acta*, **150** (1968) 666.
69. T. F. Chuang, Y. C. Awasthi and F. L. Crane, *Proc. Indiana Acad. Sci.*, **79** (1970) 110.
70. E. H. Eylar, M. A. Madoff, O. V. Brody and J. L. Oncley, *J. Biol. Chem.*, **237** (1962) 1992.
71. H. Fernandez-Moran, T. Oda, P. V. Blair and D. E. Green, *J. Cell Biol.*, **22** (1964) 63.
72. D. E. Green and R. F. Goldberger, *Molecular Insights into the Living Process*, Academic Press, New York, 1967.
73. J. P. Changeux, in: *Symmetry and Function of Biological Systems at the Macromolecular Level*, A. Engstrom and B. Strandberg (eds.), John Wiley & Sons, Inc., New York, 1969, p. 235.
74. D. E. Green and S. Ji, *Proc. Natn. Acad. Sci. USA*, **69** (1972) 726.
75. C. Howe, C. Morgan, C. de V. St. Cyr, K. C. Hsu, and H. M. Rose, *J. Virol.*, **1** (1967) 215.

76. N. H. Klein, *Fed. Proc.*, **28** (1969) 1739.
77. M. M. Burger, *Growth Control in Cell Culture*, Little Brown & Co., 1971, in press.
78. H. D. Klenk, in: *Dynamic Structure of Cell Membranes*, D. F. H. Wallach and H. Fisher (eds.), Springer-Verlag, New York, 1971, p. 97.
79. F. A. Murphy and B. N. Fields, *Virology*, **33** (1967) 625.
80. A. Klug and D. L. D. Caspar, *Adv. Virus Res.*, **7** (1960) 225.
81. A. Klug, *Fed. Proc.*, **131** (1972) 30.
82. H. Fraenkel-Conrat and R. C. Williams, *Proc. Natn. Acad. Sci. USA*, **41** (1955) 690.
83. R. R. Brown, J. A. Miller and E. C. Miller, *J. Biol. Chem.*, **209** (1954) 211.
84. S. Narasimhulu, *Arch. Biochem. Biophys.*, **147** (1971) 391.
85. L. Ernster and S. Orrenius, *Fed. Proc.*, **24** (1965) 1190.
86. E. D. Korn, *Science*, **153** (1966) 1491.
87. R. A. Capaldi, unpublished observations.
88. E. F. Korman, R. A. Harris, C. H. Williams, T. Wakabayashi and D. E. Green, *Bioenergetics*, **1** (1970) 387.
89. J. D. Robertson, T. S. Bodenheimer and D. E. Stage, *J. Cell Biol.*, **19** (1963) 159.
90. F. T. Sanel and A. A. Serpick, *Science*, **168** (1970) 1458.
91. C. A. Schnaitman, *J. Bact.*, **104** (1970) 890.
92. M. L. De Pamphilis, *J. Bact.*, **105** (1971) 1184.
93. L. Rothfield, D. Romeo and A. Hinckley, *Fed. Proc.*, **31** (1972) 12.
94. L. Rothfield, *J. Biol. Chem.*, **243** (1968) 1320.
95. E. Ribi, *J. Bact.*, **92** (1966) 1493.
96. M. J. Osborn, *Ann. Rev. Biochem.*, **38** (1969) 501.
97. R. A. Harris, D. L. Harris and D. E. Green, *Arch. Biochem. Biophys.*, **128** (1968) 219.
98. A. Tzagoloff, D. G. McConnell and D. H. MacLennan, *J. Biol. Chem.*, **243** (1968) 4117.
99. J. A. Reynolds and H. Trager, *J. Biol. Chem.*, **246** (1971) 7337.
100. R. A. Capaldi, unpublished studies.
101. A. D. Brown, *J. Mol. Biol.*, **12** (1965) 491.
102. M. Esfahani, A. R. Limbrick, S. Knutton, T. Oka and S. J. Wakil, *Proc. Natn. Acad. Sci. USA*, **68** (1971) 3180.
103. D. E. Green, T. Wakabayashi and R. F. Brucker, *Bioenergetics* (1973) in press.
103a. R. F. Brucker, T. Wakabayashi and D. E. Green, *Bioenergetics* (1973) in press.
104. C. A. Schnaitman, *J. Bact.*, **108** (1971) 553.
105. V. Braun and V. Bosch, *Proc. Natn. Acad. Sci. USA*, **69** (1971) 970.
106. G. G. Shipley, D. Atkinson and A. M. Scanu, *J. Supramol. Struc.* **1** (1972) 98.
107. B. B. Geren, *Exptl. Cell Res.*, **7** (1954) 558.
108. E. H. Eylar, J. Salk, G. C. Beveridge and L. V. Brown, *Arch. Biochem. Biophys.*, **132** (1969) 34.
109. E. H. Eylar, *Ann. N.Y. Acad. Sci.*, **195** (1972) 481.
110. G. Sherman and J. Folch-Pi, *J. Neurochem.*, **17** (1970) 567.
111. E. Racker, *Biochem. Biophys. Res. Commun.*, **10** (1963) 435.
112. J. S. O'Brien and E. L. Sampson, *J. Lipid Res.*, **6** (1965) 537.
113. J. S. O'Brien and E. L. Sampson, *J. Lipid Res.*, **6** (1965) 545.
114. F. Gonzalez-Sastre, *J. Neurochem.*, **17** (1970) 1049.
115. D. Chapman, U. B. Kanat, J. De Gier and S. A. Penkett, *J. Mol. Biol.*, **31** (1968) 101.
115a. S. Hakomori, in: *Dynamic Structure of Cell Membranes*, D. F. H. Wallach and H. Fisher (eds.), Springer-Verlag, New York, 1971, p. 65.
116. G. Vanderkooi and M. Sundaralingam, *Proc. Natn. Acad. Sci. USA*, **67** (1970) 233.
117. A. L. Lehninger, in: *Biomembranes*, Vol. 2, L. A. Manson (ed.), Plenum Press, New York, 1971, p. 147.
118. H. Baum, J. S. Rieske, H. I. Silman and S. H. Lipton, *Proc. Natn. Acad. Sci. USA*, **57** (1967) 798.
119. O. Hatase, T. Wakabayashi, H. Hayashi and D. E. Green, *Bioenergetics* (1972) in press.
120. A. Tzagoloff and P. Meagher, *J. Biol. Chem.*, **246** (1971) 7328.
120a. D. H. MacLennan and J. Asai, *Biochem. Biophys. Res. Commun.*, **33** (1968) 441.
121. D. H. MacLennan and A. Tzagoloff, *Biochemistry*, **7** (1968) 1603.
122. R. B. Beechey, *Biochem. J.*, **116** (1970) 68.
123. R. P. Poincelot and E. W. Abrahamson, *Biochemistry*, **9** (1970) 1820.
124. J. K. Blasie and M. M. Dewey, *J. Mol. Biol.*, **14** (1965) 1436.
125. N. C. Nielsen, S. Fleischer and D. G. McConnell, *Biochim. Biophys. Acta*, **211** (1970) 10.
126. P. K. Brown, *Nature New Biol.*, **236** (1972) 35.
127. J. K. Blasie and C. R. Worthington, *J. Mol. Biol.*, **39** (1969) 417.
128. W. J. Gras and C. R. Worthington, *Proc. Natn. Acad. Sci. USA*, **63** (1969) 233.
128a. R. A. Cone, *Nature New Biol.*, **236** (1972) 39.
128b. P. K. Brown, *Nature New Biol.*, **236** (1972) 35.
129. J. S. Rieske, H. Baum, C. D. Stoner and S. H. Lipton, *J. Biol. Chem.*, **242** (1967) 4854.
130. A. M. Pumphrey, *J. Biol. Chem.*, **237** (1962) 2384.

Addendum

1. Enzymes which are intrinsic to the membrane continuum may be either polymeric complexes or mobile, through-membrane monomeric proteins like the glycoproteins which are localized in the lipid phase.

2. Space limitation has made it necessary to compress our discussion of the fusion phenomenon to the point that a balanced treatment of the uncertainties still remaining in the model of the fusion membrane could not be provided. The fusion phenomenon has opened a new door to membranology and the tentative character of the molecular interpretations given in the text should be borne in mind.

The Influence of Temperature-Induced Phase Changes on the Kinetics of Respiratory and Other Membrane-Associated Enzyme Systems

John K. Raison

Plant Physiology Unit, C.S.I.R.O. Division of Food Research, and School of Biological Sciences, Macquarie University, North Ryde, Sydney, Australia, 2113

Temperature-mediated changes in the kinetics of enzyme catalysed reactions can be due to effects on a number of different parameters. If the change in temperature does not (a) inactivate the enzyme, (b) alter the affinity of the enzyme for the substrate, an activator or an inhibitor or (c) alter the pH function of the reaction components, the velocity of enzyme catalysed reactions increases with increasing temperature. The relationship between the velocity of reaction and temperature can be expressed either as the activation energy (E) or the temperature coefficient (Q_{10}). Both expressions can be derived from the empirical Arrhenius equation relating the velocity of reaction and temperature

$$\frac{\alpha \ln k}{\alpha T} = \frac{E}{RT^2} \tag{1}$$

where k is the reaction velocity constant, R the gas constant, T the absolute temperature and E a constant, subsequently called the activation energy (also written as A or μ). Integration of equation (1) gives

$$\ln \frac{k_2}{k_1} = \frac{E}{R}\left(\frac{1}{T_1} - \frac{1}{T_2}\right) \tag{2}$$

from which it can be seen that the value for E can be obtained from the slope of the straight line when $\log k$ is plotted against $1/T$

$$E = 2{\cdot}303\, R \times \text{slope}$$
$$\therefore E = 4{\cdot}576 \times \text{slope (where } R = 1{\cdot}987 \text{ cal/mole/°K)}$$

The term E is not the activation energy, ΔH^*, of the activated complex formed during an enzyme reaction as defined by the transition state theory of Eyring (cf. ref. 1) but can be related to this quantity by

$$E = \Delta H^* + RT \tag{3}$$

if the molecularity of the reaction is unity. Thus, within the physiological range of temperatures, approximately 600 cal/mole must be deduced from the calculated value of E to obtain the heat content on activation, ΔH^*. The relation between changes in heat content and other thermodynamic functions such as free energy and entropy during formation of enzyme-substrate complexes and the activation of these complexes during reactions is beyond the scope of this article. The reader is referred to the articles by Dawes;[2] Bray and White;[3] Johnson *et al.*;[1] Sizer;[4] Dixon and Webb.[5]

The other expression relating the change in reaction velocity to temperature, the temperature coefficient (Q_{10}), is defined as the factor by which the velocity increases as a result of a 10 Celsius degree rise in temperature. For most biological reactions the value of Q_{10} is between 1 and 2.

Although most enzyme catalysed reactions show a constant activation energy over a limited range of temperature, a number of reactions have been reported where a plot of the logarithm of velocity against the reciprocal of the absolute temperature (Arrhenius plot) shows two straight lines of different slope, i.e. a discontinuity or change in E with temperature.[4–7] A number of papers have also been published showing Arrhenius plots where the data is fitted to a straight line by "brute force"[8] methods. In most of these cases the realistic errors in determining the velocity have been ignored in what is apparently an attempt to make the reaction appear to conform with Arrhenius' original theory and a line of best fit applied.

The large number of examples of enzyme reactions, of both plant and animal tissues which show a distinct change in activation energy within the range of physiological temperatures, particularly reactions catalysed by enzymes associated with membrane surfaces, clearly suggests that in most cases the change is associated in some way with changes in the physical properties of the membranes. Furthermore, recent data show enzymes or multi-enzyme systems will exhibit either a varying or constant activation energy depending on the species of plant or animal from which the enzymes are derived.

This article will discuss the changes in activation energy of enzymes in terms of temperature-induced phase changes in the system, with particular reference to "crystalline" enzymes and membrane-associated enzymes of both plant and animal origin. Some of these enzymes exhibit a change in activation energy with temperature and in certain cases it is possible to relate these changes in activation energy to an alteration in the physical properties of membrane lipids. The significance of a change in activation energy is also discussed in terms of the physiological function of plants and animals at various temperatures.

Alterations in the Activation Energy of Enzymes in Terms of Changes in Physical States

The significance of abrupt changes in activation energy of some enzyme reactions and physiological processes has been the subject of considerable controversy. The similarity between the activation energy for specific enzyme reactions and complex physiological processes such as respiration and heart beat determined by Arrhenius[9] led Crozier and co-workers (for complete discussion see ref. 1) to postulate that with each physiological process, involving a multi-enzyme series of reactions, the overall process is limited by a rate-limiting or "master reaction". The change in activation energy for these processes is considered by Crozier to represent a shift from one controlling "master reaction" to another at the critical temperature (the point of discontinuity in the Arrhenius plot). Bělehrádek,[10] on the other hand, has considered that such a process would produce a smooth catenary in the region of the temperature range over which the change occurred. Thus, according to Bělehrádek,[10] the discontinuity is the result of drawing an intercept of tangents to smooth curves and denies the existence of real discontinuities. To explain the increase in activation energy at between 26° and 32°, of fumarase acting on either fumarate or malate in alkaline medium, Massey[11] has suggested the enzyme dissociates into smaller sub-units as the temperature is increased and these smaller enzyme unit have a higher activation energy. In a more recent consideration Bělehrádek[12] states that for "a real break" to be manifest in an Arrhenius plot of a series of sequential reactions the rate-limiting processes, in the two temperature zones, must have activation energy values differing from each other by at least 16 kcal/mole. Much of the available data on discontinuities in Arrhenius plots show changes in activation energy considerably less than this value.

This problem in interpretation of discontinuities in Arrhenius plots has been discussed by Kumamoto *et al.*[13] in the light of more recent data showing that in some cases the two independent straight lines of discontinuous Arrhenius plots do not intersect at the transition temperature. Kumamoto *et al.*[13] agree with Crozier and others (cf. ref. 5) that two independent processes are required to produce a discontinuity, but further stipulate that the system must provide for the exclusive functioning of these processes in their respective temperature ranges. This proposal of the exclusive operation of the independent processes within the respective temperature ranges is justified on thermodynamic grounds by considering that the system undergoes a phase change at the critical temperature. The isothermal properties of a phase change explain why the two independent processes operate exclusively immediately there is a shift from the temperature at which the phase change occurs. The phase change which allows for the exclusive functioning of the two independent processes could be in the ground state of one of the reactants or in the enzyme catalysing the reaction. An important consideration resulting from this explanation of discontinuities in Arrhenius plots is that the phase change in the reaction components may be determined by phase changes in structural components associated with the particular enzyme, but remote from the site of the catalysis. As Kumamoto *et al.*[13] have pointed out, under these conditions the compensation between enthalpy and entropy, predicted to maintain the rate constants of the two independent reactions at the

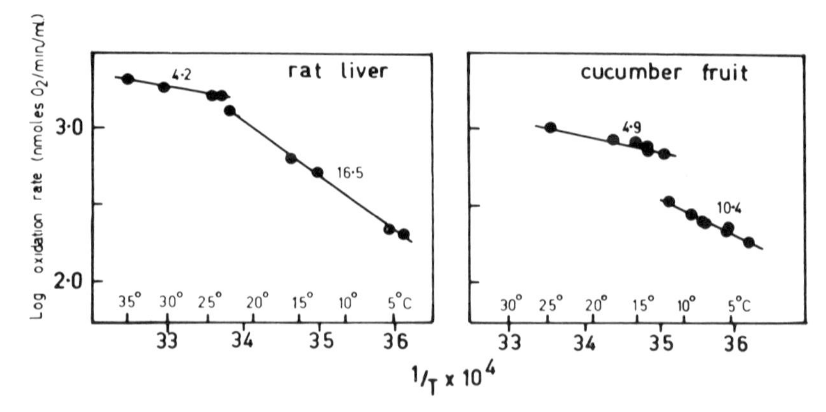

Figure 1. Arrhenius plots of state 3, succinate oxidation by rat-liver mitochondria and cucumber fruit mitochondria. The activation energies (kcal/mole) are as indicated beside each line. (Data from Kumamoto, Raison and Lyons.[13])

transition temperature constant, might not be exact. In these cases the two straight lines of an Arrhenius plot would show a non-intersecting discontinuity as typified by the examples of Arrhenius plots in Fig. 1. Many other examples of such non-intersecting discontinuities are now known (see below), but in all cases the difference in the rate constants at temperatures infinitesimally above and below the transition temperature never exceeds 0·5 log units. From thermodynamic considerations Kumamoto *et al.*[13] argue that much greater differences would be expected if the enthalpy–entropy compensation were not as exact. Their reasoning is detailed below.

By application of a simplified version of the transition state theory, neglecting partition functions, the rate constant for each of the independent reactions can be expressed as

$$k_r = \frac{Kk_\beta T}{h} e^{-\Delta F^*/RT} \tag{4}$$

where K is the transmission coefficient, k_β is Boltzmann's constant, T is the absolute temperature, h is Planck's constant, ΔF^* the change in free energy of the activated complex and R the universal gas constant. The term

$$\frac{Kk_\beta T}{h}$$

is nearly constant over the small temperature range under consideration The ratio of the two rate constants immediately above and below the transition temperature can then be expressed as

$$\frac{k_{r_1}}{k_{r_2}} = \frac{e^{-\Delta F_1^*/RT}}{e^{-\Delta F_2^*/RT}} = e^{\Delta F_2^* - \Delta F_1^*/RT} \tag{5}$$

At a point of intersection (isokinetic point) $k_{r_1} = k_{r_2}$ and

$$\Delta F_2^* - \Delta F_1^* = 0$$

However, the two independent processes on either side of the isokinetic point have different activation energies as determined experimentally (Fig. 1). The Arrhenius activation energy (E) can be related to the enthalpy by

$$E = \Delta H^* + RT \tag{3}$$

and therefore these reactions would have different enthalpy values. Since

$$\Delta F^* = \Delta H^* - T\Delta S^* \tag{6}$$

to obtain a constant ΔF^* the change in ΔH^* for these reactions requires a corresponding change in ΔS^*.

In the case of the oxidation of succinate by mitochondria from cucumber fruit shown in Fig. 1. the values for E, above and below the transition temperature at 12°, are 4·9 and 10·4 kcal/mole, i.e., ΔH^* values of 4·3 and 9·8 kcal/mole approximately 600 cal/mole less than E.

Assuming ΔF^* is equal to ΔH^* and substituting these values in equation (5), we obtain

$$\begin{aligned} k_{r_1}/k_{r_2} &= e^{(9\cdot8\times10^3 - 4\cdot3\times10^3)/566} \quad \text{at } 12°, \\ &= e^{(5\cdot5\times10^3)/566} \\ &= e^{9\cdot7} \\ \text{i.e. } k_{r_1}/k_{r_2} &= 10^{4\cdot2} \end{aligned}$$

i.e. the rate constants at the transition temperature should vary by $10^{4\cdot2}$ units if there were no simultaneous compensation of the entropy term. The fact that the rate constant at the isokinetic point for this reaction varies by only approximately $10^{0\cdot3}$ units clearly shows that the compensating change in the entropy is of the order of $10^{3\cdot9}$ entropy units.

In most of the reactions to be discussed the difference in activation energy, above and below the transition temperature, would predict even larger differences in the rate constants than that observed for succinate oxidation by cucumber mitochondria. Since the data show differences in rate constants of usually less than $10^{0\cdot2}$ units it must be assumed that almost precise enthalpy–entropy compensation is the rule rather than the exception.

Discontinuities in Arrhenius Plots as a Consequence of Phase Changes

As pointed out in the preceding section, discontinuities in Arrhenius plots can be considered as a phase change in either the enzyme or the reactants and can be dominated by events removed from the active site. It is of interest therefore to consider the events occurring in some of the classical examples of discontinuities in terms of phase changes in the systems.

A. *Reactions Involving Crystalline Enzymes*

Arrhenius plots for the activity of invertase, lipase, trypsin and pepsin[15, 16] show a discontinuity or break at 0° indicative of a phase change in the water which forms one of the reactants. Dixon and Webb[5] have objected to this interpretation on the grounds that a change in activation energy at 0° was also observed with pancreatic lipase in a reaction mixture containing 36·5% glycerol.[15] However, the water involved in the hydrolysis of tributyrin by the lipase, in the region of the active centre of the enzyme molecule, is probably a separate domain which excludes glycerol and therefore reflects the phase change of pure water at 0°.[8] It is unlikely that a phase change in the lipid substrate, tributyrin, is involved since its melting point is −75°.

Massey[11] has shown discontinuities in plots of fumarase, acting in alkaline medium on either fumarate or malate, at different temperatures depending on the pH and substrate. A discontinuity was also apparent in acid medium for the hydration of fumarate by fumarase, but not for the dehydration of malate by the same enzyme. There is no simple explanation for these observations in terms of a phase change in the substrate. In alkaline conditions the activation energy changes from a larger (14·8 kcal/mole) to a smaller value (9·3 kcal/mole) as the temperature is decreased, while in slightly acid medium an increase in activation energy (6·7 to 10·6 kcal/mole) was observed. The temperature of the change also varied depending on the pH and on the direction of the reaction; hydration of fumarate or dehydration of malate. For the alkaline conditions Massey[11] suggests that the enzyme undergoes a dissociation into smaller units as the temperature increases and the smaller units have a higher activation energy. If this is indeed the case, the two different forms of the enzyme would constitute the two phases described by Kumamoto *et al.*[13] The differences in the temperature at which the change occurs could be due to the influence of pH and substrate on the temperature-induced dissociation process. There is no satisfactory explanation of the increase in activation energy observed at 18° for the hydration of fumarate by fumarase at pH 6·35. This change in activation energy occurs at the same temperature as a change in K_m and because it occurs only in the forward reaction, was assumed to be associated with an effect on the orientation of the water involved in the reaction. In view of the observation that the affinity for inhibitors also shows a marked change in the same temperature range[17] suggests that some change occurs at this particular temperature in the structure of fumarase in the acid medium which affects the binding of substrate and inhibitors. Such changes, although inexplicable at present, obviously have the isothermal properties of the phase change discussed by Kumamto *et al.*[13] since only sharp changes are observed in the Arrhenius plots of both velocity and K_m.

The similarity between the temperature at which changes in E are observed and the temperature at which changes are observed in the structure of water adjacent to an interface, led Drost-Hansen[8] to postulate that many of the thermal anomalies in enzyme kinetics can be explained by an alteration in the physical properties of vicinal water. He stresses that these thermal anomalies occur at or near the same temperature for all systems regardless of the chemical nature of the substrates involved. The temperatures mentioned were 17°, 28°, 43° and 60°. However, as stated by Drost-Hansen,[8] changes in vicinal water must be superimposed upon the changes in the biological macromolecules, and these will undoubtedly be influenced by environmental changes. Thus for enzyme proteins an alteration in configuration reflecting structural changes in vicinal water might occur at any temperature depending upon factors such as pH and ionic strength of the medium. It is possible that changes in the structure of vicinal water might influence the conformation of the active site of an enzyme and consequently the rate constant of the reaction catalysed. This might explain the discontinuity in the Arrhenius plots of invertase, lipase, trypsin and pepsin at 0°. However, in none of the examples of discontinuities in Arrhenius plots quoted by Drost-Hansen,[8] in support of this hypothesis, has a change in vicinal water been detected.

B. *Enzymes and Multi-enzyme Systems Associated with Membranes*

(i) *Respiratory enzymes associated with mitochondrial membranes*. A major difference between the respiratory enzyme system of mitochondria derived from chilling-sensitive plants and homeothermic animals on the one hand and chilling-resistant plants and poikilothermic animals on the other is the effect of temperature on these enzyme systems.[6, 14]

Arrhenius plots of the succinate oxidase system of mitochondria from chilling-sensitive plants and homeothermic animals show a sharp discontinuity in the region of 10° to 12° and 23° respectively. A discontinuity at 17° has also been reported for succinate oxidase activity of rat-liver mitochondria[18] and at 19° for the same enzyme in a Keilin and Hartree, heart-muscle preparation from pig.[19] The same enzyme system from chilling-resistant plants[6] and poikilothermic animals[14] shows a constant activation energy over the temperature range of 0° to 25° and 0° to 37° respectively. This difference in response to temperature of the same enzyme system from the two different types of tissues is shown in Figs. 2 and 3.

Earlier studies of the effect of temperature on the kinetics of enzymes indicate that the activation energy of a particular enzyme is the same

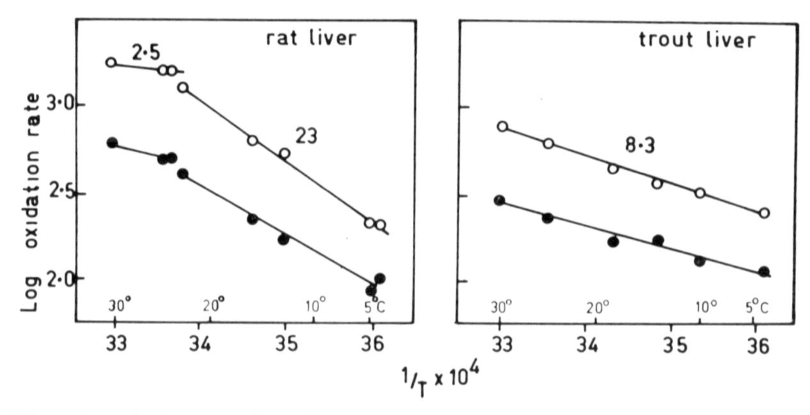

Figure 2. Arrhenius plots of state 3 (○) and state 4 (●) succinate oxidation of mitochondria from rat and trout liver. (From Lyons and Raison.[14])

regardless of the source of the enzyme and is essentially independent of environmental changes unless these factors alter the catalytic surface of the enzyme.[4] Much of the recent kinetic data on membrane-associated enzymes does, in part, confirm the earlier observations except for the significant differences between the succinate oxidase system of the two groups of the plants (chilling-resistant and chilling-sensitive) and animals (homeothermic and poikilothermic) mentioned previously.

Changes in the value of activation energy with temperature for membrane-associated systems are not confined to the succinate oxidase system. A similar discontinuity has been observed in the Arrhenius plots for the oxidation of reduced NAD (plant mitochondria only), β-hydroxybutyrate and α-oxoglutarate, the succinate: phenazine methosulphate oxidoreductase and the cytochrome *c* oxidase system measured with *N,N,N′,N′*-tetramethyl-*p*-phenylenediamine (TMPD) and ascorbate.[7] The discontinuity in the Arrhenius plots of the oxidative enzymes from the respective sensitive tissues is also apparent for both state 3 and state 4 respiration based on succinate oxidation (see Figs. 2 and 3), and with mitochondria disrupted by hypotonic swelling and by sonic disintegration.[7] Treatment of mitochondria with low concentrations of any of several different detergents abolishes the "break" or discontinuity in the Arrhenius plots resulting in an increase in the activation energy in the higher temperature range and a decrease in the low temperature range.[7]

It is of interest to note differences in the kinetic behaviour of the succinate oxidase system of liver mitochondria from the hibernating ground squirrel, *Citellus lateralis*. Mitochondria isolated from an active

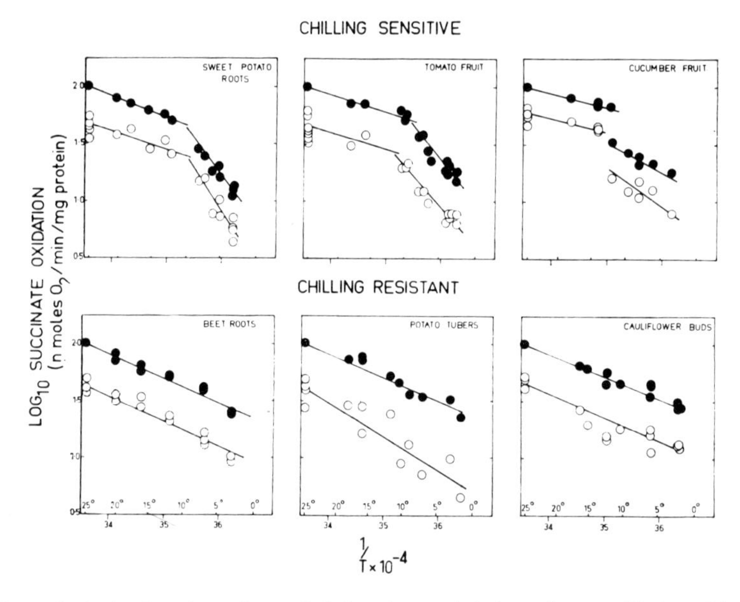

Figure 3. Arrhenius plots of state 3 (●) and state 4 (○) succinate oxidation of isolated plant mitochondria. (From Lyons and Raison.[6])

squirrel show a discontinuous Arrhenius plot for succinate oxidase activity at between 25° and 21°. However, when the mitochondria were isolated from a hibernating animal (body temperature 4°) no discontinuity is observed in the Arrhenius plot.[20] Thus the same enzyme system in the same species of animal can exhibit different temperature-dependent kinetics depending on the physiological state of the tissue. In one experiment involving one hibernating squirrel, the straight line Arrhenius plot observed with mitochondria 30 min after isolation, showed a slight discontinuity when assayed again 90 min later and a distinct discontinuity after a further 120 min.[20] This reversion from a continuous Arrhenius plot to a discontinuous plot involves a decrease in the activation energy above 25° and an increase in activation energy below this temperature. These changes in activation energy above and below the transition temperature are in the opposite direction to the changes observed in the activation energies of succinate oxidase when rat-liver mitochondria were treated with detergent.[7] However, since these changes observed with the mitochondria from the hibernating squirrel occur above and below the transition temperature it is considered to be the result of a change in the physical properties of the mitochondrial membrane; a change to a more rigid state in the low temperature range and a more flexible state in the higher temperature range.[20]

In terms of a temperature-induced phase change it is unlikely that each of the enzymes of the respiratory system of mitochondria from sensitive tissues would undergo a conformational change at the same temperature in contrast to the same enzymes of mitochondria from resistant tissues. Since mild treatment with detergent abolishes the discontinuity in the Arrhenius plots of enzymes from sensitive tissues, but does not alter the activation energy of the same enzymes from resistant tissues, the conformational change in the enzyme protein is considered to be mediated through a phase change in the membrane lipids.[7] It should be noted, however, that a soluble succinate dehydrogenase from pig heart, activated by incubation with succinate, shows a discontinuity in an Arrhenius plot at about 18° the same temperature at which a discontinuity is observed for the succinate oxidase activity of the particular heart-muscle preparation.[19] The non-activated, soluble succinate dehydrogenase shows two points of discontinuities, at 27° and 18°. The changes in activation energy are considered to be due to conformational changes in the enzyme which alter the "catalytic-centre activity" rather than the activation enthalpy.[19] In the absence of any analytical data indicating the possibility of lipid associated with the soluble dehydrogenase it is not possible to relate these changes in activation energy to a phase change in a lipid component. However, as outlined later, some soluble preparations of succinate dehydrogenase do contain associated lipid and if present in the succinate dehydrogenase from pig heart, would provide an alternative explanation for the changes in activation energy.

Phase transitions between the so-called lamella and hexagonal phases of brain phospholipid dispersed in water have been demonstrated to occur at 23°.[21] In addition, Chapman *et al.*[22] have shown that the temperature at which a particular phase can exist for a given phospholipid will depend on the melting temperature of the hydrocarbon chains. Furthermore Steim *et al.*[19, 22] have demonstrated by differential scanning calorimetry that the temperature of the phase transitions in the membranes of *Mycoplasma laidlawii*, and the lipids extracted from these membranes, is dependent on the proportions of saturated and unsaturated fatty acids present. In relating chilling sensitivity in plants to lipid composition of mitochondrial membranes, Lyons *et al.*[24]

showed a higher proportion of saturated fatty acids in the sensitive plants than in resistant plants, and also showed that a change of as little as 5% in the amount of unsaturated fatty acids in artificial mixtures of fatty acids, which approximated the composition of plant mitochondrial membranes, would change the temperature of solidification of the lipid mixtures by up to 15°.[25] A higher proportion of unsaturated fatty acids has also been reported in the mitochondria of poikilothermic animals compared to homeothermic animals.[26,27] There is thus a well defined correlation between the maintenance of a fluid state in the hydrocarbon chains in membrane lipids by increasing the degree of unsaturation, and the environmental temperature at which the mitochondria of the particular plant or animal species functions. A more direct correlation between the physical changes of membrane lipids and the change in activation energy of enzymes associated with the membrane is obtained by direct measurement of the molecular motion of spin-labelled analogues of fatty acids associated with the membrane lipids. These data clearly show a lipid phase transition at approximately 12° for mitochondria from sweet potato (chilling-sensitive) and at 23° for rat-liver mitochondria.[28] These phase transitions coincide precisely with the temperatures at which the change in activation energy of succinate oxidase is observed in the respective tissue. No phase change is observed in the membrane lipids of potato mitochondria (chilling-resistant) or trout-liver mitochondria (poikilothermic).[28] The phospholipids extracted from the membranes of the temperature sensitive mitochondria also exhibit a phase change at the same temperature as the intact mitochondria.[28]

In physical terms the phase change inferred from the abrupt change in motion of a spin-labelled fatty acid represents a sharp transition of the phospholipids of the membrane, from what Luzzati *et al.*[21] term a liquid-crystalline structure to a cogel, as the temperature is decreased below the critical temperature. Since the enzymes associated with the mitochondrial membranes undergo a phase change (as indicated by the change in activation energy) at precisely the same temperature, the perturbation associated with the liquid-crystalline to cogel transition is assumed to induce a conformational change in the active centre of the enzyme proteins. This conformational change would thus represent the phase change described by Kumamoto *et al.*[13] and allow for the exclusive functioning of the two forms of the enzyme above and below the transition temperature.

The Mg^{2+}-activated ATPase of the inner membrane of rat-liver mitochondria also shows a discontinuity in an Arrhenius plot at approximately 23°, as shown in Fig. 4. Kemp *et al.*[18] have observed a discontinuity for the same enzyme, assayed in the presence or the absence of uncouplers, at 17°, the same temperature at which these workers observed a discontinuity for succinate oxidase activity. This difference in the temperature of the discontinuity of both the ATPase and succinate oxidase activity in the two studies probably reflects slight differences in either the fatty acid composition or the degree of unsaturation of the fatty acids of the mitochondrial membrane lipids; both vary depending on the diet of the animal.[30] It is significant that the discontinuity for ATPase and for succinate oxidase occurs at the same temperature in the two independent studies. Further investigation of the ATPase system in this laboratory has shown that the discontinuity is only apparent for the oligomycin-sensitive reaction (see Fig. 4) and can be completely abolished by treating mitochondria with low

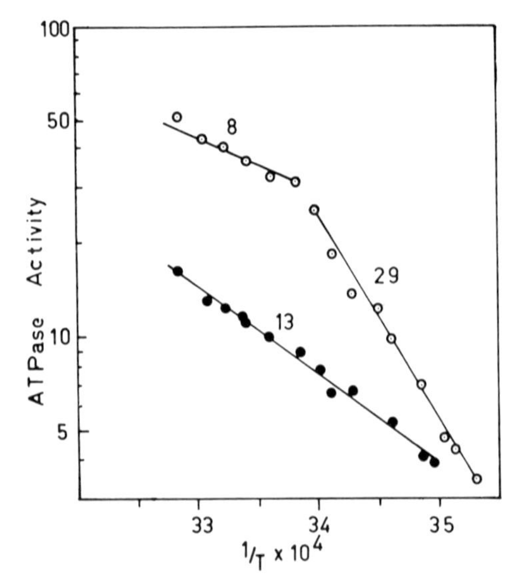

Figure 4. Arrhenius plot of the Mg^{2+}-activated ATPase of rat liver mitochondria assayed in the presence (●) and absence (○) of oligomycin. (Data of McMurchie, Raison and Cairncross.[29])

concentrations of detergent,[29] similar to the abolition of the discontinuity for succinate oxidase activity after treating mitochondria with detergent.[7] Both these observations are explicable in terms of the known structural organization of ATPase involving phospholipid and protein.[31] Oligomycin sensitivity can be abolished by even partial hydrolysis of the phospholipid with phospholipase A.[32] The coincidence of the loss of phospholipid, as a structural component of the ATPase, the loss of oligomycin sensitivity and the abolition of the discontinuity in the Arrhenius plot of activity, further enhances the concept of the induction of a conformational change in the enzyme protein through a temperature-induced phase change in the associated lipid components.

Data available regarding the temperature-dependent kinetics of the adenine nucleotide translocase of the inner membrane of rat-liver mitochondria suggest that changes in the conformation of some membrane-associated enzymes occur at temperatures differing from the temperature of the bulk lipid phase change. The temperature response of this enzyme measured over a range of 0°–20° shows a discontinuity in the Arrhenius plot at 9°, when nucleotide exchange was measured with ATP, and at 8° with ADP.[33] The respiratory enzymes,[14] the oligomycin-sensitive ATPase[29] and the lipids of the inner membrane of mitochondria[28] all exhibit major changes at 22° to 23°. It is not at present possible to relate the change in kinetics of the translocase at 8° or 9° to temperature-induced phase changes in membrane lipids as no abrupt changes in the mobility of spin-labelled fatty acids have been detected at this temperature.[28] If a phase change in a lipid component is involved in the change in activation energy of the translocase, this lipid component is apparently not influenced by changes in the physical properties of the bulk of the membrane lipids. The change in activation energy of the translocase at 8° to 9° might be an example of a temperature-induced conformational change in the enzyme protein, independent of a change in the physical properties of membrane lipids, similar to the changes observed for myosin acting on ITP or on ADP in the presence of 2,4-dinitrophenol (DNP) or *p*-chloromercuribenzoate (PCMB)[34] (see later). The slight difference in the temperature of the discontinuity in the Arrhenius plot of translocase activity, depending on the substrate, further demonstrates a similarity with the myosin system. Further studies of the physical properties of the translocase, particularly those influenced by temperature, and its association with lipids are required before a more definite explanation can be advanced to account for the discontinuity at 8° to 9°.

The nucleotide translocase is considered by some workers to be the rate limiting step in coupled-oxidation reactions by mitochondria.[18, 35] If this were so, the activation energy of state 3 respiration could never be less than the activation energy of the translocase since in any multi-enzyme process the activation energy observed must be that of the rate-limiting event. The activation energy of the nucleotide translocase of rat-liver mitochondria between 8° and 20° is 21 kcal/mole.[34] The activation energy for succinate-dependent, state 3 respiration of rat-liver mitochondria, in approximately the same temperature range, is 18·8[14] to 23[18] kcal/mole, which is not significantly different from the activation energy of the translocase. However, an Arrhenius plot of state 3 respiration of rat-liver mitochondria exhibits a sharp discontinuity at 23° (cf. Figs. 1 and 2) as determined by Lyons and Raison[14] or at 17° as determined by Kemp *et al.*[18] Above these temperatures the activation energy of succinate oxidation is considerably less, 2·5 to 8·9 kcal/mole.[13, 14, 18] Comparison of the activation energies of the translocase and succinate oxidation in state 3 would, therefore, confirm that the translocase could be the rate limiting reaction in coupled respiration in the temperature range of 0° to 17° or 0° to 23°. However, above these temperatures, that is in the physiological temperature range, the translocase must also decrease its activation energy if it is the rate-limiting step, since the activation energies reported for state 3 respiration, which involves the translocase, are considerably less in this temperature range. Because of the difficulties in estimating translocase activity of rat-liver mitochondria at elevated temperatures, clarification of a possible change in activation of the translocase activity at the temperature of the phase change in the membrane lipid could best be obtained from an investigation of this enzyme in mitochondria from chilling-sensitive plants. Mitochondria from these tissues show a lipid phase change at 10° to 12°[28] and a similar decrease in the activation energy of state 3 respiration above the transition temperatures as shown by the succinate-dependent respiration of rat-liver mitochondria.[6] In addition a comparison of the temperature dependence of the translocase with mitochondria from chilling-sensitive and resistant plants would also confirm whether the temperature-induced phase change in the lipids of the membrane influences the activation energy of the translocase.

(ii) *Enzymes associated with the endoplasmic reticulum and plasma membrane.* Similar conclusions regarding the dependence of the discontinuity in Arrhenius plots on temperature-induced changes in the physical properties of lipids associated with the enzyme can be inferred from kinetic studies of the Na^+ plus K^+-activated and the Mg^{2+}-activated ATPases of plasma and endoplasmic reticulum membranes of homeothermic animals.[18,29,36–38] Each of these enzymes exhibits a discontinuity in the Arrhenius plot between 18° and 20°, depending on the source and type of ATPase. For the Mg^{2+}-ATPase of the microsomal fraction of rat brain, an additional discontinuity is found at 6°.[36] For the ouabain-sensitive Na^+ plus K^+-activated ATPase the discontinuity is apparent for both the sensitive and insensitive reactions[38] although the sensitivity to ouabain diminishes as the temperature is reduced and there is no detectable ouabain-sensitive activity below about 5°.[38]

The association of these ATPases with membranes suggests that the configurational changes in the enzyme protein, which are assumed to occur at the transition temperature, are also induced by a phase change in the membrane lipids. This view is supported by the observation of a change in the molecular mobility of a spin-labelled fatty acid in the lipid region of a preparation of rat-liver endoplasmic reticulum at 22° to 23°, as shown in Fig. 5. The abolition of the break in the Arrhenius plot of the Na^+ plus K^+-activated ATPase by treatment with detergent[29] also suggests that changes in the physical properties of the membrane lipids are involved in the temperature-induced configurational changes of the enzyme protein.

These observations do not exclude the possibility of a direct, temperature-mediated configurational change in the enzyme protein since there is evidence which suggests such changes may occur with purified preparations of myosin[34] and the nucleotide translocase discussed above. The kinetics of hydrolysis of nucleotide triphosphates by myosin exhibits a sharp discontinuity in an Arrhenius plot at approximately 15° when the substrate is ITP or ATP plus either DNP or PCMB; ATP alone showed a constant activation energy in the same temperature range. Levy *et al.*[34] concluded from this study that portions of the myosin molecule, including the active site, can undergo a temperature-

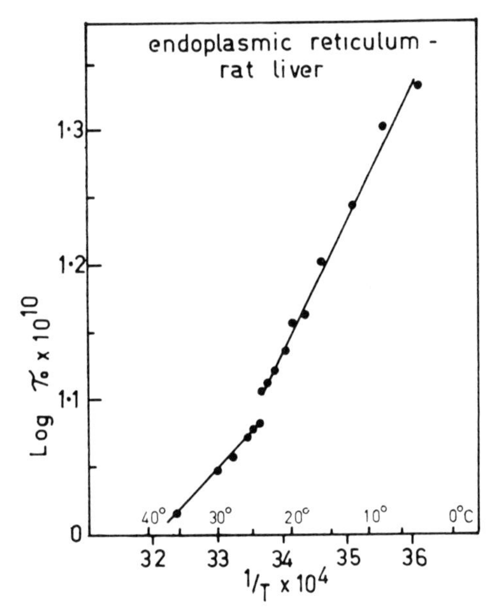

Figure 5. Arrhenius plot of the change in rotational correlation time (T_0) of the spin-label, 12 nitroxide stearic acid, in a preparation of rat-liver endoplasmic reticulum. (Data of McMurchie, Raison and Cairncross.[29])

induced conformational change, but these changes can be modified by interaction of specific groups on the protein with particular substrates or agents such as ATP, DNP or PCMB. In the binding of enzyme and substrate, the 6-amino group of the purine ring of ATP has a greater affinity for the active site than the 6-hydroxy group of ITP and evidently stabilizes the enzyme to changes in temperature since no discontinuity is observed in Arrhenius plots of activity when ATP is the substrate.[34] Cooperative transformations of enzyme protein by temperature are well known processes and the effect of the electrostatic interaction of charged groups on substrates and of compounds which either oxidize or reduce specific groups involved in cross-linking have been well documented.[1] It is of interest to note the difference between the activation energy for the reaction of myosin involving ATP alone and the reactions with ATP plus DNP or PCMB. Above the transition temperature of 15° the activation energy for hydrolysis of ATP with or without DNP or PCMB is 12 kcal/mole. Below 15° there is no change in activation energy with ATP as substrate, but with ATP plus either DNP or PCMB the activation energy increases to 25 kcal/mole.[34] This indicates that the binding of DNP and PCMB alters the tertiary structure of the ATPase below the transition temperature, but has no effect above this temperature. This is not usually the case with enzymes associated with membranes. If the activation energy of succinate oxidase or Na^+ plus K^+, activated ATPase is altered by modifying the lipid components of the membrane with detergent, the activation energy above the transition increases while the activation energy below the transition temperature decreases.[7]

A water-soluble, ouabain-sensitive, Na^+ plus K^+-dependent ATPase,[39] prepared from NaI-treated microsomal fraction by extraction with a detergent (Lubrol W) might provide an ideal system for studying the influence of membrane association on temperature-induced conformational changes of protein. The water-soluble enzyme consists of twelve subunits arranged as three tetramers. Incubation at 37° rapidly inactivates the ouabain-sensitive reaction and causes dissociation of subunits. The resulting monomers show ouabain-insensitive activity.[39] The water-soluble enzyme can also be incorporated into ATPase depleted membranes with enhancement of activity, particularly when incorporated in the presence of phosphatidyl-L-serine. It would be of interest to determine if the water soluble tetrameric form of the enzyme exhibits a discontinuity in an Arrhenius plot.

The Ca^{2+}-dependent ATPase of sarcoplasmic reticulum also shows a discontinuity in an Arrhenius plot at 10° when assayed with a relatively high concentration (5 mM) of ATP.[40] Previous studies of the temperature dependence of this enzyme, using lower concentrations of substrate (1 mM) have shown a constant activation energy over a temperature range of 5° to 20°[41] and 0° to 37°.[42] Isolated preparations of sarcoplasmic reticulum form closed vesicles, which in the presence of an energy source, such as ATP, will accumulate Ca^{2+} with a concurrent hydrolysis of ATP by the Ca^{2+}-dependent ATPase,[43] thus indicating that Ca^{2+} transport by sarcoplasmic reticulum is linked to Ca^{2+}-activated ATPase. Additional studies with pharmacological agents have shown that while both processes can be inhibited to the same extent by mersalyic acid (Salyngan), quinidine and procaine both selectively inhibit Ca^{2+} uptake to a greater extent than the Ca^{2+}-ATPase.[40] Evidence for the view that Ca^{2+} uptake and the Ca^{2+}-dependent hydrolysis of ATP are both catalysed by the one enzyme was thus contradictory. A reassessment of the effect of temperature on the activation energy of the two processes has shown significant differences which led Charnock and Frankel[40] to suggest that the sites of Ca^{2+} uptake and Ca^{2+}-ATPase are not identical and the uptake process does not necessarily utilize energy from the hydrolysis of ATP by the Ca^{2+}-dependent ATPase. Their data clearly show the Ca^{2+} uptake process to have a constant activation energy of 18·2 kcal/mole over a temperature range of 0° to 37° in contrast with the increase in activation energy shown by the Ca^{2+}-ATPase at 10°. In this reaction the activation energy above 10° was 16·3 kcal/mole, not significantly different from the Ca^{2+} uptake process, but below 10° the activation energy increased to 33·4 kcal/mole.[40] Whilst the interpretation of this data in terms of evidence in support of two separate sites of Ca^{2+} binding is open to question, in view of the fact that the activation energies of both processes, in the range of physiological temperatures, are not significantly different, if does indicate how a study of the effect of temperature may be utilized to differentiate between two apparently interdependent processes.

It is also of interest to note the different transition temperatures observed for the Ca^{2+}-dependent ATPase (10°) compared to the Na^+ plus K^+-activated ATPase (20° for rabbit sarcoplasmic reticulum[38] and 22° for rabbit heart and endoplasmic reticulum fraction[29]). If the discontinuity shown by Ca^{2+}-dependent ATPase[30] at 10°, is related to a phase change in a lipid component of the membrane, as is evident for the Na^+ plus K^+-ATPase, then the Ca^{2+}-ATPase and associated lipid must occupy a discreet region of the membrane where the lipid component is not influenced by changes in the physical properties of the bulk of membrane lipids. An alternative explanation for the change at 10° shown by the Ca^{2+}-dependent ATPase is that the protein is susceptible to a temperature-induced conformational change at 10°.

Physiological Processes Related to Enzymes Associated with Endoplasmic Reticulum and Plasma Membranes

Arrhenius plots of a variety of physiological processes involving a number of enzymes,[1] show the processes have a constant activation energy. In multicomponent processes such as these the activation energy observed from the Arrhenius plot represents the activation energy of the rate limiting process. Thus, in view of the recent data a discontinuity would be expected if a membrane-associated enzyme is the rate limiting step and this enzyme exhibits anomalous temperature behaviour. In a study of the effect of temperature on the beat rate of isolated rat heart,[29] a change in the slope of the Arrhenius plot is observed at approximately 19°. Heart beat ceases at 10°, but is detected again when the temperature is increased, thus demonstrating the reversibility of the process. The change at 19° probably corresponds with the change in activation energy of Na^+ plus K^+-activated ATPase of heart endoplasmic reticulum and plasma membrane,[29] a reaction which regulates the ion balance and thus controls the propagation of an action potential on which beat rate is dependent. The complete cessation of heart beat at 10°, which is reversible, could be a reflection of the change in activation energy of Ca^{2+}-activated ATPase reported by Charnock *et al.*[40] since muscle contraction and relaxation is dependent upon the activity of this enzyme. Although a large number of enzyme reactions are implicated in determining the rate of heart beat the large apparent activation energy of the overall process (60 to 70 kcal/mole) below 19° is the result of changes in the activation energy of the rate limiting processes. However, it is conceivable that changes in the permeability of ions due to a phase change in the membrane lipids and/or the decrease in energy production as a result of the increase in activation energy of state 3 respiration of heart mitochondria[14] could impose additional restrictions on the rate limiting step and thus increase the apparent activation energy of the overall process. Regardless of which particular enzyme reaction becomes rate-limiting below the transition temperature, it appears that the most significant event which would restrict the rate of muscular contraction of homeothermic animals is the temperature-induced phase change in the membrane lipids at about 22°. In poikilothermic animals there is no phase change in the membrane lipids, no change in the activation energy of the membrane ATPases[29] or oxidative reactions on which energy production depends[14] and heart beat in these animals show a constant apparent activation energy over a temperature range of 2° to 30°.[44] The hibernating ground squirrel, *Citellus mohavensis*, provides an interesting contrast to other homeothermic animals. Although the squirrel appears to have a number of characteristics of a homeothermic animal during the active summer months, its heart beat shows the typical behaviour of a poikilothermic animal; the heart beat rate decreases with decreasing temperature, but the apparent activation energy is constant.[45]

From the earlier observations of an association of ribosomes with membranes of the endoplasmic reticulum by Palade and Siekevitz[46] protein synthesis has been considered as a membrane-associated process. Since the lipids of the endoplasmic reticulum of liver cells of homeothermic animals have been shown to undergo a phase change at approximately 22°, it is considered likely that the activity of the associated protein synthetic system might reflect this change in the physical properties of the membrane. Preliminary data has shown that the kinetics of incorporation of amino acids by a microsomal fraction from rat liver change dramatically at 23° resulting in a substantial increase in the activation energy of the process, as shown in Fig. 6. The fact that the change in activation energy occurs at 23° is strong evidence for implying that the change in kinetic properties of the enzymes is induced by a phase change in the membrane lipids. This view is further substantiated by the observation that the incorporation of amino acids catalysed by a membrane-free preparation of ribosomes, from the cytoplasm of rat liver cells does not show a change in activation energy within the temperature range of 2° to 30°.[45] A similar increase in activation energy for the incorporation of amino acids, and hence the protein synthetic system, probably occurs in the cells of chilling-sensitive plants, at 10° to 12°, since similar phase changes have been observed in membranes of chilling-sensitive plant tissue at these temperatures. The different effects of temperature on the membrane-bound

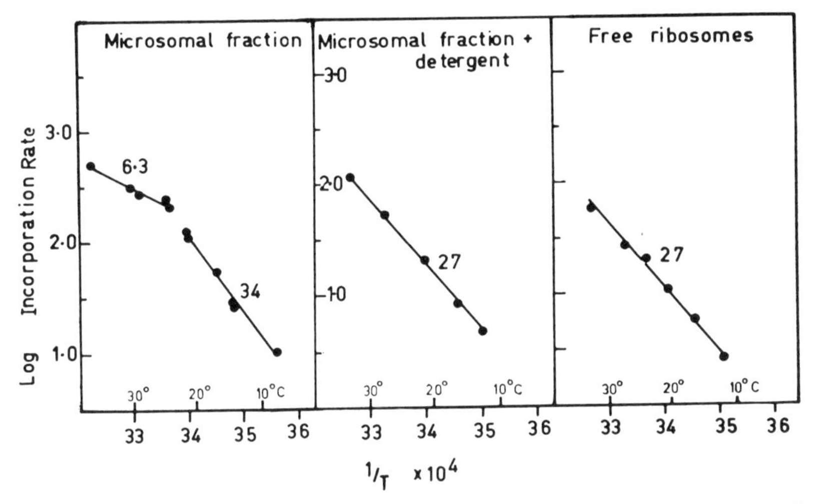

Figure 6. The effect of temperature on the activation energy of ^{14}C-leucine incorporation by a microsomal fraction and free ribosomes of rat-liver cells. (Data from Towers, Raison, Kellerman and Linnane.[47])

and free ribosome incorporating systems could be a useful method of determining whether the ribosomes of a particular incorporating system are associated with membranes, for example in organelles. The amino acid incorporation activity of rat-liver mitochondria has been found to exhibit an increase in activation energy as the temperature is decreased below 23° suggesting that the majority of the mitochondrial ribosomes are functionally associated with the mitochondrial inner membrane system.[47]

Recent evidence has shown that DNA-like RNA (D-RNA) nuclear particles, which contain messenger RNA (m-RNA), move from the nucleus to the cytoplasm of Krebs tumour cells, and become attached to cytoplasmic membranes.[48] Ribosomes attach to the bound m-RNA to form the translation complex.[48] The change in activation energy of the protein synthetic system might therefore be a reflection of a change in the conformation of m-RNA induced by changes in the surface properties of the membrane.

Enzymes Associated with the Photosynthetic-Electron Transfer System of Chloroplast Membranes

The photosynthetic capacity of some species of plants, classified by Chen *et al.*[49] as "efficient", show a sudden decrease in activity when the temperature is reduced below about 12°. The "efficient" plants were so characterized because of their high photosynthetic capacity, low CO_2 compensation concentration and the presence of the C_4-dicarboxylic acid cycle and are mostly chilling-sensitive type plants of tropical

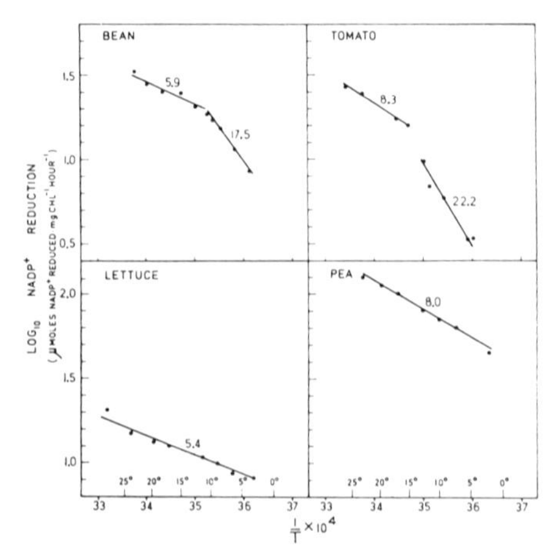

Figure 7. Arrhenius plots of the rate of NADP reduction from water for chloroplasts from chilling-sensitive (bean and tomato) and chilling-resistant (lettuce and pea) plants. The numbers beside each line are the activation energy in kcal/mole. (Data from Shneyour, Raison and Smillie.[50])

origin. Recent studies in this laboratory have shown that the decrease in the photosynthetic capacity of chilling-sensitive plants at temperatures below approximately 12° can be directly related to a change in activation energy of membrane-associated enzymes of the electron transfer system.[48]

Arrhenius plots of the rate of photoreduction of NADP from water for chloroplasts from chilling-sensitive plants (bean and tomato, Fig. 7) show a discontinuity at between 10° to 12°, similar to the discontinuity in Arrhenius plots of respiratory activity of mitochondria from some chilling-sensitive plants.[6] The increase in activation energy for the process below the transition temperature is usually in the order of three fold and is observed with chloroplasts from all the chilling-sensitive plants examined. No change in activation energy is observed with chloroplasts of chilling-resistant plants, e.g. lettuce and pea (cf.

Fig. 7). The electron transport systems of both mitochondria and chloroplasts of chilling-sensitive plants thus show similar temperature-dependent changes at the same temperature and contrast with the absence of these changes in either organelle of chilling-resistant plants.

As described previously, the distinction between the temperature response of mitochondria from chilling-sensitive and chilling-resistant plants can be directly related to the proportion of saturated and unsaturated fatty acids in the lipids of the mitochondrial membranes.[6] The same relationship, however, does not hold for chloroplast lipids of chilling-sensitive and chilling-resistant plants. Chloroplasts contain relatively high proportions of galactosyl dilinolenins (80% of total lamella lipid), and about 10% each of phosphatidy glycerol and sulphoquinovosyl diglyceride, both of which contain a linolenic or linoleic

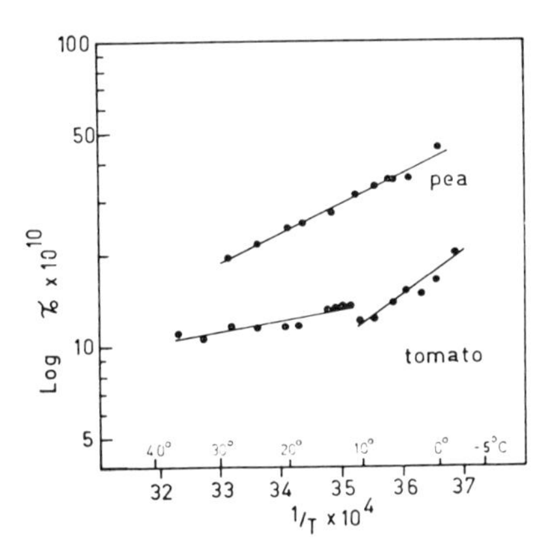

Figure 8. Arrhenius plot of the change in rotational correlation time (T_0) of the spin-label 12-nitroxide stearic acid in preparations of chloroplasts from pea and tomato leaves. (J. Raison, unpublished data.)

acid residue.[51] Based on the high content of linolenic acid in chloroplasts, the freezing point of the bulk lipids would probably be well below 10°.[25] However, physical measurements of the mobility of lipids in the chloroplast membranes by e.s.r. spectroscopy of spin-labelled fatty acids show that a phase transition occurs at approximately 10° for chloroplasts isolated from chilling-sensitive tissues (see Fig. 8). No phase change is observed in chloroplasts of chilling-resistant plants at this temperature. Thus, although a phase change in the lipids of chloroplast membranes of chilling-sensitive plants would not be predicted, considering the physical properties of the constituent fatty acids, the e.s.r. data clearly show a phase transition occurs at the same temperature as the change in activation energy of the photosynthetic electron transport enzymes and the same temperature as the phase changes in mitochondrial membrane lipids. There is little doubt that the change in activation energy of the electron transport system is a consequence of a phase change in a lipid component of the membrane, but there is insufficient data at present to predict what proportion of the total lipids is involved in the phase change and what other enzymes are affected. The increase in the activation energy of photochemical activity at the same temperature as the increase in activation energy of respiratory enzymes, provides a biochemical basis for understanding the changes in carbon metabolism,[52] photosynthetic activity[53] and morphology of chloroplasts[54] when chilling-sensitive plants are exposed to temperatures below the transition point.

Structure and Function of Membranes in Relation to Temperature-Induced Phase Changes

Various structure-function relationships have been demonstrated with membrane systems, particularly the cristae membrane of mitochondria.[55–57] The coincidence of the phase transition in the membrane lipids and the conformational change in the enzyme protein at the same temperature in a number of different membranes of plant and animal tissues provide an excellent example of this relationship, and raises important questions concerning the molecular structure of these membranes. The available data establishes membrane lipids as the locus of temperature sensitivity, but does not provide any direct evidence to indicate whether all of the lipids of the membrane undergo the phase transition or only particular or specific lipids associated with enzymes.

The correlation between the degree of unsaturation of the fatty acids of the bulk membrane lipids and the temperature at which the organisms normally function has been well established for plant and animal mitochondria,[6, 14] *Mycoplasma laidlawii*,[23] bacteria[58, 59] and fungi.[60] This would suggest that the fatty acid composition of membrane lipids is regulated to maintain the bulk membrane lipids in a similar physical state at the various growth temperatures and is understandable in view of the large increase in the activation energy of membrane-associated enzymes below the temperature of the phase change. Furthermore, both the calorimetric and spin-label method of detecting the phase change have demonstrated similar temperature-dependent, physical properties for the lipids in particulate membrane preparations and in lipid micelles formed from the lipids extracted from the membranes.[23,28] In the case of the spin-label method, the temperature of the phase change coincides precisely with the temperature at which all of the oxidative enzymes of the mitochondrial, inner membrane show a change in activation energy. The available evidence would therefore favour the view that the kinetic properties of the membrane-associated enzymes are regulated by the physical state of the bulk membrane lipids. In view of the results obtained it can be assumed that specific phospholipids which are known to be associated with succinate dehydrogenase,[61] cytochrome oxidase[62] and mitochondrial ATPase,[32] for example form an integral part of the bulk membrane lipids and are indistinguishable from the bulk lipids by the physical methods used to detect phase changes.

The structural proteins of membranes do not appear to influence the temperature at which the membrane lipids undergo a phase change which would suggest only a small proportion of the membrane lipids are involved in hydrophobic interactions with protein. This view is supported by the fact that the temperature of the phase change in mitochondrial membranes is the same regardless of whether the protein is heat denatured or in a native state and coincides with the temperature of the phase change in extracted lipids.[28] Furthermore, a comparison of the heat of transition for the lipids of *M. laidlawii* membranes with that for the extracted lipids indicates that 90% of the membrane lipids are in a bilayer configuration and only 10% are involved in hydrophobic interactions, presumably with protein.[63]

Since the change in the physical state of the membrane lipids must be transmitted to the membrane associated enzymes, to bring about the conformational change associated with the change in activation energy, there must be some structural association between the enzyme proteins and the bulk lipids. The specific phospholipids known to be associated with some membrane-associated enzymes might provide this link.

Another interesting question to arise from the relationship between the composition of membrane lipids and the minimum temperature for normal physiological functioning of the organism, is whether this temperature can be reduced by lowering the "freezing point" of the membrane lipids. The implications of success in being able to manipulate the membrane composition and reduce the minimum functional temperature would be of immense value to agriculture and medicine.

Acknowledgement

I wish to express my appreciation to Dr. J. Kumamoto, University of California, Riverside, California, for suggesting the thermodynamic implications outlined here, and to my colleagues, Dr. D. Graham and Mr. L. Campbell for comments and criticisms in the preparation of the manuscript.

References

1. F. H. Johnson, H. Eyring and M. J. Polissar, *The Kinetic Basis of Molecular Biology*, John Wiley and Sons, Inc., New York, 1954, p. 187.
2. E. A. Dawes, *Quantitative Problems in Biochemistry*, E. and S. Livingstone, Ltd., Edinburgh, 1956.
3. H. G. Bray and K. White, *Kinetics and Thermodynamics in Biochemistry*, Academic Press, New York, 1957.
4. I. W. Sizer, *Adv. in Enzymol.*, **3** (1943) 35.

5. M. Dixon and E. C. Webb, *Enzymes*, Longmans, Green and Co., London, 1958, p. 150.
6. J. M. Lyons and J. K. Raison, *Plant Physiol.*, **45** (1970) 386.
7. J. K. Raison, J. M. Lyons and W. W. Thomson, *Arch. Biochem. Biophys.*, **142** (1970) 83.
8. W. Drost-Hansen, in: *Chemistry of the Cell Interface*, Part B, H. D. Brown (ed.), Academic Press Inc., New York, 1972.
9. S. Arrhenius, *Quantitative Laws in Biological Chemistry*, G. Bell & Sons Ltd., London, 1915.
10. J. Bělehrádek, *Temperature and Living Matter Protoplasm Monographien*, Vol. 8, Gebrüder Borntraeger, Berlin, 1935, p. 50.
11. V. Massey, *Biochem. J.*, **53** (1953) 72.
12. J. Bělehrádek, *Ann. Rev. Physiol.*, **19** (1957) 59.
13. J. Kumamoto, J. K. Raison and J. M. Lyons, *J. Theoret. Biol.*, **31** (1971) 47.
14. J. M. Lyons and J. K. Raison, *Comp. Biochem. Physiol.*, **37** (1970) 405.
15. I. W. Sizer and E. S. Josephson, *Food Res.*, **7** (1942) 201.
16. A. K. Balls and H. Lineweaver, *Food Res.*, **3** (1938) 57.
17. V. Massey, *Biochem. J.*, **55** (1953) 172.
18. A. Kemp, G. S. P. Groot and H. J. Reitsma, *Biochim. Biophys. Acta*, **180** (1969) 28.
19. W. P. Zeylemaker, H. Jansen, C. Veeger and E. C. Slater, *Biochim. Biophys. Acta*, **242** (1971) 14.
20. J. K. Raison and J. M. Lyons, *Proc. Natn. Acad. Sci. USA*, **68** (1971) 2092.
21. V. Luzzati and F. Husson, *J. Cell Biol.*, **12** (1962) 207.
22. D. Chapman, P. Byrne and G. G. Shipley, *Proc. Roy. Soc. (London) A*, **290** (1966) 115.
23. J. M. Stein, M. E. Tourtellotte, J. C. Reinert, R. N. McElhaney and R. L. Rader, *Proc. Natn. Acad. Sci. USA*, **63** (1969) 104.
24. J. M. Lyons, T. A. Wheaton and H. K. Pratt, *Plant Physiol.*, **39** (1964) 262.
25. J. M. Lyons and C. M. Asmundson, *J. Am. Oil Chemist's Soc.*, **42** (1965) 1056.
26. T. Richardson and A. L. Tappel, *J. Cell Biol.*, **13** (1962) 43.
27. T. Richardson, A. L. Tappel and E. H. Grager, *Arch. Biochem. Biophys.*, **94** (1961) 1.
28. J. K. Raison, J. M. Lyons, R. J. Mehlhorn and A. D. Keith, *J. Biol. Chem.*, **246** (1971) 4036.
29. E. J. McMurchie, J. K. Raison and K. D. Cairncross, *Comp. Biochem. Physiol.* (1972), in preparation.
30. L. M. G. van Golde and L. L. M. van Deenen, *Biochim. Biophys. Acta*, **125** (1966) 496.
31. Y. Kagawa and E. Racker, *J. Biol. Chem.*, **241** (1966) 2467.
32. R. Berezney, Y. C. Awasthi and F. L. Crane, *Bioenergetics*, **1** (1970) 457.
33. E. Pfaff, H. W. Heldt and M. Klingenberg, *Europ. J. Biochem.*, **10** (1969) 484.
34. H. M. Levy, N. Sharon, E. M. Ryan and D. E. Koshland, *Biochim. Biophys. Acta*, **56** (1962) 118.
35. H. W. Heldt, in: *Regulation of Metabolic Processes in Mitochondria*, B.B.A. Lib., Vol. 7, J. M. Tager, S. Papa, E. Quagliariello and E. C. Slater (eds.), Elsevier, Amsterdam, 1966, p. 51.
36. N. Gruener and Y. Avi-Dor, *Biochem. J.*, **100** (1966) 762.
37. K. Bowler and C. J. Duncan, *Comp. Biochem. Physiol.*, **24** (1968) 1043.
38. J. S. Charnock, D. A. Cook and R. Casey, *Arch. Biochem. Biophys.* (1972), in press.
39. A. Atkinson, A. D. Gatenby and A. G. Lowe, *Nature New Biol.*, **233** (1971) 195.
40. J. S. Charnock and D. Frankel, *Abstr. 2nd Intern. Congr. Muscle Diseases*, Perth, 1971, Ab. 45.
41. G. Inesi and S. Watanabe, *Arch. Biochem. Biophys.*, **121** (1967) 665.
42. T. Yamamoto and Y. Tonomura, *J. Biochem. (Tokyo)*, **62** (1967) 558.
43. W. Hesselbach and M. Makinose, *Biochem. Z.*, **339** (1963) 94.
44. I. P. Suzdalskaya, in: *The Cell and Environmental Temperature*, Vol. 34, A. S. Troshin (ed.), Inter. Series of Monographs in Pure and Applied Biology, Pergamon Press, Oxford, 1963.
45. C. P. Lyman, *J. Mammal.*, **45** (1963) 122.
46. G. E. Palade and P. Siekevitz, *J. Biophys. Biochem. Cytol.*, **2** (1956) 671.
47. N. Towers, J. K. Raison, G. Kellerman and A. W. Linnane, unpublished data.
48. I. Faiferman, L. Cornudella, and A. O. Pogo, *Nature New Biol.*, **233** (1971) 234.
49. T. M. Chen, R. H. Brown and C. C. Black, *Weed Sci.*, **18** (1970) 399.
50. A. Shneyour, J. K. Raison and R. M. Smillie, *Biochim. Biophys. Acta* (1972) in preparation.
51. T. E. Weier and A. A. Benson in *Biochemistry of Chloroplasts*, T. W. Goodwin (ed.), Academic Press, London, 1966, p. 91.
52. T. Murata, *Physiol. Plantarum.*, **22** (1969) 401.
53. A. O. Taylor and J. A. Rowley, *Plant Physiol.*, **47** (1971) 713.
54. A. O. Taylor and A. S. Craig, *Plant Physiol.*, **47** (1971) 719.
55. C. R. Hackenbrock, *J. Cell. Biol.*, **30** (1966) 269.
56. D. E. Green, J. Asai, R. A. Harris and J. T. Penniston, *Arch. Biochem. Biophys.*, **125** (1968) 684.
57. R. W. Hendler, *Nature (London)*, **207** (1965) 1053.
58. J. Cullen, M. C. Phillips and G. G. Shipley, *Biochem. J.*, **125** (1971) 733.

59. G. Wilson, S. P. Rose and C. F. Fox, *Biochem. Biophys. Res. Commun.*, **38** (1970) 617.
60. B. L. Adams, V. McMahon and J. Seckbach, *Biochem. Biophys. Res. Commun.*, **42** (1971) 359.
61. P. Cerletti, S. Giovenco, G. Testolin and I. Binotti, *Membrane Model Form. Biol. Membranes, Proc. Meeting Intern. Conf. Biol. Membranes*, 1967, L. Bolis (ed.), North Holland Publishing. Co., Amsterdam, 1968, p. 166.
62. M. R. Lemberg, *Phys. Reviews*, **49** (1969) 48.
63. J. C. Reinert and J. M. Steim, *Science*, **168** (1970) 1580.

Interactions at the Surface of Plant Cell Protoplasts; An Electrophoretic and Freeze-etch Study†

B. W. W. Grout, J. H. M. Willison and E. C. Cocking

Department of Botany, University of Nottingham, University Park, Nottingham NG7 2RD

Introduction

The physical limitation of the cell protoplast has been demonstrated both mechanically, and microscopically, as a tangible barrier between cell cytoplasm and the external environment. In higher plant cells the isolation afforded by this limiting membrane is enhanced by the normally cellulosic cell wall, usually very close to the membrane outer surface. The work to be described is an attempt to illustrate aspects of structure and some functions of the outer protoplast membrane, or plasmalemma, and from this information, to discern something of the dynamic nature of this plasma membrane.

The experiments were performed upon spherical, naked higher plant cells (isolated protoplasts) released from tissue pieces into suspension by digestion of the cell wall with the appropriate degrading enzymes in a plasmolysing solution.

Earlier work from this laboratory on the endocytotic uptake of latex particles by tomato fruit locule protoplasts[1, 2] led to a proposed mechanism for the process which has certain similarities to the situation in animal and protistan cells.[3–5] The parameters to be met for uptake to occur were an initial surface binding of the particle to the protoplast, followed by membrane stretch over the area of particle contact. This localized stretch brought about vesiculation of the particle and the final uptake of the vesicle–particle complex into the cytoplasm. The membrane-stretch stage of this process has been well illustrated by freeze-etch replicas which supply quantitative data on changes in the number of granules per unit area of membrane.[2]

To enquire further into the phenomenon of surface adhesion, the initial stage in the uptake process, both particles and protoplasts were studied using the techniques of microelectrophoresis as applicable to whole cells and large particles. Valuable information was gained concerning electrical charges near the relevant surfaces, and also the effects of various substances believed to affect endocytosis or surface adhesion in such systems.

The techniques of freeze-etching and microelectrophoresis were also applied to aspects of membrane fusion in plant protoplast systems under study in our laboratory. From information gained by freeze-etch and thin section studies[6] it would appear that an essential pre-requisite for membrane fusion is an intimate contact, sustained for a considerable length of time.

The membrane stretch hypothesis was developed from observations of endocytosis. It has been extended, on the basis of freeze-etch observations, to a general concept applicable to other membranes.

Other theories of endocytosis are considered.[7,8]

Materials and Methods

(1) *Protoplast Isolation*

Protoplasts were prepared from tomato fruit locule tissue as has been previously described.[2]

Protoplasts from tobacco leaf mesophyll tissue were prepared using pieces cut from the leaf of 50–60 day-old *Nicotiana tabacum* var Xanthi nc plants, after the lower epidermis had been removed.[9] Peeled leaf pieces were incubated in a mixture of 5% w/v cellulase (Onozuka p. 1500), and 0·5% macerozyme (all Japan Biochemicals Co. Ltd., Nishinomiya, Japan), and 25% sucrose at 20°C for 4 h. The material was gently swirled at intervals during the treatment. The protoplasts thus released were washed twice by flotation in 25% sucrose. The plasmolyticum (sucrose) was made up in distilled water except where stated otherwise.

(2) *Incubation of Isolated Protoplasts with Latex Spheres*

Incubations of tomato fruit protoplasts with polystyrene latex spheres were carried out using methods previously described.[2]

The spheres used were Dow-Latex, 0·109 μ in diameter. The latex suspensions were added to the experimental media without prior dilution.

(3) *Induced Fusion of Petunia Leaf Protoplasts*

Protoplasts were isolated from pieces of leaf (of 50–60 day-old plant) from which the lower epidermis had been removed. These were incubated in a mixture of 0·3% macerozyme and 1% cellulase (Onozuka P 1500) in 13·5% sorbitol for approximately 18 h. Liberated protoplasts were washed once by centrifugation in fresh 13·5% sorbitol and then by flotation in 20% sucrose. The sucrose was replaced by 0·47 M $NaNO_3$ and the protoplasts were immediately centrifuged for 5 min at approximately 150 × g. Following this centrifugation, the protoplasts were quickly washed in 20% sucrose before freeze-etching.

(4) *Freeze-etching*

Specimens for freeze-etching were variously pretreated. For deep-etching, protoplasts were glutaraldehyde-fixed and washed in several changes of distilled water. In some cases, fixed protoplasts were re-suspended in 20–25% glycerol/water as described previously.[2] Otherwise protoplasts were frozen directly while suspended in the culture medium, which was generally 20% sucrose. Drops of the suspending medium, with protoplasts, were mounted on collared gold specimen holders and frozen by immersion in Freon 22 at its freezing point. Freeze-etching was performed using the now standard technique[10] utilizing a Balzers 360M freeze-etch apparatus. Platinum-carbon for shadowing was evaporated from an electron beam source (Balzers EVM 052) in most cases. Replica cleaning was as previously described.[2]

(5) *Electrophoresis*

The microelectrophoresis was carried out using a flat cell[11] with platinum black electrodes, which prevented any electrolysis or polarization. The effective inter-electrode distance was determined from measurements of the resistance of a solution of known specific conductance in the cell. The electrophoresis cell was thermostatted in an oil bath attached to the microscope, the whole apparatus being in a constant temperature room maintained at 25 ± 1°C. A known potential difference was applied across the electrodes using a Gelman Hawksley stabilized D.C. power supply. The distribution of particle mobilities across the cell was found to be parabolic and the other criteria for true electrophoresis also held. The mobility of a total of approximately 40 particles was recorded by observation of their movement against a calibrated eyepiece graticule for each current direction at each stationary level. The stationary levels were predicted using Komagata's equation.[12]

The addition of substances under investigation was made directly to the protoplast suspension in amounts to give a final concentration as stated (see Tables). Suspensions were incubated with the additives at 25°C for 20 min and then placed directly into the electrophoresis cell.

For aqueous colloidal systems the correct equation to calculate zeta potential (ζ) from electrophoretic mobility is that due to J. Th. G. Overbeek.[13] For the biological cells studied in these experiments this reduces to the Smoluckowski equation[14] as

$$u_E = \frac{\zeta D}{4\pi\eta}$$

In electrophoretic experiments with latex particles where for various ionic considerations this equation cannot be used, then the equation attributed to Henry[15] has been used

$$u_E = \frac{\zeta D}{6\pi\eta} f(ka)$$

where u_E is mobility (microns per sec per volt cm); D the dielectric constant; η the viscosity (centipoise) and $f(ka)$ is Henry's constant.

Results

Freeze-etching

1. *Changes in granule densities during endocytosis.* Our recent work on the endocytosis of latex spheres by isolated tomato fruit protoplasts[2] showed that adhering latex spheres were taken up singly in involutions of the plasmalemma which closely adhere to the latex sphere at all stages. It was proposed that the mechanism involved some form of membrane stretch. This suggestion was based on changes in the density of the granules found on frozen-etched plasmalemmae. These earlier observations were made on glutaraldehyde-fixed material and it seemed advisable to check that the observations were not induced by the fixative. Similar uptake of latex spheres were found in freeze-etched material which was unfixed and untreated with glycerol as a cryoprotective (Fig. 1). Direct comparison of granule density changes is complicated, however, by the fact that the distribution of granules between the complementary fracture faces of the membrane is altered by the medium in which the material is frozen. All the equivalent stages of endocytosis previously reported for fixed material are found in the unfixed material. The closing of the neck of the invagination may be affected to some extent by fixation since it is more usual to find invaginations with wide necks in unfixed material (arrow, Fig. 1) than in fixed material.

2. *Other changes in granule density associated with membrane "stretch".* Variations in granule density at the plasmalemma surface exposed by freeze-etching are not limited to invaginations associated with latex uptake. Occasionally evaginations of the plasmalemma of the isolated protoplast are found (Fig. 2). These have been observed light-microscopically and in unfixed freeze-etched material and cannot be considered a fixation artefact. The number of granules per unit area on the surface of the "blow-out" (Fig. 2) and on the surface of the background plasmalemma have been measured and compared. The change in granule density between plasmalemma and evaginated plasmalemma is roughly proportional to the change in membrane area which would occur if a circle of plasmalemma of the diameter of the neck of

the "blow-out" had been extended to form the "blow-out". The overall "stretch" in this case is approximately 33 times.

Extended plasmalemma of the above sort may come in contact with latex spheres (Fig. 3). Binding of the latex occurs, and it appears that the membrane folds in an attempt to accommodate the sphere. It seems probable that the membrane has reached the limit of its "stretch" capacity.

It should be noted that granule density variations are commonplace in plant cells both at the plasmalemma and in other membranes. In cytoplasmic vesicles localized granule density changes may be found which suggest that membrane "stretch" is occurring (Figs. 4 and 5). Changes of this sort may be found in both fixed and unfixed material.

3. *The fate of latex spheres following uptake.* Latex spheres enter the isolated protoplast closely bounded by plasmalemma. However, within the protoplast they are commonly found in relatively large vesicles which contain several latex spheres within their interior (Fig. 7). Most likely, these arise by fusion of the small latex-containing vesicles with pre-existing cytoplasmic vesicles. It is of interest to note that cytoplasmic vesicles can be differentiated by the freeze-etch technique. Isolated protoplasts freeze-etched while still in their native plasmolyticum usually demonstrate distinct differences in ice-crystal size between external medium, cytoplasm, and vacuole (Fig. 6). The vacuole always contains ice-crystals which are considerably larger than those in the external medium, presumably as a result of the exclusion of the plasmolyticum from the vacuole itself. These ice-crystal size differences, however, are not limited solely to the vacuole. Cytoplasmic vesicles contain ice of different sizes (Fig. 6) which tend to be characteristic of vacuole or external medium. It appears that when latex spheres are found within large cytoplasmic vesicles these vesicles contain ice which is characteristic of the external medium.

4. *The plane of fracture in freeze-etched membranes.* There is considerable evidence to suggest that frozen-fractured membranes fracture along an internal plane which is probably the hydrophobic interior of the membrane.[16] Isolated protoplasts were fixed, mounted in distilled water, and freeze-etched using the deep-etching technique.[17,18] Deep-etching involves sublimation from the surface of the ice-crystals surrounding the specimen. Thus, both the outer surface of the membrane and the fracture face are exposed. The observations (Fig. 8) of deep-etched isolated protoplast plasmalemma support the concept of internal membrane fracture. The granules of the membrane are clearly centrally located and appear to extend beyond the outer surfaces of the membrane.

5. *Additional observations of frozen-fractured membranes.* In glutaraldehyde-fixed material all membranes appear as a non-structured continuum variously covered with granules. However, occasionally in

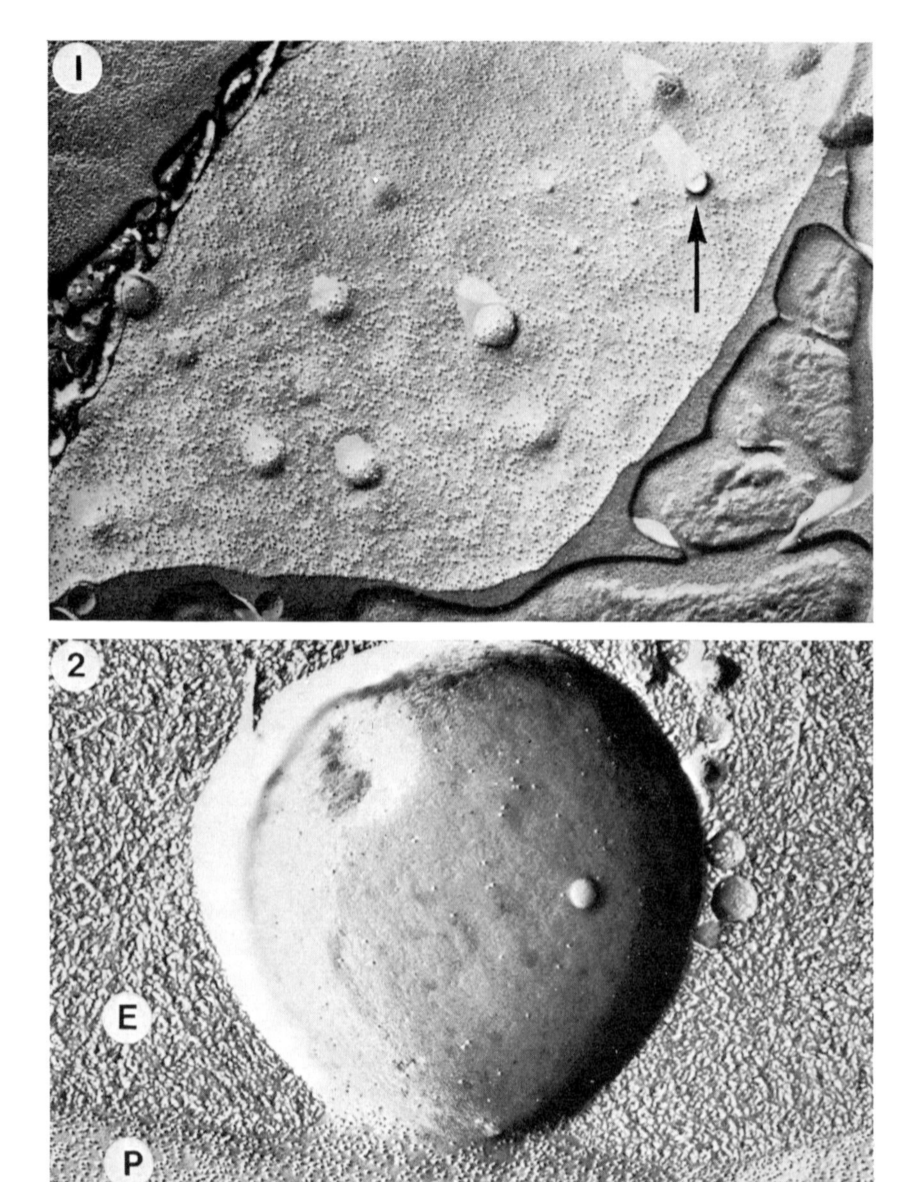

Figure 1. Endocytosis of latex spheres by an isolated tomato fruit protoplast. The plasmalemma is viewed from the inside. A wide-necked invagination has been snapped off at the neck (arrow). The specimen was prepared without chemical fixation and without the addition of a cryoprotective agent. (Magnification ×50,000.)

Figure 2. An evaginated region of tomato fruit protoplast plasmalemma. The evagination has been formed by expansion of a localized region of the plasmalemma. The specimen was glutaraldehyde-fixed and treated with glycerol. Glycerol/water eutectic (E), plasmalemma (P). (Magnification ×66,000.)

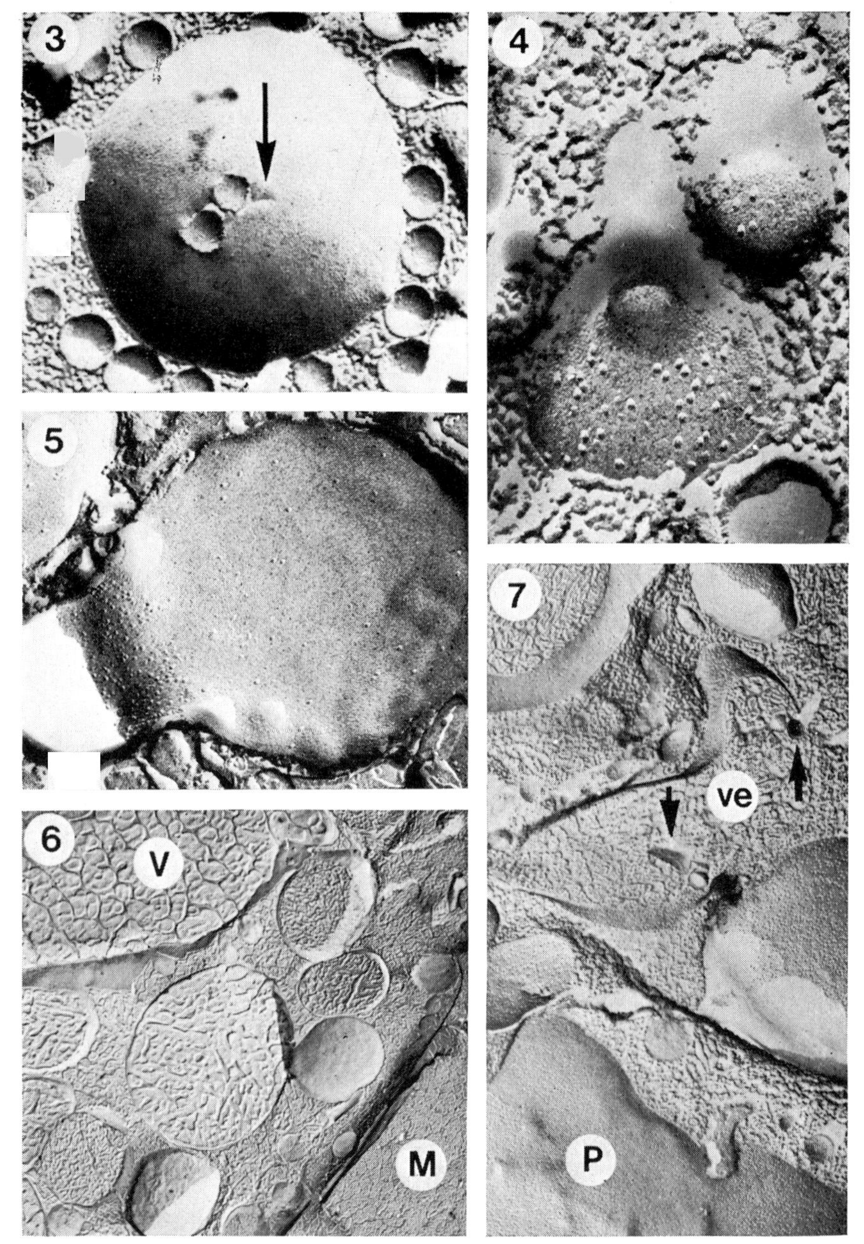

Figure 3. Binding of latex spheres to a region of expanded plasmalemma. Folding of the plasmalemma has occurred (arrow). Specimen glutaraldehyde-fixed, glycerol treated. (Magnification ×74,000.)

Figures 4 and 5. Cytoplasmic vesicles revealing membrane expansion which is indicated by changes in granule density on the fractured face of the membranes. The membrane "stretch" phenomenon. Specimens unfixed and frozen in native plasmolyticum. (Magnification ×105,000 (Fig. 4) ×64,500 (Fig. 5).)

Figure 6. Differences in ice-crystal size reveal the permeability barriers to entry of the sucrose plasmolyticum of isolated protoplasts. Cytoplasmic vesicles retain the permeability characteristics of the membranes from which they derived. Vacuole (V) frozen sucrose plasmolyticum (M). Specimen unfixed and frozen in native plasmolyticum. (Magnification ×10,500.)

Figure 7. Several latex spheres found in a relatively large cytoplasmic vesicle following endocytosis by a tomato fruit protoplast. The latex has entered the vesicle by fusion of the small latex-containing endocytotic vesicles with a pre-existing cytoplasmic vesicle. Specimen unfixed and frozen in native plasmolyticum. Cytoplasmic vesicle (Ve), plasmalemma (P), Latex spheres (arrows). (Magnification ×33,000.)

unfixed material alterations in this basic structure may be seen. These may consist of a regular pattern in the background continuum (Fig. 9); or of a series of ridges or flow-lines (Fig. 10). The latter, when found on the plasmalemma, usually seem to be associated with underlying organelles. Sometimes similar structures may be found on the fractured face of cytoplasmic vesicles (Fig. 11). Although of a similar structure these may have a different functional significance. We have seen structures of the type shown in Fig. 10 in protoplast material which has been both glycerinated and unglycerinated in sucrose plasmolyticum. It seems unlikely that these are artefacts. The tobacco leaf material (Fig. 10) was partially infiltrated with 20% glycerol for less than 1 min, and was frozen as quickly as possible after removal from the plant.

Electrophoresis

The differing mobilities of various plant protoplasts and of polystyrene latex spheres was recorded as a function of the time taken to travel a known distance as measured by a graticule. From these mobilities a value for the zeta potential has been calculated.

TABLE I. Measurement of zeta potential of tomato fruit protoplasts and polystyrene latex spheres in 20% sucrose with various added compounds

(i) Tomato fruit protoplasts

Additive	Electrophoretic mobility (U_e) μ sec^{-1} V cm^{-1}	Zeta potentials Mv
None	−0·77	−22·44
Poly-L-ornithine 2 μg/ml	Barely negative unmeasurable within error of experiment	
Poly-D-lysine 2 μg/ml	−0·44	−12·83
Sodium nitrate 8·5 mg/ml	−0·48	−13·99
DEAE Dextran 4 μg/ml	−0·47	−13·71
Glycerol monoleate 250 μg/ml	−0·79	−23·04

(ii) Polystyrene latex spheres 0·109μ diameter

Additive	Electrophoretic mobility	Zeta potential
None	−1·88	−54·82
Poly-L-ortnithine 2 μg/ml	−0·90	−26·24
Sodium nitrate 17·0 mg/ml	→0	
Glycerol monoleate 250 μg/ml	−1·78	−51·90

In experiments using sodium nitrate and poly-L-ornithine washing the protoplasts three times with 20% sucrose brought the measurements of mobility (in 20% sucrose) back to the two values for untreated material.

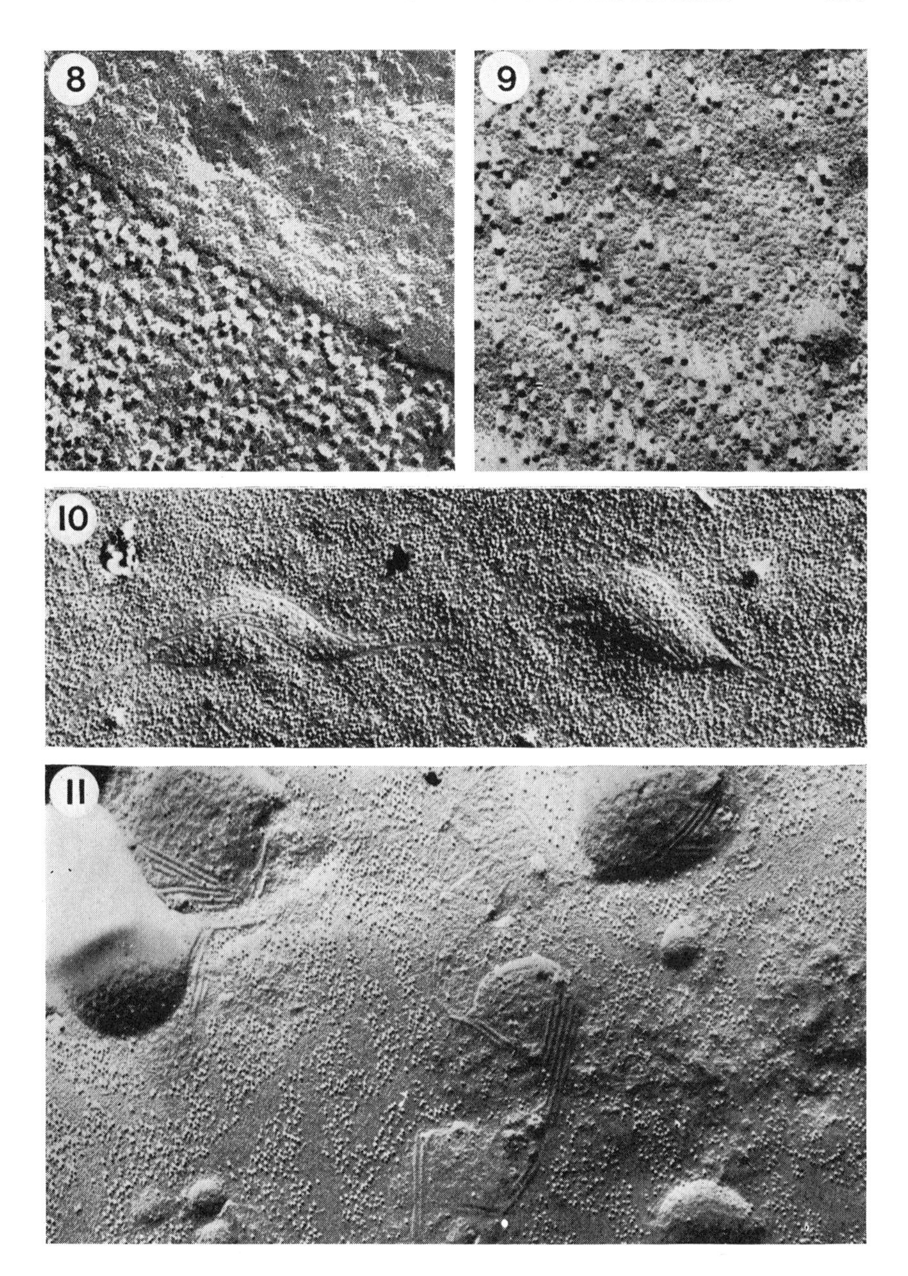

Figure 8. Deep-etched plasmalemma of tomato fruit protoplast. The fracture face of the membrane (bottom) and the real outer surface (top) are revealed. (Magnification ×130,000.)

Figure 9. A fractured membrane revealing structure in the non-granular continuum. Specimen unfixed and not treated with a cryoprotective. (Magnification ×135,000.)

Figure 10. Fractured plasmalemma of tobacco leaf cell frozen immediately after removal from the plant. 20% glycerol/water was used to mount the tissue piece but had only partially infiltrated this cell. (Magnification ×50,000.)

Figure 11. Fracture face of a cytoplasmic vesicle in a tomato fruit protoplast. No fixation or cryoprotective was used. (Magnification ×50,000.)

TABLE II. Measurement of the zeta potential of tobacco leaf protoplasts in 23% Sucrose with various added compounds

Additive	Electrophoretic mobility	Zeta potential
None	−0·90	−26·24
Poly-L-ornithine 2 μg/ml	−0·42	−12·24
Poly-L-ornithine 10 μg/ml	+0·45	+13·12
Sodium nitrate 8·5 mg/ml	→0	
Glycerol monoleate 250 μg/ml	−0·90	−26·24

Discussion

Introduction

The endocytotic uptake of polystyrene latex spheres by tomato fruit protoplasts was first observed in an electron microscope study using ultrathin sections.[1] This result has since been verified using the now standard method of freeze-etching;[2] the results of which point towards a possible mechanism for the process.

Uptake is initiated by adhesion of the latex sphere to the membrane resulting in a depression that extends solely over the area of contact of the sphere. The membrane bound particle continues to travel towards the cytoplasm and eventually comes to lie in a plasmalemma invagination closely surrounded by the membrane. The neck of the vesicle then fuses to liberate the small tightly membrane bounded vesicle into the cytoplasm. The intimacy of contact between sphere and plasmalemma at all stages of endocytosis has important implications with respect to the mechanism of endocytotic initiation (see later). Large vesicles are often observed within the cell containing several spheres, but uptake has not been seen to occur other than singly. It is probable, therefore, that the endocytotic vesicles fuse with others of similar nature, and probably also with non-sphere containing vesicles which originate from the plasmalemma (see later).

The membrane-stretch mechanism which we have proposed for the invagination process was based upon the freeze-etching observation of a reduction of granules on the membrane face in the endocytosing region. A mechanism for endocytosis by lymphocytes, macrophages and mast cells has been proposed by Allison[7] which would involve a clumping of granules. He proposed that involution then occurs by cytoplasmic microfilaments pulling on the granules. Although the two mechanisms would appear contradictory it is not improbable that the two systems are quite different, having differing mechanisms. The endocytosis of large solids by higher plant cells is an unusual process whereas the phagocytes studied by Allison are specially adapted for this process.

Freeze-etching

The phenomenon of membrane "stretch". It is not necessary to discuss at length in this paper the precise nature of freeze-etch fracture of biological membranes. This has already been treated at length by various authors, notably Branton.[19,20] Our evidence (Fig. 8) supports the view that fracture of the tomato fruit protoplast plasmalemma is along an internal plane. The similarity in appearance of other membranes suggests that internal fracture is a general phenomenon. The granules of freeze-etched membranes commonly have been presumed to have a functional significance, although this significance is often biased towards the interests of the author. The possibility that they might represent lipo-protein micelles has been proposed.[16] However, whatever their significance we have regarded them simply as markers for the state of the membrane. In the case of the endocytosis of latex spheres and of the evaginated plasmalemma (Fig. 2) there is good reason to believe that granules are neither synthesized nor do they disappear to provide extra membrane but simply that the membrane between the granules expands. The expansion of 3 to 4 times required for the endocytosis of latex spheres could, on some membrane models,[21,22] be a process for which all the additional membrane material is contained within the membrane. However, an expansion of 33 times or more (Fig. 2) is unlikely to be by such a process. More likely additional membrane material (lipo-protein or possibly lipid alone) is added by intussusception from a cytoplasmic pool.[23] The observation of folding in a highly "stretched" membrane in association with a bound latex sphere (Fig. 3) suggests that the stretching process has a limit. Presumably this limit is determined by the size of the sink of potential membrane wherever this may be.

We propose that the membrane expansions characterized by granule density changes are rapid processes which enable most cytoplasmic membranes to be highly dynamic in terms of the area which they can cover. In the case of latex sphere uptake it is probable that membrane expansion is initiated by localized changes in the electrical field impinging on the membrane (see discussion later). It seems reasonable that other membrane "stretch" phenomena are also electrically induced. In these cases, however, the motive force is not necessarily produced in the region undergoing change. A localized weakening of the membrane could enable existing forces, previously in equilibrium, to produce the change. The possibility that energetically active proteins are present in the membrane cannot be ignored. Membrane stretch mechanisms could be controlled, via the local electrical environment, by membrane bound ion pumps.

The membrane model which has the best supporting evidence is that of Robertson.[24] This is derived from the myelin sheath of nerves. Myelin forms by extension of the plasmalemma of the Schwann cell.

The membranes of freeze-etched myelin are virtually granule-free.[25,26] It seems probable, on our evidence, that myelin forms by the membrane stretch process described above. This would make myelin an untypical membrane with regard to lability and lack of variation in structure, but nevertheless having the same backbone structure as most other membranes.

Bennett's concept of membrane flow[8] is, in our view, limited by regarding the membrane as a relatively rigid sheet in which membrane turnover requires the addition, by fusion, of fully synthesized membrane. A highly labile membrane capable of considerable, rapid, localized stretch is more capable of fulfilling the multifarious transformations which typify biological membranes.[21]

Vesicle fusion. The observation that vesicles fall into distinct categories on the basis of ice crystal size suggests that the permeability properties of the tonoplast and plasmalemma from which these vesicles presumably arise are retained by the vesicles themselves. The results show clearly that the effective permeability barrier to plasmolytica is the tonoplast. Of particular interest is the observation that the latex spheres are found in large vesicles which contain ice characteristic of the external medium, not of the vacuole. These vesicles are presumably bounded by plasmalemma-type membrane. The small vesicles which contain the latex are also bounded by plasmalemma and there appears to be a recognition capacity. This specificity of recognition is probably based on electrical charge characteristics since an essential prerequisite for fusion is membrane contact and this is probably electrically determined (see discussion later).

The effect of fixation. The alterations in membrane structure revealed in Figs. 9 and 11 and the observations that glutaraldehyde may cause the closing of the necks of endocytotic invaginations show the value of the freeze-etch technique for electron-microscopic examination of membranes. The effect of glutaraldehyde on the neck of the endocytotic invaginations could account for the widespread observation of closed plasmodesmata between plant cells. From a purely teleological point of view it is in the interests of a cell to defend itself against the onslaught of fixatives by closing intercellular connections.

Other alterations in membrane structure revealed by freeze-etching. The substructure seen in the background continuum of the membrane (Fig. 9) could represent the lipoprotein micelles envisaged by Lucy.[22] Regions such as these are localized and the orientation of the substructure differs between adjacent patches. The spacing periodicity is approximately 12–13 nm. No cryoprotective was used in the instances where this substructure was revealed and it is possible that it has arisen as a result of the growth of ice-crystals close to the membrane.

Similar structures to those shown in Fig. 10 have been previously reported[27] and deserve careful assessment. They usually appear to be associated with underlying organelles closely associated with the plasmalemma and they give the impression of flow lines. The sharp angles and junctions suggest that they are not formed in association with underlying microtubules. It is tempting to believe that this is a manifestation of membrane induced cytoplasmic flow perhaps of the sort envisaged by Hejnowicz.[28] Clearly, conformational changes of a transitory nature are occurring in these regions of the membranes.

Electrophoresis

Electrical nature of latex adhesion. The most plausible explanation for the initial adhesions of latex to plasmalemma are that surface properties, probably electrical in nature, are involved. Electrophoretic

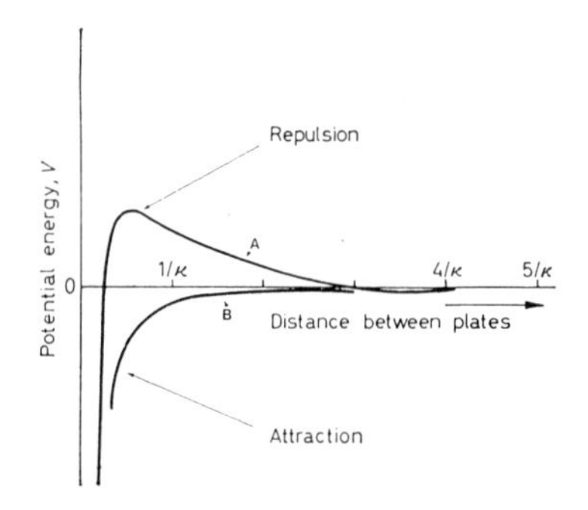

Figure 12. Total interaction energy curves for two charged surfaces, as a summation of the attraction curve and repulsion curves, showing variation with distance. (Adapted from *Introduction to Colloid and Surface Chemistry*, D. J. Shaw, Butterworth (Publ.).)

measurements were made on both protoplasts and latex spheres to investigate the nature and magnitude of these charges. Effects of various additives, all believed to promote endocytosis or cell adhesion, were studied. It was observed that both protoplasts and latex spheres had negative electrophoretic mobility and negative zeta potential, the magnitude of those of the latex being greater.

The forces which act against particle aggregation are a consequence of repulsive interactions between similarly charged electrical double layers at the surfaces. The aggregation is brought about by the universal inter-molecular London and Van der Waals attractive forces. A summation of these two forces, expressed in terms of potential energy, and their variation with the distance between the surfaces, is shown in Fig. 12(A). It can be seen that there is a peak of repulsive energy which will act as a barrier to the forces of attraction. The amount of adhesion in a mixed suspension of protoplasts and latex spheres will relate directly to the number of particles which are at a sufficiently high energy level to overcome this barrier.

The compounds under investigation appear to reduce the magnitude of the zeta potential, and in extreme cases to reverse the sign. It can be postulated, therefore, that the effects of the additives are probably due to their cationic components. Reduction of zeta potential towards zero affects the interaction of two approaching charged surfaces and will lower the repulsive "barrier" as depicted in Fig. 12(B). More particles in the suspension will therefore have sufficient kinetic and thermodynamic energy to approach within range of attractive forces. More latex spheres would, therefore, in the presence of one of these additives, adhere to the protoplast plasmalemma. This necessarily will increase the number of spheres in a membrane bound, potentially endocytotic, situation. Experiments to determine increased rates of uptake in this system are not yet complete; however, support for the tenures above is offered by experiments in which endocytotic uptake of tobacco mosaic virus by tobacco leaf protoplasts is enhanced by pre-incubation with poly-L-ornithine.[29] It is possible, however, because of their relatively large size (mol. wt. 120,000) that these polycations act as bridges between the two surfaces, part of the molecule binding to each of them. If the polycations bind flat to the surface for the whole of their length, as may well happen at low concentrations, then their mode of action will be directly comparable to that of the inorganic materials.

When considering the "membrane-stretch" of the developing endocytotic vesicle that follows the adhesion of the latex sphere (see earlier) it would seem reasonable to propose that the contact is directly responsible for initiation of the stretch. As the two charged surfaces come within a small distance of one another it is proposed that considerable energetic perturbance results, giving rise to a localized instability within the membrane architecture. The forces responsible for this disturbance would be the Van der Waals and London's attractive forces; changes in surface potential caused by electrical double-layer interaction;[30] energy redistribution resulting from elimination of charged species at the surfaces; expulsion of water, and steric molecular interferences. It is believed that the energies described above would be large enough to alter protein and lipid configurations. Our hypothesis is continued by suggesting that the resultant instability is compensated for by the rearrangement of molecules within the matrix of the membrane. This is manifested as the phenomenon of "membrane-stretch" described previously. It is possible that the instability induced in the membrane by latex sphere contact facilitates incorporation of fresh components from the cytoplasm into the membrane framework. Freeze-etch evidence would dispute the notion that stretch derives from reduction in granule number (see earlier).

As has been previously described the plasmalemma membrane at the neck of the endocytotic vesicle fuses with itself to isolate the endocytotic vesicle and repair the plasmalemma. The endocytotic vesicle is also known to fuse with various cytoplasmic vesicles. The ease and regularity with which these fusions occur led to a consideration of the events necessary for membrane/membrane fusion, and the energetic status of the membrane required for completion of the process.

Fusion. Fusion of higher plant protoplasts with one another is a manipulation under intensive investigation in our laboratories, and surface properties are of prime importance.

Electrophoretic investigation of tobacco leaf protoplasts, a system in which fusion is commonly studied, demonstrated a negative mobility and zeta potential. Compounds such as sodium nitrate, used to induce aggregation of protoplasts, reduced the value for both these properties, as did various substances used elsewhere in attempts to promote endocytosis, e.g. poly-L-ornithine. Glycerol mono-oleate, one of a group of compounds variously related to lipids, and commonly used in fusion experiments, appeared to have little or no effect on surface electrical properties.

Certain parallels between experiments aimed at inducing fusion, and concerned with endocytosis thus became apparent. In both cases one is dealing with a suspension of colloidal, or near colloidal negatively charged particles. The attractive and repulsive interactions between these particles can again be summarized by Fig. 12. For successful protoplast fusion it is necessary, as in endocytosis, to bring these protoplasts into intimate and sustained contact (Figs. 13 and 14). Substances which lower the surface charge on the particles will lower the barrier of repulsive energy represented on the interaction diagram. This increases the number of particles with sufficiently high energy to approach within the distances of London's and Van der Waals' attractive forces.

As the two membranes come within molecular distances a great deal of potential perturbing energy is available, derived from atomic and molecular interactions, steric interferences, surface potentials of the original membrane, and ionic displacements, as previously inferred for endocytosis. This again results in molecular instability within the area of membrane contact. We propose that this instability is "reduced" by fusion of the membranes at the edges of the area of contact. The membrane within the original area of contact may well re-order to form a series of vesicles, which could be considered as similar to the endocytotic vesicles with respect to their stability. The unstable membrane may, however, by dispersed into the cytoplasm. Previously it was postulated that the "inducers" of fusion acted by facilitating the initial contact phase. It is possible that these compounds may also have molecular effects on membrane instability, as under certain circumstances they appear to cause bursting of non-fused protoplasts. The very large polycationic materials, e.g. poly-L-ornithine may also have a bridging action during membrane fusion, adsorbing at different regions on the molecule to both protoplasts.

Lipid related compounds such as glycerol mono-oleate that appear not to affect surface electrical properties may well only act directly on stability of the molecular architecture of the membrane. It may, therefore, be necessary to use these compounds in combination with other means for aggregating protoplasts.

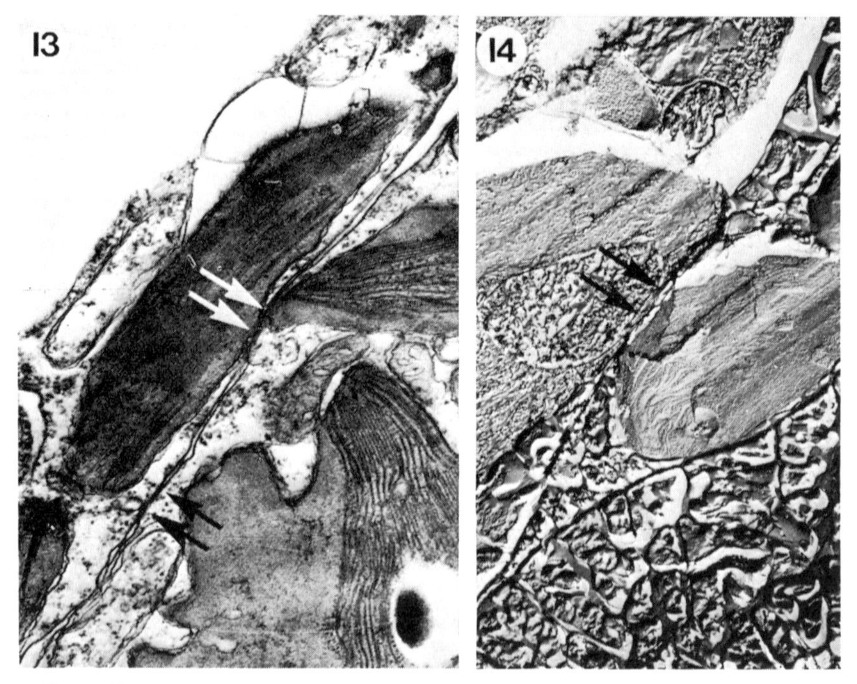

Figure 13. A thin-section micrograph of the junction between two fusing *Petunia* protoplasts. Protoplasts were induced to fuse using sodium nitrate. There are localized regions in which the two plasmalemmae come in very close contact (arrows). These are believed to be the regions of actual membrane/membrane fusion. (Magnification ×45,000.)

Figure 14. Freeze-etch micrograph of the equivalent region ot Fig. 13. This specimen was prepared without chemical fixation or prior treatment with a cryoprotective. The region of close contact (arrows) cannot be a preparative artefact. (Magnification ×8,500.)

These proposed criteria for membrane fusion can be used to derive a theory relating to membrane fusions within the cell cytoplasm. Fusion of cell organelles with other organelles, or with plasmalemma or tonoplast, may be inhibited if they are not at a sufficiently high energy level to overcome the peak of their repulsive interactions.

The plan for fusion which has been derived does not involve an external chemical agent as essential for any of its stages. The "inducers" of fusion studied were effective by decreasing the electrostatic forces of repulsion between the protoplasts. They may, however, facilitate the fusion process but are not essential to it. The polycationic compounds may well have this type of effect, as they have been reported to bring about immediate membrane contractions when applied to various protozoa.[31]

Our model for fusion relies on the physical and electrical interactions of the constituent membrane molecules when in close proximity to bring about the necessary structural rearrangements.

The fusion model of Poste and Allison shows many similarities to our own, but ATP and calcium ions are implicated as essential to the mechanisms.[32] Whilst accepting the need for ATP, our experiments would not appear to confirm a need for calcium.

Lucy has proposed a model for fusion which necessitates a high proportion of membranous lipid to be in a micellar form.[33] Our experimental plan has not been comprehensive enough to enable us to tender any valid evidence for or against this proposal.

Acknowledgements

The authors would like to thank Miss L. A. Withers for her continued advice and also the provision of Figs. 13 and 14. We should also like to express our gratitude to Mr. P. Wright, Department of Chemistry of this University, without whose continued efforts, and generous loan of electrophoretic apparatus this work could not have been completed.

References

1. M. A. Mayo and E. C. Cocking, *Protoplasma*, **68** (1969) 223.
2. J. H. M. Willison, B. W. W. Grout and E. C. Cocking, *J. Bioenergetics*, **2** (1972) 371.
3. E. D. Korn and R. A. Weismann, *J. Cell. Biol.*, **34** (1967) 219.
4. T. R. Ricketts, *Protoplasma*, **73** (1971) 387.
5. A. C. Allison, P. Davies and S. de Petris, *Nature New Biol.*, **232** (1971) 153.
6. L. A. Withers and E. C. Cocking, *J. Cell Sci.*, **11** (1972) 59.
7. A. C. Allison, in: *Symposium on Cell Interactions, Londno* 1971, Elsevier, Amsterdam.
8. H. Stanley-Bennett, in: *Handbook of Molecular Cytology*, **15**, A. Lima-de-Faria (ed.), North-Holland, Amsterdam, 1969, p. 1294.
9. J. B. Power, and E. C. Cocking, *J. Expt. Bot.*, **21** (1970) 64.
10. H. Moor and K. Muhlethaler, *J. Cell Biol.*, **17** (1963) 609.
11. G. D. Parfitt, *Journal of Oil and Colour Chemists Association*, **51** (1968) 137.
12. S. Komogata, *Researches Electrotech. Lab. Tokyo*, **348** (1933).
13. J. Th. G. Overbeek, *Adv. Colloid Sci.*, **3** (1950) 97.
14. M. Smoluckowski, in: *Handbuch der Elektrizität*, Vol. 2, B. Graetz (ed.), Leipzig, 1914, p. 333.
15. D. C. Henry, *Proc. Roy. Soc. (A)*, **133** (1931) 106.
16. D. Branton, *Proc. Natn. Acad. Sci. USA*, **55** (1968) 1048.
17. D. Branton and D. Southworth, *Exp. Cell Res.*, **47** (1967) 618.
18. P. G. Pinto da Silva and D. Branton, *J. Cell Biol.*, **45** (1970) 598.
19. D. Branton, *Ann. Rev. Pl. Physiol.*, **20** (1969) 209.
20. D. Branton and D. W. Deamer, *Membrane Structure, Protoplasmatologia II*. E. I. Springer, Wien, New York (1972) 41.
21. J. L. Kavanau, *Structure and Function in Biological Membranes I and II*, Holden-Day, San Francisco, 1965.
22. J. A. Lucy, *J. Theoret. Biol.*, **7** (1964) 360.
23. Z. A. Staehelin and O. Kiermayer, *J. Cell Sci.*, **7** (1970) 787.
24. J. D. Robertson, in: *Handbook of Molecular Cytology*, **15**, A. Lima-de-Faria (ed.), North Holland, Amsterdam, 1969, p. 1404.

25. A. Bischoff and H. Moor, *Z. Zellforsch.*, **81** (1967) 303.
26. D. Branton, *Exp. Cell Res.*, **45** (1966) 703.
27. D. H. Northcote, *Proc. Roy. Soc.* (*B*), **173** (1969) 21.
28. Z. Hejnowicz, *Protoplasma*, **71** (1970) 343.
29. R. H. A. Coutts, E. C. Cocking and B. Kassanis, *J. gen. Virol.* (1972) in press.
30. D. Gungiell, *J. Theor. Biol.*, **17** (1967) 451.
31. E. J. Ambrose and J. A. Forrester, *Symp. Soc. Exp. Biol.*, **22** (1968) 237.
32. G. Poste and A. C. Allison, *J. Theor. Biol.*, **32** (1971) 165.
33. J. A. Lucy, *Nature*, **227** (1970) 814.